Tributes
Volume 46

Relevance Logics and other Tools for Reasoning
Essays in Honor of J. Michael Dunn

Volume 36
Logic, Philosophy of Mathematics and their History.
Essays in Honor of W. W. Tait
Erich H. Reck, ed.

Volume 37
Argumentation-based Proofs of Endearment. Essays in Honor of Guillermo R. Simari on the Occasion of his 70th Birthday
Carlos I. Chesñevar, Marcelo A. Falappa, Eduardo Fermé, Alejandro J. García, Ana G. Maguitman, Diego C. Martínez, Maria Vanina Martinez, Ricardo O. Rodríguez, and Gerardo I. Simari, eds.

Volume 38
Logic, Intelligence and Artifices. Tributes to Tarcísio H. C. Pequeno
Jean-Yves Béziau, Francicleber Ferreira, Ana Teresa Martins and Marcelino Pequeno, eds.

Volume 39
Word Recognition, Morphology and Lexical Reading. Essays in Honour of Cristina Burani
Simone Sulpizio, Laura Barca, Silvia Primativo and Lisa S. Arduino, eds

Volume 40
Natural Arguments. A Tribute to John Woods
Dov Gabbay, Lorenzo Magnani, Woosuk Park and Ahti-Veikko Pietarinen, eds.

Volume 41
On Kreisel's Interests. On the Foundations of Logic and Mathematics
Paul Weingartner and Hans-Peter Leeb, eds.

Volume 42
Abstract Consequence and Logics. Essays in Honor of Edelcio G. de Souza
Alexandre Costa-Liete, ed.

Volume 43
Judgements and Truth. Essays in Honour of Jan Woleński
Andrew Schumann, ed.

Volume 44
A Question is More Illuminating than the Answer: A Festschrift for Paulo A.S. Veloso
Edward Hermann Haeusler, Luiz Carlos Pinheiro Dias Pereira and Jorge Petrucio Viana, eds.

Volume 45
Mathematical Foundations of Software Engineering. Essays in Honour of Tom Maibaum on the Occasion of his 70th Birthday and Retirement
Nazareno Aguirre, Valentin Cassano, Pablo Castro and Ramiro Demasi, eds.

Volume 46
Relevance Logics and other Tools for Reasoning.
Essays in Honor of J. Michael Dunn
Katalin Bimbó

Tributes Series Editor
Dov Gabbay   dov.gabbay@kcl.ac.uk

Relevance Logics and other Tools for Reasoning

Essays in Honor of J. Michael Dunn

edited by

Katalin Bimbó

© Individual authors and College Publications 2022. All rights reserved.

ISBN 978-1-84890-395-1

College Publications
Scientific Director: Dov Gabbay
Managing Director: Jane Spurr

http://www.collegepublications.co.uk

Cover design by Laraine Welch

All rights reserved. No part of this publication may be reproduced, stored in a retrieval system or transmitted in any form, or by any means, electronic, mechanical, photocopying, recording or otherwise without prior permission, in writing, from the publisher.

Contents

Preface vii

Gerard Allwein and William L. Harrison
Distributed Relation Algebra 1

Ofer Arieli
Four-Valued Semantics for Abstract Argumentation Frameworks
   Using (Extensions of) Dunn–Belnap Four-Valued Logic 31

Arnon Avron
On Formal Criteria for Relevance 54

Jc Beall and David Ripley
Time for Curry 65

Alex Belikov, Oleg Grigoriev and Dmitry Zaitsev
On Connegation 73

Katalin Bimbó and J. Michael Dunn
Modalities in Lattice-R 89

Ross T. Brady
Intension, Extension, Distribution and Decidability ... 128

Sergey Drobyshevich, Sergei Odintsov and Heinrich Wansing
Moisil's Modal Logic and Related Systems 150

J. Michael Dunn
Kripke's Argument for γ 178

Nicholas Ferenz
Conditional **FDE** Logics 182

Chrysafis (Takis) Hartonas
Reconciliation of Approaches to the Semantics of Logics
   without Distribution 215

Norihiro Kamide
Herbrand and Contraposition-elimination Theorems for
   Extended First-order Belnap–Dunn Logic 237

Saul A. Kripke
A Proof of Gamma 261

Edwin Mares
Models for Priorian Second-Order Logic 266

Lawrence S. Moss and Thomas F. Icard
A COMPLETENESS RESULT FOR INEQUATIONAL REASONING IN A FULL
HIGHER-ORDER SETTING 282

Hitoshi Omori and Heinrich Wansing
VARIETIES OF NEGATION AND CONTRA-CLASSICALITY IN VIEW OF DUNN
SEMANTICS 309

Gemma Robles and José M. Méndez
A CLASS OF 4-VALUED IMPLICATIVE EXPANSIONS OF FIRST-DEGREE
ENTAILMENT LOGIC (FDE) WITH THE VARIABLE-SHARING PROPERTY . 338

Yaroslav Shramko
THE DIAMOND OF MINGLE LOGICS: A FOUR-FOLD INFINITE WAY TO BE
SAFE FROM PARADOX 365

Shawn Standefer
COMPLETENESS VIA METACOMPLETENESS 394

Andrew Tedder
SITUATIONS, PROPOSITIONS, AND INFORMATION STATES 410

Alasdair Urquhart
WHO WAS SCHILLER JOE SCROGGS? 427

Yale Weiss
REVISITING CONSTRUCTIVE MINGLE: ALGEBRAIC AND OPERATIONAL
SEMANTICS 435

J. MICHAEL DUNN'S PUBLICATIONS (BOOKS AND ARTICLES) 456

Preface

J. Michael Dunn was a well-known and highly respected *logician, philosopher* and *information scientist*. Many logicians working in the area of non-classical logics knew him personally. He was highly regarded both because of his work in logic and of his role in the academic community. The idea of a collection of papers — like the present one — emerged soon after Dunn passed away in April 2021. A suggestion by Yale Weiss that I should edit such a volume led me to take on the task of editing this book.

Dunn's career spanned more than *half a century*, and while he was mainly a relevance logician, he also proved results in other areas of logic — often relying on an algebraic approach to logic. It would be unfeasible to try to sum up, in a short preface, all the theorems that Dunn proved; but it would be impossible not to mention some of the most conspicuous discoveries he made.

Dunn wrote his Ph.D. thesis about *relevance logics*. He algebraized $\mathbf{R}^t$ (and $\mathbf{E}^t$), which allowed him to apply techniques and results from universal algebra to relevance logics. Dunn invented the logic **R**-*mingle* (**RM**) — a semi-relevance logic. **RM** turned out to have an elegant algebraic semantics, relying on which Dunn proved a series of results about **RM** and its extensions. This led him (with R. K. Meyer) to a proof of the *admissibility of the rule* γ in **E**, **R** and **T**. The γ rule can be viewed as a form of disjunctive syllogism, or detachment for material conditional, and the latter immediately connects to the cut rule in a sequent calculus for 2-valued logic (such as K_1). Dunn and Meyer gave a proof of the *admissibility of the cut rule* in K_1 using a new technique inspired by the proof of the admissibility of γ in **R**. Dunn's results about sequent calculuses include the first sequent calculus formulation of $\mathbf{R}^t_+$, the introduction of *structurally free logics* (with Meyer), and using sequent calculuses to prove *decidability results* for $\mathbf{T}^t_\rightarrow$, and a group of logics near **LR** (with Bimbó).

Dunn proved results in logic that had expeditious impact on philosophical matters. In the late 1960s, Dunn and Belnap showed that the *substitutional interpretation of quantifiers* in 2-valued logic yields incompleteness in a strong sense. Dunn showed that quantified orthomodular logic with extensionality collapses into 2-valued logic; then, he gave a set of criteria for quantified non-classical logics that guaranteed a similar collapse when (a weak form of) extensionality was added. In a series of papers, Dunn developed an approach to *relevant predication* and applied this theory to the philosophical analysis of various kinds of properties.

RM was the first intensional logic (after normal modal and intuitionistic logic) to be equipped with a *relational semantics*. Dunn's *3-valued* relational semantics for **RM** did not easily generalize to other logics such as **R** or **T**; however, it allowed Dunn to develop *3-valued models of arithmetic* and of type theory. Soon after the introduction of the Meyer–Routley semantics for relevance logic, Dunn started to expand the scope of the applicability of relational semantics. This culminated in the formulation of *gaggle theory* (i.e., generalized *G*alois logics), which blossomed into a stream of papers by Dunn and by other logicians. Dunn also concerned himself with uncovering the circumstances of the emergence of the three-termed relational semantics, and he recorded this piece of history in several papers (with Bimbó, and with N. Ferenz).

Orthomodular logic became relevant again with the impending inception of the era of *quantum computers*. Dunn showed (with Hagge, Moss and Wang) that the number of *qubits* impacts the underlying logic of a quantum computer unlike the number of bits does that of a digital computer. Dunn, who was the founding dean of the School of Informatics at Indiana University, approached questions about *information* (in general), and information in practical applications using logic and philosophical ideas.

The *goal* of this collection isn't to paint a comprehensive picture of Dunn's research in logic. Rather, the volume showcases papers that were either written for this volume or they were never published before. The authors, who were invited to contribute, produced papers on a variety of topics; some papers are very directly related to Dunn's research interests whereas some others pay homage to Dunn via their approach to a question in logic. To round out the collection, two papers by Dunn, which have not appeared in print before, are also included in this volume — together with a list of his main publications. It is my hope that the variety of topics and styles will make this volume a valuable read for logicians and scholars with some interest in logic alike.

The *title of the book*, of course, was chosen to reflect its content. Several papers deal with *relevance logics* (**fde**, **R**, **EQ** and others), but not all of them do. Thus, the second part of the title — "other tools for reasoning" — aims to indicate that the topics in the collection go beyond relevance logics, and at the same time, it hearkens back to Dunn's stance on logic. Some logicians, whether classical or non-classical ones, vouch for "One true logic" — not always for the same though. Dunn, in contradistinction to this view, thought that different logics might be useful for disparate purposes. That is, logics are similar to *tools*, but their function is specific to *reasoning*.

The *sketches on the cover* depict Indiana University landmarks in Bloomington: the Sample Gates on Kirkwood Avenue and the Bicentennial Carillon in the Arboretum. Michael Dunn was a professor (later, a professor emeritus) at IU for fifty-two years.

Acknowledgments. First of all, I would like to express my gratitude to Jane Spurr, Managing Director of College Publications, for allowing me to pursue the project of editing this volume in the *Tributes* series.

The collection would not have been possible without the contributions of the authors, whom I would like to thank both for their papers and for their collaboration during the process of editing. The papers in this volume had been refereed, and I am thankful to those who acted as a referee for a paper. The referees' efforts and comments — undoubtedly — helped the authors to refine their papers.

The book has been typeset using LaTeX with a heavy reliance on the fonts and packages that sprouted from $\mathcal{A}\mathcal{M}\mathcal{S}$-TeX. The latter were developed under the auspices of the American Mathematical Society; TeX was originally designed by D. Knuth. As users, we are beholden to the architects of LaTeX and its extensions, without which it would have taken several years to complete this book — if for no other reason, merely because of the notational intricacies that are inherent in logical writings.

I am grateful for partial funding for some of my work during the editing of this volume from the *Insight Grant* #435–2019–0331 awarded by the *Social Sciences and Humanities Research Council* of Canada.

Edmonton, May 14th, 2022 KB

DISTRIBUTED RELATION ALGEBRA

Gerard Allwein and William L. Harrison

ABSTRACT. We define some extra relation operators in Relation Algebras and examine the relationship between Relation Algebras and Relevance Logic. We then extract binary relations from Kripke models of Relation Algebras and extend those to Distributed Relation Algebras. The term *distributed* refers to using a multigraph of local algebras connected by *distributed operators*. In the context of Kripke frames for Relation Algebras, the distribution refers to a multigraph of Stone spaces where each node is a *local* Kripke frame. There are then *distributed Kripke relations* connecting the local frames. Lastly, we show the relationship between the relation operators and Kan extensions and lifts from category theory.

Keywords. Distributed logic, Gaggle theory, Kan extensions and lifts, Relation algebra, Relevance logic

1. INTRODUCTION

This paper arose from considering applications of Distributed Logic in Allwein and Harrison [3] to Field Programmable Gate Arrays (FPGAs). In an FPGA application, there are several components each with its own notion of internal state. The map between the logic and FPGA application is intended to be direct. Clearly, a logic aiming to reflect this must have some notion of component as well. An FPGA device can be thought of as a sea of small circuits that are corralled by an FPGA application to form components. Every application will then have a different collection of components. Therefore, a logic reflecting the component structure must make that structure parametric to the logic. In [3], we parameterized Distributed Logic with a graph. Technically, we used *multigraphs* of graph theory with own identities: there can be multiple arcs between two nodes and multiple arcs between two nodes are different arcs; loops are permitted. The localities represented components and the arcs represented binary relations. This structure formed a distributed Kripke frame where each locality was a collection of points or states and the arcs were distributed Kripke relations, i.e., binary relations between point sets. The logic then introduced distributed modal connectives interpreted by distributed Kripke models.

We felt that we could use a different representation of FPGA applications by using a logic of distributed relations directly rather than by interpreting distributed modal connectives. One option was to formalize relation algebra (RA) as a logic, which we accomplished in Allwein et al. [4]. In this paper, we ignore the logic and work directly

---

2020 *Mathematics Subject Classification.* Primary: 03B47, Secondary: 03G15, 18N10.

Bimbó, Katalin, (ed.), *Relevance Logics and other Tools for Reasoning. Essays in Honor of J. Michael Dunn*, (Tributes, vol. 46), College Publications, London, UK, 2022, pp. 1–30.

with RA. However, the semantics of [4] does not yield a semantics as a collection of binary relations. This paper does provide such a semantics. We start from Chin and Tarski [8] and Ng [13]. We have been heavily influenced by Pratt's Action Logic [15] and the paper on the origins of the calculus of binary relations Pratt [16]. From Allwein et al. [1], we knew how to extract a calculus of relations from a three-place Kripke relation similar to that used in the semantics of Relevance Logic. In Pratt [14], there is a similar construction, but it leaves out the details. We adopt the distributed RA of [4], but then supply a new semantics using binary relations.

RA should have some connection with Relevance Logic (Anderson and Belnap [5]; Routley et al. [17]). The main composition connective is similar to Relevance Logic's fusion. The defined operators used some ideas of [15] and the fact that they can be defined in Relevance Logic from fusion and De Morgan negation. The operators are definable using the monoid of RA and De Morgan negation; the latter is manufactured from RA's converse and classical negation. Hence, there is a relationship to Classical Relevance Logic [17]. Were it not for Relevance Logic's insistence on a commutative fusion operator (a commutative monoid operator in the algebras), then the match would have been perfect. We detail the relationship here where converse, as a period-two operator on the Kripke frames, replaces the Routley–Meyer star operator and we rearrange the semantics so as not to import the commutativity; Relevance Logic's semantics builds in the commutativity.

The main issue with RAs is that in general they are not representable. In logic, this means that their completeness cannot be proved. However, from [1; 14] we can get representations for simple RAs called simple Boolean monoids. We generate these using the device used in [1], called Tabularity. This is the same notion as Tabularity in Freyd and Scedrov [11], whence the name, although we apply it to the three-place relations of Kripke frames. The result yields relations as interpretations of the elements of an RA.

We first present RA as in [8; 13] (without the induction axiom). After showing a few properties, we define some extra algebraic operators that are used in the sequel. Next, the relationship between RA and Relevance Logic is examined. We then show how to form Kripke models of the algebras in the sense of Kripke models for logics, where for us the algebras replace the logics. Using [1], we show how to extract binary relations from the Kripke models and show how to derive the algebraic operators, now as operators on actual relations. We briefly explain the mechanisms necessary to distribute RA and how the extraction of binary relations carries over to the distributed case. This yields typed relations; each relation has a source and target. These form a category of relations with two forms of composition of relations. We used the notion of Gaggle Theory from Dunn [9] and Allwein and Dunn [2] to produce intensional *nor* and *nand* operators and show how they integrate with algebras of relations. Finally, we show how the operators relate to Kan extensions and lifts, and recover the residuation rules for the relations and their operators as adjoints. The intensional *nor* and *nand* form constructions that resemble Kan extensions and lifts, however the details are a bit different and do not fit the Kan notions precisely.

The two forms of composition could have been formalized directly using the devices of enriched category theory; we do not do so here in the interest of brevity. However, a truly abstract characterization of the two forms, as well as the intensional *nor* and *nand* operations, would use enriched category theory. We expect this will have utility when discovering and proving properties about FPGA applications. Another interesting feature of the categorical notion of extensions and lifts is that the operators need not be defined over the entire distributed relation category. This notion comes in handy with respect to FPGA applications where sometimes the relations between components may not be discoverable, say, when what is known as *foreign intellectual property* (foreign IP) (components with unknown internal structure) is used. Foreign IP is used in virtually all FPGA application designs.

2. Relation Algebra (RA)

We assume the usual Boolean algebra substrate of meet, join, and Boolean negation, i.e., $\wedge, \vee, \neg$ where the Boolean $\vee$ symbol will be used in place of RA's typical $+$ symbol (which we reserve for an *intensional or* as a defined operation), and $\wedge$ is a defined operation. RA has a composition operator, $\circ$. To this we add the derived operators $\leftarrow$ and $\rightarrow$; these are the left and right residuals of relational composition, respectively, and are analogous to Relevance Logic's entailment connectives. We also add $+$, $\twoheadrightarrow$, and $\twoheadleftarrow$ for another set of residuated connectives. We use the terms *tensor* and *cotensor* to refer to $\circ$ and $+$ in the algebras. We also use these same terms in the algebras of sets that we extract from the Kripke Frames. Finally we add the self-residuated operators intensional *nor* $\downarrow$ and intensional *nand* $\uparrow$.

2.1. Axioms.

Definition 1. The RA axioms are from [8; 13].

M1. $(B, \vee, \neg, \bot, \top)$ is a Boolean algebra M2. $a^{\smile\smile} = a$ for any $a \in A$
M3. $(a \circ b) \circ c = a \circ (b \circ c)$ M4. $(a \vee b) \circ c = (a \circ c) \vee (b \circ c)$
M5. $a \circ 1 = a$ for any $a \in A$ M6. $(a \vee b)^{\smile} = a^{\smile} \vee b^{\smile}$ for any $a, b \in A$
M7. $(a \circ b)^{\smile} = b^{\smile} \circ a^{\smile}$ M8. $(a^{\smile} \circ \neg(a \circ b)) \vee \neg b = \neg b, \forall a, b \in A$

where $\bot$ and $\top$ are the bottom and top of the Boolean algebra.

There are two equivalent inequational forms for M8:

M8. $\neg(b \circ a) \circ a^{\smile} \leq \neg b$ M8'. $a^{\smile} \circ \neg(a \circ b) \leq \neg b$

It is straightforward to check that $\circ$ is a normal operator.

2.2. Defined Operators.
It is well known that $\neg(a^{\smile}) = (\neg a)^{\smile}$ and so we can elide the parentheses in the following:

Definition 2. The De Morgan operator $\sim$ is defined with $\sim a = \neg a^{\smile}$.

It is straightforward to check that $\sim$ is a De Morgan negation on the Boolean lattice of an RA. There are coentailment, intensional *nor*, and intensional *nand* operators. (We use some abbreviations in the following table such as "l." for "left," "r." for "right," "ent." for "entailment" and "int." for "intensional" in order to fit the names of the operators on each line.)

Definition 3.

Operator	Definition	Name	Operator	Definition	Name
1	1	identity	0	$\neg 1$	coidentity
$a \circ b$	$a \circ b$	tensor	$a + b$	$\sim(\sim b \circ \sim a)$	cotensor
$b \leftarrow a$	$\sim(a \circ \sim b)$	l. ent.	$b \leftarrowtail a$	$b \circ \sim a$	l. coent.
$a \rightarrow b$	$\sim(\sim b \circ a)$	r. ent.	$a \rightarrowtail b$	$\sim a \circ b$	r. coent.
$a \downarrow b$	$\sim a \circ \sim b$	int. nor	$a \uparrow b$	$\sim(a \circ b)$	int. nand

TABLE 1. Names of Intensional Operators

The following table shows the distribution properties according to their Gaggle Operator Type. The proofs are easy.

Operator Type	Distribution Property
$\circ : (\vee, \vee) \rightarrow \vee$	$a \circ (b \vee c) = (a \circ b) \vee (a \circ c)$
	$(a \vee b) \circ c = (a \circ c) \vee (b \circ c)$
$\leftarrow : (\wedge, \vee) \rightarrow \wedge$	$a \leftarrow (b \vee c) = (a \leftarrow b) \wedge (a \leftarrow c)$
	$(a \wedge b) \leftarrow c = (a \leftarrow c) \wedge (b \leftarrow c)$
$\rightarrow : (\vee, \wedge) \rightarrow \wedge$	$a \rightarrow (b \wedge c) = (a \rightarrow b) \wedge (a \rightarrow c)$
	$(a \vee b) \rightarrow c = (a \rightarrow c) \wedge (b \rightarrow c)$
$+ : (\wedge, \wedge) \rightarrow \wedge$	$a + (b \wedge c) = (a + b) \wedge (a + c)$
	$(a \wedge b) + c = (a + c) \wedge (b + c)$
$\leftarrowtail : (\vee, \wedge) \rightarrow \vee$	$a \leftarrowtail (b \wedge c) = (a \leftarrowtail b) \vee (a \leftarrowtail c)$
	$(a \vee b) \leftarrowtail c = (a \leftarrowtail c) \vee (b \leftarrowtail c)$
$\rightarrowtail : (\wedge, \vee) \rightarrow \vee$	$a \rightarrowtail (b \vee c) = (a \rightarrowtail b) \vee (a \rightarrowtail c)$
	$(a \wedge b) \rightarrowtail c = (a \rightarrowtail c) \vee (b \rightarrowtail c)$
$\downarrow : (\wedge, \wedge) \rightarrow \vee$	$a \downarrow (b \wedge c) = (a \downarrow b) \vee (a \downarrow c)$
	$(a \wedge b) \downarrow c = (a \downarrow b) \vee (b \downarrow c)$
$\uparrow : (\vee, \vee) \rightarrow \wedge$	$a \uparrow (b \vee c) = (a \uparrow b) \wedge (a \uparrow c)$
	$(a \vee b) \uparrow c = (a \uparrow b) \wedge (b \uparrow c)$

TABLE 2. Distribution Properties

The left and right arrow operators are always antitone in their source and monotone in their target. Antitone equates to flipping the lattice connective to its dual and monotone equates to preserving it. The residuation properties also follow directly from the definitions:

Residuation Property	Residuation Property
$b \leq a \rightarrow c$ iff $a \circ b \leq c$ iff $a \leq c \leftarrow b$	$a \rightarrowtail c \leq b$ iff $c \leq a + b$ iff $c \leftarrowtail b \leq a$
$a \downarrow b \leq c$ iff $c \downarrow a \leq b$	$a \leq b \uparrow c$ iff $b \leq c \uparrow a$

TABLE 3. Residuation Properties

Unwinding the definitions yields
$$a \rightarrow b = \neg(a\breve{\,} \circ \neg b) \qquad b \leftarrow a = \neg(\neg b \circ a\breve{\,}).$$
The following are all equivalent to axiom M8:

M'8. $a \circ (a \rightarrow b) \leq b$ M'8'. $(b \leftarrow a) \circ a \leq b$

M''8. $(a \rightarrow \sim b)\breve{\,} \leq b\breve{\,} \rightarrow \sim a\breve{\,}$ M''8'. $(\sim b \leftarrow a)\breve{\,} \leq \sim a\breve{\,} \leftarrow b\breve{\,}$

M'''8. $b \rightarrow \sim a \leq \sim b \leftarrow a$ M'''8'. $\sim b \leftarrow a \leq b \rightarrow \sim a$

Theorem 4. *The $+$ operator forms a monoid with 0 as its unit.*

The following theorem mirrors the rules for $\sim$ in Gentzen systems for Relevance Logic:

Theorem 5. $a \leq c + \sim b$ iff $a \circ b \leq c$ iff $b \leq \sim a + c$.

The proof follows directly from residuation and the definition of $+$. Also, $\circ$ distributes over $+$ just as in Relevance Logic.

Theorem 6. $a \circ (b + c) \leq (a \circ b) + c$.

Remark 7. The operator $+$ could be axiomatized by itself using axioms similar to those involving $\circ$. If so, then the above theorem could be taken as an axiom. We do not do so here but point it out because it features in distributed RAs when we consider two different forms of composition involving tensor and cotensor.

There are many other properties that mirror $\circ$ such as
$$(a \rightarrowtail b)\breve{\,} = b\breve{\,} \leftarrowtail a\breve{\,} \qquad (a+b)\breve{\,} = b\breve{\,} + a\breve{\,}.$$

2.3. Relation Algebra Frames (RAFs).
The Kripke frames for RA are exactly what one would expect given frames for classical relevance logics except that the Routley–Meyer ($-^*$) operator has been replaced by the weaker converse ($-\breve{\,}$) operator. That is, they are collections of points which are maximal filters when the frames arise from an RA. There is a single three-place relation which is used in evaluating the monoid operation. Also, there is a set of "zero worlds" (using the terminology of Relevance Logic) used in evaluating the unit of the monoid.

This paper will assume that an RA frame will be denoted as $\mathcal{X} = (X, \mathcal{X}, \mathbb{X})$, where $\mathcal{X} \subseteq X \times X \times X$ is the three-place relation on worlds and $\mathbb{X} \subseteq X$ is the collection of zero worlds. The symbol $\mathcal{X}$ is overloaded but since the three-place relation is so central to the frame, the reader is asked to overlook this and accept the simplicity it gives to the notation. Context will distinguish the two uses of "$\mathcal{X}$." For X and $\mathbb{X}$, this same letter in the two different fonts means different things, but both are related to the same structure.

The following definition is based on [1] where Boolean Monoid Frames are used. Here, those frames are augmented with a "converse" operator on points.

Definition 8. *A relation algebra frame*, $\mathcal{X} = (X, \mathcal{X}, \mathbb{X})$, *is a structure, where X is a set of points, $\mathcal{X} \subseteq X \times X \times X$, and $\mathbb{X} \subseteq X$ and $\mathbb{X} \neq \emptyset$. The following axioms, also called frame conditions, apply:*

FA1. $\mathcal{X}^2 uvyz$ iff $\mathcal{X}^2 u(vy)z$;[1]

---

[1] $\mathcal{X}^2$ is defined as $\mathcal{X}^2 uvyz$ iff $\exists x (\mathcal{X} uvx$ and $\mathcal{X} xyz)$ and $\mathcal{X}^2 u(vy)z$ iff $\exists w (\mathcal{X} uwz$ and $\mathcal{X} vyw)$.

FA2. there is some $z \in \mathbb{X}$ such that $\mathcal{X}xzx$ and $\mathcal{X}zxx$;
FA3. for all $y \in \mathbb{X}$, ($\mathcal{X}xyz$ or $\mathcal{X}yxz$) implies $x = z$;
FA4. $\mathcal{X}xyz$ implies $\mathcal{X}zy˘x$;
FA5. $\mathcal{X}xyz$ implies $\mathcal{X}x˘zy$.

These axioms can be augmented with the Tabularity Axiom

FA6. $\mathcal{X}xyz$ and $\mathcal{X}xy'z$ implies $y = y'$.

There are, of course, implicit universal quantifications given to the free variables in the frame conditions.

Lemma 9. $\mathcal{X}xyz$ implies $\mathcal{X}y˘x˘z˘$.

The proof follows directly from the Frame Conditions FA4 and FA5. Canonically, we let $x˘ = \{a˘ : a \in x\}$, where x is a maximal filter. Given that $(a \vee b)˘ = a˘ \vee b˘$ and that $(\neg a)˘ = \neg(a˘)$, then $x˘$ is also a maximal filter. The consequence is that $a \in x˘$ iff $a˘ \in x$. Just as in Relevance Logic,

$\mathcal{X}xyz$ iff $\forall a, b \, (a \in x$ and $b \in y$ implies $a \circ b \in z)$ and $x \in \mathbb{X}$ iff $1 \in x$.

Also as in Relevance Logic, there are the following equivalent definitions of $\mathcal{X}xyz$:

$\mathcal{X}xyz$ iff $\forall a, b \, (b \leftarrow a \in x$ and $a \in y$ implies $b \in z)$
$\mathcal{X}xyz$ iff $\forall a, b \, (a \in x$ and $a \rightarrow b \in y$ implies $b \in z)$,

where the commutativity of $\circ$ is not assumed. Just as in Relevance Logic, these definitions are equivalent because of residuation.

Theorem 10. *The modeling conditions FA1, FA2, FA3, FA4, and FA5 hold canonically in any Stone space arising as a dual space to an RA.*

Proof. That the frame conditions FA1, FA2, and FA3 hold is known from classical Relevance Logic's algebras (De Morgan monoids) that share the properties of the operator 1 and of $\circ$. Axioms FA4 and FA5 hold in the presence of the converse axioms. We show FA4:

Let $\mathcal{X}xyz$ and assume $a \in z$ and $a \rightarrow \sim b \in y˘$, then $(a \rightarrow \sim b)˘ \in y$ and so $b˘ \rightarrow \sim a˘ \in y$. Towards a reductio ad absurdum, let $\sim b \notin x$, then $b \in x˘$ and so $b˘ \in x$. From $\mathcal{X}xyz$, then $\sim a˘ \in z$. Hence $a \notin z˘˘$ and $a \notin z$, which is a contradiction. Thus, $\sim b \in x$ and $\mathcal{X}zy˘x$. ◁

Definition 11. Let A and B be sets in the power set of points of a RAF. The basic definitions are, where $\circ$ is composition in the set algebra and the unit of the monoid is t: $t = \mathbb{X}$, and

$z \in A \circ B$ iff $\exists x, y \, (\mathcal{X}xyz$ and $x \in A$ and $y \in B)$, $x \in \neg A$ iff $x \notin A$, $x \in A˘$ iff $x˘ \in A$.

Using these definitions, it is straightforward to show

Lemma 12. $(\neg A)˘ = \neg(A˘)$.

Lemma 13. *The RA axioms are valid in RAFs.*

Proof. We will only show one example using the set algebra derived from the RAFs and noting that RA is equationally defined and hence has free algebras. The axioms M1, M3 through M4 and M5 are shown in [1]. Axiom M2 is straightforward. Axiom M8 holds because defining the operator $\rightarrow$ (or $\leftarrow$) using the same relation as that for $\circ$ guarantees that the axiom holds.

As an example of the validity of an axiom, let A and B be sets of points of a RAF and $\circ$ is the set-theoretic correlate to $\circ$:

$$z \in (A \circ B)\breve{} \text{ iff } z\breve{} \in A \circ B$$
$$\text{iff } \exists x, y (\mathcal{X} xyz\breve{} \text{ and } x \in A \text{ and } y \in B)$$
$$\text{iff } \exists x, y (\mathcal{X} y\breve{} x\breve{} z \text{ and } x \in A \text{ and } y \in B)$$
$$\text{iff } \exists x, y (\mathcal{X} y\breve{} x\breve{} z \text{ and } x\breve{}\breve{} \in A \text{ and } y\breve{}\breve{} \in B)$$
$$\text{iff } \exists u, v (\mathcal{X} vuz \text{ and } u\breve{} \in A \text{ and } v\breve{} \in B)$$
$$\text{iff } \exists u, v (\mathcal{X} vuz \text{ and } u \in A\breve{} \text{ and } v \in B\breve{})$$
$$\text{iff } \exists u, v (\mathcal{X} vuz \text{ and } v \in B\breve{} \text{ and } u \in A\breve{})$$
$$\text{iff } z \in B\breve{} \circ A\breve{} \quad \triangleleft$$

A table of intensional operators can be generated by De Morgan negation and tensor. The only condition we take for granted is that of $z \in A \circ B$.

Theorem 14. *The operators of the table below are all definable in terms of the composition operator $\circ$.*

$z \in A \circ B$	$z \in A \circ B$	$\exists x, y (\mathcal{X} xyz \text{ and } x \in A \text{ and } y \in B)$
$x \in \sim(A \circ \sim B)$	$x \in B \leftarrow A$	$\forall y, z (\mathcal{X} xyz \text{ and } y \in A \text{ implies } z \in B)$
$y \in \sim(\sim B \circ A)$	$y \in A \rightarrow B$	$\forall x, z (\mathcal{X} xyz \text{ and } x \in A \text{ implies } z \in B)$
$z \in \sim(\sim B \circ \sim A)$	$z \in A + B$	$\forall x, y (\mathcal{X} xyz \text{ implies } x \in A \text{ or } y \in B)$
$x \in B \circ \sim A$	$x \in B \twoheadleftarrow A$	$\exists y, z (\mathcal{X} xyz \text{ and } y \notin A \text{ and } z \in B)$
$y \in \sim A \circ B$	$y \in A \twoheadrightarrow B$	$\exists x, z (\mathcal{X} xyz \text{ and } x \notin A \text{ and } z \in B)$
$z \in \sim A \circ \sim B$	$z \in A \downarrow B$	$\exists x, y (\mathcal{X} xyz \text{ and } x\breve{} \notin A \text{ and } y\breve{} \notin B)$
$z \in \sim(A \circ B)$	$z \in A \uparrow B$	$\forall x, y (\mathcal{X} xyz \text{ implies } x\breve{} \notin B \text{ or } y\breve{} \notin A)$

TABLE 4. Intensional Operators of Sets

The first-order logic statements follow directly from the definitions. The definitions allow us to validate the following residuation conditions:

Theorem 15.

$B \subseteq A \rightarrow C$ iff $A \circ B \subseteq C$ iff $A \subseteq C \leftarrow B$, $A \downarrow B \subseteq C$ iff $B \downarrow C \subseteq A$,

$C \twoheadleftarrow B \subseteq A$ iff $C \subseteq A + B$ iff $A \twoheadrightarrow C \subseteq B$, $A \subseteq B \uparrow C$ iff $B \subseteq C \uparrow A$.

Note that if B is the active formula in $A \circ B$ or $A + B$, then it is on the right and continues to be on the right in $C \leftarrow B$ and $C \twoheadleftarrow B$ (respectively), which is handy for recalling how the residuation rules work. That determines the direction of the arrows. Note also that the left and right shifts using $\downarrow$ or $\uparrow$ are equivalent, i.e., two lefts yield a right, two rights yield a left.

Proof. As an example, we show $C \mathrel{\leftarrow\mkern-6mu-} B \subseteq A$ implies $C \subseteq A + B$. Let $C \mathrel{\leftarrow\mkern-6mu-} B \subseteq A$ and assume (in order) $z \in C$, $\mathcal{X}xyz$, and that $y \notin B$. This gives us $\exists u, v(\mathcal{X}xuv$ and $u \notin B$ and $v \in C$), and by definition, $x \in C \mathrel{\leftarrow\mkern-6mu-} B$. Since we assumed $y \notin B$, we have $y \notin B$ implies $x \in A$, which is $x \in A$ or $y \in B$. Since we assumed $\mathcal{X}xyz$, we have $\mathcal{X}xyz$ implies ($x \in A$ or $y \in B$). By definition, $z \in A + B$. Thus, $C \subseteq A + B$. ◁

The following theorem shows cotensors as monoids in the set algebras. Cotensor will become another form of composition in the category theory.

Theorem 16. *Let* $f = X - \mathbb{X}$, *then the following formulas are sound:*

$$A + (B + C) = (A + B) + C \qquad f + A = A = A + f.$$

The formula

$$A \circ (B + C) \subseteq (A \circ B) + C$$

is validated in the set algebras. The proof uses the Frame Condition FA1 similarly to its use in showing associativity of ∘ in Relevance Logic. However, it also requires converse given the definition of $+$ on the set algebras. To remove the use of converse requires a new frame condition:

$$\exists z(\mathcal{X}xyz \text{ and } \mathcal{X}uvz) \text{ implies } \exists w(\mathcal{X}uwx \text{ and } \mathcal{X}wyv).$$

To completely separate ∘ and $+$ requires the use of two Kripke relations rather than one and is an echo of the approach in Bimbó and Dunn [7], although their work covers many more cases of operators. Thus:

Theorem 17.

$$A \circ (B + C) \subseteq (A \circ B) + C$$

holds in set algebras when the RAFs are augmented with a new three-place relation subject to a new frame condition

$$\mathcal{X}xyz \text{ and } \mathcal{Y}uvz \text{ implies } \exists w(\mathcal{X}uwx \text{ and } \mathcal{Y}wyv).$$

2.4. Relationship with Relevance Logic.

In Relevance Logic (Routley et al. [17]), there is an axiomatization of a star operator, denoted $(-^*)$. The star operator is used on worlds to interpret De Morgan negation $\sim$:

$$x \vDash \sim A \quad \text{iff} \quad x^* \nvDash A.$$

In the context of Relevance Logic, worlds canonically are prime filters. It turns out that $\sim x$, the De Morgan negation applied pointwise to the elements of the prime filter x, is a prime ideal because $\sim$ is order-reversing. The complement of a prime ideal is a prime filter. Hence canonically,

$$x^* = \mathcal{A} - \sim x,$$

where $\mathcal{A}$ is the carrier set of the De Morgan monoid we started with and where $\sim$ is applied pointwise to every element of the filter x. Relevance Logic (without classical negation) has the following in its modeling axioms:

$$x^{**} = x \qquad \mathcal{X}xyz \text{ implies } \mathcal{X}xz^*y^*.$$

This works in the presence of commutativity in the first two positions of $\mathcal{X}$ to yield the set of possibilities for Relevance Logic Frames as

$$\mathcal{X}xyz \quad \mathcal{X}yxz \quad \mathcal{X}xz^*y^* \quad \mathcal{X}z^*xy \quad \mathcal{X}yz^*x^* \quad \mathcal{X}z^*yx^*.$$

The set of possibilities using the Relational Algebra Frames is

$$\mathcal{X}xyz \quad \mathcal{X}zy\breve{}x \quad \mathcal{X}x\breve{}zy \quad \mathcal{X}z\breve{}xy\breve{} \quad \mathcal{X}yz\breve{}x \quad \mathcal{X}y\breve{}x\breve{}z\breve{}.$$

Theorem 18. *The relational formulas $\mathcal{X}zy^*x$ and $\mathcal{X}x^*zy$ which correspond to the RAF conditions $\mathcal{X}zy\breve{}x$ and $\mathcal{X}x\breve{}zy$ are not obtainable from the Relevance Logic frame conditions.*

The $(-^*)$ operator can be added to the logic with the rule

$$A \twoheadrightarrow B \text{ implies } A^* \twoheadrightarrow B^*$$

and the evaluation condition

$$x \vDash A^* \text{ iff } x^* \vDash A.$$

The rule does not add additional properties to negation not involving $(-^*)$. In Classical Relevance Logic, any one of $\sim$, $\neg$, and $(-^*)$ can be defined from the other two.

Classical Relevance Logic in general conflates $\mathcal{X}xyz$ and $\mathcal{X}yxz$, because its algebras have the axiom $a \circ b = b \circ a$. This is sometimes built into the other axioms for the frames and validates the following equality in Classical Relevance Algebra:

$$a \twoheadrightarrow b = b \twoheadleftarrow a,$$

which has the knock on effect of conflating $a \twoheadrightarrow b$ with $b \twoheadleftarrow a$. Let us separate these. In Classical Relevance Algebras one has:

CR1. $b \twoheadrightarrow \sim a \leq \sim b \twoheadleftarrow a$ \qquad CR2. $\sim b \twoheadleftarrow a \leq b \twoheadrightarrow \sim a$

and the following rules from Classical Relevance Logic [17]:

CR3. $A \twoheadrightarrow B$ implies $A^* \twoheadrightarrow B^*$ \qquad CR4. $B \twoheadleftarrow A$ implies $B^* \twoheadleftarrow A^*$

Since conditions M'''8 and M'''8' are the same as conditions CR1 and CR2, both RAs and Classical Relevance Algebras agree on these forms of contraposition. However, Classical Relevance Logic also conflates $\twoheadrightarrow$ and $\twoheadleftarrow$, so in effect Classical Relevance Algebras contain

$$a \twoheadrightarrow \sim b \leq b \twoheadrightarrow \sim a.$$

This, of course, leads to $a \circ b = b \circ a$, which is not available in RA. The inequality is not drivable in RAs either.

3. RELATION ALGEBRA OF SETS

The collection of operators fill out Gaggle Theory in [9; 2] for the relation operators. Our extraction of an algebra of relations is echoed somewhat in Dunn [10]; however, there entire three-tuples of a three-place relation were used whereas we only use two-tuples. We extract binary relations from a RAF. The extraction is in preparation for the distributed RA of the next section. The distributed RAF will then be dropped out and the relations treated as 1-arrows in a 2-category.

The paper [1] shows how to extract a Boolean monoid of relations. RAs are at least Boolean monoids. We were able to extract two entailment operators residuated with relational composition of binary relations from a Boolean monoid frame. We employ a similar strategy here for the collection of RA operators. The extracted relations are either sets or multisets of pairs of points. Without Tabularity, they are generally multisets of pairs of points; Tabularity forces out duplicate pairs.

3.1. The Extracted Relations.
The representation of Boolean monoids in [1] relied on the dual Stone space and a three-place relation induced by the monoid operator. Due to the two entailment operators being residuated with the monoid operator, the same three-place relation can be used to represent the entailment operators. Elements of an algebra of sets extracted from a frame use Roman upper case letters, i.e., $A, B, \ldots$, whereas the relations we extract use Roman upper case letters surrounded by brackets, i.e., $\langle A \rangle, \langle B \rangle, \ldots$.

Definition 19. Assume a Boolean monoid frame $\mathcal{X} = \langle X, \mathcal{X}, \mathbb{X} \rangle$ (just like a RA Frame without the $(-^\smile)$ frame conditions). Let A be an element of the set algebra, then

$$\langle A \rangle \stackrel{def}{=} \{\langle x,z \rangle : \exists y (\mathcal{X}xyz \text{ and } y \in A)\} \qquad \langle X \rangle \stackrel{def}{=} \{\langle x,z \rangle : \exists y (\mathcal{X}xyz)\}.$$

Notice that the algebra of relations does not have $X \times X$ as the top element for an ambient set X.

Definition 20. Elements of a Boolean monoid $(L, \wedge, \vee, \neg, \circ)$ are represented in the following way with β the representation function:

$$\beta(a) \stackrel{def}{=} \{x : x \text{ is a maximal filter and } a \in x\}.$$

From [1, Theorem 3.2.1], we have the following theorem:

Theorem 21. Let $\mathcal{X} = (X, \mathcal{X}, \mathbb{X})$ be a Boolean monoid frame with the additional Tabularity Axiom

$$\mathcal{X}xyz \text{ and } \mathcal{X}xy'z \text{ implies } y = y',$$

then $\langle \ldots \rangle$ is a homomorphism from the Boolean monoid of sets to a Boolean monoid of relations. Specifically,

$$\langle A \cup B \rangle = \langle A \rangle \cup \langle B \rangle \quad \langle X - A \rangle = \langle X \rangle - \langle A \rangle \quad \langle A \circ B \rangle = \langle A \rangle \odot \langle B \rangle \quad \langle \mathbb{X} \rangle = \{\langle x,x \rangle\},$$

where $\odot$ is relational composition.

The Tabularity Axiom essentially says that $\langle X \rangle$ is the largest relation and that every $\langle A \rangle$ can be "tabulated" with a monic pair of functions (cf. [11]).

3.2. Extracted Relation Operators.
The extraction is via the three-place accessibility relation of a frame. We generally assume a RAF, $\mathcal{X} = (X, \mathcal{X}, \mathbb{X})$ with Tabularity defined as in Section 2.3. A relation then uses the following prescription for A, a member of the set algebra, $\mathcal{P}(X)$, where $\mathcal{P}(-)$ is the powerset operator.

Definition 22. $\langle x,z \rangle \in \langle A \rangle$ iff $\exists y (\mathcal{X}xyz \text{ and } y \in A)$, $\langle A \rangle^\smile = \{\langle z,x \rangle : \langle x,z \rangle \in \langle A \rangle\}$, and $\neg \langle A \rangle = X - \langle A \rangle$.

In Theorem 21, we already have $\langle X - A \rangle = \langle X \rangle - \langle A \rangle$.

Theorem 23. $\neg \langle A \rangle = \langle \neg A \rangle$ and $\langle A \rangle^\smile = \langle A^\smile \rangle$.

We proved the first above; the proof of the second is straightforward.

Corollary 24. $\sim\langle A\rangle = \langle \sim A\rangle$.

In Theorem 21, we had $\langle A \circ B\rangle = \langle A\rangle \odot \langle B\rangle$. Using similar proofs and Table 4, we get the following:

Theorem 25. *The operators satisfy the following table:*

$\langle A\rangle \odot \langle B\rangle = \langle A \circ B\rangle$	$\langle A\rangle \oplus \langle B\rangle = \langle A + B\rangle$
$\langle B\rangle \leftharpoonup \langle A\rangle = \langle B \leftharpoonup A\rangle$	$\langle B\rangle \twoheadleftarrow \langle A\rangle = \langle B \twoheadleftarrow A\rangle$
$\langle A\rangle \rightharpoonup \langle B\rangle = \langle A \rightharpoonup B\rangle$	$\langle A\rangle \twoheadrightarrow \langle B\rangle = \langle A \twoheadrightarrow B\rangle$
$\langle A\rangle \downarrow \langle B\rangle = \langle A \downarrow B\rangle$	$\langle A\rangle \uparrow \langle B\rangle = \langle A \uparrow B\rangle$

TABLE 5. Relational Operators Representation

We obtain the following relational forms for the intensional operators.

Theorem 26. *The intensional operators follow straightforwardly from their definition via $\circ$ and $\sim$:*

$\langle x,z\rangle \in \langle A\rangle$	iff	$\exists y\, (\mathcal{X}xyz \text{ and } y \in A)$
$\langle x,z\rangle \in \langle A\rangle \odot \langle B\rangle$	iff	$\exists y\, (\langle x,y\rangle \in \langle A\rangle \text{ and } \langle y,z\rangle \in \langle B\rangle)$
$\langle y,z\rangle \in \langle A\rangle \rightharpoonup \langle B\rangle$	iff	$\forall x\, (\langle x,y\rangle \in \langle A\rangle \text{ implies } \langle x,z\rangle \in \langle B\rangle)$
$\langle y,z\rangle \in \langle B\rangle \leftharpoonup \langle A\rangle$	iff	$\forall x\, (\langle z,x\rangle \in \langle A\rangle \text{ implies } \langle y,x\rangle \in \langle B\rangle)$
$\langle x,z\rangle \in \langle A\rangle \oplus \langle B\rangle$	iff	$\forall y\, (\langle x,y\rangle \in \langle A\rangle \text{ or } \langle y,z\rangle \in \langle B\rangle)$
$\langle x,z\rangle \in \langle A\rangle \twoheadrightarrow \langle B\rangle$	iff	$\exists y\, (\langle y,x\rangle \notin \langle A\rangle \text{ and } \langle y,z\rangle \in \langle B\rangle)$
$\langle x,z\rangle \in \langle B\rangle \twoheadleftarrow \langle A\rangle$	iff	$\exists y\, (\langle x,y\rangle \in \langle B\rangle \text{ and } \langle z,y\rangle \notin \langle A\rangle)$
$\langle x,z\rangle \in \langle A\rangle \downarrow \langle B\rangle$	iff	$\exists y\, (\langle y,x\rangle \notin \langle A\rangle \text{ and } \langle z,y\rangle \notin \langle B\rangle)$
$\langle x,z\rangle \in \langle A\rangle \uparrow \langle B\rangle$	iff	$\forall y\, (\langle y,x\rangle \notin \langle B\rangle \text{ or } \langle z,y\rangle \notin \langle A\rangle)$

TABLE 6. Relational Operators

The proofs follow directly from the definitions.

Theorem 27. *The intensional relational operators satisfy the distribution and residuation properties of Tables 2 and 3 in this set theoretic form, e.g.,*

$$\langle A\rangle \rightharpoonup (\langle B\rangle \cap \langle C\rangle) = (\langle A\rangle \rightharpoonup \langle B\rangle) \cap (\langle A\rangle \rightharpoonup \langle C\rangle),$$
$$\langle C\rangle \subseteq \langle A\rangle \oplus \langle B\rangle \quad \text{iff} \quad \langle C\rangle \twoheadleftarrow \langle B\rangle \subseteq \langle A\rangle.$$

Proof. The proofs follow directly from the definitions. We show an example of the residuation condition above. Let $\langle C\rangle \subseteq \langle A\rangle \oplus \langle B\rangle$ and assume $\langle x,z\rangle \in \langle C\rangle \twoheadleftarrow \langle B\rangle$. From the definition of $\twoheadleftarrow$, for some y, $\langle x,y\rangle \in \langle C\rangle$ and $\langle z,y\rangle \notin \langle B\rangle$, and $\langle x,y\rangle \in \langle A\rangle \oplus \langle B\rangle$. The definition of $\oplus$ gives us $\forall u\, (\langle x,u\rangle \in \langle A\rangle \text{ or } \langle u,y\rangle \in \langle B\rangle)$. Eliminating the universal quantifier with z for u yields $\langle x,z\rangle \in \langle A\rangle$ or $\langle z,y\rangle \in \langle B\rangle$. Thus, $\langle x,z\rangle \in \langle A\rangle$.

To go in the other direction, let $\langle C\rangle \subseteq \langle A\rangle \oplus \langle B\rangle$ and assume $\langle x,z\rangle \in \langle C\rangle$. Towards a reductio ad absurdum, let $\langle x,z\rangle \notin \langle A\rangle \oplus \langle B\rangle$. From the definition of $\oplus$, we have for some y, $\langle x,y\rangle \notin \langle A\rangle$ and $\langle y,z\rangle \notin \langle B\rangle$. Therefore, $\langle x,z\rangle \in \langle C\rangle$ and $\langle y,z\rangle \notin \langle B\rangle$. By definition, $\langle x,y\rangle \in \langle C\rangle \twoheadleftarrow \langle B\rangle$ and hence $\langle x,y\rangle \in \langle A\rangle$, which is a contradiction. Therefore, $\langle x,z\rangle \in \langle A\rangle \oplus \langle B\rangle$. ◁

The following theorem is a direct consequence of the definitions.

Theorem 28. *Tensors and cotensors satisfy the following distribution law:*
$$\langle A \rangle \odot (\langle B \rangle \oplus \langle C \rangle) \subseteq (\langle A \rangle \odot \langle B \rangle) \oplus \langle C \rangle.$$

3.3. Cotensors as a Monoid of Relations.

Definition 29. $\langle t \rangle \stackrel{def}{=} \{\langle x, z \rangle : \exists u (\mathcal{X} x u z \text{ and } u \in \mathbb{X})\}$, where here t is under interpretation as a collection of "zero" worlds.

Given the axioms for frames,
$$\forall u \in \mathbb{X} (\mathcal{X} x u z \text{ implies } x = z) \quad \text{and} \quad x = z \text{ implies } \exists u \in \mathbb{X} (\mathcal{X} x u z).$$

From this and Tabularity, it is straightforward to see that
$$\langle t \rangle = \{\langle x, x \rangle\} = \mathcal{I}_X,$$

for $\mathcal{I}_X$, the identity relation on X. Hence,
$$\langle f \rangle = \sim\langle t \rangle = \neg\langle t \rangle^{\smile} = \neg\langle t \rangle = (X \times X) - \mathcal{I}_X.$$

Theorem 30. $\langle A \rangle \oplus (\langle B \rangle \oplus \langle C \rangle) = (\langle A \rangle \oplus \langle B \rangle) \oplus \langle C \rangle, \quad \langle f \rangle \oplus \langle A \rangle = \langle A \rangle = \langle A \rangle \oplus \langle f \rangle.$

4. DISTRIBUTED RELATION ALGEBRA

In [4], we presented a distributed relation logic which used a three-place relation to interpret the logic. Each local logic (i.e., logic at a locality) has a Boolean base. The interpretation used typed three-place relations much like those used for Relevance Logics. The intuitive picture is Figure 1 using the convention in Figure 2:

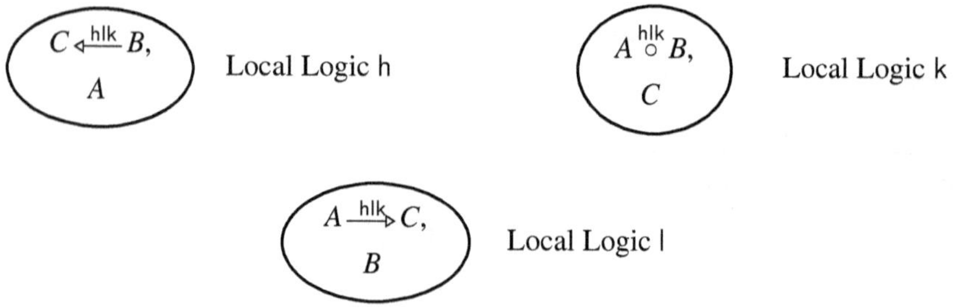

$C \xleftarrow{hlk} B,$ Local Logic h
A

$A \overset{hlk}{\circ} B,$ Local Logic k
C

$A \xrightarrow{hlk} C,$ Local Logic l
B

FIGURE 1. Distributing the Two-Place Connectives

Connective	Type h	Type l	Type k
$\overset{hlk}{\circ}$	A	B	$A \overset{hlk}{\circ} B$
$\xrightarrow{hlk}$	A	$A \xrightarrow{hlk} C$	C
$\xleftarrow{hlk}$	$C \xleftarrow{hlk} B$	B	C

FIGURE 2. Localities and their Logics

Each proposition A, B, and C is part of the local logic at a localities h, l, and k, respectively. We use the terminology of *type* h to talk about a locality h. Note the positions of the input and output types of each operator shift depending upon the operator. This matches the corresponding position of the formula containing the operator in the semantic definitions using the interpreting relations. The relation $\mathcal{R}^{hlk}$ means $\mathcal{R} \subseteq H \times L \times K$ for sets of points H, L, K at localities h, l, k, respectively:

(1) $z \models^k A \stackrel{hlk}{\circ} B$ iff $\exists x, y \, (x \models^h A$ and $y \models^l B$ and $\mathcal{R}^{hlk} xyz)$,
(2) $y \models^l A \xrightarrow{hlk} C$ iff $\forall x, z \, (x \models^h A$ and $\mathcal{R}^{hlk} xyz$ implies $z \models^k C)$,
(3) $x \models^h C \xleftarrow{hlk} B$ iff $\forall y, z \, (\mathcal{R}^{hlk} xyz$ and $y \models^l B$ implies $z \models^k C)$.

In the set algebras, $\in_k$ replaces $\models^k$, and similarly, for the rest.

4.1. Distributed Relation Algebra Axioms. The distribution structure is a hypergraph of a collection of nodes with certain three-tuples identified as *cliques* with three elements, say, hlk.[2] The diagram on the right displays a clique as a multigraph. We need a way of abstracting over cliques in a notationally convenient way. We use a string such as hlk to refer to an arbitrary clique. For any one clique hlk, we assume h refers to a node h and similarly for the rest. When we wish to restrict reference to a clique such that the clique has all the same members, we use hhh. If we will also

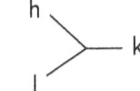

FIGURE 3. Three Place Relation $\mathcal{R}^{hlk}$

have to abstract over the members of an arbitrary clique, then we will use the variables h, l, k to range over an arbitrary clique hlk with the restriction that the variables refer to pairwise distinct positions in hlk. $h, k, l \in$ hlk denotes this. h can take on any of the values h, k, and l, and similarly, with l and k, respecting the pairwise distinct restriction. When abstracting over the clique hhh, the variables h, l, and k are still respecting the condition of no two referring to the same position in the clique. The locution $A \in h$ means that the formula A is a member of the local logic at node h.

A generic *distributed semigroup operator* has the typing structure $h \times l \to k$. Our version of heterogenous algebras is contained in the following definition.

Definition 31. A distributed RA (DRAlg) contains a distribution structure **G** of nodes called *types* and *operators* with types in a multiset of cliques **C**. A type for h is an RA with carrier set D_h, called the *local RA at* h. The *distributed operators* are of the form $\stackrel{hlk}{\circ}: D_h \times D_l \to D_k$ where we use the indices h, l, and k to index the carrier sets of the types. The phyla of heterogeneous algebras are the carrier sets. A DRAlg respects the axioms below on the right. The original axioms from [8; 13] are below on the first lines and the new typed axioms are on the second lines. For cliques hhh, hlk, hlm, mno, hno, and lkn in **C**:

M1 $(B, \vee, \neg, \bot, \top)$ is a Boolean algebra
       $(D_h, \vee, \neg, (-\breve{\ }), \bot_h, \top_h)$ is a LCAlg for each sort h

---

[2] We view three-place relations as arcs in a hypergraph, which connect more than one node — always three, in our case. Multiple arcs are permitted, so are loops.

M2 $(a \circ b) \circ c = a \circ (b \circ c)$
$(a \stackrel{hlm}{\circ} b) \stackrel{mno}{\circ} c \stackrel{o}{=} a \stackrel{hno}{\circ} (b \stackrel{lkn}{\circ} c)$

M3 $(a \vee b) \circ c = (a \circ c) \vee (b \circ c)$
$(a \vee b) \stackrel{hlk}{\circ} c \stackrel{k}{=} (a \stackrel{hlk}{\circ} c) \vee (b \stackrel{hlk}{\circ} c), \quad h, l, k \in \mathsf{hlk}$

M4 $a \circ 1 = a$ for any $a \in A$
$a \stackrel{hhh}{\circ} 1 \stackrel{h}{=} a$

M5 $a^{\smile\smile} = a$ for any $a \in A$
$a^{\smile\smile} \stackrel{h}{=} a$

M6 $(a \vee b)^{\smile} = a^{\smile} \vee b^{\smile}$ for any $a, b \in A$
$(a \vee b)^{\smile} \stackrel{h}{=} a^{\smile} \vee b^{\smile}$

M7 $(a \circ b)^{\smile} = b^{\smile} \circ a^{\smile}$, for any $a, b \in A$
$(a \stackrel{hlk}{\circ} b)^{\smile} \stackrel{k}{=} b^{\smile} \stackrel{lhk}{\circ} a^{\smile} \quad h, l, k \in \mathsf{hlk}$

M8 $(a^{\smile} \circ \neg(a \circ c)) \vee \neg c = \neg c$ for any $a, c \in A$
$(a^{\smile} \stackrel{hlk}{\circ} \neg(a \stackrel{hkl}{\circ} c)) \vee \neg c \stackrel{k}{=} \neg c \quad h, l, k \in \mathsf{hlk}$

The notation $h, l, k \in \mathsf{hlk} \in \mathbf{C}$ means that h, l, k refer to members of the set $\{h, k, l\}$ and that hlk is a clique in $\mathbf{C}$; h does not need to refer to h, and similarly, for the rest. (But h, l, k refer to distinct nodes.)

4.2. Derivation of Relational Operators. In Figure 1, each locality could support a three-place relation such as $\mathcal{R}^{\mathsf{uhv}}$, for sets H, L, K and sets U, V, and Z. Let $u \in U$, $v \in V$ and $z \in Z$ and define the relations $\langle A \rangle, \langle B \rangle, \langle C \rangle$ with

$$\langle A \rangle \stackrel{def}{=} \{\langle u, v \rangle : \exists h \in H(\mathcal{R}^{\mathsf{uhv}} uhv \text{ and } h \in_h A)\}$$

$$\langle B \rangle \stackrel{def}{=} \{\langle v, z \rangle : \exists l \in L(\mathcal{R}^{\mathsf{vlz}} vlz \text{ and } l \in_l B)\}$$

$$\langle C \rangle \stackrel{def}{=} \{\langle u, z \rangle : \exists k \in K(\mathcal{R}^{\mathsf{ukz}} ukz \text{ and } k \in_k C)\}$$

Consider each oval to be a set of points of a local Kripke frame and let H, L, K be the sets of points of the local frames at h, l, k, respectively. Similarly, U, V, Z are the sets of points of the local frames at u, v and z. In the diagram of Figure 4 (below), the cliques (like Figure 3) have been replaced by triangles.

Assuming the Tabularity Axiom now on three-place typed relations, the sets H, L, and K recede into the background. We change notation because the angle brackets are now superfluous. Let $\mathcal{T} = \langle A \rangle$, $\mathcal{S} = \langle B \rangle$, and $\mathcal{R} = \langle C \rangle$. By fiat we define these relations to be arrows with domain and codomain matching their first and second positions, respectively. Hence, $\mathcal{T}: U \to V$, $\mathcal{S}: V \to Z$, and $\mathcal{R}: U \to Z$. This yields the same collection of relation operators but now as arrows in a category of sets shown in the diagram in Figure 5 below. The puncture marks ∸ (from [11]) indicate the inner triangle and the outer pairs of arrows with common sources and targets do not commute. It is difficult to put into the diagram, but we assume none of the triangles constructed from any of the arrows commute, i.e., it is not necessarily the case that $\mathcal{S} \circ \mathcal{T} = \mathcal{R}$ (using categorical composition). In the diagram of Figure 5, one can replace ⊚, →, and ← with ⊕, ↠, and ↞, respectively, for a different notion of composition with residuals.

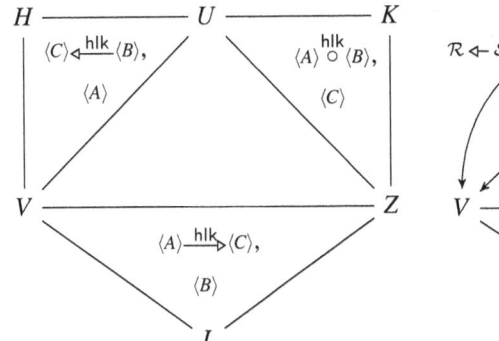
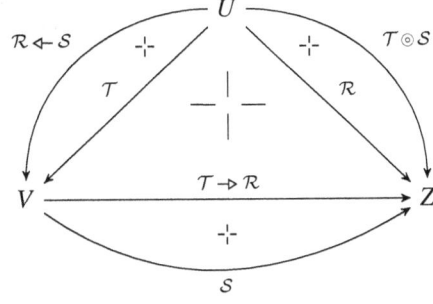

FIGURE 4. Diagram of 3-Place Relations

FIGURE 5. Diagram of Relation Operators

Definition 32.

$$\langle u,z \rangle \in \mathcal{T} \odot \mathcal{S} \quad \text{iff} \quad \exists v \in V(\langle u,v \rangle \in \mathcal{T} \text{ and } \langle v,z \rangle \in \mathcal{S})$$
$$\langle v,z \rangle \in \mathcal{T} \twoheadrightarrow \mathcal{R} \quad \text{iff} \quad \forall u \in U(\langle u,v \rangle \in \mathcal{T} \text{ implies } \langle u,z \rangle \in \mathcal{R})$$
$$\langle u,v \rangle \in \mathcal{R} \twoheadleftarrow \mathcal{S} \quad \text{iff} \quad \forall z \in Z(\langle v,z \rangle \in \mathcal{S} \text{ implies } \langle u,z \rangle \in \mathcal{R})$$

$$\langle u,z \rangle \in \mathcal{T} \oplus \mathcal{S} \quad \text{iff} \quad \forall v \in V(\langle u,v \rangle \in \mathcal{T} \text{ or } \langle v,z \rangle \in \mathcal{S})$$
$$\langle v,z \rangle \in \mathcal{T} \dashrightarrow \mathcal{R} \quad \text{iff} \quad \exists u \in U(\langle u,v \rangle \notin \mathcal{T} \text{ and } \langle u,z \rangle \in \mathcal{R})$$
$$\langle u,v \rangle \in \mathcal{R} \dashleftarrow \mathcal{S} \quad \text{iff} \quad \exists z \in Z(\langle u,z \rangle \in \mathcal{R} \text{ and } \langle v,z \rangle \notin \mathcal{S})$$

$$\langle u,z \rangle \in \mathcal{T}^{\smile} \downarrow \mathcal{S}^{\smile} \quad \text{iff} \quad \exists v \in V(\langle v,u \rangle \notin \mathcal{T}^{\smile} \text{ and } \langle z,v \rangle \notin \mathcal{S}^{\smile})$$
$$\langle u,z \rangle \in \mathcal{S}^{\smile} \uparrow \mathcal{T}^{\smile} \quad \text{iff} \quad \forall v \in V(\langle v,u \rangle \notin \mathcal{T}^{\smile} \text{ or } \langle z,v \rangle \notin \mathcal{S}^{\smile})$$

TABLE 7. Relational Operators

Note that the relations $\mathcal{T}^{\smile}$ and $\mathcal{S}^{\smile}$ used in $\downarrow$ and $\uparrow$ run in the opposite directions as $\mathcal{T}$ and $\mathcal{S}$ used in the rest. This is indicated by the converse accents of $\mathcal{T}$ and $\mathcal{S}$, i.e., $\mathcal{T}^{\smile}$ and $\mathcal{S}^{\smile}$. The use of the same letters in $\downarrow$ and $\uparrow$ as the other operations requires the relations to which they refer be turned around. We could have used new letters, but this would have prevented those arrows from being compared to the arrows in the previous operators. Hence, we use the converse $(-^{\smile})$ now as an accent. The reader can either treat the $(-^{\smile})$ as an actual converse operator on the arrow or simply treat it as an accent symbol. Note also that relational composition is the opposite of categorical composition, i.e., $\mathcal{T} \odot \mathcal{S} = \mathcal{S} \circ \mathcal{T}$.

5. KAN EXTENSIONS AND LIFTS

The definition of a right Kan lift (see below) is taken from Street [18] and the notion of 2-category from Bénabou [6]. The subsequent definitions of Kan extensions and right Kan lifts are extrapolations. Kan extensions exist in Mac Lane [12] but for functors and natural transformations; they provide the objects and arrows for the 2-categories in [6].

We make a subtle point that can be confusing if not brought out. In the extensions and lifts, we never require the underlying graph of 0-objects and 1-arrows to actually form a category. For the Kan extensions and lifts, this point is moot because the underlying graph is a category. However for intensional *nor* and *nand* this is not the case. We only require a graph with 0-arrows (objects) and 1-arrows. The 2-category has as objects the 1-arrows and as arrows the 2-arrows. We compose 1-arrows, but do not require any equations for the composition unless they involve 2-arrows. This does not affect the mathematics. We will use the locution "2-category qua category" for structure of a 1-graph with a 2-category on top when necessary. To put it another way, the lifts and extensions are statements about the 2-categories qua categories.

The upper diagrams in the extensions and lifts, e.g., Figure 13, using the diagrammatic style of [11], do not have their large triangles and the outer pairs of arrows with common sources and targets commuting (see Figure 5). Putting in the puncture marks of [11] would make the diagrams even noisier than they are currently, so we elide them with the understanding they are there. The commuting conditions are actually part of the lower diagrams on the 2-categories qua categories. This becomes more noticeable in the diagrams for ↓ and ↑ given the directions of the arrows. This is brought out in the statements of the lifts and extensions which do not involve commuting diagrams of the 1-arrows. The commutation stated is for the 2-arrows.

The proofs for the adjunctions were taken from [12] but adapted to our use of 2-categories qua categories.

5.1. 2-Diagrams.

The diagram of right whiskering in Figure 6 comes from the data for Right Kan Extensions (see below). The objects are 0-cells, the single arrows are 1-cells, and the double arrows are 2-cells. The 2-diagram on the left is equivalent to the next two diagrams, where Z^T is a functor on the 2-category qua category and merely a mapping on the 1-arrows. Using the notation of 2-categories, $Z^T \sigma = \sigma \triangleleft T$; this reads as usual in category theory: first T then σ, or if you like, σ applied to T.

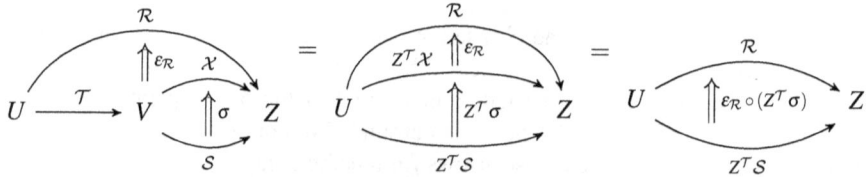

FIGURE 6. Right Whiskering of 2-Diagrams

The diagram is telling us that Z^T is a functor on the 2-category qua category, i.e., for $\sigma_1 \colon S \Rightarrow X$ and $\sigma_2 \colon X \Rightarrow Q$,

$$Z^T(\sigma_2 \circ \sigma_1) = (Z^T \sigma_2) \circ (Z^T \sigma_1).$$

Another right whiskering diagram used by Left Kan Extension with η_R in the opposite direction of and replacing ε_R is had by turning all the 2-arrows around, i.e., $\tau_2 \colon Q \Rightarrow X$ and $\tau_1 \colon X \Rightarrow S$, and

$$Z^T(\tau_1 \circ \tau_2) = (Z^T \tau_1) \circ (Z^T \tau_2),$$

with the same locution $Z^T \tau = \tau \triangleleft T$. Similarly, the diagram of Figure 7 comes from the data for right Kan lifts. U^S is a functor on the 2-category qua category and merely a mapping on the 1-arrows. Using the notation of 2-categories, $U^S \sigma = S \triangleright \sigma$, and this reads: first σ then S.

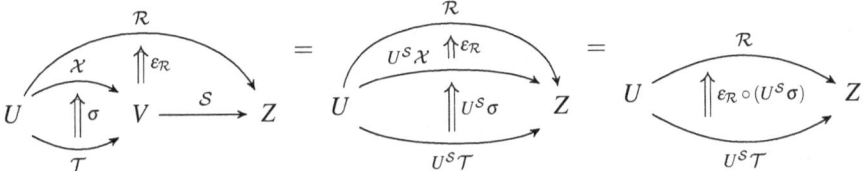

FIGURE 7. Left Whiskering of 2-Diagrams

The diagram is telling us that U^S is a functor on the 2-category qua category, i.e., for $\sigma_1 \colon T \Rightarrow X$ and $\sigma_2 \colon X \Rightarrow Q$,

$$U^S(\sigma_2 \circ \sigma_1) = (U^S \sigma_2) \circ (U^S \sigma_1).$$

A similar left whiskering diagram, used by left Kan lift with η_R in the opposite direction of ε_R and replacing ε_R, is had by turning all the 2-arrows around, i.e., $\tau_2 \colon Q \Rightarrow X$ and $\tau_1 \colon X \Rightarrow T$, and

$$U^S(\tau_1 \circ \tau_2) = (U^S \tau_1) \circ (U^S \tau_2),$$

and the locution $U^S \tau = S \triangleright \tau$.

5.2. Right and Left Kan Extensions.

Definition 33. Consider the diagrams in Figure 8 and Figure 9 below.

The diagram of Figure 8 is said to exhibit X, denoted $\mathrm{Ran}_T R$, as a *right extension of R along T* when each 2-cell $\theta \colon Z^T S \Rightarrow R$ factors as $\varepsilon_R \circ (Z^T \sigma)$ for a unique 2-cell $\sigma \colon S \Rightarrow X$. $\mathrm{Ran}_T R$ is a particular choice of right extension of R along T.

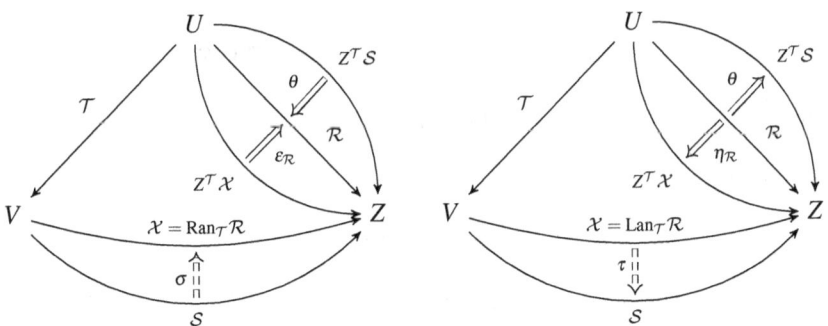

FIGURE 8. Right Kan Extension

FIGURE 9. Left Kan Extension

The dual diagram of Figure 9 is said to exhibit X, denoted $\mathrm{Lan}_T R$, as a *left extension of R along T* when each 2-cell $\theta \colon R \Rightarrow Z^T S$ factors as $(Z^T \tau) \circ \eta_R$ for a unique 2-cell $\tau \colon X \Rightarrow S$. X is a particular choice of left extension of R along T.

The Right Kan Extension defines a universal arrow from a covariant functor to an object. The Left Kan Extension defines a universal arrow from an object to a covariant functor.

Definition 34. A *universal arrow* (i.e., couniversal arrow) from Z^T to $\mathcal{R}$ is a pair $\langle \text{Ran}_T \mathcal{R}, \varepsilon_{\mathcal{R}} \rangle$ consisting of an object $\text{Ran}_T \mathcal{R} \colon V \to Z$ and an arrow $\varepsilon_{\mathcal{R}} \colon Z^T (\text{Ran}_T \mathcal{R}) \Rightarrow \mathcal{R}$ such that to every pair $\langle \mathcal{S}, \theta \rangle$ with $\mathcal{S}$ an object of Z^V and $\theta \colon Z^T \mathcal{S} \Rightarrow \mathcal{R}$, there is a unique $\sigma \colon \mathcal{S} \Rightarrow (\text{Ran}_T \mathcal{R})$ with $\theta = \varepsilon_{\mathcal{R}} \circ (Z^T \sigma)$ (where $Z^T \sigma = \sigma \triangleleft T$). In other words, every arrow θ factors uniquely through the universal arrow $\varepsilon_{\mathcal{R}}$ as in the commutative diagram of Figure 10.

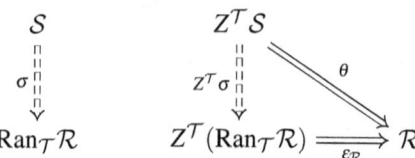

FIGURE 10. Universal Arrow from Z^T to $\mathcal{R}$

A *universal arrow* from $\mathcal{R}$ to Z^T is a pair $\langle \text{Lan}_T \mathcal{R}, \eta_{\mathcal{R}} \rangle$ consisting of an object $\text{Lan}_T \mathcal{R} \colon V \to Z$ and an arrow $\eta_{\mathcal{R}} \colon \mathcal{R} \Rightarrow Z^T (\text{Lan}_T \mathcal{R})$ such that to every pair $\langle \mathcal{S}, \theta \rangle$ with $\mathcal{S}$ an object of Z^V and $\theta \colon \mathcal{R} \Rightarrow Z^T \mathcal{S}$, there is a unique arrow $\tau \colon (\text{Lan}_T \mathcal{R}) \Rightarrow \mathcal{S}$ with $\theta = (Z^T \tau) \circ \eta_{\mathcal{R}}$ (where $Z^T \tau = \tau \triangleleft T$). In other words, every arrow θ factors uniquely through the universal arrow $\eta_{\mathcal{R}}$, as in the commutative diagram of Figure 11.

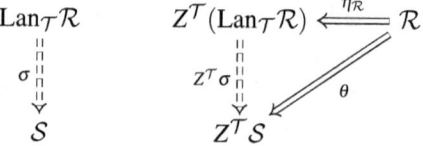

FIGURE 11. Universal Arrow from $\mathcal{R}$ to Z^T

This definition sets up the following adjunctions:

Theorem 35.

$$\varphi_{\mathcal{S},\mathcal{R}} \colon Z^U(Z^T \mathcal{S}, \mathcal{R}) \cong Z^V(\mathcal{S}, \text{Ran}_T \mathcal{R}) \qquad \varphi_{\mathcal{R},\mathcal{S}} \colon Z^V(\text{Lan}_T \mathcal{R}, \mathcal{S}) \cong Z^U(\mathcal{R}, Z^T \mathcal{S}),$$

where Z^U is a functor that returns the set of all arrows from U to Z and Z^V is a functor that returns all arrows from Z to V.

Proof. We prove the theorem for Right Kan Extensions, the proof for Left Kan Extensions is dual. We are given a universal arrow $\langle G_0 \mathcal{R}, \varepsilon_{\mathcal{R}} \rangle$ (to $\mathcal{R}$ from Z^T) for every object $\mathcal{R} \in Z^U$; there is exactly one way to make G_0 the object function of a functor G for which the transformation $\varepsilon \colon Z^T G \to I_{Z^U}$ (for I_{Z^U} the identity arrow of the 2-cell Z^U) will be natural.

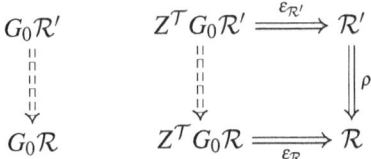

FIGURE 12. Extension of G_0 to be a Functor G

We let $G_0 = \text{Ran}_{\mathcal{T}}$ and will extend G_0 to arrows to achieve

$$(\rho\colon \mathcal{R} \Rightarrow \mathcal{R}') \mapsto G\rho\colon G_0\mathcal{R}' \Rightarrow G_0\mathcal{R}.$$

Specifically, for each $\rho\colon \mathcal{R}' \Rightarrow \mathcal{R}$, the universality of ε states that there is exactly one arrow (double dashed) which can make the diagram commute. Choose this arrow as $G\rho\colon G_0\mathcal{R}' \to G_0\mathcal{R}$; the commutativity states that ε is now natural.

We check that this choice of $G\rho$ makes G a functor. It is obvious that G preserves identity arrows because identity arrows are unique. Next, let $\rho_1\colon \mathcal{P} \Rightarrow \mathcal{R}'$ and $\rho_2\colon \mathcal{R}' \Rightarrow \mathcal{R}$, then we get $G\rho_1\colon G\mathcal{P} \Rightarrow G\mathcal{R}'$ and $G\rho_2\colon G\mathcal{R}' \Rightarrow G\mathcal{R}$. Now choose $\mu = \rho_2 \circ \rho_1\colon \mathcal{P} \Rightarrow \mathcal{R}$. There is a unique arrow $G_0\mathcal{P} \Rightarrow G_0\mathcal{R}$, which we designate $G\mu$, that causes the resulting diagram involving $\varepsilon_\mathcal{P}$, $\varepsilon_{\mathcal{R}'}$, and $\varepsilon_\mathcal{R}$ to commute. However, $G\rho_2 \circ G\rho_1$ also causes the diagram to commute, so $G\mu = G(\rho_2 \circ \rho_1) = G\rho_2 \circ G\rho_1$. Thus G is a functor.

The statement that ε is universal means that for each $\theta\colon Z^T\mathcal{S} \Rightarrow \mathcal{R}$ there is exactly one σ as in the commutative diagram in Figure 10 (above). This states that $\psi(\sigma) = \varepsilon_\mathcal{R} \circ Z^T\sigma$ defines a bijection

$$\psi_{\mathcal{S},\mathcal{R}}\colon Z^V(\mathcal{S}, G\mathcal{R}) \to Z^U(Z^T\mathcal{S}, \mathcal{R}).$$

Expanding this out a bit, by virtue of the uniqueness criterion, there is a function ψ^{-1} from $Z^U(Z^T\mathcal{S}, \mathcal{R})$ to $Z^V(\mathcal{S}, G\mathcal{R})$. The universal condition states that ψ^{-1} is a function.

To see that $\psi^{-1}_{\mathcal{S},\mathcal{R}}$ is 1-1, select $\theta', \theta \in Z^U(Z^T\mathcal{S}, \mathcal{R})$ such that $\theta' \neq \theta$. Towards a reductio ad absurdum, assume that $\psi^{-1}_{\mathcal{S},\mathcal{R}}(\theta) = \sigma$, $\theta = \varepsilon_\mathcal{R} \circ (Z^T\sigma)$, $\psi^{-1}_{\mathcal{S},\mathcal{R}}(\theta') = \sigma'$, and $\theta' = \varepsilon_\mathcal{R} \circ (Z^T\sigma')$ and that $\sigma = \sigma'$. From this latter condition, $\varepsilon_\mathcal{R} \circ (Z^T\sigma) = \varepsilon_\mathcal{R} \circ (Z^T\sigma')$, and since the preimages of θ' and θ (under $\psi_{\mathcal{S},\mathcal{R}}$) must be unique, $\theta = \theta'$, a contradiction. So $\psi^{-1}_{\mathcal{S},\mathcal{R}}$ is 1-1.

To show that $\psi^{-1}_{S\mathcal{R}}$ is surjective, let $\sigma \in V^Z(\mathcal{S}, G\mathcal{R})$, then $\psi\sigma = \varepsilon_\mathcal{R} \circ (Z^T\sigma)\colon Z^T\mathcal{S} \Rightarrow \mathcal{R}$. Therefore, there is a unique σ' such that $\sigma' \in V^Z(\mathcal{S}, G\mathcal{R})$ such that $\psi\sigma' = \varepsilon_\mathcal{R} \circ Z^T\sigma$. Since σ' is unique, then $\sigma = \sigma'$.

This bijection is natural in $\mathcal{R}$ because ε is natural, and natural in $\mathcal{S}$ because Z^T is a functor, hence gives an adjunction $\langle Z^T, G, \psi \rangle$. In this case, ε was the unit obtained from the adjunction $\langle Z^T, G, \psi \rangle$. ◁

For our situation with ⊙, $\to$ and ⊕, $\twoheadrightarrow$, we use the following definition:

Definition 36. Given an arrow $\mathcal{T}\colon U \twoheadrightarrow V$ and an object Z we consider the arrow category Z^V with the objects being the arrows $\mathcal{S}, \mathcal{X}\colon V \twoheadrightarrow Z$ and 2-arrows $\sigma\colon \mathcal{S} \Rightarrow \mathcal{X}$, then we define the covariant functors $Z_\odot^\mathcal{T}, Z_\oplus^\mathcal{T} : Z^V \to Z^U$ and the object maps $\mathrm{Ran}_\mathcal{T} : Z^U \to Z^V$ and $\mathrm{Lan}_\mathcal{T} : Z^U \to Z^V$; let $\mathcal{P}, \mathcal{Q} : V \to Z$ and $\mathcal{Y} : U \to Z$, then

$$Z_\odot^\mathcal{T} \stackrel{def}{=} \lambda \mathcal{Q}. \mathcal{T} \odot \mathcal{Q} \qquad \mathrm{Ran}_\mathcal{T} \stackrel{def}{=} \lambda \mathcal{Y}. \mathcal{T} \twoheadrightarrow \mathcal{Y}$$

$$Z_\oplus^\mathcal{T} \stackrel{def}{=} \lambda \mathcal{Q}. \mathcal{T} \oplus \mathcal{Q} \qquad \mathrm{Lan}_\mathcal{T} \stackrel{def}{=} \lambda \mathcal{Y}. \mathcal{T} \twoheadleftarrow \mathcal{Y},$$

and 2-arrows are defined by $(\nu : \mathcal{P} \Rightarrow \mathcal{Q}) \mapsto (Z_\odot^\mathcal{T} \nu : \mathcal{T} \odot \mathcal{P} \Rightarrow \mathcal{T} \odot \mathcal{Q})$ and $(\nu : \mathcal{P} \Rightarrow \mathcal{Q}) \mapsto (Z_\oplus^\mathcal{T} \nu : \mathcal{T} \oplus \mathcal{P} \Rightarrow \mathcal{T} \oplus \mathcal{Q})$.

Lemma 37. $Z_\odot^\mathcal{T}, Z_\oplus^\mathcal{T} : Z^V \to Z^U$ *are covariant functors, where in the categories Z^V and Z^U, the objects are 1-arrows and the arrows are 2-arrows.*

Proof. We show the proof for $Z_\odot^\mathcal{T}$, the proof for $Z_\oplus^\mathcal{T}$ is similar. The proof relies on the fact that arrows in the 2-category qua category compose. Assume $\nu : \mathcal{P} \Rightarrow \mathcal{Q}$ and $\mu : \mathcal{Q} \Rightarrow \mathcal{N}$, then $Z_\odot^\mathcal{T} \nu : \mathcal{T} \odot \mathcal{P} \Rightarrow \mathcal{T} \odot \mathcal{Q}$ and $Z_\odot^\mathcal{T} \mu : \mathcal{T} \odot \mathcal{Q} \Rightarrow \mathcal{T} \odot \mathcal{N}$. Hence, $(Z_\odot^\mathcal{T} \mu) \circ (Z_\odot^\mathcal{T} \nu) = Z_\odot^\mathcal{T} (\mu \circ \nu)$, where the latter follows from the properties of 2-categories qua categories from the right whiskering above. Similarly, that $Z_\odot^\mathcal{T}$ preserves the identity arrows (i.e., the equality relation) also follows from right whiskering in the 2-category qua categories.

For our context, we change notation for the ease of typesetting: $Z_\odot^\mathcal{T} \sigma = \sigma \triangleleft \mathcal{T}$ and $Z_\oplus^\mathcal{T} \tau = \tau \triangleleft \mathcal{T}$. That these functors are covariant follows from

$$\frac{\mathcal{P} \subseteq_\sigma \mathcal{Q}}{\mathcal{T} \odot \mathcal{P} \subseteq_{\sigma \triangleleft \mathcal{T}} \mathcal{T} \odot \mathcal{Q}} \qquad \frac{\mathcal{P} \subseteq_\tau \mathcal{Q}}{\mathcal{T} \oplus \mathcal{P} \subseteq_{\tau \triangleleft \mathcal{T}} \mathcal{T} \oplus \mathcal{Q}}$$

and the definitions of $\odot$ and $\oplus$. This says that $(\sigma : \mathcal{P} \Rightarrow \mathcal{Q}) \mapsto (\sigma \triangleleft \mathcal{T} : \mathcal{T} \odot \mathcal{P} \Rightarrow \mathcal{T} \odot \mathcal{Q})$ and $(\tau : \mathcal{P} \Rightarrow \mathcal{Q}) \mapsto (\tau \triangleleft \mathcal{T} : \mathcal{T} \oplus \mathcal{P} \Rightarrow \mathcal{T} \oplus \mathcal{Q})$. ◁

Theorem 38. *For our situation using Definition 36, the following diagram for $\odot$ and its dual for $\oplus$ hold: For all $\mathcal{T} : U \to V$ and $\mathcal{R} : U \to Z$,*

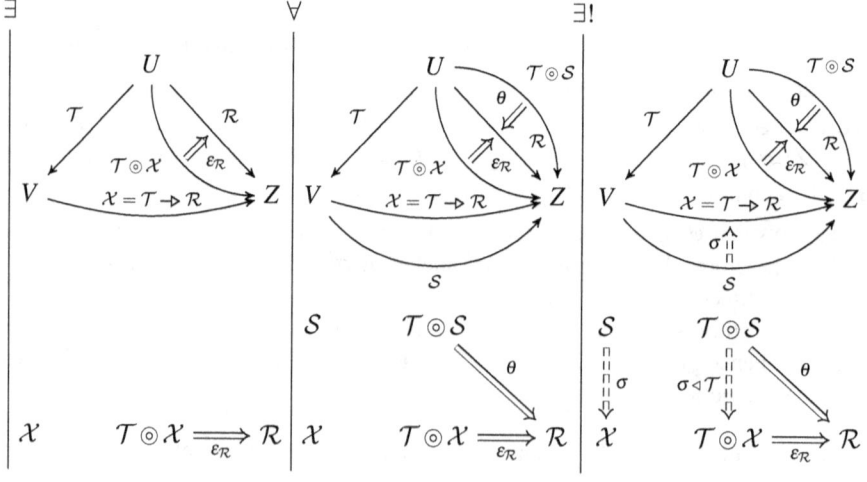

FIGURE 13. Right Kan Extension as a Right Residual

Proof. We show the proof for $\odot$; the proof for $\oplus$ is dual. We first show that σ exists. Let $\langle v,z \rangle \in \mathcal{S}$. Towards a reductio ad absurdum, assume $\langle v,z \rangle \notin \mathcal{X}$. In that case, given $\mathcal{X} = \mathcal{T} \twoheadrightarrow \mathcal{R}$, there must be some $u \in U$ such that $\langle u,v \rangle \in \mathcal{T}$ and $\langle u,z \rangle \notin \mathcal{R}$. Since $\theta : \mathcal{T} \odot \mathcal{S} \Rightarrow \mathcal{R}$ and $\langle u,z \rangle \in \mathcal{T} \odot \mathcal{S}$, then $\langle u,z \rangle \in \mathcal{R}$ and we have a contradiction. Therefore, $\langle v,z \rangle \in \mathcal{T} \twoheadrightarrow \mathcal{R}$. Hence, $\mathcal{S} \subseteq \mathcal{T} \twoheadrightarrow \mathcal{R}$ and there exists σ such that $\sigma : \mathcal{S} \Rightarrow \mathcal{X}$. Also, σ is unique since inclusions are unique in set theory.

Next we must show that $\theta = \varepsilon_{\mathcal{R}} \circ (\sigma \triangleleft \mathcal{T})$. Since $\sigma \triangleleft \mathcal{T} : \mathcal{T} \odot \mathcal{S} \Rightarrow \mathcal{T} \odot \mathcal{X}$ and $\varepsilon_{\mathcal{R}} : \mathcal{T} \odot \mathcal{X} \Rightarrow \mathcal{R}$, then $\varepsilon_{\mathcal{R}} \circ (\sigma \triangleleft \mathcal{T}) : \mathcal{T} \odot \mathcal{S} \Rightarrow \mathcal{R}$. Since $\theta : \mathcal{T} \odot \mathcal{S} \Rightarrow \mathcal{R}$ and inclusions are unique in set theory, $\theta = \varepsilon_{\mathcal{R}} \circ (\sigma \triangleleft \mathcal{T})$. ◁

Our situation corresponds to the following adjunctions:

$$\varphi_{\mathcal{S},\mathcal{R}} : Z^U(\mathcal{T} \odot \mathcal{S}, \mathcal{R}) \cong Z^V(\mathcal{S}, \mathcal{T} \twoheadrightarrow \mathcal{R}) \qquad \varphi_{\mathcal{R},\mathcal{S}} : Z^V(\mathcal{T} \twoheadrightarrow \mathcal{R}, \mathcal{S}) \cong Z^U(\mathcal{R}, \mathcal{T} \oplus \mathcal{S}).$$

We have the following logical rules for $\odot$ where $\theta = \varepsilon_{\mathcal{R}} \circ (\sigma \triangleleft \mathcal{T})$:

$$\frac{\mathcal{T} \odot \mathcal{S} \subseteq_\theta \mathcal{R} \qquad \mathcal{T} \odot (\mathcal{T} \twoheadrightarrow \mathcal{R}) \subseteq_{\varepsilon_\mathcal{R}} \mathcal{R}}{\mathcal{S} \subseteq_\sigma \mathcal{T} \twoheadrightarrow \mathcal{R}} \text{ Right Kan Extension}$$

$$\frac{\mathcal{S} \subseteq_\sigma \mathcal{T} \twoheadrightarrow \mathcal{R}}{\mathcal{T} \odot \mathcal{S} \subseteq_{\sigma \triangleleft \mathcal{T}} \mathcal{T} \odot (\mathcal{T} \twoheadrightarrow \mathcal{R})} \text{ Monotonicity}$$

The left rule is equivalent to $\mathcal{T} \odot \mathcal{S} \subseteq_\theta \mathcal{R}$ implies $\mathcal{S} \subseteq_\sigma \mathcal{T} \twoheadrightarrow \mathcal{R}$, because the second premise of the first rule always holds in our situation. This premise holds via one half of the residuation condition

$$\frac{\mathcal{T} \twoheadrightarrow \mathcal{R} \subseteq_\iota \mathcal{T} \twoheadrightarrow \mathcal{R}}{\mathcal{T} \odot (\mathcal{T} \twoheadrightarrow \mathcal{R}) \subseteq_{\varepsilon_\mathcal{R}} \mathcal{R}},$$

where ι is the equality relation and stands for the identity in the 2-category. In the Kan extension view, residuation is not available and hence that premise must be explicitly stated. The other direction is

$$\frac{\mathcal{S} \subseteq_\sigma \mathcal{T} \twoheadrightarrow \mathcal{R}}{\mathcal{T} \odot \mathcal{S} \subseteq_{\sigma \triangleleft \mathcal{T}} \mathcal{T} \odot (\mathcal{T} \twoheadrightarrow \mathcal{R}) \qquad \mathcal{T} \odot (\mathcal{T} \twoheadrightarrow \mathcal{R}) \subseteq_{\varepsilon_\mathcal{R}} \mathcal{R}}{\mathcal{T} \odot \mathcal{S} \subseteq_\theta \mathcal{R}}$$

The extra noise in the right premise of the right Kan extension rule, i.e., $\mathcal{T} \odot (\mathcal{T} \twoheadrightarrow \mathcal{R}) \subseteq_{\varepsilon_\mathcal{R}} \mathcal{R}$, is related to the ability for some right Kan extensions to be definable without necessarily having a pair of adjoint functors. In other words, propositions are considered universally quantified in logic, where here there is no such quantification. So this premise must be explicitly stated. The statement of the right Kan extension says that $\theta = \varepsilon_\mathcal{R} \circ (\sigma \triangleleft \mathcal{T})$. This holds for us, because there is only a single subset relation such that $\mathcal{T} \odot \mathcal{S} \subseteq \mathcal{R}$ whereas for Kan this must be explicitly declared, i.e., that σ is unique and hence $\varepsilon_\mathcal{R} \circ (\sigma \triangleleft \mathcal{T})$ is unique which implies that $\theta = \varepsilon_\mathcal{R} \circ (\sigma \triangleleft \mathcal{T})$.

We have the following logical rules for $\oplus$ where $\theta = (\tau \triangleleft \mathcal{T}) \circ \eta_\mathcal{R}$:

$$\frac{\mathcal{R} \subseteq_\theta \mathcal{T} \oplus \mathcal{S} \qquad \mathcal{R} \subseteq_{\eta_\mathcal{R}} \mathcal{T} \oplus (\mathcal{T} \twoheadleftarrow \mathcal{R})}{\mathcal{T} \twoheadleftarrow \mathcal{R} \subseteq_\tau \mathcal{S}} \text{ Left Kan Extension}$$

$$\frac{\mathcal{T} \twoheadleftarrow \mathcal{R} \subseteq_\tau \mathcal{S}}{\mathcal{T} \oplus (\mathcal{T} \twoheadleftarrow \mathcal{R}) \subseteq_{\tau \triangleleft \mathcal{T}} \mathcal{T} \oplus \mathcal{S}} \text{ Montonicity}$$

with similar justifications as those for $\odot$.

5.3. Right and Left Kan Lifts.

Definition 39. Consider the following diagrams, the Right Kan Lift is from [18]:

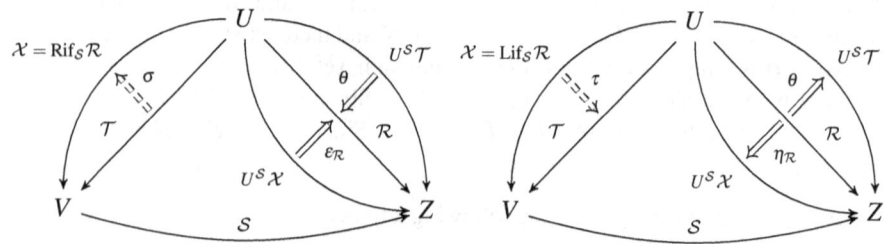

FIGURE 14. Right Kan Lift

FIGURE 15. Left Kan Lift

The diagram of Figure 14 is said to exhibit $\mathcal{X}$, denoted $\text{Rif}_S\mathcal{R}$, as a *right lifting of $\mathcal{R}$ through S* when each 2-cell $\theta\colon U^S\mathcal{T} \Rightarrow \mathcal{R}$ factors as $\varepsilon_\mathcal{R} \circ (U^S\sigma)$ for a unique 2-cell $\sigma\colon \mathcal{T} \Rightarrow \mathcal{X}$. $\text{Rif}_S\mathcal{R}$ (for which Street used $S ⋔ \mathcal{R}$) is a particular choice of right lift of $\mathcal{R}$ through S.

The dual diagram of Figure 15 is said to exhibit $\mathcal{X}$, denoted $\text{Lif}_S\mathcal{R}$, as a *left lifting of $\mathcal{R}$ through S* when each 2-cell $\theta\colon \mathcal{R} \Rightarrow U^S\mathcal{T}$ factors as $(U^S\tau) \circ \eta_\mathcal{R}$ for a unique 2-cell $\tau\colon \mathcal{X} \Rightarrow \mathcal{T}$. We write $\mathcal{X} = \text{Lif}_S\mathcal{R}$ for a particular choice of left lift of $\mathcal{R}$ through S.

Just as in Right and Left Kan Extension, the Right Kan Lift defines a universal arrow from a covariant functor to an object and the Left Kan Lift defines a universal arrow from an object to a covariant functor.

Definition 40. A *universal arrow* (i.e., couniversal arrow) from U^S to $\mathcal{R}$ is a pair $\langle \text{Rif}_S\mathcal{R}, \varepsilon_\mathcal{R} \rangle$ consisting of an object $\text{Rif}_S\mathcal{R}\colon U \to V$ and an arrow $\varepsilon_\mathcal{R}\colon U^S(\text{Rif}_S\mathcal{R}) \Rightarrow \mathcal{R}$ such that to every pair $\langle \mathcal{T}, \theta \rangle$ with $\mathcal{T}$ and object of V^U and $\theta\colon U^S\mathcal{T} \Rightarrow \mathcal{R}$, there is a unique $\sigma\colon \mathcal{T} \Rightarrow (\text{Rif}_S\mathcal{R})$ with $\theta = (U^S\sigma) \circ \varepsilon_\mathcal{R}$ (where $U^S\sigma = S \triangleright \sigma$). In other words, every arrow θ factors uniquely through the universal arrow $\varepsilon_\mathcal{R}$ as in the commutative diagram of Figure 16 (below).

A *universal arrow* from $\mathcal{R}$ to U^S is a pair $\langle \text{Lif}_S\mathcal{R}, \eta_\mathcal{R} \rangle$ consisting of an object $\text{Lif}_S\mathcal{R}\colon U \to V$ and an arrow $\eta_\mathcal{R}\colon \mathcal{R} \Rightarrow U^S(\text{Lif}_S\mathcal{R})$ such that to every pair $\langle \mathcal{T}, \theta \rangle$ with $\mathcal{T}$ an object of V^U and $\theta\colon \mathcal{R} \Rightarrow U^S\mathcal{T}$, there is a unique arrow $\tau\colon (\text{Lif}_S\mathcal{R}) \Rightarrow \mathcal{T}$ with $\theta = (U^S\tau) \circ \eta_\mathcal{R}$ (where $U^S\tau = S \triangleright \tau$). In other words, every arrow θ factors uniquely through the universal arrow $\eta_\mathcal{R}$ as in the commutative diagram of Figure 17.

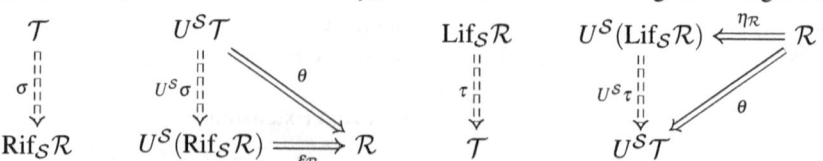

FIGURE 16. Universal Arrow from U^S to $\mathcal{R}$

FIGURE 17. Universal Arrow from $\mathcal{R}$ to U^S

Similarly to Kan Extensions, this definition sets up the following adjunctions:

Theorem 41.
$$\varphi_{\mathcal{T},\mathcal{R}}\colon Z^U(U^{\mathcal{S}}\mathcal{T},\mathcal{R}) \cong V^U(\mathcal{T},\mathrm{Rif}_{\mathcal{S}}\mathcal{R})$$
$$\varphi_{\mathcal{R},\mathcal{S}}\colon V^Z(\mathrm{Lif}_{\mathcal{S}}\mathcal{R},\mathcal{T}) \cong Z^U(\mathcal{R},U^{\mathcal{S}}\mathcal{T}),$$

where Z^U is a functor that returns the set of all arrows from U to Z and V^Z is a functor that returns all arrows from Z to V.

The proofs are similar to the adjunction proofs for Kan Extensions. For our situation with $\odot$, $\leftarrow$ and $\oplus$, $\twoheadleftarrow$, we use the following definition:

Definition 42. Given an arrow $\mathcal{S}\colon V \to Z$ and an object U we consider the arrow category V^U with objects the arrows $\mathcal{T}\colon U \to V$ and 2-arrows $\sigma\colon \mathcal{T} \Rightarrow \mathcal{X}$, then we define the covariant functors $U_\odot^{\mathcal{S}}, U_\oplus^{\mathcal{S}}\colon V^U \to Z^U$ and the object maps $\mathrm{Rif}_{\mathcal{S}}\colon Z^U \to V^U$ and $\mathrm{Lif}_{\mathcal{S}}\colon Z^U \to V^U$. Let $\mathcal{P},\mathcal{Q}\colon U \to V$; let $\mathcal{P},\mathcal{Q}\colon U \to V$ and $\mathcal{Y}\colon U \to Z$:

$$U_\odot^{\mathcal{S}} \stackrel{def}{=} \lambda \mathcal{Q}. \mathcal{Q} \odot \mathcal{S} \qquad \mathrm{Rif}_{\mathcal{S}} \stackrel{def}{=} \lambda \mathcal{Y}. \mathcal{Y} \leftarrow \mathcal{S}$$
$$U_\oplus^{\mathcal{S}} \stackrel{def}{=} \lambda \mathcal{Q}. \mathcal{Q} \oplus \mathcal{S} \qquad \mathrm{Lif}_{\mathcal{S}} \stackrel{def}{=} \lambda \mathcal{Y}. \mathcal{Y} \twoheadleftarrow \mathcal{S},$$

and 2-arrows by $(v\colon \mathcal{P}\Rightarrow \mathcal{Q}) \mapsto (U_\odot^{\mathcal{S}} v\colon \mathcal{P}\odot\mathcal{S} \Rightarrow \mathcal{Q}\odot\mathcal{S})$ and $(v\colon \mathcal{P}\Rightarrow \mathcal{Q}) \mapsto (U_\oplus^{\mathcal{S}} v\colon \mathcal{P}\oplus\mathcal{S} \Rightarrow \mathcal{Q}\oplus\mathcal{S})$.

Lemma 43. $U_\odot^{\mathcal{S}}, U_\oplus^{\mathcal{S}}\colon V^U \to Z^U$ are covariant functors where in the categories V^U, and Z^U, the objects are 1-arrows and the arrows are 2-arrows.

The proofs are similar to those for Right and Left Kan Extensions. We change notation — to ease the typesetting — as follows: $U_\odot^{\mathcal{S}}\sigma = \mathcal{S}\triangleright\sigma$ and $U_\oplus^{\mathcal{S}}\tau = \mathcal{S}\triangleright\tau$.

Theorem 44. *For our situation using Definition 42, the following diagram for $\odot$ and its dual for $\oplus$ hold: For all $\mathcal{S}\colon V \to Z$ and $\mathcal{R}\colon U \to Z$, and where $\mathcal{X} = \mathrm{Rif}_{\mathcal{S}}\mathcal{R}$ and its dual $\mathcal{X} = \mathrm{Lif}_{\mathcal{S}}\mathcal{R}$,*

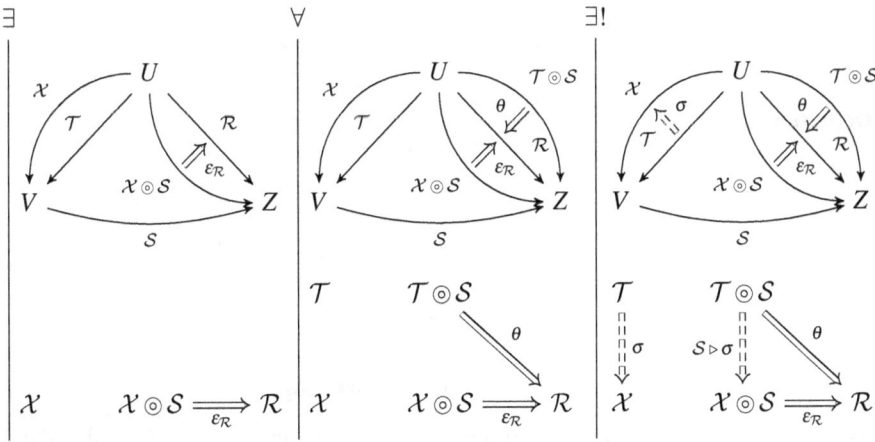

FIGURE 18. Right Kan Lift as a Left Residual

The proofs are similar to those for Right and Left Kan Extensions. Our situation corresponds to the following adjunctions:

$$Z^U(\mathcal{T} \odot \mathcal{S}, \mathcal{R}) \cong V^U(\mathcal{T}, \mathcal{R} \leftarrow \mathcal{S}) \qquad V^U(\mathcal{R} \leftarrow \mathcal{S}, \mathcal{T}) \cong Z^U(\mathcal{R}, \mathcal{T} \oplus \mathcal{S}).$$

We have the following logical rules:

$$\frac{\mathcal{T} \odot \mathcal{S} \subseteq_\theta \mathcal{R} \qquad (\mathcal{R} \leftarrow \mathcal{S}) \odot \mathcal{S} \subseteq_{\varepsilon_\mathcal{R}} \mathcal{R}}{\mathcal{T} \subseteq_\sigma \mathcal{R} \leftarrow \mathcal{S}} \text{ Right Kan Lift}$$

$$\frac{\mathcal{T} \subseteq_\sigma \mathcal{R} \leftarrow \mathcal{S}}{\mathcal{T} \odot \mathcal{S} \subseteq_{\mathcal{S} \triangleright \sigma} (\mathcal{R} \leftarrow \mathcal{S}) \odot \mathcal{S}} \text{ Montonicity}$$

The left rule is equivalent to

$$\mathcal{T} \odot \mathcal{S} \subseteq_\theta \mathcal{R} \quad \text{implies} \quad \mathcal{T} \subseteq_\sigma \mathcal{R} \leftarrow \mathcal{S},$$

because the second premise of the first rule always holds in our situation. This premise holds via one half of the residuation condition

$$\frac{\mathcal{R} \leftarrow \mathcal{S} \subseteq_\iota \mathcal{R} \leftarrow \mathcal{S}}{(\mathcal{R} \leftarrow \mathcal{S}) \odot \mathcal{S} \subseteq_{\varepsilon_\mathcal{R}} \mathcal{R}},$$

where again ι is the equality relation and stands for the identity in the 2-category. The other direction is

$$\frac{\mathcal{T} \subseteq_\sigma \mathcal{R} \leftarrow \mathcal{S}}{\mathcal{T} \odot \mathcal{S} \subseteq_{\mathcal{S} \triangleright \sigma} (\mathcal{R} \leftarrow \mathcal{S}) \odot \mathcal{S} \qquad (\mathcal{R} \leftarrow \mathcal{S}) \odot \mathcal{S} \subseteq_{\varepsilon_\mathcal{R}} \mathcal{R}}{\mathcal{T} \odot \mathcal{S} \subseteq_{\varepsilon_\mathcal{R} \circ (\mathcal{S} \triangleright \sigma)} \mathcal{R}}$$

We have the following logical rules for $\oplus$:

$$\frac{\mathcal{R} \subseteq_\theta \mathcal{T} \oplus \mathcal{S} \qquad \mathcal{R} \subseteq_{\eta_\mathcal{R}} (\mathcal{R} \leftharpoonup \mathcal{S}) \oplus \mathcal{S}}{\mathcal{R} \leftharpoonup \mathcal{S} \subseteq_\tau \mathcal{T}} \qquad \frac{\mathcal{R} \leftharpoonup \mathcal{S} \subseteq_\tau \mathcal{T}}{(\mathcal{R} \leftharpoonup \mathcal{S}) \oplus \mathcal{S} \subseteq_{\mathcal{S} \triangleright \tau} \mathcal{T} \oplus \mathcal{S}}$$

Left Kan Lift Montonicity

with similar justifications as those for $\odot$.

5.4. Intensional Nor and Nand. We use different left whiskering than the diagram in Figure 7.

Definition 45.

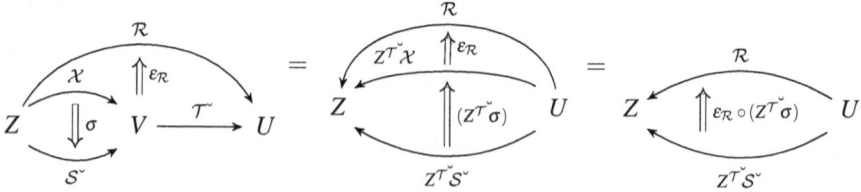

FIGURE 19. Left Whiskering

where $Z^{T^\smile}$ is a contravariant functor on the 2-category qua category. Note that the 2-arrow σ is in the opposite direction as its transform $Z^{T^\smile} \sigma$. Also note that the direction of the 1-arrows changes from the first diagram to the second and the third. That $Z^{T^\smile}$ is a contravariant functor on the 2-category qua category follows from a diagram chase, i.e., for $\sigma_1 \colon \mathcal{X} \Rightarrow \mathcal{S}^{\smile}$ and $\sigma_2 \colon \mathcal{S}^{\smile} \Rightarrow \mathcal{Q}$,

$$Z^{T^{\smile}}(\sigma_2 \circ \sigma_1) = (Z^{T^{\smile}} \sigma_1) \circ (Z^{T^{\smile}} \sigma_2).$$

Using the notation of 2-categories, $Z^{T^{\smile}} \sigma = T^{\smile} \triangleright \sigma$. A similar left whiskering diagram is used by Left Lift (see below) with $\eta_{\mathcal{R}}$ replacing $\varepsilon_{\mathcal{R}}$, τ replacing σ, and turning all the 2-arrows around.

Definition 46. Consider the following diagrams:

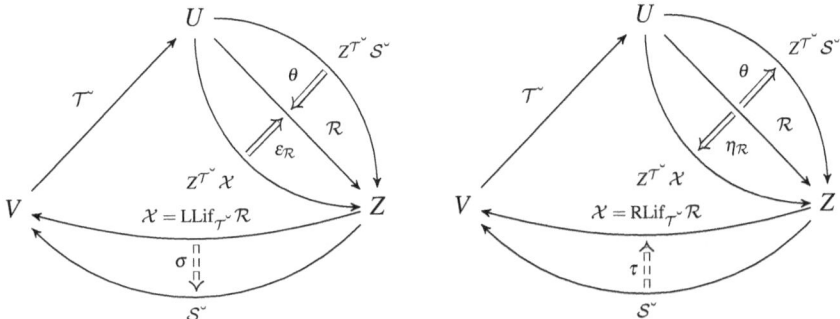

FIGURE 20. Left Lift   FIGURE 21. Right Lift

The diagram of Figure 20 is said to exhibit $\mathcal{X}$, denoted $\mathcal{X} = \text{LLif}_{T^{\smile}} \mathcal{R}$, as a *left lift of* $T^{\smile}$ *through* $\mathcal{R}$ when each 2-cell $\theta \colon Z^{T^{\smile}} \mathcal{S}^{\smile} \Rightarrow \mathcal{R}$ factors as $\varepsilon_{\mathcal{R}} \circ (Z^{T^{\smile}} \sigma)$ for a unique 2-cell $\sigma \colon \mathcal{X} \Rightarrow \mathcal{S}^{\smile}$. $\text{LLif}_{T^{\smile}} \mathcal{R}$ is a particular choice of lift of $T^{\smile}$ along $\mathcal{R}$.

The dual diagram of Figure 21 is said to exhibit $\mathcal{X}$, denoted $\text{RLif}_{T^{\smile}} \mathcal{R}$, as a *right lift of* $T^{\smile}$ *along* $\mathcal{R}$ when each 2-cell $\theta \colon \mathcal{R} \Rightarrow Z^{T^{\smile}} \mathcal{S}^{\smile}$ factors as $(Z^{T^{\smile}} \tau) \circ \eta_{\mathcal{R}}$ for a unique 2-cell $\tau \colon \mathcal{S}^{\smile} \Rightarrow \mathcal{X}$. $\text{RLif}_{T^{\smile}} \mathcal{R}$ is a particular choice of lift of $T^{\smile}$ along $\mathcal{R}$.

The Left and Right Lifts define universal arrows from and to, respectively, a contravariant functor.

Definition 47. A *universal arrow* (i.e., couniversal arrow) from $Z^{T^{\smile}}$ to $\mathcal{R}$ is a pair $\langle \text{LLif}_{T^{\smile}} \mathcal{R}, \varepsilon_{\mathcal{R}} \rangle$ consisting of an object $\text{LLif}_{T^{\smile}} \mathcal{R} \colon Z \to V$ and an arrow $\varepsilon_{\mathcal{R}} \colon Z^{T^{\smile}} (\text{LLif}_{T^{\smile}} \mathcal{R}) \Rightarrow \mathcal{R}$ such that to every pair $\langle \mathcal{S}^{\smile}, \theta \rangle$ with $\mathcal{S}^{\smile}$ an object of V^Z and $\theta \colon Z^{T^{\smile}} \mathcal{S}^{\smile} \Rightarrow \mathcal{R}$, there is a unique $\sigma \colon (\text{LLif}_{T^{\smile}} \mathcal{R}) \Rightarrow \mathcal{S}^{\smile}$ with $\theta = \varepsilon_{\mathcal{R}} \circ (Z^{T^{\smile}} \sigma)$. In other words, every arrow θ factors uniquely through the universal arrow $\varepsilon_{\mathcal{R}}$ as in the commutative diagram of Figure 22.

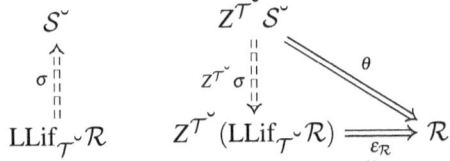

FIGURE 22. Universal from $Z^{T^{\smile}}$ to $\mathcal{R}$

A *universal arrow* from $\mathcal{R}$ to $Z^{\mathcal{T}^\vee}$ is a pair $\langle \mathrm{RLif}_{\mathcal{T}^\vee}\mathcal{R}, \eta_\mathcal{R} \rangle$ consisting of an object $\mathrm{RLif}_{\mathcal{T}^\vee}\mathcal{R}: Z \to V$ and an arrow $\eta_\mathcal{R}: \mathcal{R} \Rightarrow Z^{\mathcal{T}^\vee}(\mathrm{RLif}_{\mathcal{T}^\vee}\mathcal{R})$ such that to every pair $\langle \mathcal{S}^\vee, \theta \rangle$ with $\mathcal{S}^\vee$ an object of V^Z and $\theta: \mathcal{R} \Rightarrow Z^{\mathcal{T}^\vee}\mathcal{S}^\vee$, there is a unique $\tau: \mathcal{S}^\vee \Rightarrow (\mathrm{RLif}_{\mathcal{T}^\vee}\mathcal{R})$ with $\theta = (Z^{\mathcal{T}^\vee} \tau) \circ \eta_\mathcal{R}$. In other words, every arrow θ factors uniquely through the universal arrow $\eta_\mathcal{R}$ as in the commutative diagram of Figure 23.

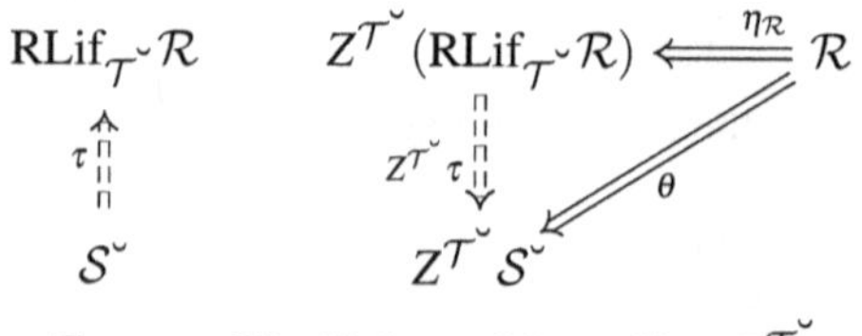

FIGURE 23. Universal from $\mathcal{R}$ to $Z^{\mathcal{T}^\vee}$

This definition sets up the following adjunctions:

Theorem 48. $\varphi_{\mathcal{S}^\vee,\mathcal{R}}: Z^U(Z^{\mathcal{T}^\vee}\mathcal{S}^\vee, \mathcal{R}) \cong V^Z(\mathrm{LLif}_{\mathcal{T}^\vee}\mathcal{R}, \mathcal{S}^\vee)$ and $\varphi_{\mathcal{R},\mathcal{S}^\vee}: Z^U(\mathcal{R}, Z^{\mathcal{T}^\vee}\mathcal{S}^\vee) \cong V^Z(\mathcal{S}^\vee, \mathrm{RLif}_{\mathcal{T}^\vee}\mathcal{R})$, where Z^U is a functor that returns the set of all arrows from U to Z and V^Z is a functor that returns all arrows from Z to V.

The proofs for the theorem are similar to that for left and right Kan lifts, respectively, remembering that $Z^{\mathcal{T}^\vee}$ is a contravariant functor. This presents no difficulties.

For our situation with $\downarrow$ and $\uparrow$, we use the following definition:

Definition 49. Given an arrow $\mathcal{T}^\vee: V \to U$ and an object Z we consider the arrow category V^Z with objects the arrows $\mathcal{S}^\vee, \mathcal{X}: Z \to V$ and 2-arrows $\sigma: \mathcal{S}^\vee \Rightarrow \mathcal{X}$, then we define the contravariant functors $Z_\downarrow^{\mathcal{T}^\vee}, Z_\uparrow^{\mathcal{T}^\vee}: V^Z \to Z^U$ and the object maps $\mathrm{LLif}_{\mathcal{T}^\vee}: Z^U \to V^Z$ and $\mathrm{RLif}_{\mathcal{T}^\vee}: Z^U \to V^Z$; let $\mathcal{P}, \mathcal{Q}: Z \to V$ and $\mathcal{Y}: U \to Z$:

$$Z_\downarrow^{\mathcal{T}^\vee} \stackrel{def}{=} \lambda \mathcal{Q}. \mathcal{T}^\vee \downarrow \mathcal{Q} \qquad \mathrm{LLif}_{\mathcal{T}^\vee} \stackrel{def}{=} \lambda \mathcal{Y}. \mathcal{Y} \downarrow \mathcal{T}^\vee,$$
$$Z_\uparrow^{\mathcal{T}^\vee} \stackrel{def}{=} \lambda \mathcal{Q}. \mathcal{Q} \uparrow \mathcal{T}^\vee \qquad \mathrm{RLif}_{\mathcal{T}^\vee} \stackrel{def}{=} \lambda \mathcal{Y}. \mathcal{T}^\vee \uparrow \mathcal{Y},$$

and 2-arrows by

$$(\nu: \mathcal{P} \Rightarrow \mathcal{Q}) \mapsto (Z_\downarrow^{\mathcal{T}^\vee} \nu: \mathcal{T}^\vee \downarrow \mathcal{Q} \Rightarrow \mathcal{T}^\vee \downarrow \mathcal{P})$$
$$(\nu: \mathcal{P} \Rightarrow \mathcal{Q}) \mapsto (Z_\uparrow^{\mathcal{T}^\vee} \nu: \mathcal{Q} \uparrow \mathcal{T}^\vee \Rightarrow \mathcal{P} \uparrow \mathcal{T}^\vee).$$

Lemma 50. $Z_\downarrow^{\mathcal{T}^\vee}, Z_\uparrow^{\mathcal{T}^\vee}: V^Z \to Z^U$ are contravariant functors where in the categories V^Z and Z^U, the objects are 1-arrows and the arrows are 2-arrows.

Proof. We show the proof for $Z_\downarrow^{\mathcal{T}^\vee}$; the proof for $Z_\uparrow^{\mathcal{T}^\vee}$ is similar. The proof relies on the fact that arrows in the 2-category qua category compose. Assume $\nu_1: \mathcal{P} \Rightarrow \mathcal{Q}$ and $\nu_2: \mathcal{Q} \Rightarrow \mathcal{N}$, then $Z_\downarrow^{\mathcal{T}^\vee} \nu_2: \mathcal{T}^\vee \downarrow \mathcal{N} \Rightarrow \mathcal{T}^\vee \downarrow \mathcal{Q}$ and $Z_\downarrow^{\mathcal{T}^\vee} \nu_1: \mathcal{T}^\vee \downarrow \mathcal{Q} \Rightarrow \mathcal{T}^\vee \downarrow \mathcal{P}$. Hence, $(Z_\downarrow^{\mathcal{T}^\vee} \nu_1) \circ (Z_\downarrow^{\mathcal{T}^\vee} \nu_2) = Z_\downarrow^{\mathcal{T}^\vee} (\nu_2 \circ \nu_1)$, where the latter follows from the properties of 2-categories qua categories from the left lift whiskering above. Similarly, that $Z_\downarrow^{\mathcal{T}^\vee}$ preserves the identity arrows (i.e., the equality relation) also follows from left lift whiskering in the 2-category qua category.

In our context, we change the notation slightly — for ease of typesetting — for functors $Z^{T^\smile}_\downarrow$ and $Z^{T^\smile}_\uparrow$, by setting $Z^{T^\smile}_\downarrow \sigma = T^\smile \triangleright \sigma$ and $Z^{T^\smile}_\uparrow \tau = T^\smile \triangleright \tau$. These functors are contravariant functors, which follows from

$$\frac{\mathcal{Q} \subseteq_\sigma \mathcal{P}}{T^\smile\downarrow\mathcal{P} \subseteq_{T^\smile\triangleright\sigma} T^\smile\downarrow\mathcal{Q}} \qquad \frac{\mathcal{Q} \subseteq_\tau \mathcal{P}}{T^\smile\uparrow\mathcal{P} \subseteq_{T^\smile\triangleright\tau} T^\smile\uparrow\mathcal{Q}},$$

and the definition of $\downarrow$ and $\uparrow$. This says that

$$(\sigma\colon \mathcal{Q} \Rightarrow \mathcal{P}) \mapsto (T^\smile \triangleright \sigma\colon T^\smile\downarrow\mathcal{Q} \Rightarrow T^\smile\downarrow\mathcal{P})$$
$$(\tau\colon \mathcal{Q} \Rightarrow \mathcal{P}) \mapsto (T^\smile \triangleright \tau\colon \mathcal{Q}\uparrow T^\smile \Rightarrow \mathcal{P}\uparrow T^\smile).$$
◁

Theorem 51. *For our situation using Definition 49, the following diagram $\downarrow$ and its dual for $\uparrow$ hold: For all $T^\smile : V \to U$ and $\mathcal{R} : U \to Z$,*

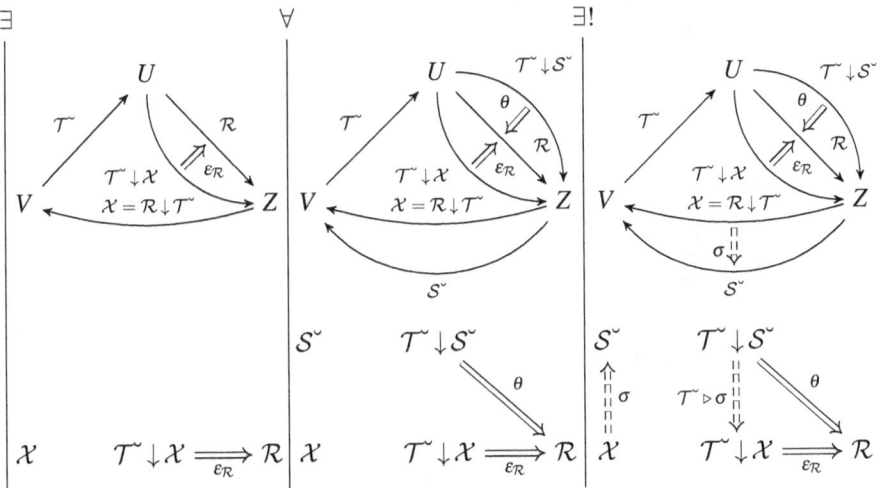

FIGURE 24. $\downarrow$ Lift of $T^\smile$ Through $\mathcal{R}$

Proof. We show the proof for $\downarrow$; the proof for $\uparrow$ is dual. We first show that σ exists. Assume $\langle z, v \rangle \in \mathcal{X}$, then $\langle z, v \rangle \in \mathcal{R}\downarrow T^\smile$ and so there is some $u \in U$ such that $\langle u, z \rangle \notin \mathcal{R}$ and $\langle v, u \rangle \notin T^\smile$. From $\theta : T^\smile\downarrow\mathcal{S}^\smile \Rightarrow \mathcal{R}$ then $\langle u, z \rangle \notin T^\smile\downarrow\mathcal{S}^\smile$. Hence, for all v', $\langle v', u \rangle \notin T^\smile$ implies $\langle z, v' \rangle \in \mathcal{S}^\smile$. Letting $v' = v$, we have $\langle v, u \rangle \notin T^\smile$ implies $\langle z, v \rangle \in \mathcal{S}^\smile$, and therefore, $\langle z, v \rangle \in \mathcal{S}^\smile$. Hence, $\mathcal{R}\downarrow T^\smile \subseteq \mathcal{S}^\smile$ and there exists σ such that $\sigma : \mathcal{R}\downarrow T^\smile \Rightarrow \mathcal{S}^\smile$. Also, σ is unique since inclusions are unique in set theory.

Next we must show that $\theta = \varepsilon_\mathcal{R} \circ (T^\smile \triangleright \sigma)$. Since $T^\smile \triangleright \sigma : T^\smile\downarrow\mathcal{S}^\smile \Rightarrow T^\smile\downarrow\mathcal{X}$ and $\varepsilon_\mathcal{R} : T^\smile\downarrow\mathcal{X} \Rightarrow \mathcal{R}$, then $\varepsilon_\mathcal{R} \circ (T^\smile \triangleright \sigma) : T^\smile\downarrow\mathcal{S}^\smile \Rightarrow \mathcal{R}$. Since $\theta : T^\smile\downarrow\mathcal{S}^\smile \Rightarrow \mathcal{R}$ and inclusions are unique in set theory, $\theta = \varepsilon_\mathcal{R} \circ (T^\smile \triangleright \sigma)$. ◁

Our situation corresponds to the following adjunctions:

$$\varphi_{\mathcal{S}^\smile,\mathcal{R}} : Z^U(T^\smile\downarrow\mathcal{S}^\smile, \mathcal{R}) \cong V^Z(\mathcal{R}\downarrow T^\smile, \mathcal{S}^\smile) \qquad \varphi_{\mathcal{S}^\smile,\mathcal{R}} : Z^U(\mathcal{R}, \mathcal{S}^\smile\uparrow T^\smile) \cong V^Z(\mathcal{S}^\smile, T^\smile\uparrow\mathcal{R}).$$

We have the following logical rules for $\downarrow$ where $\theta = \varepsilon_{\mathcal{R}} \circ (\mathcal{T}^{\smile} \triangleright \sigma)$:

$$\frac{\mathcal{T}^{\smile} \downarrow \mathcal{S}^{\smile} \subseteq_{\theta} \mathcal{R} \quad \mathcal{T}^{\smile} \downarrow (\mathcal{R} \downarrow \mathcal{T}^{\smile}) \subseteq_{\varepsilon_{\mathcal{R}}} \mathcal{R}}{\mathcal{R} \downarrow \mathcal{T}^{\smile} \subseteq_{\sigma} \mathcal{S}^{\smile}} \qquad \frac{\mathcal{R} \downarrow \mathcal{T}^{\smile} \subseteq_{\sigma} \mathcal{S}^{\smile}}{\mathcal{T}^{\smile} \downarrow \mathcal{S}^{\smile} \subseteq_{\mathcal{T}^{\smile} \triangleright \sigma} \mathcal{T}^{\smile} \downarrow (\mathcal{R} \downarrow \mathcal{T}^{\smile})}$$

Left $\downarrow$ Lift Antitonicity

The left rule is equivalent to

$$\mathcal{T}^{\smile} \downarrow \mathcal{S}^{\smile} \subseteq_{\theta} \mathcal{R} \text{ implies } \mathcal{R} \downarrow \mathcal{T}^{\smile} \subseteq_{\sigma} \mathcal{S}^{\smile},$$

because the right premise always holds in our situation. This premise holds via one half of the residuation condition

$$\frac{\mathcal{R} \downarrow \mathcal{T}^{\smile} \subseteq_{\iota} \mathcal{R} \downarrow \mathcal{T}^{\smile}}{\mathcal{T}^{\smile} \downarrow (\mathcal{R} \downarrow \mathcal{T}^{\smile}) \subseteq_{\varepsilon_{\mathcal{R}}} \mathcal{R}},$$

where ι is the equality relation and stands for the identity in the 2-category. In the left $\downarrow$ lift view, residuation is not available and hence that premise must be explicitly stated. The other direction is

$$\frac{\mathcal{R} \downarrow \mathcal{T}^{\smile} \subseteq_{\sigma} \mathcal{S}^{\smile}}{\mathcal{T}^{\smile} \downarrow \mathcal{S}^{\smile} \subseteq_{\mathcal{T}^{\smile} \triangleright \sigma} \mathcal{T}^{\smile} \downarrow (\mathcal{R} \downarrow \mathcal{T}^{\smile}) \quad \mathcal{T}^{\smile} \downarrow (\mathcal{R} \downarrow \mathcal{T}^{\smile}) \subseteq_{\varepsilon_{\mathcal{R}}} \mathcal{R}}{\mathcal{T}^{\smile} \downarrow \mathcal{S}^{\smile} \subseteq_{\theta} \mathcal{R}}$$

The following logical rules are for $\uparrow$ where $\theta = (\mathcal{T}^{\smile} \triangleright \tau) \circ \eta_{\mathcal{R}}$:

$$\frac{\mathcal{R} \subseteq_{\theta} \mathcal{S}^{\smile} \uparrow \mathcal{T}^{\smile} \quad \mathcal{R} \subseteq_{\eta_{\mathcal{R}}} (\mathcal{T}^{\smile} \uparrow \mathcal{R}) \uparrow \mathcal{T}^{\smile}}{\mathcal{S}^{\smile} \subseteq_{\tau} \mathcal{T}^{\smile} \uparrow \mathcal{R}} \qquad \frac{\mathcal{S}^{\smile} \subseteq_{\tau} \mathcal{T}^{\smile} \uparrow \mathcal{R}}{(\mathcal{T}^{\smile} \uparrow \mathcal{R}) \uparrow \mathcal{T}^{\smile} \subseteq_{\mathcal{T}^{\smile} \triangleright \tau} \mathcal{S}^{\smile} \uparrow \mathcal{T}^{\smile}}$$

Left $\uparrow$ Extension Antitonicity

with similar justifications as those for $\downarrow$.

6. Conclusion

We started with RAs as in [8; 13] and made some definitions for extra operators. We showed how to evaluate these algebras (construing them as logics) via Kripke frames using Relevance Logic [5; 17] as a template. The relationship between Relevance Logic in its classical version (from [17]) was then dissected. We showed how to retrieve an algebra of relations from the Kripke frames using the Tabularity Axiom. To get from there to a graph of relations, we recalled some of our previous work in [4] and showed how that led directly to a collection of 1-arrows where there are two categorical composition operators, tensor and cotensor. This formed a 2-category with inclusions as the 2-arrows. By not requiring the 1-category structure, we showed that intensional *nor* and *nand* also could be fit into the framework by treating the 1-arrows as merely a graph of 1-arrows and the 2-arrows forming a 2-category qua category.

This paper represents the logical evolution of our work in [4]. In that paper, the semantics relied upon the three-place Kripke relations that we borrowed from Relevance Logic (modulo a few restrictions to prevent commutativity of tensor). However, having a semantics for a distributed RA that did not use binary relations could be considered a drawback of that work. We improved on that work in this paper by showing how it can be done successfully.

The extensions and lifts were used to neatly characterize the framework from a categorical perspective. We view the extensions and lifts as prescribing those features that a realizability interpretation must meet. The extensions and lifts characterize the relational operators in a partial manner in that all the adjunctions, or residuations if you like, do need not to exist. This is an important aspect for us, because it allows the logic to capture a crucial aspect of FPGA applications, namely, that foreign IP's as black boxes exist in almost all FPGA applications thus depriving a high assurance analysis of all the relations governing the operation of the FPGA application.

The intensional *nor* and *nand* forced us to consider 2-categories qua categories and made us realize the Kan extensions and lifts do not rely upon a 1-category, only 1-arrows. As a consequence, composition was replaced by covariant bifunctors on the 2-category qua category in the case of tensor and cotensor and contravariant bifunctors in the case of intensional *nor* and *nand*.

The Kan extensions and lifts pointed out that a fully abstract algebra of relations supports two compositions, tensor and cotensor. It is possible to abstract this further into two different hom functors. The distribution of tensor over cotensor (distribution not in the sense of distributed algebra) could be formalized by using hom functors into a double monoidal category. The distribution then takes the form of certain natural transformations and their rules in the double monoidal category. One can go further and treat the lifts of intensional *nor* and *nand* as functors into that same double monoidal category. The algebraic relationships among the operators are then realized by natural transformations and rules on the double monoidal category.

REFERENCES

[1] Allwein, G., Demir, H. and Pike, L. (2004). Logics for classes of Boolean monoids, *Journal of Logic, Language and Information* **13**: 241–266.

[2] Allwein, G. and Dunn, J. M. (1993). Kripke models for linear logic, *Journal of Symbolic Logic* **58**: 514–545.

[3] Allwein, G. and Harrison, W. L. (2016). Distributed modal logic, *in* K. Bimbó (ed.), *J. Michael Dunn on Information Based Logic*, Vol. 8 of *Outstanding Contributions to Logic*, Springer International, Switzerland, pp. 331–362.

[4] Allwein, G., Harrison, W. L. and Reynolds, T. (2017). Distributed relation logic, *Logic and Logical Philosophy* **26**(1): 19–61.

[5] Anderson, A. R. and Belnap, N. D. (1975). *Entailment: The Logic of Relevance and Necessity*, Vol. I, Princeton University Press, Princeton, NJ.

[6] Bénabou, J. (1967). Introduction to bicategories, *Reports of the Midwest Category Seminar*, Springer-Verlag, Berlin, pp. 1–77.

[7] Bimbó, K. and Dunn, J. M. (2009). Symmetric generalized Galois logics, *Logica Universalis* **3**(1): 125–152.

[8] Chin, L. H. and Tarski, A. (1953). Distributive and modular laws in the arithmetic of relation algebras, *Journal of Symbolic Logic* **18**(1): 72. (Originally appeared in *University of California Publications in Mathematics*, ns. vol. 1, no. 9 (1951), pp. 341–384.).

[9] Dunn, J. M. (1991). Gaggle theory: An abstraction of Galois connections and residuation, with applications to negation, implication, and various logical operators, *in* J. van Eijck (ed.), *Logics in AI: European Workshop JELIA '90*, number 478 in *Lecture Notes in Computer Science*, Springer, Berlin, pp. 31–51.

[10] Dunn, J. M. (2001). A representation of relation algebras using Routley–Meyer frames, *in* C. A. Anderson and M. Zelëny (eds.), *Logic, Meaning and Computation. Essays in Memory of Alonzo Church*, Kluwer, Dordrecht, pp. 77–108.
[11] Freyd, P. J. and Scedrov, A. (1990). *Categories, Allegories*, North-Holland, Amsterdam.
[12] Mac Lane, S. (1998). *Categories for the Working Mathematician*, 2nd edn, Springer-Verlag, Berlin.
[13] Ng, K. C. (1984). *Relation Algebras with Transitive Closure*, PhD Thesis, University of California, Berkeley, CA.
[14] Pratt, V. (1990). Dynamic algebras as a well-behaved fragment of relation algebras, *in* C. H. Bergman, R. D. Maddux and D. L. Pigozzi (eds.), *Algebraic Logic and Universal Algebra in Computer Science*, number 425 in *Lecture Notes in Computer Science*, Springer, Amsterdam, pp. 77–110.
[15] Pratt, V. R. (1991). Action logic and pure induction, *in* J. van Eijck (ed.), *Logics in AI: European Workshop JELIA '90*, number 478 in *Lecture Notes in Computer Science*, Springer, Berlin, pp. 97–120.
[16] Pratt, V. R. (1992). Origins of the calculus of binary relations, *Proceedings of the Seventh Annual IEEE Symposium on Logic in Computer Science*, Springer-Verlag, Berlin, pp. 248–254.
[17] Routley, R., Meyer, R. K., Plumwood, V. and Brady, R. T. (1982). *Relevant Logics and Their Rivals*, Vol. 1, Ridgeview Publishing Company, Atascadero, CA.
[18] Street, R. (1983). Enriched categories and cohomology, *Questiones Mathematicae* **6**: 265–283.

NAVAL RESEARCH LABORATORY, CODE 5543, WASHINGTON, DC 20375, U.S.A.
Email: gerard.allwein@nrl.navy.mil
TWO SIX TECHNOLOGIES, INC., 901 NORTH STUART ROAD, ARLINGTON, VA 22203, U.S.A.
Email: william.lawrence.harrison@gmail.com

FOUR-VALUED SEMANTICS FOR ABSTRACT ARGUMENTATION FRAMEWORKS USING (EXTENSIONS OF) DUNN–BELNAP FOUR-VALUED LOGIC

Ofer Arieli

ABSTRACT. Four-valued labelings have been introduced as a more informative substitute to standard three-valued semantics of abstract argumentation frameworks. In this paper, we consider some 4-valued semantics for argumentation frameworks and show how they can be represented by extensions of Dunn–Belnap 4-valued logic and the corresponding 4-valued bilattice structure that naturally reflects such semantics.

Keywords. Argumentation theory, 4-valued semantics, Paraconsistent reasoning

1. INTRODUCTION

Argumentation is a cognitive process for dealing with conflicting information by generating and comparing arguments. In the seminal paper of Dung [22] this process is described by abstract argumentation frameworks (AAF for short), which can be viewed as directed graphs, where nodes represent arguments and directed edges represent attacks between the nodes. Given such a graph, a key issue is to determine set(s) of nodes, the arguments of which can be collectively accepted. Such sets determine what can be inferred from the AAF.

It is usual to evaluate AAFs in terms of 3-valued labeling functions (Baroni et al. [6; 7]): an argument can be either accepted (labeled in), rejected (labeled out), or undecided (labeled undec). However, the following example shows that such a categorization might be too crude.

Example 1. The following example is a variation of the decision-making problem presented in Pollock [40]: Suppose that a traveler has some doubts whether to take an umbrella or sunglasses to her journey. She checks two weather websites, one indicates that the weather in her destination is rainy, while the other one points out that the weather in the destination is sunny. Assuming that the web-sources are equally reliable, taking exactly one action (and so avoiding the other) is arbitrary, thus an irrational decision in this case. The traveler still has *two* further options for making a choice: she may withhold any action and wait until the weather conditions are clarified, or she may take a more practical decision and take both umbrella and sunglasses. The latter is a pragmatic approach, accepting contradictory indications whenever this doesn't cause any real risk or damage. In other situations, for instance when there are

---

2020 *Mathematics Subject Classification.* Primary: 03B70, Secondary: 03B47, 68T27.

Bimbó, Katalin, (ed.), *Relevance Logics and other Tools for Reasoning. Essays in Honor of J. Michael Dunn*, (Tributes, vol. 46), College Publications, London, UK, 2022, pp. 31–53.

conflicting symptoms obliging different medical treatments, it may be more rational to refrain from irreversible acts and look for further opinions. In both cases, though, the two neutral options (that is, those that are not biased towards the weather forecast of a particular website) have totally different consequences, so it is useful to clearly distinguish between them. This requires *two* further distinct labels, apart of in and out.

In an attempt to provide more informative and accurate descriptions to situations like those in Example 1, four-valued labelings were introduced in a number of papers (see Arieli [3], Bistarelli and Taticchi [13], Jakobovits and Vermeir [35], Riveret et al. [41]). In what follows we consider and extend some of these approaches, showing that they can be naturally embedded in (extensions of) the well-known four-valued interpretation of FDE (Anderson and Belnap [1], see also Omori and Wansing [38; 39]), introduced by J. Michael Dunn [24; 25] and Nuel Belnap [8; 9]. We thus argue that FDE-based formalisms may serve as a solid platform for representing and reasoning with argumentation frameworks.

The rest of this paper is organized as follows: in the next section we give some preliminaries on abstract argumentation frameworks and their semantics. Then, in Sections 3 and 4, we present some conflict-tolerant and conflict-free approaches (respectively) to 4-valued semantics of argumentation frameworks. In Section 5 we show how Dunn–Belnap logic and its extensions can be used for representing these semantics. Section 6 is a short conclusion of the paper.

2. ARGUMENTATION FRAMEWORKS AND THEIR SEMANTICS

We start by recalling the basic definitions behind abstract argumentation frameworks.

Definition 2 (Argumentation framework). An (abstract) *argumentation framework* (AAF) [22] is a pair $\mathcal{AF} = \langle \mathsf{Args}, \mathcal{A} \rangle$, where Args is a finite set, the elements of which are called *arguments*, and $\mathcal{A}$ is a relation on $\mathsf{Args} \times \mathsf{Args}$ whose instances are called *attacks*. When $(a,b) \in \mathcal{A}$ we say that *a attacks b* (or that *b is attacked by a*).

Given an argumentation framework $\mathcal{AF} = \langle \mathsf{Args}, \mathcal{A} \rangle$, the following notations will be useful in what follows (for an argument $a \in \mathsf{Args}$ and a set of arguments $\mathcal{E} \subseteq \mathsf{Args}$):

- The set of arguments that are attacked by a is: $a^+ = \{b \in \mathsf{Args} \mid (a,b) \in \mathcal{A}\}$.
- The set of arguments that attack a is: $a^- = \{b \in \mathsf{Args} \mid (b,a) \in \mathcal{A}\}$.
- The set of argument that are attacked by (respectively, that attack) $\mathcal{E}$ is: $\mathcal{E}^+ = \bigcup_{a \in \mathcal{E}} a^+$ (respectively, $\mathcal{E}^- = \bigcup_{a \in \mathcal{E}} a^-$).
- The set of arguments that are *defended* by $\mathcal{E}$ is: $\mathsf{Def}(\mathcal{E}) = \{a \in \mathsf{Args} \mid a^- \subseteq \mathcal{E}^+\}$.
- We say that $\mathcal{E}$ is *conflict-free*, if $\mathcal{E} \cap \mathcal{E}^+ = \emptyset$.

Thus, a set $\mathcal{E} \subseteq \mathsf{Args}$ is conflict-free if there are no attacks between arguments in $\mathcal{E}$, and an argument $a \in \mathsf{Args}$ is defended by $\mathcal{E}$ if any attacker of a is counter-attacked by (some argument in) $\mathcal{E}$.

There are two ways of giving semantics to an AAF: by extensions and by labeling. First, we describe the former.[1] The following definition lists some common types of extensions for an AAF.

---

[1] The notion of "extension" is somewhat overloaded in this paper. In the context of logics and their languages, extensions have their usual meaning of supersets (maybe with further properties like extensions of

Definition 3 (extension semantics). Let $\mathcal{AF} = \langle \text{Args}, \mathcal{A} \rangle$ be an AAF and let $\mathcal{E} \subseteq \text{Args}$ be a conflict free set of arguments. We say that $\mathcal{E}$ is:

- an *admissible* set of $\mathcal{AF}$, if $\mathcal{E} \subseteq \text{Def}(\mathcal{E})$,
- a *complete* extension of $\mathcal{AF}$, if $\mathcal{E} = \text{Def}(\mathcal{E})$,
- a *grounded* extension of $\mathcal{AF}$, if $\mathcal{E}$ is a $\subseteq$-minimal extension among the complete extensions of $\mathcal{AF}$,
- a *preferred* extension of $\mathcal{AF}$, if $\mathcal{E}$ is a $\subseteq$-maximal extension among the complete extension of $\mathcal{AF}$,
- a *stable* extension of $\mathcal{AF}$, if $\mathcal{E}$ is a complete extension of $\mathcal{AF}$ such that $\mathcal{E} \cup \mathcal{E}^+ = \text{Args}$.

Thus, a conflict-free set is admissible if it defends all of its elements, and an admissible set is complete if it defends only its elements. Also, it is well-known that the grounded extension of an AAF is unique and that stable extensions do not always exist for AAFs [22]. A discussion on the relations between these extensions, as well as definitions of further extensions, can be found in, e.g., [6; 7; 22].

Example 4. Recall Example 1 from the introduction. A corresponding argumentation framework is presented in Figure 1, abbreviating by *r s*, *u* and *g* the claims that "it is rainy," "it is sunny," "take an umbrella" and "take sunglasses" (respectively).

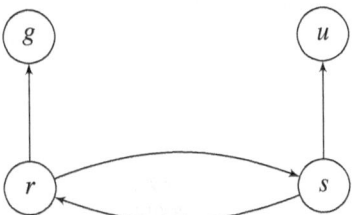

FIGURE 1. Attack diagram for Examples 1 and 4

The grounded extension in this case is the emptyset, and there are two preferred extensions, which are also the stable extensions of the framework: $\{s,g\}$ (it is sunny, take sunglasses), and $\{r,u\}$ (it is rainy, take an umbrella).

The other way of giving semantics to argumentation frameworks is by *labeling functions* (Caminada [15]; Caminada and Gabbay [16]), assigning a value from $\{\text{in}, \text{out}, \text{undec}\}$ to every element in Args. Intuitively, the label of an argument represents its status: accepted, rejected, or undecided. Given a labeling $lab: \text{Args} \to \{\text{in}, \text{out}, \text{undec}\}$, for every val $\in \{\text{in}, \text{out}, \text{undec}\}$ we denote: $\text{Val}(lab) = \{a \in \text{Args} \mid lab(a) = \text{val}\}$, and associate lab with the triplet $\langle \text{In}(lab), \text{Out}(lab), \text{Undec}(lab) \rangle$.

Common types of labeling functions are defined next.

Definition 5 (Labeling semantics). Given an abstract argumentation framework $\mathcal{AF} = \langle \text{Args}, \mathcal{A} \rangle$, let $lab : \text{Args} \to \{\text{in}, \text{out}, \text{undec}\}$ be a total function. For an argument $a \in \text{Args}$, we consider the following rules:

---

functions to larger domains). In the context of argumentation frameworks, we adopt the standard terminology, which regards to extensions as sets of arguments having certain properties, as indicated in Definition 3.

In1 If $lab(a) = \text{in}$, then there is no $b \in a^-$ such that $lab(b) = \text{in}$.
In2 If $lab(a) = \text{in}$, then for every $b \in a^-$ it holds that $lab(b) = \text{out}$.
Out If $lab(a) = \text{out}$, then there exists some $b \in a^-$ such that $lab(b) = \text{in}$.
Undec If $lab(a) = \text{undec}$, then it is not the case that for every $b \in a^-$ it holds that $lab(b) = \text{out}$ and does not exist an argument $b \in a^-$ such that $lab(b) = \text{in}$.

Now, we say that lab (w.r.t. $\mathcal{AF}$) is:

- *conflict-free*, if it satisfies conditions **In1** and **Out** for every $a \in \text{Args}$,
- *admissible*, if it satisfies conditions **In2** and **Out** for every $a \in \text{Args}$,
- *complete*, if it is admissible and satisfies condition **Undec** for every $a \in \text{Args}$.

Furthermore, if lab is complete (w.r.t. $\mathcal{AF}$), we say that it is:

- *grounded*, if for no other complete labeling lab' it holds that $\text{In}(lab') \subsetneq \text{In}(lab)$,
- *preferred*, if for no other complete labeling lab' it holds that $\text{In}(lab) \subsetneq \text{In}(lab')$,
- *stable*, if $\text{Undec}(lab) = \emptyset$.

Works on the relations between Dung-style extensions and value assignments to arguments may be traced back to Verheij [42]. The following correspondence between extension semantics and labeling semantics (adjusted to the notations of Definition 5) is shown in [16].

Proposition 6. *Let $\mathcal{AF} = \langle \text{Args}, \mathcal{A} \rangle$ be an AAF. If lab is a complete labeling w.r.t. $\mathcal{AF}$, then $\text{In}(lab)$ is a complete extension of $\mathcal{AF}$. Conversely, if $\mathcal{E}$ is a complete extension of $\mathcal{AF}$, then $\langle \mathcal{E}, \mathcal{E}^+, \text{Args} \setminus (\mathcal{E} \cup \mathcal{E}^+) \rangle$ is a complete labeling w.r.t. $\mathcal{AF}$.*

Given an argumentation framework $\mathcal{AF}$, Proposition 6 may be shown by using two mappings. One, $\text{L2E}_{\mathcal{AF}}$, from the labelings of $\mathcal{AF}$ to the extensions of $\mathcal{AF}$, is defined by $\text{L2E}_{\mathcal{AF}}(lab) = \text{In}(lab)$, and the other, $\text{E2L}_{\mathcal{AF}}$, from the extensions of $\mathcal{AF}$ to the labeling of $\mathcal{AF}$, is defined by $\text{E2L}_{\mathcal{AF}}(\mathcal{E}) = \langle \mathcal{E}, \mathcal{E}^+, \text{Args} \setminus (\mathcal{E} \cup \mathcal{E}^+) \rangle$. When the domains and ranges of these functions are restricted to complete extensions and complete labelings of $\mathcal{AF}$, the functions become bijections and each other's inverses, making complete extensions and complete labelings one-to-one related. As a consequence, one concludes that similar relations hold between grounded (respectively, preferred, stable) labelings and grounded (respectively, preferred, stable) extensions of $\mathcal{AF}$ (see [16]).

3. Conflict-Tolerant Semantics

Let's reconsider the argumentation framework in Example 4. The fact that every extension must be conflict-free dictates that one has to make an arbitrary choice between equally possible situations (rainy or sunny), as implied by the stable/preferred semantics, or abandon both of them altogether, as reflected by the grounded semantics. However, as we already argued in Example 1, there is another possibility: to be prepared for both cases. For this, conflict-freeness should be relaxed, and another label should be introduced. Intuitively, such a fourth label designates a kind of "cautious acceptance" of the arguments to which it is assigned. In our case, this means that although the conflicting situations may not occur simultaneously, both of them are still taken into account.

Another motivation for waving the conflict-freeness property (or introducing a fourth labeling value) is the following.

Example 7. Consider the attack diagram presented in Figure 2.[2] The associated argumentation framework does not have any stable extension, and the only preferred extension (which is also the grounded extension) in this case is the emptyset, meaning that none of the arguments is accepted. As we shall see in what follows, conflict-tolerant semantics enables (a weaker form of) stable extensions for every AAF, and the preferred extensions of an AAF according to this semantics are non-empty.

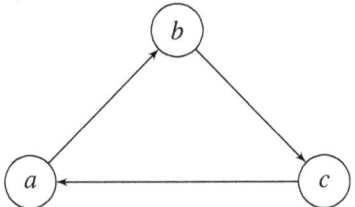

FIGURE 2. Attack diagram for Example 7

For resolving the above issues of conflict-free semantics, we now consider extension-based and labeling-based counterparts of the formalisms presented in the previous section. The conflict-free requirement is lifted from the definition of extensions, and a fourth value is added to the range of the labeling functions. This value (denoted conf, for "conflicting") intuitively designates cautious acceptance in the presence of counter-arguments.

We start with labeling semantics. The following definition is due to [3]:

Definition 8 (p-labeling). Given an abstract argumentation framework $\mathcal{AF} = \langle \text{Args}, \mathcal{A} \rangle$. A total function $lab : \text{Args} \to \{\text{in}, \text{out}, \text{conf}, \text{undec}\}$ is called *p-admissible*[3], if for every $a \in \text{Args}$ it satisfies the following rules:

pIn   If $lab(a) = \text{in}$, then for every $b \in a^-$ it holds that $lab(b) = \text{out}$.
pOut If $lab(a) = \text{out}$, then there exists some $b \in a^-$ such that $lab(b) \in \{\text{in}, \text{conf}\}$.
pConf If $lab(a) = \text{conf}$, then for every $b \in a^-$ it holds that $lab(b) \in \{\text{out}, \text{conf}\}$
         and there exists some $b \in a^-$ such that $lab(b) = \text{conf}$.
pUndec If $lab(a) = \text{undec}$, then for every $b \in a^-$ it holds that $lab(b) \in \{\text{out}, \text{undec}\}$.

We say that lab is *p-complete*, if for every $a \in \text{Args}$ it satisfies the following (stronger) rules:

pIn$^+$ $lab(a) = \text{in}$ iff for every $b \in a^-$ it holds that $lab(b) = \text{out}$.
pOut$^+$ $lab(a) = \text{out}$ iff there is some $b \in a^-$ such that $lab(b) \in \{\text{in}, \text{conf}\}$
             and there is some $b \in a^-$ such that $lab(b) \in \{\text{in}, \text{undec}\}$.
pConf$^+$ $lab(a) = \text{conf}$ iff for every $b \in a^-$ it holds that $lab(b) \in \{\text{out}, \text{conf}\}$
             and there exists some $b \in a^-$ such that $lab(b) = \text{conf}$.
pUndec$^+$ $lab(a) = \text{undec}$ iff for every $b \in a^-$ it holds that $lab(b) \in \{\text{out}, \text{undec}\}$
              and there exists some $b \in a^-$ such that $lab(b) = \text{undec}$.

Intuitively, in a four-valued labeling in and conf stand for two levels of acceptance. The former represents full acceptance of the underlying argument, while the

---
[2] In fact, any diagram with an odd-length cycle is appropriate in this case.
[3] The prefix "p" stands here and in what follows for "paraconsistent."

latter is a more cautious one, based on conflicting claims. Again, for each val $\in$ $\{\text{in}, \text{out}, \text{conf}, \text{undec}\}$ we define the set $\text{Val}(lab) = \{a \in \text{Args} \mid lab(a) = \text{val}\}$. Accordingly, a 4-valued labeling lab is associated with the quadruplet $\langle \text{In}(lab), \text{Out}(lab), \text{Conf}(lab), \text{Undec}(lab) \rangle$.

Note 9. It is not difficult to verify that Definition 8 enhances Definition 5: every admissible (respectively, complete) labeling of $\mathcal{AF}$ is also a p-admissible (respectively, p-complete) labeling of $\mathcal{AF}$ (but not the other way around).

The other types of labelings in Definition 5 can be extended to 4-valued counterparts in a similar way. For instance, a p-stable labeling lab is a p-complete labeling where $\text{Undec}(lab) = \emptyset$.

Example 10. Consider again the argumentation framework $\mathcal{AF}$ of our running example (see Example 1 and Figure 1). This framework has four p-complete labelings:

1. lab_1, in which $\text{In}(lab_1) = \text{Out}(lab_1) = \text{Conf}(lab_1) = \emptyset$ and $\text{Undec}(lab_1) = \{r, s, u, g\}$,
2. lab_2, in which $\text{In}(lab_2) = \{r, u\}$, $\text{Out}(lab_2) = \{s, g\}$, and $\text{Conf}(lab_2) = \text{Undec}(lab_2) = \emptyset$,
3. lab_3, in which $\text{In}(lab_3) = \{s, g\}$, $\text{Out}(lab_3) = \{r, u\}$, and $\text{Conf}(lab_3) = \text{Undec}(lab_3) = \emptyset$,
4. lab_4, in which $\text{In}(lab_4) = \text{Out}(lab_4) = \text{Undec}(lab_4) = \emptyset$ and $\text{Conf}(lab_4) = \{r, s, u, g\}$.

The first three labelings correspond to the three extensions discussed in Example 4. The first labeling is associated with the grounded extension of the framework. It reflects a skeptical approach, which in this case means that no action should be taken. The next two labelings meet the two preferred/stable extensions of the framework. They reflect a credulous approach in which each labeling corresponds to a different coherent assumption about the weather conditions in the destination. The fourth labeling is a new one, not available in a 3-valued semantics. Intuitively, this is a kind of an intermediate approach that takes all the weather forecasts into consideration (sunny or rainy day), and accordingly makes the required decisions. It may be understood as a "cautious" acceptance of the state of affairs, since eventually one of the forecasts will turn out to be mistaken.

We turn now to conflict-tolerant extensions. As the next definition indicates, we just give-up the conflict-freeness requirement in Definition 3.

Definition 11 (p-extensions). Let $\mathcal{AF} = \langle \text{Args}, \mathcal{A} \rangle$ be an AAF, and let $\mathcal{E} \subseteq \text{Args}$. We say that $\mathcal{E}$ is

- *paraconsistently admissible* (or: *p-admissible*) extension for $\mathcal{AF}$, if $\mathcal{E} \subseteq \text{Def}(\mathcal{E})$,
- *paraconsistently complete* (or: *p-complete*) extension for $\mathcal{AF}$, if $\mathcal{E} = \text{Def}(\mathcal{E})$.

Other counterparts of the extensions in Definition 3 can be defined just as in Definition 11, where the conflict-freeness requirement is omitted. For instance, a p-stable extension $\mathcal{E}$ of $\mathcal{AF}$ is a p-complete extension of $\mathcal{AF}$, where $\mathcal{E} \cup \mathcal{E}^+ = \text{Args}$.

Note 12. Clearly, Definition 11 enhances Definition 3: for each type of semantic sem considered in Definition 3, any sem-extension of $\mathcal{AF}$ is also a p-sem extension of $\mathcal{AF}$

(but not the other way around). For instance, the complete extensions in Example 4 are also p-complete extensions of the same framework, but there is another p-complete extension: $\{s,g,r,u\}$.

As in the 3-valued case, it is possible to define bijective mappings from labelings to extensions and from extensions to labelings. Given an argumentation framework $\mathcal{AF} = \langle \mathsf{Args}, \mathcal{A}\rangle$, we define:

- For every $\mathcal{E} \subseteq \mathsf{Args}$ and argument $a \in \mathsf{Args}$,

$$\mathsf{pE2L}_{\mathcal{AF}}(\mathcal{E})(a) = \begin{cases} \mathsf{in} & \text{if } a \in \mathcal{E} \text{ and } a \notin \mathcal{E}^+, \\ \mathsf{conf} & \text{if } a \in \mathcal{E} \text{ and } a \in \mathcal{E}^+, \\ \mathsf{out} & \text{if } a \notin \mathcal{E} \text{ and } a \in \mathcal{E}^+, \\ \mathsf{undef} & \text{if } a \notin \mathcal{E} \text{ and } a \notin \mathcal{E}^+. \end{cases}$$

- For every 4-valued labeling lab on Args,

$$\mathsf{pL2E}_{\mathcal{AF}}(lab) = \mathsf{In}(lab) \cup \mathsf{Conf}(lab).$$

The next result, shown in [3], is a counterpart of Proposition 6. Note that for the 4-valued conflict-tolerant case, unlike the 3-valued conflict-free case, this result holds also for p-admissible extensions, and not only for p-complete extensions.

Proposition 13. *Let $\mathcal{AF} = \langle \mathsf{Args}, \mathcal{A}\rangle$ be an AAF. Then:*
  (a) *If lab is a p-admissible labeling for $\mathcal{AF}$, then $\mathsf{pL2E}_{\mathcal{AF}}(lab)$ is a p-admissible extension of $\mathcal{AF}$. Conversely, if $\mathcal{E}$ is a p-admissible extension of $\mathcal{AF}$, then $\mathsf{pE2L}_{\mathcal{AF}}(\mathcal{E})$ is a p-admissible labeling for $\mathcal{AF}$.*
  (b) *If lab is a p-complete labeling for $\mathcal{AF}$, then $\mathsf{pL2E}_{\mathcal{AF}}(lab)$ is a p-complete extension of $\mathcal{AF}$. Conversely, if $\mathcal{E}$ is a p-complete extension of $\mathcal{AF}$, then $\mathsf{pE2L}_{\mathcal{AF}}(\mathcal{E})$ is a p-complete labeling for $\mathcal{AF}$.*

Moreover, we have that for every p-admissible (respectively, p-complete) labeling lab,

$$\mathsf{pE2L}_{\mathcal{AF}}(\mathsf{pL2E}_{\mathcal{AF}}(lab)) = lab,$$

and for every p-admissible (respectively, p-complete) extension $\mathcal{E}$,

$$\mathsf{pL2E}_{\mathcal{AF}}(\mathsf{pE2L}_{\mathcal{AF}}(\mathcal{E})) = \mathcal{E}.$$

Thus, the functions $\mathsf{pE2L}_{\mathcal{AF}}$ and $\mathsf{pL2E}_{\mathcal{AF}}$, restricted to the p-admissible (p-complete) labelings and the p-admissible (p-complete) extensions of $\mathcal{AF}$, are bijective, and are each other's inverse.

Note 14 (conf-free labeling). While conflict-tolerant semantics allows for tolerating conflicts among accepted arguments, it is obviously desirable to reduce such conflicts to a minimum. Consider, for instance, the argumentation framework in Figure 3.

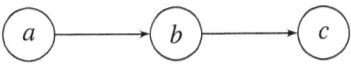

FIGURE 3. Attack diagram for Note 14

This framework has two p-stable extensions (two p-stable labelings): $\mathcal{E}_1 = \{a,b,c\}$ (lab_1, where $lab_1(a) = lab_1(b) = lab_1(c) = \mathsf{conf}$) and $\mathcal{E}_2 = \{a,c\}$ (lab_2, where $lab_2(a)$

$= lab_2(c) =$ in and $lab_2(b) =$ out). While $\mathcal{E}_1$ and lab_1 represent a degenerated interpretation to the situation at hand, according to which all the arguments are conflicting, the interpretation that is reflected by $\mathcal{E}_2$ and lab_2 is more faithful to the state of affairs: the argument that is not attacked (a), as well as the argument (c) that is defends by the unattacked argument, are accepted, while the argument (b) that is attacked by an accepted argument is rejected.

Labelings without conf-assignments are called *conf-free*. In [3] it is shown that if lab is a conf-free p-admissible (p-complete) labeling of $\mathcal{AF}$, then pL2E$_{\mathcal{AF}}(lab)$ is an admissible (complete) extension of $\mathcal{AF}$, and conversely: if $\mathcal{E}$ is an admissible (complete) extension of $\mathcal{AF}$, then pE2L$_{\mathcal{AF}}(\mathcal{E})$ is a conf-free p-admissible (p-complete) labeling of $\mathcal{AF}$. Moreover, the set of the conf-free p-admissible (p-complete) labelings of $\mathcal{AF}$ coincides with the set of the admissible (complete) labelings of $\mathcal{AF}$.

The relations between 4-valued labeling, conflict-tolerant extensions, and their conf-free variations, is summarized in Figure 4.

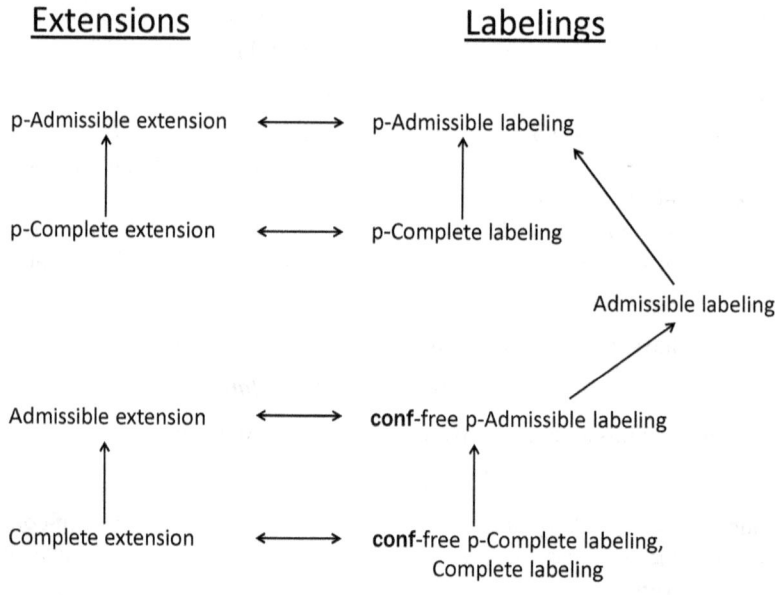

FIGURE 4. Relations between extension and labeling semantics

Note 15. Conflict-tolerant semantics turns out to be particularly useful in constraint argumentation frameworks, which are argumentation frameworks augmented with constraints about the acceptance of the arguments that must be satisfied. In such cases, the waiving of the conflict-freeness requirement allows to extract p-admissible sets of arguments that meet the constraints, while for admissible sets of arguments this is not always possible. We refer to Arieli [2] for more details on this subject.

4. Conflict-Free 4-Valued Semantics

In this section we recall some other methods of giving 4-valued semantics to abstract argumentation frameworks, all of them keep the accepted sets of arguments conflict-free (and, as before, admissible). Again, the motivation for introducing a fourth value is to have more informative labelings of the arguments, but unlike the approach described in Section 3, this time the fourth value intuitively represents "don't care" situations, i.e., a kind of "unlabeled arguments" that one does not want to consider in computing acceptability. To emphasize the difference in the intuitive meaning of the fourth value, we denote it in what follows by off (following [41]), instead of conf. The other labeling notations remain the same, namely, in, out and undec.

Definitions 16–18 describe three approaches of giving 4-valued conflict-free labeling semantics to argumentation frameworks. In all these approaches (unlike the one described in the previous section), accepted arguments are only those that are labeled in. Since none of the approaches enables attacks on in-labeled arguments by other in-labeled arguments, these approaches are indeed conflict-free.

The first labeling, due to Jakobovits and Vermeir [35], is based on four possible acceptance states obtained from the combinations of two basic labels.

Definition 16 (JV labeling). Let $\mathcal{AF} = \langle \mathsf{Args}, \mathcal{A} \rangle$ be an AAF, and let $\pm : \mathsf{Args} \to 2^{\{+,-\}}$ be a function satisfying, for every $a \in \mathsf{Args}$, the following conditions:
JV1: if $- \in \pm(a)$, there is some $b \in a^-$ such that $+ \in \pm(b)$,
JV2: if $+ \in \pm(a)$, for every $b \in a^-$ and for every $b \in a^+$, it holds that $- \in \pm(b)$.
A *JV-labeling* for $\mathcal{AF}$ is a total function $lab : \mathsf{Args} \to \{\mathsf{in}, \mathsf{out}, \mathsf{off}, \mathsf{undec}\}$, such that for every $a \in \mathsf{Args}$ it holds that $lab(a) = \mathsf{in}$ iff $\pm(a) = \{+\}$, $lab(a) = \mathsf{out}$ iff $\pm(a) = \{-\}$, $lab(a) = \mathsf{off}$ iff $\pm(a) = \{-,+\}$, and $lab(a) = \mathsf{undec}$ iff $\pm(a) = \emptyset$.[4]

The intuition behind JV-labelings is simple: to weaken an argument (i.e., to add a "$-$" to its labeling) one needs a reason, i.e., an attack from a supported argument (one whose labeling contains "$+$"). This is the first rule. The second rule indicates that a supported argument weakens the arguments that it attacks, and that one cannot support an argument unless all of its attackers are weakened. As indicated in [35], the correspondence to extension semantics is that $\mathcal{E}$ is a stable extension of $\mathcal{AF}$ iff there is a JV-labeling lab for $\mathcal{AF}$, for which $\mathsf{Undec}(lab) = \mathsf{Off}(lab) = \emptyset$ and $\mathsf{In}(lab) = \mathcal{E}$.

The next definition, due to Riveret, Oren and Sartor [41], explicates the idea that the fourth label corresponds to "don't care" situations.

Definition 17 (ROS labeling). Let $\mathcal{AF} = \langle \mathsf{Args}, \mathcal{A} \rangle$ be an AAF extending $\mathcal{AF}' = \langle \mathsf{Args}', \mathcal{A}' \rangle$ (i.e., $\mathsf{Args}' \subseteq \mathsf{Args}$ and $\mathcal{A}' \subseteq \mathcal{A}$). A total function $lab : \mathsf{Args} \to \{\mathsf{in}, \mathsf{out}, \mathsf{off}, \mathsf{undec}\}$ is a *ROS-labeling*, if there is a grounded labeling $lab' : \mathsf{Args}' \to \{\mathsf{in}, \mathsf{out}, \mathsf{undec}\}$ such that $\forall a \in \mathsf{Args}'$, $lab(a) = lab'(a)$, and $\forall a \in \mathsf{Args} \setminus \mathsf{Args}'$, $lab(a) = \mathsf{off}$.

Thus, the view of ROS-labeling is that off-labeled arguments are "ignored" and are not evaluated when computing acceptance of arguments. Note that according to ROS-labeling such computations are restricted to grounded labeling.

---

[4] In [35], arguments that are labeled off are intuitively regarded as undecided, and those that are labeled undec intuitively signify don't care policy. We keep the labeling in Definition 16 for uniformity with the other labeling methods.

In order to relate 4-valued labeling to extensions, Bistarelli and Taticchi introduce in [13] the following labeling functions:

Definition 18 (BT labeling). Let $\mathcal{AF} = \langle \text{Args}, \mathcal{A} \rangle$ be an AAF, and let lab : Args $\to$ {in, out, off, undec}. We say that lab is a:

- *conflict-free* BT-labeling, if the next conditions are satisfied for every $a \in$ Args:
  - if $lab(a) =$ in then for every $b \in a^-$ it holds that $lab(b) \neq$ in
  - if $lab(a) =$ out then there is some $b \in a^-$ such that $lab(b) =$ in
- *admissible* BT-labeling, if the next conditions are satisfied for every $a \in$ Args:
  - if $lab(a) =$ in then for every $b \in a^-$ it holds that $lab(b) \neq$ in
  - $lab(a) =$ out iff there is some $b \in a^-$ such that $lab(b) =$ in
- *complete* BT-labeling, if the next conditions are satisfied for every $a \in$ Args:
  - $lab(a) =$ in iff for every $b \in a^-$ it holds that $lab(b) \in \{$out, off$\}$
  - $lab(a) =$ out iff there is some $b \in a^-$ such that $lab(b) =$ in
- *grounded* BT-labeling, if it is a complete BT-labeling of $\mathcal{AF}$, and there is no complete BT-labeling lab' of $\mathcal{AF}$ for which $\ln(lab') \subsetneq \ln(lab)$.
- *preferred* BT-labeling, if it is an admissible BT-labeling of $\mathcal{AF}$, and there is no admissible BT-labeling lab' of $\mathcal{AF}$ for which $\ln(lab) \subsetneq \ln(lab')$.
- *stable* BT-labeling, if it is a complete BT-labeling of $\mathcal{AF}$ and Undec$(lab) = \emptyset$.

Note 19 (off-free BT labeling). Just as in the case of conflict-tolerant labeling (see Note 14), there is a one-to-one correspondence between admissible (respectively, complete, grounded preferred, stable) 4-valued BT-labelings without the fourth value and the admissible (respectively, complete, grounded preferred, stable) extensions of the same argumentation framework: As indicated in [13], lab is an off-free admissible (respectively, complete, grounded preferred, stable) BT-labeling of $\mathcal{AF}$, iff $\ln(lab)$ is an admissible (respectively, complete, grounded preferred, stable) extension of $\mathcal{AF}$. Moreover, as in the conflict-tolerant case, this implies that a complete BT-labeling lab is grounded (respectively, preferred) in the sense of Definition 5, iff it is off-free and $\ln(lab)$ is minimal (respectively, maximal) among the in-assignments of the complete BT-labelings of $\mathcal{AF}$. Likewise, a complete BT-labeling lab is stable in the sense of Definition 5, iff it is off-free and Undec$(lab) = \emptyset$.

The various approaches to 4-valued labeling defined above are based on different intuitions and are evaluated in different contexts.[5] Next, we demonstrate some of these intuitions.[6]

Example 20. Consider the following stable tennis-doubles problem, introduced in [35], and is a variation of the *stable marriage problem* (SMP, see e.g., Bistarelli and Santini [11], Iwama and Miyazaki [34]). For the doubles matches of a tennis tournament, the organizers suggest the following set of possible pairs:

$a =$ (Djokovic , Nadal) $b =$ (Nadal , Federer) $c =$ (Federer , Djokovic)
$d =$ (Djokovic , Thiem) $e =$ (Thiem , Medvedev) $f =$ (Medvedev , Zverev)

---

[5]For instance, ROS-labelings may further have a probabilistic distribution to reflect their plausibilities.
[6]Comparing the conflict-free 4-valued approaches and/or relating them to each other is beyond the scope of this paper.

Suppose further that the players express the following preferences for partners:

Nadal prefers Djokovic to Federer, Federer prefers Nadal to Djokovic,
Djokovic prefers Federer to Nadal, Djokovic prefers Federer to Thiem,
Theim prefers Djokovic to Medvedev, Medvedev prefers Thiem to Zverev.

Now, if $player_i$ prefers $player_j$ to $player_k$, then the pair ($player_i$ $player_j$) is an improvement of the pair ($player_i$ $player_k$). In terms of argumentation frameworks, this may be expressed by an attack of ($player_i$ $player_j$) on ($player_i$ $player_k$). A solution to the stable tennis-doubles problem is a choice of an acceptable set (or sets) of so-called "stable" pairs of players, a set of rejected pairs, and perhaps a set of undecided pairs. The stable pairs are those whose improvements are all rejected, and the rejected pairs are those that have an improvement that is not rejected.

The argumentation framework that is associated with the stable tennis-doubles problem in our case is shown in Figure 5.

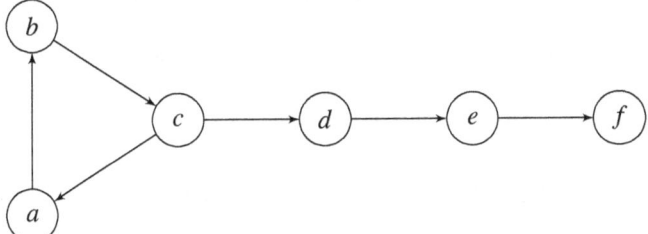

FIGURE 5. Attack diagram for Example 20

The unique admissible set in this case is the emptyset, which means that the only admissible (and so conflict-free) 3-valued labeling for this framework assigns undec to all the arguments. This means that standard (3-valued) semantics rejects all of the suggested pairs of players. This happens since the arguments a, b, and c are caught in a "preference triangle" (recall Example 7), which propagates to the other arguments (pairs) too, making all of them inadmissible.

Conflict-tolerant semantics does not really help in this case either. Its alternative solution is also degenerate, cautiously accepting all the arguments (for a reason similar to the one described above). JV-labeling, on the other hand, allows for further solutions, as listed in Table 1. For instance, according to labeling No.3 (respectively, labeling No.4) in the table, argument f (respectively, e) is accepted.

lab	a	b	c	d	e	f
1	off	off	off	off	off	off
2	off	off	off	off	off	out
3	off	off	off	off	out	in
4	off	off	off	out	in	out
5	off	off	off	off	out	undec
6	off	off	off	out	undec	undec
7	undec	undec	undec	undec	undec	undec

TABLE 1. The JV-labeling for the framework of Example 20

ROS-labelings allow for further assignments in this case. For instance, when d is ignored (and thus omitted from the graph), the graph is split to two subgraphs, consisting of the arguments $\{a,b,c\}$ and $\{e,f\}$, which yields the ROS-labeling lab, in which $\mathsf{In}(lab) = \{e\}$, $\mathsf{Out}(lab) = \{f\}$, $\mathsf{Undec}(lab) = \{a,b,c\}$ and $\mathsf{Off}(lab) = \{d\}$. Similarly, if argument e is ignored, we get a ROS-labeling lab', in which $\mathsf{In}(lab') = \{f\}$, $\mathsf{Undec}(lab') = \{a,b,c,d\}$ and $\mathsf{Off}(lab') = \{e\}$. Note that these are also complete BT-labelings of the framework in this case.

5. Representation and Reasoning with 4-Valued Logics

It is only natural to represent 4-valued labelings of an argumentation framework $\mathcal{AF} = \langle \mathsf{Args}, \mathcal{A} \rangle$ by formulas in a 4-valued logic. Indeed, for a language $\mathcal{L}$ whose atomic formulas (the elements in $\mathsf{Atoms}(\mathcal{L})$) are associated with the arguments of the framework (the elements in Args), a labeling function on Args may be viewed as a 4-valued interpretation on $\mathsf{Atoms}(\mathcal{L})$. Thus, once the labeling rules are expressed by appropriate formulas, the models of the resulting theories will correspond to the required labeling functions of the framework. This is what we show in this section, using Dunn-Belnap 4-valued logic and its extension, 4Flex.

5.1. Four-valued FDE and 4Flex.

The 4-valued variation of first-degree entailment (FDE, Anderson and Belnap [1]) was originated in the work of Dunn [23; 24] (see also [25]) and followed by papers of Belnap [8; 9]), which stimulated a lot of interest in 4-valued logics and their applications in different contexts (see, e.g., [38; 39]). In this paper we use an extension of FDE, called 4Flex (Arieli and Avron [4]), consisting of the basic FDE-connectives $\{\neg, \vee, \wedge\}$, together with D'Ottaviano and da-Costa's implication $\supset$ (da Costa [19], D'Ottaviano and da Costa [21]) and the propositional constant f, representing falsity.[7]

Given an argumentation framework $\mathcal{AF} = \langle \mathsf{Args}, \mathcal{A} \rangle$, we consider a language $\mathcal{L}_{\mathsf{Args}}$, whose atomic formulas are associated with the elements in Args.[8] A (four-valued) valuation v for $\mathcal{L}_{\mathsf{Args}}$ is a function from the atomic formulas of $\mathcal{L}_{\mathsf{Args}}$ to $\{t, f, \top, \bot\}$. These values may be arranged in a lattice structure, in which f is the minimal element, t is the maximal one, and the other two values are incomparable intermediate elements. The corresponding structure $\mathcal{FOUR} = (\{t, f, \top, \bot\}, \leq)$ is a distributive lattice with an order reversing involution $\neg$, for which $\neg t = f$, $\neg f = t$, $\neg \top = \top$ and $\neg \bot = \bot$. We shall denote the meet and the join of this lattice by $\wedge$ and $\vee$, respectively. An implication operator on $\mathcal{FOUR}$ is defined as follows: $a \supset b = t$ if $a \in \{f, \bot\}$, and $a \supset b = b$ otherwise. The truth tables of the basic connectives of $\mathcal{FOUR}$ are given below. Accordingly, a valuation is extended to complex formulas of the language $\mathcal{L}_{\mathsf{Args}}$ in the obvious way, using the truth tables of the basic lattice connectives given above: $v(\neg \psi) = \neg v(\psi)$, $v(\psi \circ \phi) = v(\psi) \circ v(\phi)$ for every $\circ \in \{\wedge, \vee, \supset\}$, and $v(\mathsf{f}) = f$.

---

[7]See [4] for a discussion on 4Flex and some of its properties.

[8]In what follows we freely exchange between an argument and the atomic formula that is associated with it.

$\vee$	t	f	$\top$	$\bot$
t	t	t	t	t
f	t	f	$\top$	$\bot$
$\top$	t	$\top$	$\top$	t
$\bot$	t	$\bot$	t	$\bot$

$\wedge$	t	f	$\top$	$\bot$
t	t	f	$\top$	$\bot$
f	f	f	f	f
$\top$	$\top$	f	$\top$	f
$\bot$	$\bot$	f	f	$\bot$

$\supset$	t	f	$\top$	$\bot$
t	t	f	$\top$	$\bot$
f	t	t	t	t
$\top$	t	f	$\top$	$\bot$
$\bot$	t	t	t	t

$\neg$	
t	f
f	t
$\top$	$\top$
$\bot$	$\bot$

Definition 21 (Models). We say that a valuation v *satisfies* an $\mathcal{L}_{\text{Args}}$-formula ψ, if $v(\psi) \in \{t, \top\}$. Thus, t and $\top$ are the *designated elements* of $\mathcal{FOUR}$ (i.e., those that represent true assertions). A valuation that satisfies every formula in a set Γ of $\mathcal{L}_{\text{Args}}$-formulas is a *model* of Γ. The set of all the models of Γ is denoted $\text{Mod}(\Gamma)$.

Note 22. The main motivation of extending 4-valued FDE to 4Flex is due to the fact that the former lacks an implication connective (namely, one that satisfies the classical deduction theorem with respect to the corresponding consequence relation; See [4]). In contrast, it is easy to verify that the implication $\supset$ is deductive in 4Flex, and that is allows for tautological formulas (those that have designated values under every valuation). Such formulas are not available in 4-valued FDE. (Indeed, every formula ψ in the language of $\{\neg, \wedge, \vee\}$ has the value $\bot$ when all its atomic formulas are assigned the value $\bot$.)

Note 23 (4-valued bilattice). There is at least one other intuitive way of ordering the elements of $\mathcal{FOUR}$. According to the alternative ordering, denoted here by $\leq_i$, $\top$ is the $\leq_i$-maximal element, $\bot$ is the $\leq_i$-minimal one, and the two "classical" values t and f are intermediate $\leq_i$-incomparable elements. This order intuitively represents differences in the amount of *information* that each element exhibits. According to this view, $\top$ represents "over" information (being the least upper bound of t and f) and $\bot$ is associated with lack of information (being the greatest lower bound of t and f).[9] The simultaneous combination of $\leq$ and $\leq_i$ in one structure forms what is known as a (four-valued) *bilattice* ([8; 9], see also Fitting [29; 30], Ginsberg [31] for overviews on bilattices and their properties), and is a cornerstone of a wealth of works on four-valued reasoning methods in AI. In particular, the information order captures the intuition behind the conflict-tolerant labeling described in Section 3, where conf and undec have the roles of $\top$ and $\bot$ (respectively).

In [23; 29; 30] it is shown that the four-valued bilattice may be viewed as a self product of a 2-valued lattice. According to this view, each value is represented by a pair of 2-valued components (x, y), and the bilattice operators are represented in terms of the operators of the 2-valued lattice (see [29] for further details). According to a dual view, the elements of the 4-valued bilattice are represented by elements of $2^{\{0,1\}}$, and so, for example, the $\leq_i$-ordering is expressed by the set-inclusion operator

---

[9]Note that all the connectives in 4-valued FDE are $\leq_i$-monotonic (thus, e.g., if $\circ$ is a binary connective in 4-valued FDE and $a_1, a_2, b_1, b_2 \in \{t, f, \top, \bot\}$ such that $a_1 \leq_i a_2$ and $b_1 \leq_i b_2$, then $a_1 \circ a_2 \leq_i b_1 \circ b_2$). It follows that only $\leq_i$-monotonic functions are expressible in 4-valued FDE. This is *not* the case when the implication $\supset$ is added to the language, which is another motivation for extending FDE to 4Flex (cf. Note 22).

on $2^{\{0,1\}}$.[10] Clearly, both methods of representing 4-valued elements by two-valued components perfectly match the idea behind the JV-labeling in Definition 16.

In the next section we show that the similarities between 4-valued valuations for giving semantics to 4-valued logics, and 4-valued labeling functions for giving semantics to abstract argumentation frameworks, can be made explicit by corresponding theories in 4Flex.

5.2. Representation of Admissible and Complete Labelings/Extensions. As already indicated above, a valuation corresponds to a labeling function and vice-versa, where the truth value t is associated with the label in, f with out, $\bot$ with undec, and $\top$ represents the additional fourth label (denoted above by conf or off). We shall denote by pL2V(lab) the valuation that corresponds to the labeling lab, and by pV2L(v) the labeling that corresponds to the valuations v.

It follows that the different states of argument that are depicted by labeling functions may be represented by corresponding $\mathcal{L}_{\text{Args}}$-formulas, as described in Table 2. In this table we use two sets of terminologies to represent different intuitions that the formulas express according to the conflict-tolerant and the conflict-free labelings. In the table (and in what follows) we denote the formula $\psi \supset f$ by not ψ.[11]

abbreviation	formula	satisfying assignments
\multicolumn{3}{c}{conflict-tolerant terminology}		
cautiously-accept(a)	a	$t, \top$
cautiously-reject(a)	$\neg a$	$f, \top$
conflicting(a)	cautiously-accept(a) $\wedge$ cautiously-reject(a)	$\top$
coherent(a)	not conflicting(a)	$t, f, \bot$
accept(a)	cautiously-accept(a) $\wedge$ coherent(a)	t
reject(a)	cautiously-reject(a) $\wedge$ coherent(a)	f
undecided(a)	not (cautiously-accept(a) $\vee$ cautiously-reject(a))	$\bot$
\multicolumn{3}{c}{conflict-free terminology}		
potentially-accept(a)	a	$t, \top$
potentially-reject(a)	$\neg a$	$f, \top$
excluded(a)	potentially-accept(a) $\wedge$ potentially-reject(a)	$\top$
included(a)	not excluded(a)	$t, f, \bot$
accept(a)	potentially-accept(a) $\wedge$ included(a)	t
reject(a)	potentially-reject(a) $\wedge$ included(a)	f
undecided(a)	not (potentially-accept(a) $\vee$ potentially-reject(a))	$\bot$

TABLE 2. Expressing different sates of arguments by $\mathcal{L}_{\text{Args}}$-formulas

---

[10]In this view, t is represented by $\{1\}$, f by $\{0\}$, $\top$ by $\{0,1\}$, and $\bot$ by $\{\}$. Thus, for instance, t is $<_i$-smaller than $\top$, since $\{1\} \subset \{0,1\}$.

[11]Thus, e.g., in Table 2, coherent(a), which abbreviates not conflicting(a), stands for the formula conflicting(a) $\supset$ f, namely: $(a \wedge \neg a) \supset f$.

Using the formulas in Table 2 it is now easy to express the conditions that define labeling functions. For instance, the conditions in Definition 8 for the p-admissible labeling of $\mathcal{AF} = \langle \mathsf{Args}, \mathcal{A} \rangle$ may be represented as follows:

pIn(x) : $\quad$ accept$(x) \supset \bigwedge_{y \in x^-}$ reject(y)
pOut(x) : $\quad$ reject$(x) \supset \bigvee_{y \in x^-}$ cautiously-accept(y)
pConf(x) : $\quad$ conflicting$(x) \supset \bigl(\bigwedge_{y \in x^-}\bigl(\text{reject}(y) \vee \text{conflicting}(y)\bigr) \wedge \bigvee_{y \in x^-} \text{conflicting}(y)\bigr)$
pUndec(x) : $\quad$ undecided$(x) \supset \bigwedge_{y \in x^-}\bigl(\text{reject}(y) \vee \text{undecided}(y)\bigr)$

The conditions for the p-complete labeling in Definition 8 may be represented in a similar way. Below, we abbreviate the formula $(\psi \supset \phi) \wedge (\phi \supset \psi)$ by $\psi \leftrightarrow \phi$:

pIn$^+(x)$: $\quad$ accept$(x) \leftrightarrow \bigwedge_{y \in x^-}$ reject(y)
pOut$^+(x)$: $\quad$ reject$(x) \leftrightarrow \bigl(\bigvee_{y \in x^-} \text{cautiously-accept}(y) \wedge \bigvee_{y \in x^-}(\text{accept}(y) \vee \text{undecided}(y))\bigr)$
pConf$^+(x)$: $\quad$ conflicting$(x) \leftrightarrow \bigl(\bigwedge_{y \in x^-}\bigl(\text{reject}(y) \vee \text{conflicting}(y)\bigr) \wedge \bigvee_{y \in x^-} \text{conflicting}(y)\bigr)$
pUndec$^+(x)$: $\quad$ undecided$(x) \leftrightarrow \bigl(\bigwedge_{y \in x^-}\bigl(\text{reject}(y) \vee \text{undecided}(y)\bigr) \wedge \bigvee_{y \in x^-} \text{undecided}(y)\bigr)$

Given an argumentation framework $\mathcal{AF} = \langle \mathsf{Args}, \mathcal{A} \rangle$, in what follows we denote by $\Psi(a, \mathcal{AF})$ the formula that is obtained from $\Psi(x)$ by substituting its variable x by an atom a that is associated with an argument $a \in \mathsf{Args}$, and where the elements in a^- (and in a^+) are determined by $\mathcal{A}$. For instance, in the argumentation framework $\mathcal{AF}$ of Figure 1 we have that $r^- = \{s\}$, thus pIn$^+(r, \mathcal{AF})$ is the formula accept$(r) \leftrightarrow$ reject(s).

The following theories may now be used for representing the p-labelings for $\mathcal{AF}$:

$$\mathsf{pADM}(\mathcal{AF}) = \bigcup_{x \in \mathsf{Args}} \mathsf{pIn}(x, \mathcal{AF}) \cup \bigcup_{x \in \mathsf{Args}} \mathsf{pOut}(x, \mathcal{AF}) \cup$$
$$\bigcup_{x \in \mathsf{Args}} \mathsf{pConf}(x, \mathcal{AF}) \cup \bigcup_{x \in \mathsf{Args}} \mathsf{pUndec}(x, \mathcal{AF})$$

$$\mathsf{pCMP}(\mathcal{AF}) = \bigcup_{x \in \mathsf{Args}} \mathsf{pIn}^+(x, \mathcal{AF}) \cup \bigcup_{x \in \mathsf{Args}} \mathsf{pOut}^+(x, \mathcal{AF}) \cup$$
$$\bigcup_{x \in \mathsf{Args}} \mathsf{pConf}^+(x, \mathcal{AF}) \cup \bigcup_{x \in \mathsf{Args}} \mathsf{pUndec}^+(x, \mathcal{AF})$$

The next proposition is easily verified (see also [2]).

Proposition 24. Let $\mathcal{AF} = \langle \mathsf{Args}, \mathcal{A} \rangle$ be an argumentation framework. Then,

(a) There is a correspondence between the 4-valued models of $\mathsf{pADM}(\mathcal{AF})$, the 4-states p-admissible labelings of $\mathcal{AF}$, and the p-admissible extensions of $\mathcal{AF}$. It holds that:
   - If ν is a model of $\mathsf{pADM}(\mathcal{AF})$ then $\mathsf{pV2L}(\nu)$ is a p-admissible labeling of $\mathcal{AF}$ and $\mathsf{pL2E}_{\mathcal{AF}}(\mathsf{pV2L}(\nu))$ is a p-admissible extension of $\mathcal{AF}$.
   - If lab is a p-admissible labeling of $\mathcal{AF}$ then $\mathsf{pL2V}(\text{lab})$ is a model of $\mathsf{pADM}(\mathcal{AF})$ and $\mathsf{pL2E}_{\mathcal{AF}}(\text{lab})$ is a p-admissible extension of $\mathcal{AF}$.
   - If $\mathcal{E}$ is a p-admissible extension of $\mathcal{AF}$ then $\mathsf{pE2L}_{\mathcal{AF}}(\mathcal{E})$ is a p-admissible labeling of $\mathcal{AF}$ and $\mathsf{pL2V}(\mathsf{pE2L}_{\mathcal{AF}}(\mathcal{E}))$ is a model of $\mathsf{pADM}(\mathcal{AF})$.

(b) There is a correspondence between the 4-valued models of $\mathsf{pCMP}(\mathcal{AF})$, the 4-states p-complete labelings of $\mathcal{AF}$, and the p-complete extensions of $\mathcal{AF}$. It holds that:

- if v is a model of $\mathsf{pCMP}(\mathcal{AF})$ then $\mathsf{pV2L}(v)$ is a p-complete labeling of $\mathcal{AF}$ and $\mathsf{pL2E}_{\mathcal{AF}}(\mathsf{pV2L}(v))$ is a p-complete extension of $\mathcal{AF}$.
- If lab is a p-complete labeling of $\mathcal{AF}$ then $\mathsf{pL2V}(lab)$ is a model of $\mathsf{pCMP}(\mathcal{AF})$ and $\mathsf{pL2E}_{\mathcal{AF}}(lab)$ is a p-complete extension of $\mathcal{AF}$.
- If $\mathcal{E}$ is a p-complete extension of $\mathcal{AF}$ then $\mathsf{pE2L}_{\mathcal{AF}}(\mathcal{E})$ is a p-complete labeling of $\mathcal{AF}$ and $\mathsf{pL2V}(\mathsf{pE2L}_{\mathcal{AF}}(\mathcal{E}))$ is a model of $\mathsf{pCMP}(\mathcal{AF})$.

Example 25. Let $\mathcal{AF}$ be the argumentation framework in Example 4, and let v be a model of $\mathsf{pCMP}(\mathcal{AF})$. If $v(r) = t$ then since $r^- = \{s\}$, $\mathsf{pIn}^+(x)$ dictates that $v(s) = f$. From a similar reason, if $v(r) = f$, by $\mathsf{pOut}^+(x)$, $v(s) = t$. Now, $\mathsf{pConf}^+(x)$ forces that if $v(r) = \top$ then $v(s) = \top$ as well, and $\mathsf{pUndec}^+(x)$ forces that if $v(r) = \bot$, so $v(s) = \bot$. Similar considerations determine the possible values of u and g, which yield the four models of $\mathsf{pCMP}(\mathcal{AF})$ shown in Table 3. As indicated in Proposition 24, these models correspond to the four p-complete labelings of $\mathcal{AF}$ (see Example 10) and the four p-complete extensions of $\mathcal{AF}$.

models				p-labelings				p-extensions
r	s	u	g	r	s	u	g	
t	f	t	f	in	out	in	out	$\{r,u\}$
f	t	f	t	out	in	out	in	$\{s,g\}$
$\top$	$\top$	$\top$	$\top$	conf	conf	conf	conf	$\{r,s,u,g\}$
$\bot$	$\bot$	$\bot$	$\bot$	undec	undec	undec	undec	$\emptyset$

TABLE 3. Models, p-labelings, and p-extensions

Conflict-free 4-valued labelings onto $\{\mathsf{in}, \mathsf{out}, \mathsf{off}, \mathsf{undec}\}$ are represented by the formulas in Table 2 in a similar way, where this time the truth value $\top$ corresponds to the label off. For instance, the following formulas represent the two conditions in Definition 16 of JV-labelings:

$\mathsf{JV1}(x):$ potentially-reject$(x) \supset \bigvee_{y \in x^-}$ potentially-accept(y),

$\mathsf{JV2}(x):$ potentially-accept$(x) \supset \bigwedge_{y \in x^+ \cup x^-}$ potentially-reject(y)

Given an argumentation framework $\mathcal{AF}$, let $\mathsf{JV1}(x, \mathcal{AF})$ and $\mathsf{JV2}(x, \mathcal{AF})$ be the formulas that are obtained from $\mathsf{JV1}(x)$ and $\mathsf{JV2}(x)$ just as formulas of the form $\Psi(x, \mathcal{AF})$ are obtained from $\Psi(x)$ in the conflict-tolerant case (as described above). Now, consider the following theory:

$$\mathsf{JV\text{-}Lab}(\mathcal{AF}) = \bigcup_{x \in \mathsf{Args}} \mathsf{JV1}(x, \mathcal{AF}) \cup \bigcup_{x \in \mathsf{Args}} \mathsf{JV2}(x, \mathcal{AF}).$$

Then, we have:
- If v is a model of $\mathsf{JV\text{-}Lab}(\mathcal{AF})$ then $\mathsf{pV2L}(v)$ is a JV-labeling of $\mathcal{AF}$.
- If lab is a JV-labeling of $\mathcal{AF}$ then $\mathsf{pL2V}(lab)$ is a model of $\mathsf{JV\text{-}Lab}(\mathcal{AF})$.

Representations in 4Flex of BT-labeling functions by formulas expressing the conditions in Definition 18 are very similar to the representations above. We leave this to the reader.

5.3. Representation of Other Labelings/Extensions.

We now turn to the other labeling functions and the corresponding extensions. Given a representation of a complete 4-valued labeling (according to either of the methods described above), representing the stable labelings (if exist) is rather easy: we just have to make sure that there are no undec-assignments. For instance, according to the conflict-tolerant approach, p-stable labelings are obtained by the models of the following theory:

$$\mathsf{pSTB}(\mathcal{AF}) = \mathsf{pCMP}(\mathcal{AF}) \cup \{\mathsf{accept}(x) \vee \mathsf{reject}(x) \vee \mathsf{conflicting}(x) \mid x \in \mathsf{Args}\}.$$

For standard 2-valued stable labeling one has to strengthen the additional conditions:[12]

$$\mathsf{pCMP}(\mathcal{AF}) \cup \{\mathsf{accept}(x) \vee \mathsf{reject}(x) \mid x \in \mathsf{Args}\}.$$

Representation of the grounded and preferred labelings is more tricky in our case, since their definitions involve minimization and maximization of in-assignments. For this, we incorporate quantified Boolean formulas (QBFs), and extend the language with three additional propositional constants t, u and c (in addition to f), representing the other truth values (and labeling), i.e., for every valuation ν, $\nu(\mathsf{t}) = t$, $\nu(\mathsf{u}) = \bot$ and $\nu(\mathsf{c}) = \top$.[13]

Quantified Boolean formulas are obtained by extending the underlying propositional language with universal and existential quantifiers $\forall, \exists$ over propositional variables. The intuitive meaning of a QBF of the form $\exists x \forall y\, \psi$, for instance, is that there exists a truth assignment for the propositional variable (the atomic formula) x such that for every truth assignment for y, the formula ψ is true. Clearly, every QBF is associated with a logically equivalent propositional formula.

Definition 26 (Models of QBFs). Let Ψ be a QBF and Γ a set of QBFs.

- An occurrence of x in Ψ is called *free*, if it is not in the scope of a quantifier Qp, for $Q \in \{\forall, \exists\}$. We denote by $\Psi[\phi_1/x_1, \ldots, \phi_n/x_n]$ the simultaneous substitution of each free occurrence of x_i in Ψ by the formula ϕ_i, for $i = 1, \ldots, n$.
- The definition of a *valuation* ν can be extended to QBFs as follows:
  - $\nu(\neg \psi) = \neg \nu(\psi)$,
  - $\nu(\psi \circ \phi) = \nu(\psi) \circ \nu(\phi)$, where $\circ \in \{\wedge, \vee, \supset\}$,
  - $\nu(\forall x\, \psi) = \nu(\psi[\mathsf{t}/x]) \wedge \nu(\psi[\mathsf{f}/x]) \wedge \nu(\psi[\mathsf{c}/x]) \wedge \nu(\psi[\mathsf{u}/x])$,
  - $\nu(\exists x\, \psi) = \nu(\psi[\mathsf{t}/x]) \vee \nu(\psi[\mathsf{f}/x]) \vee \nu(\psi[\mathsf{c}/x]) \vee \nu(\psi[\mathsf{u}/x])$.
- Again, we say that a valuation ν is a *model* of Γ, if $\nu(\Phi) \in \{t, \top\}$ for every $\Phi \in \Gamma$.

For computing grounded, respectively, preferred labelings, one has to keep only the complete labelings whose set of in-assignments is minimal, respectively maximal. For this, we introduce the following QBFs.

Definition 27. Given an abstract argumentation theory $\mathcal{AF} = \langle \mathsf{Args}, \mathcal{A} \rangle$ in which $\mathsf{Args} = \{a_1, \ldots, a_n\}$, let $\mathsf{pCMP}(\mathcal{AF})$ be the theory defined in the previous section

---

[12] Since in general stable models may not exist, this theory may not be satisfiable.

[13] As is shown in Table 2, the propositional constants t, u and c are in fact representable in the language of $\{\neg, \wedge, \vee, \supset \mathsf{f}\}$, thus their introduction here is for clarity reasons and not for extending the expressive power of the language.

for computing the p-complete extensions of $\mathcal{AF}$. We denote $\bigwedge_{a_i \in \text{Args}} \text{pCMP}(\mathcal{AF})$ the conjunction of the formulas in $\text{pCMP}(\mathcal{AF})$. Then,

- $\text{Min}_t(\text{pCMP}(\mathcal{AF}))$ is the following QBF:

$$\forall x_1,\ldots,x_n \left(\bigwedge_{a_i \in \text{Args}} \text{pCMP}(\mathcal{AF})\left[x_1/a_1,\ldots,x_n/a_n\right] \supset \right.$$
$$\left. \left(\bigwedge_{a_i \in \text{Args}, 1 \leq i \leq n} \left(\text{accept}(x_i) \supset \text{accept}(a_i)\right) \supset \bigwedge_{a_i \in \text{Args}, 1 \leq i \leq n} \left(\text{accept}(a_i) \supset \text{accept}(x_i)\right) \right) \right).$$

- $\text{Max}_t(\text{pCMP}(\mathcal{AF}))$ is the following QBF:

$$\forall x_1,\ldots,x_n \left(\bigwedge_{a_i \in \text{Args}} \text{pCMP}(\mathcal{AF})\left[x_1/a_1,\ldots,x_n/a_n\right] \supset \right.$$
$$\left. \left(\bigwedge_{a_i \in \text{Args}, 1 \leq i \leq n} \left(\text{accept}(a_i) \supset \text{accept}(x_i)\right) \supset \bigwedge_{a_i \in \text{Args}, 1 \leq i \leq n} \left(\text{accept}(x_i) \supset \text{accept}(a_i)\right) \right) \right).$$

Accordingly, we denote:

$$\text{pGRD}(\mathcal{AF}) = \text{pCMP}(\mathcal{AF}) \cup \{\text{Min}_t(\text{pCMP}(\mathcal{AF}))\},$$
$$\text{pPRF}(\mathcal{AF}) = \text{pCMP}(\mathcal{AF}) \cup \{\text{Max}_t(\text{pCMP}(\mathcal{AF}))\}.$$

Proposition 28. *Let $\mathcal{AF} = \langle \text{Args}, \mathcal{A} \rangle$ be an argumentation framework. Then:*
(a) *If v is a model of $\text{pGRD}(\mathcal{AF})$ (respectively, a model of $\text{pPRF}(\mathcal{AF})$), then $\text{pV2L}(v)$ is a p-grounded (respectively, p-preferred) labeling of $\mathcal{AF}$, and $\text{pL2E}_{\mathcal{AF}}(\text{pV2L}(v))$ is a p-grounded (respectively, p-preferred) extension of $\mathcal{AF}$.*
(b) *If lab is a p-grounded (respectively, p-preferred) labeling of $\mathcal{AF}$, then $\text{pL2V}(\text{lab})$ is a model of $\text{pGRD}(\mathcal{AF})$ (respectively, a model of $\text{pPRF}(\mathcal{AF})$), and $\text{pL2E}_{\mathcal{AF}}(\text{lab})$ is a p-grounded (respectively, p-preferred) extension of $\mathcal{AF}$.*
(c) *If $\mathcal{E}$ is a p-grounded (respectively, p-preferred) extension of $\mathcal{AF}$ then $\text{pE2L}_{\mathcal{AF}}(\mathcal{E})$ is a p-grounded (respectively, p-preferred) labeling of $\mathcal{AF}$, and $\text{pL2V}(\text{pE2L}_{\mathcal{AF}}(\mathcal{E}))$ is a model of $\text{pGRD}(\mathcal{AF})$ (respectively, a model of $\text{pPRF}(\mathcal{AF})$).*

Proof. Let v be a model of $\text{pGRD}(\mathcal{AF})$ (respectively, let v be a model of $\text{pPRF}(\mathcal{AF})$). In particular, v satisfies $\text{pCMP}(\mathcal{AF})$, thus by Proposition 24, $\text{pV2L}(v)$ is a p-complete labeling of $\mathcal{AF}$. Moreover, $\text{Min}_t(\text{pCMP}(\mathcal{AF}))$ (respectively, $\text{Max}_t(\text{pCMP}(\mathcal{AF}))$) assures that there is no other model μ of $\text{pCMP}(\mathcal{AF})$ such that $\{x \mid \mu(x) = t\} \subsetneq \{x \mid v(x) = t\}$ (respectively, $\{x \mid v(x) = t\} \subsetneq \{x \mid \mu(x) = t\}$). It follows that $\text{In}(\text{pV2L}(v))$ is $\subseteq$-minimal (respectively, $\subseteq$-maximal) among the corresponding sets of the p-complete labelings of $\mathcal{AF}$. Thus, $\text{pV2L}(v)$ is a p-grounded (respectively, p-preferred) labeling of $\mathcal{AF}$. The proofs of the other claims in the proposition are similar. ◁

Note 29 (Representation of standard 3-valued semantics of AAFs). Let $\bigwedge_{a_i \in \text{Args}} \text{SEM}(\mathcal{AF})$ be the conjunction of the formulas in a theory $\text{SEM}(\mathcal{AF})$, and let $\text{Min}_\top(\text{SEM}(\mathcal{AF}))$ be a QBF similar to $\text{Min}_t(\text{pCMP}(\mathcal{AF}))$, except that $\bigwedge_{a_i \in \text{Args}} \text{SEM}(\mathcal{AF})$ replaces $\bigwedge_{a_i \in \text{Args}} \text{pCMP}(\mathcal{AF})$ and every occurrence of $\text{accept}(\cdot)$ is replaced by $\text{conflicting}(\cdot)$. Then,

$$\text{CMP}(\mathcal{AF}) = \text{pCMP}(\mathcal{AF}) \cup \{\text{Min}_\top(\text{pCMP}(\mathcal{AF}))\}$$

represents the conf-free p-complete labelings of $\mathcal{AF}$, which by Note 14 are the (standard, 3-valued) complete labelings of $\mathcal{AF}$ in the sense of Definition 5. Similarly,

$$\mathsf{GRD}(\mathcal{AF}) = \mathsf{pGRD}(\mathcal{AF}) \cup \{\,\mathsf{Min}_\top(\mathsf{pGRD}(\mathcal{AF}))\,\}$$

represents the conf-free p-grounded labeling of $\mathcal{AF}$, which by the same note is the grounded labeling of $\mathcal{AF}$ according to Definition 5. Thus, if v is the model of $\mathsf{GRD}(\mathcal{AF})$ then $\mathsf{pV2L}(v)$ is the grounded labeling of $\mathcal{AF}$ and $\{a \in \mathsf{Args} \mid v(a) = t\}$ is the grounded extension of $\mathcal{AF}$.

Using the theory $\mathsf{CMP}(\mathcal{AF})$, another representation of the grounded semantics is obtained by the model of $\mathsf{CMP}(\mathcal{AF}) \cup \{\,\mathsf{Min}_t(\mathsf{CMP}(\mathcal{AF}))\,\}$. This immediately follows from the definition of the grounded extension and the grounded labeling as the $\subseteq$-minimal complete extension and the complete labeling with minimal in-assignments, respectively.[14] Likewise, if v is a model of

$$\mathsf{PRF}(\mathcal{AF}) = \mathsf{pPRF}(\mathcal{AF}) \cup \{\,\mathsf{Min}_\top(\mathsf{pPRF}(\mathcal{AF}))\,\}$$

then $\mathsf{pV2L}(v)$ is a preferred labeling of $\mathcal{AF}$ (again, in the sense of Definition 5) and $\{a \in \mathsf{Args} \mid v(a) = t\}$ is a preferred extension of $\mathcal{AF}$. The 3-valued preferred extensions/labelings may also be represented by the models of $\mathsf{CMP}(\mathcal{AF}) \cup \{\,\mathsf{Max}_t(\mathsf{CMP}(\mathcal{AF}))\,\}$, as follows from their definitions. Also, if v is a model of

$$\mathsf{STB}(\mathcal{AF}) = \mathsf{pSTB}(\mathcal{AF}) \cup \{\,\mathsf{Min}_\top(\mathsf{pSTB}(\mathcal{AF}))\,\}$$

then $\mathsf{pV2L}(v)$ is a stable labeling of $\mathcal{AF}$ and $\{a \in \mathsf{Args} \mid v(a) = t\}$ is a stable extension of $\mathcal{AF}$. As we have noted previously, stable labelings may also be obtained by the models of $\mathsf{pCMP}(\mathcal{AF}) \cup \{\mathsf{accept}(x) \vee \mathsf{reject}(x) \mid x \in \mathsf{Args}\}$.

In fact, the combination of formulas in extended FDE and QBFs expressing minimization and maximization requirements allows us to represent other types of standard 3-valued labelings, such as ideal labelings, eager labelings and semi-stable labelings (see [6; 7]). For instance, lab is a semi-stable labeling of $\mathcal{AF}$, if $\mathsf{Undec}(lab)$ is minimal in $\{\mathsf{Undec}(l) \mid l \text{ is a complete labeling of } \mathcal{AF}\}$. A possible representation of such a labeling would be by the theory

$$\mathsf{SSTB}(\mathcal{AF}) = \mathsf{CMP}(\mathcal{AF}) \cup \{\,\mathsf{Min}_\perp(\mathsf{CMP}(\mathcal{AF}))\,\}$$

where $\mathsf{Min}_\perp(\mathsf{CMP}(\mathcal{AF}))$ is a QBF similar to $\mathsf{Min}_\top(\mathsf{CMP}(\mathcal{AF}))$, except that every occurrence of $\mathsf{conflicting}(\cdot)$ is replaced by $\mathsf{undecided}(\cdot)$.

Different representations of these labelings using QBFs may be found, e.g., in Arieli and Caminada [5], Diller et al. [20]. A survey on other logical theories for standard, 3-valued semantics of abstract argumentation appears in Besnard et al. [10].

Representations of the conflict-free semantics described in Section 4 are similar to the representations above of conflict-tolerant semantics. For instance, given a theory $\mathsf{btCMP}(\mathcal{AF})$ for representing the complete BT-labelings of $\mathcal{AF}$, grounded, preferred, and stable BT-labelings are represented by respectively adding to $\mathsf{btCMP}(\mathcal{AF})$ the

---

[14]As shown e.g., in Dunne and Wooldridge [28], and Modgil and Caminada [37], there are simpler and computationally easier (in fact, polynomial) methods for computing the grounded extension and the grounded labeling of an AAF. Yet, our purpose in this paper is to show a uniform representation of various semantics of AAFs in various contexts. In some cases such a uniform representation is more computationally demanding.

QBFs $\text{Min}_t(\text{btCMP}(\mathcal{AF}))$, $\text{Max}_t(\text{btCMP}(\mathcal{AF}))$, and the set of formulas $\{\text{accept}(x) \vee \text{reject}(x) \vee \text{excluded}(x) \mid x \in \text{Args}\}$. In the notations of Definition 17, ROS-labelings of $\mathcal{AF} = \langle \text{Args}, \mathcal{A} \rangle$ are computed by iterating over the subsets $\mathcal{E} \subseteq \text{Args}$, assigning off to (the atomic variables that are associated with) their elements, and computing the models of $\text{GRD}(\mathcal{AF}^{\uparrow \mathcal{E}})$, where $\mathcal{AF}^{\uparrow \mathcal{E}}$ is the argumentation framework $\mathcal{AF}$, restricted to the arguments in $\text{Args} \setminus \mathcal{E}$.

Note 30. It is usually rather easy to express argumentation dynamics or domain-specific constraints by the theories described above. For instance, in the tennis-doubles problem (Example 20), the constraint that Djokovic must play either with Nadal or with Federer may be enforced by adding the formula $\text{accept}(a) \vee \text{accept}(c)$ to the theory.

6. Conclusion

In [8] it is claimed that Dunn–Belnap 4-valued logic is useful for representing *monotonic* computerized reasoning in the presence of contradictions. In turn, argumentation theory has been used, among others, for modeling *non-monotonic* reasoning for handling conflicts. This paper provides another evidence that, in fact, 4-valued FDE and its extensions are useful for representing both kinds of reasoning.[15]

Formal argumentation sometimes involves extra machinery for extending the expressive power of the frameworks. This includes, among others, additions of preferences among arguments (Kaci et al. [36]) and attacks (Bistarelli and Santini [12]), enabling of different attack relations (as in abstract dialectical frameworks, ADFs, see Brewka et al. [14]), incorporation of several interactions between arguments (Cayrol et al. [17]), and introduction of probabilities (Hunter et al. [33]).[16] Clearly, representing such extended frameworks and reasoning with them call upon corresponding enhancements of the underlying formal logics.[17] This is a subject for future work.

Acknowledgments. I would like to thank Katalin Bimbó for inviting me to contribute to this volume and for her constructive comments on this paper. I am also grateful to the other reviewer for a helpful report. This work is partially supported by the Israel Science Foundation, grant No. 550/19.

References

[1] Anderson, A. R. and Belnap, N. D. (1963). First degree entailments, *Mathematische Annalen* **149**(4): 302–319.

[2] Arieli, O. (2015). Conflict-free and conflict-tolerant semantics for constrained argumentation frameworks, *Journal of Applied Logic* **13**(4): 582–604.

[3] Arieli, O. (2016). On the acceptance of loops in argumentation frameworks, *Journal of Logic and Computation* **26**(4): 1203–1234.

---

[15]Computerized implementations of this may involve SAT-solvers or QBF-solvers, as described and illustrated e.g., in Cerutti et al. [18].

[16]Embeddings of FDE into a context of probability, like the ones described in Dunn [26], and Dunn and Kiefer [27], may be useful for this purpose.

[17]For instance, in Heyninck et al. [32] it is shown that no adequate semantics for ADFs can be formulated by a truth-functional 3-valued logic using an involutive negation.

[4] Arieli, O. and Avron, A. (2017). Four-valued paradefinite logics, *Studia Logica* **105**(6): 1087–1122.

[5] Arieli, O. and Caminada, M. W. (2013). A QBF-based formalization of abstract argumentation semantics, *Journal of Applied Logic* **11**(2): 229–252.

[6] Baroni, P., Caminada, M. W. and Giacomin, M. (2011). An introduction to argumentation semantics, *Knowledge Engineering Review* **26**(4): 365–410.

[7] Baroni, P., Caminada, M. W. and Giacomin, M. (2018). Abstract argumentation frameworks and their semantics, *in* P. Baroni, D. Gabbay, M. Giacomin and L. van der Torre (eds.), *Handbook of Formal Argumentation*, Vol. 1, College Publications, London, UK, pp. 159–236.

[8] Belnap, N. D. (1977a). How a computer should think, *in* G. Ryle (ed.), *Contemporary Aspects of Philosophy*, Oriel Press, Stocksfield.

[9] Belnap, N. D. (1977b). A useful four-valued logic, *in* J. M. Dunn and G. Epstein (eds.), *Modern Uses of Multiple-Valued Logics*, Reidel Publishing Company, Dordrecht, pp. 7–37.

[10] Besnard, P., Cayrol, C. and Lagasquie-Schiex, M. (2020). Logical theories and abstract argumentation: A survey of existing works, *Journal of Argument & Computation* **11**(1–2): 41–102.

[11] Bistarelli, S. and Santini, F. (2020). Abstract argumentation and (optimal) stable marriage problems, *Journal of Argument & Computation* **11**(1-2): 15–40.

[12] Bistarelli, S. and Santini, F. (2021). Weighted argumentation, *in* D. Gabbay, M. Giacomin, G. Simari and M. Thimm (eds.), *Handbook of Formal Argumentation*, Vol. 2, College Publications, London, UK, pp. 355–396.

[13] Bistarelli, S. and Taticchi, C. (2021). A unifying four-state labelling semantics for bridging abstract argumentation frameworks and belief revision, *in* C. S. Coen and I. Salvo (eds.), *Proceedings of the 22nd Italian Conference on Theoretical Computer Science (ICTCS'21)*, Vol. 3072 of *CEUR Workshop Proceedings*, pp. 93–106. URL: ceur-ws.org/Vol-3072.

[14] Brewka, G., Ellmauthaler, S., Strass, H., Wallner, J. P. and Woltran, S. (2017). Abstract dialectical frameworks. An overview, *IFCoLog Journal of Logics and their Applications* **4**(8): 2263–2317.

[15] Caminada, M. W. (2006). On the issue of reinstatement in argumentation, *in* M. Fischer, W. van der Hoek, B. Konev and A. Lisitsa (eds.), *Proceedings of the 10th European Conference on Logics in Artificial Intelligence (JELIA'06)*, number 4160 in *Lecture Notes in Computer Science*, Springer, Berlin, pp. 111–123.

[16] Caminada, M. W. and Gabbay, D. M. (2009). A logical account of formal argumentation, *Studia Logica* **93**(2–3): 109–145.

[17] Cayrol, C., Cohen, A. and Lagasquie-Schiex, M.-C. (2021). Higher-order interactions (bipolar or not) in abstract argumentation: A state of the art, *in* D. Gabbay, M. Giacomin, G. Simari and M. Thimm (eds.), *Handbook of Formal Argumentation*, Vol. 2, College Publications, London, UK, pp. 15–130.

[18] Cerutti, F., Gaggl, S. A., Thimm, M. and Wallner, J. P. (2017). Foundations of implementations for formal argumentation, *Journal of Applied Logics-IFCoLog Journal of Logics and their Applications* **4**(8): 2623–2706.

[19] da Costa, N. C. A. (1974). On the theory of inconsistent formal systems, *Notre Dame Journal of Formal Logic* **15**: 497–510.

[20] Diller, M., Wallner, J. P. and Woltran, S. (2015). Reasoning in abstract dialectical frameworks using quantified Boolean formulas, *Journal of Argument & Computation* **6**(2): 149–177.

[21] D'Ottaviano, I. M. and da Costa, N. C. A. (1970). Sur un problèm de Jaśkowski, *C. R. Acad Sc. Paris, Volume 270, Sèrie A* pp. 1349–1353.

[22] Dung, P. M. (1995). On the acceptability of arguments and its fundamental role in nonmonotonic reasoning, logic programming and n-person games, *Artificial Intelligence* **77**: 321–357.

[23] Dunn, J. M. (1966). *The Algebra of Intensional Logics*, PhD thesis, University of Pittsburgh. (Published as *Logic PhDs*, volume 2, College Publications, London, UK, 2019).

[24] Dunn, J. M. (1976). Intuitive semantics for first-degree entailments and 'coupled trees', *Philosophical Studies* **29**: 149–168.

[25] Dunn, J. M. (2000). Partiality and its dual, *Studia Logica* **66**(1): 5–40.

[26] Dunn, J. M. (2010). Contradictory information: Too much of a good thing, *Journal of Philosophical Logic* **39**(4): 425–452.

[27] Dunn, J. M. and Kiefer, N. M. (2019). Contradictory information: Better than nothing? The paradox of the two firefighters, *in* C. Başkent and T. M. Ferguson (eds.), *Graham Priest on Dialetheism and Paraconsistency*, Vol. 18 of *Outstanding Contributions to Logic*, Springer, Switzerland, pp. 231–247.

[28] Dunne, P. E. and Wooldridge, M. (2009). Complexity of abstract argumentation, *in* I. Rahwan and G. R. Simari (eds.), *Argumentation in Artificial Intelligence*, Springer, Berlin, pp. 85–104.

[29] Fitting, M. (1991). Bilattices and the semantics of logic programming, *Journal of Logic Programming* **11**(2): 91–116.

[30] Fitting, M. (2006). Bilattices are nice things, *in* T. Bolander, V. Hendricks and S. A. Pedersen (eds.), *Self Reference*, Vol. 178 of *CSLI Lecture Notes*, CLSI Publications, Stanford, CA, pp. 53–77.

[31] Ginsberg, M. L. (1988). Multi-valued logics: A uniform approach to reasoning in AI, *Computer Intelligence* **4**: 256–316.

[32] Heyninck, J., Thimm, M., Kern-Isberner, G., Rienstra, T. and Skiba, K. (2021). On the relation between possibilistic logic and abstract dialectical frameworks, *in* L. Amgoud and R. Booth (eds.), *Proceedings of the 19th International Workshop on Non-Monotonic Reasoning (NMR'21, November 3–5, 2021, online)*, pp. 129–138.

[33] Hunter, A., Polberg, S., Potyka, N., Rienstra, T. and Thimm, M. (2021). Probabilistic argumentation: A survey, *in* D. Gabbay, M. Giacomin, G. Simari and M. Thimm (eds.), *Handbook of Formal Argumentation*, Vol. 2, College Publications, London, UK, pp. 397–441.

[34] Iwama, K. and Miyazaki, S. (2008). A survey of the stable marriage problem and its variants, *International Conference on Informatics Education and Research for Knowledge-Circulating Society (ICKS'08, January 17, 2008, Kyoto)*, IEEE Computer Society, pp. 131–136.

[35] Jakobovits, H. and Vermeir, D. (1999). Robust semantics for argumentation frameworks, *Journal of Logic and Computation* **9**(2): 215–261.

[36] Kaci, S., van der Torre, L., Vesic, S. and Villata, S. (2021). Preference in abstract argumentation, *in* D. Gabbay, M. Giacomin, G. Simari and M. Thimm (eds.), *Handbook of Formal Argumentation*, Vol. 2, College Publications, London, UK, pp. 211–248.

[37] Modgil, S. and Caminada, M. W. (2009). Proof theories and algorithms for abstract argumentation frameworks, *in* I. Rahwan and G. R. Simari (eds.), *Argumentation in Artificial Intelligence*, Springer, Cham, pp. 105–129.

[38] Omori, H. and Wansing, H. (eds.) (2017). *Special Issue: 40 years of FDE*, Vol. 106(6) of *Studia Logica*.

[39] Omori, H. and Wansing, H. (eds.) (2019). *New Essays on Belnap–Dunn Logic*, Vol. 418 of *Synthese Library*, Springer, Switzerland.
[40] Pollock, J. L. (1994). Justification and defeat, *Artificial Intelligence* **67**(2): 377–407.
[41] Riveret, R., Oren, N. and Sartor, G. (2020). A probabilistic deontic argumentation framework, *Journal of Approximate Reasoning* **126**: 249–271.
[42] Verheij, B. (1996). Two approaches to dialectical argumentation: Admissible sets and argumentation stages, *in* J.-J. Meyer and L. van der Gaag (eds.), *Proceedings of the 8th Dutch Conference on Artificial Intelligence (NAIC'96, November 21–22, 1996, Utrecht)*, Utrecht University, Utrecht, The Netherlands, pp. 357–368.

SCHOOL OF COMPUTER SCIENCE, THE ACADEMIC COLLEGE OF TEL-AVIV, ISRAEL
Email: oarieli@mta.ac.il

On Formal Criteria for Relevance

Arnon Avron

ABSTRACT. We introduce and investigate the strong variable-sharing property, together with some other generalizations of this property. Then we show that the relevance logics **R** and **RMI**, as well as the multiplicative-additive fragment of Linear Logic, enjoy all these generalizations.

Keywords. Consequence, Relevance logics, Variable-sharing properties

1. Three Abstract Variable-sharing Properties

When does a given formal propositional logic **L** deserve to be classified as a relevance logic? In the canonical literature on the subject (like Anderson and Belnap [1]; Dunn and Restall [11]) we practically find only one clear, universally agreed upon, formal criterion:

Definition 1. A propositional logic **L** with an implication $\to$ has the *variable sharing property* (VSP) if $\mathrm{Atoms}(\varphi) \cap \mathrm{Atoms}(\psi) \neq \emptyset$ whenever $\vdash_\mathbf{L} \varphi \to \psi$. ($\mathrm{Atoms}(\varphi)$ denotes here the set of atomic formulas, or propositional variables, that occur in φ.)

Note 2. Actually, in addition to propositional variables, the atomic formulas of a given propositional logic might include propositional constants as well. However, I was unable to find in the literature an answer to the question whether a logic like $\mathbf{R}^\mathbf{t}$ (the extension of the relevance logic **R** with the propositional constant **t**) has the VSP or not. (Both Dunn and Bimbó argue in personal communications that this specific logic does. The main reason is that the standard interpretation of **t** is as the conjunction of all logical theorems. Hence, it should be viewed as involving all the variables.) For simplicity, in this paper we assume therefore that we deal only with logics which have no propositional constants.

What is the justification of the VSP criterion as given in Definition 1? In particular, why not to demand instead, e.g., that $\mathrm{Atoms}(\varphi) \cap \mathrm{Atoms}(\psi) \neq \emptyset$ whenever $\vdash_\mathbf{L} \varphi \wedge \psi$, or whenever $\vdash_\mathbf{L} \varphi \vee \psi$? The answer relies on the demand (made in passing in Definition 1) that $\to$ is an *implication* for **L**. This usually means that it satisfies at least the following (minimal) condition:

(1) $\qquad\qquad\qquad \varphi \vdash_\mathbf{L} \psi \quad \text{iff} \quad \vdash_\mathbf{L} \varphi \to \psi$

2020 *Mathematics Subject Classification.* Primary: 03B47, Secondary: 03B70.

Bimbó, Katalin, (ed.), *Relevance Logics and other Tools for Reasoning. Essays in Honor of J. Michael Dunn*, (Tributes, vol. 46), College Publications, London, UK, 2022, pp. 54–64.

The intuition behind the VSP is indeed that ψ may *logically follow* from φ only if there is some connection between the content of φ and the content of ψ. Now the only way to secure this in an abstract formal propositional logic **L** is to demand the following:

Definition 3. A propositional logic **L** has the *abstract variable sharing property* (AVSP) if $\text{Atoms}(\varphi) \cap \text{Atoms}(\psi) \neq \emptyset$ whenever $\varphi \vdash_{\mathbf{L}} \psi$.

Obviously, the AVSP is equivalent to the standard VSP in case $\to$ satisfies (1), but it is the AVSP which has the priority.[1]
The AVSP (on which the usual VSP is based) is a pure condition about $\vdash_{\mathbf{L}}$, the consequence relation of **L**. Here, however, we face a difficulty: if $\vdash_{\mathbf{L}}$ is a consequence relation according to the following standard meaning of this notion, then it cannot have the AVSP.

Definition 4. A (Tarskian) *consequence relation* (tcr) for a language $\mathcal{L}$ is a binary relation $\vdash$ between theories in $\mathcal{L}$ and formulas in $\mathcal{L}$, satisfying the following three conditions:

[R] *Reflexivity*: $\psi \vdash \psi$ (i.e., $\{\psi\} \vdash \psi$).

[M] *Monotonicity*:   If $\mathcal{T} \vdash \psi$ and $\mathcal{T} \subseteq \mathcal{T}'$, then $\mathcal{T}' \vdash \psi$.

[C] *Cut (Transitivity)*: If $\mathcal{T} \vdash \psi$ and $\mathcal{T}', \psi \vdash \varphi$ then $\mathcal{T}, \mathcal{T}' \vdash \varphi$.

Definition 4 immediately implies that if **L** has any logically valid formula ψ, then [M] implies that $\varphi \vdash_{\mathbf{L}} \psi$ for *every* φ, and so the AVSP fails. It follows that if **L** is a tcr for which (1) holds, then it cannot have the VSP. Indeed, [R] and (1) together entail that $\vdash_{\mathbf{L}} \varphi \to \varphi$ for every φ. Therefore, [M] entails that $\psi \vdash_{\mathbf{L}} \varphi \to \varphi$ for every φ and ψ. Hence $\vdash_{\mathbf{L}} \psi \to (\varphi \to \varphi)$ for every φ and ψ.

How may relevance logics overcome this difficulty, while retaining the VSP with respect to some connective $\to$ which might deserve to be called an "implication"? A very reasonable option is to replace (1) by the following weaker condition on $\to$:

(2) $\varphi \vdash_{\mathbf{L}} \psi$ iff either $\vdash_{\mathbf{L}} \varphi \to \psi$ or $\vdash_{\mathbf{L}} \psi$.

Obviously, if $\to$ satisfies (2) in **L**, and the latter has the VSP with respect to $\to$ then **L** satisfies the following general principle, whose formulation is independent of the availability in **L** of any special connective.

(3)   If $\varphi \vdash_{\mathbf{L}} \psi$ then either $\text{Atoms}(\varphi) \cap \text{Atoms}(\psi) \neq \emptyset$ or $\vdash_{\mathbf{L}} \psi$.

Since we are dealing now with conditions on the *consequence relation* of **L**, it seems unjustified to stick to the case in which there is a single premise. In fact, the following immediate generalization of (3) to the case in which there is a *(finite) set* Γ of premises seems no less intuitive:

(4)   If $\Gamma \vdash_{\mathbf{L}} \psi$ then either $\bigcup \{\text{Atoms}(\varphi) \mid \varphi \in \Gamma\} \cap \text{Atoms}(\psi) \neq \emptyset$ or $\vdash_{\mathbf{L}} \psi$.

A further obvious generalization treats cases in which some of the premises are relevant to the conclusion, and some are not.

---

[1] Note that the AVSP is applicable also for logics which have no implication, like $\mathbf{R}_{fde}$, the $\{\neg, \wedge, \vee\}$-fragment of the relevance logic **R**.

Definition 5 (Avron [7]; Avron et al. [9]). A logic $\mathbf{L} = \langle \mathcal{L}, \vdash_\mathbf{L} \rangle$ satisfies the *basic relevance criterion* if for every two theories $\mathcal{T}_1, \mathcal{T}_2$ and formula ψ, we have:

If $\mathcal{T}_1, \mathcal{T}_2 \vdash_\mathbf{L} \psi$ then either $\mathrm{Atoms}(\mathcal{T}_2) \cap \mathrm{Atoms}(\mathcal{T}_1 \cup \{\psi\}) = \emptyset$ or $\mathcal{T}_1 \vdash_\mathbf{L} \psi$, (where $\mathrm{Atoms}(\mathcal{T}) = \bigcup \{\mathrm{Atoms}(\varphi) \mid \varphi \in \mathcal{T}\}$).

2. THE STRONG VARIABLE-SHARING PROPERTY

Can the basic relevance criterion be reduced to a condition concerning a single sentence, like the VSP? It can, provided that in addition to an implication connective $\to$ which satisfies the above-mentioned condition, $\mathbf{L}$ also has a *conjunction* $\wedge$, so that $\Gamma, \theta, \varphi \vdash_\mathbf{L} \psi$ is equivalent to $\Gamma, \theta \wedge \varphi \vdash_\mathbf{L} \psi$. With $\to$ and $\wedge$ at our disposal, the basic relevance criterion can be reformulated as follows.

Definition 6. A logic $\mathbf{L} = \langle \mathcal{L}, \vdash_\mathbf{L} \rangle$ has the *strong variable sharing property* (strong VSP) with respect to the connectives $\to$ and $\wedge$ if the following condition is satisfied (where $\theta \wedge \varphi \to \psi$ abbreviates $(\theta \wedge \varphi) \to \psi$):

- If $\vdash_\mathbf{L} \theta \wedge \varphi \to \psi$, and the sentences θ and $\varphi \to \psi$ share no atomic formula, then $\vdash_\mathbf{L} \varphi \to \psi$.

Proposition 7. *Let $\mathbf{L}$ be a finitary logic. Assume that $\to$ and $\wedge$ are binary connectives of $\mathbf{L}$ that satisfy the following conditions:*

1. $\varphi \vdash_\mathbf{L} \psi$ *iff* $\vdash_\mathbf{L} \varphi \to \psi$.
2. $\Gamma, \theta, \varphi \vdash_\mathbf{L} \psi$ *iff* $\Gamma, \theta \wedge \varphi \vdash_\mathbf{L} \psi$.

If $\mathbf{L}$ has the strong VSP with respect to $\to$ and $\wedge$ then it satisfies the basic relevance criterion.

Proof. By repeated applications of item 2 we get:[2]

$(\star)$ If $\varphi_1, \ldots, \varphi_n \vdash_\mathbf{L} \psi$ then $\varphi_1 \wedge \cdots \wedge \varphi_n \vdash_\mathbf{L} \psi$.

Suppose now that $\mathcal{T}_1, \mathcal{T}_2 \vdash_\mathbf{L} \psi$, and $\mathrm{Atoms}(\mathcal{T}_2) \cap \mathrm{Atoms}(\mathcal{T}_1 \cup \{\psi\}) = \emptyset$. Since $\mathbf{L}$ is finitary, it follows that there are $\Gamma_1 \subseteq \mathcal{T}_1$ and $\Gamma_2 \subseteq \mathcal{T}_2$ such that $\Gamma_1, \Gamma_2 \vdash_\mathbf{L} \psi$. Let θ be the conjunction of all the formulas in Γ_2 and let φ be the conjunction of all the formulas in Γ_1. Then $\theta \wedge \varphi \vdash_\mathbf{L} \psi$. Hence $\vdash_\mathbf{L} \theta \wedge \varphi \to \psi$ by item 1. Since θ and $\varphi \to \psi$ share no atomic formula, it follows by the strong VSP that $\vdash_\mathbf{L} \varphi \to \psi$. By the "if" directions of items 1 and 2 it follows that $\Gamma_1 \vdash_\mathbf{L} \psi$, and so $\mathcal{T}_1 \vdash_\mathbf{L} \psi$ as well. ◁

Is the strong VSP really stronger than the VSP? It is, at least under certain rather weak conditions.

Proposition 8. *Let $\mathbf{L}$ be a logic in a language which contains $\to$, $\wedge$, and $\neg$, for which the following conditions are satisfied:*

1. *If $\vdash_\mathbf{L} \varphi \to \psi$ and $\vdash_\mathbf{L} \varphi$, then $\vdash_\mathbf{L} \psi$.*
2. $\vdash_\mathbf{L} \varphi \to \varphi$ *for every φ.*

---

[2]From item 2 it easily follows that $\varphi \wedge (\psi \wedge \theta) \vdash_\mathbf{L} (\varphi \wedge \psi) \wedge \theta$ and that $(\varphi \wedge \psi) \wedge \theta \vdash_\mathbf{L} \varphi \wedge (\psi \wedge \theta)$. This allows us to use in the sequel the notation $\varphi_1 \wedge \cdots \wedge \varphi_n$ for the conjunction of $\varphi_1, \ldots, \varphi_n$, since in the contexts in which we will use this notation, it does not matter how the brackets inside this expression are put. (Officially, we may take $\varphi_1 \wedge \varphi_2 \wedge \cdots \wedge \varphi_n$ as a short for $\varphi_1 \wedge (\varphi_2 \wedge (\cdots \wedge \varphi_n) \ldots)$.)

3. If $\vdash_L \varphi \to \psi$ then $\vdash_L \varphi \wedge \theta \to \psi$.
4. If $\vdash_L \varphi \to \psi$ then $\vdash_L \neg\psi \to \neg\varphi$.
5. There is no φ such that both $\vdash_L \varphi$ and $\vdash_L \neg\varphi$.

If **L** has the strong VSP with respect to $\to$ and $\wedge$ then it also has the VSP with respect to $\to$.

Proof. Suppose $\vdash_L \varphi \to \psi$, where φ and ψ share no atomic formula. Let p and q be two distinct variables not occurring in $\varphi \to \psi$. Then $\vdash_L \varphi \wedge (p \to p) \to \psi$ by 3, and so $\vdash_L (p \to p) \to \psi$ by strong VSP. By 4, it follows that $\vdash_L \neg\psi \to \neg(p \to p)$, and so (by 3 again) $\vdash_L \neg\psi \wedge (q \to q) \to \neg(p \to p)$. By another application of strong VSP we get $\vdash_L (q \to q) \to \neg(p \to p)$, and so $\vdash_L \neg(p \to p)$ by 1 and 2. Since $\vdash_L (p \to p)$ as well (by 2), this contradicts 5. ◁

Example 9. Dunn's semi-relevance logic **RM** satisfies all the conditions given in Proposition 8, and it does not have the VSP. (See [1] or Avron [8].) Hence it does not have the strong VSP either.[3]

Example 10. That the conditions about $\neg$ in Proposition 8 cannot be omitted is demonstrated by **CL**$^+$ and **J**$^+$, the positive fragments (i.e., the $\{\to, \wedge, \vee\}$-fragments) of classical logic (**CL**) and intuitionistic logic (**J**), respectively. It is well known, e.g., that $\vdash_{CL^+} \varphi \to \psi$ only if either $\vdash_{CL^+} \psi$ or φ and ψ share a variable. From this (using that if $\vdash_{CL^+} \varphi \wedge \theta \to \psi$ then $\vdash_{CL^+} \varphi \to (\theta \to \psi)$) it easily follows that **CL**$^+$ has the strong VSP. In contrast, it does not have the VSP, since $\vdash_{CL^+} p \to (q \to q)$. (The reasoning in the case of **J**$^+$ is similar.)

It might be worth noting that there is a simpler, but less natural, set of conditions under which the strong VSP implies VSP.

Proposition 11. *Let* **L** *be a logic in a language which contains* $\to$ *and* $\wedge$, *for which the following conditions are satisfied:*

1. *If* $\vdash_L \varphi \to \psi$ *then* $\vdash_L \varphi \wedge \theta \to \psi$.
2. *There is no ψ such that* $\vdash_L \theta \to \psi$ *for every θ.*

If **L** *has the strong VSP with respect to* $\to$ *and* $\wedge$ *then it also has the VSP with respect to* $\to$.

Proof. Suppose $\vdash_L \varphi \to \psi$, where φ and ψ share no atomic formula. Let p be a variable not occurring in $\varphi \to \psi$. Then $\vdash_L \varphi \wedge p \to \psi$ by 1, and so $\vdash_L p \to \psi$ by strong VSP. Since **L** is a logic, its consequence relation $\vdash_L$ is structural. Therefore, the facts that $\vdash_L p \to \psi$ and that p does not occur in ψ imply that $\vdash_L \theta \to \psi$ for every θ. This contradicts 2. ◁

3. WHAT IS THE CONSEQUENCE RELATION OF **R**?

The main goal of this paper is to find out whether relevance logics (and especially **R**, the logic taken in [11] as the "paradigm of a relevance logic") indeed satisfy the basic relevance criterion. But since this is a criterion about the *consequence relation*

---

[3] **RM** does have what Dunn has called the "weak relevance principle." If $\vdash_{RM} \varphi \to \psi$ then either φ and ψ share a variable, or *both* $\vdash_{RM} \neg\varphi$ and $\vdash_{RM} \psi$. (See [1, Section 29.4.].)

of the logics, not just about their sets of theorems, we have first to decide what is the consequence relation of those relevance logics. Unfortunately, the answer to this question is rather unclear. As shown in Avron [6], one can find in the relevant literature (at least) three different plausible candidates. Here they are in the case of **R**:

$\vdash_{\mathbf{R}}^{i}$: This is the tcr which results by trying to make $\to$ a full relevant implication, that is, by generalizing (2) to

$$\Gamma, \varphi \vdash_{\mathbf{L}} \psi \text{ iff either } \Gamma \vdash_{\mathbf{L}} \varphi \to \psi \text{ or } \Gamma \vdash_{\mathbf{L}} \psi.$$

For this we define: $\mathcal{T} \vdash_{\mathbf{R}}^{i} \psi$ iff there exist $\varphi_1, \varphi_2, \ldots, \varphi_n \in \mathcal{T}$ such that

$$\vdash_{\mathbf{R}} \varphi_1 \to (\varphi_2 \to (\cdots (\varphi_n \to \psi) \cdots)).$$

$\vdash_{\mathbf{R}}^{e}$: This is the relation which results by trying to reduce the multiple-premise case to the single-premise one by using $\wedge$, the official conjunction of **R**. For this we define: $\mathcal{T} \vdash_{\mathbf{R}}^{e} \psi$ iff there exist $\varphi_1, \ldots, \varphi_n \in \mathcal{T}$ such that

$$\vdash_{\mathbf{R}} \varphi_1 \wedge \cdots \wedge \varphi_n \to \psi$$

$\vdash_{\mathbf{R}}^{H}$: This is the natural relation which is induced by the Hilbert-type system HR by which **R** is defined in [1] and [11]. In other words, $\mathcal{T} \vdash_{\mathbf{R}}^{H} \psi$ iff there is finite list of formulas such that every element of which is an axiom of HR, or belongs to $\mathcal{T}$, or is derivable from two previous elements of the list by one of the inference rules of HR ([MP] for $\to$, or the adjunction rule [Ad] for $\wedge$).

It is rather easy to see that if we take $\vdash_{\mathbf{R}}$ to be $\vdash_{\mathbf{R}}^{i}$, then the basic relevance criterion for **R** follows from the VSP (that **R** is known to enjoy). However, this choice is rather unnatural (and uninteresting too). Thus the following inferences should intuitively be valid, but they fail if $\vdash_{\mathbf{R}} = \vdash_{\mathbf{R}}^{i}$.

1. $\varphi \vdash_{\mathbf{R}} (\varphi \wedge \theta) \vee (\varphi \wedge \neg \theta)$[4]
2. $\varphi, \psi \vdash_{\mathbf{R}} \varphi \wedge \psi$
3. $\varphi, \psi, \varphi \wedge \psi \to \theta \vdash_{\mathbf{R}} \theta$
4. $\varphi \to \psi \vdash_{\mathbf{R}} \varphi \to (\varphi \wedge \psi)$

In addition, the failure of the second inference in this list means that [Ad] is taken in $\vdash_{\mathbf{R}}^{i}$ only as a *rule of proof*. In view of the extensional nature of $\wedge$, I find it extremely difficult to justify this limitation.

In contrast to $\vdash_{\mathbf{R}}^{i}$, the adjunction rule is valid without any limitation for $\vdash_{\mathbf{R}}^{e}$, and so is the third inference in the above list. However, the first and forth inferences still fail. This means that $\vdash_{\mathbf{R}}^{e}$ is not really a tcr, since, e.g., $\vdash_{\mathbf{R}}^{e} \varphi \to \varphi$ and $\varphi \to \varphi, \varphi \to \psi \vdash_{\mathbf{R}}^{e} \varphi \to (\varphi \wedge \psi)$, while $\varphi \to \psi \not\vdash_{\mathbf{R}}^{e} \varphi \to (\varphi \wedge \psi)$.

This leaves us with $\vdash_{\mathbf{R}}^{H}$. This choice does not suffer from the drawbacks of the other two choices. Thus, 1–4 are all valid if we identify (as we do from now on) $\vdash_{\mathbf{R}}$ with $\vdash_{\mathbf{R}}^{H}$. There is one problem, though. The definition of $\vdash_{\mathbf{R}}^{H}$ depends on the choice of the system HR which is used for defining **R**. The next theorem provides an equivalent definition, which depends only on the set of theorems of **R**. For its proof we need the following lemma.

---

[4]This is not a valid *fde* (first-degree entailment). Hence some relevantists might deny that it is intuitively valid. I strongly believe that it is.

Lemma 12. *Let S be a finite set of propositional variables, and let $\mathbf{t}_S$ be the conjunction of all the formulas $p \to p$, where $p \in S$. Then $\vdash_\mathbf{R} \mathbf{t}_S \to (\varphi \to \varphi)$ for every φ such that $\mathsf{Atoms}(\varphi) \subseteq S$.*[5]

Proof. A straightforward induction on the structure of φ. ◁

Theorem 13. *$\mathcal{T} \vdash_\mathbf{R} \psi$ iff there exist $\varphi_1, \ldots, \varphi_n \in \mathcal{T}$ and a theorem θ of $\mathbf{R}$ s.t.*

$$\vdash_\mathbf{R} \varphi_1 \wedge \cdots \wedge \varphi_n \wedge \theta \to \psi$$

This θ can be taken as the conjunction of all formulas of the form $p \to p$, where p is an atom that occurs in $\varphi_1 \wedge \cdots \wedge \varphi_n \to \psi$.[6]

Proof. The "if" part is easy, and is left to the reader.

For the converse, assume that $\mathcal{T} \vdash_\mathbf{R} \psi$. Let $d = \psi_1, \ldots, \psi_k = \psi$ be a derivation in $\mathbf{R}$ of ψ from $\mathcal{T}$, and let $\varphi_1, \ldots, \varphi_n$ ($n \le k$) be the elements of $\mathcal{T}$ that occur in d. Without loss in generality, we may assume that the only atoms that occur in d are from $S = \mathsf{Atoms}(\varphi_1 \wedge \cdots \wedge \varphi_n \to \psi)$. (Otherwise, we can simply replace in d other atoms by atoms from this set and get another derivation in $\mathbf{R}$ of ψ from $\mathcal{T}$, which does have the required property.) Let θ be the conjunction of all formulas of the form $p \to p$, where $p \in S$. We prove by induction on i that $\vdash_\mathbf{R} \varphi_1 \wedge \cdots \wedge \varphi_n \wedge \theta \to \psi_i$. The case where $\psi_i \in \mathcal{T}$ is trivial, while the case where ψ_i is an axiom of HR easily follows from Lemma 12, using the implicational axioms of HR. The two induction steps (corresponding to the two inference rules of HR) are rather easy, using the implication-conjunction fragment of HR. Details are left for the reader. ◁

Note 14. Let HR_b^+ be the Hilbert-type system in the language of $\{\to, \wedge, \vee\}$ which is obtained from HR by deleting the contraction axiom, the distribution axiom, and the negation axioms. Let HR_b be the Hilbert-type system in the language of $\{\to, \wedge, \vee, \neg\}$ that is obtained from HR_b^+ by adding as an additional axiom the schema $(\varphi \to \psi) \to (\neg \psi \to \neg \varphi)$. By checking the proof of Theorem 13 (including the proof of Lemma 12), we see that it applies as is to any logic $\mathbf{L}$ which can be axiomatized by adding axiom schemes in their languages to HR_b^+ or to HR_b.

Corollary 15. *Any logic $\mathbf{L}$ of the type described in Note 14 (including of course $\mathbf{R}$ itself) which has the strong VSP (as we show later that $\mathbf{R}$ does) satisfies the basic relevance criterion.*

Proof. Suppose that $\mathcal{T}_1, \mathcal{T}_2 \vdash_\mathbf{L} \theta$, and $\mathsf{Atoms}(\mathcal{T}_2) \cap \mathsf{Atoms}(\mathcal{T}_1 \cup \{\theta\}) = \emptyset$. Then there are $\varphi_1, \ldots, \varphi_n \in \mathcal{T}_2$, $\psi_1, \ldots \psi_k \in \mathcal{T}_1$ such that $\varphi_1, \ldots, \varphi_n, \psi_1, \ldots \psi_k \vdash_\mathbf{L} \theta$, and $\{p_1, \ldots, p_m\} \cap \{q_1, \ldots, q_l\} = \emptyset$, where $\{p_1, \ldots, p_m\} = \mathsf{Atoms}(\{\varphi_1, \ldots, \varphi_n\})$ and $\{q_1, \ldots, q_l\} = \mathsf{Atoms}(\{\psi_1, \ldots \psi_k, \theta\})$. Therefore, Theorem 13 and Note 14 imply that $\vdash_\mathbf{L} (\varphi_1 \wedge \cdots \wedge \varphi_n \wedge \varphi^*) \wedge (\psi_1 \wedge \cdots \wedge \psi_k \wedge \psi^*) \to \theta$, where $\varphi^* = (p_1 \to p_1) \wedge \cdots \wedge (p_m \to p_m)$, and $\psi^* = (q_1 \to q_1) \wedge \cdots \wedge (q_l \to q_l)$. Hence the strong VSP of $\mathbf{R}$ entails that $\vdash_\mathbf{L} (\psi_1 \wedge \cdots \wedge \psi_k \wedge \psi^*) \to \theta$, and so $\psi_1, \ldots \psi_k \vdash_\mathbf{L} \theta$, implying that $\mathcal{T}_1 \vdash_\mathbf{L} \theta$. ◁

---

[5] This lemma is implicit in Anderson et al. [2, Section 45.2]. Essentially, it has first been proved by Ackermann. Later it was reproved in several places, like Dunn's Ph.D. thesis [10] and Maksimova's Ph.D. thesis.

[6] This is a t-free version of what is called in [11] the "enthymematic deduction theorem," originally due to Meyer et al. [12].

4. Some Important Examples

4.1. Multiplicative-additive Linear Logic.

Theorem 16. *Let $\mathbf{LL}_{ma}^{-}$ be the multiplicative-additive fragment of Linear Logic, without the propositional constants. $\mathbf{LL}_{ma}^{-}$ has the strong VSP.*

Proof. For the proof we need the following fact: If $\Gamma_1 \Rightarrow \Delta_1$ and $\Gamma_2 \Rightarrow \Delta_2$ are both non-empty (i.e., they are not the empty sequent $\Rightarrow$), and $\vdash_{\mathbf{GLL}_{ma}^{-}} \Gamma_1, \Gamma_2 \Rightarrow \Delta_1, \Delta_2$, where $\mathbf{GLL}_{ma}^{-}$ is the standard Gentzen-type system of $\mathbf{LL}_{ma}^{-}$, then $\mathrm{Atoms}(\Gamma_1 \Rightarrow \Delta_1) \cap \mathrm{Atoms}(\Gamma_2 \Rightarrow \Delta_2) \neq \emptyset$. (This well-known fact can be proved by a straightforward induction on the structure of cut-free proofs in $\mathbf{GLL}_{ma}^{-}$. Alternatively, it suffices to note that from $\Gamma_1, \Gamma_2 \Rightarrow \Delta_1, \Delta_2$ one can easily derive in $\mathbf{GLL}_{ma}^{-}$ a sequent of the form $\Rightarrow \varphi \to \psi$, where $\mathrm{Atoms}(\varphi) = \mathrm{Atoms}(\Gamma_1 \Rightarrow \Delta_1)$ and $\mathrm{Atoms}(\psi) = \mathrm{Atoms}(\Gamma_2 \Rightarrow \Delta_2)$. Hence the claim follows from the VSP for $\mathbf{LL}_{ma}^{-}$. The letter, in turn, follows from the VSP for $\mathbf{R}$.)

Next we prove by induction on the structure of cut-free proofs in $\mathbf{GLL}_{ma}^{-}$ that if $\vdash_{\mathbf{GLL}_{ma}^{-}} \varphi \wedge \psi, \Gamma \Rightarrow \Delta$, where $\Gamma \Rightarrow \Delta$ is non-empty, and $\mathrm{Atoms}(\varphi) \cap \mathrm{Atoms}(\Gamma \Rightarrow \Delta) = \emptyset$, then $\vdash_{\mathbf{GLL}_{ma}^{-}} \psi, \Gamma \Rightarrow \Delta$. Since contraction is not among the structural rules of $\mathbf{GLL}_{ma}^{-}$, the induction is rather easy. The only interesting case is the one in which the last step in the derivation of $\varphi \wedge \psi, \Gamma \Rightarrow \Delta$ introduces the indicated $\varphi \wedge \psi$. In general, the premise of this step can be in this case either $\varphi, \Gamma \Rightarrow \Delta$ or $\psi, \Gamma \Rightarrow \Delta$. However, the fact mentioned above excludes the first possibility. Hence we get that the given derivation of $\varphi \wedge \psi, \Gamma \Rightarrow \Delta$ includes a derivation of $\psi, \Gamma \Rightarrow \Delta$.

The rest the proof in now easy. Suppose that $\vdash_{\mathbf{LL}_{ma}^{-}} \theta \wedge \varphi \to \psi$. This means that $\vdash_{\mathbf{GLL}_{ma}^{-}} \Rightarrow \theta \wedge \varphi \to \psi$, and so $\vdash_{\mathbf{GLL}_{ma}^{-}} \theta \wedge \varphi \Rightarrow \psi$. If in addition the sentences θ and $\varphi \to \psi$ share no atomic formula, then the claim that we have just proved entails that $\vdash_{\mathbf{GLL}_{ma}^{-}} \varphi \Rightarrow \psi$, and so $\vdash_{\mathbf{LL}_{ma}^{-}} \varphi \to \psi$. ◁

Corollary 17. $\mathbf{LL}_{ma}^{-}$ *satisfies the basic relevance criterion.*

Proof. Immediate from Theorem 16 and Corollary 15. ◁

4.2. The Relevance Logic R.

Since no useful analytic proof system is known for the whole of $\mathbf{R}$, the proof that it has the strong VSP is far more difficult than in the case of $\mathbf{LL}_{ma}^{-}$, and uses $\mathbf{R}$'s *ternary-relation semantics* of $\mathbf{R}$-frames. We assume that the reader is acquainted with this semantics (as presented, e.g., in [11]). To apply it, we need first a definition and two easy lemmas.

Definition 18. Let F and G be $\mathbf{R}$-frames. $F \times G$ is the ternary frame in which
1. $\mathrm{Dom}(F \times G) = \mathrm{Dom}(F) \times \mathrm{Dom}(G)$.
2. $R_{F \times G}(\langle a_1, b_1 \rangle, \langle a_2, b_2 \rangle, \langle a_3, b_3 \rangle)$ iff $R_F(a_1, a_2, a_3)$ and $R_G(b_1, b_2, b_3)$.
3. $0_{F \times G} = \langle 0_F, 0_G \rangle$.
4. $\langle a, b \rangle^* = \langle a^*, b^* \rangle$.

Lemma 19. *If F and G are $\mathbf{R}$-frames, then so is $F \times G$.*

Lemma 20. *Let F and G be $\mathbf{R}$-frames, and let v and v^* be valuations in F and $F \times G$, respectively. Suppose that $\forall a \forall b (\langle a, b \rangle \vDash_{F \times G}^{v^*} p \Leftrightarrow a \vDash_F^v p)$ for every $p \in \mathrm{Atoms}(\varphi)$. Then $\forall a \forall b (\langle a, b \rangle \vDash_{F \times G}^{v^*} \varphi \Leftrightarrow a \vDash_F^v \varphi)$.*

The proofs of the two lemmas are straightforward (and left for the reader).

Theorem 21. **R** *has the strong VSP (with respect to its connectives $\to$ and $\wedge$).*

Proof. Suppose that $\mathrm{Atoms}(\theta) \cap \mathrm{Atoms}(\varphi \to \psi) = \emptyset$, while $\not\vdash_{\mathbf{R}} \varphi \to \psi$. The second assumption implies that there is an **R**-frame G, together with a set-up b and a valuation v_G in it, such that $b \vDash_G^{v_G} \varphi$, and $b \not\vDash_G^{v_G} \psi$. In addition, the fact that **R** has the VSP for $\to$ (which entails that $\not\vdash_{\mathbf{R}} \theta \to q$ in case $q \notin \mathrm{Atoms}(\theta)$) implies that there is an **R**-frame F, together with a set-up a and a valuation v_F in it, such that $a \vDash_F^{v_F} \theta$. Define a valuation v on $F \times G$ by

$$v(\langle x,y \rangle, p) = \begin{cases} v_F(x,p) & \text{if } p \in \mathrm{Atoms}(\theta) \\ v_G(y,p) & \text{otherwise.} \end{cases}$$

From Lemma 19 and Lemma 20, it follows that $F \times G$ is an **R**-frame, in which $\langle a,b \rangle \vDash_{F \times G}^{v} \theta$; $\langle a,b \rangle \vDash_{F \times G}^{v} \varphi$; $\langle a,b \rangle \not\vDash_{F \times G}^{v} \psi$. Hence $\langle a,b \rangle \not\vDash_{F \times G}^{v} \theta \wedge \varphi \to \psi$. Therefore, $\not\vdash_{\mathbf{R}} \theta \wedge \varphi \to \psi$. ◁

Note 22. The same proof would work for any other relevance logic **L** for which Lemma 19 (with **L** instead of **R**) is true.

Corollary 23. **R** *satisfies the basic relevance criterion.*

Proof. Immediate from Theorem 21 and Corollary 15. ◁

4.3. The Purely Relevance Logic RMI.

RMI is a logic in the language of **R** that was introduced in Avron [3, 4, 5]. (See also Chapter 14 of [9].) Since it is less known than **R** and **LL**, we review here some relevant information about it that can be found in the above cited sources. First, there are several ways of axiomatizing **RMI**. The simplest among them is perhaps the Hilbert-type system HRMI, which is obtained from the standard axiomatization HR of **R** by the following three changes:

1. Add the mingle axiom: $\varphi \to (\varphi \to \varphi)$.
2. Add to the adjunction rule of HR (from φ and ψ infer $\varphi \wedge \psi$) a third premise: $R(\varphi, \psi)$, where the latter is some sentence which says that φ and ψ are relevant to each other. Two such sentences are $(\varphi \to \varphi) \wedge (\psi \to \psi)$ and $(\varphi \to \varphi) + (\psi \to \psi)$ (where $\varphi + \psi$ abbreviates $\neg \varphi \to \psi$).
3. Turn the distributivity axiom $\varphi \wedge (\psi \vee \theta) \to (\varphi \vee \psi) \wedge (\varphi \vee \theta)$ of HR into a *rule* in which $R(\psi, \theta)$ is the single premise.

From the relevance point of view, **RMI** has some rather appealing properties. (See Chapters 13–14 in [9], especially Section 14.3). As argued in [6], those properties make it a prime candidate for being chosen as the most appropriate relevance logic. Here are some of them.

1. **RMI** has a very intuitive semantics. In fact, it is strongly sound and complete for a certain general family of matrices, which is based on the following rather intuitive ideas.
    (i) Propositions are divided into "domains of discourse."
    (ii) The domains are partially ordered according to certain "degrees of priority" or "degrees of dependency." This partial order induces a hierarchical, tree-like structure, in which the root has the highest "degree of priority," and a leaf the lowest one.

(iii) Each domain has its own truth-values. Usually it has two, corresponding to the classical truth-values, and classical logic is valid within it. However, domains which are leaves of the tree of domains may be degenerate, having just one truth-value I, with the intended meaning of "inconsistent," or "both true and false."
(iv) Two propositions are relevant to each other iff their associated domains are related by the grading relation. Both $\to$ and $\wedge$ produce a false proposition whenever they are applied to propositions which are not relevant to each other.

2. **RMI** has the VSP with respect to both $\to$ and $\wedge$.
3. **RMI** is *purely relevant*: it has no extensional connectives.
4. Unlike **R, E, T** and the other main relevance logics, **RMI** is *decidable*.
5. **RMI** has a corresponding cut-free hypersequential Gentzen-type system.

Among the family of matrices for **RMI**, there is one which is particularly useful, since **RMI** is strongly sound and complete relative to it. (See [9].)

Definition 24. The matrix $\mathcal{SA}$ for the language $\{\neg, \wedge, \vee, \to\}$ is $\langle SA, \mathcal{D}_{SA}, \mathcal{O}\rangle$ where:

1. $SA = [0,1] \times \{f, t, I_1, I_2, I_3, \ldots\}$.
   If $v = \langle x, a\rangle \in SA$ then x is called the *degree* of v and is denoted by $deg(v)$, while a is called the *value* of v and is denoted by $val(v)$.
2. $\mathcal{D}_{SA} = [0,1] \times \{t, I_1, I_2, I_3, \ldots\}$.
3. The operations in $\mathcal{O}$ are defined as follows:
   Negation: $\tilde{\neg}\langle x, t\rangle = \langle x, f\rangle$ $\tilde{\neg}\langle x, f\rangle = \langle x, t\rangle$ $\tilde{\neg}\langle x, I_k\rangle = \langle x, I_k\rangle$
   Implication: $deg(u\tilde{\to}v) = \min\{deg(u), deg(v)\}$

$$val(u\tilde{\to}v) = \begin{cases} I_k & u = v \text{ and } val(u) = I_k \\ t & deg(u) \leq deg(v) \text{ and } val(u) = f \\ t & deg(v) \leq deg(u) \text{ and } val(v) = t \\ f & \text{otherwise.} \end{cases}$$

Conjunction and disjunction: Let $u \preceq v$ if either $u = v$, or $dom(u) \leq dom(v)$ and $val(u) = f$, or $dom(v) \leq dom(u)$ and $val(v) = t$. Then
(i) $u \wedge v$ is u if $u \preceq v$, v if $v \preceq u$, and $\langle \inf_{\preceq}\{dom(u), dom(v)\}, f\rangle$ otherwise (i.e., if u and v are not relevant);
(ii) $u \vee v$ is v if $u \preceq v$, u if $v \preceq u$, and $\langle \inf_{\leq}\{dom(u), dom(v)\}, t\rangle$ otherwise (i.e., if u and v are not relevant).

We can now prove:

Theorem 25. **RMI** *has the strong VSP (with respect to $\to$ and $\wedge$).*

Proof. Suppose that $\mathrm{Atoms}(\theta) \cap \mathrm{Atoms}(\varphi \to \psi) = \emptyset$, while $\not\vdash_{\mathbf{RMI}} \varphi \to \psi$. The second assumption implies that there is a valuation v in $\mathcal{SA}$ such that $val(v(\varphi \to \psi)) = f$. We define a valuation v^* in $\mathcal{SA}$ such that $val(v^*(\theta \wedge \varphi \to \psi)) = f$. We let $v^*(p) = v(p)$ for each $p \notin \mathrm{Atoms}(\theta)$. This insures that $v^*(\varphi) = v(\varphi)$ and $v^*(\psi) = v(\psi)$. For $p \in \mathrm{Atoms}(\theta)$ we define $v^*(p)$ according to the reason why $val(v(\varphi \to \psi)) = f$.
Let $deg(v(\varphi)) = a$, $deg(v(\psi)) = b$.

1. $v(\varphi)$ and $v(\psi)$ are irrelevant (i.e., neither $v(\varphi) \leq v(\psi)$ nor $v(\psi) \leq v(\varphi)$). In this case, we let $v^\star(p) = v(\varphi)$ for every $p \in \mathsf{Atoms}(\theta)$. This implies that $v^\star(\theta) \in \{v(\varphi), v(\neg\varphi)\}$, and the same applies to $\theta \wedge \varphi$. It follows that $v^\star(\theta \wedge \varphi)$ and $v^\star(\psi)$ are irrelevant too. Hence $val(v^\star(\theta \wedge \varphi \to \psi)) = f$.
2. $a > b$, and $val(v(\varphi)) = t$. In this case, we let $v^\star(p) = \langle a, I_1 \rangle$ for every $p \in \mathsf{Atoms}(\theta)$. This implies that $v^\star(\theta \wedge \varphi) = v^\star(\theta) = \langle a, I_1 \rangle$. Hence, $v^\star(\theta \wedge \varphi)$ and $v^\star(\psi)$ are irrelevant, implying that $val(v^\star(\theta \wedge \varphi \to \psi)) = f$.
3. $a = b$, $val(v(\varphi)) = t$, and $val(v(\psi)) = f$. In this case, we again let $v^\star(p) = \langle a, I_1 \rangle$ for every $p \in \mathsf{Atoms}(\theta)$, and this implies that $v^\star(\theta \wedge \varphi) = v^\star(\theta) = \langle a, I_1 \rangle$. Since in this case $v^\star(\psi) = \langle a, f \rangle$, we again get that $val(v^\star(\theta \wedge \varphi \to \psi)) = f$.
4. $a < b$ and $val(v(\psi)) = f$. In this case, we let $v^\star(p) = v(\varphi)$ for every $p \in \mathsf{Atoms}(\theta)$. This implies that $v^\star(\theta) \in \{v(\varphi), v(\neg\varphi)\}$, and the same applies to $\theta \wedge \varphi$. It follows that $deg(v(\theta \wedge \varphi)) = a$, $v^\star(\psi) = \langle b, f \rangle$. Since $a < b$, it follows that $v^\star(\theta \wedge \varphi \to \psi) = \langle b, f \rangle$ too.

It follows that in all the possible cases we get that $val(v^\star(\theta \wedge \varphi \to \psi)) = f$. Hence $\not\vdash_{\mathbf{RMI}} \theta \wedge \varphi \to \psi$. ◁

Now, the variable-sharing property that **RMI** has with respect to $\wedge$ implies that Theorem 13 is not valid for it. (But note that the conditions given in Proposition 8 *are* satisfied by **RMI**.) Therefore, here Theorem 25 does not imply that **RMI** satisfies the basic relevance criterion. Nevertheless, it does.

Proposition 26 ([9, Prop. 14.57]). *RMI satisfies the basic relevance criterion.*

Proof. Suppose $\mathcal{T}_1, \mathcal{T}_2 \vdash_{\mathbf{RMI}} \psi$, and $\mathcal{T}_2$ has no variables in common with $\mathcal{T}_1 \cup \{\psi\}$. We show that $\mathcal{T}_1 \vdash_{\mathbf{L}} \psi$. Suppose otherwise. Then, there is a valuation v in $\mathcal{SA}$ which is a model of $\mathcal{T}_1$, but not of ψ. Since $\mathcal{T}_2$ has no variable in common with $\mathcal{T}_1 \cup \{\psi\}$, we may assume without loss in generality that $v(p) = \langle 1, I_1 \rangle$ for every variable p which occurs in $\mathcal{T}_2$. But then $v(\varphi) = \langle 1, I_1 \rangle$ for every $\varphi \in \mathcal{T}_2$, and so v is a model in $\mathcal{SA}$ of $\mathcal{T}_1 \cup \mathcal{T}_2$, but not of ψ. By the strong soundness and completeness of **RMI** for $\mathcal{SA}$, this contradicts our assumption that $\mathcal{T}_1, \mathcal{T}_2 \vdash_{\mathbf{RMI}} \psi$. ◁

Acknowledgments. This research was supported by The Israel Science Foundation (grant no. 550/19).

REFERENCES

[1] Anderson, A. R. and Belnap, N. D. (1975). *Entailment: The Logic of Relevance and Necessity, Vol.I*, Princeton University Press, Princeton, NJ.
[2] Anderson, A. R., Belnap, N. D. and Dunn, M. (1992). *Entailment: The Logic of Relevance and Necessity, Vol.II*, Princeton University Press, Princeton, NJ.
[3] Avron, A. (1990a). Relevance and paraconsistency – A new approach, *Journal of Symbolic Logic* **55**: 707–732.
[4] Avron, A. (1990b). Relevance and paraconsistency – A new approach. Part II: The formal systems, *Notre Dame Journal of Formal Logic* **31**: 169–202.
[5] Avron, A. (1991). Relevance and paraconsistency – A new approach. Part III: Cut-free Gentzen-type systems, *Notre Dame Journal of Formal Logic* **32**: 147–160.
[6] Avron, A. (1992). Whither relevance logic?, *Journal of Philosophical Logic* **21**: 243–281.

[7] Avron, A. (2014). What is relevance logic?, *Annals of Pure and Applied Logic* **165**: 26–48.
[8] Avron, A. (2016). RM and its nice properties, *in* K. Bimbó (ed.), *J. Michael Dunn on Information Based Logics*, Vol. 8 of *Outstanding Contributions to Logic*, Springer, Switzerland, pp. 15–43.
[9] Avron, A., Arieli, O. and Zamansky, A. (2018). *Theory of Effective Propositional Paraconsitent Logics*, Vol. 75 of *Studies in Logic (Mathematical Logic and Foundations)*, College Publications, London, UK.
[10] Dunn, J. M. (1966). *The Algebra of Intensional Logics*, Doctoral dissertation, University of Pittsburgh, Pittsburgh, PA, UMI, Ann Arbor. (Published as Vol. 2 in the *Logic PhDs* series by College Publications, London, UK, 2019.).
[11] Dunn, J. M. and Restall, G. (2002). Relevance logic, *in* D. Gabbay and F. Guenther (eds.), *Handbook of Philosophical Logic*, 2nd edn, Vol. 6, Kluwer, Dordrecht, pp. 1–136.
[12] Meyer, R. K., Dunn, J. M. and Leblanc, H. (1974). Completeness of relevant quantification theories, *Notre Dame Journal of Formal Logic* **15**: 97–121.

SCHOOL OF COMPUTER SCIENCE, TEL AVIV UNIVERSITY, TEL AVIV, ISRAEL
Email: aa@cs.tau.ac.il

TIME FOR CURRY

Jc Beall and David Ripley

ABSTRACT. Some recent approaches to Curry's paradox handle it by invoking so-called non-normal worlds: worlds at which the laws of logic fail. We present a new version of Curry paradox (a temporal Curry paradox) which seems to push on these approaches, by parity of reasoning, postulation of non-normal times: times, including times at the actual world, at which the laws of logic fail.

Keywords. Contraction, Curry's paradox, Non-normal times, Non-normal worlds, Temporal Curry paradox

1. INTRODUCTION

This paper presents a new puzzle for certain positions in the theory of truth. The relevant positions can be stated in a language including a truth predicate T and an operation $\ulcorner \ \urcorner$ that takes sentences to names of those sentences; they are positions that take the T-schema $A \leftrightarrow T(\ulcorner A \urcorner)$ to hold without restriction, for every sentence A in the language. As such, they must be based on a nonclassical logic, since paradoxes that cannot be handled classically will arise. The best-known of these paradoxes is probably the liar paradox — a sentence that says of itself (only) that it is not true — but our concern here is not with the liar. Instead, our focus is a variant of Curry's paradox [4; 6; 8; 12] — a sentence that says of itself (only) that if it is true, everything is true.

§2 is necessary stage setting; we present the standard version of Curry's paradox and the strain of response to it we wish to focus on. This strain of response crucially invokes *non-normal worlds*, that is, worlds at which the laws of logic differ from the laws that actually hold. In §3, we go on to argue that, in light of *temporal Curry paradox* (a novel version of Curry paradox that we present here), this strain of response ought also to accept *non-normal times*, that is, times at which the laws of logic *in the actual world* differ from the laws that hold now. We then consider, in §4, what this would mean for the theorists in question.

2. THE STATE OF PLAY

2.1. Curry Paradox. The conditional involved in the standard Curry sentence is the same conditional used in the T-schema, and theorists differ on just how this conditional ought to behave. For our purposes here, we focus only on *detachable* conditionals — conditionals that validate modus ponens. (By "modus ponens" we mean — throughout

---

2020 *Mathematics Subject Classification.* Primary: 03B80, Secondary: 03B44.

Bimbó, Katalin, (ed.), *Relevance Logics and other Tools for Reasoning. Essays in Honor of J. Michael Dunn*, (Tributes, vol. 46), College Publications, London, UK, 2022, pp. 65–72.

— only the so-called rule form: that A and $A \to B$ jointly imply B, that is, that the *argument* from $\{A, A \to B\}$ to B is valid.)

By the usual diagonalization methods, there is a sentence C that is equivalent to $T(\ulcorner C \urcorner) \to \bot$, where $\to$ is the (detachable) conditional in the T-schema. Here, take $\bot$ to be an "explosive sentence" from which everything follows. (It might be "everything is true," or some such.) Given these resources, a proof from the T-schema to $\bot$ threatens (where $\leftrightarrow$ is defined via $\wedge$ and $\to$ as usual):

1. $T(\ulcorner C \urcorner) \leftrightarrow C$ T-schema
2. $T(\ulcorner C \urcorner) \leftrightarrow (T(\ulcorner C \urcorner) \to \bot)$ 1, substitution
3. $T(\ulcorner C \urcorner) \to (T(\ulcorner C \urcorner) \to \bot)$ 2, $\wedge$-elim
4. $T(\ulcorner C \urcorner) \to \bot$ 3, contraction
5. C 4, substitution
6. $T(\ulcorner C \urcorner)$ 1, 5, modus ponens
7. $\bot$ 4, 6, modus ponens

Here, the step from 3 to 4 is justified by contraction: from $A \to (A \to B)$ we can conclude $A \to B$.[1] Since substituting equivalents, eliminating $\wedge$'s, and modus ponens all seem like surer steps than contraction, one plausible way to treat this paradox while retaining the T-schema is to provide a theory of $\to$ on which contraction is not a valid inference. Such theories have recently been advanced by Beall [1], Brady [3], Field [5], Priest [11], Sylvan [16], and others.

2.2. Explanation and Truth Conditions for Conditionals. Among these theorists, some think that contraction's invalidity is to be explained by $\to$'s truth conditions, typically because validity and invalidity in general are taken to be matters of truth conditions. For example, Priest argues that "validity is the relationship of truth-preservation-in-all-situations" [10, ch. 11], and this thought is also forcefully embraced in the work of Routley [15, Appendix I], and also evident in Brady's work [3]. While not universally endorsed, the thought is natural and common in philosophy.[2] Example: Why is it that (say) $A \wedge B$ implies B? The answer is that the definition of implication (validity) is "truth preservation over all conditions," and the conditions in which a conjunction is true (i.e., the truth conditions for a conjunction) have it that $A \wedge B$ is true just if both A and B are true. The explanation falls out of truth conditions and their role in validity. Likewise, the explanation for the *failure* of $A \vee B$'s implying B invokes the truth conditions for $\vee$ and the existence of situations in which $A \vee B$ is true and B is not. And the same goes, according to target theorists, for contraction.

Among such theorists, the dominant approach to truth conditions for $\to$ invokes frames involving points (worlds or world-like entities). We consider such an approach here; and we call the points "worlds" without worrying what they are. As a first approximation, let a frame be a set W of worlds, and let a model be a frame together with a relation $\Vdash$ between worlds and sentences of our language. $\Vdash$ can hold or not

---

[1]This use of "contraction" is related to, but importantly distinct from, its use to describe the structural rule that allows for repeated use of premises. For example, in the above proof, premise 1 is used twice (in the justifications of steps 2 and 6); this involves appeal to the structural rule, but not to the $\to$-related rule we call "contraction," which is involved above only in the step from 3 to 4.

[2]For example, Beall [1] and Field [5] reject the explanatory role of truth conditions.

between any world and any (non-logical) atomic sentence, but it is constrained for compound sentences. On the approach we are considering, it is these constraints that give sentential connectives their meanings. For example, here are constraints to give $\wedge$, T, and $\bot$ their meanings:

$$w \Vdash A \wedge B \quad \text{iff} \quad w \Vdash A \text{ and } w \Vdash B$$
$$w \Vdash T(\ulcorner A \urcorner) \quad \text{iff} \quad w \Vdash A$$
$$w \nVdash \bot, \text{ for any } w$$

It is sometimes useful to distinguish *extensional* connectives from *intensional* ones. Extensional connectives don't look across worlds; whether a sentence built with an extensional connective is satisfied at a world depends only on what else is satisfied at that world. Intensional connectives, on the other hand, look across worlds; whether a sentence built with an intensional connective is satisfied at a world can depend on what happens at other worlds. $\wedge$, T, and $\bot$ are all extensional. Since $\rightarrow$ will be used to express the strong T-schema connection between A and $T(\ulcorner A \urcorner)$, its constraint will take into account the relation between its antecedent and consequent across worlds, and so $\rightarrow$ is intensional:

$$w \Vdash A \rightarrow B \quad \text{iff} \quad \text{for all } w' \in W, \text{ either } w' \nVdash A, \text{ or } w' \Vdash B.$$

Now we can define validity. An argument from a set of sentences Γ to a set of sentences Δ is valid ($\Gamma \vDash \Delta$) iff in every model on every frame, at every world w such that $w \Vdash A$ for every $A \in \Gamma$, $w \Vdash B$ for some $B \in \Delta$. (This is the general, multiple-conclusion relation. Restricting to singleton conclusions reduces to the usual single-conclusion account. The general account is worth having, though nothing we say here hangs on the generality.)

2.3. Contraction Freedom and Non-normal Worlds.
But there is a problem. As things currently stand, $A \rightarrow (A \rightarrow B) \vDash A \rightarrow B$. That is, contraction holds. The other principles used in the problematic Curry argument also hold. So such truth conditions can't be the whole story; they would force us to conclude $\bot$, and thus every sentence — naked absurdity.

If it is to be contraction-free, $\rightarrow$ must derive its meaning from some constraint that doesn't force contraction on it. For ideas as to how this is to be done, we can look to frames developed for weak relevant and linear logics, in which contraction fails. Here, the usual way cuts a distinction in W, namely, *normal worlds* and *non-normal worlds*. Thus, we require our frames to be slightly more articulated, specifying a set W of worlds, and a set $N \subseteq W$ of normal worlds. For any normal world $w \in N$, we constrain $\Vdash$ as before. But for any non-normal world $w \in W \setminus N$, we treat $\rightarrow$-sentences differently; for our purposes here, we allow $\rightarrow$-sentences to be satisfied or not by non-normal worlds *arbitrarily*.[3]

If this is the only shift we make, however, we lose such important validities as modus ponens. (There will be non-normal worlds at which A holds, $A \rightarrow B$ holds, and B does not hold, since whether $A \rightarrow B$ holds there has nothing to do with where

---

[3]There are other options; the important thing for our purposes is the distinction between normal and non-normal worlds, and that this distinction matters for the constraints on $\Vdash$ when it comes to $\rightarrow$-sentences. The details of the constraints (if any) that operate at non-normal worlds are beside the point.

or whether A or B holds anywhere.) So we must also change our understanding of validity: the argument from a set of sentences Γ to a set of sentences Δ is valid ($\Gamma \vDash \Delta$) iff in every model on every frame, at every *normal* world w such that $w \Vdash A$ for every $A \in \Gamma$, $w \Vdash B$ for some $B \in \Delta$. The restriction to *normal* w in this definition ensures that, in evaluating validity, we only look at worlds in which the $\to$ is "well-behaved." In particular, modus ponens is valid, given this new understanding of validity.

Crucially, however, contraction remains invalid. Although we only look at normal worlds in checking validity, those normal worlds themselves look at *all* worlds — normal and non-normal — in the truth-conditions for $\to$-sentences. As such, it's possible for $A \to (A \to B)$ to hold at a normal world without $A \to B$ holding there; this can happen if there is a (non-normal) world at which A and $A \to B$ both hold, but where B does not hold. Counterexamples to contraction *somewhere* thus rely on counterexamples to modus ponens *somewhere else*; the distinction between normal and non-normal worlds allows us to keep these somewheres organized, so that the counterexamples to contraction are sufficient to undermine its validity, while the counterexamples to modus ponens are not.

Of course, if the invocation of non-normal worlds is meant to *explain* the failure of contraction, it is not enough simply to offer this kind of model theory. The explanation must tell us something about what non-normal worlds *are*, and why they are related to $\to$ in the way the model theory takes them to be. Indeed, such a theory is offered by Priest [9, p. 15]:

> The normal worlds are to be thought of as (logically) possible worlds. Non-normal worlds are to be thought of as (logically) impossible worlds. The idea that there can be physically impossible worlds, that is, worlds where the laws of physics are different, is a standard one. Such worlds are still logically possible. But just as some worlds have laws of physics different from the actual physical laws, so some worlds have laws of logic different from the actual logical laws.

Our approach here shall assume that this — allowing for failures of contraction, to be explained by invoking worlds at which logical laws differ — is broadly the right way to address Curry's paradox, and to explore how this approach adapts to a novel version of Curry paradox with slightly different ingredients.

3. THE MEAT

3.1. Robust Contraction-freedom. First, note that it is not enough just to avoid contraction for $\to$. Curry trouble arises if there is any connective $\Rightarrow$ meeting the following three conditions [2; 8; 14]:

$\to$-**consequence:** From $A \to B$, we can infer $A \Rightarrow B$.
$\Rightarrow$-**modus ponens:** From A and $A \Rightarrow B$, we can infer B.
$\Rightarrow$-**contraction:** From $A \Rightarrow (A \Rightarrow B)$, we can infer $A \Rightarrow B$.

In fact, we can replace the first condition, $\to$-consequence, with the condition:

$\Rightarrow$-**T-schema:** $A \Leftrightarrow T(\ulcorner A \urcorner)$ is provable.

From $\to$-consequence and the ($\to$-involving) T-schema, $\Rightarrow$-T-schema follows. And the Curry proof in §2.1 for $\to$ can simply be repeated as is for $\Rightarrow$ if $\Rightarrow$-T-schema, $\Rightarrow$-modus ponens, and $\Rightarrow$-contraction all hold.

Below, we present a connective that, at least prima facie, seems to satisfy $\Rightarrow$-T-schema, $\Rightarrow$-modus ponens, and $\Rightarrow$-contraction. We then argue that a friend of non-normal worlds explanations of the sort mentioned in §2.2 ought to acknowledge non-normal *times* to address this threat.

3.2. Temporal Curry.
Sentences are not just true at some worlds and false at others; they can also be true at some times and false at others, even within a single world. To accommodate this, we follow Kaplan [7] and expand our models. We now take our models to specify a set W of worlds and a set T of times. As before, we divide W into the normal worlds N and the others, the non-normal worlds. Now, we can specify truth conditions for our connectives almost as before. The difference is straightforward: sentences are not true at worlds, but instead true at world-and-time pairs. For all of our above connectives, the truth conditions change only slightly: they now carry an idle time parameter. For example: for any $w \in W$ and any $t \in T$,

$$\langle w,t \rangle \Vdash A \wedge B \quad \text{iff} \quad \langle w,t \rangle \Vdash A \text{ and } \langle w,t \rangle \Vdash B.$$

The conditional $\to$ is perhaps of more interest, but the same idea applies. For *normal* worlds w,

$$\langle w,t \rangle \Vdash A \to B \quad \text{iff} \quad \text{for all } w' \in W, \text{ either } \langle w',t \rangle \nVdash A, \text{ or } \langle w',t \rangle \Vdash B.$$

As before, at non-normal worlds $\Vdash$ is not constrained for $\to$-sentences; here, this lack of constraint extends to all times.

If this were all there were to temporal models, they would not be very interesting. They come into their own when we consider connectives that shift the time parameter. By analogy with the "extensional"/"intensional" terminology to describe whether a connective shifts the world parameter, we can draw a distinction between "extemporal" and "intemporal" connectives. All of our old connectives are extemporal, but intemporal connectives allow us to use the structure that temporal models provide. The most familiar intemporal connectives are unary connectives, studied by Prior [13], often written F, G, P, and H. Here, we skip these, to explore the behavior of a binary intemporal connective, which we write (only for lack of obviously better notation) as a short map arrow: $\mapsto$. Informally, $A \mapsto B$ is to be read as something like "whenever A, B." This informal reading is evident in the truth-condition $(*)$:

$$(*) \quad \langle w,t \rangle \Vdash A \mapsto B \quad \text{iff} \quad \text{for all } t' \in T, \text{ if } \langle w,t' \rangle \Vdash A, \text{ then } \langle w,t' \rangle \Vdash B.$$

Despite being intemporal, $\mapsto$ is extensional, because its range of truth values at a world w does not depend on any world beyond w.

But now trouble is brewing. Call a world-time pair $\langle w,t \rangle$ a *normal-world pair* just if $w \in N$. Presumably, we want it to be the case that $A \mapsto T(\ulcorner A \urcorner)$ and $T(\ulcorner A \urcorner) \mapsto A$ are logical truths: true at all normal-world pairs. Indeed, as far as we can see, any motivation for the T-schema's validity at *worlds* is equal motivation for its validity at *times*. (Just as it's very difficult to imagine a world at which A holds without $T(\ulcorner A \urcorner)$ holding and vice versa, so too it's very difficult for *times*.) But now the trouble bubbles to the surface. In particular, notice that, given the truth conditions for $\mapsto$, we have both $\mapsto$-*modus ponens* and $\mapsto$-*contraction*. Hence, mixed with $\mapsto$-*T-schema*, we have the ingredients for explosive Curry (see §3.1). This is a *temporal* Curry paradox that's as explosive as its standard non-temporal relative.

3.3. The Bite: Non-normal *Times*.

Since this is substantially the same problem as Curry's paradox for $\to$, we think it should receive substantially the same solution. In short, $\mapsto$ obeys the truth condition ($*$) given in §3.2 at *most* world-time pairs; however, there are world-time pairs — call them *abnormal* — at which $\mapsto$ fails to conform to the given truth conditions. (Perhaps, as with $\to$, the behavior of $\mapsto$ is arbitrary at abnormal pairs.) Such abnormal pairs involve "non-normal times," *times* at which laws of logic fail.

The bite is more than that there be *some* world-time pair $\langle w,t \rangle$ that is abnormal in having a "non-normal time." The bite is stronger: *every* world — and, hence, every *normal* world, including this (our actual) one — features in some abnormal pair. Suppose otherwise, that is, fix a world w and suppose that for all $t \in T$, the "whenever" connective $\mapsto$ obeys the given truth condition at $\langle w,t \rangle$. Then we have Curry trouble at w. Consider a Curry sentence C equivalent to $T(\ulcorner C \urcorner) \mapsto \bot$. Suppose $\langle w,t \rangle \Vdash C$. Then for all $t' \in T$, if $\langle w,t' \rangle \Vdash T(\ulcorner C \urcorner)$ then $\langle w,t' \rangle \Vdash \bot$. Since $\langle w,t' \rangle \nVdash \bot$ for all t', it must be that $\langle w,t' \rangle \nVdash T(\ulcorner C \urcorner)$ for all t'. But then we have a counterexample at $\langle w,t \rangle$ to the $\mapsto$-T-schema. This cannot be. Hence, for all t, we have $\langle w,t \rangle \nVdash C$. If we are to avoid a counterexample to the $\mapsto$-T-schema, it must be that for all t, $\langle w,t \rangle \nVdash T(\ulcorner C \urcorner)$. But, then, by $\mapsto$'s truth conditions, $\langle w,t \rangle \Vdash T(\ulcorner C \urcorner) \mapsto \bot$ for any t. So this is impossible too.

The philosophical rub comes out when we consider the actual world. As above, for any w there must be abnormal pairs $\langle w,t \rangle$ at which $\mapsto$ does not obey the given truth conditions. Consider the actual world @, and let the *non-normal times* be those times t for which $\langle @,t \rangle$ is an abnormal pair. By the argument above, there must be non-normal times. But this is philosophically awkward. It is much harder to make satisfying philosophical sense of non-normal times than it is of non-normal worlds.

Non-normal worlds, recall, are worlds where the actual laws of logic do not hold. Since worlds are unfamiliar and odd sorts of places anyhow, it is not so challenging to suppose that some of them fail laws of logic in this way. But if abnormal pairs are pairs where laws of logic do not hold — as they must be — then there must be times at which laws of logic fail *in the actual world*. This, we think, is harder to swallow. There is no modal cushion between us and the failure; it is only a matter of minutes. (It may be many minutes; maybe all of the failures are tucked away safely in the past, or far off in the future. But still, they must be there — here! — even if not now.) This failure is serious. As we saw before, for contraction to fail *here*, modus ponens must fail *somewhere*. Thus, there are times at which modus ponens fails in the actual world.

4. Possible Responses

Of course, one can simply bite the bullet and admit that there are non-normal times. Perhaps this is a discovery rather than a reductio. (Certainly, when paradoxes are in the air, one has been mistaken for the other before.) But if this is the right way to understand the situation, more needs to be said to assuage the initial awkwardness. Even those of us who were prepared to go along with non-normal worlds feel some difficulty allowing for non-normal times. A story about why logical laws might change over time, analogous to the way they can be taken to change over worlds, would be a great help to resolve this difficulty.

Another possibility would be to attack the analogy we have exploited between worlds and times. For example, perhaps there is some reason why there should be no extensional intemporal connectives like $\mapsto$, despite the presence of intensional extemporal connectives like $\to$. We don't immediately know what such a reason could be. However, if there were some reason that anything like $\mapsto$ had to be intensional, we could invalidate contraction at this world by invalidating modus ponens at some other (presumably non-normal) world. Then there would be no need to invoke non-normal times to avoid temporal Curry. Again, though, more would need to be said to make this plausible.

To sum up: if the failure of $\to$-contraction is to be explained by $\to$'s relying on worlds at which logical laws fail, then the failure of $\mapsto$-contraction ought to be explained by $\mapsto$'s relying on times at which logical laws fail. At least prima facie, however, allowing for actual times at which logical laws fail is quite awkward, more awkward than allowing for non-actual worlds at which logical laws fail. So the advocate of non-normal worlds must either 1) explain why they do not advocate non-normal times, or 2) explain why non-normal times are not as awkward as they first appear. We see no obvious way to do either of these, and so we leave this dilemma, at least for now, as a dilemma.

Acknowledgments. We are grateful to Kata Bimbó for late-stage comments. We are also grateful to many others (largely, from Alberta, Connecticut, Melbourne, New York, and St. Andrews) for discussion, too many to be individually listed here. Above all, we are grateful for the philosophical, logical and pedagogical work of Mike Dunn, in whose memory we dedicate this paper.

REFERENCES

[1] Beall, J. (2009). *Spandrels of Truth*, Oxford University Press, Oxford.
[2] Bimbó, K. (2006). Curry-type paradoxes, *Logique et Analyse* **49**(195): 227–240.
[3] Brady, R. (2006). *Universal Logic*, CSLI Publications, Stanford, CA.
[4] Curry, H. B. (1942). The inconsistency of certain formal logics, *Journal of Symbolic Logic* **7**: 115–117.
[5] Field, H. (2008). *Saving Truth from Paradox*, Oxford University Press, Oxford.
[6] Geach, P. T. (1955). On insolubilia, *Analysis* **15**(3): 71–72.
[7] Kaplan, D. (1989). Demonstratives, in J. Almog, J. Perry and H. K. Wettstein (eds.), *Themes From Kaplan*, Oxford University Press, Oxford, pp. 481–564.
[8] Meyer, R. K., Routley, R. and Dunn, J. M. (1979). Curry's paradox, *Analysis* **39**: 124–128.
[9] Priest, G. (2005). *Towards Non-Being: The Logic and Metaphysics of Intentionality*, Oxford University Press, Oxford.
[10] Priest, G. (2006a). *Doubt Truth to be a Liar*, Oxford University Press, Oxford.
[11] Priest, G. (2006b). *In Contradiction*, Oxford University Press, Oxford.
[12] Prior, A. N. (1955). Curry's paradox and 3-valued logic, *Australasian Journal of Philosophy* **33**: 177–182.
[13] Prior, A. N. (1969). *Papers on Time and Tense*, Clarendon Press, Oxford.
[14] Restall, G. (1993). How to be *really* contraction free, *Studia Logica* **52**: 381–391.
[15] Routley, R. (1980). *Exploring Meinong's Jungle and Beyond*, Ridgeview Publishing Co., Ridgeview, California.

[16] Sylvan, R. (2000). *Sociative Logics and Their Applications: Essays by the Late Richard Sylvan*, Ashgate Publishing Ltd., Oxford. Edited by D. Hyde and G. Priest.

UNIVERSITY OF NOTRE DAME, SOUTH BEND, IN, U.S.A., *Email:* jbeall@nd.edu
MONASH UNIVERSITY, VIC, AUSTRALIA, *Email:* daveripley@gmail.com

ON CONNEGATION

Alex Belikov, Oleg Grigoriev and Dmitry Zaitsev

ABSTRACT. This paper is devoted to the study of "connegation," a variant of unary truth-functional logical operation, combining the properties of conflation and negation, operations known in bilattice theory. Semantically, connegation is specified within a four-valued semantic framework, employing a particular structure of generalized truth values introduced therein. We present a logical system, **dCP**, determined by our four-valued structure, whose language is equipped with a unary propositional connective corresponding to the semantically defined connegation operation. We present axiomatizations of **dCP** in the form of Hilbert-style and Gentzen-style proof-systems and provide corresponding soundness and completeness theorems. A cut-elimination argument for the Gentzen-style calculus is presented as well.

Keywords. Conflation, Connegation, Embedding results, Generalized truth values, Gentzen-style calculi, Hilbert-style calculi, Negation

1. INTRODUCTION: MIKE DUNN AND NEGATION

In the mid 1960s, after relevance logic had began to attract more and more scholars, it had been made clear that the issues concerning relevant entailment and implication are strongly intertwined with the notion of negation. Mike Dunn was among the first to develop a framework for an analysis of relevance through a suitable semantics of negation. In his doctoral thesis Dunn [5], a class of algebraic structures, known as De Morgan lattices, were first used to semantically model Anderson and Belnap's logic of first-degree entailments, cf., e.g., [1, §15]. The approach became popular also because it received many non-algebraic implementations, e.g., in the form of many-valued, by Belnap [3; 4], and set-theoretic, by Dunn [6], semantic frameworks. Since then, philosophically oriented studies of "De Morgan negation" and its relationships with other kinds of negation-like operators constituted an independent field of research. It is worth to note that Dunn published on various negations; we mention only a few of his related publications: Dunn [7; 8; 9], Dunn and Hardegree [12].

The present article is partly motivated by similar issues. We are interested in some recently discovered negation-like operators that can be defined in the framework of generalized truth values, yet another, so to speak, independent research area, inspired by Dunn's works on "intuitive semantics" of the first-degree entailment logic.

The second source of our motivation lies in a specific view of logic as a "hybrid" system. That, again, echoes some views of Mike Dunn. As he says in his interview

---

2020 *Mathematics Subject Classification.* Primary: 03B47, Secondary: 03B50.

Bimbó, Katalin, (ed.), *Relevance Logics and other Tools for Reasoning. Essays in Honor of J. Michael Dunn*, (Tributes, vol. 46), College Publications, London, UK, 2022, pp. 73–88.

of 2015: "I don't think there will be a single post-non-classical logic. But maybe there could be something like a Swiss Army Knife of at least a number of logics, where one could choose which features to use for which purpose. Substructural logics perhaps form a paradigm here, where one can have various logics depending on what structural rules one allows, permutation, thinning, contraction, etc. And one might combine these with various modalities. And I dream of mixing probabilities into the mix somehow too" (Dunn et al. [11]). In the present article, we literally follow the same idea, however, without any connections to substructurality, modality and probability. Negation — that's our main target. We study two negation-like operators that, by themselves, can simulate the properties of other logical connectives within the same logical framework. Isn't it a Swiss Army Knife Mike Dunn was talking about?

More specifically, we are interested in unary operators introduced by N. Kamide [15] and P. Ruet [20], that can be seen as four-valued generalizations of Post's "cyclic" negation [19]. The key feature of both operations is that their double iterations can simulate the properties of classical (Boolean) negation. This observation, we guess, motivated Kamide to name a logical system with the corresponding operation as "Classical Paraconsistent Logic" **CP**. An original presentation of **CP** was given by Kamide in terms of a Gentzen-style sequent calculus. Later, **CP** has been provided with an adequate Hilbert-style formalization by H. Omori and H. Wansing [17]. Due to the fact that Kamide and Ruet's operations are perfectly dual to each other, we find it reasonable to devote the present paper basically to the development of a logical system containing Ruet's operation. By doing so, we introduce "Dual Classical Paraconsistent Logic" **dCP** and prove a bunch of proof-theoretical results about it. Firstly, we provide a sound and complete Hilbert-style calculus for **dCP**. Secondly, we present a sound and complete Gentzen-style calculus for **dCP** that enjoys the cut-elimination theorem.

The philosophical aspect of our study, which actually motivated this paper in the first place, is to propose a reasonable account of the target operations that could unite them in a single conceptual framework. To this end, we introduce the notion of "connegation," of which Kamide and Ruet's operations are examples. The notion of connegation is motivated by observation that both of these connectives can be thought of as unary truth-functional operations, combining the semantic conditions of conflation and negation, operations known in bilattice theory. In this paper, we advocate the position according to which connegation is an intuitively plausible and philosophically significant logical notion that reflects some interesting aspects of negativity, thereby calling for further philosophical and technical explorations.

2. Negation, Conflation, Kamide and Ruet's Operators

Let us briefly review some basic definitions from bilattice theory that will help us to form the context of our study and set its main problem. As the reference for bilattice theory we use Fitting [14].

A *bilattice* $\mathcal{B}$ is a tuple $\langle B, \leq_t, \leq_i \rangle$ where B is a non-empty set; $\leq_t$ and $\leq_i$ are partial orderings of B such that $\langle B, \leq_t \rangle$ and $\langle B, \leq_i \rangle$ are complete lattices.

Usually, the orderings $\leq_t$ and $\leq_i$ are interpreted as "truth" and "information" orderings, respectively. Both of them give raise to corresponding meet and join operations.

We denote the meet and join operations, associated with $\leq_t$, as $\cap$ and $\cup$, whereas their counterparts, associated with $\leq_i$, as $\otimes$ and $\oplus$, respectively.

A bilattice $\mathcal{B}$ has a *negation* if there is a mapping $\neg$ satisfying the following properties, for every $x, y \in B$: (1.1) if $x \leq_t y$, then $\neg y \leq_t \neg x$; (1.2) if $x \leq_i y$, then $\neg x \leq_i \neg y$; (1.3) $x = \neg \neg x$.

A bilattice $\mathcal{B}$ has a *conflation* if there is a mapping $-$ satisfying the following properties, for every $x, y \in B$: (2.1) if $x \leq_i y$ then $-y \leq_i -x$; (2.2) if $x \leq_t y$ then $-x \leq_t -y$; (2.3) $x = --x$.

If a bilattice $\mathcal{B}$ has both operations, they *commute* if $-\neg x = \neg -x$, for every $x \in B$.

Probably, the best known example is the bilattice $\mathcal{FOUR}_2$, which is constituted by Belnap's truth values: T ("true and not false"), B ("true and false simultaneously"), N ("neither true, nor false"), and F ("false and not true"). It is depicted in Figure 1.

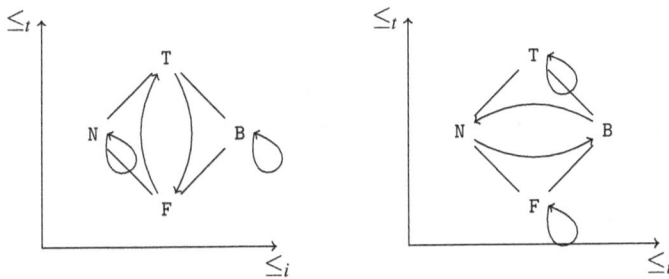

FIGURE 1. The behaviour of negation and conflation on bilattice $\mathcal{FOUR}_2$, generated by Belnap's values.

By analysing the behaviour of negation and conflation on $\mathcal{FOUR}_2$, it is possible to extract matrix definitions of these operations. Let $f^\neg$ and f^- be truth functions defined over $\{T, B, N, F\}$ that correspond to negation and conflation, respectively (see Figure 2).

x	$f^\neg(x)$	$f^-(x)$	$f^-(f^\neg(x))$	$f_K^\sim(x)$	$f_R^\sim(x)$	$f_i^\sim(f_i^\sim(x))$	$i \in \{K, R\}$
T	F	T	F	N	B	F	
B	B	N	N	T	F	N	
N	N	B	B	F	T	B	
F	T	F	T	B	N	T	

FIGURE 2. Unary operations on bilattice $\mathcal{FOUR}_2$.

The peculiarity of these functions is that their composition can, in some sense, "simulate" a function known as (classical) Boolean negation, as Figure 2 shows.

Interestingly, there are some truth functions that can simulate the properties of Boolean negation not by composition with some other functions, but through their own iteration. One of such functions was recently introduced by Kamide [15]. In the context of Belnap's framework, Kamide's operation can be represented as $f_K^\sim$ from Figure 2. Long before Kamide's work, Ruet introduced an operation which is dual to $f_K^\sim$. In the honour of Ruet, we label this operation as $f_R^\sim$. Its definition is also depicted in Figure 2.

It may seem that there is no other connection between these operations, except that they both can simulate classical negation, however, their crucial relationship is clearly visible if we graphically represent their definitions on $\mathcal{FOUR}_2$, as in Figure 3.

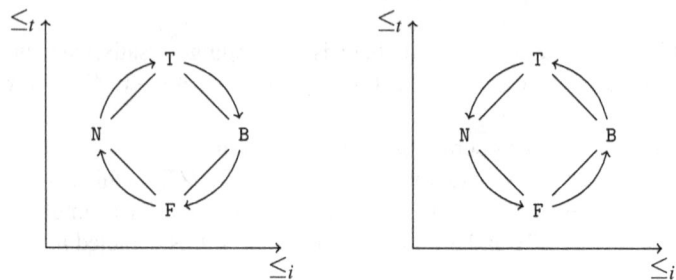

FIGURE 3. The behaviour of $f_R^\sim$ (the left one) and $f_K^\sim$ (the right one) on $\mathcal{FOUR}_2$.

Both $f_R^\sim$ and $f_K^\sim$ run through different "circles" or "cycles" over the bilattice. In this sense, they are perfectly dual to each other, because the first goes clockwise and the second goes counterclockwise. In [20], this observation motivated Ruet to call $f_R^\sim$ the "quarter turn."

As we remarked in the introductory section, Kamide introduced the logical system **CP**, containing $f_K^\sim$ as the sole unary operation. This logic can be seen as a result of replacing De Morgan negation (represented here as $f^\neg$) of A. Avron's logic **HBe** [2] (or V. M. Popov's logic **Par** [18]) with $f_K^\sim$. The proof-theoretical properties of **CP** were thoroughly studied by Kamide himself in [15], and by Omori and Wansing in [17]. In turn, Ruet was motivated by considerations of the functional completeness of the logics over $\mathcal{FOUR}_2$. In particular, he showed that a set of connectives consisting of $f_R^\sim$, $f^\neg$, and the conjunction operator (which corresponds to the truth-ordering meet on $\mathcal{FOUR}_2$) forms a functionally complete set of operations. He was not interested, as is obvious, in the comparison of Kamide's **CP** with a kind of a similar logical system, containing $f_R^\sim$ instead of $f_K^\sim$. And, of course, he was not interested in proposing a way of intuitive justification of these operations. The remaining part of our paper is intended to fill these gaps.

3. SEMANTIC FRAMEWORK: THE DUAL CLASSICAL PARACONSISTENT LOGIC

In order to put $f_K^\sim$ and $f_R^\sim$ in the unified conceptual framework, we will use the notion of "connegation," which is novel to this article. Before moving towards an explanation of the motivation for the term "connegation," we want to set out in more formal detail a version of generalized truth-value semantics for the logic containing the dual operation. Quite predictably, we denote this logic as **dCP** where **d** means "dual" and the whole name is "Dual Classical Paraconsistent Logic."

Consider, firstly, a propositional language $\mathcal{L}$, containing $\wedge$, $\vee$, $\rightarrow$, and $\sim$. The notion of a formula is defined in a standard manner. The set of all formulae of $\mathcal{L}$ and the set of all propositional variables of $\mathcal{L}$ are denoted as $\mathbb{F}$ and **Var**, respectively.

By a *positive model* is meant a structure $\mathcal{M} = \langle \mathcal{P}(2), v \rangle$ where $\mathcal{P}(2) = \{\{t,f\}, \{t\}, \{f\}, \varnothing\}$ and $v: \mathbf{Var} \mapsto \mathcal{P}(2)$. Valuational clauses concerning the positive part of the language, that is, the clauses for conjunction, disjunction, and implication, are the following:

$(\wedge_t)$ $t \in v(\varphi \wedge \psi) \Leftrightarrow t \in v(\varphi)$ and $t \in v(\psi)$
$(\wedge_f)$ $f \in v(\varphi \wedge \psi) \Leftrightarrow f \in v(\varphi)$ or $f \in v(\psi)$
$(\vee_f)$ $f \in v(\varphi \vee \psi) \Leftrightarrow f \in v(\varphi)$ and $f \in v(\psi)$
$(\vee_t)$ $t \in v(\varphi \vee \psi) \Leftrightarrow t \in v(\varphi)$ or $t \in v(\psi)$
$(\rightarrow_t)$ $t \in v(\varphi \rightarrow \psi) \Leftrightarrow$ if $t \in v(\varphi)$ then $t \in v(\psi)$
$(\rightarrow_f)$ $f \in v(\varphi \rightarrow \psi) \Leftrightarrow t \in v(\varphi)$ and $f \in v(\psi)$

At this point we interrupt the stream of definitions to consider different options for the valuational clauses for $\sim$.

Consider the following set of clauses.

(neg_t) $t \in v(\sim \varphi) \Leftrightarrow f \in v(\varphi)$ (neg_f) $f \in v(\sim \varphi) \Leftrightarrow t \in v(\varphi)$
(con_t) $t \in v(\sim \varphi) \Leftrightarrow f \notin v(\varphi)$ (con_f) $f \in v(\sim \varphi) \Leftrightarrow t \notin v(\varphi)$

Clearly, choosing conditions (neg_t) and (neg_f), we obtain the characterization of De Morgan negation, as it appears, for example, in Dunn's semantics for **FDE** in his [10]. Taking into account that Belnap's truth values can be represented as the elements of $\mathcal{P}(2)$, it is obvious that clauses (neg_t) and (neg_f) can be used as an equivalent way to define function $f^\neg$, discussed in the previous section. In turn, choosing (con_t) and (con_f) leads us to a semantic characterization of conflation, which was discussed earlier in the form of f^-.

A natural question arises: Can we use the presented clauses to characterize Kamide's and Ruet's operations? The answer is positive. Were we to admit Kamide's connective, it would be captured by clauses (neg_t) and (con_f). Let us denote a positive model augmented with them as a **CP**-*model*. The same move but in slightly different notation was actually made by Omori and Wansing in [17], who also observed that the truth condition of Kamide's $\sim$ coincides with the truth condition of negation in **FDE**. Finally, in order to characterize Ruet's operator within the generalized truth-value semantics, we can use (neg_f) and (con_t). Thus a **dCP**-*model* is a positive model augmented with (neg_f) and (con_t).

We shall define a **dCP**-*consequence relation* as follows: $\Gamma \vDash_{\mathbf{dCP}} \varphi$, if and only if, for any valuation v, if $t \in v(\gamma)$ (for every $\gamma \in \Gamma$), then $t \in v(\varphi)$. We say that a formula φ is **dCP**-*valid*, iff $t \in v(\varphi)$, for every valuation v in a **dCP**-model. The notions of **CP**-*consequence relation* and **CP**-*valid* formula are defined analogously.

Now it becomes clear that both $f_{\mathbf{K}}^{\sim}$ and $f_{\mathbf{R}}^{\sim}$ are of a hybrid nature, they share the properties of both conflation and negation. $f_{\mathbf{K}}^{\sim}$, represented via (neg_t) and (con_f), has the truth condition of negation and the falsity condition of conflation; whereas $f_{\mathbf{R}}^{\sim}$, represented via (neg_f) and (con_t), has the truth condition of conflation and the falsity condition of negation. Thus, we find it natural to label a logical connective characterized by any of these two ways as "connegation" as the derivative from "conflation" + "negation." A more detailed discussion of the notion of "connegation" is postponed to Section 5.

4. Proof-theoretical Study of Dual Classical Paraconsistent Logic

4.1. Hilbert-style Formulation. A lot of interesting proof-theoretical results regarding **CP** were obtained by Kamide [15], and Omori and Wansing [17]. In what follows, we establish some proof-theoretical results regarding **dCP**. We start with its Hilbert-style formalization.

An axiomatic calculus $\mathcal{H}$ for **dCP** contains the following list of axiomatic schemata (where $\varphi \leftrightarrow \psi$ stands for $(\varphi \to \psi) \wedge (\psi \to \varphi)$):

(A1) $\varphi \to (\psi \to \varphi)$
(A2) $(\varphi \to (\psi \to \chi)) \to ((\varphi \to \psi) \to (\varphi \to \chi))$
(A3) $(\varphi \wedge \psi) \to \varphi$
(A4) $(\varphi \wedge \psi) \to \psi$
(A5) $((\chi \to \varphi) \wedge (\chi \to \psi)) \to (\chi \to (\varphi \wedge \psi))$
(A6) $\varphi \to (\varphi \vee \psi)$
(A7) $\psi \to (\varphi \vee \psi)$
(A8) $((\varphi \to \chi) \wedge (\psi \to \chi)) \to ((\varphi \vee \psi) \to \chi)$
(A9) $(\sim\varphi \wedge \sim\psi) \leftrightarrow \sim(\varphi \wedge \psi)$
(A10) $(\sim\varphi \vee \sim\psi) \leftrightarrow \sim(\varphi \vee \psi)$
(A11) $(\varphi \wedge \sim\sim\psi) \to \sim\sim\sim(\varphi \to \psi)$
(A12) $(\sim\sim\psi \to \sim\sim\varphi) \to ((\sim\sim\psi \to \varphi) \to \psi)$

and the sole rule of inference is *modus ponens*:

(MP) $\dfrac{\varphi \to \psi, \varphi}{\psi}$.

We use standard definitions of a proof and a proof from hypotheses. We write $\Gamma \vdash_{\mathcal{H}} \varphi$ to state that there is a proof of φ from the set of formulae Γ in $\mathcal{H}$. In the light of (A1), (A2), (T7) (see Lemma 2 below) and (MP), it is clear that the pure implicative fragment of $\mathcal{H}$ is classical, and hence, the deduction theorem holds for $\mathcal{H}$.

Theorem 1. *If* $\Gamma, \varphi \vdash_{\mathcal{H}} \psi$, *then* $\Gamma \vdash_{\mathcal{H}} \varphi \to \psi$.

The following lemma reflects important features of connegations that are intrinsic to both **CP** and **dCP**. Intuitively, the first three formulas can be read as "simulated" analogues of classical De Morgan and double negation laws.

Lemma 2. *The following formulas are provable in* **dCP**:

(T1) $\varphi \leftrightarrow \sim\sim\sim\sim\varphi$
(T2) $(\sim\sim\varphi \vee \sim\sim\psi) \leftrightarrow \sim\sim(\varphi \wedge \psi)$
(T3) $(\sim\sim\varphi \wedge \sim\sim\psi) \leftrightarrow \sim\sim(\varphi \vee \psi)$
(T4) $\varphi \vee (\varphi \to \psi)$
(T5) $\sim\sim\varphi \to (\varphi \to \psi)$
(T6) $\varphi \vee \sim\sim\varphi$
(T7) $((\varphi \to \psi) \to \varphi) \to \varphi$

Proof. The proofs of these formulas are standard, keeping in mind, that $\sim\sim$ behaves like "classical" negation and the deduction theorem holds. A crucial role is played by (A12). Thus, we leave the details for an interested reader. ◁

We define a set of formulae $\mathcal{T}$ to be a theory of a logic L if it is closed under the derivability relation $\vdash_L$. A theory $\mathcal{T}$ is called non-trivial if $\mathcal{T} \neq \mathbb{F}$. A theory $\mathcal{T}$ is called prime if it satisfies the following property: if $\varphi \vee \psi \in \mathcal{T}$ then $\varphi \in \mathcal{T}$ or $\psi \in \mathcal{T}$.

Moving towards the completeness result, we will use the standard version of Lindenbaum's lemma, so we omit its proof.

Lemma 3. *If $\Gamma \nvdash_{\mathcal{H}} \varphi$, then there exists a non-trivial prime **dCP**-theory Γ', such that $\Gamma \subseteq \Gamma'$ and $\Gamma' \nvdash_{\mathcal{H}} \varphi$.*

We also need two helpful lemmas.

Lemma 4. *For every non-trivial prime **dCP**-theory $\mathcal{T}$, the following property of c-normality holds: $\varphi \in \mathcal{T}$ if and only if $\sim\sim\varphi \notin \mathcal{T}$.*

Proof. Assume $\varphi \in \mathcal{T}$ and $\sim\sim\varphi \in \mathcal{T}$. Then, by (T5), we have that for any formula ψ it holds that $\psi \in \mathcal{T}$, which is a contradiction. Assume that $\varphi \notin \mathcal{T}$ and $\sim\sim\varphi \notin \mathcal{T}$. Since $\mathcal{T}$ is prime, using (T6), we obtain a contradiction again. ◁

Lemma 5. *For every prime **dCP**-theory $\mathcal{T}$, $\varphi \to \psi \in \mathcal{T}$ iff $\varphi \notin \mathcal{T}$ or $\psi \in \mathcal{T}$.*

Proof. Assume that $\varphi \to \psi \in \mathcal{T}$ and $\varphi \in \mathcal{T}$ and $\psi \notin \mathcal{T}$. Then, by (MP), we have a contradiction. For the converse, we have two cases. Suppose that $\varphi \notin \mathcal{T}$ and $\varphi \to \psi \notin \mathcal{T}$. Then, by (T4), we have a contradiction, since $\mathcal{T}$ is prime. Now, suppose that $\psi \in \mathcal{T}$ and $\varphi \to \psi \notin \mathcal{T}$. Then, using (A1) and (MP), we obtain a contradiction. ◁

Now, for any non-trivial prime **dCP**-theory $\mathcal{T}$, let $v_\mathcal{T}$ be the **dCP**-canonical valuation, which is defined by means of the following clauses (for any $p \in \mathbf{Var}$):

$$t \in v_\mathcal{T}(p) \Rightarrow p \in \mathcal{T}, \qquad f \in v_\mathcal{T}(p) \Rightarrow \sim\sim\sim p \in \mathcal{T}.$$

Lemma 6. *dCP-canonical valuation $v_\mathcal{T}$ can be extended to an arbitrary formula π.*

Proof. By induction on the complexity of a formula π.

The case of $\pi \in \mathbf{Var}$ is covered by the definition of $v_\mathcal{T}$.

Let $\pi = \sim\varphi$. Then $t \in v_\mathcal{T}(\sim\varphi)$ implies $f \notin v_\mathcal{T}(\varphi)$. By inductive hypothesis we have $\sim\sim\sim\varphi \notin \mathcal{T}$. Using Lemma 4, we have $\sim\varphi \in \mathcal{T}$. (The other way around is identical.) Suppose $f \in v_\mathcal{T}(\sim\varphi)$. Then we obtain $t \in v_\mathcal{T}(\varphi)$, from which, using inductive hypothesis, we have $\varphi \in \mathcal{T}$. From this, using (T1), we obtain $\sim\sim\sim\sim\varphi \in \mathcal{T}$. (The other way around is identical.)

Let $\pi = \varphi \wedge \psi$. Then, $t \in v_\mathcal{T}(\varphi \wedge \psi)$ iff $t \in v_\mathcal{T}(\varphi)$ and $t \in v_\mathcal{T}(\psi)$ (by the truth condition of conjunction) iff $\varphi \in \mathcal{T}$ and $\psi \in \mathcal{T}$ (by IH) iff $\varphi \wedge \psi \in \mathcal{T}$ (by (A3), (A4), (A5)). Suppose $f \in v_\mathcal{T}(\varphi \wedge \psi)$. Then, $f \in v_\mathcal{T}(\varphi \wedge \psi)$ iff $f \in v_\mathcal{T}(\varphi)$ or $f \in v_\mathcal{T}(\psi)$ (by the falsity condition of conjunction) iff $\sim\sim\sim\varphi \in \mathcal{T}$ or $\sim\sim\sim\psi \in \mathcal{T}$ (by IH) iff $\sim\varphi \notin \mathcal{T}$ or $\sim\psi \notin \mathcal{T}$ (by Lemma 4). Then, using (A3), (A4), and (A9), both cases imply $\sim(\varphi \wedge \psi) \notin \mathcal{T}$. Applying Lemma 4, we obtain $\sim\sim\sim(\varphi \wedge \psi) \in \mathcal{T}$. The reasoning in the backward direction is slightly different but simple. We also use Lemma 4, to get $\sim(\varphi \wedge \psi) \notin \mathcal{T}$ from $\sim\sim\sim(\varphi \wedge \psi) \in \mathcal{T}$, but then, to obtain $\sim\varphi \notin \mathcal{T}$ or $\sim\psi \notin \mathcal{T}$, we deduce $\sim\varphi \wedge \sim\psi \notin \mathcal{T}$ using (A9), and $\sim\varphi, \sim\psi \vdash_\mathcal{H} \sim\varphi \wedge \sim\psi$, which is derivable in $\mathcal{H}$ in the light of (A1) and (A5).

Let $\pi = \varphi \vee \psi$. This case is analogous to the previous one; it heavily relies on a similar use of (A10).

Let $\pi = \varphi \to \psi$. In the case of $t \in v_T(\varphi \to \psi)$, we simply use Lemma 5.

Suppose $f \in v_T(\varphi \to \psi)$. Then, $f \in v_T(\varphi \to \psi)$ iff $t \in v_T(\varphi)$ and $f \in v_T(\psi)$ (by the falsity condition of implication) iff $\varphi \in \mathcal{T}$ and $\sim\sim\sim\psi \in \mathcal{T}$ (by IH) iff $\sim\sim\sim(\varphi \to \psi) \in \mathcal{T}$ (by (A11)). ◁

Now we can prove the completeness theorem.

Theorem 7. *If* $\Gamma \vDash_{\mathbf{dCP}} \varphi$ *then* $\Gamma \vdash_{\mathcal{H}} \varphi$.

Proof. Using Lemma 3 and Lemma 6. ◁

In turn, the soundness can be proved as usual.

Theorem 8. *If* $\Gamma \vdash_{\mathcal{H}} \varphi$ *then* $\Gamma \vDash_{\mathbf{dCP}} \varphi$.

4.2. Gentzen-style Formulation. Now we turn to the Gentzen-style formalization of **dCP**. A sequent calculus $\mathcal{G}$ for **dCP** contains the following set of initial sequents

$$p \Rightarrow p, \qquad \sim p \Rightarrow \sim p,$$

the following set of structural rules of inference

$$(W\Rightarrow) \frac{\Gamma \Rightarrow \Delta}{\varphi, \Gamma \Rightarrow \Delta}, \qquad (\Rightarrow W) \frac{\Gamma \Rightarrow \Delta}{\Gamma \Rightarrow \Delta, \varphi}, \qquad (Cut) \frac{\Gamma \Rightarrow \Delta, \varphi \quad \varphi, \Theta \Rightarrow \Pi}{\Gamma, \Theta \Rightarrow \Delta, \Pi},$$

and the following set of logical rules of inference

$$(\Rightarrow \sim\sim) \frac{\varphi, \Gamma \Rightarrow \Delta}{\Gamma \Rightarrow \Delta, \sim\sim \varphi} \qquad (\sim\sim \Rightarrow) \frac{\Gamma \Rightarrow \Delta, \varphi}{\sim\sim \varphi, \Gamma \Rightarrow \Delta}$$

$$(\wedge \Rightarrow) \frac{\varphi, \psi, \Gamma \Rightarrow \Delta}{\varphi \wedge \psi, \Gamma \Rightarrow \Delta} \qquad (\Rightarrow \wedge) \frac{\Gamma \Rightarrow \Delta, \varphi \quad \Gamma \Rightarrow \Delta, \psi}{\Gamma \Rightarrow \Delta, \varphi \wedge \psi}$$

$$(\vee \Rightarrow) \frac{\varphi, \Gamma \Rightarrow \Delta \quad \psi, \Gamma \Rightarrow \Delta}{\varphi \vee \psi, \Gamma \Rightarrow \Delta} \qquad (\Rightarrow \vee) \frac{\Gamma \Rightarrow \Delta, \varphi, \psi}{\Gamma \Rightarrow \Delta, \varphi \vee \psi}$$

$$(\Rightarrow \to) \frac{\varphi, \Gamma \Rightarrow \Delta, \psi}{\Gamma \Rightarrow \Delta, \varphi \to \psi} \qquad (\to \Rightarrow) \frac{\Gamma \Rightarrow \Delta, \varphi \quad \psi, \Theta \Rightarrow \Pi}{\varphi \to \psi, \Gamma, \Theta \Rightarrow \Delta, \Pi}$$

$$(\sim\wedge \Rightarrow) \frac{\sim\varphi, \sim\psi, \Gamma \Rightarrow \Delta}{\sim(\varphi \wedge \psi), \Gamma \Rightarrow \Delta} \qquad (\Rightarrow \sim\wedge) \frac{\Gamma \Rightarrow \Delta, \sim\varphi \quad \Gamma \Rightarrow \Delta, \sim\psi}{\Gamma \Rightarrow \Delta, \sim(\varphi \wedge \psi)}$$

$$(\sim\vee \Rightarrow) \frac{\sim\varphi, \Gamma \Rightarrow \Delta \quad \sim\psi, \Gamma \Rightarrow \Delta}{\sim(\varphi \vee \psi), \Gamma \Rightarrow \Delta} \qquad (\Rightarrow \sim\vee) \frac{\Gamma \Rightarrow \Delta, \sim\varphi, \sim\psi}{\Gamma \Rightarrow \Delta, \sim(\varphi \vee \psi)}$$

$$(\sim\to \Rightarrow) \frac{\varphi, \sim\sim\sim\psi, \Gamma \Rightarrow \Delta}{\sim\sim\sim(\varphi \to \psi), \Gamma \Rightarrow \Delta} \qquad (\Rightarrow \sim\to) \frac{\Gamma \Rightarrow \Delta, \varphi \quad \Gamma \Rightarrow \Delta, \sim\sim\sim\psi}{\Gamma \Rightarrow \Delta, \sim\sim\sim(\varphi \to \psi)}$$

We provide the soundness and completeness of $\mathcal{G}$ by showing the deductive equivalence of the Hilbert and Gentzen-style formulations of **dCP**.

Lemma 9. *If* $\Gamma \vdash_{\mathcal{H}} \varphi$, *then* $\mathcal{G} \vdash \Gamma \Rightarrow \varphi$.

Proof. By induction on the proof in $\mathcal{H}$. The proof is essentially straightforward, we have to show that all axioms of $\mathcal{H}$ are provable in $\mathcal{G}$, and that modus ponens is derivable in $\mathcal{G}$. We consider only one case as an example.

$$\dfrac{\psi \Rightarrow \psi}{\dfrac{\psi \Rightarrow \sim\sim\psi, \psi}{\Rightarrow \sim\sim\psi, \psi, \sim\sim\psi}(\Rightarrow\sim\sim)}(\Rightarrow W) \qquad \dfrac{\dfrac{\psi \Rightarrow \psi}{\psi, \varphi \Rightarrow \psi}(W\Rightarrow)}{\dfrac{\varphi \Rightarrow \psi, \sim\sim\psi}{(\rightarrow\Rightarrow)}}(\Rightarrow\sim\sim) \qquad \vdots$$

$$\dfrac{\sim\sim\psi \rightarrow \varphi \Rightarrow \psi, \sim\sim\psi \qquad \sim\sim\varphi, \sim\sim\psi \rightarrow \varphi \Rightarrow \psi}{\dfrac{\sim\sim\psi \rightarrow \sim\sim\varphi, \sim\sim\psi \rightarrow \varphi \Rightarrow \psi}{\dfrac{\sim\sim\psi \rightarrow \sim\sim\varphi \Rightarrow (\sim\sim\psi \rightarrow \varphi) \rightarrow \psi}{\Rightarrow (\sim\sim\psi \rightarrow \sim\sim\varphi) \rightarrow ((\sim\sim\psi \rightarrow \varphi) \rightarrow \psi)}(\Rightarrow\rightarrow)}(\Rightarrow\rightarrow)}(\rightarrow\Rightarrow)$$

$$\vdots \qquad \dfrac{\dfrac{\psi \Rightarrow \psi}{\psi, \sim\sim\varphi \Rightarrow \psi}(W\Rightarrow)}{\dfrac{\sim\sim\varphi \Rightarrow \psi, \sim\sim\psi}{(\rightarrow\Rightarrow)}}(\Rightarrow\sim\sim) \qquad \dfrac{\dfrac{\varphi \Rightarrow \varphi}{\varphi \Rightarrow \psi, \varphi}(\Rightarrow W)}{\sim\sim\varphi, \varphi \Rightarrow \psi}(\sim\sim\Rightarrow)$$

$$\dfrac{\sim\sim\psi \rightarrow \varphi \Rightarrow \psi, \sim\sim\psi \qquad \sim\sim\varphi, \sim\sim\psi \rightarrow \varphi \Rightarrow \psi}{\dfrac{\sim\sim\psi \rightarrow \sim\sim\varphi, \sim\sim\psi \rightarrow \varphi \Rightarrow \psi}{\dfrac{\sim\sim\psi \rightarrow \sim\sim\varphi \Rightarrow (\sim\sim\psi \rightarrow \varphi) \rightarrow \psi}{\Rightarrow (\sim\sim\psi \rightarrow \sim\sim\varphi) \rightarrow ((\sim\sim\psi \rightarrow \varphi) \rightarrow \psi)}(\Rightarrow\rightarrow)}(\Rightarrow\rightarrow)}(\rightarrow\Rightarrow)$$

Other cases are left for the reader. ◁

In order to prove the converse of Lemma 9, we adopt the technique used in [17, Proposition 3.14]. First, we define a formula image of a sequent $\Gamma \Rightarrow \Delta$ as follows:

$$\rho(\Gamma \Rightarrow \Delta) := \bigwedge \Gamma \rightarrow \bigvee \Delta,$$

requiring also that $\bigwedge \varnothing := (p \rightarrow p)$ and $\bigvee \varnothing := (p \wedge \sim\sim p)$, for some fixed propositional variable p.

Lemma 10. *If $\mathcal{G} \vdash \Gamma \Rightarrow \varphi$, then $\Gamma \vdash_\mathcal{H} \varphi$.*

Proof. We start by noting that $\mathcal{G} \vdash \Gamma \Rightarrow \varphi$ iff $\mathcal{G} \vdash \varnothing \Rightarrow \rho(\Gamma \Rightarrow \varphi)$. Now we have to show that for every rule of inference in $\mathcal{G}$, having the form

$$\dfrac{S_1, \ldots, S_n}{S},$$

we can prove that $\rho(S_1), \ldots, \rho(S_n) \vDash_{\mathbf{dCP}} \rho(S)$. Then, in the light of Theorem 7, this would imply that $\rho(S_1), \ldots, \rho(S_n) \vdash_\mathcal{H} \rho(S)$.

We consider only a few cases.
Case $(\Rightarrow \sim \wedge)$. Assume $t \in v(\rho(\Gamma \Rightarrow \sim\varphi))$, $t \in v(\rho(\Gamma \Rightarrow \sim\psi))$ and $t \in v(\bigwedge \Gamma)$. From the first two assumptions, applying semantic conditions, we obtain the following equivalences:

$t \in v(\rho(\Gamma \Rightarrow \sim\varphi))$ iff $t \in v(\bigwedge \Gamma \rightarrow \sim\varphi)$ iff $t \notin v(\bigwedge \Gamma)$ or $f \notin v(\varphi)$,

$t \in v(\rho(\Gamma \Rightarrow \sim\psi))$ iff $t \in v(\bigwedge \Gamma \rightarrow \sim\psi)$ iff $t \notin v(\bigwedge \Gamma)$ or $f \notin v(\psi)$.

Finally, using $t \in v(\bigwedge \Gamma)$, we have $f \notin v(\varphi)$ and $f \notin v(\psi)$, which, by the semantic conditions of connegation and conjunction, imply that $t \in v(\sim(\varphi \wedge \psi))$.

Case $(\Rightarrow \sim \rightarrow)$. Assume $t \in v(\rho(\Gamma \Rightarrow \varphi))$, $t \in v(\rho(\Gamma \Rightarrow \sim\sim\sim\psi))$ and $t \in v(\bigwedge \Gamma)$. Again, the first two assumptions imply the following equivalences:

$t \in v(\rho(\Gamma \Rightarrow \varphi))$ iff $t \in v(\bigwedge \Gamma \rightarrow \varphi)$ iff $t \notin v(\bigwedge \Gamma)$ or $t \in v(\varphi)$,

$t \in v(\rho(\Gamma \Rightarrow \sim\sim\sim\psi))$ iff $t \in v(\bigwedge \Gamma \rightarrow \sim\sim\sim\psi)$ iff $t \notin v(\bigwedge \Gamma)$ or $f \in v(\psi)$.

Using $t \in v(\bigwedge \Gamma)$, we obtain $t \in v(\varphi)$ and $f \in v(\psi)$. From this, by the semantic conditions of implication, it follows that $f \in v(\varphi \to \psi)$. Thus, using the semantic conditions of connegation, we obtain $t \in v(\sim\sim\sim(\varphi \to \psi))$. ◁

In the light of Lemma 10 and Lemma 9, the following theorems hold.

Theorem 11. $\mathcal{G} \vdash \Gamma \Rightarrow \varphi$ iff $\Gamma \vdash_{\mathcal{H}} \varphi$.

Theorem 12. $\mathcal{G} \vdash \Gamma \Rightarrow \varphi$ iff $\Gamma \vDash_{\mathbf{dCP}} \varphi$.

4.3. Cut-elimination. To prove the cut-elimination theorem for the Gentzen-style system of **dCP** we exploit an embedding technique known from the research area of multilattices and their logics, see e.g., Kamide and Shramko [16]. One of the specific points of this method is an expansion of a target language with the additional copies of the set of propositional variables. In our case we need to enrich the set of atoms of the language $\mathcal{L}_{\mathbf{CL}}$ with the set $\mathbf{Var}^* = \{p^*: p \in \mathbf{Var}\}$. Thus the set of propositional variables of $\mathcal{L}_{\mathbf{CL}}$ is $\mathbf{Var} \cup \mathbf{Var}^*$ while that of $\mathcal{L}_{\mathbf{dCP}}$ is simply $\mathbf{Var}$.

Now let us define an appropriate mapping from $\mathcal{L}_{\mathbf{dCP}}$ to $\mathcal{L}_{\mathbf{CL}}$.

Definition 13. A mapping $f: \mathcal{L}_{\mathbf{dCP}} \mapsto \mathcal{L}_{\mathbf{CL}}$ is said to be a translation from $\mathcal{L}_{\mathbf{dCP}}$ to $\mathcal{L}_{\mathbf{CL}}$ if the following equations are satisfied:

1. $f(p) = p$, $f(\sim p) = p^*$, for all $p \in \mathbf{Var}$,
2. $f(\varphi \circ \psi) = f(\varphi) \circ f(\psi)$, where $\circ \in \{\wedge, \vee, \to\}$,
3. $f(\sim\sim\varphi) = \neg f(\varphi)$,
4. $f(\sim(\varphi \wedge \psi)) = f(\sim\varphi) \wedge f(\sim\psi)$,
5. $f(\sim(\varphi \vee \psi)) = f(\sim\varphi) \vee f(\sim\psi)$,
6. $f(\sim(\varphi \to \psi)) = \neg(f(\varphi) \to f(\sim\psi))$.

For the purposes of proving the cut-elimination theorem we adopt some standard cut-free Gentzen calculus $\mathcal{G}\mathbf{CL}$ for classical logic formulated in the language $\mathcal{L}_{\mathbf{CL}}$. Additionally, we require the cut rule to be *admissible* in $\mathcal{G}\mathbf{CL}$, and in expressions like $\Gamma \Rightarrow \Delta$ used to represent sequents, letters Γ and Δ denote sets of formulas.

As usual, $f(\Gamma)$ refers to the set $\{f(\varphi): \varphi \in \Gamma\}$. We also use the notations $\vdash_{\mathcal{G}\mathbf{CL}-cut}$ and $\vdash_{\mathcal{G}\mathbf{dCP}-cut}$ to indicate the provability relation within the cut-free versions of $\mathcal{G}\mathbf{CL}$ and $\mathcal{G}\mathbf{dCP}$, correspondingly.

Lemma 14. *Let Γ and Δ be sets of formulas in $\mathcal{L}_{\mathbf{dCP}}$ and f be the translation specified in Definition 13. Then,*

(1) *If $\vdash_{\mathcal{G}\mathbf{dCP}} \Gamma \Rightarrow \Delta$, then $\vdash_{\mathcal{G}\mathbf{CL}} f(\Gamma) \Rightarrow f(\Delta)$;*
(2) *if $\vdash_{\mathcal{G}\mathbf{CL}-cut} f(\Gamma) \Rightarrow f(\Delta)$, then $\vdash_{\mathcal{G}\mathbf{dCP}-cut} \Gamma \Rightarrow \Delta$.*

Proof. (1) By induction on the length of the proof $\mathcal{P}$ of $\Gamma \Rightarrow \Delta$ in $\mathcal{G}\mathbf{dCP}$.

The case when $\Gamma \Rightarrow \Delta$ is an axiom is clear. Let us consider some cases of applying propositional connective rules.

Case ($\Rightarrow \sim\sim$). Assume that $\Gamma \Rightarrow \Delta$ has the form $\Gamma \Rightarrow \Delta', \sim\sim\varphi$ and the last inference of $\mathcal{P}$ is

$$\frac{\varphi, \Gamma \Rightarrow \Delta'}{\Gamma \Rightarrow \Delta', \sim\sim\varphi} \; (\Rightarrow \sim\sim)$$

By induction hypothesis, $\vdash_{\mathcal{GCL}} f(\varphi), f(\Gamma) \Rightarrow f(\Delta)$, so we just expand its proof using $(\Rightarrow\neg)$

$$\frac{f(\varphi), f(\Gamma) \Rightarrow f(\Delta)}{f(\Gamma) \Rightarrow f(\Delta), \neg f(\varphi)} \; (\Rightarrow\neg)$$

where $\neg f(\varphi) = f(\sim\sim\varphi)$ as required.

Case $(\sim\vee\Rightarrow)$. $\Gamma \Rightarrow \Delta$ is of the form $\sim(\varphi \vee \psi), \Gamma' \Rightarrow \Delta$ and the last inference is

$$\frac{\sim\varphi, \Gamma' \Rightarrow \Delta \quad \sim\psi, \Gamma' \Rightarrow \Delta}{\sim(\varphi \vee \psi), \Gamma' \Rightarrow \Delta} \; (\sim\vee\Rightarrow).$$

The induction hypothesis provides the proofs of the premises of the following inference:

$$\frac{f(\sim\varphi), f(\Gamma') \Rightarrow f(\Delta) \quad f(\sim\psi), f(\Gamma') \Rightarrow f(\Delta)}{f(\sim\varphi) \vee f(\sim\psi), f(\Gamma') \Rightarrow f(\Delta)} \; (\vee\Rightarrow),$$

where $f(\sim\varphi) \vee f(\sim(\psi)) = f(\sim(\varphi \vee \psi))$.

Case $(\sim\to\Rightarrow)$. $\Gamma \Rightarrow \Delta$ has the form $\sim\sim\sim(\varphi \to \psi), \Gamma' \Rightarrow \Delta$ and the last inference is of the form

$$\frac{\varphi, \sim\sim\sim\psi, \Gamma' \Rightarrow \Delta}{\sim\sim\sim(\varphi \to \psi), \Gamma' \Rightarrow \Delta} \; (\sim\to\Rightarrow).$$

Taking into account that $f(\sim\sim\sim\psi) = \neg f(\sim\psi)$, we use the known provability of the sequents on the top of the left-hand side (classically) and the right-hand side (by induction hypothesis) of the following subproof:

$$\frac{\dfrac{f(\sim\psi), f(\Gamma') \Rightarrow f(\Delta), f(\sim\psi)}{f(\Gamma') \Rightarrow f(\Delta), f(\sim\psi), \neg f(\sim\psi)} \; (\Rightarrow\neg) \quad f(\varphi), \neg f(\sim\psi), f(\Gamma') \Rightarrow f(\Delta)}{\dfrac{f(\varphi), f(\Gamma') \Rightarrow f(\Delta), f(\sim\psi)}{\dfrac{f(\Gamma') \Rightarrow f(\Delta), f(\varphi) \to f(\sim\psi)}{f(\Gamma'), \neg(f(\varphi) \to f(\sim\psi)) \Rightarrow f(\Delta)} \; (\neg\Rightarrow)} \; (\Rightarrow\to)}$$

Note that the transition to the third line requires the cut rule which is admissible in a chosen version of $\mathcal{GCL}$. Finally, $\neg(f(\varphi) \to f(\sim\psi)) = f(\sim\sim\sim(\varphi \to \psi))$.

Case $(\Rightarrow\sim\to)$. Let $\Gamma \Rightarrow \Delta$ be of the form $\Gamma \Rightarrow \Delta', \sim\sim\sim(\varphi \to \psi)$ and the last inference of $\mathcal{P}$ be of the form

$$\frac{\Gamma \Rightarrow \Delta', \varphi \quad \Gamma \Rightarrow \Delta', \sim\sim\sim\psi}{\Gamma \Rightarrow \Delta', \sim\sim\sim(\varphi \to \psi)} \; (\Rightarrow\sim\to)$$

Again, using the induction hypothesis to assert the provability of the topmost sequent (recall that $f(\sim\sim\sim\psi) = \neg f(\sim\psi)$), we obtain

$$\frac{\dfrac{f(\Gamma) \Rightarrow f(\Delta'), \neg f(\sim\psi)}{f(\Gamma), f(\sim\psi) \Rightarrow f(\Delta')} \quad f(\Gamma) \Rightarrow f(\Delta'), f(\varphi)}{\dfrac{f(\Gamma), f(\varphi) \to f(\sim\psi) \Rightarrow f(\Delta')}{f(\Gamma) \Rightarrow f(\Delta'), \neg(f(\varphi) \to f(\sim\psi))} \; (\Rightarrow\neg)} \; (\to\Rightarrow)$$

and in the last line $\neg(f(\varphi) \to f(\sim\psi)) = f(\sim\sim\sim(\varphi \to \psi))$. Additionally, an application of the cut is required to obtain the left-hand side sequent on the second line. (The details are similar to the previous case and they are omitted here.)

(2) The idea of the proof is the same as above. Suppose that we have already proved $\vdash_{\mathcal{G}\mathbf{CL}-cut} f(\Gamma) \Rightarrow f(\Delta)$ for some sequent $f(\Gamma) \Rightarrow f(\Delta)$. We inspect the last inference of the proof and activate an induction hypothesis (with respect to the premises of a rule applied in the inference) to construct a required $\mathcal{G}\mathbf{dCP}-cut$ proof.

In the axiomatic case, $f(\Gamma) \Rightarrow f(\Delta)$ is just $p \Rightarrow p$ or $p^* \Rightarrow p^*$ for some variable $p \in \mathbf{Var}$. It is clear that possible pre-images of these sequences are $p \Rightarrow p$ or $\sim p \Rightarrow \sim p$, which are axioms of $\mathcal{G}\mathbf{dCP}$ and evidently provable in $\mathcal{G}\mathbf{dCP}-cut$. Next consider some propositional cases.

Case ($\neg\Rightarrow$). Let us first assume that a sequent $f(\Gamma) \Rightarrow f(\Delta)$ is of the form $f(\sim\sim\varphi)$, $f(\Gamma') \Rightarrow f(\Delta)$ and it is the last one in some $\mathcal{G}\mathbf{CL}-cut$ proof. Since $f(\sim\sim\varphi) = \neg f(\varphi)$, the last rule applied is ($\neg\Rightarrow$)

$$\frac{f(\Gamma') \Rightarrow f(\Delta), f(\varphi)}{\neg f(\varphi), f(\Gamma') \Rightarrow f(\Delta)} \ (\neg\Rightarrow).$$

The induction hypothesis provides a $\mathcal{G}\mathbf{dCP}-cut$ proof of $\Gamma' \Rightarrow \Delta, \varphi$. Applying the rule ($\sim\sim\Rightarrow$), we obtain $\Gamma', \sim\sim\varphi \Rightarrow \Delta$ which is $\Gamma \Rightarrow \Delta$.

Case ($\wedge\Rightarrow$). Next assume that $f(\Gamma) \Rightarrow f(\Delta)$ is of the form $f(\Gamma'), f(\sim(\varphi \wedge \psi)) \Rightarrow \Delta$. Since $f(\sim(\varphi \wedge \psi)) = f(\sim\varphi) \wedge f(\sim\psi)$, the last inference of the proof is

$$\frac{f(\Gamma'), f(\sim\varphi), f(\sim\psi) \Rightarrow f(\Delta)}{f(\Gamma'), f(\sim\varphi) \wedge f(\sim\psi) \Rightarrow f(\Delta)} \ (\wedge\Rightarrow).$$

According to the induction hypothesis, $\Gamma', \sim\varphi, \sim\psi \Rightarrow \Delta$ already has a $\mathcal{G}\mathbf{dCP}-cut$ proof. Thus we extend it with the inference step

$$\frac{\Gamma', \sim\varphi, \sim\psi \Rightarrow \Delta}{\Gamma', \sim(\varphi \wedge \psi) \Rightarrow \Delta} \ (\sim\wedge\Rightarrow).$$

Other cases are similar. ◁

Theorem 15. *Any sequent $\Gamma \Rightarrow \Delta$ provable in $\mathcal{G}\mathbf{dCP}$ is provable in $\mathcal{G}\mathbf{dCP} - cut$.*

Proof. Let $\vdash_{\mathcal{G}\mathbf{dCP}} \Gamma \Rightarrow \Delta$. Then, according to Lemma 14 (1) $\vdash_{\mathcal{G}\mathbf{CL}} f(\Gamma) \Rightarrow f(\Delta)$. Since cut is redundant in the sequent calculus $\mathcal{G}\mathbf{CL}$, $\vdash_{\mathcal{G}\mathbf{CL}-cut} f(\Gamma) \Rightarrow f(\Delta)$. Using Lemma 14 (2) we conclude that $\vdash_{\mathcal{G}\mathbf{dCP}-cut} \Gamma \Rightarrow \Delta$. ◁

5. And Yet, What is Connegation?

So far we were concerned with a purely proof-theoretical study of **dCP** (and partly **CP**). Now it is time to put two different connegations represented within these logics into a more philosophical context.

Following terminology suggested by Belnap [3], we say that "φ is at least true" when $t \in v(\varphi)$, and "φ is at least false" when $f \in v(\varphi)$ (v is not specified; we keep in mind that v is extended via semantic conditions of either **dCP** or **CP**). With this notation at hand we can rewrite the semantic definition of negation and conflation.

A negation is characterized by means of

- ¬φ is at least true iff φ is at least false;
- ¬φ is at least false iff φ is at least true.

Whereas, a conflation is characterized by means of

- −φ is at least true iff φ isn't at least false;
- −φ is at least false iff φ isn't at least true.

Broadly understood, a negation is a logical term that captures a phenomenon of semantic opposition. We use negation to indicate the falsity (or untruth) of a statement;[1] that is, to indicate that a certain statement "not-φ" is opposed to the statement φ with respect to its semantic values. It is common to say that the negation is an object-language analogue of the falsity. And the opposition in question reduces to the relation of contradiction.

In this respect, the notion of conflation has been studied much less than the notion of negation. But it is clear that conflation has nothing to do with the phenomenon of semantic opposition. In [13], Melvin Fitting gives the following intuitive interpretation of conflation operator: "the conflation of a truth value is a new truth value which counts as evidence for anything that wasn't counted against originally, and counts as evidence against anything that wasn't counted for originally." Paraphrasing Fitting, the conflation transforms (or "conflates") a kind of negative information about a semantic value of a statement into a positive form. This intuition perfectly matches the semantic clauses above.

As we've suggested in the course of the present paper, connegation is of a hybrid nature, it combines properties of both negation and conflation. There are only two connegations in the four-valued framework, as the following semantic characterization shows.

A connegation from **CP** is characterized by

- ∼φ is at least true iff φ is at least false;
- ∼φ is at least false iff φ isn't at least true.

Whereas, a connegation from **dCP** is characterised by

- ∼φ is at least true iff φ isn't at least false;
- ∼φ is at least false iff φ is at least true.

It can be questioned whether these connectives appeal to any philosophically significant notion, or they are interesting only for purely formal reasons. Whether ∼'s deserve to be standing in line with such a fundamental notion as negation? The question appears to be a tricky one, but we feel obligated to present some answer. Evidently, the notion of connegation bears important and interesting relations to the logics which are representable within the framework of generalized truth values, but we believe that this notion is supposed to tell us something interesting about negativity, or at least some important kinds of negativity. We take it for granted that the role of the term corresponding to ∼ in a natural language is to form a sentence with a negative shade.

---

[1]The difference between falsity and untruth as well as between truth and non-false might be crucial in different account of negation. Much depends on the relationships between truth and falsity, whether they exclusive and exhaustive or not.

However, we stress that despite this and the fact connegations can simulate the classical negation, we tend to think that connegations themselves can hardly be understood as pure negation operators.

Here is an interpretation of $\sim$'s that motivates our position.

The interpretation relies on slightly modified but widespread treatment of truth and falsity. Now we understand an "at least true" sentence as a sentence that acquires its truth in virtue of some verification (or proof). In turn, an "at least false" sentence is now supposed to mean a sentence which is at least false in virtue of some falsification (or disproof).

Consider the case of $\sim$ from **CP**.

Two possible epistemic attitudes naturally arise, namely, a sentence φ can be at least false or at least untrue. These two situations are to be carefully distinguished because in the former case we are talking about the presence of a falsification (or disproof) and in the latter we mean only the absence of verification (or proof). What job $\sim$ does in these situations? It takes the falsity qua falsification as its primary meaning. The connegation $\sim$ indicates that the sentence $\sim \varphi$ is at least true if and only if φ is falsified. However, in a situation when φ isn't at least true, φ appears to merely lack verification. This is not enough for claiming that $\sim \varphi$ is at least true, because we admit only the presence of a falsification to justify such a claim. The connegation $\sim$ does another job here; it indicates merely that a statement about the falsification of φ, i.e., $\sim \varphi$, is falsified. In sum, $\sim$ allows one to distinguish between two different kinds of negative information, that is, the presence of falsification and the absence of verification.

Now we turn to $\sim$ from **dCP**.

In this case, the connegation behaves symmetrically, it covers the remaining epistemic attitudes in which a sentence φ is at least true or isn't at least false. But in this scenario $\sim$ expresses the failure of falsification. When would we say that the process of falsification was not successful? The falsity condition of connegation tells us that the falsification of $\sim \varphi$ breaks down if and only if φ is true in virtue of some verification. However, the situation of φ being at least non-false amounts to the mere absence of falsification of φ. Should we consider that this situation indicates the failure of φ's falsification? We think we shouldn't. So $\sim \varphi$ is at least true by default, as reflected in the truth condition of $\sim$. Summing up, we can say that $\sim$ allows one to distinguish between two different situations and decide whether a falsification of a sentence breaks down or not.

Finally, it would not be superfluous to remind ourselves that both connegations are balanced with respect to their ability to simulate classical negation. It is easy to observe that double iterations of both $\sim$'s reflect the key properties of the latter, since the following formulas are theorems in both **CP** and **dCP**.

$$(\sim\sim\psi \to \sim\sim\varphi) \to ((\sim\sim\psi \to \varphi) \to \psi)$$
$$(\sim\sim\varphi \vee \sim\sim\psi) \leftrightarrow \sim\sim(\varphi \wedge \psi)$$
$$(\sim\sim\varphi \wedge \sim\sim\psi) \leftrightarrow \sim\sim(\varphi \vee \psi)$$

$$\varphi \leftrightarrow \sim\sim\sim\sim\varphi$$
$$\sim\sim\varphi \to (\varphi \to \psi)$$
$$\varphi \vee \sim\sim\varphi$$

6. Concluding Remarks

We finish this paper by reflecting on possible implications of our study.

One of the interesting aspects of connegations is that they provide a clear route to contra-classicality. This peculiarity was already observed by Omori and Wansing in [17], so we will not dwell on this in detail. The only thing we would like to stress is that, seemingly and quite surprisingly, **dCP** exhibits more contra-classical features than **CP**. It is easy to see by comparing their Hilbert-style axiomatizations. Notice that both contra-classical formulas

(6.1) $\qquad \sim(\varphi \wedge \psi) \to (\sim\varphi \wedge \sim\psi)$

(6.2) $\qquad (\sim\varphi \vee \sim\psi) \to \sim(\varphi \vee \psi)$

are taken as initial axiomatic schemata of a Hilbert-style calculus of **dCP**.

Given the following matrices (where the *'d values are designated),

φ	$f_{\tilde{R}}$
*T	B
*B	F
N	T
F	N

$f^\to$	T	B	N	F
T	T	B	N	F
B	T	B	N	F
N	T	T	T	T
F	T	T	T	T

$f^\wedge$	T	B	N	F
T	T	B	N	F
B	B	B	N	F
N	N	N	N	F
F	F	F	F	F

$f^\vee$	T	B	N	F
T	T	T	T	T
B	T	B	B	B
N	T	B	N	N
F	T	B	N	F

it can be shown that (6.1) and (6.2) are invalid. For both cases, φ can take the value B and ψ can take the value N. The remaining axiomatic schemata of $\mathcal{H}$ and modus ponens preserve the validity in the matrices. This fact establishes the independence of (6.1) and (6.2) in the calculus of **dCP**.

Moreover, **dCP** requires taking an awkward contra-classical formula $(\varphi \wedge \sim\sim\sim\psi) \to \sim\sim\sim(\varphi \to \psi)$ as an additional axiom, which is also independent. To show its independence, the reader can use the following matrices.

φ	$f_{\tilde{R}}$
*T	B
*B	F
N	T
F	N

$f^\to$	T	B	N	F
T	T	T	N	F
B	T	T	N	F
N	B	B	B	B
F	B	B	B	B

$f^\wedge$	T	B	N	F
T	T	B	N	F
B	B	B	F	F
N	N	F	N	F
F	F	F	F	F

$f^\vee$	T	B	N	F
T	T	T	T	T
B	T	B	T	B
N	T	T	N	N
F	T	B	N	F

In turn, **CP** can be axiomatized with the two contra-classical axioms $\varphi \vee \sim\sim\varphi$ and $\varphi \to (\sim\sim\varphi \to \psi)$. Thus, we find it interesting to explore the possible implications of contra-classicality via connegation.

Last but not least, it seems to us a very tempting prospect to find a way of obtaining logical connectives that could in a similar manner simulate the behaviour of other kinds of negation-like operators, such as relevant, paraconsistent, paracomplete, and so on.

Acknowledgments. We all, the authors of this paper, were fortunate to know Mike personally. For someone, this acquaintance lasted 30 years, for another, only 5, but each of us was influenced by communication with him both scientifically and humanly. We will always keep the warmest memories of Michael Dunn — a wonderful scientist and a sympathetic, kind person.

REFERENCES

[1] Anderson, A. R. and Belnap, N. D. (1975). *Entailment: The Logic of Relevance and Neccessity*, Vol. I, Princeton University Press, Princeton, NJ.
[2] Avron, A. (1991). Natural 3-valued logics—characterization and proof theory, *Journal of Symbolic Logic* **56**(1): 276–294.
[3] Belnap, N. D. (1977). A useful four-valued logic, *in* J. M. Dunn and G. Epstein (eds.), *Modern Uses of Multiple-valued Logic*, Reidel Publishing Company, Dordrecht, pp. 8–37.
[4] Belnap, N. D. (2019). How a computer should think, *New Essays on Belnap–Dunn Logic*, Springer International Publishing, Switzerland, pp. 35–53.
[5] Dunn, J. M. (1966). *The Algebra of Intensional Logics*, Doctoral dissertation, University of Pittsburgh, Pittsburgh, PA. (Published as Vol. 2 in the *Logic PhDs* series by College Publications, London, UK, 2019.).
[6] Dunn, J. M. (1976). Intuitive semantics for first-degree entailments and 'coupled trees', *Philosophical Studies* **29**(3): 149–168.
[7] Dunn, J. M. (1993). Star and perp: Two treatments of negation, *Philosophical Perspectives* **7**: 331. (Language and Logic, 1993, J. E. Tomberlin (ed.)).
[8] Dunn, J. M. (1996). Generalized ortho negation, *in* H. Wansing (ed.), *Negation: A Notion in Focus*, W. de Gruyter, New York, NY, pp. 3–26.
[9] Dunn, J. M. (1999). A comparative study of various model-theoretic treatments of negation: A history of formal negation, *in* D. M. Gabbay and H. Wansing (eds.), *What is Negation?*, Vol. 13 of *Applied Logic*, Kluwer, Dordrecht, pp. 23–51.
[10] Dunn, J. M. (2000). Partiality and its dual, *Studia Logica* **66**(1): 5–40.
[11] Dunn, J. M., Belikov, A. A., Mertsalov, A. V., Loginov, E. V. and Iunusov, A. T. (2016). U nas vse slishkom horosho poluchaetsja (Too much of a good thing), *Finikovyi Kompot (Date Palm Compote)*, [In Russian] **10**: 58–62.
[12] Dunn, J. M. and Hardegree, G. (2001). *Algebraic Methods in Philosophical Logic*, Vol. 41 of *Oxford Logic Guides*, Oxford University Press, Oxford, UK.
[13] Fitting, M. (1989). Bilattices and the theory of truth, *Journal of Philosophical Logic* **18**(3): 225–256.
[14] Fitting, M. (2004). Bilattices are nice things, *in* V. F. Hendricks, S. A. Pedersen and T. Bolander (eds.), *Self-reference*, CSLI Publications, Stanford, CA, pp. 53–77.
[15] Kamide, N. (2017). Paraconsistent double negations as classical and intuitionistic negations, *Studia Logica* **105**(6): 1167–1191.
[16] Kamide, N. and Shramko, Y. (2016). Embedding from multilattice logic into classical logic and vice versa, *Journal of Logic and Computation* **27**: 1549–1575.
[17] Omori, H. and Wansing, H. (2018). On contra-classical variants of Nelson logic N4 and its classical extension, *Review of Symbolic Logic* **11**(4): 805–820.
[18] Popov, V. M. (1989). Sekvencial'nye formulirovki paraneprotivorechivyh logicheskih sistem, (Sequent formulations of paraconsistent logical systems), *Sintaksicheskie i semanticheskie issledovanija neekstensional'nyh logik (Syntactic and Semantic Studies of Nonextensional Logics)*, [In Russian], Moscow, Nauka, pp. 285–289.
[19] Post, E. L. (1921). Introduction to a general theory of elementary propositions, *American Journal of Mathematics* **43**(3): 163–185.
[20] Ruet, P. (1996). Complete set of connectives and complete sequent calculus for Belnap's logic, *Tech. rep., Ecole Normale Superieure. Logic Colloquium '96, Doc. LIENS-96-28*.

MODALITIES IN LATTICE-R

Katalin Bimbó and J. Michael Dunn

ABSTRACT. This paper considers modalities added to the relevance logic LR (lattice-R), which is R with the distributivity of conjunction and disjunction omitted. First, the modalities are *defined* from the Ackermann constants and the lattice connectives. Then, we introduce modalities as *primitives* equipped with some fairly usual properties. We also consider some other logics in the neighborhood. For each logic, including classical linear logic, we prove *decidability*. Lincoln, Mitchell, Scedrov, and Shankar (1992) claimed to have proved classical linear logic undecidable. We examine their work and find that their paper does not contain a proof of the admissibility of the cut rule, which would be essential for their claims to hold. Furthermore, according to their interpretation of proofs in linear logic, computations that lead to a dead-end state are not considered, unlike computations from inaccessible states that are included. The same problem with the *direction of a proof* vs the *direction of a computation* appears in all other publications that claim undecidability, including Kanovich (2016).

Keywords. Decidability, Linear logic, Modal logic, Relevance logic, Sequent calculuses

INTRODUCTION

Modality in reasoning has intrigued thinkers for millennia — at least since the time of Aristotle. Logically valid reasoning itself is often characterized in modal terms by saying that a conclusion is true necessarily, provided the premises are true. Thus it is not by chance that an attempt that aimed at tightening the connections between the notions of logical consequence and implication led to the invention of modern modal logics in the work of Clarence I. Lewis.

The *logic of entailment*, E gives a certain modal character to provable entailments. A usual definition of "$\mathcal{A}$ is necessary" in some relevance logics is by the formula $(\mathcal{A} \to \mathcal{A}) \to \mathcal{A}$. However, there are other ways to think about modality in relevance logics. In this paper, we look at an alternative definition of necessity and possibility that involves t and f, then we consider $\Box$ and $\Diamond$ as primitives.

In order to narrow our considerations, we start with the logic called *lattice*-R, which is denoted by LR. This logic was derived from the logic of *relevant implication* R by omitting the assumption that conjunction and disjunction distribute over each other; it was created by Meyer [36]. The distributivity principle does not appear to be problematic from the point of view that motivates the family of relevance logics, hence,

---

2020 *Mathematics Subject Classification.* Primary: 03B47, Secondary: 03F52, 03B45, 68T17.

Bimbó, Katalin, (ed.), *Relevance Logics and other Tools for Reasoning. Essays in Honor of J. Michael Dunn*, (Tributes, vol. 46), College Publications, London, UK, 2022, pp. 89–127.

we might wonder why to consider lattice-R at all. Lattice-R has a straightforward sequent calculus formalization that goes back to Meyer's thesis, and it was hoped way back in the 1960s, that the *decidability* of lattice-R would be a stepping stone to the decidability of R (and that of E, T, etc.). Accordingly, we will explore the question of decidability in the context of modalities and also for logics neighboring R.

Section 1 introduces lattice-R in the way it was originally defined; then we throw in some constants. We give a sequent calculus (*LLR*) and an axiomatic (*HLR*) formulation.[1] Next, in Section 2, we take up the idea of defined modalities within LLR^c, that is, lattice-R with zero-ary constants. Section 3 gives a somewhat detailed proof that LLR^c is decidable. The argument is along the standard Curry–Kripke lines, which had been successfully applied to some other logics. The next section adds $\Diamond$ and $\Box$ as new unary connectives to LLR^c. We prove that the resulting logic is decidable. In Section 5, we consider a series of logics obtained by variations on the structural rules — whether they are absent, modalized or included. Then in Section 6, we give a direct and quite detailed proof of the decidability of (classical propositional) linear logic. Finally, in Section 7, we briefly outline the argument in Lincoln et al. [35], from which they conclude a theorem that conflicts with our decidability result about linear logic in the previous sections. We pinpoint some gaps in their proof of the cut elimination theorem, and we conclude with a different interpretation of *LCLL* proofs, which dissolves the appearance of a contradiction between our result and those in [35], Kanovich [28; 27] and Forster and Larchey-Wendling [21].

1. LATTICE-R WITH CONSTANTS

The *relevant endeavor* can be quickly motivated by the desire to avoid having theorems like $\mathcal{A} \to (\mathcal{B} \to \mathcal{A})$, where $\to$ is some sort of implication. Roughly speaking, $\mathcal{B}$ gets into the theorem, although it may be completely unrelated to $\mathcal{A}$. Somewhat less obviously, $((\mathcal{A} \to \mathcal{B}) \to \mathcal{A}) \to \mathcal{A}$ is also an unwelcome theorem. It is easy to verify that the proof of these formulas in a sequent calculus for classical logic, such as Gentzen's *LK*, requires the use of some of the thinning rules. Well, then it is plain sailing to drop those rules and to see what results.

The *language* of LLR^c contains a denumerable stock of propositional variables together with a handful of logical constants.[2] The latter category is divided into three subcategories by the arity of the connectives: 0-ary, 1-ary and 2-ary. The zero-ary connectives are t ("real truth"), f ("real falsity"), T ("triviality") and F ("absurdity"). The only unary connective is $\sim$ ("De Morgan negation"). There are five binary connectives, namely, $\wedge$ ("conjunction"), $\vee$ ("disjunction"), $\circ$ ("fusion"), $\to$ ("implication" or "entailment") and $+$ ("fission"). The set of well-formed formulas is inductively defined from the base set, which comprises the propositional variables and the four zero-ary connectives, by the rest of the connectives. $\mathcal{A}, \mathcal{B}, \mathcal{C}, \ldots$ are meta-variables that range over well-formed formulas (wff's, for short).

---

[1] We would like to forewarn the reader that L in LR stands for "lattice" and not for "logistic" as in many labels for sequent calculuses, including the original *LK* and *LJ* (where we italicize *L*). Then, *LLR* is a sequent calculus formulation of LR, and so on.

[2] As we already hinted at, the superscript ^{c} in the label for the system indicates that four *zero-ary constants* are included.

Multisets constitute a datatype between sequences and sets. In a multiset, an object may have more than one occurrence, and the number of occurrences matters, but the order (of listing) of occurrences is unimportant. Here we always deal with *finite* multisets, that is, with multisets of finitely many objects, each with finitely many occurrences; we will simply talk about multisets. $\alpha, \beta, \gamma, \ldots$ are meta-variables for multisets of wff's including the empty multiset.

Definition 1. The axioms and rules of the *sequent calculus* LLR^c are as follows.

$$\alpha; F \vdash \beta \quad F\vdash \qquad \mathcal{A} \vdash \mathcal{A} \text{ id} \qquad \alpha \vdash T; \beta \quad \vdash T$$

$$f \vdash \quad f\vdash \qquad \frac{\alpha \vdash \beta}{\alpha \vdash f; \beta} \vdash f \qquad \frac{\alpha \vdash \beta}{\alpha; t \vdash \beta} \; t\vdash \qquad \vdash t \; \vdash t$$

$$\frac{\alpha; \mathcal{A} \vdash \beta}{\alpha; \mathcal{A} \wedge \mathcal{B} \vdash \beta} \wedge\vdash_1 \qquad \frac{\alpha; \mathcal{B} \vdash \beta}{\alpha; \mathcal{A} \wedge \mathcal{B} \vdash \beta} \wedge\vdash_2 \qquad \frac{\alpha \vdash \mathcal{A}; \beta \quad \alpha \vdash \mathcal{B}; \beta}{\alpha \vdash \mathcal{A} \wedge \mathcal{B}; \beta} \vdash\wedge$$

$$\frac{\alpha; \mathcal{A} \vdash \beta \quad \alpha; \mathcal{B} \vdash \beta}{\alpha; \mathcal{A} \vee \mathcal{B} \vdash \beta} \vee\vdash \qquad \frac{\alpha \vdash \mathcal{A}; \beta}{\alpha \vdash \mathcal{A} \vee \mathcal{B}; \beta} \vdash\vee_1 \qquad \frac{\alpha \vdash \mathcal{B}; \beta}{\alpha \vdash \mathcal{A} \vee \mathcal{B}; \beta} \vdash\vee_2$$

$$\frac{\alpha \vdash \mathcal{A}; \beta}{\alpha; {\sim}\mathcal{A} \vdash \beta} {\sim}\vdash \qquad \frac{\alpha; \mathcal{A} \vdash \beta}{\alpha \vdash {\sim}\mathcal{A}; \beta} \vdash{\sim}$$

$$\frac{\alpha \vdash \mathcal{A}; \beta \quad \gamma; \mathcal{B} \vdash \delta}{\alpha; \gamma; \mathcal{A} \rightarrow \mathcal{B} \vdash \beta; \delta} \rightarrow\vdash \qquad \frac{\alpha; \mathcal{A} \vdash \mathcal{B}; \beta}{\alpha \vdash \mathcal{A} \rightarrow \mathcal{B}; \beta} \vdash\rightarrow$$

$$\frac{\alpha; \mathcal{A}; \mathcal{B} \vdash \beta}{\alpha; \mathcal{A} \circ \mathcal{B} \vdash \beta} \circ\vdash \qquad \frac{\alpha \vdash \mathcal{A}; \beta \quad \gamma \vdash \mathcal{B}; \delta}{\alpha; \gamma \vdash \mathcal{A} \circ \mathcal{B}; \beta; \delta} \vdash\circ$$

$$\frac{\alpha; \mathcal{A} \vdash \beta \quad \gamma; \mathcal{B} \vdash \delta}{\alpha; \gamma; \mathcal{A} + \mathcal{B} \vdash \beta; \delta} +\vdash \qquad \frac{\alpha \vdash \mathcal{A}; \mathcal{B}; \beta}{\alpha \vdash \mathcal{A} + \mathcal{B}; \beta} \vdash+$$

$$\frac{\alpha; \mathcal{A}; \mathcal{A} \vdash \beta}{\alpha; \mathcal{A} \vdash \beta} W\vdash \qquad \frac{\alpha \vdash \mathcal{A}; \mathcal{A}; \beta}{\alpha \vdash \mathcal{A}; \beta} \vdash W$$

The notion of a *proof* in LLR^c is as usual in sequent calculuses. $\mathcal{A}$ is a *theorem* of LLR^c iff $\vdash \mathcal{A}$ is a provable sequent.

The original lattice-R does not include the constants, that is, it comprises the axiom (id) and the rules save $(t \vdash)$ and $(\vdash f)$. The last two rules, which are called *contraction*, are the only structural rules. Other commonly considered structural rules such as exchange and associativity are inherent in the datatype in the antecedent and succedent, whereas thinning is discarded both on the left and on the right — except for their special instances with t and f.

The above sequent calculus is a sensible and well-behaved sequent calculus in light of the following theorem, which involves the single cut rule.

$$\frac{\alpha \vdash \mathcal{C}; \beta \quad \gamma; \mathcal{C} \vdash \delta}{\alpha; \gamma \vdash \beta; \delta} \text{ single cut}$$

Theorem 2. (Cut theorem for LLR^c) *The cut rule is* admissible *in LLR^c.*

Proof. The cut rule formulated above is a version of the single cut rule. There are various ways to prove this rule admissible; one of them is by a triple induction on the degree of the cut formula, on the contraction measure of the cut and on the rank of the cut. We do not include the details here.[3] Here is a sample step, in which the degree of the cut formula $\mathcal{A} + \mathcal{B}$ is reduced.

$$\dfrac{\dfrac{\vdots}{\alpha \vdash \mathcal{A};\mathcal{B};\beta} \quad \dfrac{\vdots}{\gamma;\mathcal{A} \vdash \delta} \quad \dfrac{\vdots}{\varepsilon;\mathcal{B} \vdash \eta}}{\dfrac{\alpha \vdash \mathcal{A}+\mathcal{B};\beta \quad \gamma;\varepsilon;\mathcal{A}+\mathcal{B} \vdash \delta;\eta}{\alpha;\gamma;\varepsilon \vdash \beta;\delta;\eta}}$$

$$\rightsquigarrow \quad \dfrac{\dfrac{\dfrac{\vdots}{\alpha \vdash \mathcal{A};\mathcal{B};\beta} \quad \dfrac{\vdots}{\gamma;\mathcal{A} \vdash \delta}}{\alpha;\gamma \vdash \mathcal{B};\beta;\delta} \quad \dfrac{\vdots}{\varepsilon;\mathcal{B} \vdash \eta}}{\alpha;\gamma;\varepsilon \vdash \beta;\delta;\eta} \quad \triangleleft$$

The proof of the cut theorem also establishes that the addition of the zero-ary constants (one by one, or all at once) is *conservative* over the original LR.

Lattice-R can be defined by an axiom system too. We denote the Hilbert-style system by HLR^c. This calculus comprises the axiom schemas (A1)–(A17) and the rules (R1)–(R3). (Outside parentheses are omitted from wff's, as before.)

(A1) $\mathcal{A} \to \mathcal{A}$
(A2) $(\mathcal{A} \to \mathcal{B}) \to ((\mathcal{C} \to \mathcal{A}) \to (\mathcal{C} \to \mathcal{B}))$
(A3) $(\mathcal{A} \to (\mathcal{B} \to \mathcal{C})) \to (\mathcal{B} \to (\mathcal{A} \to \mathcal{C}))$
(A4) $(\mathcal{A} \to (\mathcal{A} \to \mathcal{B})) \to (\mathcal{A} \to \mathcal{B})$
(A4–5) $(\mathcal{A} \wedge \mathcal{B}) \to \mathcal{A}, \quad (\mathcal{A} \wedge \mathcal{B}) \to \mathcal{B}$
(A7) $((\mathcal{C} \to \mathcal{A}) \wedge (\mathcal{C} \to \mathcal{B})) \to (\mathcal{C} \to (\mathcal{A} \wedge \mathcal{B}))$
(A8–9) $\mathcal{A} \to (\mathcal{A} \vee \mathcal{B}), \quad \mathcal{A} \to (\mathcal{B} \vee \mathcal{A})$
(A10) $((\mathcal{A} \to \mathcal{C}) \wedge (\mathcal{B} \to \mathcal{C})) \to ((\mathcal{A} \vee \mathcal{B}) \to \mathcal{C})$
(A11–2) $({\sim}\mathcal{A} \to \mathcal{B}) \to ({\sim}\mathcal{B} \to \mathcal{A}), \quad \mathcal{A} \to {\sim}{\sim}\mathcal{A}$
(A13-4) $t, \quad (t \to {\sim}f) \wedge (f \to {\sim}t)$
(A15) $(F \to \mathcal{A}) \wedge (\mathcal{A} \to T)$
(A16) $((\mathcal{A} \circ \mathcal{B}) \to {\sim}(\mathcal{A} \to {\sim}\mathcal{B})) \wedge ({\sim}(\mathcal{A} \to {\sim}\mathcal{B}) \to (\mathcal{A} \circ \mathcal{B}))$
(A17) $((\mathcal{A} + \mathcal{B}) \to ({\sim}\mathcal{A} \to \mathcal{B})) \wedge (({\sim}\mathcal{A} \to \mathcal{B}) \to (\mathcal{A} + \mathcal{B}))$
(R1) $\mathcal{A} \to \mathcal{B}$ and $\mathcal{A}$ imply $\mathcal{B}$
(R2) $\mathcal{A}$ and $\mathcal{B}$ imply $\mathcal{A} \wedge \mathcal{B}$
(R3) $\vdash \mathcal{A}$ implies $\vdash t \to \mathcal{A}$

The notion of a *proof* is the usual one for axiom systems, and the formulas occurring in a proof are called *theorems*.

The axiom system HLR^c is equivalent to LLR^c in the sense that the two calculuses have the same set of theorems, as we state in the following theorem. (We leave the proof, which is completely routine, to the reader.)

Theorem 3. *$\mathcal{A}$ is a theorem of HLR^c iff it is a theorem of LLR^c.*

---

[3] Some details of a similar proof may be found in Bimbó [8, §2].

2. Modalities in LLR^c Defined from t and f

The symbols $\Diamond$ and $\Box$ usually stand for *unary modalities*, which are read as "diamond" and "box," or in alethic modal logics, as "possibility" and "necessity." The presence of t and f in LR^c allows us to define surrogate unary connectives.

Definition 4. $\Box \mathcal{A}$ is $t \wedge \mathcal{A}$, and $\Diamond \mathcal{A}$ is $\mathcal{A} \vee f$.

Of course, the above definition in itself is nothing more than looking at formulas with a squint. However, $\Box$ and $\Diamond$ turn out to have certain properties that are reminiscent of properties the modalities often have. The notation that we introduced was intended to prefigure this.

Lemma 5. *The formulas in* (1)–(4) *are theorems of* LLR^c, *and by* (5), *necessitation is an* admissible rule *in* LLR^c.

(1) $\Box(\mathcal{A} \to \mathcal{B}) \to (\Box \mathcal{A} \to \Box \mathcal{B})$
(2) $\Box \mathcal{A} \to \mathcal{A}$
(3) $\Box \mathcal{A} \to \Box\Box\mathcal{A}$
(4) $(\Diamond \mathcal{A} \to {\sim}\Box{\sim}\mathcal{A}) \wedge ({\sim}\Box{\sim}\mathcal{A} \to \Diamond \mathcal{A})$
(5) If $\vdash \mathcal{A}$, then $\vdash \Box \mathcal{A}$.

Proof. The proofs of the corresponding formulas are straightforward, once the defined symbols are rewritten with the primitive connectives. For instance, (1) turns into the formula $(t \wedge (\mathcal{A} \to \mathcal{B})) \to ((t \wedge \mathcal{A}) \to (t \wedge \mathcal{B}))$. (We omit the rest of the details.) ◁

The formulas in (1)–(3) resemble some well-known axioms from (normal) modal logics, when $\Box$ is viewed as $\Box$, $\wedge$ as $\wedge$, and $\to$ is taken to be $\supset$ (i.e., classical conditional). In particular, (1) looks like (K), (2) looks like (T) and (3) looks like (4).[4] It may be tempting, at first sight, to conjecture that we have found S4 in LLR^c. However, we should not forget that $\sim$ is not an orthonegation, and $\wedge$ and $\vee$ are not related to each other or to $\to$ in the way conjunction and disjunction are linked to $\supset$ (and $\neg$, orthonegation). We find another logic hidden within LLR^c though.

Linear logic, as defined in Girard [23], is sometimes called *classical linear logic*, because it shares more features with classical logic than with intuitionist logic.[5] We denote this logic by CLL. Linear logic without the modalities is called *multiplicative–additive linear logic* (or MALL). Linear logic was first defined as a one-sided sequent calculus. However, all fragments of classical linear logic that contain the negation connective may be defined equivalently as two-sided sequent calculuses.[6]

Classical linear logic can be (and has been) formulated in various ways, as in Avron [4] and [44], for instance. For our goals in this paper, it is convenient to rely on a sequent calculus formulation. Moreover, we will assume that sequents are defined as before, that is, they comprise a pair of multisets of wff's. The language of propositional CLL contains several connectives, and [23] uses unconventional notation to denote them. A translation that turns a symbol into a symbol that looks the same in another language is very manageable; hence, we list Girard's symbols together with his names for the connectives, but we immediately give our preferred notation that

---

[4]This is not a typo; (4) is the standard label for the characteristic axiom of the system S4.
[5]See, e.g., Troelstra [44], where *intuitionist linear logic* is introduced too.
[6]See Bimbó [9] for a comprehensive treatment of sequent calculuses — including calculuses for classical linear logic.

induces an identity translation between languages of logics. (In Sections 6 and 7, we turn back to using Girard's notation to facilitate comparisons.)

The zero-ary connectives are **1** (*one*, **t**), $\perp$ (*bottom*, **f**), $\top$ (*top*, **T**) and **0** (*null*, **F**). The unary connectives are $^{\perp}$ (*nil*, $\sim$), ! (*of course*, $\Diamond$ or $\square$) and ? (*why not*, $\square$ or $\Diamond$). The binary connectives are & (*with*, $\wedge$), $\oplus$ (*plus*, $\vee$), $\otimes$ (*times*, $\circ$), $\invamp$ (*par*, $+$) and $\multimap$ (*entail*, $\rightarrow$).

For the so-called exponentials (! and ?), we listed both modalities. The first modality is motivated by relational semantics, whereas the second one is based on similarities of sequent calculus rules for the punctuation marks and for modalities. For the sake of translating and comparing sequent calculuses in this paper, we use the second variant. The issue is that when the Church constants ($\top$ and **0**) are not definable from negation using the lattice operations, Kripke's rules for the modalities (or their adaptations for ! and ?) do not provide both (dual) additivity and (dual) normality for either of the two monotone operations.

In a *two-sided sequent calculus* for classical linear logic, which we denote by *LCLL*, the connective rules for the connectives that have an alter ego in LLR^c are exactly as in LLR^c. (Hence, we do no repeat those rules; rather, we simply assume that *LCLL* is formulated with standard vocabulary.) The contraction rules $(W \vdash)$ and $(\vdash W)$ are absent from *LCLL*. However, the rules below allow the introduction of ! and ? on the right- and left-hand sides of the turnstile, and they recuperate the effect of some of the contractions and thinnings in a traceable way.

Definition 6. The eight rules that involve the exponential connectives are the following. !α and ?α are multisets in which the main connective of each formula is, respectively, ! and ?.

$$\dfrac{\alpha; A \vdash \beta}{\alpha; !A \vdash \beta} \; !\vdash \qquad \dfrac{!\alpha \vdash A; ?\beta}{!\alpha \vdash !A; ?\beta} \; \vdash! \qquad \dfrac{!\alpha; A \vdash ?\beta}{!\alpha; ?A \vdash ?\beta} \; ?\vdash \qquad \dfrac{\alpha \vdash A; \beta}{\alpha \vdash ?A; \beta} \; \vdash?$$

$$\dfrac{\alpha; !A; !A \vdash \beta}{\alpha; !A \vdash \beta} \; !W\vdash \qquad \dfrac{\alpha \vdash \beta}{\alpha; !A \vdash \beta} \; !K\vdash \qquad \dfrac{\alpha \vdash \beta}{\alpha \vdash ?A; \beta} \; \vdash?K \qquad \dfrac{\alpha \vdash ?A; ?A; \beta}{\alpha \vdash ?A} \; \vdash?W$$

If we simply omit the $(W \vdash)$ and $(\vdash W)$ rules from LLR^c, then we obtain *LMALL*, a sequent calculus formalization of MALL.

Our goal now is to establish that the defined modalities in LLR^c behave sufficiently similarly to the exponentials (i.e., the modalities) of *LCLL*. Moreover, the proof of the next theorem provides a translation of wff's of CLL into LR^c, which is of special philosophical interest, given that classical linear logic's constructive character is primarily manifest via the translation of intuitionist logic into CLL. In a similar sense, LLR^c is *linear* and *constructive*.

Theorem 7. (**From LCLL to LLR^c**) *If A is a theorem of LCLL, then $\tau(A)$ is a theorem of LLR^c, where τ is defined inductively by (1)–(6).*

(1) $\tau(p)$ is p, when p is a propositional variable;
(2) $\tau(c)$ is c, where c is a zero-ary constant;
(3–5) $\tau(!A)$ is $t \wedge \tau(A)$; $\tau(?A)$ is $\tau(A) \vee f$; $\tau(A^{\perp})$ is $\sim\tau(A)$;
(6) $\tau(A * B)$ is $\tau(A) * \tau(B)$, where $*$ is a binary connective.

Proof. First, we note that τ is well-defined in the sense that it is applicable to any wff of *LCLL*, and it results in a unique wff of *LLR^c*.

The proof is by induction on χ, the height of a proof tree with root $\vdash \mathcal{A}$. We prove that if $\alpha \vdash \beta$ is provable in *LCLL*, then $\tau(\alpha) \vdash \tau(\beta)$ is provable in *LLR^c*. (τ is applied piece-wise to a multiset, and the translation of the empty multiset is itself.)

1. If $\chi = 1$, then the proof is an instance of an axiom. We note that τ is independent of the location of a formula within a sequent. Therefore, $\tau(\mathcal{A}) \vdash \tau(\mathcal{A})$ yields $\mathcal{B} \vdash \mathcal{B}$, where $\mathcal{B}$ may be $\mathcal{A}$ or may be a different formula than $\mathcal{A}$ (if there are occurrences of ! or ? in $\mathcal{A}$). Either way, $\mathcal{B} \vdash \mathcal{B}$ is an instance of (id) in *LLR^c*.

If the axiom is one of those that involve a zero-ary constant, then the claim is obviously true too.

2. If $\chi > 1$, then $\alpha \vdash \beta$ is by a rule.

2.1. The non-modal connective rules of *LCLL* turn into identical rules in *LLR^c*; furthermore, the latter rules are insensitive to the concrete shape of the parametric or subaltern wff's.[7] As an example, we consider the $(\vdash \wedge)$ rule. α is α, whereas β is $\mathcal{A} \wedge \mathcal{B}; \gamma$. On the left, we have the proof segment in *LCLL*, on the right, we have the resulting proof segment in *LLR^c*. The upper sequents are given by the hypotheses of the induction that we indicate by "i.h."

$$\frac{\alpha \vdash \mathcal{A}; \gamma \quad \alpha \vdash \mathcal{B}; \gamma}{\alpha \vdash \mathcal{A} \wedge \mathcal{B}; \gamma} \quad \overset{\text{i.h.}}{\leadsto} \quad \frac{\tau(\alpha) \vdash \tau(\mathcal{A}); \tau(\gamma) \quad \tau(\alpha) \vdash \tau(\mathcal{B}); \tau(\gamma)}{\tau(\alpha) \vdash \tau(\mathcal{A}) \wedge \tau(\mathcal{B}); \tau(\gamma)}$$

By clause (6), $\tau(\mathcal{A} \wedge \mathcal{B})$ is $\tau(\mathcal{A}) \wedge \tau(\mathcal{B})$. The other cases for the rules for non-modal connectives has the same general structure, and we omit including their details here.

2.2. The last rule may be a modal connective rule. *LCLL* has the same pleasing symmetry as *LK*, the original sequent calculus for classical logic has; hence, we consider in some detail the cases for $(!\vdash)$ and $(\vdash !)$, but leave the details of the dual cases (i.e., of $(?\vdash)$ and $(\vdash ?)$) to the reader.

We have the following subtrees.

$$\frac{\gamma; \mathcal{A} \vdash \beta}{\gamma; !\mathcal{A} \vdash \beta} \quad \overset{\text{i.h.}}{\leadsto} \quad \frac{\tau(\gamma); \tau(\mathcal{A}) \vdash \tau(\beta)}{\tau(\gamma); t \wedge \tau(\mathcal{A}) \vdash \tau(\beta)}$$

By (3), we know that $\tau(!\mathcal{A})$ is $t \wedge \tau(\mathcal{A})$, as needed.

If the sequent $\alpha \vdash \beta$ is $!\gamma \vdash !\mathcal{A}; ?\delta$ by $(\vdash !)$, then we have the following chunks of proofs.

$$\frac{!\gamma \vdash \mathcal{A}; ?\delta}{!\gamma \vdash !\mathcal{A}; ?\delta} \quad \overset{\text{i.h.}}{\leadsto} \quad \frac{\tau(!\gamma) \vdash \tau(\mathcal{A}); \tau(?\delta) \quad \dfrac{\vdash t}{\tau(!\gamma) \vdash t; \tau(?\delta)}}{\tau(!\gamma) \vdash t \wedge \tau(\mathcal{A}); \tau(?\delta)}$$

The thicker line indicates possibly several applications of rules — depending on the number of wff's in $!\gamma$ and $?\delta$. For any wff $!\mathcal{B}$ in $!\gamma$, its translation is $t \wedge \tau(\mathcal{B})$, whereas, for any wff $?\mathcal{C}$ in $?\delta$, its translation is $\tau(\mathcal{C}) \vee f$. Each $t \wedge \tau(\mathcal{B})$ can be obtained by $(t \vdash)$ and $(\wedge \vdash)$; analogously, $\tau(\mathcal{C}) \vee f$ may be gotten by $(\vdash f)$ and $(\vdash \vee)$. The last step above is justified by $(\vdash \wedge)$.

---

[7] These terms have their usual meanings following Curry [16].

2.3. There are four modalized structural rules in *LCLL*. First of all, the modalized contraction rules are special instances of their regular counterparts in *LLR^c*. That is, the claim is obviously true when the last rule is $(!W \vdash)$ or $(\vdash ?W)$.

If the last rule applied in the *LCLL* proof is $(!K \vdash)$, then α is $\gamma; !A$, and we have the following.

$$\frac{\vdots \\ \gamma \vdash \beta}{\gamma; !A \vdash \beta} \qquad \overset{\text{i.h.}}{\leadsto} \qquad \frac{\frac{\vdots \\ \tau(\gamma) \vdash \tau(\beta)}{\tau(\gamma); t \vdash \tau(\beta)}}{\tau(\gamma); t \wedge \tau(A) \vdash \tau(\beta)}$$

The rules applied in *LLR^c* are $(t \vdash)$ and $(\wedge \vdash)$. The latter rule is applicable with an arbitrary $\tau(A)$, and $\tau(!A)$ is $t \wedge \tau(A)$ by clause (3). ◁

The theorem provides a way to test wff's of *LCLL* for *non-provability*, because if their translation is not provable in *LLR^c*, then the starting formula is not provable in *LCLL*. Of course, we are using the fact, which is well known to relevance logicians, that *LLR* is decidable. We provide some details of the proof for *LLR^c* in Section 3. Provability is easily decidable if a wff of *CLL* does not contain occurrences of ! or ?, because *LLRW^c*'s decidability is an immediate consequence of the cut theorem. (*LLRW^c* is *LLR^c* without the $(\vdash W)$ or $(W \vdash)$ rules.)

An insight that we attribute to Kripke [30] is that, in relevance logics, a wff has to be introduced by a connective rule in order to be contracted. Once stated, the truth of this observation is obvious. However, a profound consequence, as Kripke realized, is that the contraction rules can be eliminated if operational rules *permit* some contraction but require none. Relying on the same observation, the amount of the permitted contractions in each operational rule may be minimized. The insight that we attribute to Dunn, is that it is sufficient to allow a formula to be contracted if it could not have been contracted in the premises.

In order to motivate the introduction of heap numbers (in Definition 8 below), we illustrate how we use heap numbers extracted from irredundant proofs in one calculus to bound the number of permitted contractions in another — but related — calculus.[8]

If we assume the usual definition of a subformula, then we may note the obvious fact that every formula has at least one subformula, but "often" it has more. Furthermore, if we count distinct occurrences of a subformula separately, then we find that some formulas have even more subformulas (in the sense of subformula occurrences). Then it is obvious that permitting as many contractions on a larger formula as we performed on some of its proper subformulas will produce at least as many or more occurrences of subformulas. Let us consider a small proof in [*LLR^c*] (cf. Definition 11).

$$\frac{\dfrac{\dfrac{\dfrac{\dfrac{\dfrac{A \vdash A \quad B \vdash B}{A, B \vdash A \circ B} [\vdash \circ] \quad \dfrac{A \vdash A \quad B \vdash B}{A, B \vdash A \circ B} [\vdash \circ]}{A, B \vdash (A \circ B) \circ (A \circ B)} [\vdash \circ]}{A \wedge B, B \vdash (A \circ B) \circ (A \circ B)} [\wedge \vdash]}{A \wedge B \vdash (A \circ B) \circ (A \circ B)} [\wedge \vdash]}{t \wedge (A \wedge B) \vdash (A \circ B) \circ (A \circ B)} [\wedge \vdash]$$

---

[8] The term "irredundant" is used in its standard sense in relevance logic; see Dunn [18, §3.6].

This proof is irredundant — with the third application of $[\vdash \circ]$ containing contractions of $\mathcal{A}$ and $\mathcal{B}$, and the second application of $[\wedge \vdash]$ containing a contraction of $\mathcal{A} \wedge \mathcal{B}$. Each of those formulas are subformulas of $!(\mathcal{A} \wedge \mathcal{B})$ (which translates via τ into $t \wedge (\mathcal{A} \wedge \mathcal{B})$). Thus the heap number of $!(\mathcal{A} \wedge \mathcal{B})$ is at least 3. (We have not generated all the irredundant proofs here, but having this proof we know that the heap number cannot be less than 3.) The following proof in [LCLL] uses three contractions as part of applications of the $[!\vdash]$ rule. This proof also happens to be irredundant, however, that is an accidental feature. In the proof search that uses the heap number as an upper bound on contractions, we do not require the resulting proof to be irredundant. Accordingly, applications of the $(!\vdash)$ and $(!W\vdash)$ rules may be separated without loss of generality.

$$\dfrac{\dfrac{\dfrac{\dfrac{\dfrac{\dfrac{\dfrac{\dfrac{\dfrac{\dfrac{\mathcal{A} \vdash \mathcal{A} \quad \mathcal{B} \vdash \mathcal{B}}{\mathcal{A}, \mathcal{B} \vdash \mathcal{A} \circ \mathcal{B}} \quad \dfrac{\mathcal{A} \vdash \mathcal{A} \quad \mathcal{B} \vdash \mathcal{B}}{\mathcal{A}, \mathcal{B} \vdash \mathcal{A} \circ \mathcal{B}}}{\mathcal{A}, \mathcal{B}, \mathcal{A}, \mathcal{B} \vdash (\mathcal{A} \circ \mathcal{B}) \circ (\mathcal{A} \circ \mathcal{B})}}{\mathcal{A}, \mathcal{B}, \mathcal{A}, \mathcal{A} \wedge \mathcal{B} \vdash (\mathcal{A} \circ \mathcal{B}) \circ (\mathcal{A} \circ \mathcal{B})}}{\mathcal{A}, \mathcal{B}, \mathcal{A} \wedge \mathcal{B}, \mathcal{A} \wedge \mathcal{B} \vdash (\mathcal{A} \circ \mathcal{B}) \circ (\mathcal{A} \circ \mathcal{B})}}{\mathcal{A}, \mathcal{B}, \mathcal{A} \wedge \mathcal{B}, !(\mathcal{A} \wedge \mathcal{B}) \vdash (\mathcal{A} \circ \mathcal{B}) \circ (\mathcal{A} \circ \mathcal{B})}}{\mathcal{A}, \mathcal{B}, !(\mathcal{A} \wedge \mathcal{B}) \vdash (\mathcal{A} \circ \mathcal{B}) \circ (\mathcal{A} \circ \mathcal{B})} \; [!\vdash]}{\mathcal{A}, \mathcal{A} \wedge \mathcal{B}, !(\mathcal{A} \wedge \mathcal{B}) \vdash (\mathcal{A} \circ \mathcal{B}) \circ (\mathcal{A} \circ \mathcal{B})}}{\mathcal{A}, !(\mathcal{A} \wedge \mathcal{B}) \vdash (\mathcal{A} \circ \mathcal{B}) \circ (\mathcal{A} \circ \mathcal{B})} \; [!\vdash]}{\mathcal{A} \wedge \mathcal{B}, !(\mathcal{A} \wedge \mathcal{B}) \vdash (\mathcal{A} \circ \mathcal{B}) \circ (\mathcal{A} \circ \mathcal{B})}}{!(\mathcal{A} \wedge \mathcal{B}) \vdash (\mathcal{A} \circ \mathcal{B}) \circ (\mathcal{A} \circ \mathcal{B})} \; [!\vdash]$$

This time, we only labeled the steps that involve a contraction.

As already hinted at by the illustration, we will rely on the theorem (proved in the next section) that LLR^c is decidable. This result is a small extension of the decidability of LLR originally proved in [36].[9]

It may be helpful to note that the decidability proof using a proof-search tree with the sequent calculus $[LLR^c]$ provides *all the irredundant proofs* of a provable sequent — unlike the example above where we only presented one irredundant proof.

Now we turn to the definition of heap numbers. Our definition uses the notion of "ancestors," which is essentially, Curry's notion (see [16, p. 199]), with some obvious modifications that are due to our calculuses being based on multisets. We briefly explain the notion ancestors in the paragraph after the definition.

---

[9]We thank Alasdair Urquhart for calling to our attention (in December 2016) the preprint paper Roorda [39], which claimed to have proved the decidability of classical linear logic. We did not know of Roorda's paper until well after we had our own proof, but his strategy is remarkably similar to ours. He uses the method of Kripke to construct a finite proof-search tree, but the problem seems to be that there is no guarantee that his tree will contain a proof of the candidate theorem if there is one. We provide such a guarantee via our *heap number* in Definition 8. As Urquhart pointed out to us, Roorda's proof does not appear in his subsequent Ph.D. thesis [40]. In fact, on p. 12 of his thesis, he mentions [35] and repeats their claim that CLL is undecidable. So, he apparently came to consider their proof to be correct and his own earlier proof to be mistaken.

Definition 8. (Heap number) Let $\vdash \mathcal{A}$ be a provable sequent. The *heap number of* $\mathcal{B}$ (where $\mathcal{B}$ is a subformula of $\mathcal{A}$) is the maximum of the total number of contractions on the ancestors of $\mathcal{B}$ in any irredundant proof of the sequent.

Given a proof, $\mathcal{B}$ may be parametric in an application of a rule, in which case, it has immediate parametric ancestors in the upper sequent. If $\mathcal{B}$ is the principal formula of a rule then it typically has subalterns in the upper sequent. (Since the calculuses from which we calculate the heap numbers have no explicit contraction rules, only the thinning rules have no subalterns in the upper sequent.) We call immediate parametric ancestors and subalterns immediate ancestors. If contraction is built into a rule, then the principal formula may encompass contractions of parametric formulas, in which case all the affected immediate parametric ancestors as well as the subalterns are immediate ancestors of the principal formula. $\mathcal{C}$ is an ancestor of $\mathcal{B}$ when $\mathcal{C}$ is in the reflexive transitive closure of the *immediate ancestor of* $\mathcal{B}$ relation.

Lemma 9. *Let* $\vdash \mathcal{A}$ *be a provable sequent of* LCLL. *The heap numbers of the subformulas of* $\mathcal{A}$ *(obtained from proofs in* [LLR^e]*) are sufficiently large as bounds on the number of contractions on each formula to construct a proof of* $\vdash \mathcal{A}$ *in* LCLL *(or* [[LCLL]]*).*

Proof. It is sufficient to consider the right-handed sequent calculus for CLL. Hence, the only contraction rule is $(\vdash ?W)$. There are two rules (beyond those for zeroary constants) that can introduce several formulas into the sequent, namely, $(\vdash \circ)$ and $(\vdash ?K)$. Clearly, a formula introduced by the latter does not need to be considered, because the rule has no subaltern. (This means that if $\mathcal{A} \vee f$ resulted from $(\vdash f)$ in [LLR^e], then $?\mathcal{A}$ can be introduced by $(\vdash ?K)$ in LCLL, if needed.) If the immediate subformula of the contracted $?\mathcal{A}$ was introduced by $(\vdash \circ)$, then the proof-search might have contained contractions on $?\mathcal{A}$ or on its subformulas. Let us assume that $n-1$ contractions on $?\mathcal{A}$ are not sufficient for the proof of the sequent in LCLL, and the proof search in [LLR^e] provided a heap number $\leq n-1$. Then there are formulas in the sequent that cannot be contracted in LCLL, which require at least n contractions on $?\mathcal{A}$. However, if a formula cannot be contracted in LCLL, then it remains in the sequent; hence, the sequent at the beginning of the proof search in [LLR^e] contains any such formula. That is, contractions on those formulas cannot reduce the number of required contractions on $?\mathcal{A}$ and its subformulas in [LLR^e]. ◁

Theorem 10. *Classical linear logic* (CLL) *is decidable.*

Proof. Given a wff $\mathcal{A}$ of CLL, τ yields a wff of LLR^e. It is decidable whether $\tau(\mathcal{A})$ is a theorem of LLR^e; if it is not, then $\mathcal{A}$ is not a theorem of CLL. If $\tau(\mathcal{A})$ is a theorem of CLL, then we can identify all the subformulas of $\tau(\mathcal{A})$, which result from the translation of an exponential subformula; we call them exceptional. The decision procedure for $\tau(\mathcal{A})$ in [LLR^e] produces all the irredundant proofs, hence, we can calculate the heap number for each exceptional formula. Then we search for a proof of $\tau(\mathcal{A})$ in a *restricted version* of [LLR^e] by building a proof search tree in which contractions on exceptional formulas are limited by their heap number.

The restrictions on the rules of [LLR^e] are the following.

1. No contraction is permitted in $[\vdash \wedge]$, $[\vee \vdash]$, $[\sim \vdash]$, $[\vdash \sim]$, $[\circ \vdash]$, $[\vdash +]$, $[\vdash \rightarrow]$.

2. A contraction is permitted in $[\wedge_1 \vdash]$ and $[\wedge_2 \vdash]$ when $\mathcal{A} \wedge \mathcal{B}$ is an exceptional wff. Dually, a contraction is permitted in $[\vdash \vee_1]$ and $[\vdash \vee_2]$, if $\mathcal{A} \vee \mathcal{B}$ is an exceptional wff.
3. A contraction is permitted in $\alpha; \gamma$ in $[\vdash \circ]$, $[+ \vdash]$ and $[\to \vdash]$ if an exceptional formula $t \wedge \mathcal{A}$ occurs both in α and γ. Dually, a contraction is permitted in $\beta; \delta$ in an application of the same rules if an exceptional formula $\mathcal{A} \vee f$ occurs both in β and δ. (The principal formulas of these rules cannot be contracted.)

The above restrictions match exactly the restricted contraction rules in CLL (see Definition 6), whereas the heap numbers provide the upper bounds on the number contractions. Therefore, the proof-search tree is finite. Since LLR^c and $LCLL$ coincide on the exponential-free fragment of CLL, and we allowed heap-number-many contractions on exceptional formulas, if $\mathcal{A}$ is a theorem of CLL, then the proof-search tree will contain a proof of $\tau(\mathcal{A})$.

Once we have proofs in the proof-search tree for the formula $\mathcal{A}$, we also check that any applications of $[\vdash \wedge]$ and $[\vee \vdash]$ with principal formulas that are translations of a !'d or ?'d formula satisfy the side conditions in Kripke's rules (i.e., of the $(\vdash !)$ and $(? \vdash)$ rules). ◁

The theorem contradicts Theorem 3.7 in [35], which states the undecidability of classical linear logic, and Corollaries 5.5 and 5.7 in [28], which state the undecidability of two Horn-fragments of linear logic. We believe that those papers do not contain proofs of the undecidability of CLL, and will provide our argument for this in Section 7. We also give another proof of the decidability of CLL below; that proof also uses in an essential way the Curry–Kripke strategy.

3. LATTICE-R'S DECIDABILITY

This section is a rather detailed presentation of the decidability of LLR^c. The decidability of LR was proved by Meyer in 1966 [36]. The addition of the zero-ary constants is not a huge extension of that result, and it has a certain resemblance to the extension of the decidability result for $R_\to$ proved by Kripke in 1959 [30] to a proof of the decidability of $R^t_\to$ in Bimbó and Dunn [13].

A core idea is to define a *contraction-free sequent calculus* that allows the proof of the same sequents as LLR^c does. This sequent calculus must be orderly, that is, the *cut theorem* has to hold for it, and cut-free proofs must have the *subformula property*. Then, we build a *proof-search tree*, which can be shown to be *finite*.

Definition 11. The *sequent calculus* $[LLR^c]$ is defined by the following axioms and rules. (Sequents are as before, and the bracket notation is explained below. This use of brackets motivates the label $[LLR^c]$ for the calculus.)

$$\alpha; F \vdash \beta \quad F\vdash \qquad \mathcal{A} \vdash \mathcal{A} \ \text{id} \qquad \alpha \vdash T; \beta \ \vdash T$$

$$f \vdash \quad f\vdash \qquad \frac{\alpha \vdash \beta}{\alpha \vdash f; \beta} \vdash f \qquad \frac{\alpha \vdash \beta}{\alpha; t \vdash \beta} \ t\vdash \qquad \vdash t \ \vdash t$$

$$\frac{\alpha; \mathcal{A} \vdash \beta}{[\alpha; \mathcal{A} \wedge \mathcal{B}] \vdash \beta} \ [\wedge\vdash_1] \qquad \frac{\alpha; \mathcal{B} \vdash \beta}{[\alpha; \mathcal{A} \wedge \mathcal{B}] \vdash \beta} \ [\wedge\vdash_2] \qquad \frac{\alpha \vdash \mathcal{A}; \beta \quad \alpha \vdash \mathcal{B}; \beta}{\alpha \vdash [\mathcal{A} \wedge \mathcal{B}; \beta]} \ [\vdash\wedge]$$

$$\frac{\alpha;A\vdash\beta \quad \alpha;B\vdash\beta}{[\alpha;A\vee B]\vdash\beta} \, [\vee\vdash] \qquad \frac{\alpha\vdash A;\beta}{\alpha\vdash[A\vee B;\beta]} \, [\vdash\vee_1] \qquad \frac{\alpha\vdash B;\beta}{\alpha\vdash[A\vee B;\beta]} \, [\vdash\vee_2]$$

$$\frac{\alpha\vdash A;\beta}{[\alpha;\sim\! A]\vdash\beta} \, [\sim\vdash] \qquad \frac{\alpha;A\vdash\beta}{\alpha\vdash[\sim\! A;\beta]} \, [\vdash\sim]$$

$$\frac{\alpha;A;B\vdash\beta}{[\alpha;A\circ B]\vdash\beta} \, [\circ\vdash] \qquad \frac{\alpha\vdash A;\beta \quad \gamma\vdash B;\delta}{[\alpha;\gamma]\vdash[A\circ B;\beta;\delta]} \, [\vdash\circ]$$

$$\frac{\alpha;A\vdash\beta \quad \gamma;B\vdash\delta}{[\alpha;\gamma;A+B]\vdash[\beta;\delta]} \, [+\vdash] \qquad \frac{\alpha\vdash A;B;\beta}{\alpha\vdash[A+B;\beta]} \, [\vdash+]$$

$$\frac{\alpha\vdash A;\beta \quad \gamma;B\vdash\delta}{[\alpha;\gamma;A\to B]\vdash[\beta;\delta]} \, [\to\vdash] \qquad \frac{\alpha;A\vdash B;\beta}{\alpha\vdash[A\to B;\beta]} \, [\vdash\to]$$

Bracketing happens in *three kinds* of situations.
1. A parametric multiset is joined with the principal wff of a rule.
2. Two parametric multisets are joined with the principal wff of a rule.
3. Two parametric multisets are joined (without the addition of the principal wff of a rule).

Situations of type 1 occur in all the $\wedge$, $\vee$ and $\sim$ rules, as well as in $[\circ\vdash]$, $[\vdash+]$ and $[\vdash\to]$. Situations of type 2 and 3 occur in the rules $[\vdash\circ]$, $[+\vdash]$ and $[\to\vdash]$ — on one or another side of the turnstile.

Definition 12. The bracketing indicates the following potential contractions — without a total loss of a wff, of course — in the respective multisets. None of the contractions is mandatory, that is, any rule can be applied without contraction, if desired.
1. The principal wff may be contracted once, if it already occurs in the parametric multiset.
2. The principal wff may be contracted once or twice, if it already occurs in one or both parametric multisets, respectively. A parametric wff may be contracted once, if it occurs in both parametric multisets.
3. A wff may be contracted once, if it already occurs in both parametric multisets.

Theorem 13. (**Cut theorem for [LLR^c]**) *The cut rule is admissible in* [LLR^c].

Proof. The proof can be carried out more or less along similar lines as the proof of Theorem 2. However, instead of dealing with the contraction rules separately, we have to verify that all the contractions, which could be carried out in a given proof, can be carried out in the transformed proof too. As a sample transformation, we give a transformation where ρ, the rank is minimal and the cut formula is a conjunction.

$$\frac{\dfrac{\vdots \qquad \vdots}{\dfrac{\alpha\vdash A;\beta \quad \alpha\vdash B;\beta}{\alpha\vdash A\wedge B;\beta} \quad \dfrac{\vdots}{\dfrac{\gamma;A\vdash\delta}{\gamma;A\wedge B\vdash\delta}}}}{[\alpha;\gamma]\vdash[\beta;\delta]} \quad\rightsquigarrow\quad \frac{\dfrac{\vdots}{\alpha\vdash A;\beta} \quad \dfrac{\vdots}{\gamma;A\vdash\delta}}{[\alpha;\gamma]\vdash[\beta;\delta]}$$

Since $A\wedge B$ does not occur in β or γ due to the assumption about the rank, any contraction must be part of the cut. All such contractions can be performed as part of the cut in the new proof. (We omit the rest of the details.) ◁

We state the obvious claim that is a consequence of the fact that any *implicit* contraction in $[LLR^c]$ is replicable by *explicit* contractions in LLR^c.

Lemma 14. *If $\mathcal{A}$ is a theorem of $[LLR^c]$, then $\mathcal{A}$ is a theorem of LLR^c.*

Next, we want to make sure that hiding the contractions in the connective rules does not diminish the capacity of the calculus with respect to proving theorems.

Lemma 15. (Curry's lemma for $[LLR^c]$) *If $\alpha \vdash \beta$ has a proof in $[LLR^c]$ with the height of the proof tree being n, and $\gamma \vdash \delta$ results from $\alpha \vdash \beta$ by one or more applications of the rules $(W \vdash)$ and $(\vdash W)$, then $\gamma \vdash \delta$ has a proof in $[LLR^c]$, where the height of the proof tree is not greater than n (i.e., it is $\leq n$).*

Proof. This is a core lemma for decidability, hence, we give more details here. The base case concerns proofs of height 1.

1. The axioms (id), $(\vdash t)$ and $(f \vdash)$ do not have instances to which a contraction rule could be applied; hence, the claim is true.

If the proof is an instance of $(F \vdash)$, then it can be the case that F is the principal formula of $(W \vdash)$ or some other wff may have multiple copies in α or in β. However, a contraction on the left cannot lead to F being dropped altogether on the left-hand side of the $\vdash$. For instance, $\alpha'; F; F; \mathcal{A}; \mathcal{A} \vdash \mathcal{B}; \mathcal{B}; \beta'$ is an instance of the axiom, but so is $\alpha'; F; \mathcal{A} \vdash \mathcal{B}; \beta'$. The case of $(\vdash T)$ is similar, modulo T occurring on the right-hand side of the $\vdash$.

2. The rest of the cases make up the inductive step. There are three kinds of rules in $[LLR^c]$. First, some rules have no contraction built in. The second group of rules has a type 1 situation on one side of the $\vdash$ and no contraction on the other side. Most rules are like this. Lastly, in three rules, there can be contraction hidden on both sides, one like type 2, the other like type 3. The concrete shape of the principal formula is really indifferent in this proof (though it is specific in each rule). Hence, we exemplify each case by detailing the step for one rule.

2.1. We will scrutinize the rules for the zero-ary constants, since, those rules (or the constants themselves) are not included in [36]. (We of course know that t does not cause a problem in $[LR^t_\to]$, as we had shown in Bimbó and Dunn [12].)

If the constant is a wff that could be contracted, then the application of the rule may be omitted. Any other contraction must involve parametric formulas, hence, the new proof is guaranteed to exist by the inductive hypothesis. Here is what happens in the case of the $(\vdash f)$ rule; the $(t \vdash)$ rule behaves dually. (We only make explicit two pairs of parametric formulas and we assume that $\mathcal{B}$ is not f. However, it should be clear that having more parametric formulas that could be contracted, or having them only on one side or having more copies of one particular formula does not change the general structure of this step in the proof.)

$$\frac{\vdots}{\dfrac{\alpha'; \mathcal{A}; \mathcal{A} \vdash f; \mathcal{B}; \mathcal{B}; \beta'}{\alpha'; \mathcal{A}; \mathcal{A} \vdash f; f; \mathcal{B}; \mathcal{B}; \beta'}} \quad \text{i.h.} \rightsquigarrow \quad \frac{\dfrac{\vdots}{\alpha'; \mathcal{A}; \mathcal{A} \vdash \mathcal{B}; \mathcal{B}; \beta'}}{\dfrac{\alpha'; \mathcal{A}; \mathcal{A} \vdash f; \mathcal{B}; \mathcal{B}; \beta'}{\alpha'; \mathcal{A} \vdash f; \mathcal{B}; \beta'}} \quad \text{i.h.} \rightsquigarrow \quad \frac{\dfrac{\vdots}{\alpha'; \mathcal{A} \vdash \mathcal{B}; \beta'}}{\alpha'; \mathcal{A} \vdash f; \mathcal{B}; \beta'}$$

2.2. In this situation, some duplicates of parametric formulas could be contracted, and additionally, the rule allows the contraction of the principal formula too, provided that it already occurred among the parametric formulas. The former sort of contraction can be dealt with by appeal to the inductive hypothesis, whereas the latter sort of contraction can result from the application of the same rule (to the new premise). As an illustration, we consider one of the $[\vdash \vee]$ rules; the other rules are similar.

$$\frac{\alpha';\mathcal{A};\mathcal{A}\vdash\mathcal{B};\mathcal{B};\mathcal{D};\mathcal{C}\vee\mathcal{D};\beta'}{\alpha';\mathcal{A};\mathcal{A}\vdash[\mathcal{B};\mathcal{B};\mathcal{C}\vee\mathcal{D};\mathcal{C}\vee\mathcal{D};\beta']} \quad \overset{\text{i.h.}}{\leadsto} \quad \frac{\alpha';\mathcal{A}\vdash\mathcal{B};\mathcal{D};\mathcal{C}\vee\mathcal{D};\beta'}{\alpha';\mathcal{A}\vdash[\mathcal{B};\mathcal{C}\vee\mathcal{D};\mathcal{C}\vee\mathcal{D};\beta']}$$

2.3. The last situation is just a notch more complicated, primarily, due to the need to keep track of where the wff's that could be contracted come from. As an illustration, we choose the $[+\vdash]$ rule and we will assume that all the contractable formulas have been made explicit — with distinct letters standing for distinct formulas. (Adding more wff's only expands the size of the sequents, but it does not alter the proof step in a crucial way.) Thus, instead of the bracket notation, we write multisets in the lower sequents.

$$\frac{\gamma;\mathcal{A};\mathcal{A}+\mathcal{B};\mathcal{C};\mathcal{E};\mathcal{E}\vdash\mathcal{D};\delta \quad \varepsilon;\mathcal{B};\mathcal{A}+\mathcal{B};\mathcal{C}\vdash\mathcal{G};\mathcal{G};\mathcal{D};\eta}{\gamma;\varepsilon;\mathcal{A}+\mathcal{B};\mathcal{A}+\mathcal{B};\mathcal{A}+\mathcal{B};\mathcal{C};\mathcal{C};\mathcal{E};\mathcal{E}\vdash\mathcal{G};\mathcal{G};\mathcal{D};\mathcal{D};\delta;\eta} \quad \overset{\text{i.h.}}{\leadsto}$$

$$\frac{\gamma;\mathcal{A};\mathcal{A}+\mathcal{B};\mathcal{C};\mathcal{E}\vdash\mathcal{D};\delta \quad \varepsilon;\mathcal{B};\mathcal{A}+\mathcal{B};\mathcal{C}\vdash\mathcal{G};\mathcal{D};\eta}{\gamma;\varepsilon;\mathcal{A}+\mathcal{B};\mathcal{C};\mathcal{E}\vdash\mathcal{G};\mathcal{D};\delta;\eta}$$

It is easy to verify that the height of the new proof tree in **2.1.–2.3.** is not greater (in some cases, strictly less) than the height of the original proof tree. ◁

Cognate sequents are often defined for sequents that comprise a pair of sequences of formulas. However, the definition straightforwardly transfers to sequents based on pairs of multisets.

Definition 16. (Cognate sequents) The sequents $\alpha \vdash \beta$ and $\gamma \vdash \delta$ are *cognate* iff (1) and (2) hold for any formula $\mathcal{A}$.
(1) $\mathcal{A}$ occurs in α iff it occurs in γ. (2) $\mathcal{A}$ occurs in β iff it occurs in δ.

The *number of occurrences* is not mentioned in the definition at all, which reflects the idea that if we would view sequents as pairs of sets, then cognation means that the set-view turns the antecedents and succedents, respectively, into the same set.

Lemma 17. (Kripke's lemma for cognate sequents) *A sequence of distinct cognate sequents, in which, if $\alpha_n \vdash \beta_n$ precedes $\alpha_m \vdash \beta_m$, then the former does not result by one or more contractions of wff's in $\alpha_m \vdash \beta_m$, is finite.*

A possibly easier-to-understand phrasing of the lemma is in terms of natural numbers. Let a finite fixed set of prime factors be given, let us say, $\{2, 5, 13\}$. Contraction is the reduction of an exponent by 1, for instance, $2^6 \cdot 5^6 \cdot 13^4$ is a contraction of $2^7 \cdot 5^6 \cdot 13^4$. Then, a sequence of distinct natural numbers over the set of fixed prime factors (all with positive integer exponents), in which earlier numbers are not (single or multiple) contractions of later numbers, is finite.[10]

---

[10]Kripke's lemma is equivalent to lemmas from other parts of mathematics, e.g., to Dickson's lemma in number theory. The truth of none of these equivalent lemmas has been questioned. The connection to Dickson's lemma was discovered by Meyer, as noted in [18] and also in its expanded version [20]. (Both

Proof. We note that Kripke's lemma is not specific to the language of a logic, that is, it does not matter what connectives occur in the formulas. The numerical illustration clearly hints toward this. A proof of Kripke's lemma may be found in Anderson and Belnap [2, §13], (and we do not repeat that proof here). ◁

Another lemma that is general, in the sense that the shape of the components in the structure is unimportant, is Kőnig's lemma about trees. *Finite branching* or *finite forking* means that no node has infinitely many children, whereas having *finite branches* means that every maximal path is finite.

Lemma 18. (Kőnig's lemma) *A tree with finite branching and with finite branches is finite.*

Proof. A detailed proof of this lemma may be found in Smullyan [41], for example, and we do not repeat that proof here. ◁

Now we can put together the latter two lemmas with some facts about $[LLR^c]$ to obtain the decidability of $[LLR^c]$, thereby, of LLR^c.

Theorem 19. (Decidability for LLR^c) *The logic LLR^c is decidable.*

Proof. To start with, we note that each formula in the language of $[LLR^c]$ has finitely many subformulas (under the usual understanding of subformulas), hence, finitely many proper subformulas. For example, if a sequent is by the $(\wedge \vdash)$ rule, then there are only two possible choices as to what the subaltern in the premise could be.

Finiteness obtains in other respects too. Each sequent contains finitely many formulas, each occurring finitely many times. Given a sequent and fixating on a rule that could have resulted in that sequent, there are finitely many contractions that could have been part of the application of that rule.

The cut theorem provides the assurance that every theorem has a cut-free proof.

Let us assume that a wff $\mathcal{A}$ is given. We construct a *proof-search tree* to determine if $\mathcal{A}$ is or is not a theorem of $[LLR^c]$. The proof-search tree has two important properties, namely, it is *a finite tree*, and if the given formula is a theorem, then the proof-search tree contains a subtree that is *a proof of the formula* in the root sequent.

The proof-search tree is built from the bottom to the top by "backward applications of the rules." The root of the tree is the sequent $\vdash \mathcal{A}$. By "backward applications of rules" we mean the consideration of potential rules (and their premises), the applications of which could result in the sequent in a given node. We may assume that the potential premises are arranged into an ordered set of leaves, and on each level we proceed from left to right — taking a node after another one, and trying to expand the tree with new nodes (forming a new level in the tree). For a node in the tree, we consider which rules could have been applied and what the premises would be. We add all those premises as children of the given node (i.e., as new leaves) to the tree as long as they do not violate the condition in Kripke's lemma. Then we move on to consider the next node.

---

contain a persuasive visualization of a concrete instance of the lemma; so does [9, §9.1].) See also Riche and Meyer [38] and Kopylov [29].

A theorem $\mathcal{A}$ has a cut-free proof in $[LLR^c]$, hence, the exhaustive search through all the possible rules and potential premises guarantees that a proof is constructed within the proof-search tree (if the formula is a theorem). On the other hand, the tree is finite, because of the already mentioned finiteness properties together with Kripke's and Kőnig's lemmas. Finally, the equivalence of LLR^c and $[LLR^c]$ with respect to provable sequents guarantees that no theorems are misclassified as unprovable when we use $[LLR^c]$ in the proof search. ◁

4. Modalities Added to LLR^c

Modalities could be added *explicitly* to LLR^c, indeed, $\Box$'s addition to LR was considered in [36]. A way to proceed is to consider some usual rules for $\Box$, and their duals for $\Diamond$ together with the connecting rules from Kripke [31], which allow us to prove versions of the so-called modal De Morgan laws for the two modalities.

Definition 20. The sequent calculus $LLR^{\Diamond\Box}$ is defined by the axioms and rules of LLR^c and the following rules.

$$\frac{\alpha; \mathcal{A} \vdash \beta}{\alpha; \Box\mathcal{A} \vdash \beta} \,\Box\vdash \qquad \frac{\Box\alpha \vdash \mathcal{A}; \Diamond\beta}{\Box\alpha \vdash \Box\mathcal{A}; \Diamond\beta} \,\vdash\Box \qquad \frac{\Box\alpha; \mathcal{A} \vdash \Diamond\beta}{\Box\alpha; \Diamond\mathcal{A} \vdash \Diamond\beta} \,\Diamond\vdash \qquad \frac{\alpha \vdash \mathcal{A}; \beta}{\alpha \vdash \Diamond\mathcal{A}; \beta} \,\vdash\Diamond$$

$\Box\alpha$ ($\Diamond\alpha$) is a multiset in which the main connective of each formula is $\Box$ ($\Diamond$). The notions of a *proof* and of a *theorem* are as for LLR^c.

There is an obvious similarity between these rules and the $(!\vdash)$, $(\vdash!)$, $(?\vdash)$ and $(\vdash?)$ rules in *LCLL* (cf. Definition 6). The analogy suggests taking ! to be $\Box$, and ? to be $\Diamond$, and this translation is very tempting. However, $\Box$ and $\Diamond$ have deeply engraved connotations in the presence of $\wedge$ and $\vee$. Especially, under the alethic reading of the connectives, it seems plausible that $\mathcal{A}$ is necessary and $\mathcal{B}$ is necessary exactly when $\mathcal{A} \wedge \mathcal{B}$ is necessary. In $LLR^{\Diamond\Box}$, it is not too difficult to prove half of this, namely, the sequent $\Box(\mathcal{A} \wedge \mathcal{B}) \vdash \Box\mathcal{A} \wedge \Box\mathcal{B}$, and dually, the sequent $\Diamond\mathcal{A} \vee \Diamond\mathcal{B} \vdash \Diamond(\mathcal{A} \vee \mathcal{B})$. Moreover, neither proof requires an application of any structural rule. In other words, if we were to omit $(W \vdash)$ and $(\vdash W)$, the sequents would remain provable. We denote by *LLRW* the contraction-less sequent calculus derived from LLR^c; its modalized version will be denoted by $LLRW^{\Diamond\Box}$.

Of course we know, though we have not yet stated it, that the cut theorem holds for $LLR^{\Diamond\Box}$; moreover, that this logic is decidable too. Nonetheless, after some proof attempts, one might convince oneself that $\Box\mathcal{A} \wedge \Box\mathcal{B} \vdash \Box(\mathcal{A} \wedge \mathcal{B})$ is not provable not only in $LLRW^{\Diamond\Box}$ but in $LLR^{\Diamond\Box}$ either. The analog sequent $!\mathcal{A} \wedge !\mathcal{B} \vdash !(\mathcal{A} \wedge \mathcal{B})$ is not provable in *LCLL*. This formula provides an example of how the proof of Theorem 10 proceeds. If $(t \wedge \mathcal{A}) \wedge (t \wedge \mathcal{B}) \vdash t \wedge (\mathcal{A} \wedge \mathcal{B})$ would not be provable in LLR^c, then we could immediately conclude that $!\mathcal{A} \wedge !\mathcal{B} \vdash !(\mathcal{A} \wedge \mathcal{B})$ is not provable in CLL. However, the translation is *provable* in LLR^c, and so a proof search has to be carried out in $[LLR^c]$ taking into account all the constraints from Theorem 10. The proof search does not produce a proof, therefore, we may conclude that the sequent is not provable in CLL. We return to the provability of $\Box\mathcal{A} \wedge \Box\mathcal{B} \vdash \Box(\mathcal{A} \wedge \mathcal{B})$ in the next section, but now we turn to what is provable in $LLR^{\Diamond\Box}$.

The following four formulas are theorems of $LLR^{\Diamond\Box}$, and (R4) is the rule of "necessitation." (We omit the details of the proofs, which are straightforward.) (A18)–(A21) look like the earlier wff's (1)–(4), in which $\Box$ and $\Diamond$ were defined connectives. By numbering these formulas and the rule consecutively, we indicate that $HLR^{\Diamond\Box}$ may be defined from HLR^c by these additions.

(A18) $\Box(\mathcal{A} \to \mathcal{B}) \to (\Box\mathcal{A} \to \Box\mathcal{B})$ (A20) $\Box\mathcal{A} \to \Box\Box\mathcal{A}$
(A19) $\Box\mathcal{A} \to \mathcal{A}$ (A21) $(\Diamond\mathcal{A} \to {\sim}\Box{\sim}\mathcal{A}) \wedge ({\sim}\Box{\sim}\mathcal{A} \to \Diamond\mathcal{A})$
(R4) $\vdash \mathcal{A}$ implies $\vdash \Box\mathcal{A}$

The cut theorem is true of $LLR^{\Diamond\Box}$, which facilitates the proof of the equivalence of the sequent and axiomatic formulations as well as the proof of decidability.

Theorem 21. (Cut theorem for $LLR^{\Diamond\Box}$) *The cut rule is* admissible *in* $LLR^{\Diamond\Box}$.

Proof. The proof proceeds as usual. An important observation is that in the transformations of proofs no other rules are used than those already used. ◁

The modalities do not appear to be too intricate — even if they do not have all the usual properties that $\Box$ and $\Diamond$ have in S4. The latter logic (more precisely, the propositional part of S4) is known to be decidable. Therefore, we may wonder whether we can adapt and extend the proof of the decidability of LLR^c to $LLR^{\Diamond\Box}$.

Definition 22. We define the *sequent calculus* denoted as $[LLR^{\Diamond\Box}]$ by taking $[LLR^c]$, and by adding the following connective rules for the modalities.

$$\frac{\alpha; \mathcal{A} \vdash \beta}{[\alpha; \Box\mathcal{A}] \vdash \beta} \; [\Box\vdash] \qquad \frac{\Box\alpha \vdash \mathcal{A}; \Diamond\beta}{\Box\alpha \vdash \Box\mathcal{A}; \Diamond\beta} \; [\vdash\Box] \qquad \frac{\Box\alpha; \mathcal{A} \vdash \Diamond\beta}{\Box\alpha; \Diamond\mathcal{A} \vdash \Diamond\beta} \; [\Diamond\vdash] \qquad \frac{\alpha \vdash \mathcal{A}; \beta}{\alpha \vdash [\Diamond\mathcal{A}; \beta]} \; [\vdash\Diamond]$$

We assume some earlier notions and notational conventions in an obvious way. Two of the rules have no bracketing at all, whereas the two others are of type 1.

Theorem 23. (Cut theorem for $[LLR^{\Diamond\Box}]$) *The cut theorem is admissible in* $[LLR^{\Diamond\Box}]$.

Proof. The proof proceeds as usual.[11] Here is a sample case from the transformation, where $\Box C$ is the cut formula. If $\rho = 2$, then $\Box C$ could not have been contracted in $[\Box\vdash]$.

$$\frac{\dfrac{\Box\alpha \vdash C; \Diamond\beta}{\Box\alpha \vdash \Box C; \Diamond\beta} \quad \dfrac{\gamma; C \vdash \delta}{\gamma; \Box C \vdash \delta}}{[\Box\alpha; \gamma] \vdash [\Diamond\beta; \delta]} \quad \rightsquigarrow \quad \frac{\Box\alpha \vdash C; \Diamond\beta \quad \gamma; C \vdash \delta}{[\Box\alpha; \gamma] \vdash [\Diamond\beta; \delta]}$$

If the right rank $\rho_r > 1$ and all the contractions in the original proof resulted from the application of the cut, then the above transformation suffices. Otherwise, that is, if $\Box C$, the principal formula of the $[\Box\vdash]$ rule, was contracted as part of the application of the rule, then we consider the number of occurrences of $\Box C$ in γ. If there are several occurrences in γ, then we permute the applications of the cut rule and of the $[\Box\vdash]$ rule. If there is only one occurrence of $\Box C$, then beyond the permutation, we also include a cut on the subaltern (which is of lower degree than $\Box C$). Here are the resulting chunks of the proof. ($n \in \mathbb{N}$ and $n > 1$.)

$$\frac{\dfrac{\Box\alpha \vdash C; \Diamond\beta}{\Box\alpha \vdash \Box C; \Diamond\beta} \quad \gamma'; (\Box C)^n; C \vdash \delta}{\dfrac{[\Box\alpha; \gamma'; (\Box C)^{n-1}; C] \vdash [\Diamond\beta; \delta]}{[\Box\alpha; \gamma'; (\Box C)^{n-1}] \vdash [\Diamond\beta; \delta]}}$$

---

[11] Some details of a related proof are given in [8, §2].

$$\cfrac{\Box\alpha \vdash C; \Diamond\beta}{[\Box\alpha;\gamma] \vdash [\Diamond\beta;\delta]} \qquad \cfrac{\cfrac{\vdots}{\Box\alpha \vdash C; \Diamond\beta}}{\cfrac{\Box\alpha \vdash \Box C; \Diamond\beta}{[\Box\alpha;\gamma;C] \vdash [\Diamond\beta;\delta]}} \cfrac{\vdots}{\gamma';\Box C; C \vdash \delta}$$

$\triangleleft$

Next, we prove Curry's lemma, which is sometimes called the height-preserving admissibility of contraction.

Lemma 24. (Curry's lemma for [$LLR^{\Diamond\Box}$]) *If $\alpha \vdash \beta$ has a proof in [$LLR^{\Diamond\Box}$] with the height of the proof tree being n, and $\gamma \vdash \delta$ results from $\alpha \vdash \beta$ by one or more applications of the rules $(W \vdash)$ and $(\vdash W)$, then $\gamma \vdash \delta$ has a proof in [$LLR^{\Diamond\Box}$], where the height of the proof tree is not greater than n (i.e., it is $\leq n$).*

Proof. The proof of this lemma seamlessly incorporates the proof of Lemma 15. We have four new rules — compared to [LLR^e]. Two of those do not permit contractions, hence, any contractions that could be applied to the lower sequent of those rules are guaranteed to exist by the hypothesis of the induction. We consider the remaining two rules, which expand case **2.2**.

2.2. If the last rule applied in the given proof is $[\Box \vdash]$, then $\Box A$ may be contracted, provided that it already occurs in the antecedent, that is, in α. We consider B and C as other wff's that potentially could be contracted. The following is an illustration of a representative case, though concretely, there might be fewer or more formulas that could be contracted.

$$\cfrac{\cfrac{\vdots}{\alpha';B;B;\Box A;A \vdash C;C;\beta'}}{[\alpha';B;B;\Box A;\Box A] \vdash C;C;\beta'} \qquad \overset{i.h.}{\leadsto} \qquad \cfrac{\cfrac{\vdots}{\alpha';B;\Box A;A \vdash C;\beta'}}{[\alpha';B;\Box A;\Box A] \vdash C;\beta'}$$

B and C can be contracted above the application of the $[\Box \vdash]$ rule, by the inductive hypothesis, and $\alpha';B;\Box A$ can be obtained using $[\Box \vdash]$.

If the last rule applied in the proof is $[\vdash \Diamond]$, then we have a dual situation. Here is the given segment, and the new chunk.

$$\cfrac{\cfrac{\vdots}{\alpha';A;A \vdash B;\Diamond B;C;C;\beta'}}{\alpha';A;A \vdash [\Diamond B;\Diamond B;C;C;\beta']} \qquad \overset{i.h.}{\leadsto} \qquad \cfrac{\cfrac{\vdots}{\alpha';A \vdash B;\Diamond B;C;\beta'}}{\alpha';A \vdash [\Diamond B;\Diamond B;C;\beta']}$$

Clearly, the height of the proof tree does not increase in either case. $\triangleleft$

Theorem 25. (Decidability for $LLR^{\Diamond\Box}$) *The logic $LLR^{\Diamond\Box}$ is decidable.*

Proof. The proof of this theorem proceeds like the proof of Theorem 19. We have two sequent calculuses, [$LLR^{\Diamond\Box}$] and $LLR^{\Diamond\Box}$, in which the same theorems are provable. Additionally, we have proved Curry's lemma for [$LLR^{\Diamond\Box}$]. The whole structure of the proof is the same as before, that is, it is through performing an exhaustive search in a finite search space. $\triangleleft$

5. LOGICS IN THE NEIGHBORHOOD OF LR$^{\Diamond\Box}$

If we keep the four connective rules for $\Diamond$ and $\Box$ fixed, then we may wonder about the effects of the inclusion of the contraction or the thinning rules, or of their modalized versions (like those in CLL). In particular, the next proof suggests the usefulness of the modalized thinning rules with $LLR^{\Diamond\Box}$. We labeled the steps where

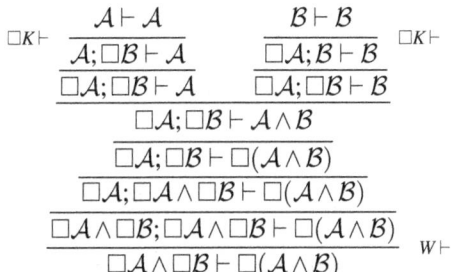

FIGURE 1. A proof of the distributivity of $\Box$ over $\wedge$

structural rules (modalized or plain ones) are applied. The proof sort of "explains" why the bottom sequent is not provable in $LLR^{\Diamond\Box}$ or in $LCLL$ (with ! instead of $\Box$ in other notation). The logic $LLR^{\Diamond\Box}$ has no thinning (except for t and f), whereas $LCLL$ does not have plain contraction. More contemplation of the proof allows us to conclude that $!\mathcal{A} \otimes !\mathcal{B} \vdash !(\mathcal{A} \& \mathcal{B})$ is provable in $LCLL$ because of $(!K\vdash)$, hence, $\Box\mathcal{A} \circ \Box\mathcal{B} \vdash \Box(\mathcal{A} \wedge \mathcal{B})$ is provable in $LBCK^{\Diamond\Box}$, for example.

$LBCK^{\Diamond\Box}$ is $LLRW^{\Diamond\Box}$ with left and right thinning rules. The letters B, C and K are motivated by the provability of the principal (simple) type schemas of the combinators B, C and K in the implicational fragment of $LBCK^{\Diamond\Box}$. If we add modalized contraction rules, then we get $LBCK^{\Diamond\Box}_{\Box\Diamond W}$, which is also known as *affine linear logic*, and it had been proved decidable by Alexei P. Kopylov, in 1995 (see [29]), using normal sequents and vector games. We give a new proof of the decidability of $LBCK^{\Diamond\Box}_{\Box\Diamond W}$ (which is conceptually different), in the second half of this section. Our proof shows that the modalization of the contraction rules does not destroy decidability. (The modalization of the thinning rules is absolutely unproblematic.)

The converse of the previously considered sequent, $!(\mathcal{A}\&\mathcal{B}) \vdash !\mathcal{A} \otimes !\mathcal{B}$ is also provable in $LCLL$, because of $(!W\vdash)$, hence, $\Box(\mathcal{A} \wedge \mathcal{B}) \vdash \Box\mathcal{A} \circ \Box\mathcal{B}$ is provable in $LLR^{\Diamond\Box}$. Thus, some of the prototypical sequents provable in $LCLL$ are the next four ones (where $\alpha \dashv\vdash \beta$ indicates that both $\alpha \vdash \beta$ and $\beta \vdash \alpha$ are provable).

$$!\mathcal{A} \otimes !\mathcal{B} \dashv\vdash !(\mathcal{A} \& \mathcal{B}) \qquad !(!\mathcal{A} \& !\mathcal{B}) \dashv\vdash !(\mathcal{A} \& \mathcal{B})$$

The proof in Figure 1 also shows that the distribution of $\Box$ over $\wedge$ is provable in $LLR^{\Diamond\Box}_{\Box\Diamond K}$, that is, $LLR^{\Diamond\Box}$ extended with a pair of modalized thinning rules. (We pointed out on page 104 that the sequent $\Box(\mathcal{A} \wedge \mathcal{B}) \vdash \Box\mathcal{A} \wedge \Box\mathcal{B}$ is always provable.)

Figure 2 (below) shows seven logics that we consider in varying details. The arrows indicate that the set of axioms and rules of a logic is a proper subset of the set of axioms and rules of another one (assuming the identity translation throughout). As a result, inclusions between sets of theorems of those logics also obtain, and they can be shown to be proper.

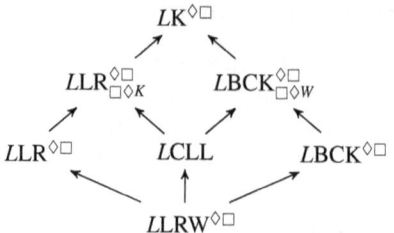

FIGURE 2. Seven logics with modalities

For the sake of clarity, the non-modalized thinning rules are the following rules:

$$\frac{\alpha \vdash \beta}{\alpha; \mathcal{A} \vdash \beta} \, K\vdash \qquad \frac{\alpha \vdash \beta}{\alpha \vdash \mathcal{A}; \beta} \, \vdash K$$

The subscripts $\Box\Diamond K$ and $\Box\Diamond W$ in the labels of some logics indicate the addition of a pair of the modalized structural rules $(\Box K \vdash)$ and $(\vdash \Diamond K)$, or $(\Box W \vdash)$ and $(\vdash \Diamond W)$.

Definition 26. *The modalized thinning and contraction rules are as follows.*

$$\frac{\alpha \vdash \beta}{\alpha; \Box \mathcal{A} \vdash \beta} \, \Box K\vdash \qquad \frac{\alpha \vdash \beta}{\alpha \vdash \Diamond \mathcal{A}; \beta} \, \vdash \Diamond K \qquad \frac{\alpha; \Box \mathcal{A}; \Box \mathcal{A} \vdash \beta}{\alpha; \Box \mathcal{A} \vdash \beta} \, \Box W\vdash \qquad \frac{\alpha \vdash \Diamond \mathcal{A}; \Diamond \mathcal{A}; \beta}{\alpha \vdash \Diamond \mathcal{A}; \beta} \, \vdash \Diamond W$$

First of all, we should note that $LK^{\Diamond\Box}$ is a *baroque logic*, because it has duplicate symbols for two of its connectives, namely, for $\wedge$ and $\vee$ (or for $\circ$ and $+$). The two pairs of the zero-ary constants also match as T and t, and F and f. Furthermore, they are definable by the lattice connectives and $\sim$. Unlike in the six other logics, $\wedge$ and $\vee$ distribute over each other. In sum, $LK^{\Diamond\Box}$ is classical logic with $\Box$ and $\Diamond$, which are S4-type modalities; that is, $LK^{\Diamond\Box}$ is a notational variant of S4.

Normality means that $\Diamond$ preserves F, or dually, $T \to \Box T$ is a theorem. It is not difficult to check that if $(\vdash K)$ and $(K\vdash)$ are in one of those logics, then both obtain, and vice versa. Thus, normality is a feature of modalities already in $LBCK^{\Diamond\Box}$. The modal operators are *monotone*, and this is their feature in all seven logics. The proof of *additivity* of $\Diamond$ requires contraction and (modalized) thinning, whereas the *normality* of $\Diamond$, as we already mentioned, requires thinning. Thus, the modalities have some S4ish properties — reflected by (A18)–(A21) and (R4) — in all seven logics, but $\Diamond$ is normal and additive, and $\Box$ has the dual of both properties only in $LK^{\Diamond\Box}$.

Propositional S4 is known to be *decidable*, and this remains true the duplicate symbols notwithstanding. The decidability of $LLRW^{\Diamond\Box}$ and of $LBCK^{\Diamond\Box}$ is immediate (because neither calculus has any contraction rules), and we have shown that $LLR^{\Diamond\Box}$ is decidable. Three other logics are left to consider. First, we focus on $LLR^{\Diamond\Box}_{\Box\Diamond K}$ and $LBCK^{\Diamond\Box}_{\Box\Diamond W}$, then we turn to $LCLL$.

The sequent calculus $[LLR^{\Diamond\Box}_{\Box\Diamond K}]$ is an extension of the sequent calculus $[LLR^{\Diamond\Box}]$ by the two modalized thinning rules. There is no contraction included in those rules. The rationale is the same as with the $(t \vdash)$ and $(\vdash f)$ rules, which may be viewed as special thinning rules. Namely, if the principal formula would be contracted, then an application of the rule may be simply omitted.

Theorem 27. (Cut theorem for $[LLR^{\Diamond\Box}_{\Box\Diamond K}]$) *The cut rule is* admissible *in* $[LLR^{\Diamond\Box}_{\Box\Diamond K}]$.

Proof. The proof extends the proof of the cut theorem for [LLR$^{\Diamond\Box}$]. We consider one new case in detail, when $\rho = 2$ and the right premise is by ($\Box K \vdash$) and $\Box C$ is the cut formula. The transformation ensures that $\Box C$ (occurring in the succedent of the left premise) disappears without an application of the cut rule.

$$\frac{\frac{\vdots}{\Box\alpha \vdash C; \Diamond\beta}}{\frac{\Box\alpha \vdash \Box C; \Diamond\beta \quad \gamma; \Box C \vdash \delta}{[\Box\alpha; \gamma] \vdash [\Diamond\beta; \delta]}} \quad \rightsquigarrow \quad \frac{\frac{\vdots}{\gamma \vdash \delta}}{\frac{\Box\alpha; \gamma \vdash \delta}{[\Box\alpha; \gamma] \vdash [\Diamond\beta; \delta]}}$$

The case, in which the pair of rules is $\langle(\vdash \Diamond K), (\Diamond \vdash)\rangle$, is the dual of this.

If the principal formula of the modalized thinning rules does not coincide with the cut formula, then the cut may be permuted upward without difficulty, because there are no side conditions on the applicability of the modalized thinning rules. (We omit the remaining details.) ◁

The cut theorem ensures the subformula property in cut-free proofs. The following lemma is preeminent for decidability.

Lemma 28. (**Curry's lemma for [LLR$^{\Diamond\Box}_{\Box\Diamond K}$]**) *If $\alpha \vdash \beta$ has a proof in [LLR$^{\Diamond\Box}_{\Box\Diamond K}$] with the height of the proof tree being n, and $\gamma \vdash \delta$ results from $\alpha \vdash \beta$ by one or more applications of the rules ($W \vdash$) and ($\vdash W$), then $\gamma \vdash \delta$ has a proof in [LLR$^{\Diamond\Box}_{\Box\Diamond K}$], where the height of the proof tree is not greater than n (i.e., it is $\leq n$).*

Proof. Once again, we suppose the proof for the logic [LLR$^{\Diamond\Box}$]. We have to extend the inductive step, namely, case **2.2**. There are two new rules, and we consider each.
2.2. Let us assume that there are some parametric wff's, $\mathcal{A}$ and $\mathcal{B}$, which have multiple occurrences that could be contracted, but $\Box C$, the principal formula of the ($\Box K \vdash$) rule is not among the contractable formulas. Then we have the following.

$$\frac{\alpha'; \mathcal{A}; \mathcal{A} \vdash \mathcal{B}; \mathcal{B}; \mathcal{B}; \beta'}{\alpha'; \mathcal{A}; \mathcal{A}; \Box C \vdash \mathcal{B}; \mathcal{B}; \mathcal{B}; \beta'} \quad \overset{i.h.}{\rightsquigarrow} \quad \frac{\alpha'; \mathcal{A} \vdash \mathcal{B}; \beta'}{\alpha'; \mathcal{A}; \Box C \vdash \mathcal{B}; \beta'}$$

It could happen that $\Box C$ already has some occurrences in the premise. Although the rule does not have any built-in contraction, the resulting sequent could be contracted. Here is an example.

$$\frac{\alpha'; \mathcal{A}; \mathcal{A}; \mathcal{A}; \Box C; \Box C \vdash \mathcal{B}; \mathcal{B}; \beta'}{\alpha'; \mathcal{A}; \mathcal{A}; \mathcal{A}; \Box C; \Box C; \Box C \vdash \mathcal{B}; \mathcal{B}; \beta'} \quad \overset{i.h.}{\rightsquigarrow} \quad \alpha'; \mathcal{A}; \Box C \vdash \mathcal{B}; \beta'$$

Dually, we have two possibilities with the ($\vdash \Diamond K$) rule. (We use two copies of $\mathcal{A}$ and $\mathcal{B}$ in these proof segments.)

$$\frac{\alpha'; \mathcal{A}; \mathcal{A} \vdash \mathcal{B}; \mathcal{B}; \beta'}{\alpha'; \mathcal{A}; \mathcal{A} \vdash \Diamond C; \mathcal{B}; \mathcal{B}; \beta'} \quad \overset{i.h.}{\rightsquigarrow} \quad \frac{\alpha'; \mathcal{A} \vdash \mathcal{B}; \beta'}{\alpha'; \mathcal{A} \vdash \Diamond C; \mathcal{B}; \beta'}$$

$$\frac{\alpha'; \mathcal{A}; \mathcal{A} \vdash \Diamond C; \mathcal{B}; \mathcal{B}; \beta'}{\alpha'; \mathcal{A}; \mathcal{A} \vdash \Diamond C; \Diamond C; \mathcal{B}; \mathcal{B}; \beta'} \quad \overset{i.h.}{\rightsquigarrow} \quad \alpha'; \mathcal{A} \vdash \Diamond C; \mathcal{B}; \beta'$$

The height of the proof does not increase in any of the cases. ◁

Theorem 29. *The logic* $[LLR^{\Diamond\Box}_{\Box\Diamond K}]$ *is decidable.*

Proof. The proof proceeds as before, hence, we skip the details here. ◁

The sequent calculus $LBCK^{\Diamond\Box}_{\Box\Diamond W}$ is defined by adding the modalized contraction rules to $LBCK^{\Diamond\Box}$, and it differs from $LK^{\Diamond\Box}$, which has non-modalized contractions. As we noted, $LK^{\Diamond\Box}$'s language could be simplified, however, for our purposes now it is useful to retain both the extensional (i.e., lattice) connectives and the intensional (including the modal) connectives.

We have noted also that $LK^{\Diamond\Box}$ is decidable. In particular, the decidability of $LK^{\Diamond\Box}$ can be proved along the lines of the decidability proofs of $LLR^{\Diamond\Box}$ and $LLR^{\Diamond\Box}_{\Box\Diamond K}$. The presence of thinning (modalized or plain) does not constitute a problem at all, because it does not even require contraction to be built into the thinning rules in the contraction-free version of the sequent calculus. $[LK^{\Diamond\Box}]$ is defined as $[LLR^{\Diamond\Box}]$ with the full left and right thinning rules added. Definition 8 does not mention (explicitly) a calculus, hence, we may use the same notion here with the assumption that the heap numbers for $LBCK^{\Diamond\Box}_{\Box\Diamond W}$ are calculated from the Curry–Kripke decision procedure for $LK^{\Diamond\Box}$.

Theorem 30. *The logic* $LBCK^{\Diamond\Box}_{\Box\Diamond W}$ *is decidable.*

Proof. Given a wff $\mathcal{A}$, we can determine if the wff is a theorem of $LK^{\Diamond\Box}$; if it is not, then $\mathcal{A}$ is not a theorem of $LBCK^{\Diamond\Box}_{\Box\Diamond W}$ either. On the other hand, we can also determine if $\mathcal{A}$ is a theorem of $LBCK^{\Diamond\Box}$; if it is, then it is a theorem of $LBCK^{\Diamond\Box}_{\Box\Diamond W}$ too. We apply a proof search procedure to the remaining wff's.

We construct a proof-search tree in $LBCK^{\Diamond\Box}_{\Box\Diamond W}$ taking into account the heap numbers for the subformulas of $\mathcal{A}$ as upper bounds on the number of applications of the $(\Box W \vdash)$ and $(\vdash \Diamond W)$ rules. The resulting tree will be finite, because there are no other contractions than those that are instances of the modalized contraction rules, and the number of their applications is bounded by the heap numbers, which are finite numbers. ◁

6. The Decidability of Linear Logic

We have already proved that classical linear logic (CLL) is decidable — as Theorem 10. CLL has a certain familiarity to many people, and it had been claimed to be undecidable in [35] (see Theorem 3.7) and in [28] (see Corollaries 5.5 and 5.7) We think though that those proofs fall short of establishing the undecidability of CLL. Since the undecidability of CLL is widely believed in the computer science community, we give a more direct proof (than the previous proof) for the decidability of CLL.

To make the reading of this proof easier for those in the linear logic community, we define a sequent calculus, which we call $[\![LCLL]\!]$, and we use Girard's notation. $[\![LCLL]\!]$ *is not* classical linear logic though. (A careful reader will recognize this logic as $[LLR^{\Diamond\Box}_{\Box\Diamond K}]$ in non-standard notation.)

The notion of a sequent is as before; a *sequent* is a pair of multisets of wff's separated by $\vdash$. We use both single and double bracketing in this calculus for permissible contractions that are built into the operational rules. For some purposes the single and the double bracketing might be treated as the same (just blur your vision). But as we shall explain after we state the rules, the double brackets sometimes mark a crucial distinction.

Definition 31. $[\![LCLL]\!]$ comprises the following axioms and rules.

$$\alpha;\mathbf{0} \vdash \beta \quad {}_{0\vdash} \qquad \mathcal{A} \vdash \mathcal{A} \quad {}_{\text{id}} \qquad \alpha \vdash \top;\beta \quad {}_{\vdash\top}$$

$$\bot \vdash \quad {}_{\bot\vdash} \qquad \frac{\alpha \vdash \beta}{\alpha \vdash \bot;\beta} \, {}_{\vdash\bot} \qquad \frac{\alpha \vdash \beta}{\alpha;\mathbf{1} \vdash \beta} \, {}_{\mathbf{1}\vdash} \qquad \vdash \mathbf{1} \quad {}_{\vdash\mathbf{1}}$$

$$\frac{\alpha;\mathcal{A} \vdash \beta}{[\alpha;\mathcal{A}\,\&\,\mathcal{B}] \vdash \beta} \, {}_{[\&\vdash_1]} \qquad \frac{\alpha;\mathcal{A} \vdash \beta}{[\alpha;\mathcal{B}\,\&\,\mathcal{A}] \vdash \beta} \, {}_{[\&\vdash_2]} \qquad \frac{\alpha \vdash \mathcal{A};\beta \quad \alpha \vdash \mathcal{B};\beta}{\alpha \vdash [\mathcal{A}\,\&\,\mathcal{B};\beta]} \, {}_{[\vdash\&]}$$

$$\frac{\alpha;\mathcal{A} \vdash \beta \quad \alpha;\mathcal{B} \vdash \beta}{[\alpha;\mathcal{A}\oplus\mathcal{B}] \vdash \beta} \, {}_{[\oplus\vdash]} \qquad \frac{\alpha \vdash \mathcal{A};\beta}{\alpha \vdash [\mathcal{A}\oplus\mathcal{B};\beta]} \, {}_{[\vdash\oplus_1]} \qquad \frac{\alpha \vdash \mathcal{A};\beta}{\alpha \vdash [\mathcal{B}\oplus\mathcal{A};\beta]} \, {}_{[\vdash\oplus_2]}$$

$$\frac{\alpha \vdash \mathcal{A};\beta}{[\alpha;\mathcal{A}^\bot] \vdash \beta} \, {}_{[{}^\bot\vdash]} \qquad \frac{\alpha;\mathcal{A} \vdash \beta}{\alpha \vdash [\mathcal{A}^\bot;\beta]} \, {}_{[\vdash{}^\bot]}$$

$$\frac{\alpha;\mathcal{A};\mathcal{B} \vdash \beta}{[\alpha;\mathcal{A}\otimes\mathcal{B}] \vdash \beta} \, {}_{[\otimes\vdash]} \qquad \frac{\alpha \vdash \mathcal{A};\beta \quad \gamma \vdash \mathcal{B};\delta}{[\alpha;\gamma] \vdash [\mathcal{A}\otimes\mathcal{B};\beta;\delta]} \, {}_{[\vdash\otimes]}$$

$$\frac{\alpha;\mathcal{A} \vdash \beta \quad \gamma;\mathcal{B} \vdash \delta}{[\alpha;\gamma;\mathcal{A}\,\mathfrak{V}\,\mathcal{B}] \vdash [\beta;\delta]} \, {}_{[\mathfrak{V}\vdash]} \qquad \frac{\alpha \vdash \mathcal{A};\mathcal{B};\beta}{\alpha \vdash [\mathcal{A}\,\mathfrak{V}\,\mathcal{B};\beta]} \, {}_{[\vdash\mathfrak{V}]}$$

$$\frac{\alpha \vdash \mathcal{A};\beta \quad \gamma;\mathcal{B} \vdash \delta}{[\alpha;\gamma;\mathcal{A}\multimap\mathcal{B}] \vdash [\beta;\delta]} \, {}_{[\multimap\vdash]} \qquad \frac{\alpha;\mathcal{A} \vdash \mathcal{B};\beta}{\alpha \vdash [\mathcal{A}\multimap\mathcal{B};\beta]} \, {}_{[\vdash\multimap]}$$

$$\frac{\alpha;\mathcal{A} \vdash \beta}{[\alpha;!\mathcal{A}] \vdash \beta} \, {}_{[!\vdash]} \qquad \frac{!\alpha \vdash \mathcal{A};?\beta}{!\alpha \vdash !\mathcal{A};?\beta} \, {}_{\vdash!}$$

$$\frac{!\alpha;\mathcal{A} \vdash ?\beta}{!\alpha;?\mathcal{A} \vdash ?\beta} \, {}_{?\vdash} \qquad \frac{\alpha \vdash \mathcal{A};\beta}{\alpha \vdash [\![?\mathcal{A};\beta]\!]} \, {}_{[\vdash?]}$$

$$\frac{\alpha \vdash \beta}{\alpha;!\mathcal{A} \vdash \beta} \, {}_{!K\vdash} \qquad \frac{\alpha \vdash \beta}{\alpha \vdash ?\mathcal{A};\beta} \, {}_{\vdash?K}$$

To start with, the brackets (whether single or double) indicate optional contractions as in Definition 12. Then $\|LCLL\|$ is equivalent to $[LLR^{\Box\Box}_{\Box\Diamond K}]$. We may weaken the logic in two different ways, each time getting CLL. First, we may forget about all the brackets and add the rules ($\Box W \vdash$) and ($\vdash \Diamond W$) (with ! for $\Box$ and ? for $\Diamond$). This is the calculus that we denote by $LCLL$. Second, we can omit the single brackets and change the meaning of the double brackets as follows.

2. If $!\mathcal{A}$ occurs both in α and γ, then it may be contracted in $[\![\alpha;\gamma]\!]$. Dually, if $?\mathcal{A}$ occurs both in β and δ, then it may be contracted in $[\![\beta;\delta]\!]$. The principal formula cannot be involved in the contraction.
3. In $[\![!\vdash]\!]$ and $[\![\vdash ?]\!]$, $!\mathcal{A}$ and $?\mathcal{A}$ may be contracted, respectively, if it occurs in α and β.

Obviously, the scope of $[\![\ ]\!]$ could be made narrower in the rules where the main connective of the principal formula of the rule is binary. We denote the logic obtained by omitting the single brackets as $[\![LCLL]\!]$.

Now we prove some useful theorems about the calculus $[\![LCLL]\!]$. Namely, every theorem of $LCLL$ is a theorem of $[\![LCLL]\!]$, and every theorem of $[\![LCLL]\!]$ has a cut-free proof (by Lemma 32). A suitable version of Curry's lemma (Lemma 33) holds too.

Lemma 32. (Cut theorem for $[\![LCLL]\!]$) *The cut rule is* admissible *in* $[\![LCLL]\!]$.

Proof. The proof is by *double induction* on the rank of the cut and the degree of the cut formula. The rank of the cut (ρ) is defined as in Gentzen [22], and the degree of the cut formula (δ) is the number of unary and binary logical connectives in the cut formula. We divide the cases within the induction into four groups, and provide some representative details.

I. Let $\delta = 0$ and $\rho = 2$. The cut formula is (1) a propositional variable (e.g., p), (2) **1**, (3) $\bot$, (4) $\top$ or (5) **0**. None of these formulas can be thinned into a sequent by the rules ($!K \vdash$) or ($\vdash ?K$), hence, both premises are by an axiom or by a rule for **1** or $\bot$. There are various ways to count the subcases; either way there are several cases, and it is straightforward to verify that the cut is directly eliminable. We give two sample cases here.

$$\cfrac{\vdash \mathbf{1} \qquad \cfrac{\vdots}{\cfrac{\alpha \vdash \beta}{\mathbf{1}; \alpha \vdash \beta}}}{\alpha \vdash \beta}$$

The proof of the premise of the application of the $(\mathbf{1} \vdash)$ rule is identical to the end sequent, hence, the cut may be omitted altogether.

$$\cfrac{\alpha \vdash \beta; \top; p \qquad p; \mathbf{0}; \gamma \vdash \delta}{\alpha; \mathbf{0}; \gamma \vdash \beta; \top; \delta}$$

The end sequent is an instance of $(\mathbf{0} \vdash)$ and also of $(\vdash \top)$, hence, both premises of the cut (and the cut itself) may be omitted.

II. Let $\delta = 0$ and $\rho > 2$, in particular, let $\rho_l > 1$. We note that the left premise cannot be the result of an application of the $(\vdash !)$ or $(? \vdash)$ rules. Furthermore, if it is by a rule for $\bot$, $\otimes$, $\invamp$, $\multimap$, $\&$, $\oplus$, **1** or $\bot$, then the principal formula cannot be contracted as part of the application of the cut. It is routine to check that the rule yielding the left premise and the cut may be permuted, and the contractions included in the given proof may be carried out after the rules have been swapped.

Let the left premise be by ($! \vdash$). The given and the transformed proof segments are as follows.

$$\cfrac{\cfrac{\vdots}{\cfrac{\mathcal{A}; \alpha \vdash \beta; p}{[\![!\mathcal{A}; \alpha]\!] \vdash \beta; p}} \qquad \cfrac{\vdots}{p; \gamma \vdash \delta}}{[\![!\mathcal{A}; \alpha; \gamma]\!] \vdash [\![\beta; \delta]\!]} \quad \leadsto \quad \cfrac{\cfrac{\mathcal{A}; \alpha \vdash \beta; p \qquad p; \gamma \vdash \delta}{[\mathcal{A}; \alpha; \gamma] \vdash [\beta; \delta]}}{[\![!\mathcal{A}; \alpha; \gamma]\!] \vdash [\![\beta; \delta]\!]}$$

If the application of the $[\![! \vdash]\!]$ rule involved a contraction of $!\mathcal{A}$, then the same contraction may be performed in the transformed proof too. (The case of $[\![\vdash ?]\!]$ is dually similar.)

Let the left premise be by $[\![\vdash ?K]\!]$. The given and the transformed proof segments are as follows.

$$\frac{\frac{\vdots}{\alpha \vdash \beta; p}\quad \vdots}{\frac{\alpha \vdash \beta; ?\mathcal{A}; p \quad p; \gamma \vdash \delta}{[\![\alpha; \gamma]\!] \vdash [\![\beta; \delta; ?\mathcal{A}]\!]}} \quad \rightsquigarrow \quad \frac{\frac{\vdots}{\alpha \vdash \beta; p} \quad \frac{\vdots}{p; \gamma \vdash \delta}}{\frac{[\![\alpha; \gamma]\!] \vdash [\![\beta; \delta]\!]}{[\![\alpha; \gamma]\!] \vdash [\![\beta; \delta; ?\mathcal{A}]\!]}}$$

If $?\mathcal{A}$ was contracted in the given proof as part of the application of the cut rule, then the last step is omitted from the transformed proof. All other contractions can be carried out as in the given proof.

$[\![LCLL]\!]$ is fully symmetric — save the $\multimap$ rules, which however, are unproblematic — when the connectives are dualized. Thus, we leave the details of the $\rho_r > 1$ case to the reader.

III. Let $\delta > 0$ and $\rho = 2$. We distinguish two groups of subcases, namely, when one of the premises is by (id) or by an axiom for $\top$ or $\mathbf{0}$, and when the two premises are by matching rules. The case when a premise is $\mathcal{A} \vdash \mathcal{A}$ is immediate. As an example, we consider $\langle (\vdash \top), [\![! \vdash]\!] \rangle$.

$$\frac{\alpha \vdash \beta; \top; !\mathcal{A} \quad \frac{\vdots}{\mathcal{A}; \gamma \vdash \delta} }{[\![\alpha; \gamma]\!] \vdash [\![\beta; \top; \delta]\!]}$$

The bottom sequent is an instance of $(\vdash \top)$, hence, the proof simplifies to that sequent.

If the principal formulas in the rules in the left and right premises have as their main connective $\bot$, $\otimes$, $\invamp$ or $\multimap$, then the transformed proof contains cuts on proper subformulas of the principal formula. The principal formula may not be contracted as part of the application of the cut rule in the given proof. Further, the parametric formulas are combined in the transformed proof in the same way as in the given proof; therefore, all the earlier contractions can be carried out. (We omit the details.)

There are four subcases with modalized cut formulas, because such formulas may be introduced by thinning too. We consider two of these cases, and leave the two others (which are duals) to the reader.

$$\frac{\frac{\vdots}{!\alpha \vdash ?\beta; \mathcal{A}}}{\frac{!\alpha \vdash ?\beta; !\mathcal{A} \quad \frac{\vdots}{!\mathcal{A}; \gamma \vdash \delta}}{[\![\,!\alpha; \gamma]\!] \vdash [\![?\beta; \delta]\!]}} \quad \rightsquigarrow \quad \frac{\frac{\vdots}{!\alpha \vdash ?\beta; \mathcal{A}} \quad \frac{\vdots}{\mathcal{A}; \gamma \vdash \delta}}{[\![\,!\alpha; \gamma]\!] \vdash [\![?\beta; \delta]\!]}$$

The transformation decreases the degree of the cut formula, and provides a possibility for the same contractions as before.

$$\frac{\frac{\vdots}{\alpha \vdash \beta}}{\frac{\alpha \vdash \beta; ?\mathcal{A} \quad \frac{\vdots}{?\mathcal{A}; !\gamma \vdash ?\delta}}{[\![\alpha; !\gamma]\!] \vdash [\![\beta; ?\delta]\!]}} \quad \rightsquigarrow \quad \frac{\frac{\vdots}{\alpha \vdash \beta}}{[\![\alpha; !\gamma]\!] \vdash [\![\beta; ?\delta]\!]}$$

In the transformed proof, the double brackets simply indicate that the $(!K \vdash)$ and $(\vdash ?K)$ steps are applied only to build up the same sequent as the bottom sequent in the given proof. (The thinning rules do not contain any contraction.) It may be useful to point out that $[\![\alpha; !\gamma]\!] \subsetneq \alpha$ and $[\![\beta; ?\delta]\!] \subsetneq \beta$ are not possible, hence, we are justified to start the transformed proof with the premise $\alpha \vdash \beta$.

IV. Let $\delta > 0$ and $\rho > 2$, in particular, let $\rho_l > 1$.

Most of the subcases in this case are similar to those in **II**. (We omit the details of those cases, where the change amounts to replacing p with $\mathcal{A}$.) Now an additional possibility is that the left premise is by $(\vdash !)$ or $(? \vdash)$. The side conditions of the rules together with $\rho_l > 1$ imply that the principal formula of either rule is not the cut formula. We consider in detail the case when the left premise is by $(\vdash !)$; the other rule may be dealt with similarly.

If $\rho_r = 1$, then the only possibility (beyond an axiom) is that the right premise is by $(? \vdash)$.

$$\frac{!\alpha \vdash ?\beta; ?\mathcal{C}; \mathcal{A} \quad \mathcal{C}; !\gamma \vdash ?\delta}{\frac{!\alpha \vdash ?\beta; ?\mathcal{C}; !\mathcal{A} \quad ?\mathcal{C}; !\gamma \vdash ?\delta}{[!\alpha; !\gamma] \vdash [?\beta; ?\delta; !\mathcal{A}]}} \quad \rightsquigarrow \quad \frac{!\alpha \vdash ?\beta; ?\mathcal{C}; \mathcal{A} \quad \frac{\mathcal{C}; !\gamma \vdash ?\delta}{?\mathcal{C}; !\gamma \vdash ?\delta}}{\frac{[!\alpha; !\gamma] \vdash [?\beta; ?\delta; \mathcal{A}]}{[!\alpha; !\gamma] \vdash [?\beta; ?\delta; !\mathcal{A}]}}$$

The transformation is justified by a decrease in ρ_l.

If $\rho_r > 1$, then the right premise cannot be by $(\vdash !)$ or $(? \vdash)$ due to the shape of the cut formula and the side conditions in those rules. If the right premise is by a rule for $\bot, \otimes, \multimap, \mathfrak{N}, \mathbf{1}$ or $\bot$, then the cut is moved upward and the transformation is justified by a decrease in ρ_r.

The remaining possibilities are that the right premise is by $[\![\,! \vdash\,]\!]$, $[\![\,\vdash\,?\,]\!]$, $(!K \vdash)$ or $(\vdash ?K)$.

$$\frac{!\alpha \vdash ?\beta; ?\mathcal{C}; \mathcal{A} \quad ?\mathcal{C}; \mathcal{B}; \gamma \vdash \delta}{\frac{!\alpha \vdash ?\beta; ?\mathcal{C}; !\mathcal{A} \quad [?\mathcal{C}; !\mathcal{B}; \gamma] \vdash \delta}{[!\alpha; !\mathcal{B}; \gamma] \vdash [?\beta; \delta; !\mathcal{A}]}} \quad \rightsquigarrow \quad \frac{\frac{!\alpha \vdash ?\beta; ?\mathcal{C}; \mathcal{A}}{!\alpha \vdash ?\beta; ?\mathcal{C}; !\mathcal{A}} \quad ?\mathcal{C}; \mathcal{B}; \gamma \vdash \delta}{\frac{[!\alpha; \mathcal{B}; \gamma] \vdash [?\beta; \delta; !\mathcal{A}]}{[!\alpha; !\mathcal{B}; \gamma] \vdash [?\beta; \delta; !\mathcal{A}]}}$$

$$\frac{!\alpha \vdash ?\beta; ?\mathcal{C}; \mathcal{A} \quad ?\mathcal{C}; \gamma \vdash \delta; \mathcal{B}}{\frac{!\alpha \vdash ?\beta; ?\mathcal{C}; !\mathcal{A} \quad ?\mathcal{C}; \gamma \vdash [\delta; ?\mathcal{B}]}{[!\alpha; \gamma] \vdash [?\beta; \delta; !\mathcal{A}; ?\mathcal{B}]}} \quad \rightsquigarrow \quad \frac{\frac{!\alpha \vdash ?\beta; ?\mathcal{C}; \mathcal{A}}{!\alpha \vdash ?\beta; ?\mathcal{C}; !\mathcal{A}} \quad ?\mathcal{C}; \gamma \vdash \delta; \mathcal{B}}{\frac{[!\alpha; \gamma] \vdash [?\beta; \delta; !\mathcal{A}; \mathcal{B}]}{[!\alpha; \gamma] \vdash [?\beta; \delta; !\mathcal{A}; ?\mathcal{B}]}}$$

The transformations are justified by a reduction in ρ_r. All the earlier contractions may be carried out in the new proof segments too. The next two cases are justified similarly.

$$\frac{!\alpha \vdash ?\beta; ?\mathcal{C}; \mathcal{A} \quad ?\mathcal{C}; \gamma \vdash \delta}{\frac{!\alpha \vdash ?\beta; ?\mathcal{C}; !\mathcal{A} \quad ?\mathcal{C}; !\mathcal{B}; \gamma \vdash \delta}{[!\alpha; !\mathcal{B}; \gamma] \vdash [?\beta; !\mathcal{A}; \delta]}} \quad \rightsquigarrow \quad \frac{\frac{!\alpha \vdash ?\beta; ?\mathcal{C}; \mathcal{A}}{!\alpha \vdash ?\beta; ?\mathcal{C}; !\mathcal{A}} \quad ?\mathcal{C}; \gamma \vdash \delta}{\frac{[!\alpha; \gamma] \vdash [?\beta; !\mathcal{A}; \delta]}{[!\alpha; !\mathcal{B}; \gamma] \vdash [?\beta; !\mathcal{A}; \delta]}}$$

$$\frac{!\alpha \vdash ?\beta; ?\mathcal{C}; \mathcal{A} \quad ?\mathcal{C}; \gamma \vdash \delta}{\frac{!\alpha \vdash ?\beta; ?\mathcal{C}; !\mathcal{A} \quad ?\mathcal{C}; \gamma \vdash \delta; ?\mathcal{B}}{[!\alpha; \gamma] \vdash [?\beta; \delta; !\mathcal{A}; ?\mathcal{B}]}} \quad \rightsquigarrow \quad \frac{\frac{!\alpha \vdash ?\beta; ?\mathcal{C}; \mathcal{A}}{!\alpha \vdash ?\beta; ?\mathcal{C}; !\mathcal{A}} \quad ?\mathcal{C}; \gamma \vdash \delta}{\frac{[!\alpha; \gamma] \vdash [?\beta; \delta; !\mathcal{A}]}{[!\alpha; \gamma] \vdash [?\beta; \delta; !\mathcal{A}; ?\mathcal{B}]}}$$

This completes the proof of the admissibility of the cut rule in $[\![LCLL]\!]$. ◁

Lemma 33. (**Curry's lemma for** $[\![LCLL]\!]$) *If $\alpha \vdash \beta$ has a proof in $[\![LCLL]\!]$ with the height of the proof tree being n, and $\gamma \vdash \delta$ results from $\alpha \vdash \beta$ by one or more applications of the rules* ($!W \vdash$) *and* ($\vdash ?W$), *then $\gamma \vdash \delta$ has a proof in $[\![LCLL]\!]$, where the height of the proof tree is not greater than n (i.e., it is $\leq n$).*

Proof. The proof of this theorem is a straightforward extension of Curry's lemma for the multiplicative–exponential fragment of CLL with six cases added. (See Theorem 14 in [8].) Namely, the basis of the induction is expanded to deal with ($\mathbf{0} \vdash$) and ($\vdash \top$), plus ($\& \vdash$), ($\vdash \&$), ($\oplus \vdash$) and ($\vdash \oplus$) are added to the inductive step. Each of these is quite routine (and we omit the details).

From another point of view, we can start with the proof of Lemma 28. We considered ($\Box K \vdash$) or ($\vdash \Diamond K$) there. Here we have to consider what happens if ($!K \vdash$) or ($\vdash ?K$) are the last rules applied in a proof. We assume that $!\mathcal{A}$, $!\mathcal{C}$, $?\mathcal{B}$ and $?\mathcal{D}$ are (pairwise) distinct, and that the former two differ from elements of α', and the latter two are not among the elements of β'. We also assume that *three* is a representative number for the general situation (and it also allows us to fit everything on a page).

Let us assume that the last rule is ($!K \vdash$). We have the following.

$$\frac{!\mathcal{C}; !\mathcal{C}; !\mathcal{C}; \alpha' \vdash \beta'; ?\mathcal{D}; ?\mathcal{D}; ?\mathcal{D}}{!\mathcal{A}; !\mathcal{C}; !\mathcal{C}; !\mathcal{C}; \alpha' \vdash \beta'; ?\mathcal{D}; ?\mathcal{D}; ?\mathcal{D}} \quad \overset{\text{i.h.}}{\leadsto} \quad \frac{!\mathcal{C}; \alpha' \vdash \beta'; ?\mathcal{D}}{!\mathcal{A}; !\mathcal{C}; \alpha' \vdash \beta'; ?\mathcal{D}}$$

If the thinned in formula is the same as $!\mathcal{C}$, then the application of ($!K \vdash$) may be simply omitted like in

$$\frac{!\mathcal{C}; !\mathcal{C}; !\mathcal{C}; \alpha' \vdash \beta'; ?\mathcal{D}; ?\mathcal{D}; ?\mathcal{D}}{!\mathcal{C}; !\mathcal{C}; !\mathcal{C}; !\mathcal{C}; \alpha' \vdash \beta'; ?\mathcal{D}; ?\mathcal{D}; ?\mathcal{D}} \quad \overset{\text{i.h.}}{\leadsto} \quad !\mathcal{C}; \alpha' \vdash \beta'; ?\mathcal{D}.$$

The case of ($\vdash ?K$) is dual to this. Here is what it looks like.

$$\frac{!\mathcal{C}; !\mathcal{C}; !\mathcal{C}; \alpha' \vdash \beta'; ?\mathcal{D}; ?\mathcal{D}; ?\mathcal{D}}{!\mathcal{C}; !\mathcal{C}; !\mathcal{C}; \alpha' \vdash \beta'; ?\mathcal{D}; ?\mathcal{D}; ?\mathcal{D}; ?\mathcal{B}} \quad \overset{\text{i.h.}}{\leadsto} \quad \frac{!\mathcal{C}; \alpha' \vdash \beta'; ?\mathcal{D}}{!\mathcal{C}; \alpha' \vdash \beta'; ?\mathcal{D}; ?\mathcal{B}}$$

$$\frac{!\mathcal{C}; !\mathcal{C}; !\mathcal{C}; \alpha' \vdash \beta'; ?\mathcal{D}; ?\mathcal{D}; ?\mathcal{D}}{!\mathcal{C}; !\mathcal{C}; !\mathcal{C}; \alpha' \vdash \beta'; ?\mathcal{D}; ?\mathcal{D}; ?\mathcal{D}; ?\mathcal{D}} \quad \overset{\text{i.h.}}{\leadsto} \quad !\mathcal{C}; \alpha' \vdash \beta'; ?\mathcal{D}$$

Next, we note that in the proof of Lemma 24, we can restrict contractions to exponential formulas. Then some of the cases disappear, whereas the others go through as before. This completes the proof. ◁

Now we turn to the decidability proof for LCLL.

Theorem 34. *Classical linear logic* (CLL) *is decidable.*

Proof. Given a wff $\mathcal{A}$, we narrow down the question whether the wff is a theorem of LCLL by ensuring that $\mathcal{A}$ is not a theorem of $LLRW^{\Diamond\Box}$, and it is a theorem of $[\![LCLL]\!]$. (If $\mathcal{A}$ is not within that range, then we already know whether it is a theorem of LCLL. Namely, if $\mathcal{A}$ is a theorem of $LLRW^{\Diamond\Box}$, then it is a theorem of LCLL, and if $\mathcal{A}$ is not a theorem of $[\![LCLL]\!]$, then it is not a theorem of LCLL.) The proof search in $[\![LCLL]\!]$ generates all the irredundant proofs of $\mathcal{A}$. By Definition 8, we calculate the heap numbers for the subformulas of $\mathcal{A}$. Then we start to build a proof-search tree in LCLL. The root is the sequent $\vdash \mathcal{A}$, and we expand the tree by scrutinizing each rule that could result in the sequent in a particular node in the tree. If there is a possibility for contractions then we add each possibility separately to the tree. However, we limit the number of contractions on each formula by its heap number. The whole tree is finite and if $\mathcal{A}$ is provable in LCLL, then the search tree will contain a proof. If $\mathcal{A}$ is not a theorem, then we will find this out in finitely many steps, namely, when the (finite) proof-search tree is completed without containing a proof. ◁

7. Remarks on "Decision Problems for Linear Logic"

Lincoln et al. [35] present what they take to be a proof of the undecidability of what we call "classical linear logic" (CLL) and they call "full propositional linear logic" (or sometimes just "linear logic"). This paper is highly original and well-motivated, exploiting the notion of linear logic as a "resource conscious logic." The proof was seemingly well-presented, and seemed to have convinced many people that CLL is undecidable. But only the most naive logician thinks that something is a proof because it is called a proof. Maybe, someday the dream will be fulfilled that all proofs will be computer checkable, but for now, and even as proofs get more and more complicated, we are largely dependent on human intelligence and a mixture of formal language, natural language, and a sometimes conventional, sometimes creative, hybrid mixture of the two. Unfortunately, and we are apologetic about this to Lincoln, Mitchell, Scedrov and Shankar (all fine logicians), but we think that there are some mistakes in their proof. We shall outline their proof both to help the reader (and ourselves) understand the virtues of their attempt, and a flaw in the proof.

The rough idea of their proof is to reduce the question of the decidability of CLL to the problem of the solvability of a question about certain finite automata, which they introduce and call *And-Branching Two-Counter Machines* **Without** *Zero-Test* (ACM for short). These are a variant of the more standard *And-Branching Two-Counter Machines* **With** *Zero-Test*. They ingeniously replace the Zero-Test with something they call "Forking." The corresponding question for the former is known to be unsolvable, and they show that the halting problem for these two is the same. They then go on to translate the question of the decidability of linear logic into the solvability of ACMs, and use the fact that ACMs are unsolvable to show that LCLL is undecidable. The "trick" is the translation between computations in ACMs and proofs in LCLL.

They start by defining (p. 261) a *theory* to be a finite set of *axioms*, and they define an *axiom* to be "a linear logic sequent of the form $\vdash C, p_{i_1}^\perp, \ldots, p_{i_n}^\perp$, where C is a MALL formula (a linear logic formula without ! or ?) and the remainder of the sequent is made up of negative literals."[12] They make it clear that the negative literals are allowed to be absent and that the restrictive form of axioms is due to their wanting to "achieve strict control over the shape of a proof."

They define that "a sequent $\vdash \Gamma$ is provable in T exactly when we are able to derive $\vdash \Gamma$ using the standard set of linear logic proof rules, in combination with axioms from

---

[12] We use ";" in the sequent calculus LCLL, but in this section we resort to "," for easy comparison with Lincoln et al. [35]. Incidentally, they use a one-sided sequent calculus, however, in the case of CLL, this affects only the presentation.

T." It is evident from context and from their Appendix B that they assume the cut rule to be in the "standard set of linear logic proof rules," just as [23] does. They go on to Lemma 3.1 that states that cut can be replaced by what they call "directed cut." They make it clear that such a derivation would be just like a proof tree in linear logic except that the leaves can be axioms from T, and not just the usual logical axioms $\vdash p_i, p_i^\perp$. Let us write $T \vdash_{LCLL} \Gamma$ for $\vdash \Gamma$ is provable from the theory T in $LCLL$. Note that they explicitly define this notion only for the case of $LCLL$, not for its multiplicative-additive fragment MALL. This is important, because later (p. 265) they say: "We have just shown how a decision problem for MALL with the addition of nonlogical axioms may be encoded in full propositional linear logic without nonlogical axioms. Thus the upcoming proof of undecidability of MALL with nonlogical axioms will yield undecidability for full propositional logic."

Notice that here they talk about "MALL with the addition of nonlogical axioms," but this has not been really defined. They actually defined provability from T in $LCLL$. We know this sounds like a picky point, and readily agree that we can make sense of MALL theories as just the obvious variant of $LCLL$ theories that does not allow applying the rules for the exponentials. But they misdescribe what they showed. What they in fact showed was how a decision problem for full propositional linear logic (not just for the MALL fragment) with the addition of nonlogical axioms may be encoded in full propositional linear logic without nonlogical axioms. However, they say (p. 260) that "We now show that if nonlogical (MALL) axioms are added to MALL, the decision problem becomes recursively unsolvable. We also show that nonlogical MALL axioms may be encoded in full propositional linear logic without nonlogical axioms, and thus we hve the result that full propositional linear logic is undecidable."

Lemmas 3.2 and 3.3 each prove different directions of the following biconditional.

For any finite set of axioms T, $\quad T \vdash_{LCLL} \Gamma \quad$ iff $\quad \vdash_{LCLL} [T], \Gamma$.

But they also seem to be saying (or tacitly implying) that

$$T \vdash_{MALL} \Gamma \quad \text{iff} \quad \vdash_{LCLL} [T], \Gamma.$$

To understand these claims we need to understand $[T]$, which translates a theory $T = \{t_1, t_2, \ldots, t_k\}$ into a multiset of linear logic formulas $?[t_1], ?[t_2], \ldots, ?[t_k]$, where $[t_i]$ is the translation of the axiom t_i into a single linear logic formula as follows. If t_i is $\vdash C, p_{i_1}^\perp, \ldots, p_{i_n}^\perp$, then $[t_i]$ is $\vdash C^\perp \otimes p_{i_1} \otimes \cdots \otimes p_{i_n}$.

Also, (p. 269) they say: "We give a translation from ACMS to linear logic with theories and show that our sequent translation of a machine in a particular state is provable in linear logic if and only if the ACM halts from that state. In fact our translation uses only MALL formulas and theories, thus with the use of our earlier encoding Lemma 3.2 and 3.3, we will have our result for propositional linear logic without nonlogical axioms. Since an instantaneous description of an ACM is given by a list of triples, it is somewhat delicate to state the induction we will use to prove soundness."

Lincoln et al. use *non-deterministic And-Branching Two-Counter Machines Without Zero-Test* (ACMs). An ACM has a set of *states* **Q**, a finite set δ of *transitions*, and *initial* and *final* states Q_I and Q_F.

Depending on the state Q_i, the ACM can do various things. Thus, where A and B are natural numbers in the first and second registers, the rules can add 1 to them, subtract 1 from them (unless they are 0 in which case the rule is not applicable), and move to the state Q_j. Or the machine can continue computation from two states Q_j and Q_k, using as inputs the values A, B in the current state Q_i.

Rule	Transition	Translation
Q_i Increment A Q_j	$\langle Q_i, A, B \rangle \mapsto \langle Q_j, A+1, B \rangle$	$\vdash q_i^\perp, (q_j \otimes a)$
Q_i Increment B Q_j	$\langle Q_i, A, B \rangle \mapsto \langle Q_j, A, B+1 \rangle$	$\vdash q_i^\perp, (q_j \otimes b)$
Q_i Decrement A Q_j	$\langle Q_i, A+1, B \rangle \mapsto \langle Q_j, A, B \rangle$	$\vdash q_i^\perp, a^\perp, q_j$
Q_i Decrement B Q_j	$\langle Q_i, A, B+1 \rangle \mapsto \langle Q_j, A, B \rangle$	$\vdash q_i^\perp, b^\perp, q_j$
Q_i Fork to Q_j and Q_k	$\langle Q_i, A, B \rangle \mapsto \langle Q_j, A, B \rangle$ and $\langle Q_k, A, B \rangle$	$\vdash q_i^\perp, (q_j \oplus q_k)$

An *instantaneous description* (ID) of an ACM M is a finite tree of ordered triples $\langle Q_i, A, B \rangle$, where $Q_i \in \mathbf{Q}$ (Q_i is a state), and A and B are natural numbers. The *accepting triple* is $\langle Q_F, 0, 0 \rangle$. An *accepting ID* is any ID where every leaf of the ID is the accepting triple. This means that no matter how the computation evolves it ends with an accepting triple, that is, in an accepting state (which is unique) and the counters containing 0.[13]

Given a triple $\langle Q_i, A, B \rangle$, its *translation* $\theta(\langle Q_i, A, B \rangle)$ is $\vdash q_i^\perp, (a^\perp)^A, (b^\perp)^B, q_F$, where the superscript ^{A} and ^{B} indicate the number of $a^\perp$'s and $b^\perp$'s in the sequent. The translation of an ID comprises the translations of the elements of the ID, that is, $\theta(E_1, E_2, \ldots, E_m) = \theta(E_1), \theta(E_2), \ldots, \theta(E_m)$.

Lincoln et al.'s main result is: "Theorem 3.7. The provability problem for propositional linear logic is recursively unsolvable." This is just a different way of saying that CLL is undecidable. Their proof consists literally of the single statement (p. 275) "From Lemmas 3.2–3.6 we obtain our main result." We shall try to construct a proof using these lemmas, and in the process end up deconstructing their proof.

As already mentioned, the first two of these lemmas can be put together as the two halves of the next biconditional.

Lemmas 3.2–3.3. *For any finite set of axioms T, $T \vdash_{LCLL} \Gamma$ iff $\vdash_{LCLL} [T], \Gamma$.*

And the last two are the two halves of the following biconditional.

Lemmas 3.5–3.6. *An ACM M accepts from the triple s iff the sequent $\theta(s)$ is provable, given the theory derived from M.*

And the middle lemma is the keystone.

Lemma 3.4. *It is undecidable whether an ACM accepts from the triple $\langle Q_I, A, B \rangle$.*

The rough idea would then be to combine these lemmas so that the undecidability of the ACM accepting from $\langle Q_I, A, B \rangle$ translates into the undecidability of provability in LCLL (without axioms).

So let us suppose that we have a method for deciding the provability of theorems in LCLL. Consider then an arbitrary ACM M, and its theory T_M that translates the instructions of the machine using the table above. Then, as a special case of Lemmas 3.2–3.3 we have:

$$T_M \vdash_{LCLL} \Gamma \quad \text{iff} \quad \vdash_{LCLL} [T_M], \Gamma.$$

Further, as a special case of Lemmas 3.5–3.6, we have:

---

[13]Lincoln et al. defined an accepting ID to be an ID each element of which is an accepting triple. They should have meant what is in this paragraph unless only one-step trivial computations are permitted.

An ACM M accepts from $\langle Q_I, A, B\rangle$ iff the sequent $\theta(\langle Q_I,A,B\rangle)$ is provable, given the theory T_M.[14]

What is the sequent $\theta(\langle Q_I,A,B\rangle)$? It is $\vdash q_I, (a^\perp)^A, (b^\perp)^B, q_F$. So the problem is to figure out whether this sequent is provable using the theory T_M, i.e., using the sequents in T_M together with applications of the cut rule.

While it is true that using the exponentials LCLL can emulate that a sequent from T_M is *not used*, *used once* or *used several times* in a MALL proof, the exponentials interact with the MALL vocabulary. In effect, the interaction implies reliance on the following claim.

$$T_M \vdash_{\text{MALL}} \Gamma \quad \text{iff} \quad \vdash_{L\text{CLL}} [T_M], \Gamma.$$

The following is an equivalent claim.

$$T_M \vdash_{\text{MALL}} \Gamma \quad \text{iff} \quad T_M \vdash_{L\text{CLL}} \Gamma.$$

From left to right, the claim is obvious and true, but the converse is less than obvious. The cut rule is *not eliminable* in the presence of proper axioms (the elements of T_M) — as Lincoln et al. [35] themselves point out on p. 262. Of course, using the cut rule in a proof is unproblematic in the sense that it is a rule and so the sequent proved is a theorem, but the cut rule causes problems for the analysis of the proof. Thus, when we try to prove that $T_M \vdash_{L\text{CLL}} \Gamma$ implies $T_M \vdash_{\text{MALL}} \Gamma$, we run into a problem, because a proof of Γ in LCLL may contain applications of the cut rule too. In other words, if the cut rule is not eliminable, then it is difficult to contemplate how the right-to-left conditional could be proved at all.

So far, we assumed that Lemmas 3.5–3.6 concerned provability in MALL. An alternative reading of the those lemmas is that they permit the use of all the rules of LCLL, but the occurrences of applications of the cut rule are limited because of Lemma 3.1, which reads as follows.

Lemma 3.1. (Cut standardization). *If there is a proof of $\vdash \Gamma$ in theory T, then there is a directed proof of $\vdash \Gamma$ in theory T.*

A *directed cut* is simply an application of the cut rule, in which at least one of the two premises is an axiom, and the cut formula is C (using the earlier notation). A *directed proof* is a derivation with only directed cuts (or no cuts at all). The cut standardization lemma holds in MALL proofs, and ensures that MALL theories in proofs in MALL can mimic the consumption of instructions in an ACM.

Before we turn to the discussion of the modeling of ACMs (and Minsky machines), we illustrate a problem with the proof of the admissibility of the cut rule in Appendix A. In the relevance logic literature, the use of a multi-cut rule is quite common, because many relevance logics contain a contraction rule (but not a thinning rule). The multi-cut rule is similar to Gentzen's mix rule in that it allows cutting out more than two formulas. On the other hand, these rules are different, because the multi-cut rule does not require the elimination of all occurrences of the cut formula. Lincoln et al. [35] opt to use both single cut and multi-cut in their elimination proof, however, the latter is only applicable to formulas that start with exponentials. In CLL, only certain

---

[14]Lemmas 3.5–3.6 do not make explicit the logic in which provability is meant. However, the first paragraph in §3.5 (p. 269) seems to suggest that it is MALL.

exponential formulas can be contracted, which explains why multi-cut is introduced for such formulas.

[35] defines the degree of the cut formula in a fairly standard manner. However, they also define the degree of a proof as the maximal degree of any cut in the proof or zero (if there is no cut). Unfortunately, the degree of the proof does not decrease in every step in the cut elimination proof.[15] Their crucial lemma reads as

Lemma A.1 (Reduce one cut). *Given a proof of the sequent $\vdash \Gamma$ in linear logic which ends in an application of* **Cut*** *of degree $d > 0$, and where the degree of the proofs of both hypotheses is less than d, we may construct a proof of $\vdash \Gamma$ in linear logic of degree less than d.*

The proof is divided into cases, and in each case a local modification of the proof is given. The transformations are similar to what is to be expected. However, it is completely obvious that several of the one-step transformations do not establish the claim of the lemma.

The following example shows an application of the single cut rule with $d = 7$. (We make explicit only the segment of the proof that is problematic.)

$$\cfrac{\cfrac{\vdash p\,\&\,(!q\oplus r)}{\vdash !(p\,\&\,(!q\oplus r))} \quad \cfrac{\cfrac{\vdots}{\vdash ?(p^\perp\oplus(?q^\perp\,\&\,r^\perp)), ?q^\perp\,\&\,r^\perp, (p\,\&\,q)\oplus r}}{\cfrac{\vdash ?(p^\perp\oplus(?q^\perp\,\&\,r^\perp)), p^\perp\oplus(?q^\perp\,\&\,r^\perp), (p\,\&\,q)\oplus r}{\cfrac{\vdash ?(p^\perp\oplus(?q^\perp\,\&\,r^\perp)), ?(p^\perp\oplus(?q^\perp\,\&\,r^\perp)), (p\,\&\,q)\oplus r}{\vdash ?(p^\perp\oplus(?q^\perp\,\&\,r^\perp)), (p\,\&\,q)\oplus r}\,?C}\,?D}}{\vdash (p\,\&\,q)\oplus r}\,\text{cut}$$

The last step in the proof is an application of the cut rule. The cut formula is principal in both premises of the cut, hence by A.2.5, the cut is moved up by a sequent in the right premise. This requires that **Cut!** be applied. However, the cut formula is the same as before, hence the degree of the proof is also the same.

Lincoln et al. [35] must have realized that they do not have an inductive proof of Lemma A.1, because they say on p. 299 that "by induction on the size of proofs, we can construct the desired proof of degree less than d." It is possible, perhaps, even plausible that one can do this. However, they do not give such a proof, indeed, they do not even define what the size of a proof is. It could be the number of propositional variables in the proof, the sum of the degrees of all formulas, the height of the proof tree, to name a few alternatives.

Another problem with the argument for Lemma A.1 is that once the above proof is transformed (as shown below), it is no longer clear which transformation is to be applied next. The cut formula is still principal in the left premise, however, it is both principal and non-principal in the right premise. There is no definition in [35] that would allow us to determine whether the cut formula in the right premise is principal or not, which is needed in order to apply (1) or (2) on p. 298. In fact, the situation is very typical, because principal formulas are (usually) unique in a rule. Then, a cut on

---

[15]Lambek [32] was able to prove a cut theorem for his calculuses by induction on one parameter that he called degree, which is however, not identical to either of the degrees just mentioned. Also, Lambek's calculuses do not contain any kind of contraction, which means that the admissibility of the single cut rule can be proved directly (without mix or multi-cut).

$$\vdots$$
$$\dfrac{\vdash p\,\&\,(!q\oplus r)}{\vdash !(p\,\&\,(!q\oplus r))} \quad \dfrac{\dfrac{\vdash ?(p^\perp\oplus(?q^\perp\,\&\,r^\perp)),\,?q^\perp\,\&\,r^\perp,(p\,\&\,q)\oplus r}{\vdash ?(p^\perp\oplus(?q^\perp\,\&\,r^\perp)),\,p^\perp\oplus(?q^\perp\,\&\,r^\perp),(p\,\&\,q)\oplus r}}{\vdash ?(p^\perp\oplus(?q^\perp\,\&\,r^\perp)),\,?(p^\perp\oplus(?q^\perp\,\&\,r^\perp)),(p\,\&\,q)\oplus r} \,?D$$
$$\dfrac{}{\vdash (p\,\&\,q)\oplus r}\ \text{cut!}$$

several formulas moved upward in a proof tree will likely come to a sequent in which the cut formula has both principal and non-principal occurrences. The usual notion of a principal formula is extended on p. 297. However, that expansion leaves one occurrence of the cut formula in the right premise of the application of the cut! rule above as a non-principal occurrence.

Presumably, we should apply the second transformation in A.2.6 now. The transformation yields two cuts in a new proof that have the same degree, and which repeat a whole branch of the proof tree. This is a point where the informal allusion to the size of the proofs would need to be made precise, because neither the height of the proof tree is decreasing nor the number of sequents or cuts does.[16]

The cut elimination proof would be the basis for the proof that directed cuts suffice. However, we believe that there is *no proof of the admissibility of the cut rule* in [35], hence, there is *no proof of Lemma 3.1* and further, of Theorem 3.7 in that paper.

The **main problem** with the alleged proofs in [35] and [27] (as well as in [21]) goes beyond what we have outlined so far. The two models, ACMs and Minsky machines, are very similar; they are both variations on what more simply are called counter machines. A particularly elegant formulation is termed *abacus machines* in Boolos and Jeffrey [15] with reference to Lambek [34].

Counter machines are "full-fledged" models of computation as proved in [15] and in [34]. However, the abacus machines *compute functions*, that is, starting with natural numbers in the counters the machine halts with some content (which may or may not be all 0's) in the counters. ACMs and Kanovich's Minsky machines do not compute any functions, rather, they *accept* a certain input. Furthermore, both models are modified to accept by a final state with *all the counters empty*.

Neither [35] nor [28] ([27]) prove that the machines that they intend to model have an undecidable halting problem. The undecidability of the halting problem for Minsky machines with a restricted halting problem was recently proved in [21, §7] via several reduction steps from the Post correspondence problem.[17]

One might wonder how the computation of one or another machine is *modeled* in propositional logic. There are well-known ways to model primitive recursive functions and computations of a Turing machine in the language of first-order arithmetic. We have explained at the beginning of this section how [35] intend to model the computations of ACMs; the rest of the authors follow a similar idea. We present in Figure 3

---

[16]Sequent calculuses are, perhaps, more difficult to understand than axiomatic systems. This may be one of the reasons behind [42], which shows that the author does not understand the proof of the cut theorem in [8]. He also seems to assume that the decision procedure for MELL should generate *all* the infinitely many proofs for a provable sequent. Of course, decision procedures, normally, do not yield all possible proofs.

[17]In all the papers that we mentioned in this paragraph, a proof that matches a computation starts with the final state. We will refer to the authors of all these papers as the authors, when we talk about this feature of the machines.

(below) a small ACM, which differs from the example in [35] in that it accepts an *infinite language* and it contains *three zero-tests*. (Their sample machine accepts the finite language $\{a^0 b^0\}$ and contains no zero-tests at all.)

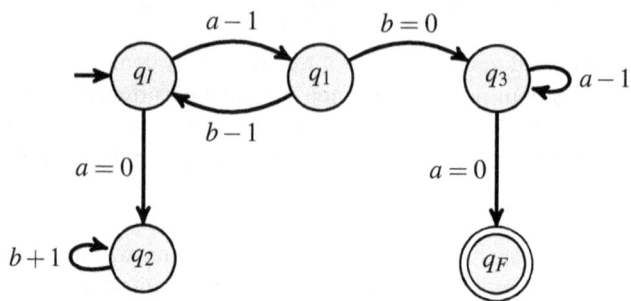

FIGURE 3. The ACM $\mathfrak{M}_1$

The picture of the ACM employs some notational conventions that are often used in visualizing finite state automata such as circles for states with the name of the state inside, and arrows with labeling for the actions of the machine. However, these similarities are somewhat superficial. The ACM receives input at the arrow pointing to q_I in the form of finitely many counters filled as desired. Then, the machine reads and occasionally *modifies* the content of the counters. The arrows labeled with $a = 0$ and $b = 0$ represent successful zero-tests. (The diagram hides the implementation of a zero-test via "and-branching.") The state q_3 is a seemingly spurious state; its function is simply to ensure that counter a is empty. The machine is so designed that if it reaches q_F, then it is guaranteed that the counters are empty, hence, the role of q_F is to indicate acceptance and halting. It is not difficult to see that $\mathfrak{M}_1$ accepts the language $\{a^m b^n : m > n\}$ (where a and b are placeholders for "first" and "second" counter). This language is not very complicated, it's easily seen to be a CFL (context-free language). Alternatively, the machine can be thought to accept when the characteristic function of the $>$ relation (on $\mathbb{N}$) evaluates to true.

For example, the full description of the computation of the machine starting with $a^3 b^1$ (i.e., 3 in the first counter, and 1 in the second counter) is the following sequence of triplets.

$$\langle q_I, a^3, b^1 \rangle, \langle q_1, a^2, b^1 \rangle, \langle q_I, a^2, b^0 \rangle, \langle q_1, a^1, b^0 \rangle, \langle q_3, a^1, b^0 \rangle, \langle q_3, a^0, b^0 \rangle, \langle q_F, a^0, b^0 \rangle$$

The set of instructions for $\mathfrak{M}_1$ encoded as axioms for a CLL theory is as follows. Here we make explicit the hidden and-branching, which we use only with the zero states z_a and z_b.

Axioms for $\mathfrak{M}_1$:

1. $\vdash q_I^\perp, a^\perp, q_1$ 2. $\vdash q_1^\perp, b^\perp, q_I$ 3. $\vdash q_3^\perp, a^\perp, q_3$ 4. $\vdash q_2^\perp, b \otimes q_2$
5. $\vdash q_I^\perp, z_a \oplus q_2$ 6. $\vdash q_1^\perp, z_b \oplus q_3$ 7. $\vdash q_3^\perp, z_a \oplus q_F$
8. $\vdash z_a^\perp, b^\perp, z_a$ 9. $\vdash z_b^\perp, a^\perp, z_b$ 10. $\vdash z_a^\perp, q_F \oplus q_F$ 11. $\vdash z_b^\perp, q_F \oplus q_F$

If we construct proofs with cuts, then these axioms cut out their negations from a sequent. If the proof is cut-free, then the same formulas have to be built-up.

Negations of axioms:

1. $q_I \otimes (a \otimes q_1^\perp)$ 2. $q_1 \otimes (b \otimes q_I^\perp)$ 3. $q_3 \otimes (a \otimes q_3^\perp)$ 4. $q_2 \otimes (b^\perp \parr q_2^\perp)$

5. $q_I \otimes (z_a^\perp \,\&\, q_2^\perp)$ 6. $q_1 \otimes (z_b^\perp \,\&\, q_3^\perp)$ 7. $q_3 \otimes (z_a^\perp \,\&\, q_F^\perp)$

8. $z_a \otimes (b \otimes z_a^\perp)$ 9. $z_b \otimes (a \otimes z_b^\perp)$ 10. $z_a \otimes (q_F^\perp \,\&\, q_F^\perp)$ 11. $z_b \otimes (q_F^\perp \,\&\, q_F^\perp)$

It is easy to see that $\vdash z_a^\perp, (b^\perp)^n, q_F$ and $\vdash z_b^\perp, (a^\perp)^n, q_F$ are provable for any $n \in \mathbb{N}$. We will omit these parts of the proof to limit the size of the tree shown. The axioms that are used in applications of cuts are listed on the left.

$$\begin{array}{l} \vdash q_3^\perp, z_a \oplus q_F \\ \quad \vdash q_3^\perp, a^\perp, q_3 \\ \\ \\ \qquad \vdash q_1^\perp, z_b \oplus q_3 \\ \qquad\quad \vdash q_I^\perp, a^\perp, q_1 \\ \qquad\quad \vdash q_1^\perp, b^\perp, q_I \\ \qquad \vdash q_I^\perp, a^\perp, q_1 \end{array}
\qquad
\begin{array}{c} \vdots \\ \vdash z_a^\perp, q_F \quad \vdash q_F^\perp, q_F \\ \hline \vdash z_a^\perp \,\&\, q_F^\perp, q_F \\ \hline \vdash q_3^\perp, q_F \qquad\qquad \vdots \\ \hline \vdash q_3^\perp, a^\perp, q_F \qquad \vdash z_b^\perp, a^\perp, q_F \\ \hline \vdash z_b^\perp \,\&\, q_3^\perp, a^\perp, q_F \\ \hline \vdash q_1^\perp, a^\perp, q_F \\ \hline q_I^\perp, a^\perp, a^\perp, q_F \\ \hline \vdash q_1^\perp, a^\perp, a^\perp, b^\perp, q_F \\ \hline \vdash q_I^\perp, a^\perp, a^\perp, a^\perp, b^\perp, q_F \end{array}$$

The proof *starts* in the *final* state. It is not accidental, because sequent calculus proofs are trees in which the root of the tree is the sequent that is proved. Hence, no tree branch in a proof can split downward.

A cut-free proof for the same sequent is the following. We indicate the negations of the axioms by their number in the listing above, and we omit the proofs leading to a z_x state from q_F together with the horizontal lines. (The two sequents that are not axioms, but easily provable, are *'d.)

$$\begin{array}{ll}
{}^*\vdash z_a^\perp, q_F & \vdash q_F^\perp, q_F \\
\vdash q_3^\perp, q_3 & \vdash z_a^\perp \,\&\, q_F^\perp, q_F, 10 \\
\vdash a^\perp, a & \vdash q_3^\perp, q_F, 10, 7 \\
\vdash q_3^\perp, q_3 & \vdash a \otimes q_3^\perp, q_F, 10, 7 \\
{}^*\vdash z_b^\perp, a^\perp, q_F & \vdash q_3^\perp, a^\perp, q_F, 10, 7, 3 \\
\vdash q_1^\perp, q_1 & \vdash z_b^\perp \,\&\, q_3^\perp, a^\perp, q_F, 10, 7, 3, 11, 9 \\
\vdash a^\perp, a & \vdash q_1^\perp, a^\perp, q_F, 10, 7, 3, 11, 9, 6 \\
\vdash q_I^\perp q_I & \vdash a \otimes q_1^\perp, a^\perp, a^\perp, q_F, 10, 7, 3, 11, 9, 6 \\
\vdash b^\perp, b & \vdash q_I^\perp, a^\perp, a^\perp, q_F, 10, 7, 3, 11, 9, 6, 1 \\
\vdash q_1^\perp, q_1 & \vdash b \otimes q_I^\perp, a^\perp, a^\perp, b^\perp, q_F, 10, 7, 3, 11, 9, 6, 1 \\
\vdash a^\perp, a & \vdash q_1^\perp, a^\perp, a^\perp, a^\perp, b^\perp, q_F, 10, 7, 3, 11, 9, 6, 1, 2 \\
\vdash q_I^\perp, q_I & \vdash a \otimes q_1^\perp, a^\perp, a^\perp, a^\perp, b^\perp, q_F, 10, 7, 3, 11, 9, 6, 1, 2 \\
 & \vdash q_I^\perp, a^\perp, a^\perp, a^\perp, b^\perp, q_F, 10, 7, 3, 11, 9, 6, 1, 2, 1
\end{array}$$

The traditional claim is that if the proof is turned upside down, then it can be seen as a modeling of the computation from q_I (with 3 in a and 1 in b) to q_F. Of course, the upside down tree is *not a proof* in CLL at all. If we try to create an interpretation from the top of the proof, then it seems that the subproofs in the whole proof tree do not have an interpretation that is independent from the whole proof tree. Another way to look at this problem is that unless the proof has a sequent of the form $\vdash q_I^\perp,\ldots,q_F$ as its root, it is not a model of (any stage of) a computation of the machine.

To further illustrate the problem, let us assume that we add a new state q_4 to $\mathfrak{M}_1$. The new state has two outgoing arrows, one pointing to q_4 itself with a label $b-1$, the other pointing to q_3 with a label $b=0$. Our new machine $\mathfrak{M}_1'$ is equivalent in terms of acceptance to $\mathfrak{M}_1$. However, there is a proof of the sequent $\vdash q_4^\perp, b^\perp, b^\perp, b^\perp, q_F$, given its theory. The state q_4 — by design — is not accessible from q_I, which means that in $\mathfrak{M}_1'$ there is no computation that involves q_4. But it is true that if we picture the machine as a special graph (like $\mathfrak{M}_1$ in Figure 3), then there is *a path* between q_4 and q_F. And starting in state q_F, and by performing the inverses of the machine's instructions, it is possible to reach state q_4 with 3 in the second counter.

[26] and [21] number the final state with 0, which gives the appearance (at first) that a proof starts at an initial state (q_0). The latter paper models computation by getting from the 0th state (called PC value) to the 1st state.

Kopylov [29] noted that provable sequents (in the normal fragment) of CLL can be given *two computational readings*. Similarly, the provable sequents of CLL may be given two computational interpretations. The emptiness of the counters at halting, and halting in a unique final state are essential for the construction of sequent calculus proofs. In other words, proofs starting with $\vdash q_F^\perp, q_F$'s that contain forking cannot be replaced by proofs that start from $\vdash q_I^\perp, q_I$ (while proving the same sequent). However, the "non-traditional" interpretation means that there are *no zero-tests* in the machine that is modeled, and the decrement and increment instructions are swapped. According to this interpretation, that we think is *the correct one*, every subtree in a proof tree is a model of a step in *reverse computation*; that is, it is a model of "running" the machine backward. (This also means that the machine may get "stuck" in a state when the subtraction cannot be performed and there is no branch that takes care of the counter's emptiness.) In view of our decidability result, we think that the machines that emerge from this interpretation — *reverse* ACMs and *reverse* Minsky machines, etc. — do not have an undecidable halting problem. In other words, our decidability result supports the conjecture that the halting problem for reverse computation in ACMs and various counter machines, in general, is decidable.

To summarize, we think that each published "proof," most prominently, that by Lincoln et al. [35] and that by Kanovich [27], has *gaps* in it. Moreover, we think that there is a real reason to believe that some of those gaps cannot be filled to complete the proofs of the undecidability of *L*CLL, because there is a conceptual mismatch between (forward/normal) computational steps and steps in a sequent calculus proof in *L*CLL. Furthermore, our proofs demonstrate that *L*CLL is *decidable*.

8. Conclusion

This paper scrutinized the issue of modalities in lattice-R. To start with, the Ackermann and Church constants (hence, modalities defined from those constants) do not interfere with the decidability of lattice-R. The addition of *primitive modal operators* with some usual rules does not lead to undecidability either. If (modalized) versions of structural rules are added (or omitted) from $LLR^{\Diamond\Box}$, then the properties of modalities vary. Nonetheless, the *decidability* of the resulting logics — no matter with however unusual modalities — stays provable. We have also proved that classical propositional linear logic is decidable, and we have explained where the proofs of earlier undecidability claims in [35], in [28] (also, [27]) and in [21] are lacking.

Acknowledgments. We are grateful to Patrick Lincoln for providing us with a summary of his and his coauthors' views on the decidability of linear logic and its computational interpretation as well as their reactions to the first draft of our paper in 2014. We thank all of them.

We would like to thank Andre Scedrov for bringing to our attention A. Kopylov's and M. Kanovich's papers. We thank Max Kanovich for providing us with a "tutorial" on his interpretation of linear logic using Minsky machines.

We also thank Alasdair Urquhart for reading a draft of our paper, and especially, for his questions about the "heap numbers" and pointing out Roorda's work.

Of course, we do not mean to imply that these researchers endorse our paper.

We would like to thank audiences at the *2015 North American Annual Meeting of the Association for Symbolic Logic* in Urbana, IL (March 2015), at the *4th CSLI Workshop on Logic, Rationality and Intelligent Interaction* in Stanford, CA (May 2015), at the *POMSIGMA* session at the *Joint Mathematics Meeting* in Atlanta, GA (January, 2017), as well as, in the *Logic Seminar* of the *Indiana University Logic Group* in Bloomington, IN (February 2015 and February 2016), where we presented talks based on parts of this paper.

AFTERWORD

The first six and a half sections of this paper were written in 2015, and they remained basically the same since then. The last several "fault-finding" pages were rewritten and expanded several times to appease referees who repeated again and again that propositional linear logic is well known to be undecidable, and first of all, we should demonstrate mistakes in published proofs. It should be noted that no referee — in all those 6–7 years of refereeing — pointed out a mistake in our paper.

REFERENCES

[1] Allwein, G. and Dunn, J. M. (1993). Kripke models for linear logic, *Journal of Symbolic Logic* **58**(2): 514–545.

[2] Anderson, A. R. and Belnap, N. D. (1975). *Entailment: The Logic of Relevance and Necessity*, Vol. I, Princeton University Press, Princeton, NJ.

[3] Anderson, A. R., Belnap, N. D. and Dunn, J. M. (1992). *Entailment: The Logic of Relevance and Necessity*, Vol. II, Princeton University Press, Princeton, NJ.

[4] Avron, A. (1988). The semantics and proof theory of linear logic, *Theoretical Computer Science* **57**: 161–184.

[5] Belnap, N. D. (1993). Life in the undistributed middle, *in* K. Došen and P. Schroeder-Heister (eds.), *Substructural Logics*, Clarendon, Oxford, UK, pp. 31–41.

[6] Belnap, N. D. and Wallace, J. R. (1961). A decision procedure for the system $E_{\bar{1}}$ of entailment with negation, *Technical Report 11, Contract No. SAR/609 (16)*, Office of Naval Research, New Haven, CT, (published in *Zeitschrift für mathematische Logik und Grundlagen der Mathematik* **11** (1965): 277–289).

[7] Bimbó, K. (2007). Relevance logics, in D. Jacquette (ed.), *Philosophy of Logic*, Vol. 5 of *Handbook of the Philosophy of Science* of *Handbook of the Philosophy of Science* (D. Gabbay, P. Thagard and J. Woods, eds.), Elsevier, Amsterdam, pp. 723–789.

[8] Bimbó, K. (2015a). The decidability of the intensional fragment of classical linear logic, *Theoretical Computer Science* **597**: 1–17.

[9] Bimbó, K. (2015b). *Proof Theory: Sequent Calculi and Related Formalisms*, CRC Press, Boca Raton, FL.

[10] Bimbó, K. (2016). Review of M. Kanovich, "The undecidability theorem for the Horn-like fragment of linear logic (Revisited)", *Mathematical Reviews* . MR3492992.

[11] Bimbó, K. (2017). On the decidability of certain semi-lattice based modal logics, in R. A. Schmidt and C. Nalon (eds.), *Automated Reasoning with Analytic Tableaux and Related Methods. Proceedings of the 26th International Conference, TABLEAUX 2017, Brasília, Brazil, September 25–28, 2017*, number 10501 in *Lecture Notes in Artificial Intelligence*, Springer Nature, Switzerland, pp. 44–61.

[12] Bimbó, K. and Dunn, J. M. (2012). New consecution calculi for $R^t_\to$, *Notre Dame Journal of Formal Logic* **53**(4): 491–509.

[13] Bimbó, K. and Dunn, J. M. (2013). On the decidability of implicational ticket entailment, *Journal of Symbolic Logic* **78**(1): 214–236.

[14] Bimbó, K. and Dunn, J. M. (2015). On the decidability of classical linear logic, (abstract), *Bulletin of Symbolic Logic* **21**(3): 358.

[15] Boolos, G. S. and Jeffrey, R. C. (1992). *Computability and Logic*, 3rd edn, Cambridge University Press, Cambridge, UK.

[16] Curry, H. B. (1963). *Foundations of Mathematical Logic*, McGraw-Hill Book Company, New York, NY. (Dover, New York, NY, 1977).

[17] Dunn, J. M. (1973). A 'Gentzen system' for positive relevant implication, (abstract), *Journal of Symbolic Logic* **38**(2): 356–357.

[18] Dunn, J. M. (1986). Relevance logic and entailment, in D. Gabbay and F. Guenthner (eds.), *Handbook of Philosophical Logic*, 1st edn, Vol. 3, D. Reidel, Dordrecht, pp. 117–224.

[19] Dunn, J. M. (1995). Positive modal logic, *Studia Logica* **55**: 301–317.

[20] Dunn, J. M. and Restall, G. (2002). Relevance logic, in D. Gabbay and F. Guenthner (eds.), *Handbook of Philosophical Logic*, 2nd edn, Vol. 6, Kluwer, Amsterdam, pp. 1–128.

[21] Forster, Y. and Larchey-Wendling, D. (2019). Certified undecidability of intuitionistic linear logic via binary stack machines and Minsky machines, *Proceedings of the 8th ACM SIGPLAN International Conference on Certified Programs and Proofs (CPP '19), January 14–15, 2019, Cascais, Portugal*, ACM, New York, NY, pp. 1–14.

[22] Gentzen, G. (1935). Untersuchungen über das logische Schließen, *Mathematische Zeitschrift* **39**: 176–210.

[23] Girard, J.-Y. (1987). Linear logic, *Theoretical Computer Science* **50**: 1–102.

[24] Girard, J.-Y. (1995). Linear logic: Its syntax and semantics, in J.-Y. Girard, Y. Lafont and L. Regnier (eds.), *Advances in Linear Logic*, number 222 in *London Mathematical Society Lecture Note Series*, Cambridge University Press, Cambridge, UK, pp. 1–42.

[25] Hopcroft, J. E., Motwani, R. and Ullman, J. D. (2007). *Automata Theory, Languages, and Computation*, 3rd edn, Pearson Education, Boston, MA.

[26] Kanovich, M. (1995). The direct simulation of Minsky machines in linear logic, in J.-Y. Girard, Y. Lafont and L. Regnier (eds.), *Advances in Linear Logic*, number 222 in *London*

Mathematical Society Lecture Note Series, Cambridge University Press, Cambridge, UK, pp. 123–145.
[27] Kanovich, M. (2016). The undecidability theorem for the Horn-like fragment of linear logic (Revisited), *Mathematical Structures in Computer Science* **26**(5): 719–744.
[28] Kanovich, M. I. (1994). Linear logic as a logic of computations, *Annals of Pure and Applied Logic* **67**: 183–212.
[29] Kopylov, A. P. (2001). Decidability of linear affine logic, *in* A. R. Meyer (ed.), *Special issue: LICS 1995*, Vol. 164 of *Information and Computation*, IEEE, pp. 173–198.
[30] Kripke, S. A. (1959). The problem of entailment, (abstract), *Journal of Symbolic Logic* **24**: 324.
[31] Kripke, S. A. (1963). Semantical analysis of modal logic I. Normal modal propositional calculi, *Zeitschrift für mathematische Logik und Grundlagen der Mathematik* **9**: 67–96.
[32] Lambek, J. (1958). The mathematics of sentence structure, *American Mathematical Monthly* **65**(3): 154–170.
[33] Lambek, J. (1961a). How to program an infinite abacus, *Canadian Mathematical Bulletin* **4**(3): 295–302.
[34] Lambek, J. (1961b). On the calculus of syntactic types, *in* R. Jacobson (ed.), *Structure of Language and its Mathematical Aspects*, American Mathematical Society, Providence, RI, pp. 166–178.
[35] Lincoln, P., Mitchell, J., Scedrov, A. and Shankar, N. (1992). Decision problems for linear logic, *Annals of Pure and Applied Logic* **56**: 239–311.
[36] Meyer, R. K. (1966). *Topics in Modal and Many-valued Logic*, PhD thesis, University of Pittsburgh, Pittsburgh, PA.
[37] Minsky, M. L. (1967). *Computation: Finite and Infinite Machines*, Automatic Computation, Prentice-Hall, Englewood Cliffs, NJ.
[38] Riche, J. and Meyer, R. K. (1999). Kripke, Belnap, Urquhart and relevant decidability & complexity, *in* G. Gottlob, E. Grandjean and K. Seyr (eds.), *Computer Science Logic (Brno, 1998)*, number 1584 in *Lecture Notes in Computer Science*, Springer, Berlin, pp. 224–240.
[39] Roorda, D. (1989). Investigations into classical linear logic, *ITLI Prepublication Series for Mathematical Logic and Foundations ML–89–08*, Institute for Language, Logic and Information, University of Amsterdam.
[40] Roorda, D. (1991). *Resource Logics: Proof-theoretical Investigations*, PhD thesis, University of Amsterdam.
[41] Smullyan, R. M. (1968). *First-order Logic*, Springer-Verlag, New York, NY.
[42] Straßburger, L. (2019). On the decision problem for MELL, *Theoretical Computer Science* **768**: 91–98.
[43] Thistlewaite, P. B., McRobbie, M. A. and Meyer, R. K. (1988). *Automated Theorem Proving in Non-classical Logics*, Pitman, London, UK.
[44] Troelstra, A. S. (1992). *Lectures on Linear Logic*, Vol. 29 of *CSLI Lecture Notes*, CSLI Publications, Stanford, CA.

DEPARTMENT OF PHILOSOPHY, UNIVERSITY OF ALBERTA, CANADA, *Email:* bimbo@ualberta.ca
LUDDY SCHOOL OF INFORMATICS, COMPUTING AND ENGINEERING, AND DEPARTMENT OF PHILOSOPHY, INDIANA UNIVERSITY, BLOOMINGTON, IN, U.S.A.

INTENSION, EXTENSION, DISTRIBUTION AND DECIDABILITY

Ross T. Brady

ABSTRACT. The cue for writing this paper is taken from a brief discussion on the reasons for the undecidability of the sentential relevant logic **R** with the participants of "The Third Workshop" at the University of Alberta, the group including Michael Dunn and the author. We start with the distinction between intension and extension, raised by the author in the above discussion. This distinction is discussed with reference to the various systems of relevant logics and classical logic, with a view to yield appropriate formalizations characterising these concepts. We follow with discussion on the role of truth in logic, and its application to rules and to Disjunctive Syllogism, in particular. In the light of the above discussions, we consider distribution in its axiom form, which was raised by Dunn. We conclude with an examination of the existing proofs of decidability of sentential relevant logics raising the respective problems for the logic R, and then with an examination of the proof of the undecidability of **R**, thus rounding out this discussion at "The Third Workshop."

Keywords. Admissible rules, Decidability, Distribution, Extension, Intension, Meaning and truth, Priming

1. INTRODUCTION

Michael Dunn will be sadly missed, not only for his significant contribution to logic (especially, relevant logic), informatics and computer science, but also for being a great friend to many, myself included. I have happy memories of his and Sally's visits to Melbourne on so many occasions. His passing has left a great hole in our lives. I am honoured to contribute to his memorial volume and very grateful that Katalin Bimbó has taken upon herself to organize it. I apologize in advance for including so much of my own work in this paper as it covers a variety of topics, which I will be addressing according to my own views on what logic ought to look like.

This paper is based on a brief discussion at The University of Alberta, Edmonton, taking place during the "The Third Workshop" on the Routley–Meyer semantics and its three-place relation R, organized by Katalin Bimbó during 2016. The discussion was initiated by Guillermo Badia who asked the question of the assembled group as to why the sentential relevant logic **R** is undecidable. In response, I started by saying that it was due to the mixing up of intension and extension and Dunn finished by saying that distribution was responsible for the undecidability. We will investigate these issues through a clarification of the respective concepts, in the process referring to some related work of Michael Dunn. Nevertheless, the author will pursue his own

2020 *Mathematics Subject Classification.* Primary: 03B47, Secondary: 03B25.

Bimbó, Katalin, (ed.), *Relevance Logics and other Tools for Reasoning. Essays in Honor of J. Michael Dunn*, (Tributes, vol. 46), College Publications, London, UK, 2022, pp. 128–149.

ideas, some of which would be at variance with those of Dunn. We initially started on the same page, focussing on strong relevant logics, but I started to deviate firstly by raising concerns over the failure of Modus Ponens to preserve truth at a base world of un-reduced Routley–Meyer semantics and then, more importantly, over the difficulties in maintaining the relevance condition when the logic **R** was applied to set theory and arithmetic. (See Brady [16] on the first point and Brady [20] on the second, with some examples in §2.)

In §2, we start with the distinction between intension and extension, these concepts giving rise to the respective formalizations: my logic of meaning containment and classical logic. We then focus on other logics, attempting to characterize them using logical concepts. In §3, we explore the concept of truth and its role in logic, in the context of analyticity and rules, this serving to round out the two key logical concepts of meaning and truth. Further, we examine the status of the disjunctive syllogism rule in this context and the classical recapture within the framework of my logic of meaning containment. In §4, we examine the similar distinction between implication and entailment, with reference to the intension/extension distinction and to the use of rules.

In §5, we consider distribution in its axiom and rule forms, evaluating these in the light of the intension/extension and implication/entailment distinctions. We drop distribution in both of these forms, whilst maintaining its rule form as an admissible rule. In §6, we explore the decidability of the various sentential relevant logics looking into their positive results together with their negative results. The methods used to prove decidability will include semantic filtration and reductio, normalized natural deduction and cut-free Gentzenization. We do consider Urquhart's undecidability proof, but conclude that the decidability of the logics is unprovable. Finally, in §7, in the light of all the above discussions, we will gain some insights as to why sentential logics such as **R** fail to be decidable, in answer to Guillermo Badia's question.

2. Intension and Extension

A useful way of starting this discussion is by examining the conceptual distinction between classes and sets, as set out in Brady [23]. Classes are generated by predicates and are thus intensionally determined in accordance with the meaning of the generating predicate. Sets are axiomatically introduced in such a way as to represent collections of individuated objects, this concept originally being expressed by Cantor. Thus, sets are extensional as they are set up in such a way to provide clarity as to what objects are in the set and what objects are not. (Note that this would apply just to recursive sets.) This contrasts with classes where the predicate determines what objects are in the class in accordance with the meaning of the particular predicate. However, there would be quite some interaction between these two concepts, as occurs in [23]. Another way of developing classes in conjunction with sets is in the class theory NBG, with its proper classes extending set theory. We need to be able to apply the intension/extension distinction to logics themselves and especially their connectives.

We start by characterizing extensional logic as classical logic. This is usually taken to be clear-cut as truth and falsity are mutually exclusive and exhaustive in line with objects either being in a set or not in the set. As mentioned above, the application of classical logic would thus be restricted to recursive sets. However, we will raise the issues of the law of excluded middle and disjunctive syllogism in §3, and classical recapture within the framework of the intensional logic about to be considered.

We next attempt to characterize intensional logic, seen through the axiomatic capturing of each of its connectives. Meaning is the core concept of logic, as all logical formalization is generally aimed at capturing the meanings of words, especially the logical words. Admittedly, meaning is harder to capture than the truth and falsity of classical logic, but I have agonized over many years to achieve the resulting system **MC** of meaning containment, with its final axiomatization in §5 below. Nevertheless, we start by considering the logics **DW** and **DJ**, and proceed in stages, much as I did over the years. We will proceed to extend **DJ** to **DJ**^{d} of [20] and [23] with the inclusion of its single-premise meta-rule in §3, and finally to the logic **MC** in §5 with the removal of distribution. A detailed account of the various connectives can be found in Brady [26]. Briefly, conjunction and disjunction satisfy their standard introduction and elimination laws, with distribution to be discussed later. Negation satisfies the De Morgan properties, making it an incomplete concept as this does not say anything about its application to atoms or to entailments. One can see in Brady [24] that this negation has a cancellation feature through the use of metavaluational trees, where negations are cancelled against each other in pairs. (However, this differs from Routley and Routley [53, page 205], where single negations are cancelled off against their unnegated forms.) This negation is also called mirror-image negation in Routley et al. [52], with the mirror acting as a line of symmetry between the negated and its corresponding unnegated formula. We note that this will not include the law of excluded middle, $A \vee \sim A$ (LEM), nor the disjunctive syllogism, $\sim A, A \vee B \Rightarrow B$ (DS), as both involve single uncancellable negations and whose main justification is based on classical truth and falsity, which are mutually inclusive and exhaustive. (See Brady [32] for a full discussion on the rejection of the LEM and see below in §3 for discussion of the LEM and the DS.) The entailment $\rightarrow$ is interpreted as a meaning containment, with more on this below and more on entailments in §4.

Looking into these two logics raises the issue of other logics that can also be given a conceptual characterization, given that logic should primarily be a conceptual study rather than just a technical study. An obvious example would be intuitionist logic which is classical logic made disjunctively constructive, as can be seen by its Gentzenization, having at most one consequent after the turnstile. However, whilst there is conceptual and technical interest in constructivity, the problem with intuitionist logic itself is not only the unjustified persistence of constructivity into applications beyond the logic itself (which would even apply beyond its mathematics), but also the remaining strength of its implicational laws such as $A \rightarrow . B \rightarrow A$, which leads to $B \rightarrow A$, if A is a theorem. (See §3 below for more on the reach of constructivity through priming.) In such a case, $B \rightarrow A$ fails the relevance condition: if $A \rightarrow B$ is a theorem, then A and B share a sentential variable. This condition is a necessary condition for a good intensional logic, as it establishes some commonality of meaning between the antecedent A and the consequent B, but it is not sufficient to characterize meanings entirely.

What we need to do is to provide a complete characterization to establish a clear concept which can then be applied not only to establish a logic but also to apply to its non-logical extensions. Indeed, there are problems in the application of logics such as **R**, as the relevance condition upon which it is based cannot be maintained in application. (See [20, p. 158], for the following examples in set theory and arithmetic: $x = y \rightarrow . p \leftrightarrow p$ and $m = n \rightarrow l = l$.) The same goes for other strong relevant logics such as **E**, **T**, **RW** and **TW**, which are all primarily based on the relevance condition without an underlying logical concept and are all too strong to characterize intensionality by themselves. (See the Appendix for these logics and all other logics mentioned in this paper.) Nevertheless, as we can see from the research work done on them, they are all technically interesting and have some degree of intensionality built into them, depending on their distancing between classical logic and my logic of intensionality, as specified below over sections §3–5.

So, we are left to consider weaker logics such as **DW** and **DJ** for appropriateness to characterize intensionality, where the meanings of all their connectives are used to determine the logic. (See the Appendix for axiomatizations of **DW** and **DJ**.) We especially need to determine the role of $\rightarrow$ as a connective capturing meaning, being aware that such a connective applies whether the constituents are true or false, without a presumption of truth for the antecedent. So, a direct analysis of the meanings of both the antecedent and consequent is required, with an appropriate relationship between them to reflect consequential certainty achieved through the analysis of the antecedent. Such a relationship would be meaning containment, with the meaning of the antecedent containing that of the consequent, used in [20; 23], and supported by a content semantics with logical contents $c(A)$ and $c(B)$ assigned to formulae A and B such that $c(A \rightarrow B) = c(c(A) \supseteq c(B))$, where contents $c(A)$ are sets of formulae and the containment between $c(A)$ and $c(B)$ is set-theoretic. Note that it is often easier to examine the meaning of the antecedent as an extension of the meaning of the consequent. In particular, the meaning of $A \vee B$ is extended to that of A and that of B, in the respective assessments of $A \rightarrow A \vee B$ and $B \rightarrow A \vee B$. This is the appropriate meaning concept for a connective, being deeper than just deductive meaning analysis, which we will consider in the next section, where we will also consider how the intensional and extensional logics should interact with each other.

We first need to tinker with a few key axioms and rules to determine a suitable logic of intensionality. We start with the conjunctive syllogism axiom, $(A \rightarrow B) \& (B \rightarrow C) \rightarrow . A \rightarrow C$, which when added to **DW** yields **DJ**. (Also, see the Appendix for the bracketing conventions.) It does seem appropriate for content containment to be transitive, especially, as it is based on set-theoretic containment, which is of course transitive. Further, we can look at the meaning of $A \rightarrow C$ as being extended to that of $(A \rightarrow B) \& (B \rightarrow C)$, since the containment of C in A is then extended to that of C in A via B, assuming transitivity. A possible problem with conjunctive syllogism is that it is a mild form of contraction in that it is deductively equivalent to $A \circ B \rightarrow A \circ (A \circ B)$, where fusion $\circ$ satisfies the two-way rule, $A \rightarrow . B \rightarrow C \Leftrightarrow A \circ B \rightarrow C$. Nevertheless, in [23], it is shown that naive set theory based on the logic **DJ** can be proved to be simply consistent, though this fails with the addition of fusion $\circ$ due to a Curry-type paradox. It should be said that fusion, in satisfying the above two-way rule, combines antecedents in a way that does not make much sense when the $\rightarrow$ is interpreted as a meaning containment.

Further tinkering will still need to be done. This will include the addition of a single-premise meta-rule, which will be discussed in §3, and the possible addition of a two-premise meta-rule together with a discussion of distribution, which will occur in §5.

3. Truth, Rules, Disjunctive Syllogism and Classical Recapture

We tend to think of truth and meaning as being the two basic semantic concepts needed to determine logical systems, but this needs some investigation. We have already considered information and necessity in Brady [33], with information being the derivative concept, true content, as argued in Brady [27], and necessity generally relying on quantification over possible worlds in a truth-theoretic semantics, which do not properly capture the meaning of disjunction as expressed in proof-theory. (On this last point, see below for further discussion and see Brady [29] for more detail.) So, we will examine the role of truth to round out our understanding of logical systems. Nevertheless, meaning is the central concept of logic in that all formalizations endeavour to capture the meanings of words in sentences. Hence, the meanings of connectives in sentential logic should be captured in all axiomatic systems.

There are two key uses of truth in deductive logic. First, analytic truth is central to logic as all logical truths are analytic, determined by analysing the meanings of their logical words. Further, the truths of key logical applications to set theory and arithmetic, and other mathematical applications as well, are all analytic as they are deduced from mathematical concepts. Towards the second usage of truth, let us first consider the definition of a valid deductive argument as an argument that requires its conclusion to be certain, given its premises, in contrast to the lack of certainty for an inductive argument, where a high probability of its conclusion suffices for a good inductive argument. This certainty would be determined by meaning analysis which can either be applied just to the constituents of the conclusion or, more commonly, to the constituents of the premises, such analysis ensuring the certainty of the conclusion. It is hard to see how else this certainty could be guaranteed. Valid logical deductions are often characterized as cases of necessary truth-preservation. However, as above, necessity is not helpful here because of its usual relationship with possible worlds in a modal context. (For a discussion of worlds, see below.) Further, the irrelevance of a certain conclusion from unrelated premises is of no concern here as relevance applies to the tighter relation between antecedent and consequent of an entailment. Having certain conclusions make a deductive argument valid does enable one to suppress analytic truths in a deductive argument. That is, $A \Rightarrow T$ is deductively equivalent with: if $A, T \Rightarrow B$ then $A \Rightarrow B$, where T is analytic. (Such suppression of analytic truths is commonly used in practice as we standardly drop off axioms that are used in a derivation.) Note that this differs from the relevant deduction of Brady [35] and Brady and Rush [38].

For the second key usage of truth, what we can say for valid deductive arguments is that truth is preserved from premises to conclusion, truth being assumed for the premises and carried through to the conclusion. The preservation of truth for the rules can be seen from the content semantics of $\mathbf{DJ}^d$ in [20] and [23, p. 63]. (We restrict ourselves here to sentential systems. However, for predicate systems, the generalization rule would need to be restricted in its application for this purpose.) The important difference between a step in a deductive argument and a formula embedded in such a step is that there is no provision for a failure to hold in the deductive step, whereas the appropriate thing to do for an embedded formula is to put its meaning into that position, with no concern about whether the embedded formula holds or not. Moreover, the truth in a deductive step is assumed truth which can even apply when the premise is false in fact (or even contradictory), but the argument proceeds on the basis of this assumed truth, determining what would follow if it were true. (A typical example of this is in the case of fictional novels.) Thus, a formula that is used as a deductive step is given a positive read, which enables this positive meaning to be applied in the deductive process in accordance with what it says. So, such a positive read is the substitute for truth in a deductive argument. This lessens the value of truth in logic to that of analytic truth, which is in turn determined by meaning. This, we believe, leaves meaning as the core concept of logic.

These are also uses of truth in standard truth-theoretic semantics, where validity of a formula requires truth in all interpretations and validity of an argument requires truth-preservation in all interpretations. However, there are problems with such semantics in that they do not provide a proper interpretation of disjunction in that, due to its formula-inductive structure, it is given the priming property (if $A \vee B$ is true at a world then either A or B is true at that world, as witnesses for the disjunction in the inductive process) at variance with its proof-theoretic meaning, with its witness-free induction on proof steps, as was discussed in [29]. Standardly, in a proof-theoretic setting, disjunctions are eliminated by assuming each disjunct separately and proving a common conclusion. There is no need to declare a particular disjunct as being the case.

Before proceeding further, we clarify the difference between meaning containment and meaning analysis used to derive conclusions of deductive arguments, expressed as rules of deduction in formal axiomatization. As discussed in §2, as a connective, meaning containment must assess the relationship between antecedent and consequent whether the antecedent is true or false. On the other hand, the meaning analysis used to validate rules of deduction assumes that the premise is true, including it as a deductive step together with any appropriate analytic truths that aid the derivation of the conclusion. Both aspects are at variance with meaning containment, where analytic truths are not always suppressible for a meaning containment to continue to hold. For example, $(A \to A) \& (B \to C) \to .A \& B \to A \& C$ is valid in **DJ** but $B \to C \to .A \& B \to A \& C$ is not. On the relationship between entailments and their corresponding rules, see [23, pp. 29–30], where the case of the formula $A \& (A \to B) \to B$ (invalid in **DJ**) is assessed in contrast to the rule, $A, A \to B \Rightarrow B$. Here, because $A \to B$ holds, what it says and means enables it to apply to the other premise A, thus enabling B to be deduced. In contrast, the two A's within the antecedent $A \& (A \to B)$ do not connect as the second A is part of a containment statement whilst the first A is on its own.

We add the following single-premise meta-rule MR1 to **DJ** to obtain the logic **DJ**^{d} of [20; 23].

MR1 If $A \Rightarrow B$ then $C \vee A \Rightarrow C \vee B$.

Initially, this was generally introduced into logics weaker than Anderson and Belnap's system **T** of ticket entailment to ensure that their reduced modellings preserved truth at the base world of a Routley–Meyer semantics. (See the Appendix for such systems.) MR1 is justified using the standard disjunction elimination rule, by first assuming C and then by assuming A. Indeed, it can be easily seen that MR1 is deductively equivalent to the rule-form of disjunction elimination: if $A \Rightarrow C$ and $B \Rightarrow C$ then $A \vee B \Rightarrow C$. For those logics which are metacomplete, no new theorems are added by MR1, as pointed out by Slaney in his [55]. (See Meyer [49], Slaney [54] and [55] and Brady [28] for an account of metacompleteness.) Nevertheless, I added MR1 so that any need for it in extensions of the logic can be guaranteed, as any logical rule persists into every application of the logic. Further, the meta-rule MR1 is very useful in a variety of applications concerning rules, including the relationship between rules and their classical $\supset$-form.

In order to proceed with the classical recapture, we first examine the rule Disjunctive Syllogism (DS), $\sim A, A \vee B \Rightarrow B$. As argued above, rules require meaning analysis to establish their conclusion, either on the assumption of the premises or within the conclusion itself. If one tries to apply this to the DS, the derivation of B would rely on the assumption of $\sim A$, the consequent rejection of A, and priming for $A \vee B$, to allow the B disjunct to be established as the conclusion. That is, together with priming for $A \vee B$, A needs to be simply consistent either just applied within itself to A and $\sim A$ or, more likely, within a broad set of formulae or, indeed, the whole system. However, these two are truly meta-theoretic properties, as they both concern provability. Thus, the DS is a metarule in the truly metatheoretic sense of the word, as opposed to the above meta-rule MR1 which is just a rule directly relating rules. So, unlike the other familiar rules, it is not derivable by meaning analysis, and thus the paraconsistent logicians do have a point in not including it in their logics, as with Ex Falso Quodlibet (EFQ), $A, \sim A \Rightarrow B$, which easily follows from the DS. The converse is also provable, making them deductively equivalent, given priming for the disjunction of the DS.[1] Recall that Belnap omitted the DS from his relevant logics as it would cause a failure in the Deduction Theorem through the absence of $\sim A \,\&\, (A \vee B) \to B$ in the logic.[2] The paraconsistentists, however, reject Ex Falso Quodlibet and hence the DS on relevance grounds.

Classical recapture for $\mathbf{DJ^d}$ needs both the LEM and the DS to apply to formulae of the system, either partially to some formulae or to all the formulae of the system, assuming of course non-triviality of the system, but also with the assumption of priming for the disjunctions occurring in the LEM and the DS. In order to prove an instance of the LEM, it must be prime in that it must be derived from one of its disjuncts, given that the LEM is not a theorem of the logic involved. As above, the disjunction in the DS must also be prime for it to be justified as holding. This will enable all classical tautologies expressed in such formulae to be derived using rules of $\mathbf{DJ^d}$ (and later the

---

[1] Given priming for $A \vee B$, we prove the DS by applying EFQ to $\sim A$ and A yielding B and hence the DS follows.

[2] Belnap made this point regarding the DS in Anderson and Belnap [1, pp. 296–300], the Deduction Theorem being called the Entailment Theorem.

system **MC** of meaning containment).³ Whatever formulae satisfy both the LEM and the DS will be termed classical formulae. Further, if the LEM holds for each member of a set of atoms then it will continue to hold for all formulae built from these atoms using &, $\vee$ and $\sim$. Similarly, if the DS, $\sim A, A \vee B \Rightarrow B$, holds for each of a set of atoms A of the DS then it continues to hold for all prime formulae built from these atoms using &, $\vee$ and $\sim$.⁴ Thus, one only needs to show that the LEM and the DS hold for atoms. However, we note that these atoms can be replaced by formulae, with the derived classical formulae built from these formulae in lieu of the atoms.

We next consider the LEM. As argued in Brady [29; 31], logic proceeds through derivations in proof and is not accurately captured by the standard truth-theoretic semantics, due to disjunction, when understood proof-theoretically, not satisfying the priming property of truth-theoretic semantics, as was discussed above. The main justification for the LEM is that sentences are either true or false. However, when truth is replaced by proof, this would require negation-completeness. However, negation-incompleteness is ubiquitous as concepts are often not fully specified, which then leads to the large-scale failure of the LEM. (See [32] for a fuller discussion of the LEM.)

In the arithmetic papers of Brady, i.e., [25] and [33], each instance of the LEM is proved via one of its disjuncts using A5 and A6 of the Appendix, whilst the DS is brought in as an admissible rule once simple consistency of the arithmetic is shown in a metacomplete system. This mode of classical recapture is ideal for systems in general, as consistency is more likely to apply generally to the whole system and it is a property one would expect a worthwhile system to satisfy. (See [31] on this point.) What we achieve here is the simple consistency of primitive recursive arithmetic in [25] and subsequently in [33] general recursive arithmetic, these theories being classically recaptured.

Let us finish by examining when the priming property should hold, as this question will often arise. Indeed, in §2 above, this question arose for intuitionist logic, which is conceptually based on constructivity. Here, the priming property would hold for applications as well as for the logic itself. However, in [32] we argue that priming should hold for logical theorem instances, with particular reference to the LEM. For **DJ**^{d} and many logics in its vicinity, priming follows from metacompleteness. (See [49] and [54; 55] for metacompleteness.) Priming still holds for primitive and general recursive arithmetic, given the form of its axiomatization and the extension of metacompleteness to these arithmetics. (See [25] and [33] on this.) Moreover, priming can fail for non-logical axioms, especially disjunctive ones. For example, it is perfectly possible to have disjunctive information, without the disjunction being resolved in favour of one disjunct or the other. It is this concept that is justifiably captured in proof-theoretic systems (as opposed to truth-theoretic semantics), each disjunct being assumed for the

---

³This is proved by using the method of normal forms, where each tautology is shown to be deductively equivalent to a conjunction of disjunctions each of which includes at least one LEM pair. Use is made of simple properties of **DJ**^{d}, including the rule-form of distribution, in these derivations. For this to apply to the system **MC** of §5 without distribution, we show that its rule-form is an admissible rule in §5. This would then hold for all metacomplete and prime extensions.

⁴To achieve this result for the DS, we not only need to use priming for each of its disjunctions but this also applies for any disjunction or negated conjunction formula occurring in this process.

purpose of entailing or deriving a common conclusion. Proof-theoretic systems must apply not only to the logic itself but also to all applications and extensions of the logic, which would include the above disjunctive information. However, if a rule is only an admissible one, as with DS above, then it holds for the logic but it needs to be checked for each and every application and extension of the logic.

4. THE IMPLICATION/ENTAILMENT DISTINCTION

Implication and entailment are traditionally related by defining entailment as necessary implication. This applied initially to classical logic, where material implication $\supset$ is taken to be the implication, and where necessity $\square$ is added in a classical modal logic to yield what was called strict implication in the form $\square(A \supset B)$ by C. I. Lewis. (This is set out in Hughes and Cresswell [45] for the Lewis modal logics **S1–S5**.) These strict implications were taken to capture entailments from a classical point of view, as appropriate to the particular modal logic. However, such strict implications were discredited by Meyer in [48] from a relevance point of view, where he shows, for example, that any sentence strictly implies a necessary truth. As a result, meaning connections, essential to a reasonable understanding of entailment are broken down. Indeed, the breaking down of these meaning connections led Brady to search for an entailment logic based on meaning rather than a necessitated implication. (See [20] for the initial work on such a logic.)

Next, let us briefly dwell on these meaning connections between antecedent and consequent of an entailment. At the connective level, as argued in [33], if the meaning connections are understood by variable-sharing, as in the relevance condition, then this, by itself, does not suffice to establish a suitable concept to base a logic on. Note that strong relevant logics such as **R** require satisfaction of the rules of Modus Ponens and Adjunction, together with the use criterion in Fitch-style natural deduction, in addition to that of the relevance condition, to provide its key characterizing features. However, it is fair to say that the relevance condition is the major one and that this would be expected to hold in applications of the logic. As discussed above and in [20] and [23], we need to move to a logic such as $\mathbf{DJ^d}$ based on meaning containment to obtain a suitable logic embracing meaning. Although satisfying the relevance condition, it is a single concept embracing meaning, which is the core semantic concept to base a logic upon. Furthermore, it is also transitive, which the relevance condition is not. (Also, see [33] on this point.) We conclude that the connective $\rightarrow$ of such a logic as $\mathbf{DJ^d}$ of meaning containment represents a good concept of entailment based on meaning, though further tightening is needed to produce our final logic **MC**.

To help characterize implication, we need to examine the other semantic concept, truth. To get started on this, we did introduce "the two inference concepts" in section 17.3 of [26], one of which was, of course, entailment, represented by the connective $\rightarrow$. The other concept was that of the primitive rules of inference and derived rules, linking the premises and conclusion of valid deductive arguments, represented by the rule $\Rightarrow$. Such rules preserve truth in sentential logics, as has been discussed above in §3 and in [26], this preservation being ensured formally in the various semantics by the inclusion of the disjunctive meta-rule MR1, which is given in the axiomatization of $\mathbf{DJ^d}$. Initially MR1 was introduced in single-premise form in Brady

[10] to give relevant logics (with distribution) a reduced Routley–Meyer semantics in which the rules preserved truth at the base world. (See [52], for these semantics.) Further, the meta-rule is used to establish truth-preservation in the algebraic-style content semantics, initially introduced in Brady [11] and Brady [12], subsequently narrowed down in [20] and [23] to the logic **DJ^d**.

With this background, we need to proceed to the determination of implication. Truth-preservation would certainly be its key property, which would differentiate it from entailment with its key property based on meaning. There is, of course, classical material implication, represented by the connective ⊃, which is truth-preserving in accordance with its truth-table. There is also relevant implication, which is understood as truth-preserving, subject to the relevance constraints of the relevance condition, its two rules, and the use requirement of its Fitch-style natural deduction system which rounds out the logic **R**. (See [1] for these two constraints on the logic **R** and its axiomatization.) As argued above, the logic **R** of relevant implication is not an appropriate logic, mainly due to the difficulty of maintaining the relevance condition in applications.

Finally, putting other systems such as **R** aside as they involve a collection of different characterizing features, let us examine the pure truth-preservation of material implication in classical logic as the obvious contender for a good concept of implication based on truth. The problem here is that its truth-preservation depends on Boolean negation. Indeed, it is the classical negation properties of the LEM and the DS (both applied to the A) that ensure that $A \supset B$ and $A \Rightarrow B$ are deductively equivalent and $A \Rightarrow B$, of course, preserves truth. (See footnote 5 for "deductive equivalence.") The trouble here is that neither the LEM nor the DS are included in the logic **DJ^d** as it is based just on De Morgan negation as a type of cancellation negation. (See above discussion in §2 and with the detail in Brady [24] on this last point.) So, the rule-form, $A \Rightarrow B$, is purer as an implicational concept in that it cuts down on extraneous elements such as negation, but not entirely as it does embody meaning analysis in getting from A to B. Given the core nature of meaning in deductive logic, this cannot be helped, as was discussed above. Further, as rules do not yield anything when A is unproven, they better capture truth-preservation than a connective such as ⊃, which would need to deal with a false antecedent. Thus, implication and entailment are separately conceptualized, being separately based on the two inferential concepts expressed using ⇒ and →, respectively, in a good meaning-based logic. We should say, however, that this applies to sentential logics only, which are the focus of this paper, due to the form of Badia's original query. At the quantificational level, the generalization rule, $A \Rightarrow \forall x A$, is not an implication and the quantificational meta-rule for ∃ is restricted to avoid instances of generalization in the application of this meta-rule, in axiomatizing the quantificational extension of **DJ^d** (and also of **MC** below).

5. Distribution

First, we briefly explain why the →-form of distribution, $A \& (B \vee C) \rightarrow (A \& B) \vee (A \& C)$, fails as a meaning containment. As set out more fully in Brady and Meinander [36], the natural deduction introduction and elimination rules for conjunction and disjunction suffice to establish their respective uniquenesses without the need for distribution. Once uniqueness is established, such conjunction and disjunction can be substituted in any context and thus distribution is not used in determining the meanings of conjunction and disjunction within the logic. Therefore, distribution in its $\to$-form is not a meaning containment and so it is dropped from the logic **DJ**^{d} in forming our final logic **MC** of meaning containment, set out in the Appendix below.

We next explain why the $\to$-form of distribution holds in truth-theoretic worlds semantics. Because of the way truth-theoretic semantics is set up by induction on formulae, each disjunction in a world requires a witness to establish its truth. Such a witness for $B \vee C$, together with A, can then be conjoined to form the respective conjunction $A \& B$ or $A \& C$ within each world, giving us $(A \& B) \vee (A \& C)$. This shows the way distribution works, that is, by enabling the respective conjunctions $A \& B$ or $A \& C$ to be formed, after assuming $A \& (B \vee C)$. Note that this form of distribution is not so easily introduced in natural deduction, as it requires a special rule $\&\vee$ to make it work, over and above the standard introduction and elimination rules for $\&$ and $\vee$. The problem here is that B and then C are assumed as further hypotheses, which cannot then be conjoined with the A as it is without such further hypothesis, making it unconjoinable as the A and the B and C will then be based on different sets of hypotheses. (See [1, pp. 273–274], regarding the difficulties using the standard introduction and elimination rules for $\&$ and $\vee$ and the introduction of a special rule, and also see [10, p. 362], for the distribution rule $\&\vee$.)

We now need to apply this in an implicational context, rather than as an entailment. This leads us to the rule-form of distribution, $A \& (B \vee C) \Rightarrow (A \& B) \vee (A \& C)$. Here, a similar problem exists at the level of rules, as B and then C still need to be assumed, making them depend on a different assumption from that of A. Indeed, the formation of conjunctions at the level of rules is borne out by the meta-rule: if $A \Rightarrow B$ and $A \Rightarrow C$ then $A \Rightarrow B \& C$, which is the conjunctive analogue of the disjunctive meta-rule: if $A \Rightarrow C$ and $B \Rightarrow C$ then $A \vee B \Rightarrow C$. This conjunctive meta-rule states that conjunctions can be formed from conjuncts which are subject to the same assumption. (Note that the rule $A, B \Rightarrow A \& B$ introduces a conjunction where both conjuncts have no assumptions.) This is at variance with what is happening in rule-distribution when the B and then the C has to be assumed, to provide a conjunction with A, which has no assumption.

Further, both the conjunction and disjunction introduction and elimination rules take the same shapes as for their respective entailments and a similar case can be made for their uniqueness at the level of rules, expressed as rule-equivalence. For this reason, we need to also reject the rule-form of distribution.

However, as for the DS, if the disjunction $B \vee C$ is prime, then the respective conjunction $A \& B$ or $A \& C$ can be obtained for either of the disjuncts B and C of the premise, yielding $A \& (B \vee C) \Rightarrow (A \& B) \vee (A \& C)$ as a metarule in the genuine sense that it requires a meta-theoretic property for it to hold. Given that **DJ**^{d}, and **MC** above, are metacomplete, which yields primeness and simple consistency, the DS and the rule-form of distribution are admissible rules for the logic and all their metacomplete extensions. It is interesting here that Dunn's quantum logic is set up in rule-form in his [41], with the consequent omission of the rule-form of distribution. We agree with this on the grounds that priming fails here, due to electrons having spin up or spin down, but neither being measurable in the context of a determined position. This serves to correct what was argued for in [36], that is, the failure of LEM. The LEM of course would only follow from one of its disjuncts.

Alternatively, the rule-form of distribution can be shown to hold if the A is an analytic truth, independently of whether the $B \vee C$ is prime or not. Then, since $B \Rightarrow A$ and $C \Rightarrow A$, $B \Rightarrow A \& B$ and $C \Rightarrow A \& C$, whereupon $A \& (B \vee C) \Rightarrow (A \& B) \vee (A \& C)$. A similar point can be made for the DS, $\sim A, A \vee B \Rightarrow B$, if $\sim A$ is an analytic truth. For then $A \Rightarrow \sim A$ and $B \Rightarrow \sim A$ and hence $A \Rightarrow A \& \sim A$ and $B \Rightarrow B \& \sim A$, yielding B on the assumption of simple consistency, without the need for $A \vee B$ to be prime.

Next, we consider the following two-premise meta-rule MR2 extending the earlier single-premise meta-rule MR1 of **DJ^d**:

MR2 If $A, B \Rightarrow C$ then $D \vee A, D \vee B \Rightarrow D \vee C$.

This also relies for its justification on the priming property to apply to its premise disjunctions $D \vee A$ and $D \vee B$, for then either D is provable or both A and B are provable, giving us $D \vee C$. It too becomes a metarule in the genuine sense that it depends on a meta-theoretic property and is an admissible rule for the logic and its metacomplete extensions. MR1 has the advantage that there is no conjunction involved and it is deductively equivalent to the justifiable rule-form of disjunction elimination. On the other hand, MR2 is deductively equivalent to the one-premise meta-rule MR1 of **DJ^d**, plus the above rule-form of distribution.[5] Note also that neither MR1 nor MR2 will increase the set of theorems of MC due to the priming property being satisfied for theorems, also following from the metacompleteness of **MC**.

In conclusion, whilst the entailment form is to be rejected, this implicational rule-form of distribution, though not being an integral part of our intensional logic **MC**, holds admissibly over **MC** and any prime extension, such as can be obtained due to metacompleteness. Note that MR2 does appear in earlier axiomatizations of **MC**, but is henceforth removed, as it embraces the rule-form of distribution.

6. Decidability

We start by examining four styles of decidability proof for various relevant logics: finite model property, semantic reductio, normalized natural deduction and Gentzenization, with a view to determine what limits their respective applications. The first two are semantic methods, both of which apply to logics with distribution. The latter two are proof-theoretic, which can apply to logics with or without distribution, though the removal of distribution creates simplification of the method and possibly a wider cast of systems to which the method would apply. This examination is done as an interesting technical exercise, without regard to the need for logics to capture appropriate concepts.

Maksimova first showed decidability for a relevant logic without the two hypothetical syllogism axioms, $A \to B \to .B \to C \to .A \to C$ and $A \to B \to .C \to A \to .C \to B$ in 1969. (See Bimbó and Dunn [8] for the details of this.) Then, Fine [42] showed

---

[5]We prove this by showing that $(D \vee A) \& (D \vee B) \Rightarrow D \vee (A \& B)$ is deductively equivalent to $A \& (B \vee C) \Rightarrow (A \& B) \vee (A \& C)$, via Belnap's form of distribution, $A \& (B \vee C) \Rightarrow (A \& B) \vee C$. Note that deductive equivalence allows substitutions to be made on formula schemes, as well as applications of the usual deductive rules.

decidability for a wide range of relevant logics without these two hypothetical syllogism axioms. This includes all the familiar strong relevant logics, including **R**, **E**, **T** and **RW**, but without these two hypothetical syllogism axioms, as well as weaker relevant logics such as **B**, **DW**, **DJ** and **DK**. He used the finite model property, which involves looking at the formulae under test in his style of semantics for relevant logics and assessing their validity or otherwise in models that are limited by the atoms and subformulae occurring in the formula, thus creating a finite model for the formula under test. However, his semantic postulates for the hypothetical syllogisms do not permit his finite model procedure to work. (See [42, p. 368], for the details.) Note that the LEM is included in his basic logic and hence all his logics under consideration. Nevertheless, we assume that the LEM can be removed, where appropriate. Note also that Fine's semantics has the advantage of simplifying the semantic postulates for $\{\rightarrow, \&\}$-formulae due to their use of theories, as opposed to the prime theories used in the Routley–Meyer semantics. As they do not require priming, Fine's theories relate more closely to the proof theory allowing the Routley–Meyer existential semantic postulates to be replaced in favour of algebraic-style postulates.

Semantic filtration is set out in [21], for the reduced Routley–Meyer semantics, which follows on from Routley's three filtration attempts in section 5.9 of [52, pp. 399–406], though only the second filtration is shown by Brady to work. The reduced semantics is further simplified by Priest and Sylvan [50] and Restall [51]. It is Restall's further simplification that we use here. The general method of filtration is to create suitable finite models which would then be used to establish the finite model property. Here, the finite models are restricted by only considering the subformulae of the formula under test, which is closed under the negation of unnegated formulae. However, the two complex postulates for the hypothetical syllogism pair are problematic in that there is difficulty in ensuring that the two z's are the same in their common conjunctive antecedent $Rabz$ and $Rzcd$ after the second filtration is applied.[6] Without these two postulates, only a finite number of models of this sort need be so generated, giving rise to a decidability argument, similar to that of Fine's for his semantics. This decidability applies to all familiar relevant logics, but without the pair of hypothetical syllogisms. However, this method is tedious and decidability is better proved using a reductio argument, still based on filtration, as given in [21, pp. 15–18]. This reductio method follows that of [45], used there for modal logics.

A reductio method is also given for contraction-less logics in [21, pp. 19–26]. This is a direct unfiltered semantic method based on truth-trees, also modelled on [45], using the reduced Routley–Meyer semantics of [51], as was used above. We start with the formula under test being assigned false at the base world and work this back to the atoms in a finite tree structure. As for the tree method, if a contradiction is proven then the formula cannot be false and must be valid. The contraction-less logics **RW**^{d}, **TW**^{d}, **DW**^{d} and **B**^{d} can be shown to be decidable by this method, with the possible addition of the LEM or the DS to **TW**^{d}, **DW**^{d} and **B**^{d}. However, any one of the axioms incorporating contraction, A10, A15, A16, and A17, when added to **TW**^{d}, **EW**^{d} or **RW**^{d}, can keep creating new worlds, producing an infinite sequence of worlds, thus

---

[6]The semantic postulates for the two hypothetical syllogism axioms are: $Rabz\,\&\,Rzcd \Rightarrow \exists x \in K.Racx\,\&\,Rbxd$, $Rabz\,\&\,Rzcd \Rightarrow \exists x \in K.Rbcx\,\&\,Raxd$.

producing a limitation for this method. The same applies to the addition of either the LEM or the DS to $\mathbf{RW^d}$, and the addition of the E-rule R5 to $\mathbf{TW^d}$. (Note that $\mathbf{TW^d}$ and $\mathbf{TW}$ have the same set of theorems, as do $\mathbf{B^d}$ and $\mathbf{B}$, $\mathbf{DW^d}$ and $\mathbf{DW}$, and $\mathbf{RW^d}$ and $\mathbf{RW}$, due to their respective metacompleteness.)

We now move on to proof-theoretic methods, starting with normalized natural deduction. As set out in Brady [22] for the logic $\mathbf{DW}$, normalization of Fitch-style natural deduction eliminates introduction and subsequent elimination rules that apply to the same connective in the same signed formula. Once normalized, one can show the usual subformula property which states that only subformulae of the formula under test can occur in its subproofs. The additional special property here is that any subformula of depth d in the formula under test can only occur in a subproof of depth d within the overall proof. This is a feature of D-level systems such as $\mathbf{DW}$, $\mathbf{DJ}$ and $\mathbf{DK}$. This puts a finite limit on the depths of subproof together with the formulae that can occur in them, ensuring the finiteness of the total number of subproofs that can be created using these subformulae. This ensures that the system is decidable and would apply to other depth relevant or D-systems. (For depth of subformulae, together with depth relevance, see Brady [9]. Depth of subproof is given by $\max(a)$, where a is the index set of a formula in the subproof, this maximum being common to all subformulae in the subproof.) However, much of this work on systems other than $\mathbf{DW}$ is still to be done and the removal of distribution would induce a much-needed simplification of a rather complicated system. (Indeed, it is hoped to use two separate subproofs in parallel in response to disjunction elimination.) As hypothetical syllogism is not depth relevant and would fail the above property, this method, as set out, does not extend to systems such as $\mathbf{TW}$ which include it.

Finally, we consider cut-free Gentzenization, which is a popular method for proving decidability, especially, when distribution is removed. The history of Gentzenization of relevant logics goes back to Kripke, who in his [46], Gentzenized the $\rightarrow$-fragment of the logics $\mathbf{R}$ and $\mathbf{E}$, $\mathbf{R}_\rightarrow$ and $\mathbf{E}_\rightarrow$, using commas to the left of the turnstile, these representing nested implications. He then went on to prove decidability by an intricate argument. Subsequently, Belnap and Wallace in [5] Gentzenized $\mathbf{E}_\rightarrow^\sim$, using commas on both sides of the turnstile, also proving its decidability by extending Kripke's work. Meyer then Gentzenized the full distribution-less logic $\mathbf{LR}$ in his [47] in similar style. He also went on to prove decidability by extending Kripke's original argument. Subsequently, Dunn Gentzenized $\mathbf{R}_+$ in 1969, first appearing as an abstract in his [40] and then in full detail in [1], using two structural connectives, commas and semicolons, to the left of the turnstile, and a single formula to the right. The comma represented extensional conjunction & and the semi-colon represented intensional fusion ∘, introduced by Belnap [3] as $\sim(A \rightarrow \sim B)$. However, Dunn, in his thesis [39] had used fusion satisfying the equivalence $A \circ B \rightarrow C. \leftrightarrow A \rightarrow .B \rightarrow C$ to introduce algebraic residuation laws. Indeed, [47] had defined the term fusion using this equivalence. Nevertheless, decidability was not provable for $\mathbf{R}_+$, Urquhart [56] and in [2], showing that it was undecidable, together with $\mathbf{E}_+$ and $\mathbf{T}_+$. (This is to be discussed below.)

Somewhat later, Grishin Gentzenized $\mathbf{LRWQ}$, i.e., $\mathbf{RW}$ without distribution but with quantifiers, and used it to prove decidability in his [44]. This was the first known proof of decidability for a full quantified logic.[7] Then, Belnap [4] and also in [2], introduced an innovative Gentzenization called display logic, which applied to **R**, **E**, **T** and other logics, centrally using a method of flipping components of the right-hand side of the turnstile onto the left so as to leave a single formula on the right. (Components can also be flipped from left to right.) Alas, this did not yield decidability. Then, Brady Gentzenized **RW**, following the work of Dunn on $\mathbf{R}_+$ in [1], and proved its decidability in [13] following Giambrone [43], who had proved decidability for $\mathbf{RW}_+$ and $\mathbf{TW}_+$. This was extended in Brady [14] with the Gentzenization and decidability of the contraction-less logics **DW**, **TW**, **EW** and **RWK**, which was then all simplified in his [15]. Subsequently, Brady [18] and [19] set out Gentzenizations of some quantified relevant logics without distribution, making use of Grishin's work in [44]. The logics **LBQ**, **LDWQ**, **LTWQ**°, **LEWQ**^{cl}, **LRWQ**, **LRWKQ** and **LRQ** were Gentzenized and all these logics except **LRQ** were shown to be decidable. Then, in Brady [17], Gentzenizations similar to that of Belnap's display logic in his [4] were given for a large range of logics with distribution, excluding **E** and **EW**. Here too, no decidability results were derived.

The general limiting factor in Gentzen systems is intensional contraction, which can be re-applied unrestrictedly, working one's way up the Gentzen proof on the antecedent side of sequents, without a complementary introduction rule, such as that for the axiom $A \to . A \to A$ of the decidable logic **RM**, that could be used to put a finite cap on the number of repetitions of contraction. (Note that $A \to . A \to A$ is inter-derivable with $A \circ A \to A$.) Such contractions provide elimination of pairs without corresponding introduction, meaning that the fusion involved is not finitely capped, which is what yields the lack of decidability. Nevertheless, more recent progress has been made in proving decidability of systems with contraction. Bimbó and Dunn [7] proved decidability for $\mathbf{T}_\to$, and Bimbó has proved decidability of the intensional fragment of classical linear logic in [6], both of which have forms of contraction in them.

We finish with undecidability, which is what Badia's question is related to. This was proved for the logics **R** and **E** by Urquhart in his [56] and reprinted in [2, pp. 348–374]. Indeed, Urquhart showed that any logic from $\mathbf{TW}_+ + A \& (A \to B) \to B$ through to **R** is undecidable. However, Brady [30] showed that the undecidability of Turing's halting problem uses the LEM in its proof, with neither of its disjuncts provable. And all undecidability arguments stem originally from Turing's problem by mapping the halting problem into the various contexts.

Similarly, in Brady and Rush [37], it was shown that Cantor's Diagonal Argument uses the LEM, with neither disjuncts provable. The LEM is similarly used in the proofs of key set-theoretic and semantic paradoxes, viz. Russell's Paradox and the Liar Paradox, as was shown in Brady [34]. These are all failures of the priming property, since the LEM is used without either of its disjuncts being provable. Since the LEM is not in the logic, the only way it can be proved is through one of its disjuncts, i.e., it should be prime. Moreover, given that the LEM only applies to recursive sets, as stated in §2, it is clear that it should fail in these cases as self-reference occurs in

---

[7] A big "thank you" to Alexander Kron and Kosta Došen for making me aware of Grishin's work, and Kron for hospitality in Serbia and Montenegro. Grishin's paper was translated from Russian with financial help from the School of Humanities of La Trobe University.

each of them. Further, Brady argued in [31] that, for the metatheory to use the LEM, decidability of the object logic must have been proved, as the LEM is understood proof-theoretically in the meta-theory. That is, the LEM would then be interpreted as: for all A, A is provable or $\sim A$ is provable, or if A is unprovable then $\sim A$ is provable.

We examine the proof of the undecidability of the halting problem, as in Brady [30]. We quote from pp. 293 and 294, with $D(D)$ being the diagonal machine applied to itself:

> We start by defining the halting set H of pairs $(M;x)$, consisting of a Turing machine M together with an input x, such that the Turing machine halts on that input.
>
> This argument has the shape: if M_H is a Turing machine that decides the set H then $D(D)$ halts iff $D(D)$ moves the cursor to the right, that is, $D(D)$ does not halt. That is, as for Cantor's diagonal argument, it takes the shape: $A \Rightarrow B \leftrightarrow \sim B$. So, by use of the LEM on $B, \ldots$, this creates the contradiction $B \,\&\sim B$, that is, that $D(D)$ both halts and does not halt, upon the assumption that H is decidable. By the classical reductio argument, there is no Turing machine M_H that decides H, and so the halting problem is undecidable.

Note that the LEM is clearly used in deriving $B \,\&\sim B$ from $B \leftrightarrow \sim B$ as $B \leftrightarrow \sim B$ is deductively equivalent to $B \vee \sim B \to B \,\&\sim B$. The "classical reductio argument" requires a derivation of the rule contraposition, if $A \Rightarrow B \,\&\sim B$ then $\sim (B \,\&\sim B) \Rightarrow \sim A$. This is achieved by applying MR1 yielding $\sim A \vee A \Rightarrow \sim A \vee (B \,\&\sim B)$, and then applying the LEM to A and the DS to $B \,\&\sim B$. Then, by De Morgan and the LEM, $\sim A$ follows, which states that the halting problem is undecidable. However, by the argument of [31], the LEM cannot be presupposed here as this would assume the decidability of the object theory which includes the halting problem, which is what is at issue. (We assume the DS here, that is the meta-theory is simply consistent, for the reasons given in [31], and $\sim A \vee (B \,\&\sim B)$ is prime, since either $\sim A$ or A is derivable, assuming the LEM.) With the failure of the LEM, the best we can do is to say that the decidability is unprovable in the expanded meta-theory, extended to include the LEM, and the decidability would still be unprovable in the meta-theory without the LEM. Projecting the halting problem to that of the theorems of **R** or **E**, the same can be said for these as well and thus their decidability would be unprovable, as a meta-meta-theoretic result, as opposed to their undecidability being provable in the meta-theory. Note that this serves to correct what was said about undecidability in [30], which mistakenly focussed on the object theory instead of the meta-theory. However, what was said in [30] about Cantor's Diagonal Argument still applies, that is the countability of the power set of the natural numbers is unprovable, as a meta-statement, rather than its uncountability being provable in the object theory.

7. CONCLUSION

So, the undecidability of the logics **R** and **E** do not matter in the general scheme of working logics, based upon appropriate concepts. And, undecidability proofs rely on the LEM which fails for Turing's halting problem, which then projects generally to the standard classical undecidability proofs. This failure of the LEM leaves their decidability or undecidability to be unproven, due to the unavailability of proofs. Nevertheless, we can still assert that such decidability arguments are not possible.

Getting back to Brady's and Dunn's response to Badia's question, is it the mixing up of intension and extension, as suggested by Brady, or is it distribution, as suggested by Dunn, that is the culprit for the undecidability of **R**? There is certainly a mixing up of entailment and implication in the formalization of distribution, though implication is only admissible for weaker (metacomplete) logics, and Dunn did introduce an extensional structural connective " , " as well as the intensional structural connective " ; " in proving distribution in his Gentzenization of $\mathbf{R}_+$ in [40]. The extensional comma is used to apply the Weakening Rule, (KE ⊢), in the proof of distribution, unavailable as an intensional rule, as this would introduce irrelevance. However, the mixture of intension and extension does present problems, as Brady suggests, in particular for strong systems with distribution, which does require such a mixture for its proof in Dunn's $\mathbf{R}_+$. The →-form of distribution does introduce a level of intensionality into a principle which is really an implication represented as an admissible rule for weaker logics, as argued in §5 above, and further convertible to an extensional principle as a ⊃-form with an application of MR1 and the LEM. So, it appears that we are both equally right because our two answers are intertwined.

However, as seen in §6, the decidability of relevant logics is limited by three components: full contraction, hypothetical syllogism and distribution. Indeed, the logic **R** without any one of these three components is decidable, given Brady's proof of decidability of **RW** in [13], Fine's proof of the decidability of **R** without hypothetical syllogism in [42] and Meyer's proof of decidability of **LR**, i.e., **R** without distribution, in his [47]. With all three present, problems in establishing a suitable cap on a decision procedure occur. The problem with full contraction, $(A \to .A \to B) \to .A \to B$, A15 below, is explained in §6 above for Gentzen systems, which also applies to weaker forms of contraction. We reiterate here that mingle $(A \to .A \to A)$ mitigates contraction and hence one should not be surprised that **RM** is decidable. Hypothetical syllogism creates complexity in Fine's semantics, as well as in the Routley–Meyer semantics, both of which were explained above in §6. Distribution creates the need for both intensional and extensional structural connectives in the Gentzen system with an uncapped intensional contraction, as observed by Dunn in [1].

Nevertheless, as argued above for Brady's intensional logic **MC**, the removal of distribution is quite justified, as it does not follow from the meanings of conjunction and disjunction. Also, the full form of contraction, is not justified in a logic of meaning containment as the antecedent $A \to .A \to B$ is virtually meaningless as its conclusion not only combines the two antecedent A's but is also self-referring. Indeed, how is it that the containment of B in A is itself contained in A? The full form of contraction only makes sense as an implication, as the two A's of the antecedent can then be combined together as an assumption of truth in a truth-preservation. However, conjunctive syllogism is a weak form of contraction, represented as $A \circ B \to A \circ (A \circ B)$ using fusion, but fusion ∘ is an inappropriate combination of premises in a logic of meaning containment. (See §2 above for some more detail on this.) Nevertheless, it should just be seen as transitivity of meaning containment. As such, conjunctive syllogism is more appropriate than hypothetical syllogism, A12 and A13 below, as we see that the hypothetical syllogisms have too much in the way of deep containment →'s involving a new formula C in their consequent for it to be contained in its antecedent $A \to B$. Hypothetical syllogism would lead to an A for B substitution into any antecedent position within a context made up of $\sim$'s and $\to$'s, and similarly, to a B for A substitution into any such consequent position. (See [23] for these substitutions and [1] for antecedent and consequent positions.) However, from the point of view of paradox solution using **TJ**^{d}, as proved in [23], both are these forms of syllogism are present. So, we view the argument against hypothetical syllogism as somewhat weaker than those against distribution and full contraction.

So, the upshot here is that the conceptual intensional system **MC** is where we should focus our efforts in proving decidability, though unproven at this moment without distribution. However, **DJ**^{d} can be shown to be decidable using semantic methods and the removal of distribution should be advantageous in proving decidability using the proof-theoretic method of normalized natural deduction.

Acknowledgments. Thank you to the referee for interesting and worthwhile comments and thank you to the editor for her dedication in providing numerous detailed and valuable comments. The paper has been considerably enhanced as a result of their efforts.

8. Appendix

We set out the systems of logic mentioned in this paper. This is especially useful to get an idea of the various tranches of decidable relevant logics, in accordance with the four methods of determination used in §6. The bracketing follows [1] with dots representing left brackets and $\to$ association to the left. Also, & and $\vee$ bind tighter than $\to$.

Primitives: $\sim, \&, \vee, \to$

Axioms:

A1. $A \to A$
A2. $A \& B \to A$
A3. $A \& B \to B$
A4. $(A \to B) \& (A \to C) \to .A \to B \& C$
A5. $A \to A \vee B$
A6. $A \to B \vee A$
A7. $(A \to C) \& (B \to C) \to .A \vee B \to C$
A8. $\sim\sim A \to A$
A9. $A \to \sim B \to .B \to \sim A$
A10. $(A \to B) \& (B \to C) \to .A \to C$
A11. $A \& (B \vee C) \to (A \& B) \vee (A \& C)$
A12. $A \to B \to .B \to C \to .A \to C$
A13. $A \to B \to .C \to A \to .C \to B$
A14. $A \to .A \to B \to B$
A15. $(A \to .A \to B) \to .A \to B$
A16. $A \to \sim A \to \sim A$
A17. $A \& (A \to B) \to B$
A18. $A \to .B \to A$
A19. $A \to .A \to A$

Rules:

R1. $A, A \to B \Rightarrow B$
R2. $A, B \Rightarrow A \& B$
R3. $A \to B, C \to D \Rightarrow B \to C \to .A \to D$
R4. $A \to \sim B \Rightarrow B \to \sim A$
R5. $A \Rightarrow A \to B \to B$

Meta-rule:

MR1. If $A \Rightarrow B$ then $C \vee A \Rightarrow C \vee B$

Systems:

B: Axioms 1–8, 11 + Rules 1–4
DW: Axioms 1–9, 11 + Rules 1–3
DJ: Axioms 1–11 + Rules 1–3
TW: DW + Axioms 12–13
TJ: TW + Axiom 10
T: TW + Axioms 15–16
EW: TW + Rule 5

E: T + Rule 5
RW: TW + Axiom 14
RWK: RW + Axiom 18
R: RW + Axiom 15, or T + Axiom 14
RM: R + Axiom 19
MC: Axioms 1–10 + Rules 1–3 + MR1

The systems with superscript ^{d}: $\mathbf{B}^d$, $\mathbf{DW}^d$, $\mathbf{DJ}^d$, $\mathbf{TW}^d$, $\mathbf{TJ}^d$ and $\mathbf{RW}^d$, are obtained by adding Meta-Rule 1 to their respective underlying system. (MR1 is derivable in **T**.) The distribution-less systems with **L** (for lattice) in front are obtained by removal of Axiom 11 from their underlying system. (This is not needed for **MC**.) The distributionless quantified systems with **Q** after them (and no ^{d}) are obtained by adding the following axioms and rules:

Primitives: $\forall, \exists$

$a, b, c, \ldots$ range over free variables. $x, y, z, \ldots$ range over bound variables. Terms $s, t, u, \ldots$ can be individual constants (when introduced) or free variables.

Quantificational Axioms:
1. $\forall x A \to A^t/x$, for any term t
2. $\forall x (A \to B) \to . A \to \forall x B$
3. $A^t/x \to \exists x A$, for any term t
4. $\forall x (A \to B) \to . \exists x A \to B$

Quantificational Rule:
1. $A^a/x \Rightarrow \forall x A$, where a is not free in A

REFERENCES

[1] Anderson, A. R. and Belnap, N. D. (1975). *Entailment: The Logic of Relevance and Necessity*, Vol. I, Princeton University Press, Princeton, NJ.
[2] Anderson, A. R., Belnap, N. D. and Dunn, J. M. (1992). *Entailment: The Logic of Relevance and Necessity*, Vol. II, Princeton University Press, Princeton, NJ.
[3] Belnap, N. D. (1960). A formal analysis of entailment, *Technical Report 7, Contract No. SAR/Nonr-609 (16)*, Office of Naval Research, New Haven, CT.
[4] Belnap, N. D. (1982). Display logic, *Journal of Philosophical Logic* **11**: 375–417.
[5] Belnap, N. D. and Wallace, J. R. (1961). A decision procedure for the system $E_{\overline{I}}$ of entailment with negation, *Technical Report 11, Contract No. SAR/609 (16)*, Office of Naval Research, New Haven, CT, (published in Zeitschrift für mathematische Logik und Grundlagen der Mathematik, vol. 11, (1965), pp. 277–289).
[6] Bimbó, K. (2015). The decidability of the intensional fragment of classical linear logic, *Theoretical Computer Science* **597**: 1–17.
[7] Bimbó, K. and Dunn, J. M. (2013). On the decidability of implicational ticket entailment, *Journal of Symbolic Logic* **78**(1): 214–236.
[8] Bimbó, K. and Dunn, J. M. (2018). Larisa Maksimova's early contributions to relevance logic, *in* S. Odintsov (ed.), *L. Maksimova on Implication, Interpolation and Definability*, Vol. 15 of *Outstanding Contributions to Logic*, Springer Nature, Switzerland, pp. 33–60.
[9] Brady, R. T. (1984a). Depth relevance of some paraconsistent logics, *Studia Logica* **43**: 63–73.
[10] Brady, R. T. (1984b). Natural deduction systems for some quantified relevant logics, *Logique et Analyse* **27**: 355–377.

[11] Brady, R. T. (1988). A content semantics for quantified relevant logics I, *Studia Logica* **47**: 111–127.
[12] Brady, R. T. (1989). A content semantics for quantified relevant logics II, *Studia Logics* **48**: 243–257.
[13] Brady, R. T. (1990). The Gentzenization and decidability of RW, *Journal of Philosophical Logic* **19**: 35–73.
[14] Brady, R. T. (1991). The Gentzenization and decidability of some contraction-less relevant logics, *Journal of Philosophical Logic* **20**: 97–117.
[15] Brady, R. T. (1992). Simplified Gentzenizations for contraction-less logics, *Logique et Analyse* **35**: 45–67.
[16] Brady, R. T. (1994). Rules in relevant logic – I: Semantic classification, *Journal of Philosophical Logic* **23**: 111–137.
[17] Brady, R. T. (1996a). Gentzenizations of relevant logics with distribution, *Journal of Symbolic Logic* **61**: 402–420.
[18] Brady, R. T. (1996b). Gentzenizations of relevant logics without distribution. I, *Journal of Symbolic Logic* **61**: 353–378.
[19] Brady, R. T. (1996c). Gentzenizations of relevant logics without distribution. II, *Journal of Symbolic Logic* **61**: 379–401.
[20] Brady, R. T. (1996d). Relevant implication and the case for a weaker logic, *Journal of Philosophical Logic* **25**: 151–183.
[21] Brady, R. T. (2003a). Semantic decision procedures for some relevant logics, *Australasian Journal of Logic* **1**: 4–27.
[22] Brady, R. T. (2006a). Normalized natural deduction systems for some relevant logics I: The logic DW, *Journal of Symbolic Logic* **71**: 35–66.
[23] Brady, R. T. (2006b). *Universal Logic*, Vol. 109 of *CSLI Lecture Notes*, CSLI Publications, Stanford, CA.
[24] Brady, R. T. (2008). Negation in metacomplete relevant logics, *Logique et Analyse* **51**: 331–354.
[25] Brady, R. T. (2012). The consistency of arithmetic, based on a logic of meaning containment, *Logique et Analyse* **55**: 353–383.
[26] Brady, R. T. (2015). Logic — The big picture, *in* J.-Y. Beziau, M. Chakraborty and S. Dutta (eds.), *New Directions in Paraconsistent Logic*, Springer, New Delhi, pp. 353–373.
[27] Brady, R. T. (2016). Comparing contents with information, *in* K. Bimbó (ed.), *J. Michael Dunn on Information Based Logics*, Vol. 8 of *Outstanding Contributions to Logic*, Springer, Switzerland, pp. 147–159.
[28] Brady, R. T. (2017a). Metavaluations, *Bulletin of Symbolic Logic* **23**: 296–323.
[29] Brady, R. T. (2017b). Some concerns regarding ternary-relation semantics and truth-theoretic semantics in general, *IFCoLog Journal of Logics and their Applications* **4**(3): 755–781. (Bimbó, K. and Dunn, J. M., (eds.), *Proceedings of the Third Workshop*, May 16–17, 2016, Edmonton, Canada).
[30] Brady, R. T. (2018). Starting the dismantling of classical mathematics, *Australasian Journal of Logic* **15**: 280–300.
[31] Brady, R. T. (2019a). The number of logical values, *in* C. Başkent and T. Ferguson (eds.), *Graham Priest on Dialetheism and Paraconsistency*, Vol. 18 of *Outstanding Contributions to Logic*, Springer, Switzerland, pp. 21–37.
[32] Brady, R. T. (2019b). On the law of excluded middle, *in* Z. Weber (ed.), *Ultralogic as Universal? The Sylvan Jungle*, Vol. 4, Springer, Switzerland, pp. 161–183.

[33] Brady, R. T. (2021a). The formalization of arithmetic in a logic of meaning containment, *Australasian Journal of Logic* **18**: 447–472.

[34] Brady, R. T. (2021b). The use of definitions and their logical representation in paradox derivation, *Synthese* **199**: 527–546.

[35] Brady, R. T. (ed.) (2003b). *Relevant Logics and Their Rivals. A Continuation of the Work of R. Sylvan, R. K. Meyer, V. Plumwood and R. Brady*, Vol. II, Ashgate, Aldershot, UK.

[36] Brady, R. T. and Meinander, A. (2013). Distribution in the logic of meaning containment and in quantum mechanics, *in* K. Tanaka, F. Berto, E. Mares and F. Paoli (eds.), *Paraconsistency: Logic and Applications*, Springer, Dordrecht, pp. 223–255.

[37] Brady, R. T. and Rush, P. A. (2008). What is wrong with Cantor's diagonal argument?, *Logique et Analyse* **51**: 185–219.

[38] Brady, R. T. and Rush, P. A. (2009). Four basic logical issues, *Review of Symbolic Logic* **2**: 488–508.

[39] Dunn, J. M. (1966). *The Algebra of Intensional Logics*, Doctoral dissertation, University of Pittsburgh, Pittsburgh, PA. Published as Vol. 2 in the *Logic PhDs* series by College Publications, London (UK), 2019.

[40] Dunn, J. M. (1973). A 'Gentzen system' for positive relevant implication, (abstract), *Journal of Symbolic Logic* **38**(2): 356–357.

[41] Dunn, J. M. (1981). Quantum mathematics, *in* P. D. Asquith and R. N. Giere (eds.), *PSA 1980: Proceedings of the 1980 Biennial Meeting of the Philosophy of Science Association*, Vol. 2, Philosophy of Science Association, East Lansing, MI, pp. 521–531.

[42] Fine, K. (1974). Models for entailment, *Journal of Philosophical Logic* **3**(4): 347–372.

[43] Giambrone, S. (1985). TW_+ and RW_+ are decidable, *Journal of Philosophical Logic* **14**: 235–254.

[44] Grishin, V. N. (1979). Herbrand's theorem for logics without contraction, (in Russian), *in* A. I. Mikhaĭlov (ed.), *Studies in Non-classical Logics and Set Theory*, Nauka, Moscow, pp. 316–329.

[45] Hughes, G. E. and Cresswell, M. J. (1996). *A New Introduction to Modal Logic*, Routledge, London, UK.

[46] Kripke, S. A. (1959). The problem of entailment, (abstract), *Journal of Symbolic Logic* **24**: 324.

[47] Meyer, R. K. (1966). *Topics in Modal and Many-valued Logic*, PhD thesis, University of Pittsburgh, Pittsburgh, PA.

[48] Meyer, R. K. (1974). Entailment is not strict implication, *Australasian Journal of Philosophy* **52**: 212–231.

[49] Meyer, R. K. (1976). Metacompleteness, *Notre Dame Journal of Formal Logic* **17**: 501–516.

[50] Priest, G. and Sylvan, R. (1992). Simplified semantics for basic relevant logics, *Journal of Philosophical Logic* **21**: 217–232.

[51] Restall, G. (1993). Simplified semantics for relevant logics (and some of their rivals), *Journal of Philosophical Logic* **22**: 481–511.

[52] Routley, R., Meyer, R. K., Plumwood, V. and Brady, R. T. (1982). *Relevant Logics and Their Rivals*, Vol. 1, Ridgeview Publishing Company, Atascadero, CA.

[53] Routley, R. and Routley, V. (1985). Negation and contradiction, *Revista Colombiana de Matemáticas* **19**: 201–230.

[54] Slaney, J. (1984). A metacompleteness theorem for contraction-free relevant logics, *Studia Logica* **43**: 159–168.

[55] Slaney, J. (1987). Reduced models for relevant logics without WI, *Notre Dame Journal of Formal Logic* **28**: 395–407.

[56] Urquhart, A. (1984). The undecidability of entailment and relevant implication, *Journal of Symbolic Logic* **49**: 1059–1073.

LA TROBE UNIVERSITY, DEPARTMENT OF POLITICS, MEDIA AND PHILOSOPHY, MELBOURNE, VIC 3086, AUSTRALIA, *Email:* ross.brady@latrobe.edu.au

Moisil's Modal Logic and Related Systems

Sergey Drobyshevich, Sergei Odintsov and Heinrich Wansing

ABSTRACT. In this paper, we outline the contents and present key logics of the 1942 paper by the Romanian logician G. Moisil entitled *Modal logic*, where Moisil attempted to develop a theory of modal operators of impossibility, contingency, necessity, and possibility based on an algebraically motivated system. Specifically, Moisil defined what is essentially the logic of lattices with two residuals: one corresponding to conjunction and a dual one corresponding to disjunction. Despite not having tools for proving the corresponding completeness result he clearly demonstrated a strong understanding of the relation between algebra and logic, which was arguably ahead of his time. Among systems which are closely related to the ones Moisil introduced in his paper are the bi-intuitionistic logic of C. Rauszer, M. Dummett's LC, first degree entailment FDE of N. D. Belnap and J. M. Dunn, and H. Leitgeb's HYPE.

Keywords. Algebraic semantics, Bi-intuitionistic logic, Coimplication, First-degree entailment, Gödel–Dummett logic, HYPE, Modal logic, Residuation

Introduction

The paper *Modal logic* [60] was published by Grigore Constantin Moisil in 1942 and later reprinted in a collection [61] of his papers entitled *Essays on non-Chrysippean Logics*. In this case, one could read "non-Chrysippean" to simply mean "non-classical." Moisil's goal was to develop an approach to modal logic based on a quite unique system among the ones that were used as foundations for modal theories at the time. Moisil himself mentions several directions of development of modal logics, namely, the strong implication logic by C. I. Lewis [52], the three- and many-valued logics by Łukasiewicz [53; 54], and the intuitionistic logics of Heyting [40], Kolmogorov [48], and Johansson [45]. It seems that, according to Moisil, these logics should be considered modal, since they differentiate between a proposition and its double negation, which, according to him, implies that the double negation can be considered as a modal operator. In fact, we will see later on that one of the modal operators in Moisil's modal logic coincides with the intuitionistic double negation. Some other important logics from the standpoint of modal theory include the logic of quantum theory by Birkhoff and von Neumann [6], Bochvar's three-valued logic [7] constructed to analyze various paradoxes, and Orlov's compatibility calculus [65].

Moisil's approach had two major sources of inspiration for [60]. The first was a recently published paper [64], where Toziro Ogasawara essentially proved that the

---

2020 *Mathematics Subject Classification.* Primary: 03–03, Secondary: 03B20, 03B45.

Bimbó, Katalin, (ed.), *Relevance Logics and other Tools for Reasoning. Essays in Honor of J. Michael Dunn*, (Tributes, vol. 46), College Publications, London, UK, 2022, pp. 150–177.

Lindenbaum algebra of intuitionistic propositional logic IPL is a bounded residuated lattice such that
$$[\varphi \to \psi] = [\psi] : [\varphi] \text{ and } [\neg \varphi] = 0 : [\varphi],$$
where ":" denotes the residual with respect to the meet operation. The starting point for Ogasawara was the axiomatization of IPL from Tarski [81], which includes eight axioms of positive intuitionistic logic together with the formulas $\neg \varphi \to (\varphi \to \psi)$ and $(\varphi \to \neg \varphi) \to \neg \varphi$. Note that [64] does not define the Lindenbaum algebra of IPL explicitly. Instead, Ogasawara first proved that a binary relation $\subset$ on the set of formulas, defined via $\varphi \subset \psi$ iff $\vdash_{\mathsf{IPL}} \varphi \to \psi$, is a preorder and then introduced the corresponding equivalence relation (denoted simply with the equality sign) via $\varphi = \psi$ iff $\varphi \subset \psi$ and $\psi \subset \varphi$. The aforementioned result is then formulated in terms of equivalence classes of this relation. Yet [64] does not establish, for instance, that this equivalence relation forms a congruence on the algebra of formulas. The second source of inspiration was the investigation of lattices with residuals by M. Ward [86] and R. Dilworth [16]. Ward and Dilworth considered not only the residual ":" with respect to the meet operation (the latter corresponding to conjunction),
$$c \leq a : b \text{ iff } cb \leq a,$$
where $\leq$ is the lattice order, and cb denotes the meet of c and b, but also a dual residual "$-$" with respect to the join operation (corresponding to disjunction),
$$a - b \leq c \text{ iff } a \leq c + b,$$
where $c + b$ is the join of c and b. According to H. B. Curry [15] lattices with residuals were considered much earlier by T. Skolem [79] (without any logical interpretation). Skolem called algebraic structures of the form $\langle L, \cdot, +, : \rangle$ *implicative lattices*, and structures of the form $\langle L, \cdot, +, - \rangle$ *subtractive lattices*. Curry used the term *Skolem lattice* to denote either an implicative or a subtractive lattice, yet he did not consider lattices with both residuals. Notably, in [60], Moisil did not refer to Skolem's work. The central idea of Moisil's paper [60] was to introduce a logic in a language which contains logical connectives corresponding to two lattice operations and two residuals, and then to investigate some modal operators in this system (hence, *modal* logic). As usual, conjunction and disjunction correspond to the two lattice operators; connectives corresponding to their residuals are implication $\to$ and coimplication $\prec$ (Moisil calls this connective *exception*).

This approach essentially implies that Moisil defined a logic via a class of algebras, which we will call here *biresiduated lattices*. This is noteworthy, given that at the time the strong connection between logics and classes of algebras that would eventually lead to the development of *abstract algebraic logic* (e.g., Font [33]) had yet to become an integral part of the field of logic. While Moisil does not establish the completeness results for his logics, he does show a quite clear understanding of the connection between his logics and algebras. For instance, at one point he refers to some theses as stating that a certain operator is an automorphism. That said, he uses biresiduated lattices explicitly only once at the very end of the paper to establish independence of some axioms.

Let us give a broad outline of [60]. The paper is written in French and with the use of bracketless (sometimes known as Polish or Łukasiewicz) notation. (Moisil makes it clear at the beginning of the paper that he is familiar with the more common notation by providing a short translation between the two.) It is divided into nineteen sections distributed between five chapters. The last chapter is untitled and the first four are called, respectively, *general modal logic*, *special modal logic*, *some specific logics* and *symmetric modal logics*. Most sections begin with a list of axioms followed by long lists of derivations from these axioms with a heavy emphasis on the laws of distributivity. There are very few designated theorems; most of those are either *uniqueness* theorems (to the effect that taking an axiomatic copy of a certain connective allows one to prove logical equivalence between the two copies) or *normal form* theorems; some important results are formulated in plain text. The last two sections are dedicated to proving the aforementioned independence of axioms (§18) and to stating some open problems (§19).

The main goal of this paper is to present some key logics Moisil has introduced in his paper and to put them into a historical perspective. In doing so we are not presenting the full scope of results in [60] and postpone the discussion of some aspects for a future occasion. The logics will be presented both axiomatically and with Kripke-style semantics. We chose to give a Kripke-style presentation both because it gives us a more natural connection with known systems and because it allows us to illuminate some interesting choices Moisil makes. To simplify the exposition, we will adopt a few conventions. Firstly, all of the presented quotes from [60] are translated, since the paper is written in French. We also translate bracketless notation into a more common one. Since Moisil does not introduce any denotations for his logics, all of the ones presented here are ours. We will present logics as Hilbert-style calculi, even though Moisil uses a distinct system he calls the *calculus of deductive schemes*, which consists of a collection of rules and meta-rules. For instance, aside from the substitution into formulas, he also has a meta-rule which allows for substitutions into what he calls *admissible* rules. Finally, we replace pairs of axioms of the form $\varphi \to \psi$ and $\psi \to \varphi$ with a single axiom $\varphi \leftrightarrow \psi := (\varphi \to \psi) \wedge (\psi \to \varphi)$ and sometime write rules as $\varphi_1, \ldots, \varphi_n / \psi$.

The structure of this paper more or less follows that of [60]. We begin every section with a short outline of relevant results presented in [60] followed by our own additions (mostly in the form of completeness results). Section 1 concerns §§1–2 of [60, Chapter 1], where Moisil first introduces positive logic and then its conservative extension with an operator $\prec$ dual to implication. This logic, which we denote as BiM, turns out to be a definitional variant of Rauszer's bi-intuitionistic logic HB [72]. In Section 2 we cover §§3–5 of [60, Chapter 1], where Moisil introduces four modal operators into BiM: impossibility, contingency, necessity, and possibility. The result of this addition is his *general modal logic*. We discuss the connection between Moisil's modal operators with the intuitionistic modal operators of Došen [9; 17]. Section 3 is dedicated to *special* extensions of the positive, intuitionistic, and general modal logics introduced in §§6–8 of [60, Chapter 2]. These logics turn out to be related to Dummett's logic [19] and to linear frames. Among the systems introduced in [60, Chapter 3] we are particularly interested in the three-valued logic (§§10–14) denoted here as GML_3, which turns out to be definitionally equivalent to Łukasiewicz's three-valued logic [53]; this is the subject of Section 3. Finally, in Section 4, we discuss the general symmetric modal logic GSML introduced in §16 [60, Chapter 4]. This system is obtained by adding an involutive and contrapositive negation inspired by that of Łukasiewicz's three-valued logic to general modal logic. It turns out that this system is closely related to two logics, which are the subject of Odintsov and Wansing [63].

There are various threads connecting Moisil's work with that of Michael Dunn. At the core of [60] are certain interests, which are heavily represented in Dunn's works. Namely, the applications of algebra to logic in general [20; 31] and residuation principles in particular [23; 24; 30; 4], as well as modal operators and the notion of negation [25; 26; 27; 28; 29; 32]. Let us also point out that the negation of Moisil's general symmetric modal logic is axiomatized via the same axioms and rules as the negation of first degree entailment FDE in Anderson and Belnap [1] (see also Dunn [20; 21]) and can be (and in fact was) equivalently characterized by means of a perp-relation and of a star-function. The relation between these two characterizations is the subject of Dunn's [25] (also [22]).

1. MOISIL'S BIRESIDUAL LOGIC BiM

In this section, we will discuss a system introduced in [60, §2] on the way of formulating Moisil's general modal logic. As such, Moisil does not give this system its own name; we will denote it here as BiM. We dedicate some space to this logic due to its connection to bi-intuitionistic logic which we will discuss shortly.

Let us first give a few definitions. For the purpose of this paper, by a *language* we understand a finite set of connectives with their arities, which contains binary *implication* $\to$, *conjunction* $\wedge$ and *disjunction* $\vee$. The smallest such language, i.e., $\{\wedge, \vee, \to\}$, we denote $\mathcal{L}_p$. *Formulas* of a given language $\mathcal{L}$ are defined as usual using propositional variables from a fixed countable set Prop and connectives in $\mathcal{L}$; the set of all formulas in a language $\mathcal{L}$ is denoted by Form$\mathcal{L}$.

As we have discussed in the introduction, Moisil formulates all of his systems via a special calculus of deductive schemes, which consists of a number of rules and meta-rules. However, it is quite easy to reformulate them as common Hilbert-style calculi, which we will do here. By a *logic* in a language $\mathcal{L}$ we will understand any set of formulas L in $\mathcal{L}$, which is closed under *modus ponens* and *substitution*. As usual, we will sometimes call elements of L its *theorems*. For technical reasons, we will associate with any logic L its multiple-conclusion consequence relation $\vdash_L$. For $\Gamma, \Delta \subseteq$ Form$\mathcal{L}$, let $\Gamma \vdash_L \Delta$ hold, if

(1) $\Delta \neq \varnothing$ and for some $\varphi_1, \ldots, \varphi_n \in \Delta$ the formula $\varphi_1 \vee \cdots \vee \varphi_n$ can be derived from Γ and theorems of L using modus ponens, or
(2) $\Delta = \varnothing$ and $\Gamma \vdash_L \varphi$, for any formula $\varphi \in$ Form$\mathcal{L}$.

It is important to point out that [60] considers no consequence relations at all.

Before introducing BiM, Moisil first formulates the *positive propositional logic* in [60, §1] as a logic in the language $\mathcal{L}_p$. It is defined by the following nine axioms:

(P1) $p \to (q \to p)$
(P2) $(p \to (p \to q)) \to (p \to q)$
(P3) $(p \to q) \to ((q \to r) \to (p \to q))$
(P6) $(r \to p) \to ((r \to q) \to (r \to (p \wedge q)))$
(P7) $p \to (p \vee q)$
(P8) $q \to (p \vee q)$

(P4) $(p \wedge q) \to p$ (P9) $(p \to r) \to ((q \to r) \to ((p \vee q) \to r))$
(P5) $(p \wedge q) \to q$

This list of axioms includes all axioms of the positive logic by Hilbert and Bernays [41, Ch. III, §3] except for three axioms for the *equivalence connective* (as it is called in [41]), which Moisil did not include in the language. Notice that this axiomatization of the positive fragment of intuitionistic logic differs from the standard one. The standard axiomatization includes the axiom

(P10) $\qquad\qquad (p \to (q \to r)) \to ((p \to q) \to (p \to r))$

from the very first axiomatization of classical propositional logic in the language $\{\to, \neg\}$ instead of (P2) and (P3), see Frege [34]. The reason for the replacement of (P10) by (P2) and (P3) is discussed in [41, Ch. III, §3.3].

Further, in [59, §2], Moisil extends the language of positive logic with a binary connective called *exception* and denoted S (from "sans" — without). We will call this connective "coimplication" and denote it as $\prec$.[1] This connective is characterized by the following axiom and rule:[2]

(M1) $p \to ((p \prec q) \vee q)$ (M2) $\chi \to (\varphi \vee \psi) / (\chi \prec \psi) \to \varphi$

In what follows we will denote this logic as BiM. Moisil's general modal logic (see Section 2) is strictly speaking an extension of BiM, but it can also be considered as simply a definitional variant of it.

Let us outline algebras which lie at the foundation of [60]. We call a lattice $\langle L, \vee, \wedge, \to, - \rangle$ with implication $\to$ and a difference operation $-$ a *biresiduated lattice*. Moisil [60, §18] defines difference as an operation satisfying two conditions:

$$a \leq (a-b) \vee b \quad \text{and} \quad (a \leq c \vee b \Rightarrow a-b \leq c),$$

which are trivially equivalent to

$$a - b \leq c \text{ iff } a \leq c \vee b,$$

and directly correspond to axiom (M1) and rule (M2) for the coimplication connective in BiM. This readily implies that the Lindenbaum algebra of BiM is a biresiduated lattice. In turn, this means that one could quite easily obtain the result that Moisil seems to have envisioned, but did not prove; namely, that BiM is sound and complete with respect to the class of all biresiduated lattices. Since it would take too much space and we have opted to characterize all logics via Kripke-style semantics, we will not prove this result here.

It turns out that the system BiM is equivalent to the propositional logic nowadays usually called *Heyting–Brouwer logic*, HB, or *bi-intuitionistic logic*, BiInt. The logic HB, alias BiInt, alias BiM has been investigated by Cecylia Rauszer in a series of papers published between 1974 and 1980 [72; 71; 73; 74; 75; 77; 76; 78], It is widely believed that this research commenced the investigation of Heyting–Brouwer logic (see, for example, Helena Rasiowa's obituary for Rauszer [70]). Neither Rauszer's

---

[1] The reasons for this choice of notation becomes clear in the next section. The symbol "$\prec$" for coimplication was introduced in Goré [35].

[2] Note that Moisil writes (M2) with propositional variables in place of formula-variables and uses a special substitution meta-rule to obtain all instances of this rule.

writings nor the careful and critical re-examination of Rauszer's work in [39] mentions Moisil's [60]. Moisil's paper was reviewed in the *Journal of Symbolic Logic* in 1948 by Atwell R. Turquette [82], but that review was not very favorable and, it seems, went largely unnoticed. In [72], Rauszer considers *semi-Boolean algebras* understood as a combination of pseudo-Boolean algebras, i.e., Heyting algebras, and their duals, Brouwerian algebras, and in [78, p. 5] she remarks that "[f]rom those investigations it appeared that an intuitionistic logic with two negations and two implications, dual to itself, would have a more elegant algebraic and model-theoretic theory than an ordinary intuitionistic logic."[3] Semi-Boolean algebras are also called *double Heyting algebras* (Beazer [3], Köhler [47]), *Heyting–Brouwer algebras* (Wolter [87]), and *bi-Heyting algebras* (Makkai and Reyes [58]), but again in those papers there are no references to Moisil's [60].

In Karl Popper's paper [69], from 1948, reprinted with corrections in [5], there is a system that extends intuitionistic logic with an "anticonditional" dual to intuitionistic implication. However, it seems that Popper as well was not aware of Moisil's work, and that it was indeed Moisil, who first defined and investigated HB. Another early paper in which coimplication was introduced is Ingebrigt Johansson's short note [46] from 1953. Remarkably, this paper does not contain any references, so it is not clear whether Johansson was aware of Moisil's or Popper's work, whereas it may be suspected that he was aware of [79].

Note that looking at HB and BiM syntactically, it is not immediately obvious that they are equivalent. There are two key differences. First, HB as defined in [72] is formulated in a language with two additional operators $\neg$ and $\mathord{\sim}$, which we will call here *(intuitionistic) negation* and *conegation*, respectively. By now, it is well known that these two negations are definable in the $\{\wedge, \vee, \rightarrow, \prec\}$-fragment of HB, and in fact oftentimes, the system is presented in the literature as one in the language without negations. This likely stems from [35] — a seminal paper on HB, where a cut-free display sequent calculus has been presented. This paper along with Crolard [13; 14] contributed to bringing HB to the attention of computer scientists. Second, each system uses one additional inference rule (aside from modus ponens), those are

$$\frac{\chi \rightarrow (\varphi \vee \psi)}{(\chi \prec \psi) \rightarrow \varphi} \quad \text{and} \quad \frac{\varphi}{\Box \varphi}$$

for BiM and HB, respectively. Note that the second rule resembles the necessitation rule of modal logics and, in fact, plays a similar role as was highlighted in [39]. As in modal logics, it makes sense to differentiate between the *local* (which does not employ the additional rule in derivations) and *global* (which does) consequence relations of HB, cf. Kracht [50]. The authors of [39] attribute to this distinction some of the errors Rauszer has made in her works. Most notably, in [71] she claims to have presented a cut-free sequent calculus for HB. It was later noted by Tarmo Uustalu that Rauszer's Gentzen-style sequent calculus for BiInt does not admit cut-elimination, see [10]. As a result, various other generalizations of ordinary Gentzen sequents have been considered and shown to be suitable for obtaining cut-free sequent calculi for HB (see,

---

[3]She also points to the fact that first-order HB validates the constant-domain axiom $\forall x(A(x) \vee B) \rightarrow (\forall x A(x) \vee B)$ (notation adjusted), but notes that this was not the immediate reason for her to introduce HB.

[37; 38; 36], [57], [66; 67], [84]). Moreover, non-standard Gentzen-style sequent systems for HB with the analytic cut property, but without the cut-rule being eliminable have been investigated in [49]. References to Moisil's work on bi-intuitionistic logic in recent literature are [43, §4.21.8] and [85], and full credit to Moisil's discovery of Leitgeb's HYPE is given in [63].

There are a few ways of establishing the equivalence between HB and BiM including a purely syntactic one. One obvious way would be through algebraic semantics since the biresiduated lattices of Moisil turn out to be exactly Heyting–Brouwer algebras. Since Moisil himself did not establish any algebraic completeness results, one would have to prove that BiM is sound and complete with respect to Heyting–Brouwer algebras. This, in fact, can be done routinely using the Lindenbaum–Tarski algebra of BiM. In this paper, we instead chose to turn to Kripke semantics, which, naturally, had not been invented yet at the time Moisil wrote his paper. This serves two purposes: first, it serves as a kind of translation for a modern logician; second, it allows us to illuminate some of the logics Moisil has introduced in his paper.

The language of BiM is $\mathcal{L}_b = \mathcal{L}_p \cup \{\prec\}$. We identify BiM with the smallest logic in the language $\mathcal{L}_b$, which contains all axioms of positive logic (P1)–(P9), axiom (M1) and is closed under (M2). By an *extension* of a logic in the language containing $\mathcal{L}_b$ (in particular, of BiM) we will understand any logic which contains the given one and is closed under (M2). Note that our definition of a consequence relation of a logic from the beginning of the section is local (in the sense discussed above): we use both modus ponens and (M2) to calculate the set of theorems of BiM, but only modus ponens is explicitly used to define $\vdash_{\text{BiM}}$.

We define a *falsity constant, (intuitionistic) negation*, a *truth constant* and *conegation*, respectively, (here p_0 is some designated propositional variable):

$$\bot := p_0 \prec p_0 \qquad \neg \varphi := \varphi \to \bot \qquad \top := p_0 \to p_0 \qquad {\sim} \varphi := \top \prec \varphi$$

The difference between negation and conegation will be clear once we will outline their semantic characterizations. Note that coimplication is sometimes interpreted in terms of exclusion. Then, whereas $\neg \varphi$ states that φ implies falsity, ${\sim}\varphi$ states that truth excludes φ. Classically, coimplication corresponds to $\varphi \wedge \neg \psi$ and both negation and conegation correspond to classical negation. There is a sense in which negation and conegation are dual to each other. One way of expressing the duality is by highlighting the fact that $\neg$ is a *paracomplete* negation insofar as $\psi \nvdash_{\text{BiM}} \varphi, \neg \varphi$ for some ψ, while ${\sim}$ is a *paraconsistent* negation because $\varphi, {\sim}\varphi \nvdash_{\text{BiM}} \psi$ for some ψ. The duality between paraconsistency and paracompleteness was studied in Michael Dunn's paper [29]. The presence of ${\sim}$ in GML makes Moisil one of the forerunners of paraconsistent logic. (Jaśkowski's discussive logic [44] is sometimes considered to be the first formal paraconsistent logic.)

We establish some properties of BiM syntactically.

Theorem 1 (Restricted deduction). *Suppose L is an extension of* BiM, *and* Γ, Δ, $\{\varphi, \psi\} \subseteq \text{Form}\mathcal{L}_b$.

(1) $\Gamma, \varphi \vdash_L \psi \iff \Gamma \vdash_L \varphi \to \psi$;
(2) $\varphi \vdash_L \psi, \Delta \iff \varphi \prec \psi \vdash_L \Delta$.

Proof. The first item is standard (see, e.g., Chagrov and Zakharyaschev [12, Theorem 1.12]).

For the second, assume $\varphi \vdash_L \psi, \Delta$. If $\Delta \neq \varnothing$, then there are $\chi_1, \ldots, \chi_n \in \Delta$ such that $\varphi \vdash_L \psi \vee \chi$, where $\chi = \chi_1 \vee \cdots \vee \chi_n$. Applying the first item, we get $\varphi \to (\psi \vee \chi) \in L$, hence, $(\varphi \prec \psi) \to \chi \in L$, by (M2), and applying the first item again, we infer $\varphi \prec \psi \vdash_L \chi$. Consequently, $\varphi \prec \psi \vdash_L \Delta$. In case $\Delta = \varnothing$, by the definition, we have to show that for an arbitrary χ, we have $\varphi \prec \psi \vdash_L \chi$. From $\varphi \vdash_L \psi$, we can clearly infer $\varphi \vdash_L \psi, \chi$ and then reason as above.

Now assume $\varphi \prec \psi \vdash_L \Delta$. By (M1), we have $\varphi \to ((\varphi \prec \psi) \vee \psi) \in L$. Then by the previous item we have $\varphi \vdash_L (\varphi \prec \psi) \vee \psi$. From $\varphi \prec \psi \vdash_L \Delta$, we infer $\varphi \prec \psi \vdash_L \psi, \Delta$ and from $\psi \vdash_L \psi$, we infer $\psi \vdash_L \psi, \Delta$. From this, using the previous item and (P9), it is easy to derive $(\varphi \prec \psi) \vee \psi \vdash_L \psi, \Delta$. Then combining $\varphi \vdash_L (\varphi \prec \psi) \vee \psi$ and $(\varphi \prec \psi) \vee \psi \vdash_L \psi, \Delta$, we obtain $\varphi \vdash_L \psi, \Delta$. ◁

Note that both deduction properties are formulated with restrictions. The first is formulated for the case of exactly one formula in the consequent, and the second one for the case of exactly one formula in the antecedent. Observe also that the second deduction property implies that $\Gamma \vdash_L \varnothing$ iff $\Gamma \vdash_L \bot$, for any extension L of BiM.

We can also establish the replacement property. It is important to note that Moisil never proves the replacement property for his logics, which makes some of his proofs incomplete.

Corollary 2 (Replacement). *For any extension L of BiM, if $\varphi \leftrightarrow \psi \in L$, then $\chi(\varphi) \leftrightarrow \chi(\psi) \in L$, where $\chi(\varphi)$ and $\chi(\psi)$ are the results of replacing some occurrence of a propositional variable in χ by φ and ψ, respectively.*

Proof. We use induction on the complexity of χ. Clearly, every theorem of the positive fragment of intuitionistic logic is a theorem of BiM, hence,

$$(p \leftrightarrow p') \to ((q \leftrightarrow q') \to ((p * q) \leftrightarrow (p' * q'))) \in \text{BiM},$$

where $* \in \{\wedge, \vee, \to\}$. This suffices to prove the induction steps corresponding to conjunction, disjunction and implication.

For the case of coimplication it is enough to show that if $\varphi \leftrightarrow \varphi', \psi \leftrightarrow \psi' \in L$, then $(\varphi \prec \psi) \leftrightarrow (\varphi' \prec \psi') \in L$. Using axioms for conjunction and symmetry considerations, it is enough to show that $(\varphi \prec \psi) \to (\varphi' \prec \psi') \in L$. Using the deduction property and the definition of a consequence relation, we get the following string of equivalences:

$$(\varphi \prec \psi) \to (\varphi' \prec \psi') \in L \iff \varphi \prec \psi \vdash_L \varphi' \prec \psi' \iff \varphi \vdash_L (\varphi' \prec \psi') \vee \psi.$$

Now, from $\psi \leftrightarrow \psi' \in L$ and the already established induction step for disjunction we infer $((\varphi' \prec \psi') \vee \psi') \leftrightarrow ((\varphi' \prec \psi') \vee \psi) \in L$, and hence, $(\varphi' \prec \psi') \vee \psi' \vdash_L (\varphi' \prec \psi') \vee \psi$. We also have $\varphi' \vdash_L \varphi$ from the assumption and $\varphi' \vdash_L (\varphi' \prec \psi') \vee \psi'$ via (M1) and the deduction property. Thus $\varphi \vdash_L \varphi'$, $\varphi' \vdash_L (\varphi' \prec \psi') \vee \psi'$ and $(\varphi' \prec \psi') \vee \psi' \vdash_L (\varphi' \prec \psi') \vee \psi$. Transitivity of derivations gives us the required $\varphi \vdash_L (\varphi' \prec \psi') \vee \psi$. ◁

We now turn to semantics. A BiM-*frame* is just a non-empty partially ordered set $\mathcal{W} = \langle W, \leq \rangle$. A BiM-model $\mathcal{M} = \langle W, \leq, v \rangle$ is a BiM-frame together with a valuation $v \colon \text{Prop} \to 2^W$, which maps propositional variables to *(upward) cones* in $\langle W, \leq \rangle$, i.e.,

$$\forall x, y \in W \left((x \in v(p) \text{ and } x \leq y) \implies y \in v(p) \right).$$

We define satisfaction clauses for formulas over elements of a model as follows:
1. $\mathcal{M}, x \vDash p \iff p \in v(x)$, where $p \in \text{Prop}$;
2. $\mathcal{M}, x \vDash \varphi \wedge \psi \iff \mathcal{M}, x \vDash \varphi$ and $\mathcal{M}, x \vDash \psi$;
3. $\mathcal{M}, x \vDash \varphi \vee \psi \iff \mathcal{M}, x \vDash \varphi$ or $\mathcal{M}, x \vDash \psi$;
4. $\mathcal{M}, x \vDash \varphi \to \psi \iff \forall y \geq x (\mathcal{M}, y \vDash \varphi \implies \mathcal{M}, y \vDash \psi)$;
5. $\mathcal{M}, x \vDash \varphi \prec \psi \iff \exists y \leq x (\mathcal{M}, y \vDash \varphi$ and $\mathcal{M}, y \nvDash \psi)$.

For a BiM-model $\mathcal{M} = \langle W, \leq, v \rangle$, we write $\mathcal{M} \vDash \varphi$ if $\mathcal{M}, x \vDash \varphi$, for all $x \in W$. For a BiM-frame $\mathcal{W} = \langle W, \leq \rangle$, we write $\mathcal{W} \vDash \varphi$ if $\mathcal{M} \vDash \varphi$, for every BiM-model $\mathcal{M}$ over $\mathcal{W}$. For a class $\mathcal{K}$ of BiM-frames, let $\Gamma \vDash_{\mathcal{K}} \Delta$, if $(\forall \varphi \in \Gamma \ \mathcal{M}, x \vDash \varphi)$ implies $(\exists \psi \in \Delta \ \mathcal{M}, x \vDash \psi)$ for every choice of $\mathcal{W} = \langle W, \leq \rangle \in \mathcal{K}$, BiM-model $\mathcal{M}$ over $\mathcal{W}$ and $x \in W$. We say that L is *sound* and *complete* with respect to a class $\mathcal{K}$ of BiM-frames if $\vdash_L = \vDash_{\mathcal{K}}$. As usual we have that

Proposition 3 (Monotonicity). *For every* BiM-*model* $\mathcal{M} = \langle W, \leq, R, v \rangle$ *and formula* φ, *we have* $\forall x, y \in W ((\mathcal{M}, x \vDash \varphi$ *and* $x \leq y) \implies \mathcal{M}, y \vDash \varphi)$.

The following is routine.

Theorem 4 (Soundness). $\mathcal{W} \vDash \varphi$, *for any* $\varphi \in$ BiM *and* BiM-*frame* $\mathcal{W}$.

We fix an extension L of BiM. We say that a pair of sets of formulas $\langle \Gamma, \Gamma' \rangle$ is *L-consistent* if $\Gamma \nvdash_L \Gamma'$, and it is *maximally L-consistent*, if it is L-consistent and $\Gamma \cup \Gamma'$ is the set of all formulas. For a set of formulas Γ, we set $\overline{\Gamma} := \text{Form}\mathcal{L} \setminus \Gamma$. Then clearly, all maximally L-consistent pairs are of the form $\langle \Gamma, \overline{\Gamma} \rangle$. The proof of the following lemma is standard (see, e.g., [12, Lemma 5.1]).

Lemma 5 (Pair extension). *If* $\Gamma \nvdash_L \Gamma'$, *then there are* $\Delta \supseteq \Gamma$ *and* $\Delta' \supseteq \Gamma'$ *such that* $\langle \Delta, \Delta' \rangle$ *is maximally L-consistent.*

Proposition 6. *Let us fix a maximally L-consistent pair* $\langle \Gamma, \Gamma' \rangle$.
 (1) *If* $\Gamma \vdash_L \varphi$, *then* $\varphi \in \Gamma$. *In particular,* $L \subseteq \Gamma$.
 (2) *If* $\varphi \vee \psi \in \Gamma$, *then* $\varphi \in \Gamma$ *or* $\psi \in \Gamma$.

Proof. (1) If $\varphi \notin \Gamma$, then $\varphi \in \overline{\Gamma}$, by the definition. Then from $\Gamma \vdash_L \varphi$, we infer $\Gamma \vdash_L \overline{\Gamma}$, which contradicts the definition of maximally L-consistent pairs $\langle \Gamma, \Gamma' \rangle$.

(2) Suppose, on the contrary, $\varphi \vee \psi \in \Gamma$, $\varphi, \psi \notin \Gamma$. Then $\varphi, \psi \in \overline{\Gamma}$. But $\varphi \vee \psi \vdash_L \varphi, \psi$, hence, $\Gamma \vdash_L \overline{\Gamma}$, which again gives us a contradiction. ◁

The *canonical L-frame* is $\mathcal{W}_L = \langle W_L, \subseteq \rangle$, where $W_L = \{\Gamma : \langle \Gamma, \overline{\Gamma} \rangle$ is maximally L-consistent$\}$. The *canonical L-model* is $\mathcal{M}_L = \langle \mathcal{W}_L, v_L \rangle$, where $v_L(p) = \{\Gamma \in W_L : p \in \Gamma\}$, for $p \in \text{Prop}$.

Lemma 7 (Canonical model). *For any* $\Gamma \in W_L$ *and for any formula* φ, *we have*

$$\mathcal{M}_L, \Gamma \vDash \varphi \iff \varphi \in \Gamma.$$

Proof. We only consider the case of coimplication. Then we have to establish the following equivalence: $\varphi \prec \psi \in \Gamma \iff \exists \Gamma' \in W_L (\Gamma' \subseteq \Gamma \ \& \ \varphi \in \Gamma' \ \& \ \psi \notin \Gamma')$.

$\impliedby$. We reason by contraposition. Suppose $\varphi \prec \psi \notin \Gamma$ and $\Gamma' \in W_L$ is such that $\Gamma' \subseteq \Gamma$ and $\varphi \in \Gamma'$. Then from $\varphi \to ((\varphi \prec \psi) \vee \psi) \in \text{BiM} \subseteq L \subseteq \Gamma'$, we conclude that $(\varphi \prec \psi) \vee \psi \in \Gamma'$ and from $\varphi \prec \psi \notin \Gamma$ we conclude $\varphi \prec \psi \notin \Gamma'$. Consequently, $\psi \in \Gamma'$, as required.

$\Longrightarrow$. Suppose that $\varphi \prec \psi \in \Gamma$. Assume additionally that $\varphi \vdash_L \psi, \overline{\Gamma}$. Then $\varphi \prec \psi \vdash_L \overline{\Gamma}$, hence, $\Gamma \vdash_L \overline{\Gamma}$, which cannot be the case. Then by the pair extension lemma, there is $\Gamma' \in W_L$ such that $\varphi \in \Gamma'$ and $\overline{\Gamma} \cup \{\psi\} \subseteq \overline{\Gamma'}$. From this, we infer $\psi \notin \Gamma'$ and $\Gamma' \subseteq \Gamma$. Indeed, if $\chi \in \Gamma'$, then $\chi \notin \overline{\Gamma'}$ and $\chi \notin \overline{\Gamma}$, which implies $\chi \in \Gamma$. ◁

We say that a BiM-frame $\mathcal{W}$ is an *L-frame* if $\mathcal{W} \vDash \varphi$ for all $\varphi \in L$. L is *canonical* if $\mathcal{W}_L$ is an *L*-frame.

Theorem 8 (Completeness). *Any canonical extension L of* BiM *is sound and complete with respect to the class* $\mathcal{K}_L$ *of all L-frames. In particular,* BiM *is sound and complete with respect to the class of all* BiM*-frames.*

Proof. If $\Gamma \nvdash_L \Delta$, then by the pair extension lemma, there is $\Gamma' \in W_L$ with $\Gamma \subseteq \Gamma'$, $\Delta \subseteq \overline{\Gamma'}$. Then by the canonical model lemma, we have $\mathcal{M}_L, \Gamma' \vDash \varphi$ for all $\varphi \in \Gamma$ and $\mathcal{M}_L \nvDash \psi$ for all $\psi \in \Delta$. Hence, $\Gamma \nvDash_{\mathcal{W}_L} \Delta$. And since $\mathcal{W}_L \in \mathcal{K}_L$, by canonicity we obtain $\Gamma \nvDash_{\mathcal{K}_L} \Delta$. ◁

Comparing this semantics to the semantics of HB, one can easily conclude that BiM is definitionally equivalent to HB as defined in [72] and coincides with HB as presented in [35].

We establish one more technical result which will be useful for us later. For a BiM-frame $\mathcal{W} = \langle W, \leq \rangle$, we denote by W^m the set of all maximal elements of W and by W_m the set of all minimal elements of W, i.e.,

$$x \in W_m \iff \forall y \in W\, (y \leq x \implies x = y);$$
$$x \in W^m \iff \forall y \in W\, (x \leq y \implies x = y).$$

We say that $\mathcal{W}$ is *bounded* if the following holds:

(bounded) $\quad \forall x \in W\, (\exists y \in W_m\, y \leq x \text{ and } \exists z \in W^m\, x \leq z)$.

Note that our notion of a bounded frame is not to be confused with other uses of the term "bounded." In particular, our notion does not presuppose the uniqueness of bounds, only that every element is contained between some minimal and some maximal element. Using Zorn's lemma it is easy to see that for any extension L of BiM its canonical frame is bounded, which implies the following.

Theorem 9. *Any canonical extension L of* BiM *is sound and complete with respect the class of all bounded L-frames.*

The semantics of BiM suggests a close relationship to temporal logic. It is clear that the following translation τ from $\mathcal{L}_b$ into the language $\{\wedge, \vee, \Box, \blacklozenge, \neg_c, \to_c\}$ of S4t, temporal S4 based on classical logic, is a straightforward generalization of the well-known Gödel–Tarski translation:

(1) $\tau(p) = \Box p$, where $p \in \mathsf{Prop}$,
(2) $\tau(\varphi \wedge \psi) = \tau(\varphi) \wedge \tau(\psi)$,
(3) $\tau(\varphi \vee \psi) = \tau(\varphi) \vee \tau(\psi)$,
(4) $\tau(\varphi \to \psi) = \Box(\tau(\varphi) \to_c \tau(\psi))$,
(5) $\tau(\varphi \prec \psi) = \blacklozenge(\tau(\varphi) \wedge \neg_c \tau(\psi))$,

where $\neg_c$ is classical negation, $\to_c$ is classical implication, and ♦ is the "sometimes in the past" modal operator, i.e., if $\mathcal{M} = \langle W, R, v \rangle$ is a Kripke model for S4t and $x \in W$, then $\mathcal{M}, x \vDash \blacklozenge \varphi$ iff $\exists y \, (yRx$ and $\mathcal{M}, y \vDash \varphi)$. In fact, the following holds:

Proposition 10 (Łukowski [56]). *For every $\mathcal{L}_b$-formula φ, $\varphi \in$ HB iff $\tau(\varphi) \in$ S4t.*

2. General Modal Logic

It is important to note that BiM plays in [60] an auxiliary role on the way to defining Moisil's *general modal logic*, which we denote here as GML, in [60, §3 and §4]. As we outlined in the introduction, Moisil was interested in introducing a new kind of modal logic and so to obtain GML he introduced four new operators into BiM in an axiomatic way. Namely, he added to BiM *impossibility*, η, and *contingency*, γ, in [60, §3] and *possibility*, μ, and *necessity*, ν, in [60, §4] via axiomatic equivalences:

(G1) $\eta p \leftrightarrow (p \to (p \prec p))$ (G3) $\mu p \leftrightarrow \eta \eta p$
(G2) $\gamma p \leftrightarrow ((p \to p) \prec p)$ (G4) $\nu p \leftrightarrow \gamma \gamma p$

Moisil sometimes refers to these four operators as *modalities*. We denote by GML the smallest logic in the language $\mathcal{L}_g = \mathcal{L}_b \cup \{\eta, \gamma, \mu, \nu\}$, which contains all axioms of BiM, (G1)–(G4) and is closed under (M2).

Moisil realized [60, §3] that his impossibility and contingency operators are kinds of negations, which are, moreover, duals of each other. Meanwhile, both positive modalities are defined as double negations [60, §4]: possibility is double impossibility and necessity is double contingency. Whereas Moisil maintains that the "impossibility of the impossibility" should be identified with the "possibility," on the latter definition he writes that "[a]lthough this interpretation seems forced, the structure of the theory as well as the example of [Łukasiewicz's three-valued logic] makes it plausible." The duality of impossibility and contingency implies in a natural way the duality of possibility and necessity. Moreover, Moisil observed in [60, §5] that his impossibility operator η is nothing more than Gentzen's negation $p \to \bot$ with $p \prec p$ substituted for $\bot$. In particular, he shows that both axioms for intuitionistic negation (axioms 4.1 and 4.11 from [40]):

(I1) $((p \to q) \wedge (p \to \eta q)) \to \eta p$ (I2) $\eta p \to (p \to q)$

are theorems of GML.

Moisil concludes §5 of [60] by stating that: "... applying I. Johansson's theorem we conclude that every theorem of general modal logic containing only the functors $\to, \vee, \wedge$ and η is a theorem of Heyting's logic." Doing so he effectively claims that GML is a conservative extension of intuitionistic logic, when formulated in the language $\{\wedge, \vee, \to, \eta\}$. Unfortunately, it is not clear what he understands by Johansson's theorem. The only reference to Johansson in the paper is to [45], which does contain a result to the effect that intuitionistic logic is a conservative extension of its positive fragment. To infer straightforwardly from this the conservativity of GML over intuitionistic logic would be a stretch. The claim itself, however, is correct as we can easily verify.

First, observe that all of the definitions and results from the previous section can be transferred directly into GML. To make a clear distinction, a GML-*frame(-model)* is the same as a BiM-frame(-model), except that we add the following satisfaction clauses for the four modalities:

6. $\mathcal{M}, x \vDash \eta\varphi \iff \forall y \in W \, (x \leq y \implies \mathcal{M}, y \nvDash \varphi)$;
7. $\mathcal{M}, x \vDash \gamma\varphi \iff \exists y \in W \, (y \leq x \text{ and } \mathcal{M}, y \nvDash \varphi)$;
8. $\mathcal{M}, x \vDash \mu\varphi \iff \forall y \in W \, (x \leq y \implies \exists z \in W \, (y \leq z \text{ and } \mathcal{M}, z \vDash \varphi))$;
9. $\mathcal{M}, x \vDash \nu\varphi \iff \exists y \in W \, (y \leq x \text{ and } \forall z \in W \, (z \leq y \implies \mathcal{M}, z \vDash \varphi))$.

Consider an *extension* of GML to be any logic in $\mathcal{L}_g$ that contains GML and is closed under (M2). Then the following is proved exactly the same way as Theorems 8 and 9:

Theorem 11. *Any canonical extension L of* GML *is sound and complete with respect to the class of all (bounded) L-frames. In particular,* GML *is sound and complete with respect to the class of all (bounded)* GML*-frames.*

Comparing this result with the completeness of intuitionistic logic with respect to its Kripke-style semantics (e.g., [12]), one immediately obtains the following.

Corollary 12. *The* $\{\wedge, \vee, \to, \eta\}$*-fragment of* GML *coincides with intuitionistic logic (with η in place of negation).*

Note that, strictly speaking, η and $\neg$ as we have defined them are two different operators, because $\eta\varphi = \varphi \to (\varphi \prec \varphi)$, whereas $\neg\varphi = \varphi \to (p_0 \prec p_0)$. Their behavior, nevertheless, is clearly the same. In fact, Moisil himself showed that $(p \prec p) \leftrightarrow (q \prec q)$ is a theorem of GML. The following statement, along with the replacement property, effectively allows us to conflate Moisil's impossibility η with $\neg$, contingency γ with $\mathord{\frown}$, possibility μ with $\neg\neg$ and necessity ν with $\mathord{\frown}\mathord{\frown}$. We will make use of these conflations throughout the remainder of the paper.

Corollary 13. *The following formulas belong to* GML*:*

$\eta p \leftrightarrow \neg p$; $\gamma p \leftrightarrow \mathord{\frown} p$; $\mu p \leftrightarrow \neg\neg p$; $\nu p \leftrightarrow \mathord{\frown}\mathord{\frown} p$.

We are now in a position to discuss these four modal operators in more detail. Whereas Moisil's impossibility is the familiar intuitionistic negation, his contingency $\mathord{\frown}$ is another negation connective that in a certain sense is dual to $\neg$. As we have seen, in the Kripke semantics for GML (and I ID), negation is a forward-looking modal operator of impossibility, while conegation $\mathord{\frown}$ is a backward-looking unnecessity operator. Note that over bounded GML-models, monotonicity allows us to simplify the satisfaction clauses for Moisil's possibility and necessity as follows.

$$\mathcal{M}, x \vDash \neg\neg\varphi \iff \forall y \in W^m \, (x \leq y \implies \mathcal{M}, y \vDash \varphi);$$
$$\mathcal{M}, x \vDash \mathord{\frown}\mathord{\frown}\varphi \iff \exists y \in W_m \, (y \leq x \text{ and } \mathcal{M}, y \vDash \varphi).$$

Thus, over bounded GML-models, Moisil's possibility can be given a satisfaction clause that follows the pattern of the satisfaction clause for $\Box\varphi$ in Kripke models for normal modal logics, with universal quantification over the set of all maximal worlds. Similarly, Moisil's necessity can be given a satisfaction clause that follows the pattern of the satisfaction clause for formulas $\Diamond\varphi$ in Kripke models for normal modal logics, with existential quantification over the set of all minimal worlds. This observation thus offers a justification for viewing Moisil's possibility $\neg\neg$ as a necessity operator and Moisil's necessity ⌐⌐ as a possibility operator.[4] We will bring more substance to this claim a little bit later, but now let us say a few words about a possible motivation for Moisil's definitions.

The key laws governing necessity and possibility for Moisil are the following theses: $\ulcorner\ulcorner \varphi \to \neg\neg \varphi$; $\ulcorner\ulcorner \varphi \to \varphi$; $\varphi \to \neg\neg \varphi$ are theorems, whereas the inverse principles are not.[5] This is in line with Łukasiewicz's view that for modal operators of necessity, $\Box$, and possibility, $\Diamond$, the formulas $\Box \varphi \to \varphi$ and $\varphi \to \Diamond \varphi$ are provable [54] (see also [55, §38]). It seems to be a natural conjecture that Moisil was heavily inspired by Łukasiewicz's works on modal logic.

On the other hand, M. Božič and K. Došen [8; 17] proved that the double intuitionistic negation is an intuitionistic necessity operator in the sense of [9]. In particular, the reason for Božič and Došen to consider double intuitionistic negation as a necessity operator is that $(\neg\neg \varphi \wedge \neg\neg \psi) \to \neg\neg(\varphi \wedge \psi)$ and $\neg\neg(\varphi \to \varphi)$ are intuitionistically valid and the following rule is validity preserving:

$$\frac{\varphi \to \psi}{\neg\neg \varphi \to \neg\neg \psi}.$$

More specifically, in [9], Božič and Došen introduced two intuitionistic modal logics, HK$\Box$ in the language with necessity ($\Box$) as the only modal operator, and HK$\Diamond$ in the language with possibility ($\Diamond$) as the only modal operator. Both of these systems were considered to be intuitionistic versions of the smallest normal modal logic K in their respective languages. The reason to consider these two systems independently is that intuitionistic negation is not strong enough to establish a duality between possibility and necessity which is available over classical logic. Axiomatically, HK$\Box$ (HK$\Diamond$) is defined as the smallest logic in the language $\{\wedge, \vee, \to, \bot, \Box\}$ ($\{\wedge, \vee, \to, \bot, \Diamond\}$), which contains all axioms of intuitionistic logic, formulas ($\Box$2) and ($\Box$3) (($\Diamond$2) and ($\Diamond$3)) and is closed under ($\Box$1) (($\Diamond$1)):

($\Box$1) $\varphi \to \psi / \Box \varphi \to \Box \psi$ ($\Diamond$1) $\varphi \to \psi / \Diamond \varphi \to \Diamond \psi$
($\Box$2) $(\Box p \wedge \Box q) \to \Box(p \to q)$ ($\Diamond$2) $\Diamond(p \vee q) \to (\Diamond p \vee \Diamond q)$
($\Box$3) $\Box \top$ ($\Diamond$3) $\neg \Diamond \bot$

In [17, p. 16], Došen specifically explains that he will (notation and reference adjusted) "not connect intuitionistic double negation with the possibility operator $\Diamond$, because $\Diamond(\varphi \vee \psi) \to (\Diamond\varphi \vee \Diamond\psi)$, which is one of the schemata characteristic for HK$\Diamond$ — the minimal normal intuitionistic modal logic with $\Diamond$ (see [9]) — does not hold when $\Diamond$ is interpreted as intuitionistic double negation." Then, in [8; 17] Božič and Došen axiomatize double intuitionistic negation in the form of the following logic:

---

[4] Note that Hughes and Cresswell [42, p. 29] argue with respect to logical necessity that an "intuitively sound principle is that whatever follows logically from a necessary truth is itself necessarily true. If we were to deny this — if, that is, we were to admit that a contingent proposition (let alone an impossible one) might follow from a necessary proposition — we should be violating a principle which has sometimes been expressed by saying that in a valid inference the conclusion runs no greater risk of falsification than the premisses do." As a convenient way of reflecting that principle they use the distribution axiom $\Box(p \to q) \to (\Box p \to \Box q)$ (notation adjusted), which is intuitionistically valid for $\Box$ as intuitionistic double negation.

[5] These three formulas can be intuitively read as "whichever is necessary is possible," "whichever is necessary is true" and "whichever is true is possible."

$\mathsf{HKdn} = \mathsf{HK}\square + p \to \square p + \square(((p \to q) \to p) \to p) + \neg\square\neg(p \to p).$[6]

For this system they establish not just $\square p \leftrightarrow \neg\neg p \in \mathsf{HKdn}$, but also that $\mathsf{HKdn} = \mathsf{HK}\square + \square p \leftrightarrow \neg\neg p$.

It turns out that we can similarly obtain an axiomatization of Moisil's necessity $\mathord{\sim}\mathord{\sim}$ as a possibility operator if we consider a natural expansion of BiM with $\lozenge$ (which coincides with a natural expansion of $\mathsf{HK}\lozenge$ with $\prec$).

Let us define $\mathsf{HBK}\lozenge$ as the smallest logic in the language $\mathcal{L}_b^\lozenge = \mathcal{L}_b \cup \{\lozenge\}$, which contains all axioms of BiM, formulas ($\lozenge$2) and ($\lozenge$3) and is additionally closed under rules (M2) and ($\lozenge$1). We start by briefly outlining the completeness result of $\mathsf{HBK}\lozenge$ with respect to the Kripke-frames from [9]. This result is a straightforward combination of the completeness proof in the previous section and the one in [9], so we skip most of the details.

An $\mathsf{HB}\lozenge$-*frame* is a structure $\mathcal{W} = \langle W, \leq, R\rangle$, where $\langle W, \leq\rangle$ is a non-empty poset and $R \subseteq W^2$ is such that the following *interplay* condition is satisfied:

$$\forall x, y, z \in W\,((x \leq y \text{ and } xRz) \implies yRz).$$

An $\mathsf{HB}\lozenge$-*model* $\mathcal{M} = \langle \mathcal{W}, v\rangle$ is an $\mathsf{HB}\lozenge$-frame together with a valuation $v\colon \text{Prop} \to 2^W$ defined as in the previous section. The satisfaction clause for possibility is as follows:

$$\mathcal{M}, x \vDash \lozenge\varphi \iff \exists y \in W\,(xRy \text{ and } \mathcal{M}, y \vDash \varphi).$$

The interplay property allows us to prove the monotonicity property in this case.

Note that this time around the completeness result is obtained for *normal extensions* of $\mathsf{HBK}\lozenge$, i.e., for logics extending $\mathsf{HBK}\lozenge$, which are closed under both (M2) and ($\lozenge$1). Then for a normal extension L of $\mathsf{HBK}\lozenge$, we define its *canonical L-frame* as $\mathcal{W}_L = \langle W_L, \subseteq, R_L\rangle$, where W_L is defined as in the previous section and for $\Gamma, \Delta \in W_L$:

$$\Gamma R_L \Delta \iff \forall\varphi\,(\varphi \in \Delta \implies \lozenge\varphi \in \Gamma).$$

The canonical L-model is defined exactly as in the previous section. Then combining the proofs from the previous section and the ones from [9], we can obtain:

Theorem 14. *Every canonical normal extension L of $\mathsf{HBK}\lozenge$ is sound and complete with respect to the class of all (bounded) L-frames.*

Consider the smallest normal extension of $\mathsf{HBK}\lozenge$ containing the following formulas (dc1) $\lozenge\top$, (dc2) $\lozenge p \to p$ and (dc3) $\neg\lozenge(p \wedge \mathord{\sim} p)$. We denote it by HBdc (*dc* stands for "double conegation"). Observe that $\neg\lozenge\bot$ is a substitution variant of $\lozenge p \to p$.

We develop some correspondence theory.

Proposition 15. *Suppose $\mathcal{W} = \langle W, \leq, R\rangle$ is a bounded $\mathsf{HBK}\lozenge$-frame. Then,*
(1) $\mathcal{W} \vDash \lozenge\top \iff \forall x \in W\, \exists y \in W\, xRy;$
(2) $\mathcal{W} \vDash \lozenge p \to p \iff \forall x \in W\, \forall y \in W\, (xRy \implies y \leq x);$
(3) $\mathcal{W} \vDash \neg\lozenge(p \wedge \mathord{\sim} p) \iff \forall x \in W\, \forall y \in W\, (xRy \implies y \in W_m).$

---

[6]Here, "+" presupposes the closure under the additional rule, e.g., ($\square$1).

Proof. For the proof of (1), see [18].

(2) $\Longleftarrow$. Assume the condition does not hold, i.e., there are $x,y \in W$ such that xRy, but $y \not\leq x$. Taking an HBK$\Diamond$-model $\mathcal{M} = \langle \mathcal{W}, v \rangle$ with $v(p) = \{z : y \leq z\}$ one can readily compute that $\mathcal{M}, x \vDash \Diamond p$ and $\mathcal{M}, x \nvDash p$. The other direction is routine.

(3) $\Longleftarrow$. Suppose the condition holds, $\mathcal{M} = \langle \mathcal{W}, v \rangle$ is an HBK$\Diamond$-model and $x \in W$. We show that $\mathcal{M}, x \nvDash \Diamond(p \wedge {\sim} p)$. To do so, assume xRy for some y. Then the condition implies that y is minimal. If $\mathcal{M}, y \vDash {\sim} p$, then for some $z \leq y$ we have $\mathcal{M}, z \nvDash p$. But y is minimal, so $z = y$ and $\mathcal{M}, y \nvDash p$. Thus either $\mathcal{M}, y \nvDash {\sim} p$ or $\mathcal{M}, y \nvDash p$, that is, $\mathcal{M}, y \nvDash p \wedge {\sim} p$. Given that y was an arbitrary accessible world, we infer $\mathcal{M}, x \nvDash \Diamond(p \wedge {\sim} p)$.

$\Longrightarrow$. Assume the condition does not hold, i.e., there are $x,y,z \in W$ such that xRy and $z < y$. Consider an HBK$\Diamond$-model $\mathcal{M} = \langle \mathcal{W}, v \rangle$ with $v(p) = \{u : y \leq u\}$. Then $z \notin v(p)$ and $\mathcal{M}, z \nvDash p$, hence, $\mathcal{M}, y \vDash {\sim} p$. On the other hand, $y \in v(p)$, hence, $\mathcal{M}, y \vDash p$ and $\mathcal{M}, y \vDash p \wedge {\sim} p$. Consequently, $\mathcal{M}, x \vDash \Diamond(p \wedge {\sim} p)$ and $\mathcal{M}, x \nvDash \neg \Diamond(p \wedge {\sim} p)$. ◁

Lemma 16. *Take a normal extension L of* HBdc, *then the canonical L-frame is an* HBdc-*frame.*

Proof. We already know that $\mathcal{W}_L$ is bounded. If $\Gamma \in \mathcal{W}_L$, then we get $\Delta \in \mathcal{W}_L$ such that $\Gamma R_L \Delta$ immediately via the canonical model lemma.

Consider $\Gamma, \Delta \in \mathcal{W}_L$ and assume $\Gamma R_L \Delta$. If $\varphi \in \Delta$, then by the definition of R_L, we have $\Diamond \varphi \in \Gamma$, hence, $\varphi \in \Gamma$ since $\Diamond \varphi \to \varphi \in$ HBdc; thus, $\Delta \subseteq \Gamma$. Suppose now there is $\Delta' \in \mathcal{W}_L$ such that $\Delta' \subset \Delta$. Then there is φ such that $\varphi \in \Delta \setminus \Delta'$, then $\varphi \wedge {\sim} \varphi \in \Delta$ by the canonical model lemma and $\Diamond(\varphi \wedge {\sim} \varphi) \in \Gamma$, by the definition of R_L, which contradicts the fact that $\neg \Diamond(\varphi \wedge {\sim} \varphi) \in$ HBdc. ◁

Proposition 17. *Suppose* $\mathcal{W} = \langle W, \leq, R \rangle$ *is an* HBdc-*frame. Then for every* $x \in W$, $\{y : xRy\} = \{z : z \leq x \text{ and } z \text{ is minimal}\}$.

Proof. One inclusion follows from the definition of an HBdc-frame. Take some minimal $z \leq x$. Then zRz, since R is serial and every accessible world from z has to be below z. Then from zRz and $z \leq x$, we derive xRz, by the interplay condition. ◁

Corollary 18. $\Diamond p \leftrightarrow {\sim}{\sim} p \in$ HBdc.

In fact, we clearly have that HBdc coincides with the smallest normal extension of HBK$\Diamond$ containing $\Diamond p \leftrightarrow {\sim}{\sim} p$.

Note that we can easily expand the translation at the end of last section to accommodate HBdc. We inductively define the translation $\tau^\Diamond$ from $\mathcal{L}_b^\Diamond$ into $\mathcal{L}_b$ as follows:

1. $\tau^\Diamond(p) = p$, $p \in$ Prop;
2. $\tau^\Diamond(\varphi \circ \psi) = \tau^\Diamond(\varphi) \circ \tau^\Diamond(\psi)$, for $\circ \in \{\wedge, \vee, \to, \prec\}$;
3. $\tau^\Diamond(\Diamond \varphi) = {\sim}{\sim} \tau^\Diamond(\varphi))$.

Proposition 19. *For every* $\mathcal{L}^\Diamond$-*formula* φ, *we have* $\varphi \in$ HBdc *iff* $\tau(\tau^\Diamond(\varphi)) \in$ S4t.

Proof. By Proposition 10 and Corollary 18. ◁

3. SPECIAL MODAL LOGIC

Chapter II of [60] is dedicated to developing *special modal logic* — an extension of general modal logic with additional axioms. This system is introduced in [60, §8], but along the way Moisil also introduces special extensions of positive logic and of intuitionistic logic in §6 and §7, respectively. We introduce these logics and outline some of Moisil's results before giving a Kripke-style characterization and making some further remarks.

The key formulas for formulating the special extensions are

(S1) $((p \wedge q) \to r) \to ((p \to r) \vee (q \to r))$ (S2) $((r \prec p) \wedge (r \prec q)) \to (r \prec (p \vee q))$.

One can see that these two formulas seem to be dual to each other in some sense.

The *special positive logic* SPL and the *special intuitionistic logic* SIL are extensions of positive and intuitionistic logic, respectively, with (S1). The *special modal logic* SML is an extension of GML with both (S1) and (S2).

In [60, §6], Moisil only gives some preliminary derivations using (S1). The key fact for us here is that he establishes that the formula

(S3) $\qquad\qquad\qquad (p \to q) \vee (q \to p)$,

which one might recognize as an axiom of Dummett's logic LC [19], is a theorem of SPL. In fact, we will show later that (S1) is equivalent to Dummett's axiom over positive logic.

The section [60, §6] on special intuitionistic logic begins with the remark that (notation adjusted) "the introduction of the distributivity axiom (S1) to intuitionist logic profoundly changes its structure." As Moisil points out, one way in which it does this is that the following "modal principle of excluded middle, weaker than that of classical logic"

(WEM) $\qquad\qquad\qquad \neg p \vee \neg \neg p$,

is a theorem of SIL. Observe that under Moisil's reading of modalities this principle can be intuitively read as stating that every statement is either impossible or possible. Further, he essentially establishes that the following are theorems of SIL:

(AM1) $\neg\neg(p \vee q) \leftrightarrow (\neg\neg p \vee \neg\neg q)$ (AM2) $\neg\neg(p \wedge q) \leftrightarrow (\neg\neg p \wedge \neg\neg q)$.

What is interesting here is that he comments on these two formulas by saying that possibility $\neg\neg$ "is an automorphism." This exhibits a good understanding of the connection between his logical derivations and corresponding algebraic facts since $\neg\neg$ would indeed induce a lattice automorphism on Heyting algebras validating (S1). A similar result is obtained in [60, §8] for necessity and special modal logic, namely, that the following are theorems of SML:

(AM3) $\llcorner\llcorner(p \vee q) \leftrightarrow (\llcorner\llcorner p \vee \llcorner\llcorner q)$ (AM4) $\llcorner\llcorner(p \wedge q) \leftrightarrow (\llcorner\llcorner p \wedge \llcorner\llcorner q)$.

Next, Moisil observes that the triple negation laws $\neg\neg\neg p \leftrightarrow \neg p$ and $\llcorner\llcorner\llcorner p \leftrightarrow \llcorner p$, which were previously established in GML, along with the following theorems of SML, $\neg\llcorner p \leftrightarrow \llcorner\llcorner p$ and $\llcorner\neg p \leftrightarrow \neg\neg p$, allow one to show that (notation adjusted) "modalities can be reduced to four ($\neg, \llcorner, \neg\neg, \llcorner\llcorner$)." It is fairly clear here that by modalities he understands strings of $\neg$ and $\llcorner$. Notice that, strictly speaking, one would also require the replacement property to be able to fully reduce any string of negations to one of the four listed strings, which Moisil himself does not establish, but which does indeed hold in all extensions of GML and so in SML as well.

Finally, Moisil shows that the following *law of excluded fourth* is a theorem of SML, (EF) $\llcorner\llcorner p \vee \neg p \vee (\llcorner p \wedge \neg\neg p)$. Moisil reads this formula as stating that every statement is either necessary, or impossible, or problematic. It is unclear why he considers statements which are both contingent and possible ($\sim p \wedge \neg\neg p$) to be problematic.

We proceed to develop a correspondence theory for (S1) and (S2). Along the way we establish that the result of adding (S1) is equivalent to the result of adding (S3) and the result of adding (S2) is equivalent to the result of adding

(S4) $\qquad\qquad\qquad \neg((p \prec q) \wedge (q \prec p)).$

Notice that the duality between (S3) and (S4) is even more clear than that between (S1) and (S2), if one looks at (S4) as saying that $(p \prec q) \wedge (q \prec p)$ is a counter-theorem of the logic, i.e., $(p \prec q) \wedge (q \prec p) \vdash_{\mathsf{SML}} \varnothing$.

Proposition 20. *Suppose* $\mathcal{W} = \langle W, \leq \rangle$ *is a GML-frame.*
(1) *The following are equivalent:* (i) $\mathcal{W} \models$ (S1), (ii) $\mathcal{W} \models$ (S3) *and*
     (iii) $\forall x,y,z \in W\, ((x \leq y \text{ and } x \leq z) \implies (y \leq z \text{ or } z \leq y))$.
(2) *The following are equivalent:* (i) $\mathcal{W} \models$ (S2), (ii) $\mathcal{W} \models$ (S4) *and*
     (iii) $\forall x,y,z \in W\, ((x \geq y \text{ and } x \geq z) \implies (y \leq z \text{ or } z \leq y))$.

Proof. We prove (2); (1) is proved similarly.

(iii) $\implies$ (i). Take some GML-model $\mathcal{M} = \langle \mathcal{W}, v \rangle$ and $x \in W$. Suppose $\mathcal{M}, x \models (r \prec p) \wedge (r \prec q)$, then there is $y \leq x$ such that $\mathcal{M}, y \models r$ and $\mathcal{M}, y \not\models p$ and $z \leq x$ such that $\mathcal{M}, z \models r$ and $\mathcal{M}, z \not\models q$. By the assumption, we have $y \leq z$ or $z \leq y$. We consider the former case, the latter is similar. By monotonicity, we have $\mathcal{M}, y \not\leq q$, hence, $\mathcal{M}, y \not\models p \vee q$. We conclude that $\mathcal{M}, x \models r \prec (p \vee q)$.

(iii) $\implies$ (ii) is similar.

(i) $\implies$ (ii) and (ii) $\implies$ (iii) are established by contraposition. Suppose the condition does not hold, that is, there are $x, y, z \in W$ such that $x \geq y$, $x \geq z$, but y, z are incomparable with respect to $\leq$. Define a GML-model $\mathcal{M} = \langle \mathcal{W}, v \rangle$ such that $v(p) = \{u : y \leq u\}$, $v(q) = \{u : z \leq u\}$ and $v(r) = v(p) \cup v(q)$. Then we clearly have $\mathcal{M}, y \models p$, $\mathcal{M}, y \not\models q$, $\mathcal{M}, y \models r$, $\mathcal{M}, z \not\models p$, $\mathcal{M}, z \models q$ and $\mathcal{M}, z \models r$. From this we infer $\mathcal{M}, x \models p \prec q$ and $\mathcal{M}, x \models q \prec p$, hence, $\mathcal{M}, x \not\models \neg((p \prec q) \wedge (q \prec p))$ and (ii) does not hold. To establish that (i) does not hold, first observe that $\mathcal{M}, x \models r \prec q$. Now, suppose $u \leq x$ is such that $\mathcal{M}, u \models r$. Then $u \geq y$ or $u \geq z$, hence, $\mathcal{M}, u \models p$ or $\mathcal{M}, u \models q$, respectively, and $\mathcal{M}, u \models p \vee q$ in either case. Consequently, $\mathcal{M}, x \not\models r \prec (p \vee q)$ and $\mathcal{W} \not\models$ (S2). ◁

Lemma 21. *For* $i \in \{1, 2, 3, 4\}$, *if* (Si) $\in L$ *then* $\mathcal{W}_L \models$ (Si), *where* $\mathcal{W}_L$ *is the canonical L-frame.*

Proof. Given the previous proposition it is enough to establish that the condition corresponding to (S1) and (S3) holds for $\mathcal{W}_L$ in case (S3) $\in L$, and that the condition corresponding to (S2) and (S4) holds for $\mathcal{W}_L$ in case (S4) $\in L$. We establish only the latter.

Suppose $\Gamma, \Delta, \Theta \in W_L$ are such that $\Delta, \Theta \subseteq \Gamma$. Suppose also neither $\Delta \subseteq \Theta$, nor $\Theta \subseteq \Delta$. Then there are φ, ψ such that $\varphi \in \Delta \setminus \Theta$, $\psi \in \Theta \setminus \Delta$. Then, using the canonical model lemma, we obtain $\varphi \prec \psi \in \Gamma$, $\psi \prec \varphi \in \Gamma$, and consequently, $(\varphi \prec \psi) \wedge (\psi \prec \varphi) \in \Gamma$. On the other hand, from (S4) $\in L \subseteq \Gamma$ we infer $\neg((\varphi \prec \psi) \wedge (\psi \prec \varphi)) \in \Gamma$, which contradicts the choice of Γ. ◁

Notice that the results of the last two statements can clearly be transferred to positive and intuitionistic logic. An immediate consequence of that is that the results of adding (S1) and (S3) to intuitionistic logic coincide.

Corollary 22. *Dummett's logic* LC *coincides with special intuitionistic logic* SIL.

To summarize, in his special intuitionistic logic Moisil has introduced LC 17 years before it was done by Dummett. Note that according to von Plato [68], this system was already introduced in 1913 by Skolem.[7] It is also worth pointing out that we are not the first ones to make the connection between Moisil's work and LC — it was already established by Lloyd Humberstone in [43, p. 312], albeit without a proof.

It is well known (cf. [12]) that LC is sound and complete with respect to the class of linearly ordered sets (chains). Would it not imply that the result of adding (S1) to GML will already give us a logic sound and complete with respect to linear orders? The answer to this is negative. The way one can obtain this result for LC is by employing generated subframes (e.g., [12]). In the intuitionistic case generated subframes are just (upward) cones and thus taking a generated subframe of a frame satisfying the condition corresponding to (S1) will yield a linear order. This is not the case for GML and HB since the notion of a generated subframe has to change due to the presence of backward-looking coimplication. More specifically, one can see that restricting a GML-model to a cone does not necessarily preserve validity of formulas. Consider $\mathcal{M} = \langle W, \leq, v \rangle$, where $W = \{x, y, z\}$ with $y, z \leq x$ and y, z being incomparable, $v(p) = \{x, y\}$, $v(q) = \{x, z\}$ and $v(r) = \{x, y, z\}$. Then, following the proof of proposition 20, we would have $\mathcal{M}, x \nvDash$ (S2), hence, $\mathcal{M}, x \vDash \neg$(S2). Yet we would not have $\mathcal{M}', x \vDash \neg$(S2), where $\mathcal{M}'$ is the natural restriction of $\mathcal{M}$ to $\{x\}$, which is the intuitionistic submodel of $\mathcal{M}$ generated by x.

Suppose $\mathcal{W} = \langle W, \leq \rangle$ is a GML-frame and $x \in W$. Denote by W_x the smallest subset of W containing x and closed both upwards and downwards. The latter condition means that $\forall y \in W_x \forall z \in W ((y \leq z \text{ or } z \leq y) \implies z \in W_x)$. It is easy to see that

$$W_x = \{y \in W : \exists n \in \mathbb{N} \exists u_1, \ldots, u_n \in W \, x = u_1 (\leq \circ \leq^{-1}) u_2 \cdots u_{n-1} (\leq \circ \leq^{-1}) u_n = y\}.$$

Denote by $\mathcal{W}_x = \langle W_x, \leq_x \rangle$ the restriction of $\mathcal{W}$ to W_x, i.e., $y \leq z \iff y \leq_x z$ for $y, z \in W_x$. If, additionally, $\mathcal{M} = \langle \mathcal{W}, v \rangle$ is a GML model, then let $\mathcal{M}_x := \langle \mathcal{W}_x, v_x \rangle$, where $v_x(p) = v(x) \cap W_x$. Then we call $\mathcal{W}_x$ and $\mathcal{M}_x$ the *x-generated subframe* and the *x-generated submodel* of $\mathcal{W}$ and $\mathcal{M}$, respectively.

Lemma 23 (Submodel). *Suppose* $\mathcal{M} = \langle W, \leq, v \rangle$ *is a GML-model,* $x \in W$ *and* $\mathcal{M}_x = \langle W_x, \leq_x, v_x \rangle$ *is its x-generated submodel. Then for all* $y \in W_x$,
$$\mathcal{M}, y \vDash \varphi \iff \mathcal{M}_x, y \vDash \varphi.$$
In particular, $\mathcal{W} \vDash \varphi$ *implies* $\mathcal{W}_x \vDash \varphi$, *where* $\mathcal{W} = \langle W, \leq \rangle$.

Proof. Routine induction on the complexity of φ. ◁

Corollary 24. SML *is sound and complete with respect to the class of all (bounded)* GML*-frames* $\mathcal{W} = \langle W, \leq \rangle$ *such that* $\leq$ *is a linear order.*

---

[7]Since [19] does not reference any of Skolem's works, it is safe to assume that Dummett was not aware of the connection.

Proof. If $\varphi \notin$ SML, then there is SML-model $\mathcal{M} = \langle W, \leq, v \rangle$ and $x \in W$ such that $\mathcal{M}, x \nvDash \varphi$. Then $\mathcal{M}_x, x \nvDash p$ and clearly the underlying frame of the x-generated submodel $\mathcal{M}_x$ is a linearly ordered set. ◁

As a result, the special extensions of Moisil are exactly those extensions of positive, intuitionistic, and general modal logic that are sound and complete with respect to linear orders. This is especially noteworthy since it means that despite not having access to Kripke semantics which makes this connection transparent, Moisil felt that one needs to add the new axiom (S2) to make GML behave in a similar way to the special positive and intuitionistic logics.

4. THREE-VALUED LOGIC

The third chapter of [60] is entitled "some specific logics" and is dedicated to building on top of special modal logic to express classical logic (§9), Łukasiewicz's [53] three-valued logic L_3 (§§10–14) and, finally, Słupecki's [80] logic (§15). We postpone an in-depth discussion of this chapter for a future paper, but will dedicate some space to an extension of SML introduced in [60, §10], which we denote here as GML_3. This system is obtained from SML by adding the following formula as an axiom:

(T1) $\qquad ((\ulcorner p \to \ulcorner q) \wedge (\neg p \to \neg q)) \to (q \to p).$

The main result concerning this logic is that it is equivalent to Łukasiewicz's three-valued logic. As Moisil points out, the key property of this logic is the that the following *determinism principle* is among its theorems.

(T2) $\qquad ((\ulcorner\ulcorner p \to \ulcorner\ulcorner q) \wedge (\neg\neg p \to \neg\neg q)) \to (p \to q).$

This formula can be intuitively read as "if the necessity of p implies the necessity of q and the possibility of p implies the possibility of q, then p implies q."

To obtain the result on the equivalence between GML_3 and L_3, Moisil first adds Łukasiewicz's negation $\sim$ in [59, §11] and implication $\to_L$ in [59, §12] (N and C_L in his notation, respectively) axiomatically to GML_3. The negation $\sim$ is characterized by (LN) $\sim \varphi \leftrightarrow (\neg \varphi \vee (\varphi \wedge \ulcorner \varphi))$. Regarding this negation he observes that all of the De Morgan laws and both double negation laws hold with respect to it. Moreover, he points out that coimplication is definable via this negation and implication, since $(\varphi \prec \psi) \leftrightarrow \sim(\sim \psi \to \sim \varphi)$ would be a theorem of the resulting system.

Similarly, Łukasiewicz's implication is introduced via the following axiom

(LI) $\qquad (\varphi \to_L \psi) \leftrightarrow ((\varphi \to \psi) \wedge (\sim \psi \to \sim \varphi)).$

For brevity we conflate the resulting expansion with GML_3 here.

The next two sections [59, §§13–14] are dedicated to establishing the completeness result of GML_3 with respect to three-valued truth tables which extend those of Łukasiewicz's three-valued logic.[8]

---

[8] Moisil uses values v, f and p instead of 1, 0 and i, which we use here.

$\vee$	0	i	1
0	0	i	1
i	i	i	1
1	1	1	1

$\wedge$	0	i	1
0	0	0	0
i	0	i	i
1	0	i	1

$\rightarrow$	0	i	1
0	1	1	1
i	0	1	1
1	0	i	1

$\prec$	0	i	1
0	0	0	0
i	i	0	0
1	1	1	0

$\rightarrow_L$	0	i	1
0	1	1	1
i	i	1	1
1	0	i	1

φ	$\sim\varphi$	$\neg\varphi$	$\ulcorner\varphi$	$\neg\neg\varphi$	$\ulcorner\ulcorner\varphi$
0	1	1	1	0	0
i	i	0	1	1	0
1	0	0	0	1	1

Note that taking the $\{\wedge, \vee, \rightarrow_L, \sim\}$-fragment of these truth tables gives us exactly the truth tables for Łukasiewicz's three-valued logic and that 1 is considered as the only designated value when the tables are read as a matrix.

The soundness part is routine; for the completeness Moisil first develops conjunctive and disjunctive normal forms, which are exactly the same as for classical logic, except that all formulas of the following forms are allowed as atoms: $p, \sim p, \neg p, \ulcorner p, \neg\neg p$ and $\ulcorner\ulcorner p$ (where $p \in$ Prop). He then gives a criterion of theoremhood in terms of conjunctive normal forms to the effect that $\varphi \in \text{GML}_3$ iff its conjuctive normal form $\varphi_1 \wedge \cdots \wedge \varphi_n$ (with $\varphi_i = \psi_{i1} \vee \cdots \vee \psi_{is_i}$ and ψ_{ij} being atomic in the above sense) is such that for every $1 \leq i \leq n$, there is $p \in$ Prop such that φ_i contains as its disjunctive members either (i) p and $\ulcorner p$, or (ii) $\ulcorner p$ and $\ulcorner\ulcorner p$, or (iii) $\ulcorner p$ and $\neg\neg p$, or (iv) $\neg p$ and $\neg\neg p$.

To establish the equivalence of GML_3 and Łukasiewicz's three-valued logic, Moisil first shows that all of the axioms and rules of Wajsberg's axiom system [83] hold for GML_3 and then remarks that one could reduce every formula of GML to a formula in the language $\rightarrow_L, \sim$ according to the following previously obtained equivalences:

(i)–(iii) $\neg\neg\varphi \leftrightarrow (\sim\varphi \rightarrow_L \varphi)$ $\neg\varphi \leftrightarrow \sim\neg\neg\varphi$ $\ulcorner\varphi \leftrightarrow (\varphi \rightarrow_L \sim\varphi)$

(iv)–(v) $(\varphi \vee \psi) \leftrightarrow \sim(\sim\varphi \vee \sim\psi)$ $(\varphi \wedge \psi) \leftrightarrow \sim(\sim\varphi \vee \sim\psi)$

(vi) $(\varphi \prec \psi) \leftrightarrow ((\ulcorner\ulcorner\varphi \wedge \ulcorner\psi) \vee (\varphi \wedge \ulcorner\varphi \wedge \neg\psi))$

(vii) $(\varphi \rightarrow \psi) \leftrightarrow (\neg\varphi \vee \ulcorner\ulcorner\psi \vee (\ulcorner\varphi \wedge \neg\neg\psi) \vee (\ulcorner\ulcorner\varphi \wedge \psi \wedge \ulcorner\psi))$

Then employing completeness results of GML_3 and Łukasiewicz's logic with respect to the truth tables above establishes what Moisil claims is "an identity between the three-valued logic and the logic of Mr. Łukasiewicz."

A few remarks are in order. First, Moisil writes regarding the third and the first equivalences above that "these are the definitions of operators γ, η due to Mr. Tarski." We remind the reader that $\gamma = \ulcorner$ denotes Moisil's contingency and $\eta = \neg$ denotes Moisil's impossibility operator. Note that the first equivalence is not, in fact, a definition of impossibility $\neg$, but of double impossibility $\neg\neg$, which is Moisil's possibility operator. We plan to investigate this claim at a future occasion. The definitions for implication and coimplication are given in the shape of disjunctive normal forms. Finally, strictly speaking, the two systems cannot coincide since they are formulated in different languages, which suggests that Moisil employs a different notion of identity;

what he effectively shows is that GML_3 is definitionally equivalent to Łukasiewicz's three-valued logic.

We now proceed to investigate the frame condition corresponding to (T1).

Proposition 25. *Suppose* $\mathcal{W} = \langle W, \leq \rangle$ *is an* SML-*frame. Then,*

$$\mathcal{W} \vDash (T1) \iff \forall x, y, z \in W \, (x \leq y \leq z \implies (x = y \text{ or } y = z)).$$

Proof. $\Longleftarrow$. We reason by contraposition. Take some SML-model $\mathcal{M} = \langle \mathcal{W}, v \rangle$ over $\mathcal{W}$ and assume that $\mathcal{M}, x \vDash \mathbin{\sim} p \to \mathbin{\sim} q$, $\mathcal{M}, x \vDash \neg p \to \neg q$, $\mathcal{M}, x \vDash q$ and $\mathcal{M}, x \nvDash p$. From $\mathcal{M}, x \vDash q$ we infer $\mathcal{M}, x \nvDash \neg q$, hence, $\mathcal{M}, x \nvDash \neg p$ and there is some $y \geq x$ such that $\mathcal{M}, y \vDash p$. Then from $x \leq y$ and $\mathcal{M}, x \nvDash p$, we infer $\mathcal{M}, y \vDash \mathbin{\sim} p$ and hence $\mathcal{M}, y \vDash \mathbin{\sim} q$ and there exists $z \leq y$ such that $\mathcal{M}, z \nvDash q$. Given frame conditions corresponding to SML and monotonicity we infer that $z \leq x$. Finally, we see that x and y cannot coincide since they do not agree on p, and x and z cannot coincide since they do not agree on q. Consequently, our frame condition cannot hold.

$\Longrightarrow$. Again, we use contraposition. Assume that $x, y, z \in W$ are such that $x \leq y \leq z$, $x \neq y$ and $y \neq z$. Consider an SML-model $\mathcal{M} = \langle \mathcal{W}, v \rangle$ such that $v(q) = \{u \in W : z \leq u\}$ and $v(p) = W \setminus \{u \in W : u \leq x\}$. It is routine to verify that $\mathcal{M}, y \nvDash (T1)$. ◁

Lemma 26. *If L is an extension of* SML *that contains* (T1), *then* $\mathcal{W}_L \vDash (T1)$.

Proof. It is enough to show that the frame condition of the previous proposition holds for the canonical L-frame. We make heavy use of the canonical model lemma (Lemma 7). Suppose on the contrary that $\Gamma \subset \Delta \subset \Theta$, for some $\Gamma, \Delta, \Theta \in W_L$. Then there are $\varphi \in \Theta \setminus \Delta$ and $\psi \in \Delta \setminus \Gamma$. Then $\psi \to \varphi \notin \Delta$, and hence, by (T1) $\neg \varphi \to \neg \psi \notin \Delta$ or $\mathbin{\sim} \varphi \to \mathbin{\sim} \psi \notin \Delta$. Consider the former; then, there is $\Delta' \supseteq \Delta$ with $\neg \varphi \in \Delta'$ and $\neg \psi \notin \Delta'$. Given the canonicity of SML, we have $\Delta' \subseteq \Theta$ or $\Theta \subseteq \Delta'$. In both cases, we obtain a contradiction because of $\varphi \in \Theta$ and $\neg \varphi \in \Delta'$. For the latter possibility, there is, again, $\Delta' \supseteq \Delta$ with $\mathbin{\sim} \varphi \in \Delta'$ and $\mathbin{\sim} \psi \notin \Delta'$. Then $\psi \in \Delta$ and $\mathbin{\sim} \psi \notin \Delta'$ give us a contradiction. ◁

Theorem 27. GML_3 *is sound and complete with respect to a single two-element chain.*

Proof. We sketch the proof. Using generated subframes we immediately conclude that every formula not belonging to GML_3 can be refuted on a chain of no more than two elements. Thus it remains to see that every formula which is refuted on a singleton frame can be refuted on a two-element chain. Consider φ and $\mathcal{M} = \langle \{x\}, \leq, v \rangle$ such that $\mathcal{M}, x \nvDash \varphi$. Then define $\mathcal{M}' = \langle \{x, y\}, \leq', v' \rangle$, where $x \leq' y$ and

$$v'(p) = \begin{cases} \{x, y\}, & \text{if } x \in v(p); \\ \varnothing, & \text{if } x \notin v(p). \end{cases}$$

Then a routine induction on the complexity of ψ shows that $\mathcal{M}, x \vDash \psi$ is equivalent to $\mathcal{M}', x \vDash \psi$, which concludes the proof. ◁

Note that this result essentially implies that GML_3 is a conservative extension of *here-and-there* — an important logic for logical programming (see, e.g., Balbiani et al. [2] and references therein). Observe that the logic of a two-element chain will naturally be three-valued as monotonicity guarantees that any formula will be satisfied either on both elements of the chain, or on the top element, or on none of the elements.

5. General Symmetric Modal Logic

Chapter 4 of [60] is titled "Symmetric modal logic" and is dedicated to adding a new negation to general modal logic; this negation is clearly motivated by that of Łukasiewicz's three-valued logic. The resulting *general symmetric modal logic* GSML is obtained from GML by stating that negation is involutive and contrapositive:

(GS1) $p \leftrightarrow \sim\sim p$ \hspace{2cm} (GS2) $\varphi \to \psi / \sim\psi \to \sim\varphi$.

Note that Humberstone [43, p. 1240] calls this negation *Moisil negation*. As far as we can tell, Moisil considers $\sim$ to be the only negation of GSML. While we would consider his contingency and impossibility operators to be negations, he considers them only as modal operators. Observe also that taking (GS2) along with the two conjuncts of (GS1) gives one exactly the same axioms and rule that characterize negation in first degree entailment FDE in Anderson and Belnap [1] (see also Dunn [21]).

The key properties of this negation that Moisil establishes are that it satisfies all the De Morgan laws and that once again coimplication can be recovered from implication and negation as before. That is, $(\varphi \prec \psi) \leftrightarrow \sim(\sim\psi \to \sim\varphi) \in$ GSML.

As was outlined in [63] this implies that GSML is definitionally equivalent to the propositional version of Hannes Leitgeb's logic HYPE [51] and to the involutive extension of logic N* Cabalar et al. [11] (see also Odintsov [62]). In particular, all three logics share the same $\{\wedge, \vee, \to, \sim\}$-fragment.

We only give here a semantic characterization of GSML, for technical details one can consult [62; 51; 63]. The one we will present here is in terms of a *star-function*, but it is important to note that Leitgeb also gives a characterization in terms of a *perp-relation*; these two types of semantics are the subject of Dunn [25] (see also [20; 22]).

By a GSML-*frame* we mean a triple $\mathcal{W} = \langle W, \leq, * \rangle$, where $\langle W, \leq \rangle$ is a partially ordered set and $*\colon W \to W$ is an involutive anti-monotone function, i.e., (i) $x \leq y$ implies $y^* \leq x^*$ and (ii) $x^{**} = x$, for all $x, y \in W$. A GSML-*model* $\mathcal{M} = \langle \mathcal{W}, v \rangle$ is defined exactly as a GML-model with the following additional satisfaction clause for negation $\mathcal{M}, x \vDash \sim\varphi \iff \mathcal{M}, x^* \nvDash \varphi$.

Adopting all of the definitions of the first section and reasoning exactly as in the proofs of Theorems 8 and 9 we obtain:

Theorem 28. *Any canonical extension L of GSML is sound and complete with respect to the class of all (bounded) L-frames. In particular, GSML is sound and complete with respect to the class of all (bounded) GSML-frames.*

6. Summary

We considered the very remarkable paper [60] by the Romanian mathematician, logician, and computer scientist Grigore Constantin Moisil (1906–1973). The aim of our paper was to shed some light on what Moisil was doing in that paper, to outline the connections of his contributions in [60] with more recent developments, and to supplement formal proofs to some facts Moisil seemed to have envisioned.

For several reasons Moisil's article remained largely unnoticed. It was published in the middle of World War II in a not widely known Romanian journal, it was written in French rather than the modern *lingua franca* of science, English, and it made use of the — at the time still popular but nowadays hardly ever used — Łukasiewicz notation.

Moreover, the paper received a not very favorable review in the *Journal of Symbolic Logic* ([82]). As we have tried to make clear, Moisil's paper is, however, a gem and bears witness to the great originality of its author.

It deserves to be highlighted that Moisil, who was influenced by Łukasiewicz's modal logic, presented highly innovative results in his [60] insofar as he

(1) motivated logics by algebras and the notion of residuation;
(2) made use of algebraic semantics (even if without completeness proofs), and thereby, made an early contribution to algebraic logic;
(3) introduced coimplication six years before Karl Popper and 11 years earlier than Ingebrigt Johansson;
(4) introduced the bi-intuitionistic logic BiM 32 years earlier than Cecylia Rauszer introduced the definitionally equivalent Heyting–Brouwer logic HB;
(5) introduced Michael Dummett's superintuitionistic logic LC (also know as Gödel logic or Gödel–Dummett logic) 17 years before Dummett, and independently from Skolem (cf. [68]);
(6) introduced the logic HYPE already 77 years earlier than Hannes Leitgeb.

In addition to outlining Moisil's achievements in [60], in the present paper, a number of results related to Moisil's work have been obtained.

In Section 1, the Restricted Deduction Theorem and the Replacement Theorem for BiM were proved. Moreover, it was shown that BiM is sound and complete with respect to the class of all BiM-frames and that any canonical extension L of BiM is sound and complete with respect the class of all bounded L-frames. This result turned out to be useful in the discussion of Moisil's general modal logic, GML. We observed that any canonical extension L of GML is sound and complete with respect to the class of all (bounded) L-frames, and thus GML is sound and complete with respect to the class of all (bounded) GML-frames. This characterization allows one to give a satisfaction clause for formulas $\neg\neg \varphi$ ($\mathbin{\rule[0.5ex]{0.6em}{0.4pt}}\mathbin{\rule[0.5ex]{0.6em}{0.4pt}} \varphi$) in bounded GML-models that follows the pattern of the satisfaction clause for $\Box \varphi$ ($\Diamond \varphi$) in Kripke models for normal modal logics, thereby, connecting Moisil's reading of $\neg\neg$ as expressing possibility with Došen's understanding of intuitionistic double negation as a necessity operator.

In relation with Božič and Došen's work on intuitionistic modal logics, in Section 2, the bi-intuitionistic modal logic HBK$\Diamond$ was defined, and it was shown that every canonical normal extension L of HBK$\Diamond$ is sound and complete with respect to the class of all (bounded) L-frames. Moreover, the axiom system HBdc was defined and it was shown that HBdc is the smallest normal extension of HBK$\Diamond$ that contains $\Diamond p \leftrightarrow \mathbin{\rule[0.5ex]{0.6em}{0.4pt}}\mathbin{\rule[0.5ex]{0.6em}{0.4pt}} p$.

The coincidence of Moisil's special intuitionistic logic with the superintuitionistic logic LC was already stated by Humberstone. In Section 3, this result was proved by observing that Moisil's axiom (S1) is equivalent with Dummett's linearity axiom and that the result of adding this axiom to GML is canonical. Something similar was shown for Moisil's axiom S2 and a dual of Dummett's axiom. Moreover, by introducing suitable notions of a generated subframe and a generated submodel, it was shown that SML, the extension of GML by (S1) and (S2), is sound and complete with respect to the class of all (bounded) GML-frames $\mathcal{W} = \langle W, \leq \rangle$, where $\leq$ is a linear order.

In Section 4, after explaining Moisil's demonstration that his axiomatic extension GML_3 of SML is definitionally equivalent with Łukasiewicz's three-valued logic, corresponding frame conditions were presented for the axioms (T1) and (T2) that are added to SML to obtain GML_3, and a proof was sketched of the fact that GML_3 is sound and complete with respect to a single two-element chain.

Finally, in Section 5 on Moisil's general symmetric logic, GSML, the system that is definitionally equivalent with HYPE, it was observed that any canonical extension L of GSML is sound and complete with respect to the class of all (bounded) L-frames.

We intend to discuss the historical aspects of Moisil's [60] in greater detail in a sequel to the present paper.

Acknowledgments. We would like to thank a number of people: Graham Priest and Mircea Dumitru for helping us to get hold of a copy of Moisil's *Logique modale*; Andre Scedrov for an invitation to present the results of this paper at the Penn Logic and Computation Seminar; an anonymous referee and Katalin Bimbó for their helpful comments on an earlier version of the paper. Finally, we would like to thank numerous people who expressed their interest and participated in discussions during one of the presentations on the topic.

REFERENCES

[1] Anderson, A. R. and Belnap, N. D. (1975). *Entailment: The logic of relevance and necessity*, Vol. I, Princeton University Press, Princeton, NJ.

[2] Balbiani, P., Dieguez, M. and Fariñas del Cerro, L. (2019). Setting the basis for Here and There modal logic, *Journal of Applied Logics: The IFCoLog Journal of Logics and their Applications* **6**(7): 1475–1500.

[3] Beazer, R. (1980). Subdirectly irreducible double Heyting algebras, *Algebra Universalis* **10**(1): 220–224.

[4] Bimbó, K. and Dunn, J. M. (2008). *Generalized Galois Logics: Relational Semantics of Nonclassical Logical Calculi*, Vol. 188 of *CSLI Lecture Notes*, CSLI Publications, Stanford, CA.

[5] Binder, D., Piecha, T. and Schroeder-Heister, P. (eds.) (2022). *The Logical Writings of Karl Popper*, Vol. 58 of *Trends in Logic*, Springer, Cham.

[6] Birkhoff, G. and von Neumann, J. (1975). The logic of quantum mechanics, *The Logico-Algebraic Approach to Quantum Mechanics*, Vol. 5a of *The University of Western Ontario Series in Philosophy of Science*, Springer, Dordrecht, pp. 1–26.

[7] Bochvar, D. (1938). Ob odnom trehznachnom ischislenii i ego primenenii k analizu paradoksov klassicheskogo rasshirennogo funkcional'nogo ischislenija, *Matematicheskij sbornik* **4**(46): 287–308. (English translation: Bochvar, D. & Bergmann, M. (1981). On a three-valued logical calculus and its application to the analysis of the paradoxes of the classical extended functional calculus, *History and Philosophy of Logic* **2**: 87–112.).

[8] Božić, M. and Došen, K. (1983). Axiomatizations of intuitionistic double negation, *Bulletin of the Section of Logic* **12**(2): 99–102.

[9] Božić, M. and Došen, K. (1984). Models for normal intuitionistic modal logics, *Studia Logica* **43**(3): 217–245.

[10] Buisman, L. and Goré, R. (2007). A cut-free sequent calculus for bi-intuitionistic logic, in N. Olivetti (ed.), *Lecture Notes in Computer Science*, number 4548, Springer, Berlin, pp. 90–106.

[11] Cabalar, P., Odintsov, S. and Pearce, D. (2006). Logical foundations of well-founded semantics, *in* P. Doherty, J. Mylopoulos and C. Welty (eds.), *Principles of Knowledge Representation and Reasoning: Proceedings of the 10th International Conference (KR2006)*, AAAI Press, Menlo Park, CA, pp. 25–36.

[12] Chagrov, A. and Zakharyaschev, M. (1997). *Modal Logic*, Oxford University Press, Oxford.

[13] Crolard, T. (2001). Subtractive logic, *Theoretical Computer Science* **254**(1–2): 151–185.

[14] Crolard, T. (2004). A formulae-as-types interpretation of subtractive logic, *Journal of Logic and Computation* **14**(4): 529–570.

[15] Curry, H. B. (1963). *Foundations of Mathematical Logic*, McGraw-Hill Book Company, New York.

[16] Dilworth, R. P. (1990). Abstract residuation over lattices, *in* K. P. Bogart, R. Freese and J. Kung (eds.), *The Dilworth Theorems: Selected Papers of Robert P. Dilworth*, Birkhäuser, Boston, MA, pp. 309–315.

[17] Došen, K. (1984). Intuitionistic double negation as a necessity operator, *Publications de l'Institut Mathématique, Nouvelle série* **35**(49): 15–20.

[18] Došen, K. (1985). Models for stronger normal intuitionistic modal logics, *Studia Logica* **44**(1): 39–70.

[19] Dummett, M. (1959). A propositional calculus with denumerable matrix, *Journal of Symbolic Logic* **24**(2): 97–106.

[20] Dunn, J. M. (1966). *The Algebra of Intensional Logics*, Doctoral dissertation, University of Pittsburgh, Pittsburgh, PA. (Published as Vol. 2 in the *Logic PhDs* series by College Publications, London (UK), 2019.).

[21] Dunn, J. M. (1976). Intuitive semantics for first-degree entailments and 'coupled trees', *Philosophical Studies* **29**(3): 149–168.

[22] Dunn, J. M. (1986). Relevance logic and entailment, *in* D. M. Gabbay and F. Guenthner (eds.), *Handbook of Philosophical Logic*, Vol. 3, D. Reidel, Dordrecht, pp. 117–224.

[23] Dunn, J. M. (1991a). Gaggle theory: An abstraction of Galois connections and residuation, with applications to negation, implication, and various logical operators, *in* J. van Eijck (ed.), *Logics in AI: European Workshop on Logics in Artificial Intelligence (JELIA '90)*, number 478 in *Lecture Notes in Computer Science*, Springer, Berlin, pp. 31–51.

[24] Dunn, J. M. (1991b). Partial-gaggles applied to logics with restricted structural rules, *in* K. Došen and P. Schroeder-Heister (eds.), *Substructural logics*, Clarendon Press, Oxford, pp. 63–108.

[25] Dunn, J. M. (1993). Star and perp: Two treatments of negation, *Philosophical Perspectives* **7**: 331–357. (Language and Logic, 1993, J. E. Tomberlin (ed.)).

[26] Dunn, J. M. (1995). Positive modal logic, *Studia Logica* **55**(2): 301–317.

[27] Dunn, J. M. (1996). Generalized ortho negation, *in* H. Wansing (ed.), *Negation. A Notion in Focus*, W. de Gruyter, Berlin, pp. 3–26.

[28] Dunn, J. M. (1999). A comparative study of various model-theoretic treatments of negation: A history of formal negation, *in* D. M. Gabbay and H. Wansing (eds.), *What is Negation?*, Vol. 13 of *Applied Logic Series*, Kluwer, Dordrecht, pp. 23–51.

[29] Dunn, J. M. (2000). Partiality and its dual, *Studia Logica* **66**(1): 5–40.

[30] Dunn, J. M., Gehrke, M. and Palmigiano, A. (2005). Canonical extensions and relational completeness of some substructural logics, *Journal of Symbolic Logic* **70**(3): 713–740.

[31] Dunn, J. M. and Hardegree, G. (2001). *Algebraic Methods in Philosophical Logic*, Vol. 41 of *Oxford Logic Guides*, Oxford University Press, Oxford.

[32] Dunn, J. M. and Zhou, C. (2005). Negation in the context of gaggle theory, *Studia Logica* **80**(2–3): 235–264.

[33] Font, J. M. (2016). *Abstract Algebraic Logic: An Introductory Textbook*, College Publications, Norcross, GA.
[34] Frege, G. (1879). *Begriffsschrift, eine der arithmetischen nachgebildete Formelsprache des reinen Denkens*, Verlag von Louis Nebert, Halle.
[35] Goré, R. (2000). Dual intuitionistic logic revisited, *in* R. Dyckhoff (ed.), *Proceedings of Automated Reasoning with Analytic Tableaux and Related Methods (TABLEAUX 2000)*, number 1847 in *Lecture Notes in Computer Science*, Springer, Berlin, pp. 252–267.
[36] Goré, R. and Lellmann, B. (2019). Syntactic cut-elimination and backward proof-search for tense logic via linear nested sequents, *in* S. Cerrito and A. Popescu (eds.), *Automated Reasoning with Analytic Tableaux and Related Methods (TABLEAUX 2019)*, number 11714 in *Lecture Notes in Computer Science*, Springer, Cham, pp. 185–202.
[37] Goré, R., Postniece, L. and Tiu, A. (2008). Cut-elimination and proof-search for bi-intuitionistic logic using nested sequents, *in* C. Areces and R. Goldblatt (eds.), *Advances in Modal Logic*, Vol. 7, College Publications, London, pp. 43–66.
[38] Goré, R., Postniece, L. and Tiu, A. (2010). Cut-elimination and proof search for bi-intuitionistic tense logic, *in* L. Beklemishev, V. Goranko and V. Shehtman (eds.), *Advances in Modal Logic*, Vol. 8, College Publications, London, pp. 156–177.
[39] Goré, R. and Shillito, I. (2020). Bi-intuitionistic logics: a new instance of an old problem, *in* N. Olivetti, R. Verbrugge, S. Negri and G. Sandu (eds.), *Advances in Modal Logic*, Vol. 13, College Publications, London, pp. 269–288.
[40] Heyting, A. (1930). Die formalen Regeln der intuitionistischen Logik, *Sitzungsberichte der preussischen Akademie der Wissenschaften, Physikalisch-Mathematische Klasse*, pp. 42–56, 57–71, 158–169.
[41] Hilbert, D. and Bernays, P. (1934). *Grundlagen der Mathematik I*, Springer, Berlin.
[42] Hughes, G. E. and Cresswell, M. J. (1968). *An Introduction to Modal Logic*, Methuen, London.
[43] Humberstone, L. (2001). *The Connectives*, MIT Press, Cambbridge, MA.
[44] Jaśkowski, S. (1948). Rachunek zdań dla systemów dedukcyjnych sprzecznych [in Polish], *Studia Societatis Scientiarum Torunensis, Sectio A* **1**: 55–77. (English translation: Jaśkowski, L. (1969). Propositional calculus for contradictory deductive systems, *Studia Logica* **24**: 143–157).
[45] Johansson, I. (1930). Der Minimalkalkül, ein reduzierter intuitionistischer Formalismus, *Compositio Mathematica* **4**: 119–136.
[46] Johansson, I. (1953). On the possibility to use the subtractive calculus for the formalization of constructive theories, *Proceedings of 11th International Congress of Philosophy, Brussels*, North-Holland, Amsterdam, pp. 60–64.
[47] Köhler, P. (1980). A subdirectly irreducible double Heyting algebra which is not simple, *Algebra Universalis* **10**(1): 189–194.
[48] Kolmogorov, A. N. (1925). O principe tertium non datur, *Matematicheskij sbornik* **32**: 646–667. (English translation: Kolmogorov, A. N. (1967). On the principle of excluded middle, *in* J. van Heijenoort (ed.), *From Frege to Gödel: A Source-book in Mathematical Logic*, Harvard University Press, Cambridge, MA, pp. 416–437).
[49] Kowalski, T. and Ono, H. (2017). Analytic cut and interpolation for bi-intuitionistic logic, *Review of Symbolic Logic* **10**(2): 259–283.
[50] Kracht, M. (2007). Modal consequence relations, *in* P. Blackburn, J. van Benthem and F. Wolter (eds.), *Handbook of Modal Logic*, Vol. 3 of *Studies in Logic and Practical Reasoning*, Elsevier, Amsterdam, pp. 491–545.
[51] Leitgeb, H. (2019). HYPE: A system of hyperintensional logic (with an application to semantic paradoxes), *Journal of Philosophical Logic* **48**(2): 305–405.

[52] Lewis, C. I. (1918). *A survey of symbolic logic*, University of California Press, Berkeley.
[53] Łukasiewicz, J. (1920). O logice trójwartościowej [in Polish], *Ruch filosoficzny* **5**: 169–170. (English translation: Łukasiewicz, J. (1970). On three-valued logic, in L. Borkowski (ed.), *Selected Works*, North-Holland, Amsterdam, pp. 87–88).
[54] Łukasiewicz, J. (1930). Philosophische Bemerkungen zu mehrwertigen Systemen des Aussagenkalküls, *Comptes rendue des seauces de la Societe des Sciences et des Lettres de Varsovie, Classe III* **23**: 51–77. (English translation: Łukasiewicz, J. (1970). Philosophical remarks on many-valued systems of propositional logic, in L. Borkowski (ed.), *Selected Works*, North-Holland, Amsterdam, pp. 153–178).
[55] Łukasiewicz, J. (1951). *Aristotle's Syllogistic from the Standpoint of Modern Formal Logic*, 2nd edn, Clarendon Press, Oxford.
[56] Łukowski, P. (1996). Modal interpretation of Heyting–Brouwer logic, *Bulletin of the Section of Logic* **25**: 80–83.
[57] Lyon, T., Tiu, A., Goré, R. and Clouston, R. (2020). Syntactic interpolation for tense logics and bi-intuitionistic logic via nested sequents, in M. Fernández and A. Muscholl (eds.), *28th EACSL Annual Conference on Computer Science Logic (CSL 2020)*, Leibniz-Zentrum für Informatik, Schloss Dagstuhl, pp. 28:1–28:16.
[58] Makkai, M. and Reyes, G. E. (1995). Completeness results for intuitionistic and modal logic in a categorical setting, *Annals of Pure and Applied Logic* **72**(1): 25–101.
[59] Moisil, G. C. (1941). Remarques sur la logique modale du concept, *Annales de l'Académie Roumaine, Mémoires de la section scientifique, ser. 3* **16**: 975–1012.
[60] Moisil, G. C. (1942). Logique modale, *Disquisitiones Mathematicae et Physicae* **2**: 3–98.
[61] Moisil, G. C. (1972). *Essais sur les logiques non chrysippiennes*, Éditions de l'Académie de la République Socialiste de Roumanie, Bucharest.
[62] Odintsov, S. P. (2010). Combining intuitionistic connectives and Routley negation, *Siberian Electronic Mathematical Reports* **7**: 21–41.
[63] Odintsov, S. and Wansing, H. (2020). Routley star and hyperintensionality, *Journal of Philosophical Logic* **50**(1): 33–56.
[64] Ogasawara, T. (1939). Relations between intuitionistic logic and lattices, *Journal of Science of the Hiroshima University, Series A* **9**: 157–164.
[65] Orlov, I. E. (1928). Ischislenie sovmestnosti predlozhenij, *Matematicheskij sbornik* **34**: 263–286.
[66] Pinto, L. and Uustalu, T. (2009). Proof search and counter-model construction for bi-intuitionistic propositional logic with labelled sequents, *Proceedings of Automated Reasoning with Analytic Tableaux and Related Methods (TABLEAUX 2009)*, number 5607 in Lecture Notes in Computer Science, Springer, Berlin, pp. 295–309.
[67] Pinto, L. and Uustalu, T. (2011). Relating sequent calculi for bi-intuitionistic propositional logic, in S. van Bakel, S. Berardi and U. Berger (eds.), *Proceedings of Third International Workshop on Classical Logic and Computation*, Vol. 47 of *Electronic Proceedings in Theoretical Computer Science*, pp. 57–72.
[68] von Plato, J. (2003). Skolem's discovery of Gödel–Dummett logic, *Studia Logica* **73**(1): 153–157.
[69] Popper, K. R. (1948). On the theory of deduction, Part II. The definitions of classical and intuitionist negation, *Koninklijke Nederlandse Akademie van Wetenschappen, Proceedings of the Section of Sciences* **51**: 173–183. (Reprinted in *Indagationes Mathematicae* **10** (1948): 44–54).
[70] Rasiowa, H. (1994). Cecylia rauszer, *Studia Logica* **53**(4): 467–471.
[71] Rauszer, C. (1974a). A formalization of the propositional calculus of H–B logic, *Studia Logica* **33**(1): 23–34.

[72] Rauszer, C. (1974b). Semi-Boolean algebras and their applications to intuitionistic logic with dual operations, *Fundamenta Mathematicae* **83**(3): 219–249.
[73] Rauszer, C. (1976). On the strong semantical completeness of any extension of the intuitionistic predicate logic, *Bulletin de l'Academie Polonaise des Sciences* **24**: 81–87.
[74] Rauszer, C. (1977a). An algebraic approach to the Heyting–Brouwer predicate calculus, *Fundamenta Mathematicae* **96**: 127–135.
[75] Rauszer, C. (1977b). Applications of Kripke models to Heyting–Brouwer logic, *Studia Logica* **36**(1-2): 61–71.
[76] Rauszer, C. (1977c). Craig interpolation theorem for an extension of intuitionistic logic, *Bulletin de l'Academie Polonaise des Sciences* **25**: 127–135.
[77] Rauszer, C. (1977d). Model theory for an extension of intuitionistic logic, *Studia Logica* **36**(1–2): 73–87.
[78] Rauszer, C. (1980). *An algebraic and Kripke-style approach to a certain extension of intuitionistic logic*, Vol. 167 of *Dissertationes Mathematicae*, Institute of Mathematics, Polish Academy of Sciences, Warsaw.
[79] Skolem, T. (1919). *Untersuchungen über die Axiome des Klassenkalküls und über Produktations- und Summations-Probleme, welche gewisse Klassen von Aussagen betreffen*, Vol. 3 of *Videnskapsselskapet Skrifter, I. Matematisk-naturvidenskabelig Klasse*, In Kommission bei Jacob Dybwad, Kristiania.
[80] Słupecki, J. (1936). Der volle dreiwertige Aussagenkalkül, *Comptes rendus de la Société des sciences et des lettres de Varsovie* **29**: 9–11.
[81] Tarski, A. (1938). Aussagenkalkül und Topologie, *Fundamenta Mathematicae* **31**: 103–134.
[82] Turquette, A. R. (1948). Review of Gr. C. Moisil, Logique modale, *Journal of Symbolic Logic* **13**(3): 162–163.
[83] Wajsberg, M. (1931). Aksjomatyzacja trójwartościowego rachunku zdań [in Polish], *Comptes rendue des seauces de la Societe des Sciences et des Lettres de Varsovie, Classe III* **24**: 259–262.
[84] Wansing, H. (2008). Constructive negation, implication, and co-implication, *Journal of Applied Non-Classical Logics* **18**(2–3): 341–364.
[85] Wansing, H. (2016). Falsification, natural deduction and bi-intuitionistic logic, *Journal of Logic and Computation* **26**(1): 425–450.
[86] Ward, M. (1938). Structure residuation, *Annals of Mathematics, Second Series* **39**: 558–568.
[87] Wolter, F. (1998). On logics with co-implication, *Journal of Philosophical Logic* **27**(4): 353–387.

RUHR UNIVERSITY, BOCHUM, *Email:* Heinrich.Wansing@rub.de

Kripke's Argument for γ

J. Michael Dunn

The following argument is entirely due to Kripke,[*] and was intended by him to be dual to the usual Meyer–Dunn style of argument.[#] Kripke's argument is modeled on completeness proofs for tableaux systems, wherein (in effect) a partial valuation is extended to a total valuation. It thus (as Kripke remarked) avoids the apparatus of inconsistent theories that is so characteristic of the Meyer–Dunn strategy, wherein an inconsistent complete theory is cut down to a consistent complete theory.

I will run the argument for R without quantifiers out of laziness, but Kripke's proof works for quantifiers as well, surely by analogy with completeness results for tableau systems for first order logic.

The formalism. Let's work with signed formulas TA, FA à la Smullyan, and write down "truth trees" using the following rules of Smullyan,

```
 TA∧B        FA∧B         FA∨B        TA∨B         T~A      F~A
   |         /   \           |         /  \          |        |
  TA        FA   FB          FA       TA   TB        FA       TA
  TB                         FB
```

together with two additional rules,

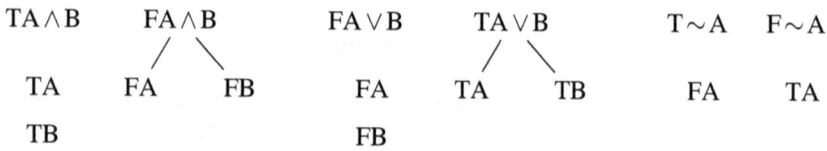

All the rules except the very last are "analytic" (have the subformula property). After we have defined "closure" for the trees, it is easy to see that the formalism without the very last rule is equivalent to the formalism with it in the sense that the same sentences will have closed trees.

A branch of a tree is *closed* iff *either* both TA, FA are in it for some A, or FA appears in it for some axiom of A of R (could just say "theorems" here, but it is interesting that "axiom" will do). A tree is *closed* iff all its branches are closed.

2020 *Mathematics Subject Classification.* Primary: 03B47.

[*] Though the write up is my own, and any mistakes or infelicities should be charged to me.
[#] In particular, to that version due entirely to Meyer using "(quasi-)metavaluations."

Bimbó, Katalin, (ed.), *Relevance Logics and other Tools for Reasoning. Essays in Honor of J. Michael Dunn*, (Tributes, vol. 4{#}), College Publications, London, UK, 2022, pp. 178–181.

The argument. Suppose $\vdash_R A$, $\vdash_R \sim A \vee B$, yet $\dashv_R B$. Run a tableau for FB. If the tableau closes, it is easy to see that $\vdash_R B$, contrary to hypothesis. Run a tableau in an "efficient" way, so that rules are applied as often as possible. Because of the very last rule, this means (since, the tableau doesn't close) that some branch will run on forever. Let S be the set of signed sentences in that branch. It is clear that S has the following properties:

(1) no TA, FA $\in$ S

(2) no FA $\in$ S, where A is an R axiom

$\alpha \begin{cases} (3)\ TA \wedge B \in S \Rightarrow TA, TB \in S, \quad FA \vee B \in S \Rightarrow FA, FB \in S, \\ T \sim A \in S \Rightarrow FA \in S, \quad F \sim A \in S \Rightarrow TA \in S \end{cases}$

$\beta \begin{cases} (4)\ TA \vee B \in S \Rightarrow TA \in S \text{ or } TB \in S \\ (5)\ TA \to B \in S \Rightarrow FA \in S \text{ or } TB \in S \end{cases}$

(6) FB $\in$ S $\Rightarrow$ FA $\to$ B $\in$ S or FA $\in$ S

Further, (2) can be changed to

(2′) no FA $\in$ S, where, A is an R theorem

without loss, as we now verify.

Proof (by induction on length of proof of A). If A is an axiom, immediate. Suppose A comes from B, and B $\to$ A. If FA $\in$ S, then by (6) FB or FB $\to$ A $\in$ S, which by inductive hypothesis, they are not.

Define V: Sentences $\to \{0,1\}$ as follows: If Tp $\in$ S, V(p) = 1; if Fp $\in$ S, V(p) = 0; and otherwise V(p) is arbitrary (for explicitness, set then V(p) = 0). This is for p a propositional variable. Let V(A $\wedge$ B), V(A $\vee$ B), V($\sim$A) be *Boolean*, i.e., defined in the usual truth-table way. Set

$$V(A \to B) = 1 \quad \Leftrightarrow \quad (1)\ V(A) = 0 \text{ or } V(B) = 1, \text{ and } (2)\ FA \to B \notin S.$$

Note that (1) is the extensional condition and (2) the intensional condition.

Lemma 0. *For arbitrary sentences A, V(A) is uniquely defined as 0 or 1.*
(Proof by trivial induction; tacitly used in sequel.)

Lemma 1. *For arbitrary sentences A,*

(i) TA $\in$ S $\Rightarrow$ V(A) = 1
(ii) FA $\in$ S $\Rightarrow$ V(A) = 0

Proof (by induction on complexity of A). Immediate when A is a propositional variable. Cases when A = B $\wedge$ C, B $\vee$ C, $\sim$B fall by routine considerations. So we consider case where A = B $\to$ C. *Ad* (i). Let TB $\to$ C $\in$ S. Then FB $\in$ S or TC $\in$ S. So by inductive hypothesis, V(B) = 0 or V(C) = 1. So extensional condition is satisfied. Further, since S is consistent, FB $\to$ C $\notin$ S. So intensional condition is satisfied. So V(B $\to$ C) = 1. *Ad* (ii). Suppose FB $\to$ C $\in$ S. It cannot be that V(B $\to$ C) = 1 since this requires (intensional condition) that FB $\to$ C $\notin$ S. So V(B $\to$ C) = 0.

Lemma 2. $\vdash_R A \Rightarrow V(A) = 1$.

Lemma 2 follows by routine induction from following:

Sublemma 1. *If* A *is an* R-*axiom,* $V(A) = 1$.

Sublemma 2. (i) $V(A) = V(A \to B) = 1 \Rightarrow V(B) = 1$;
(ii) $V(A) = V(B) = 1 \Rightarrow V(A \wedge B) = 1$.

Part (ii) of Sublemma 2 is trivial (since V is Boolean).
Part (i) comes from extensional condition.

Proof of Sublemma 1 is more involved and proceeds by examining axioms one by one. We do only two by way of example.

Assertion. We wish to show $V(A \to .A \to B \to B) = 1$. Suppose not. Then since we have an axiom, it must be the case that $FA \to .A \to B \to B \notin S$. So intensional condition is satisfied. So extensional condition must fail. So $V(A) = 1$ and $V(A \to B \to B) = 0$. Clearly, $FA \to B \to B \notin S$, for if otherwise, by property 6 of S, then, either $FA \in S$ or $FA \to .A \to B \to B \in S$. The first disjunct fails since $V(A) = 1$ and so by Lemma 1 (ii) $FA \notin S$. The second disjunct fails as noted above. So since intensional condition thus holds for $A \to B \to B$, it must be extensional condition that fouls up. So $V(A \to B) = 1$ and $V(B) = 0$. But those, together with $V(A) = 1$ lead (extensional condition) to $V(B) = 1$, a contradiction.

Contraposition. We wish to show $V(A \to \sim B \to .B \to \sim A) = 1$. Suppose not. Then since we have an axiom, it must be the case that $FA \to \sim B \to .B \to \sim A \notin S$. But then we must have $V(A \to \sim B) = 1$ and $V(B \to \sim A) = 0$. We cannot have $FB \to \sim A \in S$ since then by property 6, we would have $FA \to \sim B \in S$ or $F(A \to \sim B \to .B \to \sim A) \in S$. The latter, we have noted above does not hold, but we cannot have $FA \to \sim B \in S$ either since $V(A \to \sim B) = 1$, which by Lemma 1 (ii) means $FA \to \sim B \notin S$.

The proof of the admissibility of γ now falls quickly from the lemmas. By Lemma 2, $V(A) = V(\sim A \vee B) = 1$, since both $\vdash_R A$ and $\vdash_R \sim A \vee B$. But by "truth-tables," then $V(B) = 1$. Yet by Lemma 2, $V(B) = 0$, since $FB \in S$. Contradiction. Q.E.D.

AFTERWORD

The original hand-written manuscript bears Dunn's signature and the date 23rd July 1978. Sometime later, Dunn had the manuscript typed up. Both the handwritten manuscript and its typed version are in the Archives of Indiana University, Bloomington. The present typeset version is based on a photocopy of the typescript (that Dunn gave to KB in 2016) and a photocopy of the handwritten original. We publish this short paper in this volume with permission from Sarah J. Dunn.

Obviously, this is not a full-length paper. Dunn might have intended this to be a theorem (perhaps, the main theorem or the initial theorem in a series of theorems) in a longer paper. Dunn gave a talk, entitled "Some proof-theoretic aspects of relevance logic," in the Logic Seminar at the Mathematical Institute, University of Oxford, in April 1978, where he might have mentioned insights from [13], [14] and [12]. But the paper contains no "context," and it does not introduce the problem of the admissibility of γ in R or even the logic R. The paper has no references, although Dunn alludes to other proofs of the admissibility of γ. The list below aims to compensate for the lack of references, and also to situate the admissibility of γ. The references contain books and articles on relevance logics (including R), and a paper that motivated the quest to prove γ admissible, which was first accomplished in Meyer and Dunn [13]. Further papers deal with other proofs of the same, the history of the problem, the admissibility of γ in logics closely related to R and E, for example, in extensions of Dunn's RM.

REFERENCES

[1] Anderson, A. R. (1963). Some open problems concerning the system E of entailment, *Acta Philosophica Fennica* **16**: 9–18.

[2] Anderson, A. R. and Belnap, N. D. (1975). *Entailment: The Logic of Relevance and Necessity*, Vol. I, Princeton University Press, Princeton, NJ.

[3] Anderson, A. R., Belnap, N. D. and Dunn, J. M. (1992). *Entailment: The Logic of Relevance and Necessity*, Vol. II, Princeton University Press, Princeton, NJ.

[4] Bimbó, K. (2007). Relevance logics, *in* D. Jacquette (ed.), *Philosophy of Logic*, Vol. 5 of *Handbook of the Philosophy of Science* of *Handbook of the Philosophy of Science* (D. Gabbay, P. Thagard and J. Woods, eds.), Elsevier, Amsterdam, pp. 723–789.

[5] Brady, R. T. (ed.) (2003). *Relevant Logics and Their Rivals. A Continuation of the Work of R. Sylvan, R. K. Meyer, V. Plumwood and R. Brady*, Vol. II, Ashgate, Aldershot, UK.

[6] Dunn, J. M. (1970). Algebraic completeness results for R-mingle and its extensions, *Journal of Symbolic Logic* **35**(1): 1–13.

[7] Dunn, J. M. (1986). Relevance logic and entailment, *in* D. Gabbay and F. Guenthner (eds.), *Handbook of Philosophical Logic*, 1st edn, Vol. 3, D. Reidel, Dordrecht, pp. 117–224.

[8] Dunn, J. M. and Meyer, R. K. (1989). Gentzen's cut and Ackermann's gamma, *in* J. Norman and R. Sylvan (eds.), *Directions in Relevant Logic*, Kluwer, Dordrecht, pp. 229–240.

[9] Friedman, H. and Meyer, R. K. (1992). Whither relevant arithmetic?, *Journal of Symbolic Logic* pp. 824–831.

[10] Mares, E. D. and Meyer, R. K. (1992). The admissibility of γ in **R4**, *Notre Dame Journal of Formal Logic* **33**(2): 197–206.

[11] Meyer, R. K. (1976a). Ackermann, Takeuti and Schnitt: γ for higher-order relevant logics, *Bulletin of the Section of Logic of the Polish Academy of Sciences* **5**: 138–144.

[12] Meyer, R. K. (1976b). Metacompleteness, *Notre Dame Journal of Formal Logic* **17**: 501–516.

[13] Meyer, R. K. and Dunn, J. M. (1969). E, R and γ, *Journal of Symbolic Logic* **34**(3): 460–474.

[14] Meyer, R. K., Dunn, J. M. and Leblanc, H. (1974). Completeness of relevant quantification theories, *Notre Dame Journal of Formal Logic* **15**(1): 97–121.

[15] Meyer, R. K., Giambrone, S. and Brady, R. T. (1984). Where gamma fails, *Studia Logica* **43**: 247–256.

[16] Routley, R. and Meyer, R. K. (1973). The semantics of entailment, *in* H. Leblanc (ed.), *Truth, Syntax and Modality. Proceedings of the Temple University Conference on Alternative Semantics*, North-Holland, Amsterdam, pp. 199–243.

[17] Routley, R., Meyer, R. K., Plumwood, V. and Brady, R. T. (1982). *Relevant Logics and Their Rivals*, Vol. 1, Ridgeview Publishing Company, Atascadero, CA.

[18] Smullyan, R. M. (1968). *First-order Logic*, Springer-Verlag, New York, NY.

[19] Urquhart, A. (2016). The story of γ, *in* K. Bimbó (ed.), *J. Michael Dunn on Information Based Logic*, Vol. 8 of *Outstanding Contributions to Logic*, Springer Nature, Switzerland, pp. 93–105.

CONDITIONAL FDE LOGICS

Nicholas Ferenz

ABSTRACT. The present work adds a conditional connective to the logic **FDE**. The conditional is defined as in Chellas' conditional logics, and two base logics are constructed (based on Chellas' *minimal* and *standard* conditional logics). I further consider (i) propositional extensions of these logics in which the behavior of the conditional can be largely engineered to fit one's purpose, and (ii) first-order extensions which are modeled using the Mares–Goldblatt interpretation of the quantifiers. The main results are soundness and completeness for a wide range of (first-order) conditional **FDE** logics.

Keywords. Conditional logic, Conditionals, FDE, Relevant logic

1. INTRODUCTION

The logic **FDE**, especially in its four-valued presentation, and sometimes referred to as the Belnap–Dunn logic due to its origins in the work of both Nuel Belnap [3; 4] and J. Michael Dunn [9], is typically presented without a conditional connective. Moreover, adding a conditional to **FDE** is not always straightforward, and sometimes presents interesting problems. An infamous problem shared by many paraconsistent logics is that the standard definitions of the material conditional (via ¬ and either ∨ or ∧) result in a conditional for which modus ponens is not valid. While there are several extensions and expansions of **FDE** with a conditional connective, the present work adds to this list by combining Dunn's two-valued relational approach to **FDE** with conditionals, where the conditionals are defined (semantically) as in the conditional logics presented by Chellas in [6] and in Chapter 10 of [7]. In addition, we present a general frame semantics that, strictly speaking, is only necessary for some of the extensions (quantified or propositional) of the base logics constructed here.

Recently, Ma and Wong [17] and Wansing and Unterhuber [40] have also defined Chellas-style conditional logics based on **FDE**. Both approaches combine the *standard models* of Chellas with the logic **FDE**. The major differences in the approach taken in this work consist of the generalization to *minimal models* and our treatment of first-order extensions.

There are many extant ways of adding a conditional to **FDE**, and we will briefly summarize a few approaches. The interested reader is directed to Omori and Wansing [19], for an overview with more depth. The first, and often troubling approach, is the truth-functional approach. Here a conditional's truth value is determined only

---

2020 *Mathematics Subject Classification.* Primary: 03B47, Secondary: 03B45, 03B62.

Bimbó, Katalin, (ed.), *Relevance Logics and other Tools for Reasoning. Essays in Honor of J. Michael Dunn*, (Tributes, vol. 46), College Publications, London, UK, 2022, pp. 182–214.

by the truth values of the antecedent and consequent. The prominent downside to this approach is that the resulting conditional often lacks certain desirable properties (e.g., modus ponens and contraposition). Recently, Pelletier, Hazen, and sometimes Sutcliff [13; 35] have defined a conditional they call *cmi* (for classical material implication) in **FDE** and related logics. This conditional detaches (i.e., is modus ponensable), and can be used to create a useful conditional for which contraposition is valid.

Relevant logics represent an approach capable of capturing (to some approximation, and in contrast with truth-functional approaches) the intensional nature of natural language conditionals. In the Routley–Meyer ternary relational semantics (e.g., see [26; 27; 30]), a three-place relation between points (of **FDE** situations, but with a conditional) can model a highly intensional conditional. A related approach adds a strict implication between those situations. On the strict conditional approach (with a good falsity condition and a binary relation) we can obtain the logics sometimes called K_4 and (with non-normal situations) N_4 [20; 19].

Here we will define conditional **FDE**-logics. These logics extend **FDE** with an intensional conditional. Indeed, the semantics of these logics share a close connection to neighborhood ternary relational semantics.[1] The base conditional **FDE**-logics can be extended to recapture many desirable properties of the conditional, including modus ponens. Moreover, the conditional **FDE**-logics defined here also have comparisons with dynamic modal logics (with formula-indexed modalities). Notably, the logics we develop below are distinct from the neighborhood relevant logics by their treatment of negation (for which no star-worlds are required), and by presenting essentially as a set of formula-formula sequents.

The paper is divided as follows. First, we set out the preliminaries in Section 2 (notations, logical vocabulary, etc.), 2.1 (the logic **FDE**), and 2.2 (Chellas' conditional logics). Notably, in Section 2.1, the logic of **FDE** is given by a contraposition-less axiomatization, and a first order extension is defined. (The Appendix contains the proofs of soundness and completeness for quantified **FDE** with respect to a Mares–Goldblatt style semantics.) The basis and theme of the semantics given is Dunn's two-valued relational semantics for **FDE**. The conditional logics of [7] are given in Section 2.2, where the *minimal* and *standard* conditional logics are defined.

Then, in Section 3 we combine propositional **FDE** with minimal conditional logics. We develop these logics without (at least as primitive) the rule of contraposition in order to construct the minimal conditional **FDE**-logics that will play the star role in the quantified extensions below. This section ends with discussions of the possible admissibility of the rule of contraposition, and semantic characterizations of numerous extensions. Section 4 mirrors the previous section, but with standard conditional logics in place of the minimal conditional logics.

The remainder of the paper, Section 6, is devoted to quantified extensions of the minimal conditional **FDE**-logics. Logics are defined and proved to be sound and complete with respect to a Mares–Goldblatt style semantics. Extensions of these logics are likely desirable for many applications, as the base quantified minimal conditional logics do not validate the sequent $\forall x(\mathcal{A} \Rightarrow \mathcal{B}) \vdash \mathcal{A} \Rightarrow \forall x\mathcal{B}$, where x is not free in $\mathcal{A}$. A

---

[1]For neighborhood ternary relational semantics, see Routley and Meyer [28; 29], Lavers [15], Goble [11], and more recently Standefer [34] and Tedder [37] and Tedder and Ferenz [38].

countermodel is given to this formula, which is essentially a Barcan Formula (with conditionals instead of boxes). We end by giving some characterization results for axioms where the conditional and quantifiers interact.

2. Preliminaries

Propositional and first-order languages are defined as follows. For propositional logics, we assume a denumerable set of a atomic propositions or propositional variables, denoted as lowercase letters from p through s, with or without subscripts. For any propositional logic, the set of (well-formed) formulas is defined in the usual way, restricted to the set of connectives of that logic. The particular set of connectives will be clear from the context, and will be a selection from $\neg$ (negation), $\wedge$ (truth-functional conjunction), $\vee$ (truth-functional disjunction), $\Rightarrow$ (conditional), $\circ$ (fusion), $\Leftarrow$ (left-conditional). A truth-functional implication $\rightarrow$, and bi-implication $\leftrightarrow$, are taken to be defined in the usual way.

For first-order logics, we assume a denumerable set of variables, which will be denoted by lowercase letters near the end of the Latin alphabet, sometimes with integer decoration (e.g., x, y, z, y_4). A *signature* is a set $\mathcal{L}$ consisting of a non-empty but at most denumerable set of predicate symbols and an at most denumerable set of individual constant symbols. Each predicate symbol is of the form P^n, where n is the arity of the predicate. The arity is often omitted. I shall denote individual constants by c, with or without subscripts. A *term* is denoted by τ, with or without integer decoration. An $\mathcal{L}$-*term*, for signature $\mathcal{L}$, is the union of the variables and constants of $\mathcal{L}$. A term is *closed* when it contains no variables, otherwise it is open.

For a given signature $\mathcal{L}$, the atomic formulas (atomic $\mathcal{L}$-formulas) are those of the form $P^n(\tau_1, \ldots, \tau_n)$, where $P^n \in \mathcal{L}$ and $\tau_1, \ldots, \tau_n$ are $\mathcal{L}$-terms. The set of well-formed formulas of a first-order logic with signature $\mathcal{L}$ is defined in the usual way, extending the propositional connectives with the cases for $\forall x$ (universal quantification) and $\exists x$ (existential quantification), for each variable x. We will use calligraphic, uppercase Latin letters to range over the set of well-formed formulas, for both propositional and first-order logics.

An instance of a variable x is *bound* in the formula $\mathcal{A}$ if either (1) the instance is the x of an expressions $\forall x$ or $\exists x$ occurring in $\mathcal{A}$, or (2) the instance of x occurs within the scope of a quantifier, $\forall x$ or $\exists x$. An instance is *free* when it is not bound, and a formula with no free variables is called a *sentence*. A term τ is *free for* (or freely substitutable for) x in $\mathcal{A}$ if, for every variable y occurring in τ, there are no free occurrences of x in $\mathcal{A}$ that are in the scope of a quantifier $\forall y$ or $\exists y$.

We shall write $\mathcal{A}[\tau/x]$ for the result of replacing every free occurrence of x in $\mathcal{A}$ with the term τ. Similarly, we will use $\mathcal{A}[\tau_0/v_0, \ldots, \tau_n/v_n]$ for the result of simultaneously replacing v_0 through v_n with τ_0 through τ_n, respectively. Similar substitution notation is adopted for sequents (a pair of formulas separated by "$\vdash$"), where $(\mathcal{A} \vdash \mathcal{B})[\tau/x] = \mathcal{A}[\tau/x] \vdash \mathcal{B}[\tau/x]$.

A variable assignment, $g \in U^\omega$, assigns an element of the domain U to each variable. In detail, we order the variables and associate each position in that ordering with an element of the domain, using gn to denote the nth element of the ordering. That is, a variable assignment is a denumerable list of elements in the domain. An x-variant

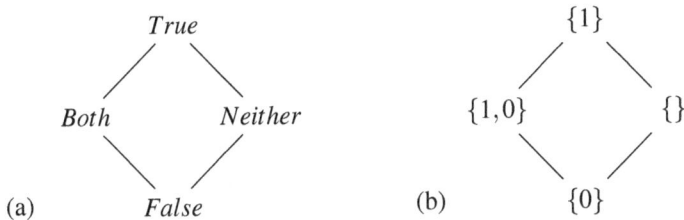

FIGURE 1. Hasse Diagrams

of a variable assignment g differs from g in at most the assignment to the variable x, and the set of all x-variants of g is denoted xg. We write $g[j/n]$ (or $g[j/x]$) to represent the variable assignment just like g, except that the n-th element in the order g (or x) is replaced by the element j.

2.1. First-Degree Entailment. **FDE** is the first-degree fragment of the many relevant logics, including **B**, **T**, **E** and **R**. That is, it was originally defined in Belnap's dissertation, and soon after in Anderson and Belnap [1], as the set of theorems of **E** of the form $\mathcal{A} \to \mathcal{B}$, where $\mathcal{A}$ and $\mathcal{B}$ can only contain connectives from the set $\{\neg, \wedge, \vee\}$. Given this relation to relevant logics, there are at least as many semantic approaches to **FDE** as there are relevant logics. A short, non-exhaustive list includes the star-world approaches (cf. Australian Plan semantics for relevant logics; e.g., see Sylvan (né Routley) and Meyer [26; 27], Sylvan and Plumwood (né Routley) [31], and Restall [21]), two-valued relational and four-valued approaches (cf. the American Plan semantics; e.g., see Restall [22] and Sylvan [24]), approaches combining both the American and the Australian plan (Logan [16]), and the algebraic approach (that is, with De Morgan lattices; e.g., see Dunn [8]).

In the logic of First-Degree Entailment, the four-valued interpretation employs valuations that map propositions to one of 4 truth values: **True**, **False**, **Both**, and **Neither**. These truth values are often put into a Hasse diagram, as in Figure 1(a).

With the truth values of Both and Neither, **FDE** is a logic that allows for both truth value gaps and gluts. In fact, by removing elements of the truth value set we can obtain the paraconsistent logic **LP**, Kleene's paracomplete logic **K3**, and classical logic (as all the connectives preserve classical values). On the other hand, the relational semantics of **FDE** captures the idea of gaps and gluts in an interesting way. In Figure 1(b), the truth values are the elements of the powerset of $\{True, False\}$ (hereafter $\{\mathbf{t}, \mathbf{f}\}$). The set $\{\}$ is a *gap* or lack of truth value, and the $\{\mathbf{t}, \mathbf{f}\}$ is a glut or overabundance of truth values. This relational semantics was developed by Dunn in [9], and is the basis for some of the semantic constructions of the present work. Following [9], and generalizing to a frame of points, we present a semantics for **FDE**.

Definition 1. An **FDE**-situation-model is a pair $\mathfrak{M} = \langle K, V \rangle$, where K is a non-empty set of points (situations), and V is a valuation function that assigns to each world-atom(ic sentence) pair a subset of $\{\mathbf{t}, \mathbf{f}\}$.[2] V is extended to every sentence by the following. For every $\alpha \in K$,

---

[2]From this function we can easily obtain the more familiar function (in most presentations of modal logic) from atomic sentences to the set of situations at which they are true, and the less familiar function

(i) $\mathbf{t} \in V(\alpha, \neg \mathcal{A})$ iff $\mathbf{f} \in V(\alpha, \mathcal{A})$
(ii) $\mathbf{f} \in V(\alpha, \neg \mathcal{A})$ iff $\mathbf{t} \in V(\alpha, \mathcal{A})$
(iii) $\mathbf{t} \in V(\alpha, \mathcal{A} \wedge \mathcal{B})$ iff $\mathbf{t} \in V(\alpha, \mathcal{A})$ and $\mathbf{t} \in V(\alpha, \mathcal{B})$
(iv) $\mathbf{f} \in V(\alpha, \mathcal{A} \wedge \mathcal{B})$ iff $\mathbf{f} \in V(\alpha, \mathcal{A})$ or $\mathbf{f} \in V(\alpha, \mathcal{B})$
(v) $\mathbf{t} \in V(\alpha, \mathcal{A} \vee \mathcal{B})$ iff $\mathbf{t} \in V(\alpha, \mathcal{A})$ or $\mathbf{t} \in V(\alpha, \mathcal{B})$
(vi) $\mathbf{f} \in V(\alpha, \mathcal{A} \vee \mathcal{B})$ iff $\mathbf{f} \in V(\alpha, \mathcal{A})$ and $\mathbf{f} \in V(\alpha, \mathcal{B})$

Definition 2. Consequence relations are defined as follows:

(i) $\mathcal{A} \vDash^{\mathfrak{M}}_{FDE} \mathcal{B}$ iff, for every $\alpha \in K$, if $\mathbf{t} \in V(\alpha, \mathcal{A})$, then $\mathbf{t} \in V(\alpha, \mathcal{B})$.
(ii) $\mathcal{A} \vDash_{FDE} \mathcal{B}$ iff, for every model $\mathfrak{M}$, $\mathcal{A} \vDash^{\mathfrak{M}}_{FDE} \mathcal{B}$.

This single-premise — or FMLA–FMLA, in the terminology of [14] — consequence relation, is exactly the first-degree fragment of the Logic of Entailment **E**. In other words, as shown in [1], for $\mathcal{A}$ and $\mathcal{B}$ in the language of **FDE**, $\mathcal{A} \vDash^{\mathfrak{M}}_{FDE} \mathcal{B}$ iff $\vdash_\mathbf{E} \mathcal{A} \Rightarrow \mathcal{B}$. However, **FDE** can be generalized to allow premise sets.[3] We will often remove the sub- and super-scripts from $\vDash^{\mathfrak{M}}_{FDE}$ ($\vDash_{FDE}$ or simply $\vDash$), when the intent is clear from the context.

Here we are considering **FDE**-situation-models (built on frames of points) instead of the simpler models lacking a set of situations. This is primarily due to a better fit with the remainder of this paper. However, it also provides some additional utility. We will often want to talk about the truth set of a sentence. The truth set of a sentence $\mathcal{A}$, sometimes referred to as a proposition or UCLA proposition, is the set of situations at which $\mathcal{A}$ is (at least) true. We record the following definition and fact.

Definition 3. The truth set of a proposition $\mathcal{A}$ in model $\mathfrak{M}$, denoted by $||\mathcal{A}||^{+}_{\mathfrak{M}}$ is the set $\{\alpha \in K: \mathbf{t} \in V(\alpha, \mathcal{A})\}$. The falsity set of a proposition, denoted with a minus sign decoration as in $||\mathcal{A}||^{-}_{\mathfrak{M}}$ is the set $\{\alpha \in K: \mathbf{f} \in V(\alpha, \mathcal{A})\}$.

Throughout the paper, we will drop the $\mathfrak{M}$, when it is implied by context. In addition, we will sometimes drop the "+" of a truth set, when doing so produces no ambiguity.

Fact 4. *Some facts about the consequence relation are equivalently stated using the truth sets:*

(1) $\mathcal{A} \vDash \mathcal{B}$ *iff* $||\mathcal{A}||^{+}_{\mathfrak{M}} \subseteq ||\mathcal{B}||^{+}_{\mathfrak{M}}$.
(2) $\mathcal{A} \vDash \mathcal{B}$ *and* $\mathcal{B} \vDash \mathcal{A}$ *iff* $||\mathcal{A}||^{+}_{\mathfrak{M}} = ||\mathcal{B}||^{+}_{\mathfrak{M}}$.

2.1.1. *Axiomatic Proof Theory.* The logic of **FDE** can be axiomatized in a number of ways. Here, following [33], we will use a sequent-based axiomatization which we

---

from atomic sentences to the set of situations at which they are false. For **FDE**, these are commonly represented as positive and negative valuation functions V^+ and V^-.

[3]The conditionals of **E** have single sentence antecedents, and so we are presented with multiple ways of connecting **E** to a multi-premise **FDE**. As a first option, we could take the premises as conjoined, and conjoin the premises into a single antecedent in a conditional of **E**. The other option is to take the (non-commutative) "fusion" of the premises by right-nesting conditionals. That is where $\mathcal{A}, \mathcal{B} \vdash_{FDE} \mathcal{C}$, we take its corresponding conditional in **E** to be $\mathcal{A} \Rightarrow (\mathcal{B} \Rightarrow \mathcal{C})$. On this route, the order of premises matters, as **E** does not have permutation, so we must give up the commutativity of premises in **FDE**.

will call the De Morgan-based **FDE** axiomatization. The key feature of this axiom system is that there is no rule of contraposition.

(∧-El) $\mathcal{A} \land \mathcal{B} \vdash \mathcal{A}$ (∧-Er) $\mathcal{A} \land \mathcal{B} \vdash \mathcal{B}$
(∨-Il) $\mathcal{A} \vdash \mathcal{A} \lor \mathcal{B}$ (∨-Ir) $\mathcal{B} \vdash \mathcal{A} \lor \mathcal{B}$
(∧∨-D) $\mathcal{A} \land (\mathcal{B} \lor \mathcal{C}) \vdash (\mathcal{A} \land \mathcal{B}) \lor \mathcal{C}$
(DNI) $\mathcal{A} \vdash \neg\neg\mathcal{A}$ (DNE) $\neg\neg\mathcal{A} \vdash \mathcal{A}$
(DM1) $\neg(\mathcal{A} \lor \mathcal{B}) \vdash \neg\mathcal{A} \land \neg\mathcal{B}$ (DM2) $\neg\mathcal{A} \land \neg\mathcal{B} \vdash \neg(\mathcal{A} \lor \mathcal{B})$
(DM3) $\neg(\mathcal{A} \land \mathcal{B}) \vdash \neg\mathcal{A} \lor \neg\mathcal{B}$ (DM4) $\neg\mathcal{A} \lor \neg\mathcal{B} \vdash \neg(\mathcal{A} \land \mathcal{B})$

(RT) $\mathcal{A} \vdash \mathcal{B}$ and $\mathcal{B} \vdash \mathcal{C} \Rightarrow \mathcal{A} \vdash \mathcal{C}$
(R∧-I) $\mathcal{A} \vdash \mathcal{B}$ and $\mathcal{A} \vdash \mathcal{C} \Rightarrow \mathcal{A} \vdash (\mathcal{B} \land \mathcal{C})$
(R∨-E) $\mathcal{A} \vdash \mathcal{B}$ and $\mathcal{C} \vdash \mathcal{B} \Rightarrow (\mathcal{A} \lor \mathcal{C}) \vdash \mathcal{B}$

Rules are written in the form $\Gamma \Rightarrow \mathcal{A} \vdash \mathcal{B}$, where Γ is a set of sequents. As in the rules given above, we write "$\mathcal{A} \vdash \mathcal{B}$ and $\mathcal{C} \vdash \mathcal{B}$" for "$\{\mathcal{A} \vdash \mathcal{B}, \mathcal{C} \vdash \mathcal{B}\}$." A rule of the form $\Gamma \Rightarrow \mathcal{A} \vdash \mathcal{B}$ means that if $\mathcal{C} \vdash \mathcal{D}$, for each sequent $\mathcal{C} \vdash \mathcal{D} \in \Gamma$, then you may infer the sequent $\mathcal{A} \vdash \mathcal{B}$. If a sequent $\mathcal{A} \vdash \mathcal{B}$ is derivable in an axiom system $\mathfrak{L}$, we will call the sequent a theorem of $\mathfrak{L}$.

Soundness and completeness are straightforward to prove for this system with respect to the models defined above.

2.1.2. *Quantified FDE.* There are numerous first-order extensions of **FDE**. We define a number of systems from the following list of axioms and rules:

(∀E) $\forall x \mathcal{A} \vdash \mathcal{A}[\tau/x]$, where τ is free for x in $\mathcal{A}$
(∃I) $\mathcal{A}[\tau/x] \vdash \exists x \mathcal{A}$, where τ is free for x in $\mathcal{A}$

(R∀I) $\mathcal{A} \vdash \mathcal{B} \Rightarrow \mathcal{A} \vdash \forall x \mathcal{B}$, where x is not free in $\mathcal{A}$
(R∃E) $\mathcal{A} \vdash \mathcal{B} \Rightarrow \exists x \mathcal{A} \vdash \mathcal{B}$, where x is not free in $\mathcal{B}$

(EC1) $\forall x(\mathcal{A} \lor \mathcal{B}) \vdash \mathcal{A} \lor \forall x \mathcal{B}$, where x is not free in $\mathcal{A}$
(EC2) $\mathcal{A} \land \exists x \mathcal{B} \vdash \exists x(\mathcal{A} \land \mathcal{B})$, where x is not free in $\mathcal{A}$

(Dual1) $\forall x \neg \mathcal{A} \dashv\vdash \neg \exists x \mathcal{A}$
(Dual2) $\exists x \neg \mathcal{A} \dashv\vdash \neg \forall x \mathcal{A}$

We define a couple logics as follows:[4]

QFDE = **FDE** + (∀E) + (∃I) + (R∀I) + (R∃E) + (Dual1) + (Dual2)
FDEQ = **QFDE** + (EC1) + (EC2)

The next lemma states that two other statements of the duality of the quantifiers are provable sequents. However, note that without contraposition (Dual1) and (Dual2) are not provable from (Dual3) and (Dual4).

Lemma 5. *The following are theorems of* **QFDE** *(and a fortiori of* **FDEQ***).*

(Dual3) $\neg \forall x \neg \mathcal{A} \dashv\vdash \exists x \mathcal{A}$
(Dual4) $\neg \exists x \neg \mathcal{A} \dashv\vdash \forall x \mathcal{A}$

Proof. For (Dual4), the right-to-left direction is given by

---

[4] The permutation of **Q** with a logic's name to indicate the presence of extensional confinement is fairly standard in relevant logic, especially, in more recent works. This convention is adopted, for example, in [10; 12; 18; 38].

(1) $\forall x \mathcal{A} \vdash \mathcal{A}[\tau/x]$ ($\forall$E)
(2) $\mathcal{A}[\tau/x] \vdash \neg\neg\mathcal{A}[\tau/x]$ (DNI)
(3) $\forall x \mathcal{A} \vdash \forall x \neg\neg \mathcal{A}$ (RT), (R$\forall$I)
(4) $\forall x \mathcal{A} \vdash \neg\exists x \neg \mathcal{A}$ (Dual1), (RT)

while the left-to-right direction is by

(1) $\neg\exists x \neg \mathcal{A} \vdash \forall x \neg\neg \mathcal{A}$ (Dual1)
(2) $\forall x \neg\neg \mathcal{A} \vdash \forall x \mathcal{A}$ ($\forall$E), (DNE), (RT)
(3) $\neg\exists x \neg \mathcal{A} \vdash \forall x \mathcal{A}$ (RT)

The proof of (Dual3) is similar, but dual. ◁

The following lemma is used in the completeness proof.

Lemma 6. *If sequent $(\mathcal{A} \vdash \mathcal{B})[c/x]$ is a theorem, where c does not occur in $\mathcal{A} \vdash \mathcal{B}$, then $(\mathcal{A} \vdash \mathcal{B})[\tau/x]$ is also a theorem, where τ is a term nor occurring in $\mathcal{A} \vdash \mathcal{B}$.*

Proof. Given a proof of $(\mathcal{A} \vdash \mathcal{B})[c/x]$, where c does not occur in $\mathcal{A} \vdash \mathcal{B}$, let τ be a term not occurring in this sequent. Replace c throughout the proof with τ. As τ is new, and the axioms and rules do not rely on constants in their statements, the result is a proof of $(\mathcal{A} \vdash \mathcal{B})[\tau/x]$. ◁

2.1.3. QFDE- and FDEQ-models.

For the semantics of first-order extensions of **FDE**, we employ the general frame Mares–Goldblatt semantics. This requires some additional preliminaries. First, we add a set of admissible propositions (*Prop*), and a set of admissible propositonal functions (*PropFun*) to our models. The former is the subset of the sets of situations K that are to be treated as admissible propositions; that is, as sets of situations that together can coherently serve as the truth set of a sentence. The latter are functions from value assignments for variables to the admissible propositions. We will represent a truth set by a propositional function φ applied to a variable assignment g: φg is the truth set of the formula φ under the variable assignment g.

In the first-order setting, we will break from the use of Dunn's two-valued, relational semantics in form but not in spirit. That is, the same motivations and interpretations are present. However, we will focus on the propositions (truth sets) and propositional functions.

A proposition is the truth set of a formula. On these sets of situations, we use the set theoretic $\cap$ and $\cup$ operators, and the operators $\neg$, $\sqcap$, and $\sqcup$ defined as follows. For $\sqcap$, and $\sqcup$, for every set of sets of situations S: $\sqcap S =_{df} \bigcup \{X \in Prop : X \subseteq \cap S\}$; $\sqcup S =_{df} \bigcap \{X \in Prop : \cup S \subseteq X\}$. We state the definition of $\neg$ using a model's valuation condition, although we could push negation into the frame proper (but we don't due to the fact that our giving negative (falsity) propositional functions to each atomic formula guarantees closure under this defined negation operator). For *every truth and falsity set $||\mathcal{A}||^+$ and $||\mathcal{A}||^-$ in Prop*, for some formula $\mathcal{A}$ — in the models below, $|\mathcal{A}|^+g$ and $|\mathcal{A}|^-g$ for some variable assignment g — we define $\neg ||\mathcal{A}||^+ =_{df} ||\mathcal{A}||^-$ and $\neg ||\mathcal{A}||^- =_{df} ||\mathcal{A}||^+$.

A consequence of this is that the set of admissible propositions *Prop* will be closed only under $\neg$ with respect to truth and falsity sets of formulas relative to a particular model, and not arbitrary elements of *Prop*; however, this does no harm as (i) soundness only requires this weaker closure, and (ii) this closure is found in the canonical model.

Essentially, the defined $\sqcap$ and $\sqcup$ enable the restriction of the intersections and unions to admissible propositions with the same behaviour, as *Prop* is not required to be closed under infinite intersections and unions. Goldblatt describes this definition of $\sqcap$ as "Motivated by the intuition that the sentence $\forall x \varphi$ expresses the conjunction of all the sentences $\varphi[a/x]$" [12, p. 17]. A similar, but dual, motivation underlies $\sqcup$. Goldblatt adds to this a "more semantic" interpretation. A proposition X entails Y when $X \subseteq Y$, and we call Y *weaker* and X *stronger* than the other. The operation $\sqcap S$, therefore, gives the weakest member of *Prop* that entails every member in S. If the (infinite) conjunction of the members of S is in *Prop*, then $\sqcap S = \cap S$. The reader is directed to Mares and Goldblatt [18] and Goldblatt [12] for additional explication of these operators.

Using these operators, we can define operations on propositional functions. For the truth functional connectives, we have for every $g \in U^\omega$ and $\varphi, \psi \in PropFun$, $(\neg \varphi)g = \neg(\varphi g)$; $(\varphi \cap \psi)g = \varphi g \cap \psi g$; and $(\varphi \cup \psi)g = \varphi g \cup \psi g$. Then, for the quantifiers, we define $\forall_n$ and $\exists_n$ by, for every $g \in U^\omega$, every $n \in \omega$, and every $\varphi \in PropFun$,

$$(\forall_n \varphi)g = \bigsqcap_{j \in U} \varphi g[j/n] = \bigcup \{X \in Prop : X \subseteq \bigcap_{j \in U} \varphi g[j/n]\},$$

$$(\exists_n \varphi)g = \bigsqcup_{j \in U} \varphi g[j/n] = \bigcap \{X \in Prop : \bigcup_{j \in U} \varphi g[j/n] \subseteq X\}.$$

Definition 7. A **QFDE**-model is a tuple $\langle K, U, Prop, PropFun, |-|\rangle$, where K is a non-empty set of points, U is a non-empty set of individuals, *Prop* and *PropFun* satisfy conditions (c1)–(c3), and the valuation function $|-|$ is defined as below.

(c1) *Prop* is closed under $\cap, \cup, \neg$, as defined above.
(c2) *PropFun* is closed under $\cap, \cup, \neg$.[5]
(c3) *PropFun* is closed under $\exists_n$ and $\forall_n$, for every $n \in \omega$.

The valuation function $|-|$ is then defined by assigning

(1) an element $|c| \in U$ to each constant symbol c;
(2) a positive function $|P|^+ : U^n \longrightarrow \wp(K)$ to each n-ary predicate symbol P;
(3) a negative function $|P|^- : U^n \longrightarrow \wp(K)$ to each n-ary predicate symbol P;
(4) a positive and negative admissible propositional function $|\mathcal{A}|^+$ and $|\mathcal{A}|^-$ (each of type $U^\omega \longrightarrow Prop$) to each formula $\mathcal{A}$ such that, when $\mathcal{A}$ is the atomic $P\tau_1, \ldots, \tau_n$, the propositional function is given by, for every $g \in U^\omega$,

$$|P\tau_1 \ldots \tau_n|^+ g = |P|^+(|\tau_1|g, \ldots, |\tau_n|g)$$
$$|P\tau_1 \ldots \tau_n|^- g = |P|^-(|\tau_1|g, \ldots, |\tau_n|g),$$

where $|\tau|g \in U$ is $|c|$ if τ is the constant c, and gn if τ is the variable x_n. When $\mathcal{A}$ is not atomic, the function assigned to the formula is given by, for

---

[5] Closure under $\neg$, strictly speaking, follows from the definition of the valuation function and the closure under $\cap, \cup, \forall_n$, and $\exists_n$. A model's valuation function assigns negative (falsity) functions for atomic propositions. Furthermore, all negative functions *reduce* via the extension of the valuation function to the base cases, namely, positive and negative functions for atomics.

every $g \in U^\omega$,

$|\mathcal{A} \wedge \mathcal{B}|^+ g = |\mathcal{A}|^+ g \cap |\mathcal{B}|^+ g$ $\qquad$ $|\mathcal{A} \wedge \mathcal{B}|^- g = |\mathcal{A}|^- g \cup |\mathcal{B}|^- g$
$|\mathcal{A} \vee \mathcal{B}|^+ g = |\mathcal{A}|^+ g \cup |\mathcal{B}|^+ g$ $\qquad$ $|\mathcal{A} \vee \mathcal{B}|^- g = |\mathcal{A}|^- g \cap |\mathcal{B}|^- g$
$|\neg \mathcal{A}|^+ g = |\mathcal{A}|^- g$ $\qquad$ $|\neg \mathcal{A}|^- g = |\mathcal{A}|^+ g$
$|\forall x_n \mathcal{A}|^+ g = (\forall_n |\mathcal{A}|^+) g$ $\qquad$ $|\forall x_n \mathcal{A}|^- g = (\exists_n |\mathcal{A}|^-) g$
$|\exists x_n \mathcal{A}|^+ g = (\exists_n |\mathcal{A}|^+) g$ $\qquad$ $|\exists x_n \mathcal{A}|^- g = (\forall_n |\mathcal{A}|^-) g$

We may also define the valuation using the equality $|\forall x_n \mathcal{A}|^+ g = \bigcap_{h \in xg} |\mathcal{A}|^+ h$, and similarly for the other quantifier cases. This is because $\bigcap_{h \in xg} |\mathcal{A}|^+ h = \bigcap_{j \in U} |\mathcal{A}|^+ g[j/n]$, as in [18].

The functions $|P|^+$ and $|P|^-$ give what can be called the extension and the anti-extension of a predicate. That is, the propositional functions that return truth and falsity sets, respectively, when given a variable assignment.

The formula $\forall x \mathcal{A}$ is true at a situation (given a variable assignment) just when it is a member of the weakest admissible proposition that entails every instance of $\mathcal{A}$. The formula $\forall x \mathcal{A}$ is false at a situation when it is a member of the strongest admissible proposition that is entailed by every falsity set of the instances of $\mathcal{A}$. We may explicate the existential quantifier using a dual interpretation. Note also that this guarantees the duality of the quantifiers via negation — that is, the validity of (Dual1) and (Dual2) — by the following identity:

$$|\neg \forall x_n \mathcal{A}|^+ = |\forall x_n \mathcal{A}|^- = \exists_n |\mathcal{A}|^- = \exists_n |\neg \mathcal{A}|^+ = |\exists x_n \neg \mathcal{A}|^+$$

Definition 8. Consequence relations are defined as follows:

(i) $\mathcal{A} \vDash_{\mathbf{QFDE}}^{\mathfrak{M}} \mathcal{B}$ iff, for every $g \in U^\omega$, $|\mathcal{A}|^+ g \subseteq |\mathcal{B}|^+ g$.

(ii) $\mathcal{A} \vDash_{\mathbf{QFDE}} \mathcal{B}$ iff, for every **QFDE**-model $\mathfrak{M}$, $\mathcal{A} \vDash_{\mathbf{QFDE}}^{\mathfrak{M}} \mathcal{B}$.

For the logic **FDEQ**, the conditions added to the models for the validity of the EC axioms are, for every $\varphi \in PropFun$, $X, Y \in Prop$, $n \in \omega$, and $g \in U^\omega$:

(cEC1) $X - Y \subseteq \bigcap_{j \in U} \varphi(g[j/n])$ only if $X - Y \subseteq (\forall_n \varphi) g$

(cEC2) $\bigcup_{j \in U} \varphi(g[j/n]) \subseteq X \cup \overline{Y}$ only if $|\exists_n \varphi| g \subseteq X \cup \overline{Y}$.

Without the rule of contraposition, it appears (though is yet to be proved) that (EC1) and (EC2) are independent in the background of **QFDE**. Semantically, the **QFDE**-models that satisfy (cEC1) will validate (EC2) if it is the case that $\mathcal{A} \dashv\vdash \mathcal{B}$ implies $|\mathcal{A}|^- = |\mathcal{B}|^-$. This would be the case if the rule of contraposition was an admissible rule. This has not been proved either way for **QFDE** with or without either of the extensional confinement axioms. Thus, for now we opt to give semantic conditions for both axioms, whether or not they are independent.

Theorem 9 (Soundness and Completeness). *The logics defined are sound and complete with respect to their class of models. That is,*

(1) $\mathcal{A} \vdash_{\mathbf{QFDE}} \mathcal{B}$ *iff* $\mathcal{A} \vDash_{\mathbf{QFDE}} \mathcal{B}$, $\qquad$ and $\qquad$ (2) $\mathcal{A} \vdash_{\mathbf{FDEQ}} \mathcal{B}$ *iff* $\mathcal{A} \vDash_{\mathbf{FDEQ}} \mathcal{B}$.

The proof of this theorem is tangential to the main theme of conditional **FDE**-logics, and is, therefore, relegated to the Appendix.

2.2. Conditional Logics. Conditional logics, as presented in [7], add a non-truth-functional conditional based on neighborhood semantics. In [7], these logics are used to construct deontic logics; however, conditional logics offer an alternative construction for conditionals on top of a truth-functional logic, which presents the opportunity to add a solid conditional to **FDE**. Chellas defines a class of *standard conditional models* (corresponding, in a rough sense, with normal modal logics) and *minimal conditional models* (corresponding roughly with classical modal logics).[6]

2.2.1. *Minimal Conditional Models.* Here we restate Chellas' model definitions, the axiom systems corresponding to these semantics, and some extensions of these logics. Most of the results are given (or stated) in [7], with some minor exceptions we will note. We begin with the minimal conditional models.

Definition 10. A *minimal conditional model* is a tuple $\langle K, f, V \rangle$ where, K is a non-empty set of situations, f is a function from a situation and *proposition* to a *set of propositions* (i.e., $f \colon (K \times \wp(K)) \longrightarrow \wp(\wp(K))$), and V is a valuation function that assigns a set of worlds to each atomic formula. V is then extended to every sentence by the following. For every $\alpha \in K$,

(i) $\alpha \vDash p$ iff $\alpha \in V(p)$
(ii) $\alpha \vDash \neg \mathcal{A}$ iff $\alpha \nvDash \mathcal{A}$
(iii) $\alpha \vDash \mathcal{A} \wedge \mathcal{B}$ iff $\alpha \vDash \mathcal{A}$ and $\alpha \vDash \mathcal{B}$
(iv) $\alpha \vDash \mathcal{A} \vee \mathcal{B}$ iff $\alpha \vDash \mathcal{A}$ or $\alpha \vDash \mathcal{B}$
(v) $\alpha \vDash \mathcal{A} \to \mathcal{B}$ iff $\alpha \nvDash \mathcal{A}$ or $\alpha \vDash \mathcal{B}$
(vi) $\alpha \vDash \mathcal{A} \Rightarrow \mathcal{B}$ iff $\|\mathcal{B}\| \in f(\alpha, \|\mathcal{A}\|)$

The conditional $\Rightarrow$ is the only non-straightforward case, and it can use a bit of explanation. Chellas explains that $f(\alpha, \|\mathcal{A}\|)$ returns a set of all the propositions that are "necessary relative to the condition expressed by $\mathcal{A}$ at that world $[\alpha]$" [7, p. 270]. The semantics here is essentially neighborhood semantics, as the focus is on propositions (sets of situations) instead of worlds, unlike the *standard conditional models* below. In fact, the relation to Classical, Monotonic, and Regular logics can be seen quite clearly if we rewrite $\mathcal{A} \Rightarrow \mathcal{B}$ as $[\mathcal{A}]\mathcal{B}$ (as in [7]), where "$[\mathcal{A}]$" is treated as a modal operator, and we have a denumerable number of modal operators indexed to formulas.

2.2.2. *Soundness, Completeness, and Some Extensions.* Here we first define all of the two-valued conditional logics we will consider using a Hilbert-style axiom system. A list of axioms and rules is given, then we define logics using that list. Our list is as follows:

(PL) All the theorems of propositional, two-valued logic.
(ID) $\mathcal{A} \Rightarrow \mathcal{A}$
(CMP) $(\mathcal{A} \Rightarrow \mathcal{B}) \to (\mathcal{A} \to \mathcal{B})$
(RCEA) $\mathcal{A} \leftrightarrow \mathcal{B} \Rightarrow \mathcal{A} \Rightarrow \mathcal{C} \leftrightarrow \mathcal{B} \Rightarrow \mathcal{C}$
(RCEC) $\mathcal{B} \leftrightarrow \mathcal{C} \Rightarrow \mathcal{A} \Rightarrow \mathcal{B} \leftrightarrow \mathcal{A} \Rightarrow \mathcal{C}$
(RCM) $\mathcal{B} \to \mathcal{C} \Rightarrow \mathcal{A} \Rightarrow \mathcal{B} \to \mathcal{A} \Rightarrow \mathcal{C}$
(RCR) $(\mathcal{B} \wedge \mathcal{D}) \to \mathcal{C} \Rightarrow ((\mathcal{A} \Rightarrow \mathcal{B}) \wedge (\mathcal{A} \Rightarrow \mathcal{D})) \to \mathcal{A} \Rightarrow \mathcal{C}$
(RCN) $\mathcal{B} \Rightarrow \mathcal{A} \Rightarrow \mathcal{B}$

---

[6]We use the terms *classical, regular, monotonic* as in Segerberg [32], where they were introduced.

The minimal conditional logic (or least/smallest classical conditional logic), denoted by **CE**, is defined by (PL) + (RCEA) + (RCEC). This logic is sound and complete for the minimal models. The least monotonic conditional logic, **CM**, is defined as **CE** + (RCM). The least regular conditional logic, **CR**, is defined as **CM** + (RCR). Finally, the least normal conditional logic, **CK** is defined as **CR** + (RCN).

The semantic conditions for these axioms are given here by correspondence in the background of **CE**. Chellas [7] gives semantic conditions for (ID) and (CMP) in standard models. The conditions given by Chellas for the rules (RCM), (RCR), and (RCN), characterize those axioms in the background of **CE**, although the proof is not given explicitly by Chellas. For (cID) and (cCMP), Chellas only gives conditions in the standard models, but the conditions below are easy to check.[7]

The conditions are as follows:

(cID) $X \in f(\alpha, X)$;
(cCMP) if $\alpha \in X$ and $Y \in f(\alpha, X)$, then $\alpha \in Y$;
(cRCM) if $Y \cap Y' \in f(\alpha, X)$, then $Y \in f(\alpha, X)$ and $Y' \in f(\alpha, X)$;
(cRCR) if $Y \in f(\alpha, X)$ and $Y' \in f(\alpha, X)$, then $Y \cap Y' \in f(\alpha, X)$;
(cRCN) $K \in f(\alpha, X)$.

Some characterization results are recorded in the following lemmas.

Fact 11. *For minimal conditional models:*
(1) **CE** *is sound and complete with respect to the class of minimal conditional models.*
(2) *The logics extending* **CE** *with the axioms and rules listed above are sound and complete with respect the the class of models determined by the conditions corresponding to the additional axioms and rules.*

2.2.3. *Standard Conditional Models.* In two-valued modal logic, the neighborhood and relational models are related. A very informative result concerning their relation is that the relational models are point-wise equivalent to the *augmented* neighborhood models. In particular, this result says that augmented neighborhood models are just relational models under a different description; you can transform the models back and forth, keeping the set of situations and the atomic valuations the same, and only replacing neighborhoods with relations or vice versa. For conditional logics, there is similarly a set of conditions such that, when a minimal model satisfies these conditions, it is sound and complete with respect to the least normal conditional logic characterized by the standard conditional models. Unlike the two-valued case relating neighborhood and relational frames, here, we have a relation between two distinct kinds of neighborhood-based frames.

While Chellas [7] does not provide proofs of these results, the related equivalence in terms of soundness and completeness is given. That is, the logic **CK** is sound and complete with respect to the standard conditional models defined below, and with respect to the minimal conditional models that satisfy (cRCM), (cRCR), and (cRCN). For the logic **CK** and its extensions, we can give what Chellas calls standard models, where the f function returns a single set of worlds instead of a set of sets of worlds.

---

[7]Note that (RCR) implies (RCM), but the condition (cRCR) does not imply (cRCM). Logics with (RCM) satisfy both conditions. Note also that I have renamed Chellas' conditions to conform to the conventions in this paper, where (cXYZ) is the condition for the axiom or rules named (XYZ).

Definition 12. A *standard conditional model* is a tuple $\langle K, f, V \rangle$ where, K is a non-empty set of situations, f is a function from a situation and *proposition* to a *proposition* (i.e., $f: (K \times \wp(K)) \longrightarrow \wp(K)$), and V is a valuation function that assigns a set of worlds to each atomic formula. V then is extended to every sentence as in the minimal conditional models, with the exception of the case for conditionals, which is

(i) $\alpha \vDash \mathcal{A} \Rightarrow \mathcal{B}$ iff $f(\alpha, ||\mathcal{A}||) \subseteq ||\mathcal{B}||$.

The difference between minimal and standard models is similar to the difference between neighborhood and relational models, as a set of sets of worlds is reduced to just a single set (their intersection). The standard models can be seen as a multi-relational model, where the relations are indexed to formulas as in $\beta \in f(\alpha, ||\mathcal{A}||)$ iff $\alpha R_\mathcal{A} \beta$.

The logic **CK** can be axiomatized as above, or equivalently, by adding to classical logic, (RCEA) and the rule

(RCK) $(\mathcal{B}_1 \wedge \cdots \wedge \mathcal{B}_n) \to C \Rightarrow ((\mathcal{A} \Rightarrow \mathcal{B}_1) \wedge \cdots \wedge (\mathcal{A} \Rightarrow \mathcal{B}_n)) \to (\mathcal{A} \Rightarrow C)$.

For further extensions of these logics, particularly in the deontic setting, the reader is directed to the aforementioned chapter of [7].

3. Minimal Conditional **FDE** Logics

As conditional logics can be distinguished based on whether or not they can be given standard or minimal conditional models, we will separate our discussion of Conditional **FDE** Logics based on this divide. First, we will examine the logics defined by combining **FDE** with the minimal conditional models. The logics characterized by these models may be called classical in the Segerberg-like sense of the conditional being congruent in both the antecedent and consequent. That is, provably equivalent formulas are substitutable in both the antecedent and consequent position of a conditional. The term "minimal," although merely the name of a collection of semantic structures, is often used to label the logics characterized by said structures. Because the term "classical" can refer to types of modal logics as well as two-valued truth-functional logic, we will tend to (ab)use the term "minimal" when referring to the logics of this section.

The approach taken here is semantics-first. That is, we start with the semantic idea of conditional logics (as in [7, Ch. 10]), but replace the "classical" foundation with **FDE**. As in Routley's "American Plan Completed" [24], there are several ways we may go about defining the semantics in terms of the falsity condition for the conditionals. Here we opt for reducing the negated conditionals to non-negated conditionals by pushing the negation into the consequent of the conditional. That is, by the equivalence of $\neg(\mathcal{A} \Rightarrow \mathcal{B})$ with $\mathcal{A} \Rightarrow \neg\mathcal{B}$. This approach is taken, for example, by Wansing in connexive logic (e.g., see [39]). In addition, this approach enables a conditional to have all four combinations of truth values. Another method which would ensure this is where a conditional is false when some of the antecedent-f-selected situations are such that the negation of the consequent is true. We call this the *overlap* method and the former method the *container* method.

Wansing and Unterhuber [40] combine these methods for standard (and not minimal) conditional logics. In particular, there is the connective $\Box\!\!\rightarrow$ of their logic **cCL**, which employs the container method for both negated and non-negated conditionals.

Our approach differs from their $\Box\!\to$ in **cCL** in that we, additionally, generalize in this section to *minimal* conditional logics.

Note the contrast with Ma and Wong [17], whose paraconsistent conditional logic employs a positive selection function f^+ and a negative selection function f^-. Moreover, we will only construct logics based on the **FDE** axiomatization without contraposition (again, in contrast to [17]).[8] We will say more about contraposition, and its potential admissibility, in Section 3.2.

Here we will define general frames — that is, admissible semantics that restrict the possible truth sets to a special subset of sets of points we call *Prop*. We have already defined the operator $\neg$ on (the truth and falsity sets of formulas in) *Prop*. A conditional operator is defined by the following; for every $X, Y \in Prop$,

$$X \Rightarrow Y = \{a \in K : Y \in f(a, X)\}.$$

Definition 13. A $\mathbf{CE_{FDE}}$-model is a tuple $\mathfrak{M} = \langle K, f, Prop, V \rangle$ where, K is a nonempty set of situations, $Prop \subseteq \wp(K)$, f is a function from a situation and *proposition* to a set of *propositions* (i.e., $f: (K \times Prop) \longrightarrow \wp(Prop)$) and the condition (c1a) is satisfied.

(c1a) *Prop* is closed under $\cap, \cup, \neg,$ and $\Rightarrow$.

Moreover, V is a valuation function that assigns to each world-atom(ic sentence) pair an element of $\wp(\{\mathbf{t}, \mathbf{f}\})$ such that for every atomic sentence p, $||p||^+, ||p||^- \in Prop$. V is then extended to every sentence using (i)–(vi) of Definition 1 for $\neg, \wedge,$ and $\vee$, and using (vii) and (viii) below for $\Rightarrow$. For every $\alpha \in K$,

(vii) $\mathbf{t} \in V(\alpha, \mathcal{A} \Rightarrow \mathcal{B})$ iff $||\mathcal{B}||^+_{\mathfrak{M}} \in f(\alpha, ||\mathcal{A}||^+_{\mathfrak{M}})$;
(viii) $\mathbf{f} \in V(\alpha, \mathcal{A} \Rightarrow \mathcal{B})$ iff $||\neg\mathcal{B}||^+_{\mathfrak{M}} \in f(\alpha, ||\mathcal{A}||^+_{\mathfrak{M}})$.[9]

The truth condition for the conditional is the expected condition for minimal conditional logics. The falsity condition, referred to above as the container approach, reduces the negated conditionals to bare conditionals. The f-function is fairly unrestricted, so the reader may easily confirm that a conditional can be just true, just false, neither, and both truth and false under this definition.

Lemma 14. *In every $\mathbf{CE_{FDE}}$-model $\mathfrak{M}$, every sentence $\mathcal{A}$ is mapped onto an element of Prop by the valuation function.*

Proof. The proof is by induction on the complexity of $\mathcal{A}$. The atomic case is by definition. We will only show the conditional case. Assume that $\mathcal{A} = \mathcal{B} \Rightarrow \mathcal{C}$. We have

$$||\mathcal{B} \Rightarrow \mathcal{C}|| = \{a \in K : \mathbf{t} \in V(\alpha, \mathcal{B} \Rightarrow \mathcal{C})\}$$
$$= \{a \in K : ||\mathcal{C}|| \in f(\alpha, ||\mathcal{B}||)\}$$
$$= ||\mathcal{B}|| \Rightarrow ||\mathcal{C}||$$

By the induction hypothesis, both $||\mathcal{B}||$ and $||\mathcal{C}||$ are elements of *Prop*, and so, by (c1a), $||\mathcal{B} \Rightarrow \mathcal{C}|| \in Prop$. ◁

---

[8] An earlier version of this paper additionally contained the logic defined by adding contraposition to the minimal conditional **FDE**-logic defined here. To capture these logics, the minimal models need only be modified by defining the consequent relation as preserving truth in the left-to-right direction, and falsity in the right-to-left direction.

[9] The reader is reminded of the convention to drop the "+" superscript for truth sets in what follows.

The consequence relation is then defined as only left-to-right truth preservation.

Definition 15. The consequence relations $\vDash^{\mathfrak{M}}_{\mathbf{CE_{FDE}}}$ and $\vDash_{\mathbf{CE_{FDE}}}$ are defined as in Definition 2, but with respect to $\mathbf{CE_{FDE}}$.

We say that *Prop* (or a model) is *full* when $Prop = \wp(K)$. Several extensions of $\mathbf{CE_{FDE}}$ are not complete with respect to full models. This will be shown later, and is partially due to the fact that no sentence is contained in every prime theory.

First we will provide an axiom system for this logic, and then prove soundness and completeness. Then, we will address whether or not the rule of contraposition is admissible.

The De Morgan-based axiom system for **FDE**, on which the system below is defined, is a more flexible axiomatic system than the one with the rule of contraposition. For example, this axiomatisation is known to be more amenable to extensions with new connectives [33, p. 313].

Definition 16. The logic $\mathbf{CE_{FDE}}$ is defined axiomatically by extending the De Morgan-based axiomatization of **FDE** with the following:

($\neg$-$\Rightarrow$) $\neg(\mathcal{A} \Rightarrow \mathcal{B}) \dashv\vdash \mathcal{A} \Rightarrow \neg\mathcal{B}$
(RCEA) $\mathcal{A} \dashv\vdash \mathcal{B} \Rightarrow \mathcal{A} \Rightarrow \mathcal{C} \dashv\vdash \mathcal{B} \Rightarrow \mathcal{C}$
(RCEC) $\mathcal{B} \dashv\vdash \mathcal{C} \Rightarrow \mathcal{A} \Rightarrow \mathcal{B} \dashv\vdash \mathcal{A} \Rightarrow \mathcal{C}$

A proof is defined in the usual way, but let us briefly note how to use the occurrences of "$\dashv\vdash$" in the new rules. Because $\mathcal{A} \dashv\vdash \mathcal{B}$ is shorthand for the metalogical conjunction of $\mathcal{A} \vdash \mathcal{B}$ and $\mathcal{B} \vdash \mathcal{A}$. Citing an axiom such as ($\neg$-$\Rightarrow$) will justify one of the metalogical conjuncts in a proof. Following this, the rules will cite both metalogical conjuncts to produce a single metalogical conjunct per line. For example,

(1) $\mathcal{A} \vdash \neg\neg\mathcal{A}$		(DNI)
(2) $\neg\neg\mathcal{A} \vdash \mathcal{A}$		(DNE)
(3) $\mathcal{A} \Rightarrow \mathcal{C} \vdash \neg\neg\mathcal{A} \Rightarrow \mathcal{C}$		(1), (2), (RCEA)
(4) $\neg\neg\mathcal{A} \Rightarrow \mathcal{C} \vdash \mathcal{A} \Rightarrow \mathcal{C}$		(1), (2), (RCEA)

We take this to not only be a proof of $\neg\neg\mathcal{A} \Rightarrow \mathcal{C} \vdash \mathcal{A} \Rightarrow \mathcal{C}$, using the last line of the proof, but also of $\neg\neg\mathcal{A} \Rightarrow \mathcal{C} \dashv\vdash \mathcal{A} \Rightarrow \mathcal{C}$, using the last two lines of the proof. (Although we could always introduce "$\dashv\vdash$" to the proof system, with suitable introduction and elimination rules to achieve the same result.)

3.1.1. Soundness.

Lemma 17. *The axioms* ($\wedge$-*El*), ($\wedge$-*Er*), ($\vee$-*Il*), ($\vee$-*Ir*), ($\wedge\vee$-*D*), (*DNI*), (*DNE*), *and the De Morgan axioms are all valid in the class of* $\mathbf{CE_{FDE}}$-*models.*

The proof is similar to those already in the literature.

Lemma 18. *The axiom* ($\neg$-$\Rightarrow$) *is valid in the class of* $\mathbf{CE_{FDE}}$-*models.*

Proof. Note that
$\mathbf{t} \in V(\alpha, \neg(\mathcal{A} \Rightarrow \mathcal{B}))$ iff $\mathbf{f} \in V(\alpha, \mathcal{A} \Rightarrow \mathcal{B})$
iff $||\neg\mathcal{B}|| \in f(\alpha, ||\mathcal{A}||)$
iff $\mathbf{t} \in V(\alpha, \mathcal{A} \Rightarrow \neg\mathcal{B})$ ◁

Lemma 19. *The rules (RT), (R∧-I), (R∨-E), (RCEA), (RCEC) preserve validity in the class of* $\mathbf{CE_{FDE}}$*-models.*

Proof. The rules of **FDE** can be shown to be valid using the usual arguments. For (RCEA) and (RCEC), suppose that $\mathcal{A} \vDash \mathcal{B}$ and $\mathcal{B} \vDash \mathcal{A}$. Then $||\mathcal{A}|| = ||\mathcal{B}||$. Assume that $||\mathcal{C}|| \in f(\alpha, ||\mathcal{A}||)$. Immediately, we get $||\mathcal{C}|| \in f(\alpha, ||\mathcal{B}||)$, as required, using the equality above. For (RCEC), a similar argument may be applied. ◁

Theorem 20 (Soundness). *The logic* $\mathbf{CE_{FDE}}$ *is sound with respect to the class of* $\mathbf{CE_{FDE}}$*-models. That is, if* $\mathcal{A} \vdash \mathcal{B}$*, then* $\mathcal{A} \vDash \mathcal{B}$*.*

3.1.2. Completeness. Let $\Gamma \gg_{\mathbf{CE_{FDE}}} \Delta$ mean that there are some $\mathcal{A}_1, \ldots, \mathcal{A}_n \in \Gamma$ and $\mathcal{B}_1, \ldots, \mathcal{B}_m \in \Delta$ such that $(\mathcal{A}_1 \land \cdots \land \mathcal{A}_n) \vdash (\mathcal{B}_1 \lor \cdots \lor \mathcal{B}_m)$ is a theorem of $\mathbf{CE_{FDE}}$, where Γ and Δ are sets of formulas. When $\Delta = \{\mathcal{A}\}$, we'll just write $\Gamma \gg_{\mathbf{CE_{FDE}}} \mathcal{A}$ for $\Gamma \gg_{\mathbf{CE_{FDE}}} \{\mathcal{A}\}$, and similarly where Γ is a singleton.

Definition 21. An L-theory is a set of formulas Γ such that if $\Gamma \gg_{\mathbf{CE_{FDE}}} \mathcal{A}$, then $\mathcal{A} \in \Gamma$. A theory Γ is *prime* if and only if, if $\mathcal{A} \lor \mathcal{B} \in \Gamma$, then either $\mathcal{A} \in \Gamma$ or $\mathcal{B} \in \Gamma$.

Definition 22 (Canonical Model for $\mathbf{CE_{FDE}}$). The *canonical model* for $\mathbf{CE_{FDE}}$ is $\mathfrak{M}^{\mathbb{C}}_{\mathbf{CE_{FDE}}} = \langle K^{\mathbb{C}}, f^{\mathbb{C}}, Prop^{\mathbb{C}}, V^{\mathbb{C}} \rangle$, where

  (i) $K^{\mathbb{C}}$ is the set of prime $\mathbf{CE_{FDE}}$-theories;
 (ii) $||\mathcal{A}|| = \{\alpha \in K^{\mathbb{C}} : \mathcal{A} \in \alpha\}$;
(iii) $f^{\mathbb{C}}$ is defined by $||\mathcal{B}|| \in f^{\mathbb{C}}(\alpha, ||\mathcal{A}||)$ iff $\mathcal{A} \Rightarrow \mathcal{B} \in \alpha$;
 (iv) $Prop^{\mathbb{C}}$ is defined to be the set $\{X \in \wp(K^{\mathbb{C}}) : X = ||\mathcal{A}||$ for some formula $\mathcal{A}\}$;
  (v) $V^{\mathbb{C}}$ is defined, for every $\alpha \in K^{\mathbb{C}}$, and every atomic sentence p, by
      (a) $\mathbf{t} \in V^{\mathbb{C}}(\alpha, p)$ iff $p \in \alpha$;
      (b) $\mathbf{f} \in V^{\mathbb{C}}(\alpha, p)$ iff $\neg p \in \alpha$.
 (vi) The valuation is extended to all well-formed formulas as usual.

Lemma 23 (Pair Extension). *If* $\Gamma \not\gg_{\mathbf{CE_{FDE}}} \Delta$*, then there is some prime* $\mathbf{CE_{FDE}}$*-theory* Γ' *such that* $\Gamma \subseteq \Gamma'$ *and* $\Gamma' \cap \Delta = \emptyset$*.*

For the proof, the reader is directed to Restall [23, 5.1–5.2].

Lemma 24. $\mathcal{A} \vdash \mathcal{B}$ *is a theorem iff* $||\mathcal{A}|| \subseteq ||\mathcal{B}||$*.*

Proof. The left-to-right direction is immediate by definition. For the right-to-left direction, by contraposition assume that $\mathcal{A} \vdash \mathcal{B}$ is not a theorem. Then $\mathcal{A} \not\gg_{\mathbf{CE_{FDE}}} \mathcal{B}$, and so by the previous lemma we have a prime theory α such that $\mathcal{A} \in \alpha$, but $\mathcal{B} \notin \alpha$. Therefore $||\mathcal{A}|| \not\subseteq ||\mathcal{B}||$. ◁

Lemma 25. *The canonical model is a* $\mathbf{CE_{FDE}}$*-model.*

Proof. This is fairly straightforward to check. Note that the set of prime theories is non-empty, so K is non-empty. The function f is of the right type; by definition only truth sets (elements of *Prop*) appear in the second place of the function, and only sets of truth sets (sets of elements of *Prop*) are returned by the function. Further, (c1a) is satisfied. This is shown by the following equations.

$$||\mathcal{A} \land \mathcal{B}|| = ||\mathcal{A}|| \cap ||\mathcal{B}|| \qquad ||\mathcal{A} \lor \mathcal{B}|| = ||\mathcal{A}|| \cup ||\mathcal{B}||$$
$$||\mathcal{A} \Rightarrow \mathcal{B}|| = ||\mathcal{A}|| \Rightarrow ||\mathcal{B}|| \qquad ||\neg \mathcal{A}|| = \neg ||\mathcal{A}||$$

The equations are straightforward to prove. We will show the details of only the conditional. Suppose that $\alpha \in ||\mathcal{A} \Rightarrow \mathcal{B}||$. This is if and only if $\mathcal{A} \Rightarrow \mathcal{B} \in \alpha$. That is, if and only if $||\mathcal{B}|| \in f(\alpha, ||\mathcal{A}||)$, which is also if and only if $\alpha \in ||\mathcal{A}|| \Rightarrow ||\mathcal{B}||$. Moreover, for each atomic proposition p, $||p||^+, ||p||^- \in Prop$, as can seen by the definition of the canonical valuation. ◁

Lemma 26 (Truth). $\mathbf{t} \in V^{\mathbb{C}}(\alpha, \mathcal{A})$ iff $\mathcal{A} \in \alpha$ (i.e., $\alpha \in ||\mathcal{A}||$).

Proof. The proof is by induction on the complexity of $\mathcal{A}$. The truth-functional connective cases are straightforward. We will show the case for the conditional.

First we must show that $\{\alpha \in K^{\mathbb{C}} : \mathcal{A} \Rightarrow \mathcal{B} \in \alpha\} = \{\alpha \in K^{\mathbb{C}} : \mathbf{t} \in V(\alpha, \mathcal{A} \Rightarrow \mathcal{B})\}$. For the left-to-right direction, suppose that $\mathcal{A} \Rightarrow \mathcal{B} \in \alpha$. Then, by the definition of $f^{\mathbb{C}}$, $||\mathcal{B}|| \in f^{\mathbb{C}}(\alpha, ||\mathcal{A}||)$, which gives us that $\mathbf{t} \in V(\alpha, \mathcal{A} \Rightarrow \mathcal{B})$. For the other direction, suppose that $\mathbf{t} \in V(\alpha, \mathcal{A} \Rightarrow \mathcal{B})$. Then $||\mathcal{B}|| \in f^{\mathbb{C}}(\alpha, ||\mathcal{A}||)$, and so $\mathcal{A} \Rightarrow \mathcal{B} \in \alpha$. ◁

Theorem 27 (Completeness). *The logic* $\mathbf{CE_{FDE}}$ *is complete with respect to the class of* $\mathbf{CE_{FDE}}$*-models. That is, if* $\mathcal{A} \vDash \mathcal{B}$*, then* $\mathcal{A} \vdash \mathcal{B}$.

Completeness may be shown by the usual contrapositive method. Suppose that $\mathcal{A} \nvdash \mathcal{B}$. Then there is a prime theory α, obtained by the extension lemma on the closure of $\mathcal{A}$ under $\vdash$, such that $\mathcal{A} \in \alpha$, but $\mathcal{B} \notin \alpha$. By the truth lemma, $\mathbf{t} \in V^{\mathbb{C}}(\alpha, \mathcal{A})$ and $\mathbf{t} \notin V^{\mathbb{C}}(\alpha, \mathcal{B})$, and therefore, $\mathcal{A} \nvDash \mathcal{B}$.

In summary, we have proved soundness and completeness for the minimal conditional **FDE** logic defined without contraposition. The chosen falsity condition opted for here works well with minimal models. Roughly, $f(\alpha, ||\mathcal{A}||)$ is the set of propositions that are $\mathcal{A}$-possible at α. So if $||\mathcal{B}||$ is one of those propositions, then the conditional $\mathcal{A} \Rightarrow \mathcal{B}$ is true at α. However, having one of those accessible propositions overlap $||\neg \mathcal{B}||$ is insufficient to give us the negation of the conditional, for $||\neg \mathcal{B}||$ is not accessible in this way. Thus, the overlap method is not motivated.

3.2. Rule Contraposition. Contraposition is not truth-preserving in every $\mathbf{CE_{FDE}}$-model. We can show this by constructing a single model in which the rule does not preserve truth. This is not surprising. In logics where γ is admissible, we can sometimes identify models in which $\mathcal{A}$ and $\neg \mathcal{A} \vee \mathcal{B}$ are true, but in which $\mathcal{B}$ is not true. We have to look at the set of theorems, and not merely the sentences made true by any old model, in order to address rule admissibility. Nonetheless, here is a quick demonstration.

Let $K = \{e\}$, $Prop = \wp(K)$, and $f(e, ||\mathcal{A}||) = \emptyset$. Set $f(e, ||\mathcal{C}||) = \{\{e\}\} = \{||\neg \mathcal{D}||\}$, and let $||\mathcal{A}|| = ||\mathcal{B}|| = ||\mathcal{C}|| = ||\mathcal{D}|| = ||\neg \mathcal{B}|| = \emptyset$.

We have that $\mathcal{A} \Rightarrow \mathcal{B} \vdash \mathcal{C} \Rightarrow \mathcal{D}$ is valid in this model, for it satisfies "if $\emptyset = ||\mathcal{B}|| \in f(e, ||\mathcal{A}||) = \emptyset$, then $\emptyset = ||\mathcal{D}|| \in f(e, ||\mathcal{C}||) = \emptyset$." In fact, this is satisfied vacuously, as the antecedent is false: the empty set is not a member of itself. Additionally, we have that $\mathcal{C} \Rightarrow \neg \mathcal{D}$, which is $\neg(\mathcal{C} \Rightarrow \mathcal{D})$, is true at e, by definition of f. But also that $\mathcal{A} \Rightarrow \neg \mathcal{B}$, which is $\neg(\mathcal{A} \Rightarrow \mathcal{B})$, is not true at e. So $\neg(\mathcal{A} \Rightarrow \mathcal{B}) \nvDash \neg(\mathcal{C} \Rightarrow \mathcal{D})$.

The question remains as to whether the rule is admissible by being valid over the whole class of $\mathbf{CE_{FDE}}$-models. That is, if *every* model makes $A \vdash B$ true, then do they all also make $\neg B \vdash \neg A$ true? I strongly expect a positive answer to this question.

The conditionals introduced by the rules (RCEA) and (RCEC) (i) require the inter-provability of a sentence with another, and (ii) add the same sentence in the same position in the introduced conditional on each side of the sequent. Semantically, then, we can observe that a sentence is modeled by the same truth set as itself; moreover, inter-derivable sentences have the same truth set. Combined with the semantics for conditionals and their negations, this strongly suggests that contraposition is an admissible rule.

Whether or not contraposition is admissible, this is quite a fragile admissibility. Let's briefly make a few observations on extensions in which the rule of contraposition is inadmissible.

First, adding truth value constants to these logics can result in contraposition inadmissibility.

Lemma 28. *The logic* CE_{FDE} *extended with the constants* **T** *and* **B** — *where* $V(\alpha, \mathbf{T}) = \{\mathbf{t}\}$ *and* $V(\alpha, \mathbf{B}) = \{\mathbf{t}, \mathbf{f}\}$, *for all* α — *does not admit the rule of contraposition.*

Proof. It is easy to check that $T \vDash B$. However, it is not the case that $\neg B \vDash \neg T$, as $\mathbf{f} \in V(\alpha, \mathbf{B})$, but $\mathbf{f} \notin V(\alpha, \mathbf{T})$. ◁

It may be desirable to have a theorem of identity, $(ID_\Rightarrow) \mathcal{A} \Rightarrow \mathcal{A}$. As we are working with ⊢-sequents only, we can either be happy with $\mathcal{A} \vdash \mathcal{A}$, or add the irrelevant axiom that the identity formula $(ID_\Rightarrow)$ follows from anything. This also results in contraposition inadmissibility.

Lemma 29. *The logic* CE_{FDE} *extended with the semantic condition if* $\alpha \in Y$, *for some* Y, *then* $X \in f(\alpha, X)$, *for every* X — *which corresponds to the axiom* $\mathcal{A} \vdash \mathcal{B} \Rightarrow \mathcal{B}$ — *does not admit the rule of contraposition.*

Proof. In this logic, $\mathcal{A} \Rightarrow \mathcal{A} \vDash \mathcal{B} \Rightarrow \mathcal{B}$, but not necessarily $\neg(\mathcal{B} \Rightarrow \mathcal{B}) \vDash \neg(\mathcal{A} \Rightarrow \mathcal{A})$. The reader may easily construct a model to demonstrate this fact (where $||\neg \mathcal{B}|| \in f(\alpha, ||\mathcal{B}||)$, but $||\neg \mathcal{A}|| \notin f(\alpha, ||\mathcal{A}||)$). ◁

Following Ross Brady, we may define the *depth* of a conditional subformula of a formula to be "the degree of nestings of '⇒''s required to 'reach' the occurrence of the subformula" [5, p. 64]. Some desirable axioms (and rules) may include (or introduce) sequents where the maximum depth of a conditional subformula on one side of the sequent is not the same as the maximum on the other side. For example, take the contraction sequent $\mathcal{A} \Rightarrow (\mathcal{A} \Rightarrow \mathcal{B}) \vdash \mathcal{A} \Rightarrow \mathcal{B}$. Sequents such as this will ensure the inadmissibility of contraposition, as the truth set of a conditional need not be equivalent to the truth set of a non-conditional. (Of course, that is only if we do not add additional axioms or rules to regain contraposition.) We will not take the time here to formally examine the concept of depth, or to develop the sufficient and necessary conditions for the inadmissibility of contraposition when asymmetric conditional-depth sequents are included.

We thus conclude that, even if the rule of contraposition is admissible in CE_{FDE}, it is a fragile admissibility that does not survive some fairly minor extensions.

3.3. Extensions. The conditional of CE_{FDE} is quite "weak," in that many sequents with conditional formula occurrences are, perhaps surprisingly, not theorems. A short list of example sequents is presented in the following lemma.

Lemma 30. *The follow statements of non-theorems of* $\mathbf{CE_{FDE}}$ *hold.*

(1) $\mathcal{A} \Rightarrow (\mathcal{A} \Rightarrow \mathcal{B}) \nvdash \mathcal{A} \Rightarrow \mathcal{B}$
(2) $\mathcal{A} \Rightarrow \mathcal{B} \nvdash \mathcal{A} \Rightarrow (\mathcal{A} \Rightarrow \mathcal{B})$
(3) $\mathcal{A} \Rightarrow \mathcal{B} \nvdash (\mathcal{B} \Rightarrow \mathcal{C}) \Rightarrow (\mathcal{A} \Rightarrow \mathcal{C})$
(4) $\mathcal{A} \Rightarrow \mathcal{B} \nvdash (\mathcal{C} \Rightarrow \mathcal{A}) \Rightarrow (\mathcal{C} \Rightarrow \mathcal{B})$

Proof. For (1), consider the model $\langle \{e\}, f, Prop, V \rangle$, with f defined by $f(e,X) = \{\|\mathcal{A} \Rightarrow \mathcal{B}\|\}$, and V defined by $\|\mathcal{A}\| = \|\mathcal{B}\| = \emptyset$, and $\|\mathcal{A} \Rightarrow \mathcal{B}\| = \{e\}$. Let *Prop* be full. It is easy to verify (1) on this model. For (2), a similar construction may be defined.

For (3), consider the model $\langle \{e\}, f, V \rangle$ with, for all X, $f(e,X) = \{\{e\}\} = \{\|\mathcal{B}\|\}$, and where $\|\mathcal{A} \Rightarrow \mathcal{C}\| = \emptyset$, and *Prop* is full. It is straightforward to show this is a countermodel for (3). A countermodel for (4) is left to the reader to construct. ◁

The logic may be interestingly extended by various axioms and rules. Consider the following list:

(W) $\mathcal{A} \Rightarrow (\mathcal{A} \Rightarrow \mathcal{B}) \vdash \mathcal{A} \Rightarrow \mathcal{B}$
(W^c) $\mathcal{A} \Rightarrow \mathcal{B} \vdash \mathcal{A} \Rightarrow (\mathcal{A} \Rightarrow \mathcal{B})$
(B') $\mathcal{A} \Rightarrow \mathcal{B} \vdash (\mathcal{B} \Rightarrow \mathcal{C}) \Rightarrow (\mathcal{A} \Rightarrow \mathcal{C})$
(B) $\mathcal{A} \Rightarrow \mathcal{B} \vdash (\mathcal{C} \Rightarrow \mathcal{A}) \Rightarrow (\mathcal{C} \Rightarrow \mathcal{B})$
(IDr) $\mathcal{A} \vdash \mathcal{B} \Rightarrow \mathcal{B}$

(RCT) $\mathcal{A} \vdash \mathcal{B} \Rightarrow \mathcal{B} \Rightarrow \mathcal{C} \vdash \mathcal{A} \Rightarrow \mathcal{C}$
(RCM) $\mathcal{B} \vdash \mathcal{C} \Rightarrow \mathcal{A} \Rightarrow \mathcal{B} \vdash \mathcal{A} \Rightarrow \mathcal{C}$
(RCR) $\mathcal{B} \wedge \mathcal{D} \vdash \mathcal{C} \Rightarrow (\mathcal{A} \Rightarrow \mathcal{B}) \wedge (\mathcal{A} \Rightarrow \mathcal{D}) \vdash \mathcal{A} \Rightarrow \mathcal{C}$

These axioms and rules (with the rules denoted with an "R" prefix) are characterized by conditions below. Note that these conditions are restricted to elements of *Prop*. For the $\mathbf{CE_{FDE}}$-models, *Prop* was non-essential, but for these extensions *Prop* appears to be necessary. The reason is that the completeness proofs only seem to work (for these conditions) with the restriction to *Prop*. For example, without *Prop* the condition (cRCM) would imply that the set of prime theories $K^{\mathbb{C}}$ is an element of every set $f(\alpha, X)$, but (even with the definition of f) there is no $\mathbf{CE_{FDE}}$ sentence that is contained in every $\mathbf{CE_{FDE}}$-prime theory. The following conditions are each to be appended by "for every X, Y, Y' and $Z \in Prop$."

(cW) If $X \Rightarrow Y \in f(\alpha, X)$, then $Y \in f(\alpha, X)$.
(cW^c) If $Y \in f(\alpha, X)$, then $X \Rightarrow Y \in f(\alpha, X)$.
(cB') If $Y \in f(\alpha, X)$, then $X \Rightarrow Z \in f(\alpha, Y \Rightarrow Z)$.
(cB) If $Y \in f(\alpha, X)$, then $Z \Rightarrow Y \in f(\alpha, Z \Rightarrow X)$.
(cIDr) $X \in f(\alpha, X)$, if $\alpha \in Y$ for some Y.

(cRCT) $Z \in f(\alpha, Y)$ and $X \subseteq Y$ imply $Z \in f(\alpha, X)$.
(cRCM) $Y \in f(\alpha, X)$ and $Y \subseteq Z$ imply $Z \in f(\alpha, X)$.
(cRCR) $Y \in f(\alpha, X)$, $Y' \in f(\alpha, X)$, and $Y \cap Y' \subseteq Z$ imply $Y \cap Y' \in f(\alpha, X)$.

Lemma 31. *In any logic extending* $\mathbf{CE_{FDE}}$, *the axioms* (W), (W^c), (B'), (B), *and* (IDr) *and the rules* (RCT), (RCM), (RCR) *are characterized by models satisfying the corresponding conditions above.*

Proof. The cases for the axioms are straightforward. We include only the cases for the rules.

Case (RCT). Soundness: Assume that $\mathcal{A} \vdash \mathcal{B}$. That is $||\mathcal{A}|| \subseteq ||\mathcal{B}||$. Further suppose that $||\mathcal{C}|| \in f(\alpha, ||\mathcal{B}||)$. So $||\mathcal{C}|| \in f(\alpha, ||\mathcal{A}||)$ as required. Completeness: Suppose that $Z \in f(\alpha, Y)$ and $X \subseteq Y$. By definition, $X = ||\mathcal{A}||$, $Y = ||\mathcal{B}||$ and $Z = ||\mathcal{C}||$, for some sentences $\mathcal{A}, \mathcal{B}, \mathcal{C}$, and $\mathcal{B} \Rightarrow \mathcal{C} \in \alpha$. Since α is a theory, and by the rule (RCM), $\mathcal{A} \Rightarrow \mathcal{C} \in \alpha$ for every $\mathcal{A}$. Thus, $||\mathcal{C}|| \in f(\alpha, ||\mathcal{A}||)$; that is, $Z \in f(\alpha, X)$, as required. Therefore, (cRCT) is satisfied by the canonical model.

Case (RCM). Soundness: Suppose that $\mathcal{A} \vdash \mathcal{B}$, that is, $||\mathcal{A}|| \subseteq ||\mathcal{B}||$. Further suppose that $||\mathcal{A}|| \in f(\alpha, X)$. Applying the condition gives $||\mathcal{B}|| \in f(\alpha, X)$, as required. Completeness: Suppose that $Y \in f(\alpha, X)$ and $Y \subseteq Z$. From the former, we have that $\mathcal{A} \Rightarrow \mathcal{B} \in \alpha$, where $Y = ||\mathcal{B}||$ and $X = ||\mathcal{A}||$. From the latter, we have that $\mathcal{B} \vdash \mathcal{C}$ for $Z = ||\mathcal{C}||$. Thus, by (RCM) and the fact that α is a theory, we have that $\mathcal{A} \Rightarrow \mathcal{C} \in \alpha$, which gives $Z \in f(\alpha, X)$.

Case (RCR). Soundness is straightforward. For completeness, assume that $Y \in f(\alpha, X)$, $Y' \in f(\alpha, X)$, and $Y \cap Y' \subseteq Z$. Then $X = ||\mathcal{A}||$, $Y = ||\mathcal{B}||$, $Y' = ||\mathcal{D}||$, and $Z = ||\mathcal{C}||$ for some formulas $\mathcal{A}, \mathcal{B}, \mathcal{D}$ and $\mathcal{C}$. It follows that $\mathcal{A} \Rightarrow \mathcal{B}, \mathcal{A} \Rightarrow \mathcal{D} \in \alpha$. Thus, since theories are closed under conjunction, $(\mathcal{A} \Rightarrow \mathcal{B}) \wedge (\mathcal{A} \Rightarrow \mathcal{D}) \in \alpha$. Moreover, we have that $(\mathcal{A} \Rightarrow \mathcal{B}) \wedge (\mathcal{A} \Rightarrow \mathcal{D}) \vdash (\mathcal{A} \Rightarrow \mathcal{C})$ is a theorem, by (RCR), Lemma 24 and $Y \cap Y' \subseteq Z$. Thus $\mathcal{A} \Rightarrow \mathcal{C} \in \alpha$, which is our desired result that $Z \in f(\alpha, X)$. Therefore, (cRCR) is satisfied by the canonical model. ◁

Note that the rule of (RCN) is missing. This is due to the logic containing no formula-theorems, but only sequents.

Some of the extensions defined so far can be equivalently specified with corresponding axioms. This is noted by Chellas in [6] for (CM) and (CR) in the classical setting, additionally shown here for (CT).[10]

We define the following axioms:

(CM) $\mathcal{A} \Rightarrow (\mathcal{B} \wedge \mathcal{C}) \vdash (\mathcal{A} \Rightarrow \mathcal{B}) \wedge (\mathcal{A} \Rightarrow \mathcal{C})$
(CR) $(\mathcal{A} \Rightarrow \mathcal{B}) \wedge (\mathcal{A} \Rightarrow \mathcal{C}) \vdash \mathcal{A} \Rightarrow (\mathcal{B} \wedge \mathcal{C})$
(CT) $\mathcal{B} \Rightarrow \mathcal{C} \vdash (\mathcal{B} \wedge \mathcal{D}) \Rightarrow \mathcal{C}$

Lemma 32. *Where (CM), (CR), and (CT) are defined above,*
  (i) $\mathbf{CE_{FDE}} + (CM)$ *is equivalent to* $\mathbf{CE_{FDE}} + (RCM)$;
  (ii) $\mathbf{CE_{FDE}} + (CR)$ *is equivalent to* $\mathbf{CE_{FDE}} + (RCR) + (RCM)$;
  (iii) $\mathbf{CE_{FDE}} + (CT)$ *is equivalent to* $\mathbf{CE_{FDE}} + (RCT)$.

Proof. We show only the case for (CT).

(1) $\mathcal{B} \wedge \mathcal{D} \vdash \mathcal{B}$ ($\wedge$-El)
(2) $\mathcal{B} \Rightarrow \mathcal{C} \vdash (\mathcal{B} \wedge \mathcal{D}) \Rightarrow \mathcal{C}$ 1, (RCT)

(1) $\mathcal{A} \vdash \mathcal{B}$ Assumption
(2) $\mathcal{A} \dashv\vdash \mathcal{A} \wedge \mathcal{B}$ 1, (R$\wedge$-I), ($\wedge$-El)
(3) $\mathcal{B} \Rightarrow \mathcal{C} \vdash (\mathcal{A} \wedge \mathcal{B}) \Rightarrow \mathcal{C}$ (CT)
(4) $(\mathcal{A} \wedge \mathcal{B}) \Rightarrow \mathcal{C} \dashv\vdash \mathcal{A} \Rightarrow \mathcal{C}$ 2, (RCEA)

---

[10] Note that (CR) here is what Chellas calls (CC).

(5) $\mathcal{B} \Rightarrow C \vdash \mathcal{A} \Rightarrow C$ 3, 4, (RT)

◁

4. Standard Conditional FDE-Logics

The standard conditional **FDE**-logics are defined similarly, by combining two-valued standard conditional logics with **FDE**.

Definition 33. A $\mathbf{CK_{FDE}}$-model is a tuple $\mathfrak{M} = \langle K, f, Prop, V \rangle$ where, K is a non-empty set of situations, f is a function from a situation and *proposition* to a *proposition* (i.e., $f: (K \times Prop) \longrightarrow Prop$), $Prop \subseteq \wp(K)$, the condition (c1a) defined above in Definition 13 is satisfied, and V is a valuation function that assigns to each world-atom(ic sentence) pair an element of $\{\mathbf{t}, \mathbf{f}\}$ such that the truth and falsity sets of every atomic proposition are elements of *Prop*. V is extended to every sentence using (i)–(vi) from Definition 1 for $\neg, \wedge$ and $\vee$, and using (ix) and (x) below for $\Rightarrow$. For every $\alpha \in K$,

(ix) $\mathbf{t} \in V(\alpha, \mathcal{A} \Rightarrow \mathcal{B})$ iff $f(\alpha, \|\mathcal{A}\|_{\mathfrak{M}}^+) \subseteq \|\mathcal{B}\|_{\mathfrak{M}}^+$;
(x) $\mathbf{f} \in V(\alpha, \mathcal{A} \Rightarrow \mathcal{B})$ iff $f(\alpha, \|\mathcal{A}\|_{\mathfrak{M}}^+) \subseteq \|\neg\mathcal{B}\|_{\mathfrak{M}}^+$.

For the valuations of conditionals, the truth condition is the same as the truth condition for Chellas' conditional logics. A conditional is (at least) true under a valuation at a situation α — that is, $\mathbf{t} \in V(\alpha, \mathcal{A} \Rightarrow \mathcal{B})$ — when the accessibility relation relative to the antecedent and α results in a set of situations at which the consequent is (at least) true. Intuitively, $\mathcal{A} \Rightarrow \mathcal{B}$ is true at α if we look at the set of situations that α takes to be such that $\mathcal{A}$ is at least true, all of those situations are also $\mathcal{B}$ situations. The falsity condition again reduces negated conditionals to non-negated conditionals.

Lemma 34. *In every* $\mathbf{CK_{FDE}}$-*model* $\mathfrak{M}$, *every sentence* $\mathcal{A}$ *is mapped onto an element of Prop by the valuation function.*

The proof is as in Lemma 14.

Definition 35. *The consequence relations* $\vDash_{\mathbf{CK_{FDE}}}^{\mathfrak{M}}$ *and* $\vDash_{\mathbf{CK_{FDE}}}$ *are defined as in Definition 2, but with respect to* $\mathbf{CK_{FDE}}$.

4.1.1. Axiom system.
The logic $\mathbf{CK_{FDE}}$ is defined axiomatically, and somewhat redundantly, as the logic $\mathbf{CE_{FDE}}$ (that is, with De Morgan axioms and without the rule of contraposition) plus the following rule (RCK).

(RCK) $(\mathcal{B}_1 \wedge \cdots \wedge \mathcal{B}_n) \vdash C \Rightarrow ((\mathcal{A} \Rightarrow \mathcal{B}_1) \wedge \cdots \wedge (\mathcal{A} \Rightarrow \mathcal{B}_n)) \vdash \mathcal{A} \Rightarrow C$, for $n \geq 1$.

4.1.2. Soundness and completeness.

Theorem 36 (Soundness). *The logic* $\mathbf{CK_{FDE}}$ *is sound with respect to the class of* $\mathbf{CK_{FDE}}$-*models. That is, if* $\mathcal{A} \vdash \mathcal{B}$, *then* $\mathcal{A} \vDash \mathcal{B}$.

This theorem is easy to check.

For completeness, we first define theories in the usual way. The canonical model is defined as follows.

Definition 37 (Canonical Model for **CK_FDE**). The *canonical model* for **CK_FDE** is $\mathfrak{M}^{\mathbb{C}}_{\mathbf{CK_{FDE}}} = \langle K^{\mathbb{C}}, f^{\mathbb{C}}, Prop^{\mathbb{C}}, V^{\mathbb{C}} \rangle$, where $K^{\mathbb{C}}$, $Prop^{\mathbb{C}}$ and $V^{\mathbb{C}}$ are defined by (i), (ii), (iv)–(vi) of Definition 22 (using **CK_FDE**-theories in place of **CE_FDE**-theories), and $f^{\mathbb{C}}$ is defined by

(iii′) $\beta \in f^{\mathbb{C}}(\alpha, ||\mathcal{A}||)$ iff, if $\mathcal{A} \Rightarrow \mathcal{B} \in \alpha$ then $\mathcal{B} \in \beta$.

Lemma 38 (Extension). *If $\Gamma \not\succeq_{\mathbf{CK_{FDE}}} \Delta$, then there is some prime **CK_FDE**-theory Γ' such that $\Gamma \subseteq \Gamma'$ and $\Gamma' \cap \Delta = \emptyset$.*

Lemma 39 (f-Squeeze). *If β is a theory, α is a prime theory, $\beta \in f'(\alpha, X)$ and $\mathcal{E} \notin \beta$, then there is a prime theory $\beta' \supseteq \beta$ such that $\beta' \in f^{\mathbb{C}}(\alpha, X)$ and $\mathcal{E} \notin \beta'$.*[11]

Proof. Given a theory β such that $\beta \in f'(\alpha, X)$, $\mathcal{E} \notin \beta$, and α is a prime theory, by Lemma 38 there is a prime extension of $\beta' \supseteq \beta$ such that $\mathcal{E} \notin \beta'$. In addition, given the definition of f' (and $f^{\mathbb{C}}$), $\beta' \in f'$, and since β' is prime, $\beta' \in f^{\mathbb{C}}(\alpha, X)$. ◁

Lemma 40. *The canonical **CK_FDE**-model is an **CK_FDE**-model.*

This lemma is straightforward to check.

Lemma 41 (Truth Lemma). $\mathbf{t} \in V^{\mathbb{C}}(\alpha, \mathcal{A})$ *iff* $\mathcal{A} \in \alpha$.

Proof. The proof is by strong induction on the complexity of $\mathcal{A}$ defined as follows. (i) An atomic has a complexity of 1. (ii) The complexity of $\neg \mathcal{A}$ is the complexity of $\mathcal{A}$ plus 1. The complexity of $\mathcal{A} \otimes \mathcal{B}$ (for $\otimes \in \{\Rightarrow, \wedge, \vee\}$) is the complexity of $\mathcal{A}$, plus the complexity of $\mathcal{B}$, plus 1.

The only interesting case is that of a conditional and its negation, corresponding to the truth and falsity conditions of a conditional. Consider $\mathcal{A} \Rightarrow \mathcal{B}$. For the left-to-right direction, assume that $\mathbf{t} \in V^{\mathbb{C}}(\alpha, \mathcal{A} \Rightarrow \mathcal{B})$ and that $\mathcal{A} \Rightarrow \mathcal{B} \notin \alpha$. By the former, $f(\alpha, ||\mathcal{A}||) \subseteq ||\mathcal{B}||$. Now consider the set $\gamma = \{\mathcal{C} : \mathcal{A} \Rightarrow \mathcal{C} \in \alpha\}$. By the rule (RCK) and the fact that α is a prime theory, γ is a theory, and by assumption, $\mathcal{B} \notin \gamma$. By the squeeze lemma above, there is a prime $\gamma' \supseteq \gamma$ such that $\mathcal{B} \notin \gamma'$. Moreover, by construction $\gamma' \in f(\alpha, ||\mathcal{A}||)$, and so given the implication from the original supposition (after the induction hypothesis) $\mathcal{B} \in \gamma'$, a contradiction. Thus $\mathcal{A} \Rightarrow \mathcal{B} \in \alpha$. For the other direction, assume that $\mathcal{A} \Rightarrow \mathcal{B} \in \alpha$. Next suppose that $\mathbf{t} \notin V(\alpha, \mathcal{A} \Rightarrow \mathcal{B})$. Then $\exists \beta \in K$ such that $\beta \in f(\alpha, ||\mathcal{A}||)$, but $\beta \notin ||\mathcal{B}||$. Thus, $\mathcal{B} \notin \beta$, by the induction hypothesis. But by the definition of f, $\mathcal{B} \in \beta$, a contradiction. Therefore $\mathbf{t} \in V(\alpha, \mathcal{A} \Rightarrow \mathcal{B})$, as required.

The case for a negated conditional is similar, except that the induction hypothesis applies so that $\neg \mathcal{B} \in \gamma'$ iff $\gamma' \in ||\neg \mathcal{B}||$. This works using strong induction, as the complexity of $\neg \mathcal{B}$ is strictly less than the complexity of the negated conditional. ◁

Theorem 42 (Completeness). *The logic **CK_FDE** is complete with respect to the class of **CK_FDE**-models. That is, if $\mathcal{A} \vDash \mathcal{B}$, then $\mathcal{A} \vdash \mathcal{B}$.*

Completeness may be shown by the usual contrapositive method. Suppose that $\mathcal{A} \nvdash \mathcal{B}$. Then there is a prime theory α, obtained by the extension lemma on the closure of $\mathcal{A}$ under $\vdash$, such that $\mathcal{A} \in \alpha$, but $\mathcal{B} \notin \alpha$. By the truth lemma, $\mathbf{t} \in V^{\mathbb{C}}(\alpha, \mathcal{A})$ and $\mathbf{t} \notin V^{\mathbb{C}}(\alpha, \mathcal{B})$, and therefore, $\mathcal{A} \nvDash \mathcal{B}$.

---

[11] Note that f' is the relaxation of $f^{\mathbb{C}}$ to theories.

5. MODUS PONENS

Many extensions of **FDE** with a truth-functional conditional fail to validate the rule of modus ponens. However, there are several formulations of rules and axioms that could be considered a kind of modus ponens in the systems constructed so far.

The rule form of modus ponens as $\mathcal{A}, \mathcal{A} \Rightarrow \mathcal{B} \Rightarrow \mathcal{B}$ is inappropriate because $\mathcal{A}$ would have to be a theorem. That is, it would have to be a sequent, as all of our theorems are sequents of the form $X \vdash Y$. As something of the form "$(\mathcal{C} \vdash \mathcal{D}) \Rightarrow \mathcal{B}$" is undefined, we could opt for the defined "$\mathcal{C} \vdash \mathcal{D} \Rightarrow \mathcal{A} \vdash \mathcal{B}$," resulting in the following meta-rule formulation.

(MP^m) If $\mathcal{C} \vdash \mathcal{D}$ is a theorem and $\mathcal{C} \vdash \mathcal{D} \Rightarrow \mathcal{A} \vdash \mathcal{B}$ is admissible, then infer that $\mathcal{A} \vdash \mathcal{B}$ is a theorem.

However, (MP^m) is just a restatement of how the proof system works, and is, therefore, derivable. If you prove that $\mathcal{C} \vdash \mathcal{D} \Rightarrow \mathcal{A} \vdash \mathcal{B}$ is admissible, then if you also have a proof of the sequent $\mathcal{C} \vdash \mathcal{D}$, then you can apply the admissible rule without citing (MP^m). Even though this rule does not add anything to our logics, we will reject it as formalizing modus ponens for $\Rightarrow$ primarily because $\Rightarrow$ does not have any occurrences in its formulation. So much for the classical multi-premise rule approach. Thus, we are left with a couple options. First, the sequent

(MP1) $\mathcal{A} \Rightarrow \mathcal{B} \vdash \mathcal{A} \Rightarrow \mathcal{B}$

is a decent option for modus ponens. This is because (i) it is a valid sequent, and (ii) it roughly states that if a conditional is true, then its antecedent will imply (via that conditional) the consequent. The downsides of this formulation are that (i) it does not guarantee closure of theories under $\Rightarrow$, and consequently (ii) it does not quite capture a more robust sense of modus ponens in which, if you have both the antecedent of a conditional and the conditional, then you also have the consequent.

While (MP1) was decent, it failed what we will consider the be the prime desideratum for modus ponens, namely, *that it guarantees the closure of theories under the conditional*. With this in mind, let's continue exploring the options.

The next axiom is the one adopted by Chellas (as evidenced by his naming of the axiom MP). In the language of **FDE** — since it involves a truth-functional conditional — the axiom is as follows:

(MP2) $\mathcal{A} \Rightarrow \mathcal{B} \vdash \neg \mathcal{A} \vee \mathcal{B}$.

Unfortunately, this also does not close theories under the conditional. The reason is that, when $\mathcal{A}$ is both true and false, the material conditional does not detach, and thus we are back to the same problem, namely, the lack of modus ponens for truth-functional conditionals in **FDE**. However, this axiom does give us a sort of reflexivity, where the truth of a conditional in a theory has truth-functional consequences in that very theory. Nonetheless, let's move on.

The final option we will consider is the axiom

(MP3) $(\mathcal{A} \Rightarrow \mathcal{B}) \wedge \mathcal{A} \vdash \mathcal{B}$.

This axiom, which is often called weak contraction, does close theories under the conditional.[12] The semantic condition is as follows:

---

[12]In non-sequent form, namely $((\mathcal{A} \Rightarrow \mathcal{B}) \wedge \mathcal{A}) \Rightarrow \mathcal{B}$, it is also known as the *modus ponens axiom*.

(cMP3) $Y \in f(\alpha, X)$ and $\alpha \in X$ imply $\alpha \in Y$.

We have seen multiple ways of expressing modus ponens in conditional **FDE**-logics. The latter two, (MP2) and (MP3) may be added non-trivially to the minimal and standard logics. The logics **CE**$_{\mathbf{FDE}}$ and **CK**$_{\mathbf{FDE}}$ lack these forms of modus ponens, a rule of contraposition, contraposition for the conditional, and other often desirable properties for conditionals, but can be extended to logics that have these properties by construction. We conclude the conditional **FDE**-logics in fact add a robust and useful conditional which may be customized with ease (in contrast to the truth-functional conditionals, for example).

6. Quantified Conditional **FDE**-Logics

The most immediate quantified extensions of Conditional **FDE**-Logics — that is, those that just adjoin the axioms and rules of **CE**$_{\mathbf{FDE}}$ with **QFDE** (in an appropriately combined language) — do not validate some fairly desirable properties of the interaction between a conditional and quantifiers. For example, this logic does not validate the sequent $\forall x(\mathcal{A} \Rightarrow \mathcal{B}) \vdash (\mathcal{A} \Rightarrow \forall x \mathcal{B})$, where x is not free in $\mathcal{A}$. Here, we first develop the most-straightforward quantified conditional **FDE**-logic as a base for exploring the relationship between conditionals and quantifiers.

Definition 43. The logic **CE**$_{\mathbf{QFDE}}$ (**CE**$_{\mathbf{FDEQ}}$) is defined by extending the logic **QFDE** (**FDEQ**) by adding a conditional to the language and adding the axioms and rules that define **CE**$_{\mathbf{FDE}}$.

Lemma 44. *The following sequents are theorems of* **CE**$_{\mathbf{QFDE}}$.

(i) $\mathcal{A} \Rightarrow \forall x \mathcal{B} \vdash \mathcal{A} \Rightarrow \neg \exists x \neg \mathcal{B}$
(ii) $\forall x \mathcal{A} \Rightarrow \mathcal{B} \vdash \neg \exists x \neg \mathcal{A} \Rightarrow \mathcal{B}$
(iii) $\forall x \mathcal{A} \wedge \forall x \mathcal{B} \vdash \forall x(\mathcal{A} \wedge \mathcal{B})$
(iv) $\forall x(\mathcal{A} \wedge \mathcal{B}) \vdash \forall x \mathcal{A} \wedge \forall x \mathcal{B}$

Proofs for these sequents can be easily constructed.

Lemma 45. *If the sequent* $(\mathcal{A} \vdash \mathcal{B})[c/x]$ *is a theorem of* **CE**$_{\mathbf{QFDE}}$, *where c does not occur in* $\mathcal{A} \vdash \mathcal{B}$, *then* $(\mathcal{A} \vdash \mathcal{B})[\tau/x]$ *is also a theorem, where τ is a term not occurring in* $\mathcal{A} \vdash \mathcal{B}$.

The proof of the previous lemma is as above, in the similar Lemma 6.

6.1. Semantics. The semantics will be much as it was for quantified **FDE**, with the necessary modifications for the conditional. As in the models for quantified logics above, we will also lift the operations onto propositions and propositional functions. To this end, let use define a conditional operation on elements of *Prop*. We have already defined the $\Rightarrow$ operation on *Prop* in Section 3. This is further lifted to the propositional functions as you would expect, namely, by

$$(\varphi \Rightarrow \psi)g = \varphi g \Rightarrow \psi g,$$

for every $\varphi, \psi \in PropFun$ and $g \in U^\omega$.

Definition 46. A **CE$_{QFDE}$**-model is a tuple $\mathfrak{M} = \langle K, f, U, Prop, PropFun, |-|\rangle$ defined by extending Definition 7. The additional element f is a function from a situation and *proposition* to a set of *propositions* (i.e., $f\colon (K \times Prop) \longrightarrow \wp(Prop)$). We add to conditions (c1)–(c3) the following condition (c4).

(c4) *Prop* is closed under $\Rightarrow$; *PropFun* is closed under $\Rightarrow$.

Finally, the valuation function $|-|$ is defined as in **QFDE**-models, but with the following additions for conditionals:

$$|\mathcal{A} \Rightarrow \mathcal{B}|^+ g = |\mathcal{A}|^+ g \Rightarrow |\mathcal{B}|^+ g \qquad |\mathcal{A} \Rightarrow \mathcal{B}|^- g = |\mathcal{A}|^+ g \Rightarrow |\mathcal{B}|^- g$$

A **CE$_{FDEQ}$**-model is a **CE$_{QFDE}$**-model that satisfies both (cEC1) and (cEC2).

Definition 47. The consequence relations $\vDash^{\mathfrak{M}}_{CE_{QFDE}}$ and $\vDash_{CE_{QFDE}}$ are defined as in Definition 8, but with respect to **CE$_{QFDE}$**.

Unlike the case of quantified **FDE**, the proofs of soundness and completeness are not left to the Appendix, for we aim to explore these models with greater focus.

6.2. Soundness. For soundness, we can extend the arguments in the Appendix by the cases for the conditional. The proof is mostly left to the reader, but we will provide some arguments.

The axiom ($\neg\text{-}\Rightarrow$) is valid in every **CE$_{QFDE}$**-model. To see this, consider the following chain of equalities:

$$\begin{aligned}|\neg(\mathcal{A} \Rightarrow \mathcal{B})|^+ g &= |\mathcal{A} \Rightarrow \mathcal{B}|^- g \\ &= |\mathcal{A}|^+ g \Rightarrow |\mathcal{B}|^- g \\ &= |\mathcal{A}|^+ g \Rightarrow |\neg\mathcal{B}|^+ g \\ &= |\mathcal{A} \Rightarrow \neg\mathcal{B}|^+ g\end{aligned}$$

This chain of equalities holds for any $g \in U^\omega$, so the axiom is valid (given the definition of the consequence relation).

Consider the rule (RCEA). The arguments for **CE$_{FDE}$** are easily adapted. Suppose that $\mathcal{A} \vDash \mathcal{B}$ and $\mathcal{B} \vDash \mathcal{A}$. Then, for every model and every $g \in U^\omega$, we have that $|\mathcal{A}|^+ g = |\mathcal{B}|^+ g$. Thus, for any point α in any these models, for any $g \in U^\omega$, $f(\alpha, |\mathcal{A}|^+ g) = f(\alpha, |\mathcal{B}|^+ g)$. This gives us that $\mathcal{A} \Rightarrow \mathcal{C} \vDash \mathcal{B} \Rightarrow \mathcal{C}$ and $\mathcal{B} \Rightarrow \mathcal{C} \vDash \mathcal{A} \Rightarrow \mathcal{C}$, as required.

6.3. Completeness. For completeness, first, we define theories and prime theories as before, but with respect to the **CE$_{QFDE}$** and **CE$_{FDEQ}$**. Again, here we shall use $\mathbb{L}$ to refer to both **CE$_{QFDE}$** and **CE$_{FDEQ}$**.

Lemma 48 (Pair Extension). *If $\Gamma \not\gg_\mathbb{L} \Delta$, then there is a prime $\mathbb{L}$-theory Γ' such that $\Gamma \subseteq \Gamma'$ and $\Gamma' \cap \Delta = \emptyset$.*

Proof. The proof is fairly standard, and the reader is directed to Anderson, Belnap, and Dunn [2, pp. 123–126] and again to [23, 5.1–5.2], and is invited to check that the $\vdash$ relation is indeed pair extension acceptable. Note well that the theories are not required to be ω-complete. The lack of ω-completeness as a requirement means that we can have theories which contain every instance of $\forall x \mathcal{A}$, but yet do not contain the universally quantified formula $\forall x \mathcal{A}$. ◁

Corollary 49. *If $\mathcal{A} \nvdash \mathcal{B}$, then there is a prime theory Γ such that $\mathcal{A} \in \Gamma$, but $\mathcal{B} \notin \Gamma$.*

Definition 50 (Canonical Model for **CE**$_{\mathbf{QFDE}}$). The *canonical model* for **CE**$_{\mathbf{QFDE}}$, $\mathfrak{M}^{\mathbb{C}}_{\mathbf{QFDE}}$, is a tuple $\langle K^{\mathbb{C}}, f^{\mathbb{C}}, U^{\mathbb{C}}, Prop^{\mathbb{C}}, PropFun^{\mathbb{C}}, |-|^{\mathbb{C}} \rangle$, where $K^{\mathbb{C}}$ and $f^{\mathbb{C}}$ are defined as in Definition 22 (but with respect to **CE**$_{\mathbf{QFDE}}$-theories), and

(i) $U^{\mathbb{C}}$ is the set of individual constants.
(ii) For every closed formula $\mathcal{A}$, $||\mathcal{A}|| =_{df} \{\alpha \in K \colon \mathcal{A} \in \alpha\}$.
(iii) $Prop^{\mathbb{C}} =_{df} \{||\mathcal{A}|| \colon \mathcal{A} \text{ is a closed formula}\}$.
(iv) For any $g \in U^{\omega}$, each variable is mapped to a constant, and so we may define the substitution of all variables in a formula with their respective constants. We obtain the closed formula $\mathcal{A}^g$ by $\mathcal{A}[g0/x_0, \ldots, gn/x_n, \ldots]$.
(v) For each formula $\mathcal{A}$, the function $\varphi_{\mathcal{A}} \colon U^{\omega} \longrightarrow Prop$ is given by $(\varphi_{\mathcal{A}})g = ||\mathcal{A}^g||$.
(vi) $PropFun$ is the set of all functions $\varphi_{\mathcal{A}}$, for every formula $\mathcal{A}$.
(vii) The canonical valuation is defined by
  (a) $|c| = c$;
  (b) $|P|^+ c_1 \ldots c_n = ||Pc_1 \ldots c_n||$;
  (c) $|P|^- c_1 \ldots c_n = ||\neg Pc_1 \ldots c_n||$.[13]
  (d) The valuation is extended to all formulas as before.

Lemma 51. *The canonical models for* **CE**$_{\mathbf{QFDE}}$ *and* **CE**$_{\mathbf{FDEQ}}$ *satisfy (c1), (c2), (c3), and (c4).*

Proof. The cases for (c1)–(c3) can be proven using the arguments for the non-conditional case in the Appendix. For (c4), note the equalities, for closed $\mathcal{A}$ and $\mathcal{B}$:

$$||\mathcal{A} \Rightarrow \mathcal{B}|| = \{\alpha \in K \colon \mathcal{A} \Rightarrow \mathcal{B} \in \alpha\} \qquad \text{Df: } ||-||$$
$$= \{\alpha \in K \colon ||\mathcal{B}|| \in f(\alpha, ||\mathcal{A}||)\} \qquad \text{Df: } f$$
$$= ||\mathcal{A}|| \Rightarrow ||\mathcal{B}|| \qquad \text{Df: } X \Rightarrow Y$$

This shows that *Prop* is closed under $\Rightarrow$, for every element in prop is of the form $||\mathcal{C}||$ for a closed formula $\mathcal{C}$. The argument to show that *PropFun* is closed under $\Rightarrow$ is similar to the other cases for binary operators for (c2). In particular, we obtain the equality $\varphi_{\mathcal{A}} \Rightarrow \varphi_{\mathcal{B}} = \varphi_{\mathcal{A} \Rightarrow \mathcal{B}}$, which is used in the Truth Lemma below. ◁

Lemma 52. *The canonical model for* **CE**$_{\mathbf{FDEQ}}$ *satisfies (cEC1) and (cEC2).*

The proof is as for **FDEQ** in the Appendix.

Lemma 53. *For every n-ary predicate P, terms $\tau_1, \ldots, \tau_n$, and $g \in U^{\omega}$,*

(i) $(P\tau_1, \ldots, \tau_n)^g = P(|\tau_1|g, \ldots, |\tau_n|g)$
(ii) $|P\tau_n, \ldots, \tau_n|^+ = \varphi_{P\tau_n, \ldots, \tau_n}$
(iii) $|P\tau_n, \ldots, \tau_n|^- = \varphi_{\neg P\tau_n, \ldots, \tau_n}$

The proof is as for **FDEQ**.

Lemma 54 (Truth Lemma). *For every formula $\mathcal{A}$, $|\mathcal{A}|^+ = \varphi_{\mathcal{A}}$. That is, for every $g \in U^{\omega}$, $|\mathcal{A}|^+ g = ||\mathcal{A}^g||$.*

---

[13]Equivalently, we could define falsity sets for closed formulas as primitive, but this definition is sufficient.

Proof. The proof is by induction on the complexity of $\mathcal{A}$. The new case is for conditionals. Suppose that $\mathcal{A} = \mathcal{B} \Rightarrow \mathcal{C}$. It follows that

$$\begin{aligned}
|\mathcal{B} \Rightarrow \mathcal{C}|^+ &= |\mathcal{B}|^+ \Rightarrow |\mathcal{C}|^+ && \text{Df: } |-|^+ \\
&= \varphi_{\mathcal{B}} \Rightarrow \varphi_{\mathcal{C}} && \text{Inductive Hypothesis} \\
&= \varphi_{\mathcal{B} \Rightarrow \mathcal{C}} && \text{(c4) Equality of Lemma 51}
\end{aligned}$$
◁

Theorem 55 (Completeness). *The logic* **CE**$_{\mathbf{QFDE}}$ *(and* **CE**$_{\mathbf{FDEQ}}$*) is complete with respect to the class of* **CE**$_{\mathbf{QFDE}}$*-models (***CE**$_{\mathbf{FDEQ}}$*-models). That is, if* $\mathcal{A} \vDash_{\mathbf{CE}_{\mathbf{QFDE}}} \mathcal{B}$, *then* $\mathcal{A} \vdash_{\mathbf{CE}_{\mathbf{QFDE}}} \mathcal{B}$, *and similarly, for* **CE**$_{\mathbf{FDEQ}}$.

Proof. The proof is similar to the case for **QFDE**. Suppose that $\mathcal{A} \vDash_{\mathbf{CE}_{\mathbf{QFDE}}} \mathcal{B}$. By the Truth Lemma, for every $g \in U^{\omega}$, we have that $||\mathcal{A}^g|| \subseteq ||\mathcal{B}^g||$. By Corollary 49, we have that $\mathcal{A}^g \vdash \mathcal{B}^g$ — that is, $(\mathcal{A} \vdash \mathcal{B})[\ldots gn/x_n \ldots]$ for the free variables x_n in $\mathcal{A}$ and $\mathcal{B}$ — as every prime filter is an element of the canonical model. By repeated applications of Lemma 45, the proof of $\mathcal{A}^g \vdash \mathcal{B}^g$ may be turned into a proof of $\mathcal{A} \vdash \mathcal{B}$, as required, since the variables occurring in $\mathcal{A} \vdash \mathcal{B}$ do not occur in $\mathcal{A}^g \vdash \mathcal{B}^g$. The proof is similar for **CE**$_{\mathbf{FDEQ}}$.
◁

6.4. Quantifier Distribution (Countermodels).

The sequent $\forall x(\mathcal{A} \Rightarrow \mathcal{B}) \vdash (\mathcal{A} \Rightarrow \forall x \mathcal{B})$, where x is not free in $\mathcal{A}$, and the similar distribution of the universal quantifier into the antecedent (as an existential quantifier) are not theorems of **CE**$_{\mathbf{QFDE}}$ or **CE**$_{\mathbf{FDEQ}}$. To show this, consider the following model.

Suppose we have a tuple $\langle K, f, U, Prop, PropFun, |-|\rangle$ defined by $K = \{a, b, c\}$, $U = \{j, k\}$, $Prop = \wp(K)$, $PropFun$ is the set of all propositional functions $\{j, k\}^{\omega} \longrightarrow Prop$. Further suppose that $f(a, \{a\}) = f(b, \{a\}) = f(c, \{a\}) = \{\{b\}, \{c\}\}$; for every other $X \in Prop$, $f(a, X) = f(b, X) = f(c, X) = \emptyset$. Further, suppose we have a signature with a zero place P_1^0, and a one-place P_2^1 (which we will often write without a term), and that the valuation $|-|$ is defined so that, for every $g \in U^{\omega}$, $|P_1^0|^+ g = \{a\}$, $|P_2^1 x_n|^+ g[j/x_n] = \{b\}$ for all variables x_n, and $|P_2^1 x_n|^+ g[k/x_n] = \{c\}$ for all variables x_n. We will show that $a, b, c \in |\forall x(P_1^0 \to P_2^1)|^+$, but also that $a, b, c \notin |(P_1^0 \to \forall x P_2^1)|^+$, where x is, obviously, not free in P_1^0.

First, we show that $a, b, c \in |\forall x(P_1^0 \to P_2^1)|^+$. Since Prop is finite $\bigcap S = \bigcap S$. So then

$$\begin{aligned}
|\forall x_n(P_1^0 \to P_2^1)|^+ g &= \forall_n |(P_1^0 \to P_2^1)|^+ g \\
&= \bigcap_{i \in U} |(P_1^0 \to P_2^1)|^+ g[i/n] \\
&= \bigcap_{i \in U} |(P_1^0 \to P_2^1)|^+ g[i/n]
\end{aligned}$$

Now, since $|P_2^1 x_n|^+ g[j/x_n] = \{b\} \in f(a, |P_1^0|^+ g[j/x_n])$ and that $f(a, |P_1^0|^+ g[j/x_n]) = f(b, |P_1^0|^+ g[j/x_n]) = f(c, |P_1^0|^+ g[j/x_n])$, we have that a, b and c are all in $|(P_1^0 \to P_2^1)|^+ g[j/n]$. Similarly, we can show this for $|(P_1^0 \to P_2^1)|^+ g[k/n]$. The intersection of $\{a, b, c\}$ with itself is itself. Thus, given the equality established above, $a, b, c \in |\forall x(P_1^0 \to P_2^1)|^+$.

To show that $a, b, c \notin |(P_1^0 \to \forall x P_2^1)|^+$, it suffices to show that $|\forall x P_2^1|^+ g \notin f(a, |P_1^0|^+)$, for all $g \in U^\omega$, and similarly, with b and c. We easily obtain the equalities

$$|\forall x_n P_2^1|^+ g = \forall_n |P_2^1|^+ g = \bigcap_{i \in U} |P_2^1|^+ g[i/n]$$
$$= \bigcap_{i \in U} |P_2^1|^+ g[i/n] = |P_2^1|^+ g[j/n] \cap |P_2^1|^+ g[k/n] = \emptyset$$

Note that the empty set is not an *element* of $f(\gamma, X)$, for any $\gamma \in K$ and $X \in Prop$, so in particular $|\forall x P_2^1|^+ g \notin f(a, |P_1^0|^+)$. Similar results can be shown for b and c to obtain our claim. Thus, $\forall x(P_1^0 \Rightarrow P_2^1) \vdash (P_1^0 \Rightarrow \forall x P_2^1)$ is not a theorem of $\mathbf{CE_{QFDE}}$.

Note that not only does the model fail to satisfy the sequent, since $|\forall x(P_1^0 \Rightarrow P_2^1)|^+ \not\subseteq |(P_1^0 \Rightarrow \forall x P_2^1)|^+$, but it is also a model of the formula $\forall x(P_1^0 \Rightarrow P_2^1)$, since every point of the frame is in $|\forall x(P_1^0 \Rightarrow P_2^1)|^+$.[14] (We can easily modify the model by $f(b, \{a\}) = f(c, \{a\}) = \emptyset$ to produce a model that fails to satisfy the sequent, but isn't also a model of the $\forall x(P_1^0 \Rightarrow P_2^1)$.)

Let us label some quantifier distribution (over the conditional) sequents:

($\forall$-$\forall$) $\forall x(\mathcal{A} \Rightarrow \mathcal{B}) \vdash \mathcal{A} \Rightarrow \forall x \mathcal{B}$, where x is not free in $\mathcal{A}$
($\forall$-$\exists$) $\forall x(\mathcal{A} \Rightarrow \mathcal{B}) \vdash \exists x \mathcal{A} \Rightarrow \mathcal{B}$, where x is not free in $\mathcal{B}$
($\forall$-$\forall^c$) $\mathcal{A} \Rightarrow \forall x \mathcal{B} \vdash \forall x(\mathcal{A} \Rightarrow \mathcal{B})$, where x is not free in $\mathcal{A}$
($\forall$-$\exists^c$) $\exists x \mathcal{A} \Rightarrow \mathcal{B} \vdash \forall x(\mathcal{A} \Rightarrow \mathcal{B})$, where x is not free in $\mathcal{B}$

For the sequent ($\forall$-$\exists$), a model can be more easily constructed. Clearly, the sets $f(\alpha, ||\mathcal{A}||)$ and $f(\alpha, ||\exists x \mathcal{A}||)$ are not constricted to be related by the definition of the models so far. Given this and the model above, we record the following fact.

Fact 56. *The sequents ($\forall$-$\forall$) and ($\forall$-$\exists$) are not theorems of $\mathbf{CE_{QFDE}}$ or $\mathbf{CE_{FDEQ}}$.*

What about the converse sequents ($\forall$-$\forall^c$) and ($\forall$-$\exists^c$)? The sequents ($\forall$-$\forall$) and ($\forall$-$\exists$) are conditional translations of Barcan formulas, and similarly, their converses are translations of converse Barcan Formulas. As in the case of quantified relevant logics, these translations of Converse Barcan Formulas are provable when the (formula-indexed) modality is monotonic in the right way — here by (RCM) and (RCT).

Fact 57. *The sequents ($\forall$-$\forall^c$) and ($\forall$-$\exists^c$) are theorems of $\mathbf{CE_{QFDE}}$ + (RCM) and $\mathbf{CE_{FDEQ}}$ + (RCT), respectively.*

Proof. The derivations below suffice.

(1) $\forall x \mathcal{B} \vdash \mathcal{B}$ ($\forall$E)
(2) $\mathcal{A} \Rightarrow \forall x \mathcal{B} \vdash \mathcal{A} \Rightarrow \mathcal{B}$ 1, (RCM)
(3) $\mathcal{A} \Rightarrow \forall x \mathcal{B} \vdash \forall x(\mathcal{A} \Rightarrow \mathcal{B})$ 2, (R$\forall$I)

(1) $\mathcal{A} \vdash \exists x \mathcal{A}$ ($\exists$I)
(2) $\exists x \mathcal{A} \Rightarrow \mathcal{B} \vdash \mathcal{A} \Rightarrow \mathcal{B}$ 1, (RCT)
(3) $\exists x \mathcal{A} \Rightarrow \mathcal{B} \vdash \forall x(\mathcal{A} \Rightarrow \mathcal{B})$ 2, (R$\forall$I)

◁

---

[14]Being a model of a formula has not been explicitly defined, since we never have an empty left-hand-side of a turnstile. However, we can say that a model is *a model of a formula* $\mathcal{A}$ when every point in that model makes the formula $\mathcal{A}$ at least true.

For the sequents ($\forall$-$\forall^c$) and ($\forall$-$\exists^c$), it is easy to see — from a semantic point of view — why (cRCT) and (cRCM) are sufficient.

Our Barcan formulas ($\forall$-$\forall$) and ($\forall$-$\exists$) are difficult to characterize in the general frame setting, even when expressed using a $\square$ modality (e.g., see [12]). So the behavior of the conditional and the quantifier observed so far is not surprising. The formula $\forall x(\mathcal{A} \Rightarrow \mathcal{B}) \vdash (\mathcal{A} \Rightarrow \forall x\mathcal{B})$ is similar to the Barcan Formula. That is, when translating arrows to formula-indexed boxes, we obtain the formula $\forall x \square_\mathcal{A} \mathcal{B} \vdash \square_\mathcal{A} \forall x \mathcal{B}$, which is the Barcan Formula expressed as a $\vdash$-sequent. In Goldblatt [12], Ferenz [10] and Tedder and Ferenz [38], the best (and indeed only) characterization for the Barcan Formula in the Mares–Goldblatt semantics is a straightforward transliteration into a semantic (algebraic) dialect. That is, the *semantic* condition that $\forall_n \square \varphi \subseteq \square \forall_n \varphi$. Analogously, we can use the following conditions obtained by applying a similar approach, making the requisite changes for the restriction that x is not free in $\mathcal{A}$.

(c$\forall$-$\forall$) $\forall_n(\varphi \Rightarrow \psi) \subseteq \varphi \Rightarrow \forall_n \psi$, where $\varphi g = \varphi g[j/x_n]$ for every $j \in U$ and $g \in U^\omega$.
(c$\forall$-$\exists$) $\forall_n(\psi \Rightarrow \varphi) \subseteq \exists_n \psi \Rightarrow \varphi$, where $\varphi g = \varphi g[j/x_n]$ for every $j \in U$ and $g \in U^\omega$.

The restriction on these conditions ensures that φ does not "have x free."

Lemma 58. **CE$_{\text{QFDE}}$**-*models satisfying the conditions above are adequate for extensions of* **CE$_{\text{QFDE}}$** *with the corresponding axioms. That is,*

(i) **CE$_{\text{QFDE}}$** + ($\forall$-$\forall$) *is sound and complete with respect to the* **CE$_{\text{QFDE}}$**-*models satisfying (c$\forall$-$\forall$);*

(ii) **CE$_{\text{QFDE}}$** + ($\forall$-$\exists$) *is sound and complete with respect to the* **CE$_{\text{QFDE}}$**-*models satisfying (c$\forall$-$\exists$).*

Proof. Goldblatt's arguments in [12] for a similar condition for the Barcan formula in classical settings are used here. The proof of soundness is straightforward. For completeness, we first show that if $\mathcal{A} \vdash \mathcal{B}$ is a theorem, then in the canonical model we have $\varphi_\mathcal{A} \subseteq \varphi_\mathcal{B}$. For any $\alpha \in K$ and $g \in U^\omega$, by definition $\alpha \in \varphi_\mathcal{A} g$ implies $\alpha \in ||\mathcal{A}^g||$, which is $\mathcal{A}^g \in \alpha$. But since α is a theorem and $\mathcal{A} \vdash \mathcal{B}$ is a theorem, $\mathcal{B}^g \in \alpha$, and so $\alpha \in \varphi_\mathcal{B}$. Combined with the fact that ($\forall$-$\forall$) (or ($\forall$-$\exists$)) is a theorem, the canonical model is indeed a model satisfying the corresponding constraint. ◁

Model conditions that are merely axioms rewritten into a semantic dialect are great for obtaining completeness in general frame settings, given that theories are closed under the consequence relation. Nonetheless, it is desirable to have a less transliterated semantic condition. For example, the property of reflexivity of a binary relation as characterizing the T axiom in modal logics offers, in many respects, a much better semantic explanation of the axiom T compared to the transliterated condition that $\square \varphi \subseteq \varphi$. So far, the efforts to produce a nice semantic condition for the Barcan formula (or rule) in quantified modal (relevant) logics, and for the formulas ($\forall$-$\forall$) and ($\forall$-$\exists$) in this paper, have not been fruitful. The transliterated condition is what [12] (in the classical case) and [10] (in the relevant case) give for the Barcan Formula. So far, every other condition that I have demonstrated to work is (i) lengthy, and not surprisingly, (ii) just an unnecessarily wordy reformulation of (c$\forall$-$\forall$). Moreover, these conditions contain phrases like "S is the range of a propositional function φ

restricted to a domain of $x_n g$ for some variable x_n," which needlessly complicates matters without providing the desired insight.[15]

7. Conclusion and Further Work

In conclusion, we have constructed two classes of propositional **FDE**-logics, and extended the most general of these classes with quantifiers. This work opens up several areas of further research. Some pioneering work of interest to the relevant logician on the relation between functional models and relational models is found in Routley [25] and Sylvan, Meyer, and Plumwood [36]. There, the relation between $R\alpha XY$, $R\alpha X\beta$ and $Y \in f(\alpha, X)$ and $f(\alpha, X) \subseteq Y$ is introduced. In terms of furthering that development and the current work, future directions of research include tying down the exact relation between the sequent systems given here and the usual bunch of relevant logics (with ternary relational semantics), especially as the latter are often presented as sets of theorems. A particularly interesting line of research in this direction is determining what the intensional conjunction $\circ$ and intensional truth t can (or should) look like in these sequent systems.

Acknowledgments. The author wishes to thank Katalin Bimbó, Andrew Tedder, attendees of talks based on the paper, and an anonymous referee for helpful discussions and comments. This paper was supported by RVO 67985807 and by the Czech Science Foundation project GA22-01137S.

Appendix. Metatheory for FDE

A.1.1. *Soundness.* Soundness can be proved using the arguments of [18] and [10], with modifications for the existential quantifier (and the presentation as a sequent system). The only interesting new case in the proof is recorded in the following lemma.

Lemma 59. *The axiom (EC2) is valid in the class of* **FDEQ**-*frames.*

Proof. Suppose in an arbitrary model, and arbitrary assignment g, that $\alpha \in |\mathcal{A} \wedge \exists x_n \mathcal{B}|^+ g$, with x not occurring free in $\mathcal{A}$. Then $\alpha \in |\exists x_n \mathcal{B}|^+ g$, and $\alpha \in |\mathcal{A}|^+ g$. For reductio, assume that

$$\alpha \notin |\exists x_n (\mathcal{A} \wedge \mathcal{B})|^+ g = \exists_n |\mathcal{A} \wedge \mathcal{B}|^+ g = \bigsqcup_{h \in xg} |\mathcal{A} \wedge \mathcal{B}|^+ h.$$

Then, by the definition of $\bigsqcup$, there is an $X \in Prop$ such that $\alpha \notin X$ and $\bigcup_{h \in xg} |\mathcal{A} \wedge \mathcal{B}| h \subseteq X$. But then, since x does not occur free in $\mathcal{A}$, we have that $\bigcup_{h \in xg} |\mathcal{B}|^+ h \subseteq X \cup \overline{|\mathcal{A}|^+ g}$. By (cEC2), we have $|\exists_n \mathcal{B}|^+ g \subseteq X \cup \overline{|\mathcal{A}|^+ g}$. But $\alpha \in |\exists_n \mathcal{B}|^+ g$ and $\alpha \notin X \cup \overline{|\mathcal{A}|^+ g}$, whence we obtain contradiction. ◁

---

[15]There is some insight to be gained here. We want the $f(\alpha, X)$ to be closed under $\bigcap$, but just for those sets that correspond to the right kind of propositional function. Otherwise, it would close it under finite intersection, and thus entail the rule (RCR).

A.1.2. *Completeness.* Theories, prime theories and $\gg_\mathbb{L}$ are defined as in Section 3.1.2, but with respect to **QFDE** and **FDEQ**. For this section, we shall use $\mathbb{L}$ to refer to both **QFDE** and **FDEQ**. The proof in [2] may be used to show the following:

Lemma 60 (Extension). *If $\Gamma \not\gg_\mathbb{L} \Delta$, then there is a prime $\mathbb{L}$-theory Γ' such that $\Gamma \subseteq \Gamma'$ and $\Gamma' \cap \Delta = \emptyset$.*

Corollary 61. *If $\mathcal{A} \not\vdash \mathcal{B}$, then there is a prime theory Γ, where $\mathcal{A} \in \Gamma$ but $\mathcal{B} \notin \Gamma$.*

Definition 62 (Canonical Model for **QFDE**). The *canonical model* for **QFDE** is $\mathfrak{M}^\mathbb{C}_{\mathbf{QFDE}} = \langle K^\mathbb{C}, U^\mathbb{C}, Prop^\mathbb{C}, PropFun^\mathbb{C}, |-|^\mathbb{C}\rangle$, where each element is defined as in Definition 50 (without $f^\mathbb{C}$) with respect to **QFDE**-theories.

Lemma 63. *The canonical model for **QFDE** (**FDEQ**) satisfies (c1)–(c3).*

The proof is from [18], with the existential quantifier cases from [10].

Lemma 64. *Where $\exists x \mathcal{A}$ is closed, $\exists x \mathcal{A} \in \alpha$ iff for every $X \in Prop$, $\bigcup_{c \in U} \mathcal{A}[c/x] \subseteq X$ implies $\alpha \in X$.*

Proof. Left-to-right: Suppose first that $\exists x \mathcal{A} \in \alpha$. For reductio, suppose that there is an $X \in Prop$ such that $\bigcup_{c \in U} \mathcal{A}[c/x] \subseteq X$ and $\alpha \notin X$. However, $X = ||\mathcal{B}||$ for some closed $\mathcal{B}$. Further, it must the case that $\exists x \mathcal{A} \vdash \mathcal{B}$ is a theorem. If it were not, then $\mathcal{A}[c/x] \not\vdash \mathcal{B}$. But then, by the extension lemma, we can extend α to a prime theory δ such that (i) $\mathcal{A}[c/x] \in \delta$ and (ii) $\mathcal{B} \notin \delta$. But then by assumption we have $\delta \in \bigcup_{c \in U} \mathcal{A}[c/x]$, and so $\delta \in ||\mathcal{B}||$, a contradiction. So $\exists x \mathcal{A} \vdash \mathcal{B}$ is a theorem. But α is a theory, and so $\mathcal{B} \in \alpha$, which means $\alpha \in X$, our contradiction.

Right-to-left: Suppose that for every $X \in Prop$, $\bigcup_{c \in U} \mathcal{A}[c/x] \subseteq X$ implies $\alpha \in X$. Take $X = ||\exists x \mathcal{A}||$. By the axiom ($\exists$I), it follows that $||\mathcal{A}[c/x]|| \subseteq ||\exists x \mathcal{A}||$, for every $c \in U$. So in particular, we obtain $\bigcup_{c \in U} \mathcal{A}[c/x] \subseteq ||\exists x \mathcal{A}||$, and so $\alpha \in ||\exists x \mathcal{A}||$, which is $\exists x \mathcal{A} \in \alpha$. ◁

Lemma 65. *The canonical model for **FDEQ** satisfies (cEC1) and (cEC2).*

Proof. For (cEC1), see [18, Theorem 10.3]. For (cEC2), suppose condition is applicable to some φ, X, Y, n, and g. By the definition of the canonical model, $X = ||\mathcal{A}||$, for some sentence $\mathcal{A}$, and $\varphi = \varphi_\mathcal{B}$ for some formula $\mathcal{B}$.

Suppose that $\bigcup_{c \in U} \varphi(g[c/n]) \subseteq X \cup \overline{||\mathcal{A}||}$. Thus, for every $c \in U$, we have that

$$\varphi_\mathcal{B} g[c/n] \subseteq X \cup \overline{||\mathcal{A}||}$$

$$||\mathcal{A}|| \cap ||\mathcal{B}^{g[c/n]}|| \subseteq X$$

$$||\mathcal{A} \wedge (\mathcal{B}^{g[c/n]})|| \subseteq X$$

Because $\mathcal{A}$ is closed, and the identity $\mathcal{A}^{g\backslash n}[c/n] = \mathcal{A}^{g[c/n]}$,[16]

$$\mathcal{A} \wedge (\mathcal{B}^{g[c/n]}) = \mathcal{A} \wedge (\mathcal{B}^{g\backslash n}[c/n])$$
$$= (\mathcal{A} \wedge \mathcal{B})^{g\backslash n}[c/n]$$

---

[16] The formula $\mathcal{A}^{g\backslash n}$, the result of applying the substitution of the assignment g but leaving the variable x_n alone, is defined as $\mathcal{A}[g0/x_0, \ldots, g(n-1)/x_{n-1}, x_n/x_n, g(n+1)/x_{n+1}, \ldots]$. Below we will freely use the identity (for both quantifiers) $\exists x_n(\mathcal{A}^{g\backslash n}) = (\exists x_n \mathcal{A})^g$.

So, in particular we have that $||\mathcal{A} \wedge (\mathcal{B}^{g[c/n]})|| = ||(\mathcal{A} \wedge \mathcal{B})^{g\backslash n}[c/n]||$. Now, suppose $a \notin X \cup \overline{||\mathcal{A}||}$. Then $\alpha \notin X$, which means that $\alpha \notin ||(\mathcal{A} \wedge \mathcal{B})^{g\backslash n}[c/n]||$, for every $c \in U$. By Lemma 64, we have $\exists x_n ((\mathcal{A} \wedge \mathcal{B})^{g\backslash n}) \notin \alpha$. By (EC2) and α being a theory, it follows that $\mathcal{A} \wedge \exists x_n \mathcal{B}^{g\backslash n} \notin \alpha$. But $\alpha \notin \overline{||\mathcal{A}||}$ entails $\alpha \in ||\mathcal{A}||$, which gives $\mathcal{A} \in \alpha$. Thus, $\exists x_n (\mathcal{B})^{g\backslash n} \notin \alpha$. Thus, $\alpha \notin ||(\exists x_n \mathcal{B})^g|| = (\varphi_{\exists x_n \mathcal{B}})g = (\exists_n \varphi_\mathcal{B})g$. As α was arbitrary, this gives the desired result that $|\exists_n \varphi|g \subseteq X \cup \overline{Y}$. ◁

The next theorem shows that all of the atomic formulas are given both positive and negative functions within *PropFun*.

Lemma 66. *For every n-ary predicate P, terms* $\tau_1, \ldots, \tau_n$, *and* $g \in U^\omega$,

(i) $(P\tau_1, \ldots, \tau_n)^g = P(|\tau_1|g, \ldots, |\tau_n|g)$
(ii) $|P\tau_n, \ldots, \tau_n|^+ = \varphi_{P\tau_n, \ldots, \tau_n}$
(iii) $|P\tau_n, \ldots, \tau_n|^- = \varphi_{\neg P\tau_n, \ldots, \tau_n}$

The proof is similar to that in [18].

Lemma 67 (Truth Lemma). *For every formula* $\mathcal{A}$, $|\mathcal{A}|^+ = \varphi_\mathcal{A}$. *That is, for every* $g \in U^\omega$, $|\mathcal{A}|^+ g = ||\mathcal{A}^g||$.

Proof. The proof is straightforward, as in [18], given the equalities shown in the proof of Lemma 63. ◁

Theorem 68 (Completeness). *The logic* **QFDE** *(and* **FDEQ***) is complete with respect to the class of* **QFDE**-*models* (**FDEQ**-*models*). *That is, if* $\mathcal{A} \vDash_{\mathbf{QFDE}} \mathcal{B}$, *then* $\mathcal{A} \vdash_{\mathbf{QFDE}} \mathcal{B}$, *and similarly, for* **FDEQ**.

The proof is as in the proof of Theorem 55.

REFERENCES

[1] Anderson, A. R. and Belnap, N. D. (1962). Tautological entailments, *Philosophical Studies* **13**: 9–24.
[2] Anderson, A. R., Belnap, N. D. and Dunn, J. M. (1992). *Entailment: The Logic of Relevance and Necessity*, Vol. II, Princeton University Press, Princeton.
[3] Belnap, N. D. (1977a). How a computer should think, in G. Ryle (ed.), *Contemporary Aspects of Philosophy*, Oriel Press, Stocksfield, pp. 30–35.
[4] Belnap, N. D. (1977b). A useful four-valued logic, in J. M. Dunn and G. Epstein (eds.), *Modern Uses of Multiple-Valued Logics*, D. Reidel Publishing Co., Dordrecht, pp. 8–37.
[5] Brady, R. T. (1984). Depth relevance of some paraconsistent logics, *Studia Logica* **43**: 63–73.
[6] Chellas, B. (1975). Basic conditional logic, *Journal of Philosophical Logic* **4**(2): 133–153.
[7] Chellas, B. (1980). *Modal Logic: An Introduction*, Cambridge University Press, Cambridge.
[8] Dunn, J. M. (1966). *The Algebra of Intensional Logics*, Doctoral dissertation, University of Pittsburgh, Pittsburgh, PA. (Published as Vol. 2 in the *Logic PhDs* series by College Publications, London, UK, 2019.).
[9] Dunn, J. M. (1976). Intuitive semantics for first-degree entailments and 'coupled trees', *Philosophical Studies: An International Journal for Philosophy in the Analytic Tradition* **29**(3): 149–168.

[10] Ferenz, N. (n.d.). Quantified modal relevant logics, *Review of Symbolic Logic* . (Forthcoming).
[11] Goble, L. (2003). Neighborhoods for entailment, *Journal of Philosophical Logic* **32**: 483–529.
[12] Goldblatt, R. (2011). *Quantifiers, Propositions and Identity: Admissible Semantics for Quantified Modal and Substructural Logics*, Cambridge University Press, Cambridge.
[13] Hazen, A. P. and Pelletier, F. J. (2020). K3, Ł3, LP, RM3, A3, FDE, M: How to make many-valued logics work for you, *in* H. Omori and H. Wansing (eds.), *New Essays on Belnap–Dunn Logic*, Vol. 418 of *Synthese Library*, Springer, New York, pp. 155–190.
[14] Humberstone, L. (2011). *The Connectives*, MIT Press, Cambridge.
[15] Lavers, P. (1985). *Generating Intensional Logics*, Master's thesis, University of Adelaide.
[16] Logan, S. A. (2020). Putting the stars in their places, *Thought: A Journal of Philosophy* **9**: 188–197.
[17] Ma, M. and Wong, C.-T. (2020). A paraconsistent conditional logic, *Journal of Philosophical Logic* **49**: 883–903.
[18] Mares, E. D. and Goldblatt, R. (2006). An alternative semantics for quantified relevant logic, *Journal of Symbolic Logic* **71**(1): 163–187.
[19] Omori, H. and Wansing, H. (2017). 40 years of FDE: An introductory overview, *Studia Logica* **105**: 1021–1049. (Special issue: *40 Years of FDE*).
[20] Priest, G. (2008). *An Introduction to Non-lassical Logic*, 2nd edn, Cambridge University Press, Cambridge.
[21] Restall, G. (1993). Simplified semantics for relevant logics (and some of their rivals), *Journal of Philosophical Logic* **22**: 481–511.
[22] Restall, G. (1995). Four-valued semantics for relevant logics (and some of their rivals), *Journal of Philosophical Logic* **24**: 139–160.
[23] Restall, G. (2000). *An Introduction to Substructural Logics*, Routledge, New York.
[24] Routley, R. (1984). The American plan completed: Alternative classical-style semantics, without stars, for relevant and paraconsistent logics, *Studia Logica* **43**(1): 199–243.
[25] Routley, R. (1989). Philosophical and linguistic inroads: Multiply intensional relevant logics, *in* J. Norman and R. Sylvan (eds.), *Directions in Relevant Logic*, Kluwer Academic Publishers, Dordrecht, pp. 269–304.
[26] Routley, R. and Meyer, R. K. (1972). The semantics of entailment, II, *Journal of Philosophical Logic* **1**: 53–73.
[27] Routley, R. and Meyer, R. K. (1973). The semantics of entailment, *in* H. Leblanc (ed.), *Truth, Syntax, and Modality*, North-Holland, Amsterdam, pp. 199–243.
[28] Routley, R. and Meyer, R. K. (1975). Towards a general semantical theory of implication and conditionals. I. Systems with normal conjunctions and disjunctions and aberrant and normal negations, *Reports on Mathematical Logic* **4**: 67–89.
[29] Routley, R. and Meyer, R. K. (1976). Towards a general semantical theory of implication and conditions. II. Improved negation theory and propositional identity, *Reports on Mathematical Logic* **9**: 47–62.
[30] Routley, R., Meyer, R. K., Plumwood, V. and Brady, R. T. (1982). *Relevant Logics and Their Rivals*, Vol. 1, Ridgeview Publishing Company, Atascadero.
[31] Routley, R. and Routley, V. (1972). The semantics of first-degree entailment, *Noûs* **6**: 335–359.
[32] Segerberg, K. (1971). *An Essay in Classical Modal Logic*, PhD thesis, Stanford.
[33] Shramko, Y. (2020). First-degree entailment and structural reasoning, *in* H. Omori and H. Wansing (eds.), *New Essays on Belnap–Dunn Logic*, Vol. 418 of *Synthese Library*, Springer, pp. 311–324.

[34] Standefer, S. (2019). Tracking reasons with extensions of relevant logics, *Logic Journal of the IGPL* **27**: 543–569.
[35] Sutcliffe, G., Hazen, A. P. and Pelletier, F. J. (2018). Making Belnap's "useful four-valued logic" useful, *in* K. Brawner and V. Rus (eds.), *Proceedings of the 31st International Florida Artificial Intelligence Research Society Conference, FLAIRS*, AAAI Press, pp. 116–121.
[36] Sylvan, R., Meyer, R. K. and Plumwood, V. (2003). Multiplying connectives and multiply intensional logics, *in* R. Brady (ed.), *Relevant Logics and their Rivals,* Vol. 2, Ashgate, Burlington, pp. 17–37.
[37] Tedder, A. (2021). Information flow in logics in the vicinity of **BB**, *Australasian Journal of Logic* **18**: 1–24.
[38] Tedder, A. and Ferenz, N. (n.d.). Neighbourhood semantics for quantified relevant logics, *Journal of Philosophical Logic* . (Forthcoming).
[39] Wansing, H. (2005). Connexive modal logic, *in* R. Schmidt, I. Pratt-Hartmann, M. Reynolds and H. Wansing (eds.), *Advances in Modal Logic, Volume 5*, King's College Publications, London, pp. 367–383.
[40] Wansing, H. and Unterhuber, M. (2019). Connexive conditional logic, I, *Logic and Logical Philosophy* **28**: 567–610.

INSTITUTE OF COMPUTER SCIENCES OF THE CZECH ACADEMY OF SCIENCES, PRAGUE, CZECHIA
Email: ferenz@cs.cas.cz

Reconciliation of Approaches to the Semantics of Logics without Distribution

Chrysafis (Takis) Hartonas

In memoriam Jon Michael Dunn

ABSTRACT. This article clarifies and indeed completes an approach to the relational semantics of logics that may lack distribution (Dunn's non-distributive gaggles). The approach was initiated by Dunn and this author several years ago and again pursued by the present author over the last three years or so. It uses sorted frames with an incidence relation on sorts (polarities), equipped with additional sorted relations, but, in the spirit of Occam's razor principle, it drops the extra assumptions made in the generalized Kripke frames approach, initiated by Gehrke, that the frames be separated and reduced (RS-frames). We show in this article that, despite rejecting the additional frame restrictions, all the main ideas and results of the RS-frames approach relating to the semantics of non-distributive logics are captured in this simpler framework. This contributes in unifying the research field, and, in an important sense, it complements and completes Dunn's gaggle theory project for the particular case of logics that may drop distribution.

Keywords. Canonical lattice extensions, Gaggle theory, Relational semantics for non-distributive logics, RS-frames

1. Preliminaries

1.1. A Note on Motivation. The motivation for this article is twofold. First, it aims at complementing Dunn's gaggle theory project pursued in Dunn [11; 12]; Allwein and Dunn [1]; Dunn and Hardegree [14]; Dunn and Zhou [15]; Dunn et al. [13]; Bimbó and Dunn [2], by addressing the case of logics that may lack distribution. This has been the topic of recent research both by the present author and by researchers in the RS-frames approach, initiated by Gehrke [18]. The second motivation relates to clarifying points of convergence and divergence between our approach and that of RS-frames. Dunn himself seems to have stood at the junction of the two, as he has contributed, with this author, the lattice representation result Hartonas and Dunn [33] on which our approach is based, while having also contributed in applying the RS-frames approach to his gaggle theory project, with Gehrke and Palmigiano [13]. Indeed we conclude that apart from dropping the "RS" from the "RS-frames" approach, which is to say the assumptions that frames (polarities with relations) are *Separated and Reduced*, the two approaches have nearly identical objectives and nearly identical techniques,

---

2020 *Mathematics Subject Classification.* Primary: 03G10, Secondary: 03B47, 06B15.

Bimbó, Katalin, (ed.), *Relevance Logics and other Tools for Reasoning. Essays in Honor of J. Michael Dunn*, (Tributes, vol. 46), College Publications, London, UK, 2022, pp. 215–236.

though they have developed separately. In our concluding remarks we point at some issues that seem to indicate that the approach we have taken may be better suited for some purposes (though no doubt the same can be said for the RS-approach, for some other purposes).

1.2. Normal Lattice Expansions.

In [11], Dunn introduced the notion of a distributoid, as a distributive lattice with various operations that in each argument place either distribute or co-distribute over either meets or joins, always returning the same type of lattice operation (always a meet, or always a join). To define technically the extended notion, to which we refer as a *normal lattice expansion*, let $\{1,\partial\}$ be a 2-element set, $\mathcal{L}^1 = \mathcal{L}$ and $\mathcal{L}^\partial = \mathcal{L}^{op}$ (the opposite lattice, order reversed). Extending the Jónsson–Tarski terminology [36], a function $f\colon \mathcal{L}_1 \times \cdots \times \mathcal{L}_n \longrightarrow \mathcal{L}_{n+1}$ is *additive* and *normal*, or a *normal operator*, if it distributes over finite joins of the lattice $\mathcal{L}_i$, for each $i = 1, \ldots, n$, delivering a join in $\mathcal{L}_{n+1}$.

Definition 1. An n-ary operation f on a bounded lattice $\mathcal{L}$ is *a normal lattice operator of distribution type* $\delta(f) = (i_1, \ldots, i_n; i_{n+1}) \in \{1,\partial\}^{n+1}$ if it is a normal additive function $f\colon \mathcal{L}^{i_1} \times \cdots \times \mathcal{L}^{i_n} \longrightarrow \mathcal{L}^{i_{n+1}}$ (distributing over finite joins in each argument place), where each i_j, for $j = 1, \ldots, n+1$, is in the set $\{1,\partial\}$, i.e., $\mathcal{L}^{i_j}$ is either $\mathcal{L}$, or $\mathcal{L}^\partial$.

Example 2. A normal diamond operator $\Diamond$ is a normal lattice operator of distribution type $\delta(\Diamond) = (1;1)$, i.e., $\Diamond\colon \mathcal{L} \longrightarrow \mathcal{L}$, distributing over finite joins of $\mathcal{L}$. A normal box operator $\Box$ is also a normal lattice operator in the sense of Definition 1, of distribution type $\delta(\Box) = (\partial;\partial)$, i.e., $\Box\colon \mathcal{L}^\partial \longrightarrow \mathcal{L}^\partial$ distributes over finite joins of $\mathcal{L}^\partial$, which are then just meets of $\mathcal{L}$.

An **FL**$_{ew}$-algebra (also referred to as a full **BCK**-algebra, or a commutative integral residuated lattice) $\mathcal{A} = (L, \wedge, \vee, 0, 1, \circ, \rightarrow)$ is a normal lattice expansion, where $\delta(\circ) = (1,1;1)$, $\delta(\rightarrow) = (1,\partial;\partial)$, i.e., $\circ\colon \mathcal{L} \times \mathcal{L} \longrightarrow \mathcal{L}$ and $\rightarrow\colon \mathcal{L} \times \mathcal{L}^\partial \longrightarrow \mathcal{L}^\partial$ are both normal lattice operators with the familiar distribution properties.

Dropping exchange, $\circ$ may have two residuals $\leftarrow, \rightarrow$, one in each argument place, where $\delta(\leftarrow) = (\partial,1;\partial)$, i.e., $\leftarrow\colon \mathcal{L}^\partial \times \mathcal{L} \longrightarrow \mathcal{L}^\partial$.

De Morgan Negation $\neg$ is a normal lattice operator and it has both the distribution type $\delta_1(\neg) = (1;\partial)$ and $\delta_2(\neg) = (\partial;1)$, as it switches both joins to meets and meets to joins.

The Grishin operators [23] $\leftsmallarrow, \star, \rightsmallarrow$, satisfying the familiar co-residuation conditions $a \geq c \leftsmallarrow b$ iff $a \star b \geq c$ iff $b \geq a \rightsmallarrow c$ have the respective distribution properties, which are exactly captured by assigning to them the distribution types $\delta(\star) = (\partial,\partial;\partial)$ ($\star$ behaves as a binary box operator and it is known as *fission* in the relevance logic literature), $\delta(\leftsmallarrow) = (1,\partial;1)$ and $\delta(\rightsmallarrow) = (\partial,1;1)$.

Dunn's distributoids as defined in [11] are the special case of a normal lattice expansion where the underlying lattice is distributive. BAO's (Boolean Algebras with Operators) [36; 37] are the special case where the underlying lattice is a Boolean algebra and all normal operators distribute over finite joins of the Boolean algebra, i.e., they are all of distribution types of the form $(1,\ldots,1;1)$. For BAO's there is no need to consider operators of other distribution types, as they can be obtained by composition of operators with Boolean complementation. For example, in studying residuated

Boolean algebras [38], Jónsson and Tsinakis introduce a notion of *conjugate operators* and they show that intensional implications (division operations) $\backslash, /$ (the residuals of the product operator $\circ$) are interdefinable with the conjugates (at each argument place) $\triangleleft, \triangleright$ of $\circ$, i.e., $a \backslash b = (a \triangleright b^-)^-$ and $a \triangleright b = (a \backslash b^-)^-$ (and similarly for $/$ and $\triangleleft$, see [38] for details). Note that $\backslash, /$ are not operators in the sense of [36], whereas $\triangleleft, \triangleright$ are.

The relational representation of BAO's in [36] extended Stone's representation [45] of Boolean algebras, using the space of ultrafilters of the algebra. To quote Copeland [9], the results of [36] "can be viewed as in effect a treatment of all the basic modal axioms and corresponding properties of the accessibility relation. Kripke described this paper by Jónsson and Tarski as the 'most surprising anticipation' of his own work." It appears fair to say that, in retrospect, the relational representation of [36] forms, logically speaking, the technical basis of the subsequently introduced by Kripke [39; 40; 41] possible worlds semantics (relational semantics), with its well-known impact on the development of normal modal logics.

Dunn's approach and objective in [11] has been to achieve the same unified semantic treatment for the logics of distributive lattices with various quasioperators, now based on the Priestley representation [43] of distributive lattices in ordered Stone spaces (simplifying Stone's original representation [44] of distributive lattices), using the space of prime filters, and abstracting over various specific results in the semantics of distributive, non-classical logics, notably Relevance Logics.

For non-distributive lattices, Urquhart pioneered a representation theorem [47], using the space of maximally disjoint filter-ideal pairs. Over the years, Urquhart's representation has proven notoriously difficult to work with, though some authors, including Allwein and Dunn [1], as well as Düntsch, Orłowska, Radzikowska and Vakarelov [16] have based a semantic treatment of specific systems on it. Hartonas and Dunn [32], published in 1997 as [33], provide a lattice representation and duality result based on the representation of semilattices and of Galois connections and abstracting over Goldblatt's [22] representation of ortholattices, replacing orthocomplementation with the trivial Galois connection (the identity map $\iota : \mathcal{L} \longrightarrow (\mathcal{L}^\partial)^\partial$). Hartonas [24] presents another lattice representation, extended to include various lattice expansions. Both [33; 24] form the background of a representation and duality result for normal lattice expansions Hartonas [25; 31], extending the representation of [33].

The bulk of Dunn's work on gaggles predates the extension of the theory of canonical extensions to bounded lattices advanced by Gehrke and Harding in [19], extending the Jónsson–Tarski results for perfect extensions of Boolean algebras [36] and the Gehrke–Jónsson following extension to distributive lattice expansions [20]. Subsequently, Gehrke [18] proposed generalized Kripke frames (RS-frames), based on Hartung's lattice representation [35], as a suitable framework for the relational semantics of logics lacking distribution Chernilovskaya et al. [4], including full Linear Logic Coumans et al. [10]. The RS-frames approach to the semantics of logics without distribution was further developed by Palmigiano and co-workers, notably Conradie [8; 5; 7].

2. POLARITIES WITH RELATIONS

2.1. Definitions, Notational Conventions and Basic Facts.
Let $\{1,\partial\}$ be a set of sorts and (Z_1, Z_∂) a sorted set. Base sorted frames $\mathfrak{F} = (X, \perp, Y) = (Z_1, \perp, Z_\partial)$ are triples consisting of nonempty sets $Z_1 = X$, $Z_\partial = Y$ and a binary relation $\perp \subseteq X \times Y$, which will be referred to as the *Galois relation* of the frame. It generates a Galois connection $(\ )^\perp : \mathcal{P}(X) \leftrightarrows \mathcal{P}(Y)^\partial : {}^\perp(\ )$ ($V \subseteq U^\perp$ iff $U \subseteq {}^\perp V$), defined by

$$U^\perp = \{y \in Y : \forall x \in U \; x \perp y\} = \{y \in Y : U \perp y\}$$
$$^\perp V = \{x \in X : \forall y \in V \; x \perp y\} = \{x \in X : x \perp V\}$$

Triples (X, R, Y), $R \subseteq X \times Y$, where R is treated as the Galois relation of the frame, are variously referred to in the literature as *polarities*, after Birkhoff [3], as *formal contexts*, in the Formal Concept Analysis (FCA) tradition Ganter and Wille [17], or as *object-attribute (categorization, classification, or information) systems* Orłowska [42], Vakarelov [48], or as *generalized Kripke frames* [18], or as *polarity frames* in the bi-approximation semantics of Suzuki [46].

A subset $A \subseteq X$ will be called *stable* if $A = {}^\perp(A^\perp)$. Similarly, a subset $B \subseteq Y$ will be called *co-stable* if $B = ({}^\perp B)^\perp$. Stable and co-stable sets will be referred to as *Galois sets*, disambiguating to *Galois stable* or *Galois co-stable* when needed and as appropriate.

$\mathcal{G}(X), \mathcal{G}(Y)$ designate the families (complete lattices) of stable and co-stable sets, respectively. Note that the Galois connection restricts to a duality of the complete lattices of Galois stable and co-stable sets $(\ )^\perp : \mathcal{G}(X) \simeq \mathcal{G}(Y)^\partial : {}^\perp(\ )$.

Preorder relations are induced on each of the sorts, by setting for $x, z \in X$, $x \preceq z$ iff $\{x\}^\perp \subseteq \{z\}^\perp$ and, similarly, for $y, v \in Y$, $y \preceq v$ iff ${}^\perp\{y\} \subseteq {}^\perp\{v\}$. We use Γ as the upper closure operator, and simplify $\Gamma(\{x\})$ to Γx. A (sorted) frame is called *separated* if the preorders $\preceq$ (on X and on Y) are in fact partial orders $\leq$. Note that if the frame is separated, then $\Gamma x = \Gamma z$ iff $x = z$ and ${}^\perp\{y\} = {}^\perp\{v\}$ iff $y = v$. Thus we can identify X and Y with the corresponding subsets of $\mathcal{G}(X), \mathcal{G}(Y)$. Moreover, we can identify $Z = X \cup Y$ with the family of sets $\{\Gamma x : x \in X\} \cup \{{}^\perp\{y\} : y \in Y\} \subseteq \mathcal{G}(X)$. The following result is due to Gehrke [18, Proposition 2.7, Corollary 2.11].

Proposition 3 (Gehrke [18]). *In a separated frame $(X, \perp, Y)$ the set $Z = X \cup Y$ is partially ordered by the relation $\leq$ defined for $x, z \in X$ and $y, v \in Y$ by*

$$
\begin{array}{llll}
x \leq y & \text{iff} & \Gamma x \subseteq {}^\perp\{y\} & \text{iff} \quad x \perp y \\
x \leq z & \text{iff} & \Gamma x \subseteq \Gamma z & \text{iff} \quad z \leq x \\
y \leq v & \text{iff} & {}^\perp\{y\} \subseteq {}^\perp\{v\} & \text{iff} \quad y \leq v \\
y \leq x & \text{iff} & {}^\perp\{y\} \subseteq \Gamma x & \text{iff} \quad \forall u \in X \forall w \in Y (u \perp y \wedge x \perp w \longrightarrow u \perp w)
\end{array}
$$

Moreover, $\mathcal{G}(X)$ is the Dedekind–MacNeille completion $\overline{Z}$ of $(Z, \leq)$.

We caution the reader not to confuse the relation $\leq$ on $X \cup Y$ with the relation $\leq$ induced by the Galois connection on each of X and Y in a separated frame. Generalized Kripke frames, introduced by Gehrke [18], are separated and *reduced* polarities (RS-frames), where the latter is defined by the conditions

1. $\forall x \in X \exists y \in Y (x \not\leq y \wedge \forall z \in X (z \leq x \wedge z \neq x \longrightarrow z \leq y))$
2. $\forall y \in Y \exists x \in X (x \not\leq y \wedge \forall v \in Y (y \leq v \wedge y \neq v \longrightarrow x \leq v))$

The concept goes back to Wille's Formal Concept Analysis (FCA) framework [17] and Hartung's lattice representation theorem [35]. Gehrke observes that being reduced means that all the elements of X are completely join irreducible in X (equivalently, in Z, equivalently in $\overline{Z} = \mathcal{G}(X)$)) and, dually, that all the elements of Y are completely meet irreducible in Y (equivalently in Z, equivalently in $\overline{Z} = \mathcal{G}(X)$). This turns the poset $(Z, \leqslant)$ to what is called a *perfect* poset in [13] (join-generated by the set $J^\infty(Z)$ of its join irreducibles and meet-generated by the set $M^\infty(Z)$ of its meet irreducibles), which can be represented as the two-sorted frame $(J^\infty(Z), M^\infty(Z), \leqslant)$.

To model additional logical operators, RS-frames are equipped with relations subject to the requirement that all their sections be stable sets, where a section of a relation is the set obtained by leaving one argument place unfilled. For example, to model the Lambek calculus product operator ∘, RS-frames are equipped with a relation $S \subseteq Y \times (X \times X)$, all sections of which are required to be stable, and an operation $\otimes$ is generated on stable sets by defining (see [18])

$$A \otimes C = {}^\perp\{y \in Y : \forall x \in A \, \forall z \in C \; ySxz\}$$
$$= \{u \in X : \forall y \in Y \, (\forall x, z \in X (x \in A \wedge z \in C \longrightarrow ySxz) \longrightarrow u \perp y)\}$$

In the canonical frame (which is precisely Hartung's representation [35] of the Lindenbaum–Tarski algebra $\mathcal{L}$ of the logic), the relation S is defined by the condition $ySxz$ iff $\forall a, c \in \mathcal{L} (a \in x \wedge c \in z \longrightarrow a \circ c \in y)$.

In our own approach, initiated with Dunn and the lattice representation result of [32; 33] and gradually developed in the last three years or so by this author [25; 34; 30; 27; 26; 28; 31], we have preferred to apply Occam's razor principle and reject any property of frames that can be rejected while still allowing for the derivation of the needed results. Therefore we work with polarities that need not be separated, or reduced. Section stability for the additional relations, however, is retained as a requirement.[1]

In the rest of this section, we specify the dual objects (polarities with relations) of normal lattice expansions, thereby a class of frames for logics that may lack distribution is described. Representation issues (for completeness arguments) is discussed in the next section and it amounts to an extension of the lattice representation published by this author and Dunn [33]. We do not specify any particular logical signature (except for assuming that conjunction, disjunction and logical constants for truth and falsity are present). The logical setting is very much the same as that detailed by Conradie and Palmigiano in [6] and the reader is referred to this article for a syntactic description of the logics. We save some space by working here with the algebraic structures corresponding to logics that may lack distribution (LE-logics, in the terminology of Conradie et al. [8]), i.e., with normal lattice expansions.

Remark 4 (Notational Conventions). For a sorted relation $R \subseteq \prod_{j=1}^{n+1} Z_{i_j}$, where $i_j \in \{1, \partial\}$ for each j (and thus $Z_{i_j} = X$ if $i_j = 1$, and $Z_{i_j} = Y$ when $i_j = \partial$), we make the convention to regard it as a relation $R \subseteq Z_{i_{n+1}} \times \prod_{j=1}^{n} Z_{i_j}$. We agree to write its sort type as $\sigma(R) = (i_{n+1}; i_1 \cdots i_n)$ and for a tuple of points of suitable sort we write

---

[1]The usefulness of this property has not been fully acknowledged in some of our previous writings, as we did not make section stability explicit. We fill in this gap in this article.

$uRu_1\cdots u_n$ for $(u,u_1,\ldots,u_n)\in R$. We often display the sort type as a superscript, as in R^σ. Thus, for example, $R^{\partial 1\partial}$ is a subset of $Y\times(X\times Y)$. In writing then $yR^{\partial 1\partial}xv$ it is understood that $x\in X=Z_1$ and $y,v\in Y=Z_\partial$. The sort superscript is understood as part of the name designation of the relation, so that, for example, $R^{111}, R^{\partial\partial 1}$ name two different relations.

We use Γ to designate upper closure $\Gamma U=\{z\in X:\exists x\in U\ x\preceq z\}$, for $U\subseteq X$, and similarly for $U\subseteq Y$. U is *increasing* (an upset) iff $U=\Gamma U$. For a singleton set $\{x\}\subseteq X$, we write Γx rather than $\Gamma(\{x\})$, and similarly for $\{y\}\subseteq Y$.

We typically use the standard FCA [17] priming notation for each of the two Galois maps $^\perp(\ ), (\ )^\perp$. This allows for stating and proving results for each of $\mathcal{G}(X), \mathcal{G}(Y)$ without either repeating definitions and proofs, or making constant appeals to duality. Thus for a Galois set G, $G'=G^\perp$, if $G\in\mathcal{G}(X)$ (G is a Galois stable set), and otherwise $G'=\ ^\perp G$, if $G\in\mathcal{G}(Y)$ (G is a Galois co-stable set).

For an element u in either X or Y and a subset W, respectively of Y or X, we write $u|W$, under a well-sorting assumption, to stand for either $u\perp W$ (which stands for $u\perp w$, for all $w\in W$), or $W\perp u$ (which stands for $w\perp u$, for all $w\in W$). Well-sorting means that either $u\in X, W\subseteq Y$, or $W\subseteq X$ and $u\in Y$, respectively. Similarly, for the notation $u|v$, where u,v are elements of different sorts.

We designate n-tuples (of sets, or elements) using a vectorial notation, setting $(G_1,\ldots,G_n)=\vec{G}\in\prod_{j=1}^n\mathcal{G}(Z_{i_j})$, $\vec{U}\in\prod_{j=1}^n\wp(Z_{i_j})$, $\vec{u}\in\prod_{j=1}^n Z_{i_j}$ (where $i_j\in\{1,\partial\}$). Most of the time we are interested in some particular argument place $1\leq k\leq n$ and we write $\vec{G}[F]_k$ for the tuple $\vec{G}$ where $G_k=F$ (or G_k is replaced by F). Similarly $\vec{u}[x]_k$ is $(u_1,\ldots,u_{k-1},x,u_{k+1},\ldots,u_n)$. For brevity, we write $\vec{u}\preceq\vec{v}$ for the pointwise ordering statements $u_1\preceq v_1,\ldots,u_n\preceq v_n$. We also let $\vec{u}\in\vec{W}$ stand for the conjunction of componentwise membership statements $u_j\in W_j$, for all $j=1,\ldots,n$.

To refer to sections of relations (the sets obtained by leaving one argument place unfilled) we make use of the notation $\vec{u}[\_]_k$ which stands for the $(n-1)$-tuple $(u_1,\ldots,u_{k-1}, [\_]_k, u_{k+1},\ldots,u_n)$ and similarly for tuples of sets, extending the membership convention for tuples to cases such as $\vec{u}[\_]_k\in\vec{F}[\_]_k$ and similarly for ordering relations $\vec{u}[\_]_k\preceq\vec{v}[\_]_k$. We also quantify over tuples (with, or without a hole in them), instead of resorting to an iterated quantification over the elements of the tuple, as for example in $\exists\vec{u}[\_]_k\in\vec{F}[\_]_k\exists v,w\in G\ wR\vec{u}[v]_k$. Quantification as in the example just given is always understood under a well-sorting assumption, which however is typically omitted, though always tacitly assumed.

We extend the vectorial notation to distribution types, summarily writing $\delta=(\vec{i_j};i_{n+1})$ for $(i_1,\ldots,i_n;i_{n+1})$. Then, for example, $\vec{i_j}[\partial]_k$ is the tuple with $i_k=\partial$. Furthermore, we let $\overline{i_j}=\partial$, if $i_j=1$ and $\overline{i_j}=1$, when $i_j=\partial$.

Lemma 5. *Let $\mathfrak{F}=(X,\perp,Y)$ be a polarity and $u\in Z=X\cup Y$.*

1. *$\perp$ is increasing in each argument place.*
2. *$(\Gamma u)'=\{u\}'$, and $\Gamma u=\{u\}''$ is a Galois set.*
3. *Galois sets are increasing, i.e., $u\in G$ implies $\Gamma u\subseteq G$.*
4. *For a Galois set G, $G=\bigcup_{u\in G}\Gamma u$.*
5. *For a Galois set G, $G=\bigvee_{u\in G}\Gamma u=\bigcap_{v|G}\{v\}'$.*
6. *For a Galois set G and any set W, $W''\subseteq G$ iff $W\subseteq G$.*

Proof. By simple calculation. Proof details are included in [25], Lemma 2.2. For claim 4, $\bigcup_{u \in G} \Gamma u \subseteq G$ by claim 3 (Galois sets are upsets). In claim 5, given our notational conventions, the claim is that if $G \in \mathcal{G}(X)$, then $G = \bigcap_{G \perp y} {}^{\perp}\{y\}$ and if $G \in \mathcal{G}(Y)$, then $G = \bigcap_{x \perp G}\{x\}^{\perp}$. ◁

For the purposes of this article, the following definition of closed and open elements suffices.

Definition 6 (Closed and Open Elements). The principal upper sets of the form Γx, with $x \in X$, will be called *closed*, or *filter* elements of $\mathcal{G}(X)$, while sets of the form ${}^{\perp}\{y\}$, with $y \in Y$, will be referred to as *open*, or *ideal* elements of $\mathcal{G}(X)$. Similarly for $\mathcal{G}(Y)$. A closed element Γu is *clopen* iff there exists an element v, with $u|v$, such that $\Gamma u = \{v\}'$.

By Lemma 5, the closed elements of $\mathcal{G}(X)$ join-generate $\mathcal{G}(X)$, while the open elements meet-generate $\mathcal{G}(X)$ (similarly for $\mathcal{G}(Y)$).

Definition 7 (Galois Dual Relation). For a relation R, of sort type σ, its *Galois dual relation* R' is the relation defined by $uR'\vec{v}$ iff $\forall w (wR\vec{v} \longrightarrow w|u)$. In other words, $R'\vec{v} = (R\vec{v})'$.

For example, given a relation R^{111} its Galois dual is the relation $R^{\partial 11}$ where for any $x, z \in X$, $R^{\partial 11}xz = (R^{111}xz)^{\perp} = \{y \in Y : \forall u \in X\, (uR^{111}xz \longrightarrow u \perp y)\}$ and, similarly, for a relation $S^{\partial 1 \partial}$ its Galois dual is the relation $S^{11\partial}$ where for any $z \in X, v \in Y$ we have $S^{11\partial}zv = {}^{\perp}(S^{\partial 1 \partial}zv)$, i.e., $xS^{11\partial}zv$ holds iff for all $y \in Y$, if $yS^{\partial 1 \partial}zv$ obtains, then $x \perp y$.

Definition 8 (Sections of Relations). For an $(n+1)$-ary relation R^{σ} and an n-tuple $\vec{u}$, $R^{\sigma}\vec{u} = \{w : wR^{\sigma}\vec{u}\}$ is the *section* of R^{σ} determined by $\vec{u}$. To designate a section of the relation at the k-th argument place we let $\vec{u}[\_]_k$ be the tuple with a hole at the k-th argument place. Then $wR^{\sigma}\vec{u}[\_]_k = \{v : wR^{\sigma}\vec{u}[v]_k\} \subseteq Z_{i_k}$ is the k-th section of R^{σ}. Note that $R^{\sigma}\vec{u}$ is the $(n+1)$-th section of the relation.

2.2. Image Operators, Conjugates and Residuals. If R^{σ} is a relation on a sorted frame $\mathfrak{F}$, of some sort type $\sigma = (i_{n+1}; i_1 \cdots i_n)$, then as in the unsorted case, R^{σ} (but we shall drop the displayed sort type when clear from context) generates a (sorted) *image operator* α_R, defined by (1), of sort $\sigma(\alpha_R) = (i_1, \ldots, i_n; i_{n+1})$, defined by the obvious generalization of the Jónsson–Tarski image operators in [36]

$$(1) \quad \alpha_R(\vec{W}) = \{w \in Z_{i_{n+1}} : \exists \vec{w}\,(wR\vec{w} \wedge \bigwedge_{j=1}^{n}(w_j \in W_j))\} = \bigcup_{\vec{w} \in \vec{W}} R\vec{w},$$

where for each j, $W_j \subseteq Z_{i_j}$.

Thus α_R is a normal and completely additive function in each argument place, therefore, it is residuated, i.e., for each k there is a set-operator β_R^k satisfying the condition:

$$(2) \quad \alpha_R(\vec{W}[V]_k) \subseteq U \text{ iff } V \subseteq \beta_R^k(\vec{W}[U]_k).$$

Hence $\beta_R^k(\vec{W}[U]_k)$ is the largest set V s.t. $\alpha_R(\vec{W}[V]_k) \subseteq U$, and thus, it is definable by

$$(3) \quad \beta_R^k(\vec{W}[U]_k) = \bigcup \{V : \alpha_R(\vec{W}[V]_k) \subseteq U\}.$$

Let $\overline{\alpha}_R$ be the closure of the restriction of α_R to Galois sets $\vec{F}$,

(4) $$\overline{\alpha}_R(\vec{F}) = (\alpha_R(\vec{F}))'' = \left(\bigcup_{j=1,\ldots,n}^{w_j \in F_j} R\vec{w} \right)'' = \bigvee_{\vec{w} \in \vec{F}} (R\vec{w})'',$$

where $F_j \in \mathcal{G}(Z_{i_j})$, for each $j \in \{1,\ldots,n\}$. The operator $\overline{\alpha}_R$ is sorted and its sorting is inherited from the sort type of R. For example, if $\sigma(R) = (\partial; 11)$, $\alpha_R \colon \wp(X) \times \wp(X) \longrightarrow \wp(Y)$, hence, $\overline{\alpha}_R \colon \mathcal{G}(X) \times \mathcal{G}(X) \longrightarrow \mathcal{G}(Y)$. Single sorted operations $\overline{\alpha}_R^1 \colon \mathcal{G}(X) \times \mathcal{G}(X) \longrightarrow \mathcal{G}(X)$ and $\overline{\alpha}_R^\partial \colon \mathcal{G}(Y) \times \mathcal{G}(Y) \longrightarrow \mathcal{G}(Y)$ can be then extracted by composing appropriately with the Galois connection (which is a duality of Galois stable and co-stable sets): $\overline{\alpha}_R^1(A,C) = (\overline{\alpha}_R(A,C))'$ (where $A,C \in \mathcal{G}(X)$) and, similarly, $\overline{\alpha}_R^\partial(B,D) = \overline{\alpha}_R(B',D')$ (where $B,D \in \mathcal{G}(Y)$). Similarly, for the n-ary case.

Definition 9 (Complex Algebra). Let $\mathfrak{F} = (X, \bot, Y, R)$ be a polarity with a relation R of some sort $\sigma(R) = (i_{n+1}; i_1 \cdots i_n)$. The *full complex algebra of* $\mathfrak{F}$ is the structure $\mathfrak{F}^+ = (\mathcal{G}(X), \overline{\alpha}_R^1)$ and its *dual full complex algebra* is the structure $\mathfrak{F}^\partial = (\mathcal{G}(Y), \overline{\alpha}_R^\partial)$.

Most of the time we work with the dual sorted algebra $(\ )^{\pm} \colon \mathcal{G}(X) \simeq \mathcal{G}(Y)^\partial \colon {}^{\pm}(\ )$, as it allows for considering sorted operations that distribute over joins in each argument place (which are either joins of $\mathcal{G}(X)$, or of $\mathcal{G}(Y)$, depending on the sort type of the operation). Single-sorted normal operators are then extracted in the complex algebra by composition with the Galois maps, as indicated above.

Remark 10 (Objective). The primary objective of the current section is to specify conditions under which the residuation structure $\alpha_R \dashv \beta_R^k$ is preserved under the restriction and closure operation described above so that the sorted operator $\overline{\alpha}_R$ on Galois sets is residuated, hence it distributes over arbitrary joins of Galois sets. The notion of conjugate operators we next introduce is useful in this context. Conjugates were introduced in [38] for residuated Boolean algebras. We generalize here to the sorted case, using the duality provided by the Galois connection rather than by classical complementation.

Definition 11 (Conjugates). Let α be an image operator (generated by some relation R) of sort type $\sigma(\alpha) = (\vec{i}_j; i_{n+1})$ and $\overline{\alpha}$ the closure of its restriction to Galois sets in each argument place, as defined above. A function $\overline{\gamma}^k$ on Galois sets, of sort type $\sigma(\overline{\gamma}^k) = (\vec{i}_j [\overline{i_{n+1}}]_k; \overline{i}_k) = (i_1,\ldots,i_{k-1}, \overline{i_{n+1}}, i_{k+1}, \ldots, i_n; \overline{i}_k)$ (where recall that $\overline{i}_j = \partial$ if $i_j = 1$ and $\overline{i}_j = 1$ when $i_j = \partial$) is a *conjugate* of $\overline{\alpha}$ at the k-th argument place (or a k-conjugate) iff the following condition holds

(5) $$\overline{\alpha}(\vec{F}) \subseteq G \text{ iff } \overline{\gamma}^k(\vec{F}[G']_k) \subseteq F_k',$$

for all Galois sets $F_j \in \mathcal{G}(Z_{i_j})$ and $G \in \mathcal{G}(Z_{i_{n+1}})$.

It follows from the definition of a conjugate function that $\overline{\gamma}$ is a k-conjugate of $\overline{\alpha}$ iff $\overline{\alpha}$ is one of $\overline{\gamma}$ and we thus call $\overline{\alpha}, \overline{\gamma}$ k-conjugates. Note that the priming notation for both maps of the duality $(\ )^{\pm} \colon \mathcal{G}(X) \simeq \mathcal{G}(Y)^\partial \colon {}^{\pm}(\ )$ packs together, in one form, four distinct cases (due to sorting) of conjugacy.

Example 12. In the case of a ternary relation R^{111} of the indicated sort type, an image operator $\alpha_R = \odot \colon \mathcal{P}(X) \times \mathcal{P}(X) \longrightarrow \mathcal{P}(X)$ is generated. Designate the closure of its restriction to Galois stable sets by $\overline{\odot} \colon \mathcal{G}(X) \times \mathcal{G}(X) \longrightarrow \mathcal{G}(X)$. Then $\overline{\alpha} = \overline{\odot}$ is of sort type $\sigma(\overline{\odot}) = (1,1;1)$. If $\overline{\gamma}_R^2 = \triangleright \colon \mathcal{G}(X) \times \mathcal{G}(Y) \longrightarrow \mathcal{G}(Y)$, with $\sigma(\triangleright) = (1,\partial;\partial)$, then $\overline{\odot}, \triangleright$ are *conjugates* iff for any Galois stable sets $A, F, C \in \mathcal{G}(X)$ it holds that $A \overline{\odot} F \subseteq C$ iff $A \triangleright C' \subseteq F'$.

We point out that, given an operator $\triangleright \colon \mathcal{G}(X) \times \mathcal{G}(Y) \longrightarrow \mathcal{G}(Y)$, if we now define $\Rightarrow \colon \mathcal{G}(X) \times \mathcal{G}(X) \longrightarrow \mathcal{G}(X)$ by $A \Rightarrow C = (A \triangleright C')' = {}^{\perp}(A \triangleright C^{\perp})$, it is immediate that $\overline{\odot}, \triangleright$ are conjugates iff $\overline{\odot}, \Rightarrow$ are residuated. In other words

$$A \overline{\odot} F \subseteq C \text{ iff } A \triangleright C' \subseteq F' \text{ iff } F \subseteq A \Rightarrow C.$$

Lemma 13. *The following are equivalent.*
1. $\overline{\alpha}_R$ *distributes over any joins of Galois sets at the k-th argument place.*
2. $\overline{\alpha}_R$ *has a k-conjugate $\overline{\gamma}_R^k$ defined on Galois sets by*

$$\overline{\gamma}_R^k(\vec{F}) = \bigcap \{G \colon \overline{\alpha}_R(\vec{F}[G']_k) \subseteq F_k'\}.$$

3. $\overline{\alpha}_R$ *has a k-residual $\overline{\beta}_R^k$ defined on Galois sets by*

$$\overline{\beta}_R^k(\vec{F}[G]_k) = (\overline{\gamma}_R^k(\vec{F}[G']_k))' = \bigvee \{G' \colon \overline{\alpha}_R(\vec{F}[G']_k) \subseteq F_k'\}.$$

Proof. Existence of a k-residual is equivalent to distribution over arbitrary joins and the residual is defined by

$$\overline{\beta}_R^k(\ldots, F_{k-1}, H, F_{k+1}, \ldots) = \bigvee \{G \colon \overline{\alpha}_R(\ldots, F_{k-1}, G, F_{k+1}, \ldots) \subseteq H\}.$$

We show that the distributivity assumption (1) implies that (2) and (3) are equivalent, i.e., that

$$\overline{\alpha}_R(\vec{F}[G]_k) \subseteq H \text{ iff } \overline{\gamma}_R^k(\vec{F}[H']_k) \subseteq G' \text{ iff } G \subseteq \overline{\beta}_R^k(\vec{F}[H]_k).$$

We illustrate the proof for the unary case only, as the other parameters remain idle in the argument.

Assume $\overline{\alpha}_R(G) \subseteq H$ and let $\overline{\gamma}_R(H') = \bigcap \{E \colon \overline{\alpha}_R(E') \subseteq H\}$, a Galois set by definition, given that G, H, E are assumed to be Galois sets. Then G' is in the set whose intersection is taken. Hence, $\overline{\gamma}_R(H') \subseteq G'$ follows from the definition of $\overline{\gamma}_R$. It also follows by definition that $G \subseteq \overline{\beta}_R(H) = (\overline{\gamma}_R(H'))'$.

Assuming $G \subseteq \overline{\beta}_R(H)$ we obtain by definition that $G \subseteq (\overline{\gamma}_R(H'))'$, hence, $G \subseteq \bigvee \{E' \colon \overline{\alpha}_R(E') \subseteq H\}$, using the definition of $\overline{\gamma}_R$ and duality. Hence by the distributivity assumption $\overline{\alpha}_R(G) \subseteq \bigvee \{\overline{\alpha}_R(E') \colon \overline{\alpha}_R(E') \subseteq H\} \subseteq H$. This establishes that $\overline{\alpha}_R(G) \subseteq H$ iff $\overline{\gamma}_R(H') \subseteq G'$ iff $G \subseteq \overline{\beta}_R(H)$, as desired. ◁

Definition 14. We let $\beta_{R/}^k$ be the restriction of $\overline{\beta}_R^k$ of equation (3) to Galois sets, according to its sort type, explicitly defined by (6):

(6) $\qquad \beta_{R/}^k(\vec{E}[G]_k) = \bigcup \{F \in \mathcal{G}(Z_{i_k}) \colon \alpha_R(\vec{E}[F]_k) \subseteq G\}.$

Theorem 15. *If $\overline{\alpha}_R$ is residuated in the k-th argument place, then $\beta_{R/}^k$ is its residual and $\beta_{R/}^k(\vec{E}[G]_k)$ is a Galois set, i.e., the union in equation (6) is actually a join in $\mathcal{G}(Z_{i_k})$.*

Proof. It suffices to argue the unary case only. We have that $\beta_{R/}(G) = \bigcup\{F : \alpha_R(F) \subseteq G\}$, for Galois sets F, G.

Note first that $\overline{\alpha}_R(F) \subseteq G$ iff $F \subseteq \beta_{R/}(G)$. Left-to-right is obvious by definition and by the fact that for a Galois set G and any set U, $U'' \subseteq G$ iff $U \subseteq G$. If $F \subseteq \beta_{R/}(G) \subseteq \beta_R(G)$, then by residuation $\alpha_R(F) \subseteq G$. Given that G is a Galois set, it follows that $\overline{\alpha}_R(F) \subseteq G$.

If indeed $\overline{\alpha}_R$ is residuated on Galois sets with a map $\overline{\beta}_R$, then the residual is defined by $\overline{\beta}_R(G) = \bigvee\{F : \overline{\alpha}_R(F) \subseteq G\} = \bigvee\{F : \alpha_R(F) \subseteq G\}$ and this is precisely the closure of $\beta_{R/}(G) = \bigcup\{F : \alpha_R(F) \subseteq G\}$. But in that case we obtain $F \subseteq \overline{\beta}_R(G)$ iff $\overline{\alpha}_R(F) \subseteq G$ iff $\alpha_R(F) \subseteq G$ iff $F \subseteq \beta_{R/}(G)$ and setting $F = \overline{\beta}_R(G)$ it follows that $\overline{\beta}_R(G) \subseteq \beta_{R/}(G) \subseteq \overline{\beta}_R(G)$. ◁

Lemma 16. $\beta_{R/}^k$ *is defined equivalently by (7) and by (8).*

(7) $\quad\quad\quad \beta_{R/}^k(\vec{E}[G]_k) = \bigcup\{\Gamma u \in \mathcal{G}(Z_{i_k}) : \alpha_R(\vec{E}[\Gamma u]_k) \subseteq G\}$

(8) $\quad\quad\quad \beta_{R/}^k(\vec{E}[G]_k) = \{u \in Z_{i_k} : \alpha_R(\vec{E}[\Gamma u]_k) \subseteq G\}$

Proof. $\beta_{R/}^k$ is defined by equation (6), so if $u \in \beta_{R/}^k(\vec{E}[G]_k)$, let $F \in \mathcal{G}(Z_{i_k})$ be such that $u \in F$ and $\alpha_R(\vec{E}[F]_k) \subseteq G$. Then $\Gamma u \subseteq F$ and by monotonicity of α_R we have $\alpha_R(\vec{E}[\Gamma u]_k) \subseteq \alpha_R(\vec{E}[F]_k) \subseteq G$ and this establishes the left-to-right inclusion for the first identity of the lemma. The converse inclusion is obvious since Γu is a Galois set.

For the second identity, the right-to-left inclusion is obvious. Now if u is such that $\alpha_R(\vec{E}[\Gamma u]_k) \subseteq G$ and $u \preceq w$, then $\Gamma w \subseteq \Gamma u$ and then by monotonicity of α_R it follows that $\alpha_R(\vec{E}[\Gamma w]_k) \subseteq \alpha_R(\vec{E}[\Gamma u]_k) \subseteq G$.

This shows that $\bigcup\{\Gamma u \in \mathcal{G}(Z_{i_k}) : \alpha_R(\vec{E}[\Gamma u]_k) \subseteq G\}$ is contained in the set $\{u \in Z_{i_k} : \alpha_R(\vec{E}[\Gamma u]_k) \subseteq G\}$, and given the first part of the lemma, the second identity obtains as well. ◁

Definition 17 (Conjugate Relations). Let $\mathfrak{F} = (X, \bot, Y, R, S)$, where $\sigma(R) = (i_{n+1}; i_1 \cdots i_k \cdots i_n)$, $\sigma(S) = (t_{n+1}; t_1 \cdots t_k \cdots t_n)$, where $t_{n+1} = \overline{i_k}, t_k = \overline{i_{n+1}}$ and for $j \notin \{k, n+1\}$, $t_j = i_j$. Let α_R and η_S be the generated image operators and $\overline{\alpha}_R, \overline{\eta}_S$ be the closures of their respective restrictions to Galois sets.

The relations R, S will be called *k-conjugate relations* iff the Galois set operators $\overline{\alpha}_R, \overline{\eta}_S$ are *k*-conjugates (Definition 11), i.e., just in case (given that G, F'_k are Galois sets) $\alpha_R(\vec{F}) \subseteq G$ iff $\eta_S(\vec{F}[G']_k) \subseteq F'_k$.

Lemma 18. *Let* $\mathfrak{F} = (X, \bot, Y, R, S)$, *where* $\sigma(R) = (i_{n+1}; i_1 \cdots i_k \cdots i_n)$, $\sigma(S) = (t_{n+1}; t_1 \cdots t_k \cdots t_n)$, *where* $t_{n+1} = \overline{i_k}, t_k = \overline{i_{n+1}}$ *and for* $j \notin \{k, n+1\}$, $t_j = i_j$. *Assume that the k-th sections of the Galois dual relation* R' *of* R *are Galois sets. Let* T *be the relation defined, for* $w \in Z_{\overline{i_k}}$, *by* $vT\vec{p}[w]_k$ *iff* $w \in (vR'\vec{p}[\_]_k)'$ *iff* $\forall u \in F_k (vR'\vec{p}[u]_k \longrightarrow u|w)$.

If the constraint (9) below holds in the frame, then R and S are k-conjugates.

(9) $\quad\quad\quad \forall v \in Z_{\overline{i_{n+1}}} \forall \vec{p}[\_]_k \in \vec{Z}_{i_j}[\_]_k \forall w \in Z_{\overline{i_k}} (vT\vec{p}[w]_k \leftrightarrow wS\vec{p}[v]_k)$

Proof. We have

$\alpha_R(\vec{F}) \subseteq G \quad$ iff $\quad \bigcup_{\vec{p} \in \vec{F}} R\vec{p} \subseteq G \quad$ iff $\quad \forall \vec{p}\,(\vec{p} \in \vec{F} \longrightarrow (R\vec{p} \subseteq G))$

iff $\forall \vec{p} \, (\vec{p} \in \vec{F} \longrightarrow (G' \subseteq R'\vec{p}))$
iff $\forall \vec{p} \, (\vec{p} \in \vec{F} \longrightarrow \forall v \in Z_{\overline{i_{n+1}}} (G|v \longrightarrow vR'\vec{p}))$
iff $\forall \vec{p} \, \forall v \in Z_{\overline{i_{n+1}}} (\vec{p}[\_]_k \in \vec{F}[\_]_k \land p_k \in F_k \land G|v \longrightarrow vR'\vec{p}[p_k]_k)$
iff $\forall \vec{p} \, \forall v \in Z_{\overline{i_{n+1}}} (\vec{p}[\_]_k \in \vec{F}[\_]_k \land G|v \longrightarrow (p_k \in F_k \longrightarrow vR'\vec{p}[p_k]_k))$
iff $\forall \vec{p}[\_]_k \forall v \in Z_{\overline{i_{n+1}}} (\vec{p}[\_]_k \in \vec{F}[\_]_k \land G|v \longrightarrow (F_k \subseteq vR'\vec{p}[\_]_k))$
(using the hypothesis that the k-th sections of R' are Galois sets)
iff $\forall \vec{p}[\_]_k \forall v \in Z_{\overline{i_{n+1}}} (\vec{p}[\_]_k \in \vec{F}[\_]_k \land G|v \longrightarrow ((vR'\vec{p}[\_]_k)' \subseteq F'_k))$
iff $\forall \vec{p}[\_]_k \forall v \in Z_{\overline{i_{n+1}}} \left(\vec{p}[\_]_k \in \vec{F}[\_]_k \land G|v \longrightarrow \forall w \in Z_{\overline{i_k}} (vT\vec{p}[w]_k \longrightarrow F_k|w)\right)$
iff $\forall \vec{p}[\_]_k \forall v \in Z_{\overline{i_{n+1}}} \forall w \in Z_{\overline{i_k}} \left(vT\vec{p}[w]_k \land \vec{p}[\_]_k \in \vec{F}[\_]_k \land G|v \longrightarrow F_k|w\right)$

On the other hand, we have

$\eta_S(\vec{F}[G']_k) \subseteq F'_k$ iff $\bigcup_{\vec{p}[v]_k \in \vec{F}[G']_k} S\vec{p}[v]_k \subseteq F'_k$
iff $\forall \vec{p}[\_]_k \forall v \in Z_{\overline{i_{n+1}}} \forall w \in Z_{\overline{i_k}} \left(wS\vec{p}[v]_k \land \vec{p}[\_]_k \in \vec{F}[\_]_k \land G|v \longrightarrow F_k|w\right)$

and thus the claim of the lemma is proved. ◁

Theorem 19. *Let $\mathfrak{F} = (X, \bot, Y, R)$ be a frame with an $(n+1)$-ary sorted relation, of some sort $\sigma(R) = (i_{n+1}; i_1 \cdots i_n)$. If for any $w \in Z_{\overline{i_{n+1}}}$ and any $(n-1)$-tuple $\vec{p}[\_]_k$ with $p_j \in Z_{i_j}$, for each $j \in \{1, \ldots, n\} \setminus \{k\}$, the sections $wR'\vec{p}[\_]_k$ of the Galois dual relation R' of R are Galois sets, then $\overline{\alpha}_R$ distributes at the k-th argument place over arbitrary joins in $\mathcal{G}(Z_{i_k})$.*

Proof. Define the relation T from R as in the statement of Lemma 18,

$$vT\vec{p}[w]_k \text{ iff } w \in (vR'\vec{p}[\_]_k)'.$$

Then use equation (9), repeated below, as a definition for a relation S

$$\forall v \in Z_{\overline{i_{n+1}}} \forall \vec{p}[\_]_k \in \vec{Z}_{i_j}[\_]_k \forall w \in Z_{\overline{i_k}} (vT\vec{p}[w]_k \leftrightarrow wS\vec{p}[v]_k).$$

Note that the sort type of S, as defined, is $\sigma(S) = (t_{n+1}; t_1 \cdots t_k \cdots t_n)$, where $t_{n+1} = \overline{i_k}, t_k = \overline{i_{n+1}}$ and for $j \notin \{k, n+1\}$, $t_j = i_j$. By the proof of Lemma 18, the relations R and S are k-conjugates. Consequently, by Lemma 13, $\overline{\alpha}_R$ distributes at the k-th argument place over arbitrary joins in $\mathcal{G}(Z_{i_k})$ and it has a k-residual which, by Theorem 15, is precisely the restriction to Galois sets $\beta_{R/}^k$ (defined by equation (6), equivalently by Lemma 16) of the k-residual β_R^k of the image operator α_R. ◁

By composition with the Galois connection, single-sorted operators $\overline{\alpha}_R^1, \overline{\alpha}_R^\partial$ can be obtained on $\mathcal{G}(X)$ and $\mathcal{G}(Y)$, respectively. Given that the Galois connection is a duality between Galois stable and co-stable sets, completely normal lattice operators (dual to each other) are obtained on $\mathcal{G}(X)$ and $\mathcal{G}(Y)$, respectively. Therefore we have proven the following result.

Corollary 20. *Let $\mathfrak{F} = (X, \bot, Y, (R_p)_{p \in P})$ be a polarity with relations indexed in some set P, of sort types σ_p $(p \in P)$ and such that every section of the Galois dual relations R'_p $(p \in P)$ is a Galois set. Then the full complex algebra $\mathfrak{F}^+$ of $\mathfrak{F}$ is a normal lattice expansion where a relation of sort type $(i_{n+1}; \vec{i}_j)$ determines a completely normal lattice operator of distribution type $(\vec{i}_j; i_{n+1})$.*

3. Representation of Normal Lattice Expansions

A bounded lattice expansion is a structure $\mathcal{L} = (L, \leq, \wedge, \vee, 0, 1, \mathcal{F}_1, \mathcal{F}_\partial)$, where $\mathcal{F}_1$ consists of normal lattice operators f of distribution type $\delta(f) = (\vec{i_j}; 1)$ (i.e., of output type 1), while $\mathcal{F}_\partial$ consists of normal lattice operators h of distribution type $\delta(h) = (\vec{i_j}; \partial)$ (i.e., of output type ∂). For representation purposes, nothing depends on the size of the operator families $\mathcal{F}_1$ and $\mathcal{F}_\partial$ and we may as well assume that they contain a single member, say $\mathcal{F}_1 = \{f\}$ and $\mathcal{F}_\partial = \{h\}$. In addition, the representation argument is uniform for operators of any arity, so we may assume they are both n-ary.

3.1. Canonical Frame Construction.
The canonical frame is constructed as follows, based on [32; 33; 24; 25].

First, the base polarity $\mathfrak{F} = (\text{Filt}(\mathcal{L}), \perp, \text{Idl}(\mathcal{L}))$ consists of the sets $X = \text{Filt}(\mathcal{L})$ of filters and $Y = \text{Idl}(\mathcal{L})$ of ideals of the lattice and the relation $\perp \subseteq \text{Filt}(\mathcal{L}) \times \text{Idl}(\mathcal{L})$ is defined by $x \perp y$ iff $x \cap y \neq \emptyset$, while the representation map ζ_1 sends a lattice element $a \in L$ to the set of filters that contain it, $\zeta_1(a) = \{x \in X : a \in x\} = \{x \in X : x_a \subseteq x\} = \Gamma x_a$. Similarly, a co-represenation map ζ_∂ is defined by $\zeta_\partial(a) = \{y \in Y : a \in y\} = \{y \in Y : y_a \subseteq y\} = \Gamma y_a$. It is easily seen that $(\zeta_1(a))' = \zeta_\partial(a)$ and, similarly, $(\zeta_\partial(a))' = \zeta_1(a)$. The images of ζ_1, ζ_∂ are precisely the families (sublattices of $\mathcal{G}(X), \mathcal{G}(Y)$, respectively) of clopen elements of $\mathcal{G}(X), \mathcal{G}(Y)$, since clearly $\Gamma x_a = {}^\perp\{y_a\}$ and $\Gamma y_a = \{x_a\}^\perp$. For further details the reader is referred to [32; 33].

Second, for each normal lattice operator a relation is defined, such that if $\delta = (\vec{i_j}; i_{n+1})$ is the distribution type of the operator, then $\sigma = (i_{n+1}; \vec{i_j})$ is the sort type of the relation. Without loss of generality, we have restricted to the families of operators $\mathcal{F}_1 = \{f\}$ and $\mathcal{F}_\partial = \{h\}$, so that we shall define two corresponding relations R, S of respective sort types $\sigma(R) = (1; i_1 \cdots i_n)$ and $\sigma(S) = (\partial; t_1 \cdots t_n)$, where for each j, i_j and t_j are in $\{1, \partial\}$. In other words,

$$R \subseteq X \times \prod_{j=1}^{n} Z_{i_j} \qquad S \subseteq Y \times \prod_{j=1}^{n} Z_{t_j}.$$

To define the relations, we use the point operators introduced in [24] (see also [25]). In the generic, case we examine, we need to define two sorted operators

$$\widehat{f} : \prod_{j=1}^{n} Z_{i_j} \longrightarrow Z_1 \qquad \widehat{h} : \prod_{j=1}^{n} Z_{t_j} \longrightarrow Z_\partial \qquad \text{(recall that } Z_1 = X, Z_\partial = Y\text{)}.$$

Assuming for the moment that the point operators have been defined, the canonical relations R, S are defined by

(10) $\qquad xR\vec{u} \quad \text{iff} \quad \widehat{f}(\vec{u}) \subseteq x \quad \text{(for } x \in X \text{ and } \vec{u} \in \prod_{j=1}^{n} Z_{i_j})$

$\qquad yS\vec{v} \quad \text{iff} \quad \widehat{h}(\vec{v}) \subseteq y \quad \text{(for } y \in Y \text{ and } \vec{v} \in \prod_{j=1}^{n} Z_{t_j})$

Returning to the point operators and letting x_e, y_e be the principal filter and principal ideal, respectively, generated by a lattice element e, these are uniformly defined as follows, for $\vec{u} \in \prod_{j=1}^{n} Z_{i_j}$ and $\vec{v} \in \prod_{j=1}^{n} Z_{t_j}$,

$$\widehat{f}(u_1,\ldots,u_n) = \bigvee\{x_{f(a_1,\ldots,a_n)}: \bigwedge_{j=1}^{n}(a_j \in u_j)\} = \bigvee\{x_{f(\vec{a})}: \vec{a} \in \vec{u}\}$$

(11) $\quad \widehat{h}(v_1,\ldots,v_n) = \bigvee\{y_{h(a_1,\ldots,a_n)}: \bigwedge_{j=1}^{n}(a_j \in v_j)\} = \bigvee\{y_{h(\vec{a})}: \vec{a} \in \vec{v}\}$

In other words, $\widehat{f}(\vec{u})$ is the filter generated by the set $\{f(\vec{a}): \vec{a} \in \vec{u}\}$, and similarly, $\widehat{h}(\vec{v})$ is the ideal generated by the set $\{h(\vec{a}): \vec{a} \in \vec{v}\}$.

Example 21 (FL$_{ew}$). We consider as an example the case of associative, commutative, integral residuated lattices $\mathcal{L} = (L, \leq, \wedge, \vee, 0, 1, \circ, \rightarrow)$, the algebraic models of **FL$_{ew}$** (the associative full Lambek calculus with exchange and weakening), also referred to in the literature as full **BCK**. By residuation of $\circ, \rightarrow$, the distribution types of the operators are $\delta(\circ) = (1,1;1)$ and $\delta(\rightarrow) = (1,\partial;\partial)$. Let $(\text{Filt}(\mathcal{L}), \pm, \text{Idl}(\mathcal{L}))$ be the canonical frame of the bounded lattice $(L, \leq, \wedge, \vee, 0, 1)$. Designate the corresponding canonical point operators by $\oplus$ and $\leadsto$, respectively. They are defined by (11)

$$x \oplus z = \bigvee\{x_{a \circ c}: a \in x \wedge c \in z\} \in \text{Filt}(\mathcal{L}) \qquad (x, z \in \text{Filt}(\mathcal{L}))$$

$$x \leadsto v = \bigvee\{y_{a \rightarrow c}: a \in x \wedge c \in v\} \in \text{Idl}(\mathcal{L}) \qquad (x \in \text{Filt}(\mathcal{L}), v \in \text{Idl}(\mathcal{L}))$$

where recall that we write x_e, y_e for the principal filter and ideal, respectively, generated by the lattice element e, so that $x \oplus z \in \text{Filt}(\mathcal{L})$, while $(x \leadsto v) \in \text{Idl}(\mathcal{L})$.

The relations $R^{111}, S^{\partial 1 \partial}$ are then defined by

$$uR^{111}xz \text{ iff } x \oplus z \subseteq u \qquad\qquad yS^{\partial 1 \partial}xv \text{ iff } (x \leadsto v) \subseteq y$$

of sort types $\sigma(R) = (1;11)$ and $\sigma(S) = (\partial;1\partial)$. The canonical **FL$_{ew}$**-frame is therefore the structure $\mathfrak{F} = (\text{Filt}(\mathcal{L}), \pm, \text{Idl}(\mathcal{L}), R^{111}, S^{\partial 1 \partial})$.

3.2. Properties of the Canonical Frame.

Lemma 22. *The following hold for the canonical frame.*

1. *The frame is separated.*
2. *For $\vec{u} \in \prod_{j=1}^{n} Z_{i_j}$ and $\vec{v} \in \prod_{j=1}^{n} Z_{t_j}$, the sections $R\vec{u}$ and $S\vec{v}$ are closed elements of $\mathcal{G}(X)$ and $\mathcal{G}(Y)$, respectively.*
3. *For $x \in X, y \in Y$, the n-ary relations xR, yS are decreasing in every argument place.*

Proof. For 1, just note that the ordering $\preceq$ is set-theoretic inclusion (of filters, and of ideals, respectively), hence separation of the frame is immediate.

For 2, by the definition of the relations, $R\vec{u} = \{x: \widehat{f}(\vec{u}) \subseteq x\} = \Gamma(\widehat{f}(\vec{u}))$ is a closed element of $\mathcal{G}(X)$ and similarly for $S\vec{v}$.

For 3, if $w \subseteq u_k$, then $\{x_{f(a_1,\ldots,a_n)}: a_k \in w \wedge \bigwedge_{j \neq k}(a_j \in u_j)\}$ is a subset of the set $\{x_{f(a_1,\ldots,a_n)}: \bigwedge_j (a_j \in u_j)\}$, hence taking joins it follows that $\widehat{f}(\vec{u}[w]_k) \subseteq \widehat{f}(\vec{u})$. By definition, if $xR\vec{u}$ holds, then we obtain $\widehat{f}(\vec{u}[w]_k) \subseteq \widehat{f}(\vec{u}) \subseteq x$, hence $xR\vec{u}[w]_k$ holds as well. Similarly, for the relation S. ◁

Lemma 23. *In the canonical frame, $xR\vec{u}$ holds iff $\forall \vec{a} \in L^n (\vec{a} \in \vec{u} \longrightarrow f(\vec{a}) \in x)$. Similarly, $yS\vec{v}$ holds iff $\forall \vec{a} \in L^n (\vec{a} \in \vec{v} \longrightarrow h(\vec{a}) \in y)$.*

Proof. By definition $xR\vec{u}$ holds iff $\widehat{f}(\vec{u}) \subseteq x$, where $\widehat{f}(\vec{u})$, by its definition (11) is the filter generated by the elements $f(\vec{a})$, for $\vec{a} \in \vec{u}$, hence, clearly $\vec{a} \in \vec{u}$ implies $f(\vec{a}) \in x$. Similarly, for the relation S. ◁

Lemma 24. *Where R', S' are the Galois dual relations of the canonical relations R, S, $yR'\vec{u}$ holds iff $\widehat{f}(\vec{u}) \perp y$ iff $\exists \vec{b} \, (\vec{b} \in \vec{u} \wedge f(\vec{b}) \in y)$. Similarly, $xS'\vec{v}$ holds iff $x \perp \widehat{h}(\vec{v})$ iff $\exists \vec{e} \, (\vec{e} \in \vec{v} \wedge h(\vec{e}) \in x)$.*

Proof. By definition of the Galois dual relation, $yR'\vec{u}$ holds iff for all $x \in X$, if $xR\vec{u}$ obtains, then $x \perp y$. By definition of the canonical relation R, for any $x \in X$, $xR\vec{u}$ holds iff $\widehat{f}(\vec{u}) \subseteq x$ and thereby $\widehat{f}(\vec{u})R\vec{u}$ always obtains. Hence, $yR'\vec{u}$ is equivalent to $\forall x \in X \, (\widehat{f}(\vec{u}) \subseteq x \longrightarrow x \cap y \neq \emptyset)$, from which it follows that $\widehat{f}(\vec{u}) \perp y$ iff $yR'\vec{u}$ obtains.

To show that $yR'\vec{u}$ holds iff $\exists \vec{a} \, (\vec{a} \in \vec{u} \wedge f(\vec{a}) \in y)$, since the direction from right to left is trivially true, assume $yR'\vec{u}$, or equivalently, by the argument given above, assume that $\widehat{f}(\vec{u}) \perp y$, i.e., $\widehat{f}(\vec{u}) \cap y \neq \emptyset$ and let $e \in \widehat{f}(\vec{u}) \cap y$. By $e \in \widehat{f}(\vec{u})$ and definition of $\widehat{f}(\vec{u})$ as the filter generated by the set $\{f(\vec{a}) : \vec{a} \in \vec{u}\}$, let $\vec{a}^1, \ldots, \vec{a}^s$, for some positive integer s, be n-tuples of lattice elements (where $\vec{a}^r = (a_1^r, \ldots, a_n^r)$, for $1 \leq r \leq s$) such that $f(\vec{a}^1) \wedge \cdots \wedge f(\vec{a}^s) \leq e$ and $a_j^r \in u_j$ for each $1 \leq r \leq s$ and $1 \leq j \leq n$. Recall that the distribution type of f is $\delta(f) = (i_1, \ldots, i_n; 1)$, where for $j = 1, \ldots, n$ we have $i_j \in \{1, \partial\}$ and define elements $b_1, \ldots, b_n$ as follows.

$$b_j = \begin{cases} a_j^1 \wedge \cdots \wedge a_j^s & \text{if } i_j = 1; \\ a_j^1 \vee \cdots \vee a_j^s & \text{if } i_j = \partial. \end{cases}$$

When $i_j = 1$, f is monotone at the j-th argument place, u_j is a filter and $b_j \leq a_j^r \in u_j$, for all $r = 1, \ldots, s$, so that $b_j = a_j^1 \wedge \cdots \wedge a_j^s \in u_j$. Similarly, when $i_{j'} = \partial$, f is antitone at the j'-th argument place, while $u_{j'}$ is an ideal, so that $b_{j'} = a_{j'}^1 \vee \cdots \vee a_{j'}^s \in u_{j'}$. This shows that $\vec{b} \in \vec{u}$ and it remains to show that $f(\vec{b}) \in y$. We argue that $f(\vec{b}) \leq f(\vec{a}^1) \wedge \cdots \wedge f(\vec{a}^s) \leq e$ and the desired conclusion follows by the fact that $e \in y$, an ideal.

For any $1 \leq r \leq s$, let $\vec{a}^r[b_j]_j^{i_j=1}$ be the result of replacing a_j^r by b_j in the tuple $\vec{a}^r$ and in every position j from 1 to n such that $i_j = 1$ in the distribution type of f. Since $b_j \leq a_j^r$ and f is monotone at any such j-th argument place, it follows that $f(\vec{a}^r[b_j]_j^{i_j=1}) \leq f(\vec{a}^r)$, for all $1 \leq r \leq s$.

In addition, for any $1 \leq r \leq s$, let $\vec{a}^r[b_j]_j^{i_j=1}[b_{j'}]_{j'}^{i_{j'}=\partial}$ be the result of replacing $a_{j'}^r$ by $b_{j'}$ in the tuple $\vec{a}^r[b_j]_j^{i_j=1}$ and in every position j' from 1 to n such that $i_{j'} = \partial$ in the distribution type of f. Since $b_{j'} \geq a_{j'}^r$ and f is antitone at any such j'-th argument place, it follows that $f(\vec{a}^r[b_j]_j^{i_j=1}[b_{j'}]_{j'}^{i_{j'}=\partial}) \leq f(\vec{a}^r[b_j]_j^{i_j=1}) \leq f(\vec{a}^r)$, for all $1 \leq r \leq s$. Since $\vec{a}^r[b_j]_j^{i_j=1}[b_{j'}]_{j'}^{i_{j'}=\partial} = \vec{b}$ we obtain that

$$f(\vec{b}) = f(\vec{a}^r[b_j]_j^{i_j=1}[b_{j'}]_{j'}^{i_{j'}=\partial}) \leq f(\vec{a}^r[b_j]_j^{i_j=1}) \leq f(\vec{a}^1) \wedge \cdots \wedge f(\vec{a}^s) \leq e,$$

hence, $f(\vec{b}) \in y$ and this completes the proof, as far as the relation R is concerned.
The argument for the relation S is similar, and can be safely left to the reader. ◁

Lemma 25. *In the canonical frame, all sections of the Galois dual relations R', S' of the canonical relations R, S are Galois sets.*

Proof. There are two cases to handle, one for each of the relations R', S', with two subcases for each one, depending on whether i_k is 1, or ∂.

Case of the relation R': We have $R'\vec{u} = (R\vec{u})'$, by definition, so the section $R'\vec{u}$ is a Galois (co-stable) set. It remains to be shown that for any $y \in Y$ and $\vec{u}[\_]_k$, the k-th section $yR'\vec{u}[\_]_k$ is a Galois set, for any $1 \leq k \leq n$. There are two subcases to consider, accordingly as $i_k = 1$, or $i_k = \partial$ and recall that $\delta(f) = (i_1,\ldots,i_n;1)$. Hence if $i_k = 1$, then f is monotone and it distributes over finite joins at the k-th argument place and if $i_k = \partial$, then f is antitone and it co-distributes at the k-th argument place over finite meets (turning them into joins).

Subcase $i_k = 1$: Then $yR'\vec{u}[\_]_k \subseteq X = \text{Filt}(\mathcal{L})$.

Let v be the ideal generated by the set $V = \{b \in L : \exists \vec{a}[\_]_k \ f(\vec{a}[b]_k) \in y\}$. For any $x \in X$ such that $yR'\vec{u}[x]_k$ holds, by Lemma 24, we have $\widehat{f}(\vec{u}[x]_k) \perp y$, equivalently, $\exists \vec{a}[\_]_k \exists b \, (\vec{a}[b]_k \in \vec{u}[x]_k \land f(\vec{a}[b]_k) \in y)$. Thus, $b \in x \cap v$ and so $yR'\vec{u}[\_]_k \perp v$.

We assume $z \in (yR'\vec{u}[\_]_k)''$ and we need to show that $yR'\vec{u}[z]_k$. The assumption implies that $z \perp v$, i.e., for some lattice element e we have $e \in z \cap v$. By $e \in v$, let $b_1,\ldots,b_s \in V \subseteq v$, for some positive integer s, be elements such that $e \leq b_1 \vee \cdots \vee b_s \in v$, since v is an ideal.

Since $b_1,\ldots,b_s \in V$, there are tuples of lattice elements $\vec{c}^r[\_]_k$ such that $f(\vec{c}^r[b_r]_k) \in y$, for each $1 \leq r \leq s$. Considering the distribution type of f and as in the proof of Lemma 24, define the tuple of elements $\vec{a} = (a_1,\ldots,a_n)$ by

$$a_j = \begin{cases} c_j^1 \wedge \cdots \wedge c_j^s & \text{if } i_j = 1; \\ c_j^1 \vee \cdots \vee c_j^s & \text{if } i_j = \partial. \end{cases}$$

For each $1 \leq r \leq s$, let $\vec{c}^r[b_r]_k[a_j]_j^{i_j=1}$ be the result of replacing in $\vec{c}^r[b_r]_k$ all c_j^r by a_j whenever $i_j = 1$ in the distribution type of f. Then for each r as above and by monotonicity of f at the j-th argument place whenever $i_j = 1$ we have $f(\vec{c}^r[b_r]_k[a_j]_j^{i_j=1}) \leq f(\vec{c}^r[b_r]_k) \in y$, an ideal, hence $f(\vec{c}^r[b_r]_k[a_j]_j^{i_j=1}) \in y$. Let also $\vec{c}^r[b_r]_k[a_j]_j^{i_j=1}[a_{j'}]_{j'}^{i_{j'}=\partial}$ be the result of further replacing in $\vec{c}^r[b_r]_k[a_j]_j^{i_j=1}$ all $c_{j'}^r$ by $a_{j'}$ whenever $i_{j'} = \partial$ in the distribution type of f. Since f is antitone at the j'-th position when $i_{j'} = \partial$ (given that the output type of f is assumed to be $i_{n+1} = 1$), we obtain $f(\vec{c}^r[b_r]_k[a_j]_j^{i_j=1}[a_{j'}]_{j'}^{i_{j'}=\partial}) \leq f(\vec{c}^r[b_r]_k[a_j]_j^{i_j=1}) \leq f(\vec{c}^r[b_r]_k) \in y$ and so $f(\vec{c}^r[b_r]_k[a_j]_j^{i_j=1}[a_{j'}]_{j'}^{i_{j'}=\partial}) \in y$. But, having performed substitutions in all places (except for the k-th) $\vec{c}^r[b_r]_k[a_j]_j^{i_j=1}[a_{j'}]_{j'}^{i_{j'}=\partial} = \vec{a}[b_r]_k$, for each $1 \leq r \leq s$.

It follows from the above that, for each $1 \leq r \leq s$ we have $f(\vec{a}[b_r]_k) \in y$, an ideal, hence, also $\bigvee_{r=1}^s f(\vec{a}[b_r]_k) \in y$. By case assumption, $i_k = 1$, hence, f distributes over finite joins in the k-th argument place and then we obtain that $f(\vec{a}[b_1 \vee \cdots \vee b_s]_k) =$

$\bigvee_{r=1}^{s} f(\vec{a}[b_r]_k) \in y$. By $e \leq b_1 \vee \cdots \vee b_s$ and monotonicity of f at the k-th argument place we obtain $f(\vec{a}[e]_k) \leq f(\vec{a}[b_1 \vee \cdots \vee b_s]_k) \in y$, hence, also $f(\vec{a}[e]_k) \in y$.

Therefore, there exists $\vec{a}[e]_k \in \vec{u}[z]_k$ such that $f(\vec{a}[e]_k) \in y$ which, by Lemma 24 is equivalent to $yR'\vec{u}[z]_k$. This shows that $(yR'\vec{u}[\_]_k)'' \subseteq yR'\vec{u}[\_]_k$, so the section has been shown to be a Galois (stable) set.

The subcase $i_k = \partial$ and the case of the relation S are treated similarly, and we leave details to the interested reader. ◁

The canonical frame for a lattice expansion $\mathcal{L} = (L, \leq, \wedge, \vee, 0, 1, f, h)$, where $\delta(f) = (i_1, \ldots, i_n; 1)$ and $\delta(h) = (t_1, \ldots, t_n; \partial)$ $(i_j, t_j \in \{1, \partial\})$ is the structure $\mathcal{L}_+$, that is, $\mathfrak{F} = (\text{Filt}(\mathcal{L}), \perp, \text{Idl}(\mathcal{L}), R, S)$. By Proposition 25, the canonical relations R, S are compatible with the Galois connection generated by $\perp \subseteq X \times Y$, in the sense that all sections of their Galois dual relations are Galois sets. Set operators α_R, η_S are defined as in Section 2.2 and we let $\overline{\alpha}_R, \overline{\eta}_S$ be the closures of their restrictions to Galois sets (according to their distribution types). Note that $\overline{\alpha}_R(\vec{F}) \in \mathcal{G}(X)$, while $\overline{\eta}_S(\vec{G}) \in \mathcal{G}(Y)$, given the output types of f, h (alternatively, given the sort types of R, S).

It follows from Theorem 19 and Lemma 25, that the sorted operators $\overline{\alpha}_R, \overline{\eta}_S$ on Galois sets distribute over arbitrary joins of Galois sets (stable or co-stable, according to the sort types of R, S) in each argument place.

Note that $\overline{\alpha}_R, \overline{\eta}_S$ are sorted maps, taking their values in $\mathcal{G}(X)$ and $\mathcal{G}(Y)$, respectively. We define single-sorted maps on $\mathcal{G}(X)$ (analogously for $\mathcal{G}(Y)$) by composition with the Galois connection

(12) $\quad \overline{\alpha}_f(A_1, \ldots, A_n) \;=\; \overline{\alpha}_R(\ldots, \underbrace{A_j}_{i_j=1}, \ldots, \underbrace{A'_r}_{i_r=\partial}, \ldots) \quad (A_1, \ldots, A_n \in \mathcal{G}(X))$,

(13) $\quad \overline{\eta}_h(B_1, \ldots, B_n) \;=\; \overline{\eta}_S(\ldots, \underbrace{B_r}_{i_r=\partial}, \ldots, \underbrace{B'_j}_{i_j=1}, \ldots) \quad (B_1, \ldots, B_n \in \mathcal{G}(Y))$.

Given that the Galois connection is a duality of Galois stable and Galois co-stable sets, it follows that the distribution type of $\overline{\alpha}_f$ is that of f and that $\overline{\alpha}_f$ distributes, or co-distributes, over arbitrary joins and meets in each argument place, according to its distribution type, returning joins in $\mathcal{G}(X)$. Similarly, for $\overline{\eta}_h$. Thus, the lattice representation maps $\zeta_1 : (L, \leq, \wedge, \vee, 0, 1) \longrightarrow \mathcal{G}(X)$ and $\zeta_\partial : (L, \leq, \wedge, \vee, 0, 1) \longrightarrow \mathcal{G}(Y)$ are extended to maps $\zeta_1 : \mathcal{L} \longrightarrow \mathcal{G}(X)$ and $\zeta_\partial : \mathcal{L} \longrightarrow \mathcal{G}(Y)$ by setting

$$\zeta_1(f(a_1, \ldots, a_n)) \;=\; \overline{\alpha}_f(\zeta_1(a_1), \ldots, \zeta_1(a_n)) \;=\; \overline{\alpha}_R(\ldots, \underbrace{\zeta_1(a_j)}_{i_j=1}, \ldots, \underbrace{\zeta_\partial(a_r)}_{i_r=\partial}, \ldots)$$

(14) $\quad \zeta_\partial(f(a_1, \ldots, a_n)) \;=\; \left(\overline{\alpha}_f(\zeta_1(a_1), \ldots, \zeta_1(a_n))\right)'$

$\quad\quad\;\; \zeta_1(h(a_1, \ldots, a_n)) \;=\; \left(\overline{\eta}_h(\zeta_\partial(a_1), \ldots, \zeta_\partial(a_n))\right)'$

(15) $\quad \zeta_\partial(h(a_1, \ldots, a_n)) \;=\; \overline{\eta}_h(\zeta_\partial(a_1), \ldots, \zeta_\partial(a_n))$

It has been therefore established that there exists a map from normal lattice expansions to polarities with relations, as specified in the following concluding result.

Corollary 26. *Given a normal lattice expansion $\mathcal{L} = (L, \leq, \wedge, \vee, 0, 1, \mathcal{F}_1, \mathcal{F}_\partial)$, where $\mathcal{F}_1, \mathcal{F}_\partial$ are families of normal lattice operators of output types 1 and ∂, respectively, the dual frame $\mathcal{L}_+$ of the lattice expansion $\mathcal{L}$ is a polarity with additional relations,* where for a normal lattice operator f of distribution type $(\vec{i}_j; i_{n+1})$ the corresponding frame relation R_f is of sort type $(i_{n+1}; \vec{i}_j)$ and where all sections of its Galois dual relation R'_f are Galois sets.

3.3. Representation, Canonical Extensions and RS-Frames. A canonical lattice extension is defined in [19] as a pair (α, C) where C is a complete lattice and α is an embedding of a lattice $\mathcal{L}$ into C such that the following density and compactness requirements are satisfied.

- (density) $\alpha[\mathcal{L}]$ is *dense* in C, where the latter means that every element of C can be expressed both as a meet of joins and as a join of meets of elements in $\alpha[\mathcal{L}]$;
- (compactness) for any set A of closed elements and any set B of open elements of C, $\bigwedge A \leq \bigvee B$ iff there exist finite subcollections $A_1 \subseteq A, B_1 \subseteq B$ such that $\bigwedge A_1 \leq \bigvee B_1$,

where the *closed elements* of C are defined in [19] as the elements in the meet-closure of the representation map α and the *open elements* of C are defined dually as the join-closure of the image of α.

Proposition 27. $\mathcal{G}(X)$ *(the lattice of Galois stable subsets of the set of filters) is a canonical extension of the (bounded) lattice* $(L, \leq, \wedge, \vee, 0, 1)$.

Proof. This was shown by Gehrke and Harding in [19]. More precisely, existence of canonical extensions is proven in [19] by demonstrating that the compactness and density requirements are satisfied in the representation due to Hartonas and Dunn [33], which is precisely the representation presented in Section 3.1. ◁

Proposition 28. *The canonical representations of the normal lattice operators f, h, of respective output types* 1, ∂, *as defined by the equations* (14) *and* (15), *are the σ and π-extension (in the terminology of* [19]), *respectively, of f, h.*

Proof. In the representation of Section 3.1, the closed and open elements of $\mathcal{G}(X)$ are the sets of the form $\Gamma x (x \in X)$ and $\doteq \{y\}$ $(y \in Y)$, respectively. For a unary lattice operator $f: \mathcal{L} \longrightarrow \mathcal{L}$, its σ-extension in a canonical extension C of the lattice $\mathcal{L}$ is defined in [19] by equation (16), where K is the set of closed elements of C and O is its set of open elements.

(16) $f_\sigma(k) = \bigwedge \{f(a): k \leq a \in L\}$ $\quad f_\sigma(u) = \bigvee \{f_\sigma(k): K \ni k \leq u\}$

(17) $f_\pi(o) = \bigvee \{f(a): L \ni a \leq o\}$ $\quad f_\pi(u) = \bigwedge \{f_\pi(o): u \leq o \in O\}$

where in these definitions $\mathcal{L}$ is identified with its isomorphic image in C and $a \in L$ is then identified with its representation image.

Working concretely with the canonical extension of [33], the σ extension f_σ: $\mathcal{L}_\sigma \longrightarrow \mathcal{L}_\sigma$ of a monotone map f as in equation (16) and the dual σ-extension f_σ^∂: $\mathcal{L}_\sigma^\partial \longrightarrow \mathcal{L}_\sigma^\partial$ (not used in [19]) are defined by instantiating equation (16) in the concrete canonical extension of [33]. For $x \in X$ and $y \in Y$ and where x_e is a principal filter, y_e a principal ideal and the closed elements are precisely the principal upper sets Γu (for each of X, Y) we have, by instantiating (16),

$$\begin{aligned}
f_\sigma(\Gamma x) &= \bigwedge\{\alpha_X(fa): a \in \mathcal{L}, \Gamma x \leq \alpha_X(a)\} = \bigwedge\{\Gamma x_{fa}: \Gamma x \subseteq \Gamma x_a\} \\
&= \bigwedge\{\Gamma x_{fa}: a \in x\} = \Gamma(\bigvee\{x_{fa}: a \in x\}) \\
&= \Gamma(\widehat{f}(x)) \\
f_\sigma^\partial(\Gamma y) &= \bigwedge\{\alpha_Y(fa): a \in \mathcal{L}, \Gamma y \leq \alpha_Y(a)\} = \bigwedge\{\Gamma y_{fa}: \Gamma y \subseteq \Gamma y_a\} \\
&= \bigwedge\{\Gamma y_{fa}: a \in y\} = \Gamma(\bigvee\{y_{fa}: a \in y\})
\end{aligned}$$

Hence $f_\sigma(\Gamma x) = \Gamma(\widehat{f}(x))$, where $\widehat{f}$ is the point operator we defined, after [24], by equation (11). For an n-ary operator f, first observe that in a product $\prod_{j=1}^{n} \mathcal{G}(Z_{i_j})$, where when $i_j = \partial$ then $Z_{i_j} = Y$ and $\mathcal{G}(Z_{i_j}) = \mathcal{G}(Y)$, closed elements are n-tuples of closed elements of the factors $(\Gamma u_1, \ldots, \Gamma u_n)$. Then $f_\sigma(\Gamma u_1, \ldots, \Gamma u_n) = \Gamma(\widehat{f}(\vec{u}))$, by the same analysis. This is a sorted operator and by composition with the Galois connection we obtain (in the case examined the output type is 1) the single-sorted σ-extension $f^\sigma(\ldots, \underbrace{\Gamma u_j}_{i_j = 1}, \ldots, \underbrace{{}^\pm\{u_r\}}_{i_r = \partial}, \ldots) = \Gamma(\widehat{f}(\vec{u}))$.

For an arbitrary Galois stable set A and unary monotone f, $f^\sigma(A)$ is defined in (16) using join-density of closed elements. Hence we obtain $f^\sigma(A) = \bigvee_{x \in A} f_\sigma(\Gamma x) = \bigvee_{x \in A} \Gamma(\widehat{f}(x))$.

For an n-ary monotone map we similarly obtain that $f^\sigma(\vec{F}) = \bigvee_{\vec{u} \in \vec{F}} \Gamma(\widehat{f}(\vec{u}))$. Since $w \in \Gamma(\widehat{f}(\vec{u}))$ iff $\widehat{f}(\vec{u}) \subseteq w$ iff $wR\vec{u}$, by the way the canonical relation R was defined in equation (10), so that $\Gamma(\widehat{f}(\vec{u})) = R\vec{u}$, we obtain that (observing that $R\vec{u}$ is a Galois set, indeed a closed element of $\mathcal{G}(X)$)

$$f^\sigma(\vec{F}) = \bigvee_{\vec{u} \in \vec{F}} R\vec{u} = \left(\bigcup_{\vec{u} \in \vec{F}} R\vec{u}\right)'' = \overline{\alpha}_R(\vec{F}).$$

Note that σ-extensions, as defined in [19], are sorted maps and then a single-sorted map is obtained by composing with the Galois connection, as shown in equations (14) and (15).

The π-extension is simply the Galois image of the dual σ-extension, so there is nothing new to discuss and the proof is complete. We only note further that the way we have canonically proceeded is to represent a lattice operator with output type 1 by its σ-extension and one of output type ∂ by its π-extension. ◁

Remark 29 (Canonical Relations in the RS-Frames Approach). In modeling the Lambek calculus product operator, $\circ$, of distribution type $(1,1;1)$ (see Example 21), the canonical relation R^{111} was defined by (using the point operators) $uRxz$ iff $x \oplus z \subseteq u$, where $x \oplus z$ is the filter generated by the elements $a \circ c$, with $a \in x$ and $c \in z$. By Lemma 23, specialized to this case, this amounts to the classical definition of a canonical relation, familiar from the Boolean and distributive case, by the clause

$$uRxz \text{ iff } \forall a, c \in \mathcal{L} (a \in x \wedge c \in z \longrightarrow a \circ c \in u).$$

By Lemma 24, specialized to the particular case, $yR'xz$ holds iff $x \oplus z \perp y$ iff $\exists a, c \in \mathcal{L}(a \in x \wedge c \in z \wedge a \circ c \in y)$, which is precisely Gehrke's [18] canonical relation definition for the Lambek product operator.

This is no isolated matter, as in the RS-frames approach the choice is made to work directly with a relation that can be nevertheless defined as the Galois dual of a classically defined accessibility relation. This is also witnessed by the way Goldblatt [21] proceeds, generalizing on Gehrke's [18], to define relations and set-operators in a frame. Indeed, examining (for his case of interest) additive operators F on stable sets he defines a relation S_F by setting $yS_F\vec{z}$ iff $F(\Gamma z_1,\ldots,\Gamma z_n) \perp y$ (which in the case of a binary, completely additive operator F, is equivalent in the canonical frame to $z_1 \odot z_2 \perp y$). It is merely a matter of choice and convenience, given the purpose at hand, which relation to decide to work with. Gehrke [18] does indeed point out that instead of using the relation $S \subseteq Y \times (X \times X)$ defined as above (for the Lambek product operator), one could use a relation $R \subseteq X \times (X \times X)$, which is actually the Galois dual of S, but she does not dwell much on the matter.

Though, to the best of this author's knowledge, it has not been made explicit in the RS-frames approach how relations are to be defined corresponding to arbitrary normal lattice operators in general (but only in cases of particular examples), the relations on an RS-frame corresponding to normal lattice operators of some distribution type $(\vec{i}_j; i_{n+1})$ are the Galois duals of our canonical accessibility relations, hence they are systematically of sort type $(\overline{i_{n+1}}; \vec{i}_j)$, and operators are defined from them. For example, Goldblatt [21] (generalizing Gehrke's [18] definition for the Lambek product operator) defines from a relation $S \subseteq Y \times X^n$ an operator F_S by setting

$$(18) \quad F_S^{\bullet}(\vec{F}) = \bigcap \{S\vec{z}: \vec{z} \in \vec{F}\}$$

$$(19) \quad F_S(\vec{F}) = {}^{\perp}F_S^{\bullet}(\vec{F}) = \bigvee \{{}^{\perp}(S\vec{z}): \vec{z} \in \vec{F}\} = \bigvee_{\vec{z} \in \vec{F}} {}^{\perp}(S\vec{z}).$$

The relation R defined as the Galois dual of S, i.e., by $R\vec{z} = {}^{\perp}(S\vec{z})$ is precisely a relation of sort type $(i_{n+1}; \vec{i}_j)$ and, assuming section stability, R and S are each other's Galois dual. Therefore, we obtain

$$(20) \quad F_S(\vec{F}) = \bigvee_{\vec{z} \in \vec{F}} {}^{\perp}(S\vec{z}) = \bigvee_{\vec{z} \in \vec{F}} R\vec{z} = \left(\bigcup_{\vec{z} \in \vec{F}} R\vec{z} \right)''$$

A comparison of equations (4) and (20) reveals then that the two definitions are variants of each other.

4. CONCLUSIONS

We have argued that the two approaches, the one developed in this article (concluding and completing our previous recent work on the subject) and the RS-frames approach really only differ in whether the polarity is assumed to be separated and reduced or not. The results of this article have shown that nothing is lost by dropping these additional assumptions, as far as the semantics of logics without distribution is concerned.

There are three points of interest, however, that are worth making.

First, a Stone type duality for RS-frames (essentially for Hartung's lattice representation) has encountered difficulties, similar to these encountered in extending Urquhart's representation to a full Stone duality. In [25], we have developed a duality result for normal lattice expansions, extending the representation of [33]. In view of Goldblatt's recent proposal [21] of a notion of bounded morphisms for polarities, this result, combined with the results of this article, can be improved to a Stone duality for normal lattice expansions with bounded morphisms as the morphisms in the dual category of polarities. This project has been carried out in the sequel [31] of the present article.

Second, the approach we have presented in this article allows for relating the logic of non-distributive lattices to the sorted, residuated (poly)modal logic of polarities with relations, where the residuated pair of modal operators is generated by the complement of the Galois relation of the frame. Preliminary results in this direction have been reported in [29; 30] by this author, but the area is far from fully explored. Regarding non-distributive logics as fragments of sorted, residuated (poly)modal logics allows for importing techniques and results from modal logic in the field of logics lacking distribution.

Finally, we believe that the semantic framework presented in this article fully complements Dunn's gaggle theory project, and in an important sense it completes the project for the case of non-distributive logical calculi.

REFERENCES

[1] Allwein, G. and Dunn, J. M. (1993). Kripke models for linear logic, *Journal Symbolic Logic* **58**(2): 514–545.

[2] Bimbó, K. and Dunn, J. M. (2008). *Generalized Galois Logics: Relational Semantics of Nonclassical Logical Calculi*, Vol. 188 of *CSLI Lecture Notes*, CSLI Publications, Stanford, CA.

[3] Birkhoff, G. (1979). *Lattice Theory*, American Mathematical Society Colloquium Publications, Vol. 25, 3rd edn, American Mathematical Society, Providence, RI.

[4] Chernilovskaya, A., Gehrke, M. and Van Rooijen, L. (2012). Generalized Kripke semantics for the Lambek–Grishin calculus, *Logic Journal of the IGPL* **20**(6): 1110–1132.

[5] Conradie, W., Frittella, S., Palmigiano, A., Piazzai, M., Tzimoulis, A. and Wijnberg, N. M. (2016). Categories: How I learned to stop worrying and love two sorts, *Logic, Language, Information, and Computation*, number 9803 in *Lecture Notes in Computer Science*, Springer, Berlin, pp. 145–164.

[6] Conradie, W. and Palmigiano, A. (2019). Algorithmic correspondence and canonicity for non-distributive logics, *Annals of Pure and Applied Logic* **170**(9): 923–974.

[7] Conradie, W., Palmigiano, A., Robinson, C. and Wijnberg, N. (2020). Non-distributive logics: From semantics to meaning, in A. Rezus (ed.), *Contemporary Logic and Computing*, Vol. 1 of *Landscapes in Logic*, College Publications, London, UK, pp. 38–86.

[8] Conradie, W., Palmigiano, A. and Tzimoulis, A. (2018). Goldblatt–Thomason for LE-logics.

[9] Copeland, B. J. (2002). The genesis of possible worlds semantics, *Journal of Philosophical Logic* **31**(2): 99–137.

[10] Coumans, D., Gehrke, M. and van Rooijen, L. (2014). Relational semantics for full linear logic, *Journal of Applied Logic* **12**(1): 50–66.

[11] Dunn, J. M. (1991). Gaggle theory: An abstraction of Galois connections and residuation, with applications to negation, implication, and various logical operators, in J. van Eijck (ed.), *Logics in AI: European Workshop JELIA '90*, number 478 in *Lecture Notes in Computer Science*, Springer, Berlin, pp. 31–51.

[12] Dunn, J. M. (1993). Partial gaggles applied to logics with restricted structural rules, in K. Došen and P. Schroeder-Heister (eds.), *Substructural Logics (Tübingen, 1990)*, Vol. 2 of *Studies in Logic and Computation*, Oxford University Press, New York, pp. 63–108.

[13] Dunn, J. M., Gehrke, M. and Palmigiano, A. (2005). Canonical extensions and relational completeness of some substructural logics, *Journal of Symbolic Logic* **70**(3): 713–740.

[14] Dunn, J. M. and Hardegree, G. M. (2001). *Algebraic Methods in Philosophical Logic*, Vol. 41 of *Oxford Logic Guides*, Clarendon Press, New York.

[15] Dunn, J. M. and Zhou, C. (2005). Negation in the context of gaggle theory, *Studia Logica* **80**(2-3): 235–264.

[16] Düntsch, I., Orłowska, E., Radzikowska, A. M. and Vakarelov, D. (2004). Relational representation theorems for some lattice-based structures, *Journal of Relational Methods in Computer Science (JORMICS)* **1**: 132–160.

[17] Ganter, B. and Wille, R. (1999). *Formal Concept Analysis*, Springer-Verlag, Berlin.

[18] Gehrke, M. (2006). Generalized Kripke frames, *Studia Logica* **84**(2): 241–275.

[19] Gehrke, M. and Harding, J. (2001). Bounded lattice expansions, *Journal of Algebra* **238**(1): 345–371.

[20] Gehrke, M. and Jónsson, B. (2004). Bounded distributive lattice expansions, *Mathematica Scandinavica* **94**(1): 13–45.

[21] Goldblatt, R. (2020). Morphisms and duality for polarities and lattices with operators, *FLAP* **7**(6): 1017–1070.

[22] Goldblatt, R. I. (1974). Semantic analysis of orthologic, *Journal of Philosophical Logic* **3**(1–2): 19–35.

[23] Grishin, V. N. (1983). A generalization of a system of Ajdukiewicz and Lambek, *Studies in Nonclassical Logics and Formal Systems*, Nauka, Moscow, pp. 315–334.

[24] Hartonas, C. (1997). Duality for lattice-ordered algebras and for normal algebraizable logics, *Studia Logica* **58**(3): 403–450.

[25] Hartonas, C. (2018). Stone duality for lattice expansions, *Logic Journal of the IGPL* **26**(5): 475–504.

[26] Hartonas, C. (2019a). Discrete duality for lattices with modal operators, *Journal of Logic and Computation* **29**(1): 71–89.

[27] Hartonas, C. (2019b). Duality results for (co)residuated lattices, *Logica Universalis* **13**(1): 77–99.

[28] Hartonas, C. (2019c). Game-theoretic semantics for non-distributive logics, *Logic Journal of the IGPL* **27**(5): 718–742.

[29] Hartonas, C. (2019d). Lattice logic as a fragment of (2-sorted) residuated modal logic, *Journal of Applied Non-Classical Logics* **29**(2): 152–170.

[30] Hartonas, C. (2020). Modal translation of substructural logics, *Journal of Applied Non-Classical Logics* **30**(1): 16–49.

[31] Hartonas, C. (2021). Duality for normal lattice expansions and sorted, residuated frames with relations, *CoRR*, abs/2110.06924 .

[32] Hartonas, C. and Dunn, J. M. (1993). Duality theorems for partial orders, semilattices, Galois connections and lattices, *Technical report*, Indiana University Logic Group Technical Report IULG-93-26.

[33] Hartonas, C. and Dunn, J. M. (1997). Stone duality for lattices, *Algebra Universalis* **37**(3): 391–401.

[34] Hartonas, C. and Orłowska, E. (2019). Representation of lattices with modal operators in two-sorted frames, *Fundamenta Informaticae* **166**(1): 29–56.

[35] Hartung, G. (1992). A topological representation of lattices, *Algebra Universalis* **29**(2): 273–299.

[36] Jónsson, B. and Tarski, A. (1951). Boolean algebras with operators. I, *American Journal of Mathematics* **73**: 891–939.

[37] Jónsson, B. and Tarski, A. (1952). Boolean algebras with operators. II, *American Journal of Mathematics* **74**: 127–162.

[38] Jónsson, B. and Tsinakis, C. (1993). Relation algebras as residuated Boolean algebras, *Algebra Universalis* **30**(4): 469–478.

[39] Kripke, S. A. (1959). A completeness theorem in modal logic, *Journal of Symbolic Logic* **24**: 1–14.

[40] Kripke, S. A. (1963a). Semantical analysis of modal logic. I. Normal modal propositional calculi, *Zeitschrift für Mathematische Logik und Grundlagen der Mathematik* **9**: 67–96.

[41] Kripke, S. A. (1963b). Semantical considerations on modal logic, *Acta Philosophica Fennica* **16**: 83–94.

[42] Orłowska, E. (1984). Modal logics in the theory of information systems, *Zeitschrift für mathematische Logik und Grundlagen der Mathematik* **30**(13-16): 213–222.

[43] Priestley, H. A. (1970). Representation of distributive lattices by means of ordered stone spaces, *Bulletin of the London Mathematical Society* **2**: 186–190.

[44] Stone, M. H. (1937). Topological representation of distributive lattices and Brouwerian logics, *Časopis pro pěstování matematiky a fysiky, Část matematická* **67**: 1–25.

[45] Stone, M. H. (1938). The representation of Boolean algebras, *Bulletin of the American Mathematical Society* **44**(12): 807–816.

[46] Suzuki, T. (2014). On polarity frames: Applications to substructural and lattice-based logics, *Advances in Modal Logic, 10*, College Publications, London, UK, pp. 533–552.

[47] Urquhart, A. (1978). A topological representation theory for lattices, *Algebra Universalis* **8**(1): 45–58.

[48] Vakarelov, D. (1998). Information systems, similarity relations and modal logics, *Incomplete Information: Rough Set Analysis*, Vol. 13 of *Studies in Fuzziness and Soft Computing*, Physica, Heidelberg, pp. 492–550.

UNIVERSITY OF THESSALY, GREECE, *Email:* hartonas@uth.gr

HERBRAND AND CONTRAPOSITION-ELIMINATION THEOREMS FOR EXTENDED FIRST-ORDER BELNAP–DUNN LOGIC

Norihiro Kamide

Dedicated to the memory of professor J. Michael Dunn

ABSTRACT. Belnap–Dunn logic is known to be useful in broad areas of computer science. A useful first-order extension of Belnap–Dunn logic is required to develop an expressive inconsistency-tolerant automated theorem proving framework that can simultaneously handle indefinite and definite information. In this study, Herbrand and contraposition-elimination theorems (and other theorems) are proved for a Gentzen-type sequent calculus FBD+ for a first-order extension of De and Omori's axiomatic propositional extension BD+ of Belnap–Dunn logic, in which classical negation and classical implication are added. These fundamental theorems provide a proof-theoretic justification for developing an FBD+-based inconsistency-tolerant automated theorem proving framework that can simultaneously handle indefinite and definite information.

Keywords. Belnap–Dunn logic, Contraposition-elimination theorem, Gentzen-type sequent calculus, Herbrand theorem

1. INTRODUCTION

Belnap–Dunn logic (also known as *Belnap and Dunn's four-valued logic, first-degree entailment logic*, or *Dunn–Belnap logic*) [5; 4; 12; 11] and its extensions and generalizations [2; 13; 40; 46; 45; 39] are considered useful in broad areas of computer science, including inconsistency-tolerant reasoning, logic programming, and knowledge representation. Additional information on Belnap–Dunn logic and its applications can be found in [14; 5; 4; 12; 11; 34]. A useful first-order extension of Belnap–Dunn logic is required to develop an expressive inconsistency-tolerant automated theorem proving framework that can simultaneously handle indefinite and definite information. In this study, we prove the Herbrand and contraposition-elimination theorems (and other theorems) for the Gentzen-type sequent calculus FBD+ introduced by Kamide and Omori in [25] for an extended first-order Belnap–Dunn logic with both paraconsistent and classical negations. The target first-order logic discussed in this study is a first-order extension of De and Omori's axiomatic propositional extension BD+ [10] of Belnap–Dunn logic, in which classical negation and classical implication are added.

2020 *Mathematics Subject Classification.* Primary: 03B47, Secondary: 03B50, 03F05.

Bimbó, Katalin, (ed.), *Relevance Logics and other Tools for Reasoning. Essays in Honor of J. Michael Dunn*, (Tributes, vol. 46), College Publications, London, UK, 2022, pp. 237–260.

The Herbrand and contraposition-elimination theorems proved in this study provide a proof-theoretic justification for developing an FBD+-based automated theorem proving framework that can simultaneously handle indefinite and definite information. The Herbrand theorem for first-order classical logic is considered a fundamental theorem for realizing automated theorem proving frameworks [8; 37]. By contrast, the contraposition-elimination theorem, which has been considered, for example, in [27] and [3], provides an alternative "compact" Gentzen-type sequent calculus (i.e., it uses only a few logical inference rules), which is theorem-equivalent to FBD+. The alternative sequent calculus derived from the contraposition-elimination theorem is expected to be useful for generating simple negated proofs of automated theorem proving frameworks (e.g., it is easy to generate proofs of $\sim\alpha$ from a proof of α). In addition to these fundamental results, the existing standard algorithms for automated theorem proving frameworks based on first-order classical logic can also be applied to an FBD+-based automated theorem proving framework using an embedding theorem that was previously proved in [25]. This embedding theorem proved in [25] also plays a central and important role in proving the cut-elimination, completeness, Herbrand, and Craig interpolation theorems for FBD+.

The propositional logic BD+, originally introduced as a Hilbert-style axiomatic system in [10], was obtained from a Hilbert-style axiomatic system for propositional classical logic with the standard language $\{\wedge, \vee, \rightarrow, \neg\}$ by adding the following axiom schemes with a paraconsistent negation connective $\sim$:

1. $\sim\sim\alpha \leftrightarrow \alpha$,
2. $\sim(\alpha \wedge \beta) \leftrightarrow (\sim\alpha \vee \sim\beta)$,
3. $\sim(\alpha \vee \beta) \leftrightarrow (\sim\alpha \wedge \sim\beta)$,
4. $\sim(\alpha \rightarrow \beta) \leftrightarrow (\neg\sim\alpha \wedge \sim\beta)$,
5. $\sim\neg\alpha \leftrightarrow \neg\sim\alpha$.

We note that the characteristic axiom schemes of BD+ are $\sim(\alpha \rightarrow \beta) \leftrightarrow (\neg\sim\alpha \wedge \sim\beta)$ and $\sim\neg\alpha \leftrightarrow \neg\sim\alpha$, which were considered quite natural and plausible in terms of many-valued semantics in [10]. It was also shown in [10] that BD+ is essentially equivalent to *Béziau's four-valued modal logic* PM4N [6] and *Zaitsev's paraconsistent logic* FDEP [47]. Another system that is essentially equivalent to BD+ is that of Méndez and Robles PŁ4 [32]. In addition, BD+ was observed in [24] to be a conservative extension of *Avron's self-extensional four-valued paradefinite logic* (SE4) [3] (i.e., SE4 is a classical-negation-free fragment of BD+). Furthermore, some modal and intuitionistic variants of BD+ have recently been studied by Kamide in [23]. In [38], a similar (but different) first-order extension of Belnap–Dunn logic with an additional unary connective $\triangle$, referred to as the *Baaz's delta operator*, was studied by Sano and Omori based on a Gentzen-type natural deduction system. This extended logic with $\triangle$ is considered equivalent to an extended first-order Belnap-Dunn logic with *exclusion negation* [10].

Gentzen-type sequent calculi for Belnap–Dunn logic have been extensively studied. See, for example, [28; 20] for a survey of Gentzen-type sequent calculi for Belnap–Dunn logic. Several Gentzen-type sequent calculi for BD+ were introduced by Kamide in [22]. It was shown in [22] that completeness (with respect to a valuation semantics) and cut-elimination theorems hold for various Gentzen-type sequent calculi for BD+ and its neighbors. Using the cut-elimination theorem for one of these calculi for BD+, we can derive the fact that BD+ is a conservative extension of Belnap–Dunn logic and propositional classical logic. In [25], the Gentzen-type sequent calculus FBD+ for a first-order extension of BD+, which is also used and investigated in the present study, was introduced by Kamide and Omori. The first-order system FBD+ is obtained from a Gentzen-type sequent calculus for BD+ by adding the standard logical inference rules for the universal and existential quantifiers $\forall$ and $\exists$ in first-order classical logic as well as the special negated logical inference rules that correspond to the following axiom schemes:

1. $\sim \forall x \alpha \leftrightarrow \exists x \sim \alpha$,
2. $\sim \exists x \alpha \leftrightarrow \forall x \sim \alpha$.

It was shown in [25] that (syntactical and semantical) embedding, cut-elimination, and completeness (with respect to valuation and many-valued semantics) theorems hold for FBD+. Thus, this study extends and refines the results of the previous work [25] to obtain the foundations of FBD+-based inconsistency-tolerant automated theorem proving framework that can simultaneously handle indefinite and definite information.

We remark that FBD+, BD+, SE4, and Belnap–Dunn logic are *paraconsistent logics* [35], which are suitable for handling inconsistency-tolerant (or paraconsistent) reasoning with indefinite information. In general, paraconsistent logics that employ a paraconsistent negation connective $\sim$ are logics with the property of paraconsistency with respect to $\sim$, which rejects the law $(\alpha \wedge \sim \alpha) \rightarrow \beta$ of explosion. See, for example, [35] for more information on paraconsistency. In the following, we show that FBD+ is useful for representing both paraconsistent (indefinite) and non-paraconsistent (definite) situations (information) simultaneously, as it uses both paraconsistent and classical negation connectives. We now consider an illustrative example of clinical reasoning, which is regarded as inconsistency-tolerant reasoning. In clinical reasoning, we do not want a description $(s(x) \wedge \sim s(x)) \rightarrow d(x)$ to be satisfied for any symptom s and disease d, where $\sim s(x)$ states that "person x does not have symptom s" and $d(x)$ states that "person x suffers from disease d." We do not want this because situations may exist that support the truth of both $s(a)$ and $\sim s(a)$ for some individual a but not the truth of $d(a)$. However, we also require the classical negation connective $\neg$ for clinical reasoning. For example, we do not want to describe $inHospital(x) \wedge \neg inHospital(x)$, where $\neg inHospital(x)$ states that "person x is not in a hospital." Although $s(x)$ and $d(x)$ are vague predicates with indefinite information, $inHospital(x)$ is a crisp and complete predicate with definite information. Thus, $\neg$ is required to represent the negation of this type of crisp and complete situation with definite information. We can then appropriately represent and handle the following situation by FBD+: $\neg inHospital(John) \wedge healthy(John) \wedge \sim health(John)$, which states that "John is not in a hospital, and he is healthy and not healthy."

The remainder of this paper is structured as follows. In Section 2, along the lines of [25], we define FBD+ and present a theorem for embedding FBD+ into a Gentzen-type sequent calculus FLK for first-order classical logic. This embedding theorem plays a crucial role in proving the Craig interpolation and Herbrand theorems for FBD+. In Section 3, we prove the contraposition-elimination theorem for FBD+. This contraposition-elimination theorem shows that the following (global) contraposition rule is admissible in cut-free FBD+:

$$\frac{\Delta \Rightarrow \Gamma}{\sim\Gamma \Rightarrow \sim\Delta} \quad \text{(contraposition)}$$

Using this theorem, we construct an alternative compact Gentzen-type sequent calculus, FBD+*, which is theorem-equivalent to FBD+. The contraposition-elimination theorem is regarded as a characteristic theorem for FBD+, as this theorem or its variants do not hold for Gentzen-type sequent calculi for typical paraconsistent logics such as *Nelson's paraconsistent four-valued logic* N4 [1; 33; 44; 28; 29], which is regarded as an extension of Belnap–Dunn logic. The contraposition-elimination theorem for a Gentzen-type sequent calculus *GSE*4 for SE4 (i.e., the classical-negation-free fragment of BD+) was proved by Avron in [3]. In Section 4, we first prove the strong-equivalence replacement (or substitution) theorem for FBD+, which was addressed in [25] (but where its proof was not given). Next, we prove the Herbrand theorem for FBD+ using this strong-equivalence replacement theorem and the theorem for embedding FBD+ into FLK. The strong-equivalence replacement theorem is a particularly novel property, as it is lacking in some typical paraconsistent logics such as N4. In Section 5, we show the Craig interpolation and Maksimova separation theorems for FBD+ using the theorem for embedding FBD+ into FLK. Furthermore, we introduce an alternative cut-free Gentzen-type sequent calculus FBD+° and prove the theorem equivalence between FBD+° and FBD+. In Section 6, we conclude the study and address some remarks.

2. Preliminary: Embedding Theorem

First, we introduce the first-order language $\mathcal{L}_{\text{FBD+}}$ of an extended first-order *Belnap-Dunn logic* with classical negation. This language is also denoted as $\mathcal{L}$ when we have no confusion. *Formulas* of $\mathcal{L}$ are constructed from countably many predicate and function symbols and countably many individual variables and constants by the following logical connectives: $\wedge$ (conjunction), $\vee$ (disjunction), $\rightarrow$ (implication), $\neg$ (classical negation), $\sim$ (paraconsistent negation), $\exists$ (existential quantifier), and $\forall$ (universal quantifier). We use small letters $p, q, \ldots$ to denote predicate symbols or atomic formulas, small letters $x, y, \ldots$ to denote individual variables, small letters $t, t_1, \ldots$ to denote terms, small Greek letters $\alpha, \beta, \ldots$ to denote formulas, and Greek capital letters $\Gamma, \Delta, \ldots$ to denote finite (possibly empty) sets of formulas. We use an expression $\alpha[t/x]$ to denote the formula that is obtained from the formula α by replacing all free occurrences of the individual variable x in α with the term t, but avoiding a clash of variables by an appropriate renaming of bound variables. We consider a 0-ary function and a 0-ary predicate to be an individual constant and a propositional variable, respectively. If Φ is the set of all atomic formulas of $\mathcal{L}$, then it is said that $\mathcal{L}$ is based on Φ. We use expressions $\sim\Gamma$ and $\neg\Gamma$ to denote the sets $\{\sim\gamma : \gamma \in \Gamma\}$ and $\{\neg\gamma : \gamma \in \Gamma\}$, respectively. We use the symbol $\equiv$ to denote the equality of expressions symbol by symbol. We call an expression of the form $\Gamma \Rightarrow \Delta$ a *sequent*. We use an expression $\alpha \Leftrightarrow \beta$ to represent the abbreviation of the sequents $\alpha \Rightarrow \beta$ and $\beta \Rightarrow \alpha$. We use an expression $L \vdash S$ to denote the fact that a sequent S is provable in a sequent calculus L. If L of $L \vdash S$ is clear from the context, L may be omitted. A rule R of inference

is called *admissible* in a sequent calculus L if the following condition is satisfied: For any instance
$$\frac{S_1 \cdots S_n}{S}$$
of R, if $L \vdash S_i$ for all i, then $L \vdash S$. Furthermore, R is called *derivable* in L if there is a derivation from $S_1, \cdots, S_n$ to S in L. It is remarked that a rule R of inference is admissible in a sequent calculus L if and only if two sequent calculi L and $L+R$ are theorem-equivalent.

We now introduce a Gentzen-type sequent calculus FBD+ for the first-order extension of De and Omori's extended Belnap–Dunn logic BD+ with classical negation, which was introduced in [25].

Definition 1 (FBD+ [25]). In the following definition, we use a symbol t to denote an arbitrary term and a symbol z to denote an individual variable which obeys the eigenvariable condition (i.e., z does not occur as a free individual variable in the lower sequent of the rule).

The initial sequents of FBD+ are of the form: For any atomic formula p,
$$p \Rightarrow p \qquad \sim p \Rightarrow \sim p.$$

The structural inference rules of FBD+ are of the form:
$$\frac{\Gamma \Rightarrow \Delta, \alpha \quad \alpha, \Sigma \Rightarrow \Pi}{\Gamma, \Sigma \Rightarrow \Delta, \Pi} \text{ (cut)}$$

$$\frac{\Gamma \Rightarrow \Delta}{\alpha, \Gamma \Rightarrow \Delta} \text{ (we-left)} \qquad \frac{\Gamma \Rightarrow \Delta}{\Gamma \Rightarrow \Delta, \alpha} \text{ (we-right)}$$

The non-negated logical inference rules of FBD+ are of the form:

$$\frac{\alpha, \beta, \Gamma \Rightarrow \Delta}{\alpha \wedge \beta, \Gamma \Rightarrow \Delta} \text{ ($\wedge$left)} \qquad \frac{\Gamma \Rightarrow \Delta, \alpha \quad \Gamma \Rightarrow \Delta, \beta}{\Gamma \Rightarrow \Delta, \alpha \wedge \beta} \text{ ($\wedge$right)}$$

$$\frac{\alpha, \Gamma \Rightarrow \Delta \quad \beta, \Gamma \Rightarrow \Delta}{\alpha \vee \beta, \Gamma \Rightarrow \Delta} \text{ ($\vee$left)} \qquad \frac{\Gamma \Rightarrow \Delta, \alpha, \beta}{\Gamma \Rightarrow \Delta, \alpha \vee \beta} \text{ ($\vee$right)}$$

$$\frac{\Gamma \Rightarrow \Delta, \alpha \quad \beta, \Gamma \Rightarrow \Delta}{\alpha \rightarrow \beta, \Gamma \Rightarrow \Delta} \text{ ($\rightarrow$left)} \qquad \frac{\alpha, \Gamma \Rightarrow \Delta, \beta}{\Gamma \Rightarrow \Delta, \alpha \rightarrow \beta} \text{ ($\rightarrow$right)}$$

$$\frac{\Gamma \Rightarrow \Delta, \alpha}{\neg \alpha, \Gamma \Rightarrow \Delta} \text{ ($\neg$left)} \qquad \frac{\alpha, \Gamma \Rightarrow \Delta}{\Gamma \Rightarrow \Delta, \neg \alpha} \text{ ($\neg$right)}$$

$$\frac{\alpha[t/x], \Gamma \Rightarrow \Delta}{\forall x \alpha, \Gamma \Rightarrow \Delta} \text{ ($\forall$left)} \qquad \frac{\Gamma \Rightarrow \Delta, \alpha[z/x]}{\Gamma \Rightarrow \Delta, \forall x \alpha} \text{ ($\forall$right)}$$

$$\frac{\alpha[z/x], \Gamma \Rightarrow \Delta}{\exists x \alpha, \Gamma \Rightarrow \Delta} \text{ ($\exists$left)} \qquad \frac{\Gamma \Rightarrow \Delta, \alpha[t/x]}{\Gamma \Rightarrow \Delta, \exists x \alpha} \text{ ($\exists$right)}$$

The negated logical inference rules of FBD+ are of the form:

$$\frac{\alpha, \Gamma \Rightarrow \Delta}{\sim\sim\alpha, \Gamma \Rightarrow \Delta} \text{ ($\sim\sim$left)} \qquad \frac{\Gamma \Rightarrow \Delta, \alpha}{\Gamma \Rightarrow \Delta, \sim\sim\alpha} \text{ ($\sim\sim$right)}$$

$$\frac{\sim\alpha, \Gamma \Rightarrow \Delta \quad \sim\beta, \Gamma \Rightarrow \Delta}{\sim(\alpha \wedge \beta), \Gamma \Rightarrow \Delta} \text{ ($\sim\wedge$left)} \qquad \frac{\Gamma \Rightarrow \Delta, \sim\alpha, \sim\beta}{\Gamma \Rightarrow \Delta, \sim(\alpha \wedge \beta)} \text{ ($\sim\wedge$right)}$$

$$\frac{\sim\alpha,\sim\beta,\Gamma\Rightarrow\Delta}{\sim(\alpha\vee\beta),\Gamma\Rightarrow\Delta}\;(\sim\vee\text{left})\qquad\frac{\Gamma\Rightarrow\Delta,\sim\alpha\quad\Gamma\Rightarrow\Delta,\sim\beta}{\Gamma\Rightarrow\Delta,\sim(\alpha\vee\beta)}\;(\sim\vee\text{right})$$

$$\frac{\sim\beta,\Gamma\Rightarrow\Delta,\sim\alpha}{\sim(\alpha\to\beta),\Gamma\Rightarrow\Delta}\;(\sim\to\text{left})\qquad\frac{\sim\alpha,\Gamma\Rightarrow\Delta\quad\Gamma\Rightarrow\Delta,\sim\beta}{\Gamma\Rightarrow\Delta,\sim(\alpha\to\beta)}\;(\sim\to\text{right})$$

$$\frac{\Gamma\Rightarrow\Delta,\sim\alpha}{\sim\neg\alpha,\Gamma\Rightarrow\Delta}\;(\sim\neg\text{left})\qquad\frac{\sim\alpha,\Gamma\Rightarrow\Delta}{\Gamma\Rightarrow\Delta,\sim\neg\alpha}\;(\sim\neg\text{right})$$

$$\frac{\sim\alpha[z/x],\Gamma\Rightarrow\Delta}{\sim\forall x\alpha,\Gamma\Rightarrow\Delta}\;(\sim\forall\text{left})\qquad\frac{\Gamma\Rightarrow\Delta,\sim\alpha[t/x]}{\Gamma\Rightarrow\Delta,\sim\forall x\alpha}\;(\sim\forall\text{right})$$

$$\frac{\sim\alpha[t/x],\Gamma\Rightarrow\Delta}{\sim\exists x\alpha,\Gamma\Rightarrow\Delta}\;(\sim\exists\text{left})\qquad\frac{\Gamma\Rightarrow\Delta,\sim\alpha[z/x]}{\Gamma\Rightarrow\Delta,\sim\exists x\alpha}\;(\sim\exists\text{right})$$

To address an embedding theorem, we introduce a Gentzen-type sequent calculus FLK for first-order classical logic. The language $\mathcal{L}_{\text{FLK}}$ of FLK is obtained from $\mathcal{L}_{\text{FBD+}}$ by deleting $\sim$. This is also simply denoted as $\mathcal{L}$ when we have no confusion.

Definition 2 (FLK). The system FLK is the $\sim$-free fragment of FBD+. Namely, FLK is obtained from FBD+ by deleting the negated logical inference rules and the initial sequents of the form $\sim p \Rightarrow \sim p$.

Remark 3. We make the following remarks.

1. Let L be FLK or FBD+. For any formula α, the sequent $\alpha \Rightarrow \alpha$ is provable in cut-free L. This can be shown by induction on α.
2. The following sequents are provable in cut-free FBD+:
   (a) $\sim\sim\alpha \Leftrightarrow \alpha$,
   (b) $\sim(\alpha\wedge\beta) \Leftrightarrow \sim\alpha\vee\sim\beta$,
   (c) $\sim(\alpha\vee\beta) \Leftrightarrow \sim\alpha\wedge\sim\beta$,
   (d) $\sim(\alpha\to\beta) \Leftrightarrow \neg\sim\alpha\wedge\sim\beta$,
   (e) $\sim\neg\alpha \Leftrightarrow \neg\sim\alpha$,
   (f) $\sim\exists x\alpha \Leftrightarrow \forall x\sim\alpha$,
   (g) $\sim\forall x\alpha \Leftrightarrow \exists x\sim\alpha$.
3. The cut-elimination theorem for FLK is well known (see e.g., [15; 41]).
4. The inference rules ($\sim\to$left) and ($\sim\to$right) correspond to the Hilbert-style axiom scheme $\sim(\alpha\to\beta) \leftrightarrow \neg\sim\alpha\wedge\sim\beta$. The inference rules ($\sim\neg$left) and ($\sim\neg$right) correspond to the Hilbert-style axiom scheme $\sim\neg\alpha \leftrightarrow \sim\neg\alpha$. These axiom schemes were originally introduced by De and Omori [10] to axiomatize the extended propositional Belnap–Dunn logic BD+ with classical negation.
5. The $\{\wedge,\vee,\to,\sim\}$-fragment of FBD+, which fragment was called A4 in [24], was introduced in [24] as an alternative Gentzen-type sequent calculus for Avron's self-extensional four-valued paradefinite logic SE4 [3]. The original Gentzen-type sequent calculus GSE4, which was introduced in [3], is obtained from A4 by replacing the initial sequents of the form $p \Rightarrow p$ and $\sim p \Rightarrow \sim p$ for any propositional variable p with the initial sequents of the form $\alpha \Rightarrow \alpha$ for any formulas. Since the sequents of the form $\alpha \Rightarrow \alpha$ for any formula α are provable in cut-free A4, the systems GSE4 and A4 are theorem-equivalent. The cut-elimination-theorem for GSE4 was shown in [3].

Next, we introduce a translation from FBD+ to FLK, and by using this translation, we address a theorem for embedding FBD+ into FLK.

Definition 4. We fix a set Φ of atomic formulas, and define the set $\Phi' := \{p' : p \in \Phi\}$ of atomic formulas. Let the languages $\mathcal{L}_{\text{FBD+}}$ and $\mathcal{L}_{\text{FLK}}$ be defined as above based on sets Φ and $\Phi \cup \Phi'$, respectively. A mapping f from $\mathcal{L}_{\text{FBD+}}$ to $\mathcal{L}_{\text{FLK}}$ is inductively defined by the following clauses:

1. For any $p \in \Phi$, $f(p) := p$, $f(\sim p) := p' \in \Phi'$,
2. $f(\alpha \wedge \beta) := f(\alpha) \wedge f(\beta)$,
3. $f(\alpha \vee \beta) := f(\alpha) \vee f(\beta)$,
4. $f(\alpha \to \beta) := f(\alpha) \to f(\beta)$,
5. $f(\neg \alpha) := \neg f(\alpha)$,
6. $f(\forall x \alpha) := \forall x f(\alpha)$,
7. $f(\exists x \alpha) := \exists x f(\alpha)$,
8. $f(\sim(\alpha \wedge \beta)) := f(\sim \alpha) \vee f(\sim \beta)$,
9. $f(\sim(\alpha \vee \beta)) := f(\sim \alpha) \wedge f(\sim \beta)$,
10. $f(\sim(\alpha \to \beta)) := \neg f(\sim \alpha) \wedge f(\sim \beta)$,
11. $f(\sim \sim \alpha) := f(\alpha)$,
12. $f(\sim \neg \alpha) := \neg f(\sim \alpha)$,
13. $f(\sim \forall x \alpha) := \exists x f(\sim \alpha)$,
14. $f(\sim \exists x \alpha) := \forall x f(\sim \alpha)$.

We use an expression $f(\Gamma)$ to denote the result of replacing every occurrence of a formula α in Γ by an occurrence of $f(\alpha)$.

Remark 5. We make the following remarks on the translation function f.

1. The translation function f is independent of terms. Thus, f can be used based on an arbitrary language with or without individual constant and function symbols.
2. The expressions p and p' in Definition 4 include, for example, $p(x_1, x_2)$ and $p'(x_1, x_2)$, respectively, with some individual variables x_1 and x_2. Also, the expression $f(\sim p(x_1, x_2))$ with f coincides with $p'(x_1, x_2)$.
3. A similar translation function has been used by Vorob'ev [42], Rautenberg [36], and Gurevich [16] to embed Nelson's constructive logic [1; 33] into intuitionistic logic. A similar translation was used by Burgess [7] (pp. 107–108) to show the relationship between Belnap–Dunn logic and classical logic. Some similar translation functions have also been used, for example, in [26; 21; 30; 23] to embed some paraconsistent logics into classical logic.

We have the following syntactical embedding theorem, which was proved in [25] as Theorem 10. This theorem will be used for proving the Craig interpolation and Herbrand theorems for FBD+.

Theorem 6 (Embedding from FBD+ into FLK [25]). *Let Γ, Δ be sets of formulas in $\mathcal{L}_{\text{FBD+}}$, and f be the mapping defined in Definition 4.*

1. FBD+ $\vdash \Gamma \Rightarrow \Delta$ *iff* FLK $\vdash f(\Gamma) \Rightarrow f(\Delta)$.
2. FBD+ $-$ (cut) $\vdash \Gamma \Rightarrow \Delta$ *iff* FLK $-$ (cut) $\vdash f(\Gamma) \Rightarrow f(\Delta)$.

Using Theorem 6, we can obtain the following cut-elimination theorem, which was also proved in [25].

Theorem 7 (Cut-elimination for FBD+ [25]). *The rule* (cut) *is admissible in cut-free* FBD+.

Remark 8. Theorem 6 plays a critical role for proving some desired properties of FBD+. Actually, the cut-elimination (Theorem 7) and completeness theorems for FBD+ were proved in [25] using Theorem 6. In this study, Theorem 26 (Existence of quasi-Skolem normal form), Theorem 28 (Herbrand), and Theorem 32 (Craig interpolation) will be proved using Theorem 6.

3. Contraposition-elimination Theorem

In this section, we prove the contraposition-elimination theorem for FBD+. Prior to proving it, we have to show the following proposition.

Proposition 9. *The following rules are admissible in cut-free* FBD+:

$$\frac{\sim\sim\alpha, \Gamma \Rightarrow \Delta}{\alpha, \Gamma \Rightarrow \Delta} \,(\sim\sim \text{left}^{-1}) \qquad \frac{\Gamma \Rightarrow \Delta, \sim\sim\alpha}{\Gamma \Rightarrow \Delta, \alpha} \,(\sim\sim \text{right}^{-1})$$

Proof. For ($\sim\sim$left^{-1}), this proposition can be proved by induction on the proofs P of the upper sequent $\sim\sim\alpha, \Gamma \Rightarrow \Delta$ of ($\sim\sim$left^{-1}) in cut-free FBD+. For ($\sim\sim$right^{-1}), it can be proved similarly. ◁

Remark 10. Similar to Proposition 9, we can show that the following rules are admissible in cut-free FBD+:

$$\frac{\sim\neg\alpha, \Gamma \Rightarrow \Delta}{\Gamma \Rightarrow \Delta, \sim\alpha} \,(\sim\neg \text{left}^{-1}) \qquad \frac{\Gamma \Rightarrow \Delta, \sim\neg\alpha}{\sim\alpha, \Gamma \Rightarrow \Delta} \,(\sim\neg \text{right}^{-1})$$

Using Proposition 9, we show the following contraposition-elimination theorem for FBD+.

Theorem 11 (Contraposition-elimination for FBD+). *The following rule is admissible in cut-free* FBD+:

$$\frac{\Delta \Rightarrow \Gamma}{\sim\Gamma \Rightarrow \sim\Delta} \,(contraposition).$$

Proof. By induction on the proofs P of the upper sequent $\Delta \Rightarrow \Gamma$ of (contraposition) in cut-free FBD+. We show some of the critical cases.

1. Case ($\rightarrow$left): The last inference of P is of the form:

$$\frac{\Gamma \Rightarrow \Delta, \alpha \quad \beta, \Gamma \Rightarrow \Delta}{\alpha \rightarrow \beta, \Gamma \Rightarrow \Delta} \,(\rightarrow \text{left})$$

By induction hypothesis, we have FBD+ $-$ (cut) $\vdash \sim\alpha, \sim\Delta \Rightarrow \sim\Gamma$ and FBD+ $-$ (cut) $\vdash \sim\Delta \Rightarrow \sim\Gamma, \sim\beta$. Then, we obtain the required fact:

$$\frac{\begin{array}{c} \vdots \\ \sim\alpha, \sim\Delta \Rightarrow \sim\Gamma \end{array} \quad \begin{array}{c} \vdots \\ \sim\Delta \Rightarrow \sim\Gamma, \sim\beta \end{array}}{\sim\Delta \Rightarrow \sim\Gamma, \sim(\alpha \rightarrow \beta)} \,(\sim\rightarrow \text{left})$$

2. Case ($\rightarrow$right): The last inference of P is of the form:

$$\frac{\alpha, \Gamma \Rightarrow \Delta, \beta}{\Gamma \Rightarrow \Delta, \alpha \rightarrow \beta} \,(\rightarrow \text{right})$$

By induction hypothesis, we have FBD+ $-$ (cut) $\vdash {\sim}\beta, {\sim}\Delta \Rightarrow {\sim}\Gamma, {\sim}\alpha$. Then, we obtain the required fact:

$$\vdots$$
$$\frac{{\sim}\beta, {\sim}\Delta \Rightarrow {\sim}\Gamma, {\sim}\alpha}{{\sim}(\alpha \to \beta), {\sim}\Delta \Rightarrow {\sim}\Gamma} \ ({\sim}{\to}\text{left})$$

3. Case (${\sim}\neg$right): The last inference of P is of the form:

$$\frac{{\sim}\alpha, \Gamma \Rightarrow \Delta}{\Gamma \Rightarrow \Delta, {\sim}\neg\alpha} \ ({\sim}\neg\text{right})$$

By induction hypothesis, we have FBD+ $-$ (cut) $\vdash {\sim}\Delta \Rightarrow {\sim}\Gamma, {\sim}{\sim}\alpha$. Then, we obtain the required fact:

$$\vdots$$
$$\frac{{\sim}\Delta \Rightarrow {\sim}\Gamma, {\sim}{\sim}\alpha}{{\sim}\Delta \Rightarrow {\sim}\Gamma, \alpha} \ ({\sim}{\sim}\text{right}^{-1})$$
$$\frac{}{\neg\alpha, {\sim}\Delta \Rightarrow {\sim}\Gamma} \ (\neg\text{left})$$
$$\frac{}{{\sim}{\sim}\neg\alpha, {\sim}\Delta \Rightarrow {\sim}\Gamma} \ ({\sim}{\sim}\text{left})$$

where (${\sim}{\sim}\text{right}^{-1}$) is admissible in cut-free FBD+ by Proposition 9.

4. Case (${\sim}{\to}$left): The last inference of P is of the form:

$$\frac{{\sim}\beta, \Gamma \Rightarrow \Delta, {\sim}\alpha}{{\sim}(\alpha \to \beta), \Gamma \Rightarrow \Delta} \ ({\sim}{\to}\text{left})$$

By induction hypothesis, we have FBD+ $-$ (cut) $\vdash {\sim}{\sim}\alpha, {\sim}\Delta \Rightarrow {\sim}\Gamma, {\sim}{\sim}\beta$. Then, we obtain the required fact:

$$\vdots$$
$$\frac{{\sim}{\sim}\alpha, {\sim}\Delta \Rightarrow {\sim}\Gamma, {\sim}{\sim}\beta}{{\sim}{\sim}\alpha, {\sim}\Delta \Rightarrow {\sim}\Gamma, \beta} \ ({\sim}{\sim}\text{right}^{-1})$$
$$\frac{}{\alpha, {\sim}\Delta \Rightarrow {\sim}\Gamma, \beta} \ ({\sim}{\sim}\text{left}^{-1})$$
$$\frac{}{{\sim}\Delta \Rightarrow {\sim}\Gamma, \alpha \to \beta} \ ({\to}\text{right})$$
$$\frac{}{{\sim}\Delta \Rightarrow {\sim}\Gamma, {\sim}{\sim}(\alpha \to \beta)} \ ({\sim}{\sim}\text{right})$$

where (${\sim}{\sim}\text{left}^{-1}$) and (${\sim}{\sim}\text{right}^{-1}$) are admissible in cut-free FBD+ by Proposition 9.

5. Case (${\sim}{\to}$right): The last inference of P is of the form:

$$\frac{{\sim}\alpha, \Gamma \Rightarrow \Delta \quad \Gamma \Rightarrow \Delta, {\sim}\beta}{\Gamma \Rightarrow \Delta, {\sim}(\alpha \to \beta)} \ ({\sim}{\to}\text{right})$$

By induction hypothesis, we have FBD+ $-$ (cut) $\vdash {\sim}\Delta \Rightarrow {\sim}\Gamma, {\sim}{\sim}\alpha$ and FBD+ $-$ (cut) $\vdash {\sim}{\sim}\beta, {\sim}\Delta \Rightarrow {\sim}\Gamma$. Then, we obtain the required fact:

$$\vdots \qquad \vdots$$

$$({\sim}{\sim}\text{right}^{-1}) \ \frac{{\sim}\Delta \Rightarrow {\sim}\Gamma, {\sim}{\sim}\alpha}{{\sim}\Delta \Rightarrow {\sim}\Gamma, \alpha} \quad \frac{{\sim}{\sim}\beta, {\sim}\Delta \Rightarrow {\sim}\Gamma}{\beta, {\sim}\Delta \Rightarrow {\sim}\Gamma} \ ({\sim}{\sim}\text{left}^{-1})$$
$$\frac{\alpha \to \beta, {\sim}\Delta \Rightarrow {\sim}\Gamma}{} \ ({\to}\text{left})$$
$$\frac{}{{\sim}{\sim}(\alpha \to \beta), {\sim}\Delta \Rightarrow {\sim}\Gamma} \ ({\sim}{\sim}\text{left})$$

where (${\sim}{\sim}\text{left}^{-1}$) and (${\sim}{\sim}\text{right}^{-1}$) are admissible in cut-free FBD+ by Proposition 9.

6. Case ($\sim \exists$right): The last inference of P is of the form:

$$\frac{\Gamma \Rightarrow \Delta, \sim \alpha[z/x]}{\Gamma \Rightarrow \Delta, \sim \exists x \alpha} \ (\sim \exists\text{right})$$

By induction hypothesis, we have FBD+ $-$ (cut) $\vdash \sim\sim\alpha[z/x], \sim\Delta \Rightarrow \sim\Gamma$. Then, we obtain the required fact:

$$\vdots$$
$$\sim\sim\alpha[z/x], \sim\Delta \Rightarrow \sim\Gamma$$
$$\vdots \ (\sim\sim\text{left}^{-1})$$
$$\frac{\alpha[z/x], \sim\Delta \Rightarrow \sim\Gamma}{\exists x\alpha, \sim\Delta \Rightarrow \sim\Gamma} \ (\exists\text{left})$$
$$\frac{}{\sim\sim\exists x\alpha, \sim\Delta \Rightarrow \sim\Gamma} \ (\sim\sim\text{left})$$

where ($\sim\sim$left^{-1}) is admissible in cut-free FBD+ by Proposition 9. ◁

Remark 12. We make the following remarks.

1. Theorem 11 is considered to be a novel property of FBD+, because this theorem does not hold for some typical paraconsistent logics with an implication connective (e.g., it does not hold for Nelson's paraconsistent four-valued logic N4).
2. The contraposition-elimination theorem for a Gentzen-type sequent calculus *GSE*4 for Avron's self-extensional four-valued paradefinite logic SE4, which was observed in [24] to be the classical-negation-free fragment of BD+, was proved by Avron in [3].
3. The contraposition-elimination theorems for other systems have been studied, for example in [27], wherein the contraposition-elimination theorem was proved by Kamide and Wansing for a Gentzen-type sequent calculus for symmetric praraconsistent logic that has both implication and co-implication connectives.

Next, we introduce an alternative Gentzen-type sequent calculus FBD+* for the first-order extension of BD+. This system FBD+* will be shown to be theorem-equivalent to FBD+ by using Theorem 11.

Definition 13 (FBD+*). The system FBD+* is obtained from FBD+ by replacing the negated initial sequents of the form $\sim p \Rightarrow \sim p$ and the negated logical inference rules ($\sim\wedge$left), ($\sim\wedge$right), ($\sim\vee$left), ($\sim\vee$right), ($\sim\to$left), ($\sim\to$right), ($\sim\neg$left), ($\sim\neg$right), ($\sim\forall$left), ($\sim\forall$right), ($\sim\exists$left), and ($\sim\exists$right) with the logical inference rule (contraposition).

Remark 14. We make the following remarks.

1. For any formula α, the sequent $\alpha \Rightarrow \alpha$ is provable in FBD+* using (cut). This can be shown by induction on α.
2. However, it cannot be shown that this sequent is provable in cut-free FBD+*, because as will be shown, cut-elimination theorem does not hold for FBD+*.

3. Let p and q be distinct propositional variables. Then, the following are examples of FBD+*-proofs using (contraposition).

$$
\begin{array}{c}
\text{(we-right)} \dfrac{p \Rightarrow p}{p \Rightarrow p, q} \\
\text{($\rightarrow$right)} \dfrac{\Rightarrow p, p \rightarrow q}{\Rightarrow p, p \rightarrow q} \\
\text{(contraposition)} \dfrac{\sim p, \sim(p \rightarrow q) \Rightarrow}{\sim(p \rightarrow q) \Rightarrow \neg \sim p} \\
\text{($\neg$right)} \dfrac{\sim(p \rightarrow q) \Rightarrow \neg \sim p \wedge \sim q}{\sim(p \rightarrow q) \Rightarrow \neg \sim p \wedge \sim q}
\qquad
\dfrac{q \Rightarrow q}{q, p \Rightarrow q} \text{ (we-left)} \\
\dfrac{q \Rightarrow p \rightarrow q}{\sim(p \rightarrow q) \Rightarrow \sim q} \text{ ($\rightarrow$right)} \\
\text{(contraposition)} \\
\text{($\wedge$right)}
\end{array}
$$

$$
\dfrac{\dfrac{p \Rightarrow p \quad q \Rightarrow q}{p \rightarrow q, p \Rightarrow q} \text{ ($\rightarrow$left)}}{\dfrac{\sim q \Rightarrow \sim(p \rightarrow q), \sim p}{\dfrac{\neg \sim p, \sim q \Rightarrow \sim(p \rightarrow q)}{\neg \sim p \wedge \sim q \Rightarrow \sim(p \rightarrow q)} \text{ ($\wedge$left)}} \text{ ($\neg$left)}} \text{ (contraposition)}
$$

4. The proofs displayed just above and the fact presented in the first item of this remark imply that the sequents of the form $\sim(\alpha \rightarrow \beta) \Leftrightarrow \neg \sim \alpha \wedge \sim \beta$ for any formulas α and β, which correspond to one of the characteristic axiom schemes of BD+, are provable in cut-free FBD+*.

Prior to smoothly proving the equivalence between FBD+ and FBD+*, we have to show the following proposition.

Proposition 15. *The rules* $(\sim\sim\text{left}^{-1})$ *and* $(\sim\sim\text{right}^{-1})$ *are derivable in* FBD+*.

Proof. We show only the following case. The other case can be shown similarly.

Case $(\sim\sim\text{left}^{-1})$: We have the following proof.

$$
\vdots
$$

$$
(\sim\sim\text{right}) \dfrac{\alpha \Rightarrow \alpha}{\dfrac{\alpha \Rightarrow \sim\sim\alpha \quad \sim\sim\alpha, \Gamma \Rightarrow \Delta}{\alpha, \Gamma \Rightarrow \Delta} \text{ (cut)}}
$$
◁

Using Theorem 11 and Proposition 15, we can obtain the following theorem.

Theorem 16 (Equivalence between FBD+ and FBD+*). *The systems* FBD+ *and* FBD+* *are theorem-equivalent.*

Proof. We have the following proofs.

First, we show FBD+* $\vdash \Gamma \Rightarrow \Delta$ implies FBD+* $\vdash \Gamma \Rightarrow \Delta$ for any sets Γ and Δ of formulas. This fact is proved by induction on the proofs P of $\Gamma \Rightarrow \Delta$ in FBD+. We distinguish the cases according to the last inference of P, and show some cases. The other cases can be shown similarly or straightforwardly. For example, the proof of Case $(\sim\rightarrow\text{right})$ can be shown in a similar manner as that for Case $(\sim\wedge\text{left})$ using Proposition 15.

(1) Case $\sim p \Rightarrow \sim p$: The last inference of P is of the form $\sim p \Rightarrow \sim p$. In this case, we obtain the required fact:

$$
\dfrac{p \Rightarrow p}{\sim p \Rightarrow \sim p} \text{ (contraposition)}
$$

(2) Case ($\sim\wedge$left): We have the following proof:

$$\text{(contraposition)} \quad \cfrac{\cfrac{\sim\alpha,\Gamma\Rightarrow\Delta}{\sim\Delta\Rightarrow\sim\Gamma,\sim\sim\alpha}}{\sim\Delta\Rightarrow\sim\Gamma,\alpha} \quad \cfrac{\cfrac{\sim\beta,\Gamma\Rightarrow\Delta}{\sim\Delta\Rightarrow\sim\Gamma,\sim\sim\beta}}{\sim\Delta\Rightarrow\sim\Gamma,\beta} \text{ (contraposition)} \text{ ($\sim\sim$right^{-1})}$$
$$\cfrac{\sim\Delta\Rightarrow\sim\Gamma,\alpha\wedge\beta}{\sim(\alpha\wedge\beta),\sim\sim\Gamma\Rightarrow\sim\sim\Delta} \text{ ($\wedge$right)}$$
$$\vdots \quad (\sim\sim\text{left}^{-1})\, (\sim\sim\text{right}^{-1})$$
$$\sim(\alpha\wedge\beta),\Gamma\Rightarrow\Delta$$

where ($\sim\sim$left^{-1}) and ($\sim\sim$right^{-1}) are derivable in FBD+* by Proposition 15.

Second, we show FBD+* $\vdash \Gamma\Rightarrow\Delta$ implies FBD+ $\vdash \Gamma\Rightarrow\Delta$ for any sets Γ and Δ of formulas. This fact is proved by induction on the proofs Q of $\Gamma\Rightarrow\Delta$ in FBD+*. We distinguish the cases according to the last inference of Q. It is sufficient to show the case when the last inference of Q is (contraposition). This case is obtained by Theorem 11. ◁

Remark 17. We can obtain another simple proof for the first direction of the proof of Theorem 16 without using Proposition 15. But, we intend to obtain a systematic and unified proof using Proposition 15. As an example of simple proof, we show the following proof for Case ($\sim\wedge$left) without using Proposition 15 but using (cut):

$$\text{(contraposition)} \quad \cfrac{\cfrac{(\wedge\text{right})\cfrac{\alpha\Rightarrow\alpha \quad \beta\Rightarrow\beta}{\alpha,\beta\Rightarrow\alpha\wedge\beta}}{\sim(\alpha\wedge\beta)\Rightarrow\sim\alpha,\sim\beta} \quad \sim\alpha,\Gamma\Rightarrow\Delta}{\cfrac{\sim(\alpha\wedge\beta),\Gamma\Rightarrow\Delta,\sim\beta}{\sim(\alpha\wedge\beta),\Gamma\Rightarrow\Delta}} \quad \sim\beta,\Gamma\Rightarrow\Delta \text{ (cut)}$$

We have the following negative result on the cut-elimination theorem for FBD+*.

Theorem 18 (Failure of cut-elimination for FBD+*). *Cut-elimination theorem does not hold for* FBD+*.

Proof. A counterexample sequent is $p\Rightarrow\sim(p\wedge\sim p)$, where p is a propositional variable. Obviously, this sequent cannot be proved in cut-free FBD+*. However, this sequent can be proved in FBD+* using (cut) by:

$$(\sim\sim\text{right})\,\cfrac{p\Rightarrow p}{p\Rightarrow\sim\sim p} \quad \cfrac{\cfrac{\cfrac{\cfrac{p\Rightarrow p}{\sim p\Rightarrow\sim p}\text{ (contraposition)}}{p,\sim p\Rightarrow\sim p}\text{ (we-left)}}{p\wedge\sim p\Rightarrow\sim p}\text{ ($\wedge$left)}}{\sim\sim p\Rightarrow\sim(p\wedge\sim p)} \text{ (contraposition)}$$
$$\cfrac{}{p\Rightarrow\sim(p\wedge\sim p)} \text{ (cut)}$$

◁

Remark 19. The counterexample sequent $p\Rightarrow\sim(p\wedge\sim p)$ displayed in the proof of Theorem 18 for the failure of the cut-elimination for FBD+* is provable in cut-free FBD+ by:

$$\frac{p\Rightarrow p}{p\Rightarrow \sim\sim p} \; (\sim\sim\text{right})$$
$$\frac{p\Rightarrow \sim\sim p}{p\Rightarrow \sim p, \sim\sim p} \; (\text{we-right})$$
$$\frac{p\Rightarrow \sim p, \sim\sim p}{p\Rightarrow \sim(p\wedge\sim p)} \; (\sim\wedge\text{right})$$

4. Herbrand Theorem

In this section, we prove the Herbrand theorem for FBD+. Prior to proving it, we have to prove the strong equivalence replacement theorem.

Definition 20 (Strong equivalence for FBD+-formulas). An expression $\alpha \leftrightarrow_s \beta$ for any formulas α and β of FBD+, which is called a *strong equivalence* between α and β, is defined by FBD+ $\vdash \alpha \Leftrightarrow \beta$ and FBD+ $\vdash \sim\alpha \Leftrightarrow \sim\beta$.

Proposition 21. *We have the following list of strong equivalences:*

1. $\sim(\alpha\wedge\beta) \leftrightarrow_s \sim\alpha \vee \sim\beta,$
2. $\sim(\alpha\vee\beta) \leftrightarrow_s \sim\alpha \wedge \sim\beta,$
3. $\sim(\alpha\to\beta) \leftrightarrow_s \neg\sim\alpha \wedge \sim\beta,$
4. $\sim\sim\alpha \leftrightarrow_s \alpha,$
5. $\sim\neg\alpha \leftrightarrow_s \neg\sim\alpha,$
6. $\sim(\forall x\alpha) \leftrightarrow_s \exists x\sim\alpha,$
7. $\sim(\exists x\alpha) \leftrightarrow_s \forall x\sim\alpha.$

Proof. The proof is straightforward. We show some cases.

(1) Case (3): In this case, we show only the following cases:

$$\frac{\dfrac{\alpha\Rightarrow\alpha \quad \beta\Rightarrow\beta}{\alpha,\alpha\to\beta\Rightarrow\beta}}{\dfrac{\sim\sim\alpha,\alpha\to\beta\Rightarrow\beta}{\dfrac{\sim\sim\alpha,\alpha\to\beta\Rightarrow\sim\sim\beta}{\dfrac{\alpha\to\beta\Rightarrow\sim\neg\sim\alpha,\sim\sim\beta}{\dfrac{\alpha\to\beta\Rightarrow\sim(\neg\sim\alpha\wedge\sim\beta)}{\sim\sim(\alpha\to\beta)\Rightarrow\sim(\neg\sim\alpha\wedge\sim\beta)}}}}}$$

$$\frac{\dfrac{\alpha\Rightarrow\alpha}{\dfrac{\alpha\Rightarrow\sim\sim\alpha}{\dfrac{\alpha\Rightarrow\beta,\sim\sim\alpha}{\dfrac{\alpha,\sim\neg\sim\alpha\Rightarrow\beta}{\sim\neg\sim\alpha\Rightarrow\alpha\to\beta}}}} \quad \dfrac{\dfrac{\beta\Rightarrow\beta}{\dfrac{\sim\sim\beta\Rightarrow\beta}{\dfrac{\alpha,\sim\sim\beta\Rightarrow\beta}{\sim\sim\beta\Rightarrow\alpha\to\beta}}}}{}}{\dfrac{\sim(\neg\sim\alpha\wedge\sim\beta)\Rightarrow\alpha\to\beta}{\sim(\neg\sim\alpha\wedge\sim\beta)\Rightarrow\sim\sim(\alpha\to\beta)}}$$

(2) Case (6): In this case, we show only the following cases:

$$\frac{\dfrac{\alpha[z/x]\Rightarrow\alpha[z/x]}{\dfrac{\alpha[z/x]\Rightarrow\sim\sim\alpha[z/x]}{\dfrac{\forall x\alpha\Rightarrow\sim\sim\alpha[z/x]}{\dfrac{\forall x\alpha\Rightarrow\sim\exists x\sim\alpha}{\sim\sim(\forall x\alpha)\Rightarrow\sim\exists x\sim\alpha}}}}}{}$$

$$\frac{\dfrac{\alpha[z/x]\Rightarrow\alpha[z/x]}{\dfrac{\sim\sim\alpha[z/x]\Rightarrow\alpha[z/x]}{\dfrac{\sim\exists x\sim\alpha\Rightarrow\alpha[z/x]}{\dfrac{\sim\exists x\sim\alpha\Rightarrow\forall x\alpha}{\sim\exists x\sim\alpha\Rightarrow\sim\sim(\forall x\alpha)}}}}}{} \quad\triangleleft$$

Remark 22. We make the following remarks on the strong equivalence. It is known that $\sim(\alpha\to\beta) \leftrightarrow_s \alpha\wedge\sim\beta$ does not hold for the standard Gentzen-type sequent calculus for Nelson's paraconsistent four-valued logic N4. Actually, $\sim(\alpha\to\beta) \leftrightarrow \alpha\wedge\sim\beta$ is a characteristic axiom scheme of N4, but $\sim\sim(\alpha\to\beta) \leftrightarrow \sim(\alpha\wedge\sim\beta)$ is not a theorem of N4. See e.g., [44; 43] for more information on strong equivalence in some variants of N4.

Theorem 23 (Strong equivalence replacement for FBD+). *Let α be a subformula of a formula γ, and γ^* be the formula obtained from γ by replacing an occurrence of α with that of β. Then, we have:*

$$\text{If } \alpha \leftrightarrow_s \beta, \text{ then } \gamma \leftrightarrow_s \gamma^*.$$

Proof. We use induction on γ, and we show some cases.

1. Case $\gamma \equiv \gamma_1 \to \gamma_2$: It is sufficient to show $\vdash \gamma_1 \to \gamma_2 \Leftrightarrow (\gamma_1 \to \gamma_2)^*$ and $\vdash \sim(\gamma_1 \to \gamma_2) \Leftrightarrow (\sim(\gamma_1 \to \gamma_2))^*$ where $(\gamma_1 \to \gamma_2)^*$ and $(\sim(\gamma_1 \to \gamma_2))^*$ coincide with $\gamma_1^* \to \gamma_2^*$ and $\sim(\gamma_1^* \to \gamma_2^*)$, respectively. We show only the latter case. By induction hypothesis, we have $\gamma_i \leftrightarrow_s \gamma_i^*$ ($i \in \{1,2\}$), i.e., $\vdash \gamma_i \Leftrightarrow \gamma_i^*$ and $\vdash \sim\gamma_i \Leftrightarrow \sim\gamma_i^*$. We thus obtain the required fact:

$$\cfrac{\cfrac{\text{ind. hyp.} \vdots}{\sim\gamma_1^* \Rightarrow \sim\gamma_1}\quad \cfrac{\text{ind. hyp.} \vdots}{\sim\gamma_2 \Rightarrow \sim\gamma_2^*}}{\cfrac{\sim\gamma_2, \sim\gamma_1^* \Rightarrow \sim\gamma_1 \quad \sim\gamma_2 \Rightarrow \sim\gamma_2^*, \sim\gamma_1}{\cfrac{\sim\gamma_1^*, \sim(\gamma_1 \to \gamma_2) \Rightarrow \quad \sim(\gamma_1 \to \gamma_2) \Rightarrow \sim\gamma_2^*}{\sim(\gamma_1 \to \gamma_2) \Rightarrow (\sim(\gamma_1 \to \gamma_2))^*}}}$$

$$\cfrac{\cfrac{\text{ind. hyp.} \vdots}{\sim\gamma_1 \Rightarrow \sim\gamma_1^*}\quad \cfrac{\text{ind. hyp.} \vdots}{\sim\gamma_2^* \Rightarrow \sim\gamma_2}}{\cfrac{\sim\gamma_2^*, \sim\gamma_1 \Rightarrow \sim\gamma_1^* \quad \sim\gamma_2^* \Rightarrow \sim\gamma_2, \sim\gamma_1^*}{\cfrac{\sim\gamma_1, (\sim(\gamma_1 \to \gamma_2))^* \Rightarrow \quad (\sim(\gamma_1 \to \gamma_2))^* \Rightarrow \sim\gamma_2}{(\sim(\gamma_1 \to \gamma_2))^* \Rightarrow \sim(\gamma_1 \to \gamma_2)}}}$$

2. Case $\gamma \equiv \neg\delta$: It is sufficient to show $\vdash \neg\delta \Leftrightarrow (\neg\delta)^*$ and $\vdash \sim\neg\delta \Leftrightarrow (\sim\neg\delta)^*$ where $(\neg\delta)^*$ and $(\sim\neg\delta)^*$ coincide with $\neg\delta^*$ and $\sim\neg\delta^*$, respectively. We show only the latter case. By induction hypothesis, we have $\delta \leftrightarrow_s \delta^*$, i.e., $\vdash \delta \Leftrightarrow \delta^*$ and $\vdash \sim\delta \Leftrightarrow \sim\delta^*$. We thus obtain the required fact:

$$\cfrac{\cfrac{\text{ind. hyp.} \vdots}{\sim\delta^* \Rightarrow \sim\delta}}{\cfrac{\Rightarrow (\sim\neg\delta)^*, \sim\delta}{\sim\neg\delta \Rightarrow (\sim\neg\delta)^*}} \qquad \cfrac{\cfrac{\text{ind. hyp.} \vdots}{\sim\delta \Rightarrow \sim\delta^*}}{\cfrac{\Rightarrow \sim\neg\delta, \sim\delta^*}{(\sim\neg\delta)^* \Rightarrow \sim\neg\delta}}$$

3. Case $\gamma \equiv \sim\delta$: It is sufficient to show $\vdash \sim\delta \Leftrightarrow (\sim\delta)^*$ and $\vdash \sim\sim\delta \Leftrightarrow (\sim\sim\delta)^*$ where $(\sim\delta)^*$ and $(\sim\sim\delta)^*$ coincide with $\sim\delta^*$ and $\sim\sim\delta^*$, respectively. We show both cases below. By induction hypothesis, we have $\delta \leftrightarrow_s \delta^*$, i.e., $\vdash \delta \Leftrightarrow \delta^*$ and $\vdash \sim\delta \Leftrightarrow \sim\delta^*$. Thus, the former case is immediately obtained from the induction hypothesis. For the latter case, we obtain the required fact:

$$\cfrac{\cfrac{\text{ind. hyp.} \vdots}{\delta \Rightarrow \delta^*}}{\cfrac{\delta \Rightarrow (\sim\sim\delta)^*}{\sim\sim\delta \Rightarrow (\sim\sim\delta)^*}} \qquad \cfrac{\cfrac{\text{ind. hyp.} \vdots}{\delta^* \Rightarrow \delta}}{\cfrac{\delta^* \Rightarrow \sim\sim\delta}{(\sim\sim\delta)^* \Rightarrow \sim\sim\delta}}$$

4. Case $\gamma \equiv \forall x \delta$: It is sufficient to show $\vdash \forall x\delta \Leftrightarrow (\forall x\delta)^*$ and $\vdash \sim\forall x\delta \Leftrightarrow (\sim\forall x\delta)^*$ where $(\forall x\delta)^*$ and $(\sim\forall x\delta)^*$ coincide with $\forall x\delta^*$ and $\sim\forall x\delta^*$, respectively. We

show both cases below. By induction hypothesis, we have $\delta[z/x] \leftrightarrow_s \delta[z/x]^*$, i.e., $\vdash \delta[z/x] \Leftrightarrow \delta[z/x]^*$ and $\vdash \sim\delta[z/x] \Leftrightarrow \sim\delta[z/x]^*$. Thus, the former case immediately obtained from the induction hypothesis. For the latter case, we obtain the required fact:

$$\begin{array}{cc}
\text{ind. hyp.} \vdots & \vdots \text{ ind. hyp.} \\
\sim\delta[z/x] \Rightarrow \sim\delta[z/x]^* & \sim\delta[z/x]^* \Rightarrow \sim\delta[z/x] \\
\hline
\sim\delta[z/x] \Rightarrow (\sim\forall x\delta)^* & \sim\delta[z/x]^* \Rightarrow \sim\forall x\delta \\
\hline
\sim\forall x\delta \Rightarrow (\sim\forall x\delta)^* & (\sim\forall x\delta)^* \Rightarrow \sim\forall x\delta
\end{array}$$

◁

Prior to showing the Herbrand theorem for FBD+, we need to introduce some notions.

Definition 24. A formula of the form p or $\sim p$ where p is an atomic formula is called a *quasi-atomic formula*. A formula α is called a *quasi-literal* if α is a quasi-atomic formula or a formula of the form $\neg\beta$ where β is a quasi-atomic formula. A quasi-literal α is called the complement of a quasi-literal β if $\alpha \equiv \neg\beta$ or $\beta \equiv \neg\alpha$. The complement of a quasi-literal α is denoted as α^*, (i.e., $\alpha^* \equiv \neg\gamma$ if $\alpha \equiv \gamma$, and $\alpha^* \equiv \gamma$ if $\alpha \equiv \neg\gamma$, where γ is a quasi-atomic formula).

Definition 25. A *quasi-disjunctive normal form* of a $\{\forall, \exists\}$-free formula (of FBD+) is obtained from a usual disjunctive normal form of a formula (of FLK) by replacing (the part of) "literal" with "quasi-literal."

A *quasi-Skolem normal form* of a formula (of FBD+) is obtained from a usual Skolem normal form by replacing "literal" with "quasi-literal," i.e., a formula of the form $\exists x_1 \cdots \exists x_m (\alpha_1 \vee \cdots \vee \alpha_n)$ ($0 \leq m$, $1 \leq n$) is a quasi-Skolem normal form if $\alpha_1 \vee \cdots \vee \alpha_n$ is a quasi-disjunctive normal form with some usual Skolem functions and $x_1, \ldots, x_m$ are the mutually distinct free individual variables occurring in $\alpha_1 \vee \cdots \vee \alpha_n$.

A formula β is called an *instance* of a quasi-Skolem normal form $\exists x_1 \cdots \exists x_m (\alpha_1 \vee \cdots \vee \alpha_n)$ if β is obtained from $\alpha_1 \vee \cdots \vee \alpha_n$ by replacing $x_1, \ldots, x_m$ with any terms, i.e., β is a substitution instance of $\alpha_1 \vee \cdots \vee \alpha_n$.

Using Theorem 6, Proposition 21 and Theorem 23, we obtain the following theorem.

Theorem 26 (Existence of quasi-Skolem normal form for FBD+). *For any formula α, we can construct a quasi-Skolem normal form α^s such that* FBD+ $\vdash \Rightarrow \alpha$ *if and only if* FBD+ $\vdash \Rightarrow \alpha^s$.

Proof. Let Φ be a set of atomic formulas, and Φ' be the set $\{p' : p \in \Phi\}$ of atomic formulas. Consider a Φ-based language of FBD+ and a $\{\Phi, \Phi'\}$-based language of FLK. We show the way of constructing a quasi-Skolem normal form α^s from an arbitrary formula in FBD+. First, each occurrence of the connective $\sim$ occurring in α moves to the most inner position of all the other connectives by the virtue of Proposition 21 and Theorem 23. Then, we obtain the formula in which all occurrences of $\sim$ are of the form $\overbrace{\sim\cdots\sim}^{n} p$ where p is an atomic formula occurring in α. By using Proposition 21 and Theorem 23, we reduce such occurrences of $\sim$ by the following way. If n is odd, then we replace $\overbrace{\sim\cdots\sim}^{n} p$ with $\sim p$, and if n is even, then we replace $\overbrace{\sim\cdots\sim}^{n} p$ with p. The formula β obtained in such a way is logically equivalent to α by Proposition 21 and Theorem 23. Hence, we have FBD+ $\vdash \Rightarrow \alpha$ iff FBD+ $\vdash \Rightarrow \beta$. Next, we transform each formula of the form $\sim p$ appearing in β into $p' \in \Phi'$. This transformation is justified by Theorem 6. Then, we construct a Skolem normal form β^s of FLK in the standard way. Finally, a quasi-Skolem normal form α^s in FBD+ is obtained from β^s by replacing all the occurrences of atomic formulas of the form p' ($p' \in \Phi'$) with $\sim p$. This replacement is also justified by Theorem 6. ◁

Remark 27. We illustrate a quasi-Skolem normal form by an example. Consider a formula
$$\alpha \equiv \forall x \exists y \exists z \forall w (\sim p(x,y,z,w) \to p(x,y,z,w)),$$
where p is a predicate symbol and x, y, z, w are mutually distinct individual variables. Then, we can obtain the following quasi-Skolem normal form α^s of α
$$\alpha^s \equiv \exists y \exists z (\neg \sim p(c,y,z,f_s(y,z)) \vee p(c,y,z,f_s(y,z))),$$
where c is a new individual constant and f_s is a Skolem function.

The following Herbrand theorem, which is a purely syntactic formulation, is well known.

Theorem 28 (Herbrand theorem for FLK). *For any Skolem normal form α, FLK $\vdash \Rightarrow \alpha$ if and only if there is a finite set Δ of instances of α such that FLK $\vdash \Rightarrow \Delta$.*

Remark 29. We make the following remarks on the Herbrand theorem.
1. The direct syntactic proof of Theorem 28 is presented, for example, in [17].
2. An alternative semantic formulation of the Herbrand theorem also holds for the standard semantics for FLK (see e.g., [8]).
3. Assuming the completeness theorem for FLK, these two syntactic and semantic formulations represent the same thing.
4. The present syntactic formulation can also be obtained as a corollary of Gentzen's mid-sequent theorem (see e.g., [41]).

Using Theorems 6 and 28, we obtain the following Herbrand theorem for FBD+.

Theorem 30 (Herbrand theorem for FBD+). *For any quasi-Skolem normal form α, FBD+ $\vdash \Rightarrow \alpha$ if and only if there is a finite set Δ of instances of α such that FBD+ $\vdash \Rightarrow \Delta$.*

Proof. Let α be a quasi-Skolem normal form $\exists x_1 \cdots \exists x_m (\alpha_1 \vee \cdots \vee \alpha_n)$ ($0 \leq m$, $1 \leq n$), and FBD+ $\vdash \Rightarrow \alpha$. By Theorems 6 and 28, we have: FBD+ $\vdash \Rightarrow \alpha$ iff FLK $\vdash \Rightarrow f(\alpha)$ iff FLK $\vdash \Rightarrow f(\Delta)$, where $f(\alpha)$ is obtained from α by replacing all the occurrences of the negated atomic formulas of the form $\sim p$ with p' (i.e., $f(\sim p)$), and $f(\Delta)$ denotes a finite set of instances of $f(\alpha)$. Note that $f(\Delta)$ is of the form $\{f(\alpha_1^1 \vee \cdots \vee \alpha_n^1), f(\alpha_1^2 \vee \cdots \vee \alpha_n^2), \ldots, f(\alpha_1^l \vee \cdots \vee \alpha_n^l)\}$, where $f(\alpha_1^j \vee \cdots \vee \alpha_n^j)$ ($j \in \{1, 2, \ldots, l\}$) is obtained from $f(\alpha_1 \vee \cdots \vee \alpha_n)$ by replacing $x_1, \ldots, x_m$ with some terms. By Theorem 6, we then obtain FLK $\vdash \Rightarrow f(\Delta)$ iff FBD+ $\vdash \Rightarrow \Delta$, where Δ is obtained from $f(\Delta)$ by replacing all the occurrences of $f(\sim p)$ (i.e., p') with $\sim p$, and hence, Δ is a finite set of instances of α. Note that Δ is of the form $\{\alpha_1^1 \vee \cdots \vee \alpha_n^1, \alpha_1^2 \vee \cdots \vee \alpha_n^2, \ldots, \alpha_1^l \vee \cdots \vee \alpha_n^l\}$, where $\alpha_1^j \vee \cdots \vee \alpha_n^j$ ($j \in \{1, 2, \ldots, l\}$) is obtained from $\alpha_1 \vee \cdots \vee \alpha_n$ by replacing $x_1, \ldots, x_m$ with some terms. ◁

5. OTHER THEOREMS

5.1. Craig Interpolation and Maksimova Separation Theorems.
In what follows, we show the Craig interpolation and Maksimova separation (Maksimova principle of variable separation) theorems for FBD+ by using a similar embedding-based proof method proposed and studied in [18; 19; 26].

Remark 31. We make the following remarks on Craig interpolation and Maksimova separation theorems.

1. The Craig interpolation theorem for FLK is well known [9; 41].
2. The Maksimova separation theorem for FLK is well known. This theorem can be derived from the Craig interpolation theorem for FLK.
3. Maksimova separation theorem was originally proved in [31] for some relevant logics, wherein an example of a relevant logic for which Maksimova separation theorem doesn't hold was also shown.

To show the theorems for FBD+, we now assume a slightly simplified first-order language without individual constants and function symbols. An expression $V(\alpha)$ is used to denote the set of all predicate symbols occurring in α.

Theorem 32 (Craig interpolation for FBD+). *Suppose* FBD+ $\vdash \alpha \Rightarrow \beta$ *for any formulas α and β. If $V(\alpha) \cap V(\beta) \neq \emptyset$, then there exists a formula γ such that*

1. FBD+ $\vdash \alpha \Rightarrow \gamma$ *and* FBD+ $\vdash \gamma \Rightarrow \beta$;
2. $V(\gamma) \subseteq V(\alpha) \cap V(\beta)$.

If $V(\alpha) \cap V(\beta) = \emptyset$, then

3. FBD+ $\vdash \Rightarrow \neg \alpha$ *or* FBD+ $\vdash \Rightarrow \beta$.

Proof. (Sketch) We give a sketch of the proof. Prior to proving the theorem, we need to prove the following statement (*):

Let I_p be $\{p, p'\}$ where $p \in \Phi$ and $p' \in \Phi'$. Let f be the mapping defined in Definition 4. For any atomic formula p in $\mathcal{L}_{\text{FBD+}}$ and any formula α in $\mathcal{L}_{\text{FBD+}}$,

1. $p \in V(\alpha)$ iff $q \in V(f(\alpha))$ for some $q \in I_p$,
2. $p \in V(\sim \alpha)$ iff $q \in V(f(\sim \alpha))$ for some $q \in I_p$.

This statement is proved by (simultaneous) induction on α. In the following discussion, the subscript p of I_p is omitted for the sake of brevity. The base step is proved as follows. For 1, we have that $p \in V(p)$ iff $p = f(p) \in V(f(p))$ by the definition of f. For 2, we have that $p \in V(\sim p)$ iff $p' = f(\sim p) \in V(f(\sim p))$, by the definition of f. For the induction step, we show only the case $\alpha \equiv \beta \rightarrow \gamma$ as follows. For 1, we obtain that $p \in V(\beta \rightarrow \gamma)$ iff $p \in V(\beta)$ or $p \in V(\gamma)$ iff $[r \in V(f(\beta))$ for some $r \in I]$ or $[s \in V(f(\gamma))$ for some $s \in I]$ (by induction hypothesis for 1) iff $q \in V(f(\beta) \rightarrow f(\gamma))$ for some $q \in I$ iff $q \in V(f(\beta \rightarrow \gamma))$ for some $q \in I$ (by the definition of f). For 2, we obtain that $p \in V(\sim(\beta \rightarrow \gamma))$ iff $p \in V(\sim \beta)$ or $p \in V(\sim \gamma)$ iff $[r \in V(f(\sim \beta))$ for some $r \in I]$ or $[s \in V(f(\sim \gamma))$ for some $s \in I]$ (by induction hypothesis for 2) iff

$q \in V(f(\sim\beta) \wedge f(\sim\gamma))$ for some $q \in I$ iff $q \in V(\neg f(\sim\beta) \wedge f(\sim\gamma))$ for some $q \in I$ iff $q \in V(f(\sim(\beta \rightarrow \gamma)))$ for some $q \in I$ (by the definition of f).

Next, using the statement (*), we show the following statement (**):

Let f be the mapping defined in Definition 4. For any formulas α and β in $\mathcal{L}_{\text{FBD+}}$, if $V(f(\alpha)) \subseteq V(f(\beta))$, then $V(\alpha) \subseteq V(\beta)$.

This statement can be proved as follows. Assume $V(f(\alpha)) \subseteq V(f(\beta))$, and let $p \in V(\alpha)$. Then, by the statement (*), we obtain $q \in V(f(\alpha))$ for some $q \in I$. By the assumption, $q \in V(f(\beta))$ for some $q \in I$, and hence, $p \in V(\beta)$, by the statement (*).

Using the statement (**), we can obtain the required theorem as follows.

First, we prove the case $V(\alpha) \cap V(\beta) \neq \emptyset$ as follows. Suppose FBD+ $\vdash \alpha \Rightarrow \beta$ and $V(\alpha) \cap V(\beta) \neq \emptyset$. Then, we have FLK $\vdash f(\alpha) \Rightarrow f(\beta)$, by Theorem 6. By the Craig interpolation theorem for FLK, we have the following: There exists a formula γ in $\mathcal{L}_{\text{FLK}}$ such that

1. FLK $\vdash f(\alpha) \Rightarrow \gamma$ and FLK $\vdash \gamma \Rightarrow f(\beta)$,
2. $V(\gamma) \subseteq V(f(\alpha)) \cap V(f(\beta))$.

Since γ is a formula of FLK, γ is regarded as in $\mathcal{L}^* = \mathcal{L}_{\text{FLK}} - \Phi'$ ($\subseteq \mathcal{L}_{\text{FBD+}}$). Then, we have the fact $\gamma = f(\gamma)$ for any $\gamma \in \mathcal{L}^*$. This fact can be shown by induction on γ. By Theorem 6, we thus obtain that there exists a formula γ such that

1. FBD+ $\vdash \alpha \Rightarrow \gamma$ and FBD+ $\vdash \gamma \Rightarrow \beta$,
2. $V(f(\gamma)) \subseteq V(f(\alpha)) \cap V(f(\beta))$.

Now it is sufficient to show that $V(f(\gamma)) \subseteq V(f(\alpha)) \cap V(f(\beta))$ implies $V(\gamma) \subseteq V(\alpha) \cap V(\beta)$. This can be shown by the statement (**).

Second, we prove the case $V(\alpha) \cap V(\beta) = \emptyset$ as follows. Suppose FBD+ $\vdash \alpha \Rightarrow \beta$ and $V(\alpha) \cap V(\beta) = \emptyset$. Then, we have FLK $\vdash f(\alpha) \Rightarrow f(\beta)$ by Theorem 6. We also have ($\star$): $V(f(\alpha)) \cap V(f(\beta)) = \emptyset$. To show this, it is sufficient to prove that $V(\alpha) \cap V(\beta) = \emptyset$ implies $V(f(\alpha)) \cap V(f(\beta)) = \emptyset$. We now show the contrapositive, i.e., $V(f(\alpha)) \cap V(f(\beta)) \neq \emptyset$ implies $V(\alpha) \cap V(\beta) \neq \emptyset$. Suppose $q \in V(f(\alpha)) \cap V(f(\beta))$ with $q \in \Phi \cup \Phi'$. If q is of the form $f(p) = p \in \Phi$, then we obviously have $p \in V(\alpha) \cap V(\beta)$. If q is of the form $f(\sim p) = p' \in \Phi'$, then we have $p \in V(\alpha) \cap V(\beta)$. Therefore we obtain ($\star$). Thus, by the Craig interpolation theorem for FLK, we have: FLK $\vdash \Rightarrow \neg f(\alpha)$ or FLK $\vdash \Rightarrow f(\beta)$, where $\neg f(\alpha)$ coincides with $f(\neg\alpha)$ by the definition of f. By Theorem 6, we thus obtain the required fact, that is, FBD+ $\vdash \Rightarrow \neg\alpha$ or FBD+ $\vdash \Rightarrow \beta$. ◁

Theorem 33 (Maksimova separation for FBD+). *If* FBD+ $\vdash \alpha_1 \wedge \beta_1 \Rightarrow \alpha_2 \vee \beta_2$ *for any formulas* $\alpha_1, \alpha_2, \beta_1$ *and* β_2 *with* $V(\alpha_1, \alpha_2) \cap V(\beta_1, \beta_2) \neq \emptyset$, *then either* FBD+ $\vdash \alpha_1 \Rightarrow \alpha_2$ *or* FBD+ $\vdash \beta_1 \Rightarrow \beta_2$.

Proof. Suppose $V(\alpha_1, \alpha_2) \cap V(\beta_1, \beta_2) \neq \emptyset$ and FBD+ $\vdash \alpha_1 \wedge \beta_1 \Rightarrow \alpha_2 \vee \beta_2$. Then, we have that FBD+ $\vdash \alpha_1, \beta_1 \Rightarrow \alpha_2, \beta_2$. We remark that to show this fact, we need to prove the invertibility of the logical inference rules concerning $\wedge$ and $\vee$, but this invertibility can straightforwardly be shown. Then, by this fact, we obtain FBD+ $\vdash \alpha_1, \neg\alpha_2 \Rightarrow \neg\beta_1, \beta_2$. Thus, we obtain FBD+ $\vdash \alpha_1 \wedge \neg\alpha_2 \Rightarrow \neg\beta_1 \vee \beta_2$. By Theorem 32, we obtain FBD+ $\vdash \Rightarrow \neg(\alpha_1 \wedge \neg\alpha_2)$ or FBD+ $\vdash \Rightarrow \neg\beta_1 \vee \beta_2$. We thus obtain the required fact FBD+ $\vdash \alpha_1 \Rightarrow \alpha_2$ or FBD+ $\vdash \beta_1 \Rightarrow \beta_2$ by:

or
$$\Rightarrow \neg(\alpha_1 \wedge \neg\alpha_2) \quad \cfrac{\cfrac{\cfrac{\alpha_1 \Rightarrow \alpha_1}{\alpha_1 \Rightarrow \alpha_2, \alpha_1} \quad \cfrac{\vdots}{\cfrac{\alpha_2 \Rightarrow \alpha_2}{\alpha_2, \alpha_1 \Rightarrow \alpha_2}}}{\cfrac{\alpha_1 \Rightarrow \alpha_2, \alpha_1 \wedge \neg\alpha_2}{\neg(\alpha_1 \wedge \neg\alpha_2), \alpha_1 \Rightarrow \alpha_2}}}{\alpha_1 \Rightarrow \alpha_2} \text{ (cut)}$$

$$\Rightarrow \neg\beta_1 \vee \beta_2 \quad \cfrac{\cfrac{\cfrac{\beta_1 \Rightarrow \beta_1}{\beta_1 \Rightarrow \beta_1, \beta_2}}{\cfrac{\neg\beta_1, \beta_1 \Rightarrow \beta_2}{\neg\beta_1 \vee \beta_2, \beta_1 \Rightarrow \beta_2}} \quad \cfrac{\vdots}{\cfrac{\beta_2 \Rightarrow \beta_2}{\beta_2, \beta_1 \Rightarrow \beta_2}}}{\beta_1 \Rightarrow \beta_2} \text{ (cut)}$$

respectively. ◁

5.2. Equivalence and Cut-elimination Theorems for another Sequent Calculus.

Next, we introduce an alternative cut-free Gentzen-type sequent calculus FBD+° and prove the theorem-equivalence between cut-free FBD+° and cut-free FBD+.

Definition 34 (FBD+°). The system FBD+° is obtained from FBD+ by replacing the negated logical inference rules ($\sim\rightarrow$left) and ($\sim\rightarrow$right) with the following negated logical inference rules:

$$\cfrac{\sim\neg\alpha, \sim\beta, \Gamma \Rightarrow \Delta}{\sim(\alpha \rightarrow \beta), \Gamma \Rightarrow \Delta} \;(\sim\rightarrow\text{left}^\star) \qquad \cfrac{\Gamma \Rightarrow \Delta, \sim\neg\alpha \quad \Gamma \Rightarrow \Delta, \sim\beta}{\Gamma \Rightarrow \Delta, \sim(\alpha \rightarrow \beta)} \;(\sim\rightarrow\text{right}^\star)$$

Remark 35. In [22], the negated logical inference rules ($\sim\rightarrow$left*) and ($\sim\rightarrow$right*) were introduced for the propositional system BD+. These negated logical inference rules correspond to the slightly modified Hilbert-style axiom scheme $\sim(\alpha \rightarrow \beta) \leftrightarrow (\sim\neg\alpha \wedge \sim\beta)$ from the original axiom scheme $\sim(\alpha \rightarrow \beta) \leftrightarrow (\neg\sim\alpha \wedge \sim\beta)$.

By contrast to FBD+*, we obtain the following nice property for FBD+°, which implies the cut-elimination theorem for FBD+°.

Theorem 36 (Cut-free equivalence between FBD+ and FBD+°). *The systems* FBD+ $-$ (cut) *and* FBD+° $-$ (cut) *are theorem-equivalent.*

Proof. (Sketch) We give a sketch of the proof. Prior to showing the theorem equivalence, we remark that the following rules are admissible in cut-free FBD+:

$$\cfrac{\sim\neg\alpha, \Gamma \Rightarrow \Delta}{\Gamma \Rightarrow \Delta, \sim\alpha} \;(\sim\neg\text{left}^{-1}) \qquad \cfrac{\Gamma \Rightarrow \Delta, \sim\neg\alpha}{\sim\alpha, \Gamma \Rightarrow \Delta} \;(\sim\neg\text{right}^{-1})$$

This fact can be proved in a similar way as for Proposition 9.

Then, the admissibility of ($\sim\rightarrow$left*) and ($\sim\rightarrow$right*) in cut-free FBD+ is proved as follows:

$$\dfrac{\dfrac{\vdots}{\sim\neg\alpha,\sim\beta,\Gamma\Rightarrow\Delta}\ (\sim\neg\text{left}^{-1})}{\dfrac{\sim\beta,\Gamma\Rightarrow\Delta,\sim\alpha}{\sim(\alpha\rightarrow\beta),\Gamma\Rightarrow\Delta}\ (\sim\rightarrow\text{left})}$$

$$(\sim\neg\text{right}^{-1})\ \dfrac{\dfrac{\Gamma\Rightarrow\Delta,\sim\neg\alpha}{\sim\alpha,\Gamma\Rightarrow\Delta}\quad \dfrac{\vdots}{\Gamma\Rightarrow\Delta,\sim\beta}}{\Gamma\Rightarrow\Delta,\sim(\alpha\rightarrow\beta)}\ (\sim\rightarrow\text{right})$$

where $(\sim\neg\text{left}^{-1})$ and $(\sim\neg\text{right}^{-1})$ are admissible in cut-free FBD+.

The derivability of $(\sim\rightarrow\text{left})$ and $(\sim\rightarrow\text{right})$ in cut-free FBD+° is proved as follows:

$$\dfrac{\dfrac{\vdots}{\sim\beta,\Gamma\Rightarrow\Delta,\sim\alpha}\ (\sim\neg\text{left})}{\dfrac{\sim\neg\alpha,\sim\beta,\Gamma\Rightarrow\Delta}{\sim(\alpha\rightarrow\beta),\Gamma\Rightarrow\Delta}\ (\sim\rightarrow\text{left}^\star)}$$

$$(\sim\neg\text{right})\ \dfrac{\dfrac{\sim\alpha,\Gamma\Rightarrow\Delta}{\Gamma\Rightarrow\Delta,\sim\neg\alpha}\quad \dfrac{\vdots}{\Gamma\Rightarrow\Delta,\sim\beta}}{\Gamma\Rightarrow\Delta,\sim(\alpha\rightarrow\beta)}\ (\sim\rightarrow\text{right}^\star)$$

Therefore, FBD+ − (cut) and FBD+° − (cut) are theorem-equivalent. ◁

Theorem 37 (Cut elimination for FBD+°)**.** *The rule* (cut) *is admissible in cut-free* FBD+°.

Proof. By Theorems 36 and 7. ◁

Remark 38. Using Theorems 36 and 37, we can obtain the embedding, Herbrand and contraposition-elimination theorems for FBD+° as well as the Craig interpolation and Maksimova separation theorems for it.

6. Concluding Remarks

Belnap–Dunn logic is considered useful for handling inconsistency-tolerant reasoning with indefinite (vague and incomplete) information by the paraconsistent negation connective $\sim$. De and Omori's extended propositional Belnap–Dunn logic with classical negation (BD+) is also useful for handling classical logic-based normal reasoning with definite (crisp and complete) information by the classical negation connective $\neg$. A useful first-order extension of Belnap–Dunn logic and its neighbors is required for developing an expressive automated theorem proving framework that can appropriately handle both inconsistency-tolerant and classical logic-based normal reasoning with both indefinite and definite information. The Gentzen-type sequent calculus FBD+ for a first-order extension of De and Omori's BD+ is regarded as a first-order extension of this type. However, the Herbrand and contraposition-elimination theorems (and other fundamental theorems) for FBD+ have yet to be proved. In this study, we proved the Herbrand, contraposition-elimination, Craig interpolation, and Maksimova separation theorems for FBD+. The previously obtained embedding theorem [25] of FBD+ into a Gentzen-type sequent calculus FLK for first-order classical logic was effectively used for proving these theorems for FBD+. We also introduced two alternative Gentzen-type sequent calculi FBD+* and FBD+°, which are theorem-equivalent to FBD+. These theorems and Gentzen-type sequent calculi are intended to provide a proof-theoretic justification for developing FBD+-based automated theorem proving framework for appropriately handling and combining both inconsistency-tolerant and classical logic-based normal reasoning with both indefinite and definite information.

We next address a remark on a purely paraconsistent (or paradefinite) subsystem FA4 of FBD+. The subsystem FA4 is the $\neg$-free fragment of FBD+, which is regarded as a first-order extension of a Gentzen-type sequent calculus (A4) [24] for Avron's self-extensional four-valued paradefinite logic (SE4) [3]. We can show the cut-elimination, contraposition-elimination, and strong-equivalence replacement theorems for FA4. However, the Herbrand and Craig interpolation theorems with the same formulations as those for FBD+ cannot be shown, because these theorems are formulated using $\neg$ (i.e., $\neg$ plays a critical role in these theorems). We can also show some theorems for syntactically and semantically embedding FA4 into FLK. In these embedding theorems, we cannot replace FLK with the $\neg$-free fragment (i.e., positive fragment) of FLK because we require the condition $f(\sim(\alpha \to \beta)) := \neg f(\sim \alpha) \wedge f(\sim \beta)$ of the translation function, as given in Definition 4. Using these embedding theorems, we can also obtain the completeness theorem (with respect to a valuation semantics) for FA4. The valuation semantics of FA4 can be obtained from the valuation semantics [25] of FBD+ by deleting the clauses concerning $\neg$.

Next, we address a remark on a Gentzen-type sequent calculus FBDe for a first-order extension of De and Omori's propositional extension BDe [10] of Belnap–Dunn logic. Some Gentzen-type sequent calculi for BDe were also investigated in [22]. The system FBDe is obtained from FBD+ by replacing the negated logical inference rules ($\sim\to$left), ($\sim\to$right), ($\sim\neg$left), and ($\sim\neg$right) with the following negated logical inference rules:

$$\frac{\alpha, \sim\beta, \Gamma \Rightarrow \Delta}{\sim(\alpha \to \beta), \Gamma \Rightarrow \Delta} \; (\sim\to\text{left}^{\sharp}) \qquad \frac{\Gamma \Rightarrow \Delta, \alpha \quad \Gamma \Rightarrow \Delta, \sim\beta}{\Gamma \Rightarrow \Delta, \sim(\alpha \to \beta)} \; (\sim\to\text{right}^{\sharp})$$

$$\frac{\alpha, \Gamma \Rightarrow \Delta}{\sim\neg\alpha, \Gamma \Rightarrow \Delta} \; (\sim\neg\text{left}^{\sharp}) \qquad \frac{\Gamma \Rightarrow \Delta, \alpha}{\Gamma \Rightarrow \Delta, \sim\neg\alpha} \; (\sim\neg\text{right}^{\sharp})$$

These negated logical inference rules correspond to the Hilbert-style axiom schemes $\sim(\alpha \to \beta) \leftrightarrow (\alpha \wedge \sim\beta)$ and $\sim\neg\alpha \leftrightarrow \alpha$. On the one hand, we cannot prove the Herbrand, contraposition-elimination, and strong-equivalence replacement theorems for FBDe. This fact implies that the Herbrand and contraposition-elimination theorems for FBD+ are regarded as novel properties. However, we can prove the cut-elimination, completeness (with respect to a standard valuation semantics), Craig interpolation, and Maksimova separation theorems for FBDe and some theorems for syntactically and semantically embedding FBDe into the $\neg$-free fragment of FLK (i.e., a Gentzen-type sequent calculus for positive first-order classical logic).

Finally, we provide an interesting direction for future research. In this study, we examined the Herbrand and contraposition-elimination theorems for FBD+. However, we have yet to introduce some temporal and modal extensions of FBD+ and prove the Herbrand and contraposition-elimination theorems for these temporal and modal extensions. These systems and theorems are required for developing inconsistency-tolerant temporal, epistemic, etc. automated theorem proving frameworks with both indefinite and definite information. Based on these systems and theorems, we can obtain the foundations of these expressive theorem proving frameworks with some realistic applications.

Acknowledgments. We would like to thank the anonymous referee and Prof. Katalin Bimbó for their valuable comments and suggestions. This research was supported by JSPS KAKENHI Grant Numbers JP18K11171 and JP16KK0007 and Grant-in-Aid for Takahashi Industrial and Economic Research Foundation.

REFERENCES

[1] Almukdad, A. and Nelson, D. (1984). Constructible falsity and inexact predicates, *Journal of Symbolic Logic* **49**: 231–233.
[2] Arieli, O. and Avron, A. (1998). The value of the four values, *Artificial Intelligence* **102**(1): 97–141.
[3] Avron, A. (2020). The normal and self-extensional extension of Dunn–Belnap logic, *Logica Universalis* **14**(3): 281–296.
[4] Belnap, N. D. (1977a). How a computer should think, *in* G. Ryle (ed.), *Contemporary Aspects of Philosophy*, Oriel Press Ltd., Stocksfield, pp. 30–55.
[5] Belnap, N. D. (1977b). A useful four-valued logic, *in* J. M. Dunn and G. Epstein (eds.), *Modern Uses of Multiple-valued Logic*, Reidel Publishing Company, Dordrecht, pp. 8–37.
[6] Béziau, J.-Y. (2011). A new four-valued approach to modal logic, *Logique et Analyse* **54**(213): 109–121.
[7] Burgess, J. P. (2009). *Philosophical Logic*, Princeton Foundations of Contemporary Philosophy, Princeton University Press, Princeton, NJ.
[8] Buss, S. R. (1998). An introduction to proof theory, *in* S. R. Buss (ed.), *Handbook of Proof Theory*, Elsevier, pp. 1–78.
[9] Craig, W. (1957). Three uses of the Herbrand–Gentzen theorem in relating model theory and proof theory, *Journal of Symbolic Logic* **22**(3): 269–285.
[10] De, M. and Omori, H. (2015). Classical negation and expansions of Belnap–Dunn logic, *Studia Logica* **103**(4): 825–851.
[11] Dunn, J. M. (1966). *The Algebra of Intensional Logics*, Doctoral dissertation, University of Pittsburgh, Pittsburgh, PA. Published as Vol. 2 in the *Logic PhDs* series by College Publications, London (UK), 2019.
[12] Dunn, J. M. (1976). Intuitive semantics for first-degree entailment and 'coupled trees', *Philosophical Studies* **29**(3): 149–168.
[13] Dunn, J. M. (2000). Partiality and its dual, *Studia Logica* **66**(1): 5–40.
[14] Dunn, J. M. and Restall, G. (2002). Relevance logic, *in* D. Gabbay and F. Guenthner (eds.), *Handbook of Philosophical Logic*, 2nd edn, Vol. 6, Kluwer, Amsterdam, pp. 1–128.
[15] Gentzen, G. (1969). *Collected Papers of Gerhard Gentzen, M. E. Szabo, ed.*, Studies in Logic and the Foundations of Mathematics, North-Holland, Amsterdam.
[16] Gurevich, Y. (1977). Intuitionistic logic with strong negation, *Studia Logica* **36**: 49–59.

[17] Hayashi, S. (1989). *Mathematical Logic* [in Japanese], Corona Publishing Co. LTD., Tokyo, Japan.
[18] Kamide, N. (2011). Notes on Craig interpolation for LJ with strong negation, *Mathematical Logic Quarterly* **57**(4): 395–399.
[19] Kamide, N. (2015). Trilattice logic: An embedding-based approach, *Journal of Logic and Computation* **25**(3): 581–611.
[20] Kamide, N. (2018). Proof theory of paraconsistent quantum logic, *Journal of Philosophical Logic* **47**(2): 301–324.
[21] Kamide, N. (2019a). An extended paradefinite Belnap–Dunn logic that is embeddable into classical logic and vice versa, *in* A. P. Rocha, L. Steels and H. J. van den Herick (eds.), *Proceedings of the 11th International Conference on Agents and Artificial Intelligence (ICAART 2019)*, Vol. 2, SciTePress, pp. 377–387.
[22] Kamide, N. (2019b). Gentzen-type sequent calculi for extended Belnap–Dunn logics with classical negation: A general framework, *Logica Universalis* **13**(1): 37–63.
[23] Kamide, N. (2021a). Modal and intuitionistic variants of extended Belnap–Dunn logic with classical negation, *Journal of Logic, Language and Information* **30**(3): 491–531.
[24] Kamide, N. (2021b). Notes on Avron's self-extensional four-valued paradefinite logic, *Proceedings of the 51st IEEE International Symposium on Multiple-Valued Logic (ISMVL 2021)*, pp. 43–49.
[25] Kamide, N. and Omori, H. (2017). An extended first-order Belnap–Dunn logic with classical negation, *in* A. Baltag, J. Seligman and T. Yamada (eds.), *Proceedings of the 6th International Workshop on Logic, Rationality, and Interaction (LORI 2017)*, number 10455 in *Lecture Notes in Computer Science*, pp. 11–15.
[26] Kamide, N. and Shramko, Y. (2017). Embedding from multilattice logic into classical logic and vice versa, *Journal of Logic and Computation* **27**(5): 1549–1575.
[27] Kamide, N. and Wansing, H. (2010). Symmetric and dual paraconsistent logics, *Logic and Logical Philosophy* **19**(1–2): 7–30.
[28] Kamide, N. and Wansing, H. (2012). Proof theory of Nelson's paraconsistent logic: A uniform perspective, *Theoretical Computer Science* **415**: 1–38.
[29] Kamide, N. and Wansing, H. (2015). *Proof theory of N4-related paraconsistent logics*, Vol. 54 of *Studies in Logic*, College Publications, London, UK.
[30] Kamide, N. and Zohar, Y. (2020). Completeness and cut-elimination for first-order ideal paraconsistent four-valued logic, *Studia Logica* **108**(3): 549–571.
[31] Maksimova, L. (1976). The principle of separation of variables in propositional logics, *Algebra i Logika* **15**: 168–184.
[32] Méndez, J. M. and Robles, G. (2015). A strong and rich 4-valued modal logic without Łukasiewicz-type paradoxes, *Logica Universalis* **9**(4): 501–522.
[33] Nelson, D. (1949). Constructible falsity, *Journal of Symbolic Logic* **14**: 16–26.
[34] Omori, H. and H. Wansing (eds.) (2019). *New essays on Belnap–Dunn Logic*, number 418 in *Synthese Library*, Springer Nature.
[35] Priest, G. (2002). Paraconsistent logic, *in* D. Gabbay and F. Guenthner (eds.), *Handbook of Philosophical Logic*, 2nd edn, Vol. 6, Kluwer, Dordrecht, pp. 287–393.
[36] Rautenberg, W. (1979). *Klassische und nicht-klassische Aussagenlogik*, Vieweg, Braunschweig.
[37] Robinson, J. A. (1965). A machine-oriented logic based on the resolution principle, *Journal of the ACM* **12**(1): 23–41.
[38] Sano, K. and Omori, H. (2014). An expansion of first-order Belnap–Dunn logic, *Logic Journal of the IGPL* **22**(3): 458–481.

[39] Shramko, Y. (2016). Truth, falsehood, information and beyond: The American plan generalized, *in* K. Bimbó (ed.), *J. Michael Dunn on Information Based Logics*, Vol. 8 of *Outstanding Contributions to Logic*, Springer Nature, Switzerland, pp. 191–212.

[40] Shramko, Y. and Wansing, H. (2005). Some useful sixteen-valued logics: How a computer network should think, *Journal of Philosophical Logic* **34**: 121–153.

[41] Takeuti, G. (2013). *Proof Theory*, 2nd edn, Dover Publications, Inc., Mineola, NY.

[42] Vorob'ev, N. N. (1952). A constructive propositional calculus with strong negation [in Russian], *Doklady Akademii Nauk, USSR* **85**: 465–468.

[43] Wansing, H. (1993a). Informational interpretation of substructural propositional logics, *Journal of Logic, Language and Information* **2**(4): 285–308.

[44] Wansing, H. (1993b). *The Logic of Information Structures*, number 681 in *Lecture Notes in AI*, Springer, Berlin.

[45] Wansing, H. (2010). The power of Belnap: Sequent systems for $SIXTEEN_3$, *Journal of Philosophical Logic* **39**(4): 369–393.

[46] Zaitsev, D. (2009). A few more useful 8-valued logics for reasoning with tetralattice $EIGHT_4$, *Studia Logica* **92**(2): 265–280.

[47] Zaitsev, D. (2012). *Generalized Relevant Logic and Models of Reasoning*, Doctoral dissertation, Moscow State Lomonosov University, Moscow.

TEIKYO UNIVERSITY, DEPARTMENT OF INFORMATION AND ELECTRONIC ENGINEERING, TOYO-SATODAI 1–1, UTSUNOMIYA, TOCHIGI 320–8551, JAPAN, *Email:* drnkamide08@kpd.biglobe.ne.jp

A Proof of Gamma

Saul A. Kripke

ABSTRACT. This paper is dedicated to the memory of Mike Dunn. His untimely death is a loss not only to logic, computer science, and philosophy, but to all of us who knew and loved him. The paper gives an argument for closure under γ in standard systems of relevance logic (first proved by Meyer and Dunn [3]). For definiteness, I chose the example of **R**. The proof also applies to **E** and to the quantified systems **RQ** and **EQ**. The argument uses semantic tableaux (with one exceptional rule not satisfying the subformula property). It avoids the previous arguments' use of cutting down inconsistent sets of formulas to consistent sets. Like all tableau arguments, it extends partial valuations to total valuations.

Keywords. Completeness, Partial valuation, Relevance logic, Rule γ, Semantic tableau

This note gives a new proof of the closure of such systems as **R**, **E**, **RQ**, and **EQ** under Ackermann's rule γ, based on the idea of a semantic tableau.* The usual proofs of γ, beginning with Meyer and Dunn [3], all "cut down" an "inconsistent valuation" to a "consistent" one. The present proof proceeds dually: no inconsistent valuation is used, but rather a partial valuation is extended to a total valuation, as is usual with tableau completeness proofs. The proof is in fact very similar in flavor to the usual completeness proofs of tableau procedures.

For convenience, we fix our attention on **R**. We assume a usual axiomatization of **R**, with *modus ponens* for the relevant conditional and adjunction as the only rules. The system obtained by adjoining γ as an additional rule is called $\mathbf{R}_\gamma$.

We assume the reader is thoroughly familiar with semantic tableaux, originally introduced by Beth [1]. However, we will informally sketch the idea of a tableau construction in the present context. Following Smullyan [4], we use signed formulae: ordered pairs $\langle A, T \rangle$ and $\langle A, F \rangle$, where A is a formula of **R**, representing in Beth's terminology that A appears on the left or the right, respectively. A tableau is then simply a set S of signed formulae. A rule extends a tableau S to one (or two) immediate descendants, S' (S' and S'') such that $S \subseteq S'$ (and $S \subseteq S''$).

The rules for conjunction and negation are usual:

$\wedge T$ Set $S' = S \cup \{\langle A, T \rangle, \langle B, T \rangle\}$, where $\langle A \wedge B, T \rangle \in S$.
$\wedge F$ Set $S' = S \cup \{\langle A, F \rangle\}$ and $S'' = S \cup \{\langle B, F \rangle\}$, where $\langle A \wedge B, F \rangle \in S$.
$\sim T$ Set $S' = S \cup \{\langle A, F \rangle\}$, where $\langle \sim A, T \rangle \in S$.
$\sim F$ Set $S' = S \cup \{\langle A, T \rangle\}$, where $\langle \sim A, F \rangle \in S$.

The rules for disjunction are dual to those for conjunction. As for $\to$:

$\to T$ Set $S' = S \cup \{\langle A,F\rangle\}$ and $S'' = S \cup \{\langle B,T\rangle\}$, where $\langle A \to B, T\rangle \in S$.
Mpon Set $S' = S \cup \{\langle A,F\rangle\}$ and $S'' = S \cup \{\langle A \to B, F\rangle\}$, where $\langle B,F\rangle \in S$.

Mpon is rather different from the usual tableau rules in that it does *not* decompose a formula into subformulas.

A *construction* proceeds in *stages*. The initial stage is the unit set of tableaux $\{S\}$. Each stage consists in a finite set of tableaux $\{S_1,\ldots,S_n\}$. The $n+1$th stage comes from the nth by replacing some set S_i by its immediate descendant or its two immediate descendants according to one of the rules. As usual in tableau constructions, the procedure can be diagrammed as a tree, where binary branching occurs in connection with the rules $\wedge F$, $\vee T$, $\to T$, and Mpon. A tableau S is *closed* if and only if for some formula A, $\langle A,F\rangle \in S$ and *either* $\langle A,T\rangle \in S$ or A is an axiom of **R**. A stage $\{S_1,\ldots,S_n\}$ is closed if and only if each S_i is closed.

We can stipulate a fixed priority ordering for applying rules if we wish. The point is to make the stages of a construction determinate, given the initial stage. We assume that the ordering is such that every applicable rule is eventually applied. The construction for A is the construction whose initial stage is $\{\{\langle A,F\rangle\}\}$. A construction is *closed* if some one of its stages is closed.

A *valuation* is a map v whose domain is the set of formulae of **R**, and whose range is $\{T,F\}$. A valuation is *admissible* if and only if:

(i) It respects the usual conditions for truth functions.
(ii) If $v(A \to B) = v(A) = T$, then $v(B) = T$ (equivalently, given (i): if $v(A \to B) = T$, $v(A \supset B) = T$).
(iii) If A is an axiom of **R**, $v(A) = T$.

A formula A is *valid* if and only if for every admissible valuation v, $v(A) = T$.

Theorem 1. *If A is a theorem of $\mathbf{R}_\gamma$, A is a theorem of $\mathbf{R}$.*

Proof. The theorem follows from Lemmas 2–4. The crucial step is Lemma 3. ◁

Lemma 2. *If A is a theorem of $\mathbf{R}_\gamma$, A is valid.*

Proof. The axioms of $\mathbf{R}_\gamma$ are valid, and the rules preserve validity. ◁

Lemma 3. *If A is valid, the construction for A is closed.*

Proof. We prove the contrapositive. Suppose the construction for A is not closed. Then by the usual argument from König's Lemma, there is an infinite set S of signed formulae such that:[1]

(i) S is closed under the rules (e.g., for $\wedge F$, if $\langle B \wedge C, F\rangle \in S$, either $\langle B,F\rangle \in S$ or $\langle C,F\rangle \in S$; for $\wedge T$, if $\langle B \wedge C, T\rangle \in S$, $\langle B,T\rangle \in S$ and $\langle C,T\rangle \in S$; for Mpon, if $\langle B,F\rangle \in S$, then for any C, either $\langle C \to B, F\rangle \in S$ or $\langle C,F\rangle \in S$; etc.).
(ii) $\langle A,F\rangle \in S$.
(iii) For no formula B are both $\langle B,T\rangle$ and $\langle B,F\rangle \in S$.
(iv) If B is an axiom of **R**, $\langle B,F\rangle \notin S$.

Define a valuation $v(B)$ by induction on the complexity of B. If B is atomic, set $v(B) = T$ (F) if and only if $\langle B,T\rangle \in S$ ($\langle B,T\rangle \notin S$). For truth-functional formulas, define v so as to respect the truth-functions. $v(B \to C) = T$ if and only if $v(B) = F$ or $v(C) = T$, and $\langle B \to C, F\rangle \notin S$; otherwise, $v(B \to C) = F$.

We needed to show that A is not valid. This will follow if $v(A) = F$ and v is an admissible valuation. That $v(A) = F$ follows from Sublemma 3.1, given that $\langle A, F\rangle \in S$. v obviously satisfies conditions (i) and (ii) for admissibility. Condition (iii) is Sublemma 3.2.

Sublemma 3.1. For any formula B, if $\langle B, T\rangle \in S$, then $v(B) = T$; if $\langle B, F\rangle \in S$, then $v(B) = F$.

Proof. This is the usual lemma for the completeness of a tableau procedure. It is proved by induction on the number of connectives in A. If A is atomic and $\langle A, T\rangle \in S$, the result follows by the definition of v. If $\langle A, F\rangle \in S$, then $\langle A, T\rangle \notin S$. So by the definition of v, $v(A) = F$. Suppose B is $C \wedge D$, and the lemma holds for C and D. Then if $\langle B, T\rangle \in S$, then by the closure of S under the rules, $\langle C, T\rangle \in S$ and $\langle D, T\rangle \in S$, so $v(C) = v(D) = T$. So $v(B) = v(C \wedge D) = T$. If $\langle B, F\rangle \in S$, then either $\langle C, F\rangle \in S$ or $\langle D, F\rangle \in S$, so by inductive hypothesis $v(C) = F$ or $v(D) = F$, so $v(C \wedge D) = F$. Similarly, for the other truth functional formulas. If B is $C \to D$ and $\langle B, T\rangle \in S$, then either $\langle C, F\rangle \in S$ or $\langle D, T\rangle \in S$. So, by inductive hypothesis, either $v(C) = F$ or $v(D) = T$. Also, since $\langle C \to D, T\rangle \in S$, $\langle C \to D, F\rangle \notin S$, so by definition of v, $v(B) = v(C \to D) = T$. If $\langle C \to D, F\rangle \in S$, then by definition of v, $v(C \to D) = F$.

Sublemma 3.2. If B is an axiom of **R**, $v(B) = T$.

Proof. This goes case by case. We give two sample cases. The reader can verify the others.

Suppose B is $(C \to (C \to D)) \to (C \to D)$. To show $v(B) = T$, suppose for *reductio* that $v(B) = F$. Then by definition of v, since B is an implicational formula, either $\langle B, F\rangle \in S$, or $v(C \to (C \to D)) = T$ and $v(C \to D) = F$. Since B is an axiom, $\langle B, F\rangle \in S$ is impossible, so $v(C \to (C \to D)) = T$ and $v(C \to D) = F$. Since $v(C \to D) = F$, either $\langle C \to D, F\rangle \in S$, or $v(C) = T$ and $v(D) = F$. Suppose $\langle C \to D, F\rangle \in S$. Then by closure of S under *Mpon*, either $\langle (C \to (C \to D)) \to (C \to D), F\rangle \in S$ or $\langle C \to (C \to D), F\rangle \in S$. But the first alternative is impossible, as already observed, and the second is impossible, since $v(C \to (C \to D)) = T$. So $v(C) = T$ and $v(D) = F$. But then, by definition of v, $v(C \to (C \to D)) = F$. This is a contradiction.

Suppose B is $(C \to (D \to E)) \to (D \to (C \to E))$. Suppose $v(B) = F$. $\langle B, F\rangle \in S$ is impossible, so $v(C \to (D \to E)) = T$ and $v(D \to (C \to E)) = F$. So either $\langle D \to (C \to E), F\rangle \in S$ or $v(D) = T$ and $v(C \to E) = F$. In the former case, by *Mpon*, either $\langle (C \to (D \to E)) \to (D \to (C \to E)), F\rangle \in S$, or $\langle C \to (D \to E), F\rangle \in S$. Both are already ruled out, since the first is an axiom and the second contradicts $v(C \to (D \to E)) = T$. So $\langle D \to (C \to E), F\rangle \notin S$, so $v(D) = T$ and $v(C \to E) = F$. Since $v(C \to E) = F$, either $\langle C \to E, F\rangle \in S$, or $v(C) = T$ and $v(E) = F$. If $\langle C \to E, F\rangle \in S$, by *Mpon*, either $\langle D \to (C \to E), F\rangle \in S$ or $\langle D, F\rangle \in S$. But $\langle D \to (C \to E), F\rangle \in S$ has already been ruled out, and $\langle D, F\rangle \in S$ is impossible, since then, by Sublemma 2.1, $v(D) = F$, which has already been ruled out. So, $\langle C \to E, F\rangle \notin S$, hence $v(C) = T$ and $v(E) = F$. Also we already have $v(D) = T$. Hence by definition of v, $v(C \to (D \to E)) = F$. This is a contradiction. ◁

Lemma 4. *If the construction for A is closed, A is a theorem of* **R**.

Proof. Any of the usual methods of proving that a tableau procedure is contained in a corresponding axiomatic system will do. For example, let a tableau $S = \{\langle B_1, T\rangle, \ldots, \langle B_m, T\rangle, \langle C_1, F\rangle, \ldots, \langle C_n, F\rangle\}$ (m or n may $= 0$). Then define the *characteristic formula* of S as $\neg B_1 \vee \cdots \vee \neg B_m \vee C_1 \vee \cdots \vee C_n$. Note that if S is closed, its characteristic formula is provable in **R**, since either it has two disjuncts of the forms B and $\neg B$, or some disjunct is an axiom of **R**. If a stage of a construction is $\{S_1, \ldots, S_q\}$ with characteristic formulae $D_1, \ldots, D_q$, let the characteristic formula of the stage be $D_1 \wedge \cdots \wedge D_q$. Then the characteristic formula of a closed stage is provable in **R**. By inspection of the various tableau rules, if C is the characteristic formula of a non-initial stage of a construction and C' is the characteristic formula of the preceding stage, $C \to C'$ is provable in **R**. Hence by transitivity of $\to$, if C is the characteristic formula of any stage of a construction and D is the characteristic formula of the initial stage, $C \to D$ is provable in **R**. Note that the characteristic formula of the initial stage of the construction for A is A itself. So, if the construction for A is closed, and C is the characteristic formula of the closed stage, $C \to A$ and C are both theorems of **R**. Thus, A is. ◁

The proof above, except for its treatment of $\to$ and the axioms of **R**, is very close to the usual completeness proofs of tableau procedures. It is shown that the theorems of **R**, of $\mathbf{R}_\gamma$, and the valid formulae are all coextensive, though the semantical notion of validity used is of little independent interest. In Lemma 3, the partial function defined by $v(A) = T$ if $\langle A, T\rangle \in S$, and $v(A) = F$ if $\langle A, F\rangle \in S$, and undefined otherwise, is shown to extend to an admissible valuation defined on all formulae.

Although for definiteness the theorem was stated for **R**, the proof applies equally well, for example, to **E**. If quantifiers are added, as in **RQ** or **EQ**, the proof extends readily. Here we define an admissible valuation over a nonempty domain D, and the quantifiers are evaluated in the usual way. For the tableaux, quantifier rules of the usual kind are added.

Acknowledgments. My thanks to Oliver Marshall, Yale Weiss, and especially to Romina Padró for their help in producing this version. I would like to thank Katalin Bimbó for some technical help. This paper has been completed with support from the Saul A. Kripke Center at the City University of New York, Graduate Center.

Notes

[*] Unfortunately, the original handwritten manuscript of this paper was undated. Mike Dunn, however, recalled that I verbally reported this result to him in the summer of 1978 at Oxford. In an email to the Saul Kripke Center, dated February 6th, 2017, Dunn said: "Saul's communication to me was verbal without much detail. We were both visitors at Oxford in the spring of 1978 and [I] know that I at least stayed through mid-summer. I think that early summer/late spring at Oxford might be when/where he told me about his proof, but my memory is not clear on this. Anyway, I wrote up his proof, probably within a couple of months after he told me about it, and sent him a copy. I attach what I sent him. It is dated July 23, 1978, and I think was sent to him shortly after then." (See also Dunn and Meyer [2, §5].)

I originally thought that I was influenced by the result in Dunn and Meyer [2] connecting the proofs of γ to a method of proving Gentzen's cut elimination, but in the email mentioned above, Dunn said: "I sent him [me] a copy of 'Gentzen's cut ...' Feb. 12, 1980. It wasn't published until 1989, because of delay in publication of the Norman–Sylvan volume." So unless I had seen another copy or heard them give a talk, it is unlikely that I was influenced by their paper.

Other proofs of γ in relevance logic had long been around since the original paper by Meyer and Dunn [3], some of them making it much simpler. However, the present proof was novel in that it was based on the usual completeness proofs of tableau (cut-free) methods, and it is not based on cutting down an inconsistent set of statements to a consistent set.

[1] S is not a tableau of the construction but is the union of an increasing sequence of non-closed tableaux, each of which is an immediate descendant of its predecessor in the sequence.

REFERENCES

[1] Beth, E. W. (1955). Semantic entailment and formal derivability, *Mededelingen der koninklijke, Nederlandse Akademie van Wetenschappen, Afdeling Letterkunde* (Nieuwe Reeks) **18**(13): 309–342.
[2] Dunn, J. M. and Meyer, R. K. (1989). Gentzen's cut and Ackermann's gamma, *in* J. Norman and R. Sylvan (eds.), *Directions in Relevant Logic*, Kluwer, Dordrecht, pp. 229–240.
[3] Meyer, R. K. and Dunn, J. M. (1969). E, R and γ, *Journal of Symbolic Logic* **34**(3): 460–474.
[4] Smullyan, R. M. (1968). *First-order Logic*, Springer-Verlag, New York, NY.

SAUL KRIPKE CENTER, THE GRADUATE CENTER, CITY UNIVERSITY OF NEW YORK, NEW YORK, U.S.A.

MODELS FOR PRIORIAN SECOND-ORDER LOGIC

Edwin Mares

ABSTRACT. This paper is both a historical examination of Arthur Prior's approach to second-order logic in *Objects of Thought* and a formal reconstruction of it. I claim that Prior has in mind a form of supervenience thesis — that truths formulable in a second-order language supervene on first-order truths. I take this thesis, along with certain of Prior's other claims, and construct a substitutional semantics that vindicates the supervenience claims. I then argue that the logic of this semantics is ramified second-order logic.

This paper is dedicated to the memory of J. Michael Dunn. Mike was my thesis supervisor and in recent years he had become my friend. Among a great many other things, I learned from him a respect for the history of logic.

Keywords. Model theory, Arthur Prior, W. V. O. Quine, Second-order logic, Substitutional quantification

1. Introduction

Objects of Thought [11] is a book that Arthur Prior was writing at the time of his death in 1969. Although some sections are not complete, it is clear what his aim and approach is in the book: Prior describes and defends a sweeping nominalism. In particular, Prior objects to the view that the very structure of our language and the logic underlying it commit us to the existence of abstract objects, such as properties, propositions or sets. Prior adumbrates, but does not develop, a view of quantification over predicates that supposedly does not have any untoward metaphysical consequences.

My purpose in this paper is both to provide a historical account of what Prior's view in *Objects of Thought* and to formalise this view. In providing the formalisation, I do not attempt to vindicate Prior's attempt to avoid ontological commitments. I use model theory, which is based on set theory, and hence is committed to some of those entities that Prior rejects. Rather, I wish to make his view clear and precise. The use of model theory *might* be understood by readers as a scaffolding that can be discarded after it is built — and treated as a mere fiction. But I am suspicious of any such moves

---

2020 *Mathematics Subject Classification.* Primary: 03–03, Secondary: 03B16, 03C30.

Bimbó, Katalin, (ed.), *Relevance Logics and other Tools for Reasoning. Essays in Honor of J. Michael Dunn*, (Tributes, vol. 46), College Publications, London, UK, 2022, pp. 266–281.

— I find model theory genuinely explanatory in treating semantic matters and find that the explanation disappears when we claim that it is a mere fiction.[1]

The key aspect that I wish to concentrate on is what I think of as Prior's supervenience claim. A set of propositions X *supervenes* over a set of propositions Y (the supervenience base) if and only if any two worlds (or models) which differ in terms of the X propositions that they make true must also differ in terms of the Y propositions that they make true. In Prior's theory, X is the class of second-order truths and Y the class of first-order truths. He thinks of all true statements of second-order logic as being made true by the sorts of features of the world that can be expressed in first-order logic (which are, in turn, made true by truths of zeroth-order logic, i.e., predicate logic without quantifiers). We find this sort of view also in the *Tractatus*, but Wittgenstein appeals to atomic facts in making clear the sort of supervenience that he accepts, and Prior rejects facts. Rather, Prior employs John Wisdom's notion of a logical construction to state his view. The result, unlike Wittgenstein's theory, is a theory that is *essentially informal* — by its very nature it resists formalisation.

Thus, I cannot vindicate the view by formalising it. In fact, I do violence to it. But what I do is extract an important aspect of the theory and show that it has an interesting property. Contrary to what Prior says, the logic behind the sort of supervenience that Prior wants is a form of the ramified theory of types. Prior claims to reject the ramified theory in favour of the simple theory of types, but I think he displays a confusion in doing so. Moreover, the fact that this logic is ramified is not a mere artefact of my formalisation, but rather integral to the notion of supervenience.

The plan of the paper is as follows. I begin with a historical exposition of Prior's comments regarding second-order logic in *Objects of Thought*. I then describe a model theory in which every individual in the domain has a name. I use this model theory in section 6 to describe a class of "Prior models" for second-order logic. I show in section 7 that a straightforward version of the supervenience thesis holds of Prior models. In section 8, I define classes of "overhang models" and supervaluational models and show that supervenience holds of the former but not of the latter. I end by pointing out that the logic of Prior models is ramified second-order logic, contrary to Prior's explicit rejections of ramified type theory.

2. PRIOR ON SECOND-ORDER LOGIC

One of the goals of *Objects of Thought* is to present a *nominalistic* analysis of natural language, using the tools of formal logic. Prior wishes to avoid any commitment to abstract objects, in particular, to propositions, facts, sets, and properties. In the first and second chapters, he presents a strategy to avoid commitment to such entities by use of John Wisdom's theory of logical constructions, which I describe below. In the third chapter, he attempts to undermine Quine's thesis of ontological commitment. On Prior's reading, Quine thinks that bound variables in the statements that we accept indicate the entities that we thereby must accept. On Quine's view, second-order quantification commits us to the existence of sets. Thus, for example, if we accept

---

[1] In an interview, Peter Hacker told me that Prior viewed logic as a "grid" that he could lay over natural language, and the usefulness of logic is only in its ability to illuminate features of natural language, not for example to expose weaknesses of natural language.

$\forall P \forall x(Px \supset Rx)$ then we must accept both individuals (as values of "x") and sets of individuals (as values of "P"). Prior claims, to the contrary, that the acceptance of statements that include second-order quantifiers do not commit us to the existence of sets. In fact, he thinks that, for example, the sentence $\forall P \forall x(Px \supset Gx)$ does not commit us to anything other than the truth of its universal instantiations, such as $\forall x(Fx \supset Gx)$, $\forall x(Gx \supset Gx)$, $\forall x(Hx \supset Gx)$, and so on. Moreover, none of these instantiations is committed to the existence of sets.

I think that Prior misreads Quine. Quine thinks of theories rather than individual sentences as carrying ontological commitment. Our choice of a logic in which to regiment a theory carries with it ontological commitments [12]. Thus, to formulate a theory of physics, say, then if we use second-order logic that theory will not only be committed to the existence of electrons, but to the existence of sets of electrons as well. Quine prefers to use a first-order set theory to talk about sets, as opposed to a second-order logic, since the first talks about sets directly and the second does so in a rather obscure and indirect manner [13, §35]. In addition, Quine, unlike Prior, is not interested in giving a formal logical analysis of natural language. Rather, he thinks that the central philosophical purpose of formal logic is to provide a framework in which scientific theories can be regimented.[2]

Despite the fact that Prior's attack on Quine misses its mark by a wide margin, I think that Prior's position with regard to higher-order logic is extremely interesting. I read Prior as proposing a form of *supervenience* thesis about higher-order truths. A statement of the form $\forall PA(P)$, is made true by its instantiations $A(F)$, $A(G)$, and so on. These Fs and Gs are predicates of a supposedly unproblematic first-order language. Prior's view is that higher-order truths *supervene* on first-order (or perhaps zeroth-order) truths. At first, this theory might seem like a version of the view of Wittgenstein's *Tractatus* or Russell's *Lectures on Logical Atomism*, according to which higher-order truths are all reducible to first-order facts.[3] But Prior's view is interestingly different from these theories in that Prior rejects the existence of facts in chapter one of *Objects of Thought*. Prior thinks that we can talk about facts, but facts are logical fictions of a sort and are useful only as shorthand to describe features of the world that should properly not be thought of as entities in their own right.

In adopting this latter view, Prior relies on a theory of *logical constructs* and *logical fictions* developed by John Wisdom [18]. A logical fiction is a locution that seems to refer to things but may not really refer at all. For example, in "the average New Zealand family has three children" the phrase "the average New Zealand family" need not refer to any particular family and there may not be any family in New Zealand that has three children for this sentence to be true. The average New Zealand family is a logical construct and it is used to refer in a derivative sense to all New Zealand families, but need not directly refer to anything.

---

[2]Quine expresses this view extremely clearly in his unpublished paper, "Reflections on Models and Truth" [14], but it comes through also in his published works. Since I am neither interested in Quine exposition or whether Prior's attack on Quine is successful in this paper, I will leave this topic to the side now.

[3]This is not entirely true for Russell, who thought that relations might sometimes hold between relations, and are not reducible to the sorts of things that one finds in first-order facts [16, pp. 102–103].

For Prior, facts and propositions are logical fictions. Consider a situation in which Susan says "Inflation will surpass 4% next year" and Jane says "That certainly fits the facts." Here "facts" does not stand for a collection of entities of some kind, but its use allows us to indicate a much longer statement such as "the government has had to put too much money in circulation due to the COVID lockdown, the supply chain has been disrupted for a wide variety of goods, ..." This longer statement might not even be possible to be stated. It might be infinitely long. The word "facts" is a useful tool to say things of this sort that we might not want or be able to make otherwise. Propositions are likewise to be understood as façons de parler. To say that Susan believes the proposition that eating animals is wrong is to say no more or less than that Susan believes that eating animals is wrong.

Prior treats truth in a similar manner. He accepts a form of the redundancy theory of truth. On this theory, we can remove the phrase "it is true that" in sentences like "it is true that the sky is blue" without loss. "It is true that the sky is blue" means the same as "the sky is blue." The ability to rephrase sentences containing "it is true that" to sentence that do not contain that locution shows that the original sentences are not really about anything that we call "truth." Ascribing truth or falsehood to sentences is merely to agree or disagree with what they say [11, p. 11].[4] Like logical fictions such as "the average New Zealand family," "it is true that" is useful in representing our thoughts in an abbreviated manner, but it does not express a real component of those thoughts.

In doing semantics, the use of the notion of truth is often thought to be essential. We usually set out truth conditions for each connective to give a recursive theory of truth for a formal language and that this recursive theory tells us the meaning of the connective. Prior rejects the theory in which truth conditions for the connectives are represented in a metalanguage [11, ch. 7]. In fact, Prior rejects the Tarskian semantic notion of truth altogether [11, pp. 100–101].

Without the notion of truth, and without the appeal to abstract objects, Prior's semantics becomes extremely informal. Consider a sentence of the form "$\forall x A(x)$." This sentence is true in the Tarskian semantics if and only if $A(x)$ is true of everything in the domain. On Prior's view, we say that $\forall x A(x)$ (note the disappearance of the quotation marks and "is true") if and only if $A(john), A(susan), A(jane), \ldots$. I will discuss later the problem of this ellipsis. But now I point out that Prior accepted the commitment to things like John, Susan, and Jane, which can be represented by names — which can take argument position in formal sentences.

With regard to second-order logic Prior has two aims. First, he wants to hold that predicate expressions do not refer to anything. He says that verbs in general, do not have "the job of designating objects" [11, p. 35]. Second, he wants to vindicate second-order (and other higher-order) quantification as legitimate. A second-order variable for Prior stands in for a verb of some sort. For example, in the open sentence, "Peter φs Paul," the variable "φ" is a place-holder for a transitive verb. Thus, "Peter visits Paul," "Peter insults Paul," "Peter is quite rude to Paul," "Peter slanders Paul," "Peter hits Paul," "Peter murders Paul," and so on, are all values of "Peter φs Paul."

---

[4]In this, Prior's theory is very much like the prosentential theory of truth of Dorothy Grover, Joseph Camp, and Nuel Belnap [8; 7].

But neither "φ" nor any of the verbs that can replace it in a sentence refer to anything by themselves, nor do the sentences in which they occur refer to anything (since propositions and facts are just logical fictions).

This view of predicate expressions gives Prior's view of second-order quantification a rather substitutional feel. In reading it, I feel pulled to say that, for Prior, $\forall PA(P)$ is true if and only if, for all predicates F, $A(F)$ is true. But recall that truth is a logical construct for Prior. What we should say is something closer to $\forall PA(P)$ is true if and only if for each predicate F, $A(F)$. But there's a problem with this reading. Prior wants to avoid recourse to abstract objects in order to understand the semantics of ordinary language. This sort of substitutional semantics is Platonist. In its models, it contains sentences that no one has ever or will ever utter. The only way to understand these things, if we want to treat the theory realistically, is to take them to be abstract objects: sentence types.

I think it is partly for this reason that Prior shies away from giving a semantics that gives us biconditionals of some sort for the application of sentences. The biconditionals that state truth-conditions, say, typically force abstract objects (sets, sentence types, properties, propositions, ...) on us, especially when treating higher-order quantification. I find Prior's refusal to give a theory that contains biconditionals a refusal to give a real theory of meaning for natural language, which I would have thought was his central goal. In order to deal with this tension, I present a formal theory, that does contain biconditional truth conditions. But I do not pretend that it is Prior's theory. If we want, we can think of the theory as a fiction, in the sense of contemporary fictionalist theories, that captures Prior's ideas in a more rigorous manner than he was willing to present them.

Prior appeals to his substitutional-like semantics in order to circumvent the appeal to sets to explain quantification over predicates:

> Quine would argue, I think, that the quantified forms $\forall x \varphi x$ and $\exists x \varphi x$ do not commit us to the existence of any other sorts of entities than do the corresponding singular forms φa, φb, etc., which follow from the former and entail the latter. Why, then, should he suppose that the quantified forms $\exists \varphi \varphi a$, $\exists \varphi \exists \varphi x$, etc., commit us to the existence of entities which we are not committed to by the forms φa, ψa, $\exists x \varphi x$ from which they follow? Or that the form $\exists p \delta p$ commits us to the existence of kinds of entities to which we are not committed by specific 'δq's from which it follows? The alleged emergence of these new ontological commitments has an almost magical air about it. [11, p. 43]

Prior's argument appeals to a general principle: If A entails B, then at most B can carry the ontological commitments already carried by A. A normal zeroth order sentence φa is committed only to whatever a is. φa entails $\exists x \varphi x$ and this is likewise committed only to there being things that are φ (like a). But φa also entails $\exists \varphi \varphi a$ as well. By the general principle, $\exists \varphi \varphi a$ is committed only to, at most, the things that φa is committed. So, $\exists \varphi \varphi a$ is committed only to a and not to things like properties or sets.

I don't want to discuss the entailment view of ontological commitment to which Prior appeals.[5] This passage is interesting to me for another reason. It gives us a fairly

---

[5] Phillip Bricker [2] presents a clear taxonomy of theories of ontological commitment, including entailment-based ones like Prior's.

clear picture of Prior's view of higher-order quantification. A sentence $\forall PA(P)$ is true if and only if for all verbs F, $A(F)$. Here the schema $A(F)$ only represents sentences in a Pickwickean sense. It is a logical fiction that allows us to express features of the world that we take to be of the form $A(F)$ (here too, the notion of form is only being used as a logical fiction).

I suggest that Prior is giving us a picture of truth in natural language according to which there are some basic truths that are captured by zeroth order sentences and that all the indicative sentences of natural language are determined in their truth or falsity by those basic truths. But there is a deeper problem here. Even talking about basic truths is to engage in a logical fiction. There is no class of entities that are basic truths, according to Prior. Such things would be facts (or near enough to make Prior reject them). Prior's nominalism is a form of what David Armstrong calls "ostrich nominalism" [1]. It constitutes a refusal to give an explanation. I find this refusal frustrating, although I do not have an argument that it is illegitimate.[6]

3. Languages

My constructions use a variety of related languages. The base first-order language, $\mathcal{L}$, contains the connectives $\neg$ (negation) and $\wedge$ (conjunction) and the quantifier $\forall$ (universal quantifier), as well as variables, $x_1, x_2, \ldots$, individuals constants, and predicate constants (and parentheses). The defined connectives are disjunction, implication, equivalence, and the existential quantifier:

$$A \vee B =_{df} \neg(\neg A \wedge \neg B); \qquad A \supset B =_{df} \neg(A \wedge \neg B);$$
$$A \equiv B =_{df} (A \supset B) \wedge (B \supset A); \qquad \exists x A =_{df} \neg \forall x \neg A.$$

The models that I use carry with them names for all the objects in their domains. For each set of individual constants C, the formulae of $\mathcal{L}_C$ is the set of first-order formulae that can be constructed in the usual way with names from C and from the language $\mathcal{L}$.

The second-order language $\mathcal{L}^2$ is an extension of $\mathcal{L}$ with predicate variables for each finite arity n, $P^n, Q^n, R^n, \ldots$ and quantifiers binding such variables. Note that the formation rules of $\mathcal{L}^2$ bar predicates, including predicate variables, from being logical subjects. This means that we cannot have formulas like PQ, where P is a second-order predicate and Q is a first-order predicate. The language also contains sets of *predicate parameters*. I use "Pr" to denote an arbitrary set of predicate parameters. The language $\mathcal{L}_{Pr}$ is the language $\mathcal{L}$ extended with formulae constructed from the vocabulary in $\mathcal{L}$ together with the parameters of Pr. Similar definitions hold when the subscript C and the superscript 2 are added.

4. Robinson First-Order Models

In this section, I introduce first-order models in which every member of the domain has a name, which I will use in later sections in order to formulate metaphysical supervenience. This way of treating model theory seems to have its origin in Abraham

---

[6]Wisdom's method of logical constructions, especially in the way that Prior employs it, turns handwaving into a philosophical art. When the ellipses come out, I sometimes want to say that I just don't know how the series is supposed to continue. But that may be just a feeling of frustration from someone who loves philosophical theorising.

Robinson's PhD thesis [15], in which he uses these sorts of models to prove compactness.[7] I have chosen Robinson models to make clear and precise the supervenience thesis. In a Robinson model we can represent every fact as a true formula.

Here is a more formal definition of a Robinson model:

Definition 1. A first-order Robinson model, $\mathfrak{M}$, is a triple $\langle D, C, v \rangle$ such that D is a non-empty set (the domain of individuals of $\mathfrak{M}$), C is a non-empty set (of names), and v is a function from n-place predicates into the power set of D^n, and from C **onto** D.

To make the truth clauses slightly simpler, I assume that C always includes all the names from $\mathcal{L}$. Note that the assumption that every individual has a name makes the language a very non-standard form of first-order language. Standard first-order languages have $\aleph_0$ many well-formed formulae. If the domain of a model contains uncountably many individuals, then the language will be uncountable as well.

The use of Robinson models makes the theory of truth slightly simpler than for the more common Tarski models. There is no need for the assignments to variables, nor any talk of "x-variants" of value assignments. Open formulae are not given values at all. Here are the recursive truth conditions. For $\mathfrak{M} = \langle D, C, v \rangle$:

1. Where F is an n-ary predicate, $\mathfrak{M} \models F c_1 \ldots c_n$ iff $\langle v_f(c_1), \ldots, v_f(c_n) \rangle \in v(F)$;
2. $\mathfrak{M} \models A \wedge B$ if and only if $\mathfrak{M} \models A$ and $\mathfrak{M} \models B$;
3. $\mathfrak{M} \models \neg A$ if and only if $\mathfrak{M} \not\models A$;
4. $\mathfrak{M} \models \forall x A$ if and only if for all $c \in C$, $\mathfrak{M} \models A[c/x]$,

where $A[c/x]$ is the result of replacing every free occurrence of x in A with c.

If we fix C, then we get a model theory that is equivalent to a substitutional theory of the first-order quantifiers and the compactness theorem would fail. For we would have the set of formulas Γ, the set of all Pc, for $c \in C$, that entails $\forall x P x$, whereas no finite subset of Γ entails $\forall x P x$.

It seems quite plausible to attribute a Robinson-like treatment of quantification to Prior. In the standard referential treatment of the universal quantifier, $\forall x A(x)$ is true if and only if $A(x)$ is true of every individual. But Prior thinks that we can only think of an open sentence's being true of an individual in an indirect manner. Prior says,

> If we start from an open sentence such as 'x is red-haired' and ask what the variable 'x' stands for here, the answer depends on what we mean by 'stands for.' The variable may be said, in the first place, to stand for a name (or to keep the place of a name) in the sense that we obtain an ordinary closed sentence by replacing it with a name, i.e., by *any* genuine name of an individual object or person, say 'Peter.' The name 'Peter' itself stands for a person, viz. the man Peter, in the sense of referring to or designating the man; and the variable 'x' may be said, in a secondary sense, to stand for individual objects or persons such as Peter. It 'stands for' any such object or person in the sense that it stands for (keeps a place for) any name that stands for (refers to) an object or person. [11, p. 35]

Variables, according to Prior, only refer to those things that are their values by means of standing in for names that actually refer to things. This strongly suggests that Prior

---

[7]Leigh Steinhardt had slightly earlier formulated a similar semantical theory — in which it is assumed that every member of the domain has a name — but her formulation was not, properly speaking, model theoretic, so I chose to name the models after Robinson instead of Steinhardt [17].

does not think that variables can refer to things that do not have names. But Prior denies that:

> I do not say 'Something is red-haired' or 'For some x, x is red-haired' is true *only* if there is some true sentence which specifies it, since its truth may be due to the red-hairdness of some object for which our language has no name or one which no one is in a position to point to while saying '*This* is red-haired'. If we want to bring an 'only if' into it the best we can do, ultimately is to say that 'For some x, x is red-haired' is true if and only there is some red-haired object or person, but this is only to say that it is true if and only if for some x, x is red-haired. [11, p. 36]

This is an admission that the basic truths in the world outstrip our language, at least in terms of the objects that they can be about. In effect, in this passage Prior rejects substitutional semantics for the first-order fragment of natural language. Despite this rejection of substitutional semantics, I think we can understand his view in terms of Robinson models in the following manner. A Robinson model may represent the language of a speaker or community of speakers by analysing it *as if* there were names in that language for all the things that they consider individuals. Thus, the use of Robinson models seems reasonable.

5. Substitutional Semantics for Second-Order Quantification

In this section, I present a substitutional semantics for second-order quantification. In section 6 below, I combine this semantics with the theory of Robinson models to produce a Priorian theory of quantification. In this section, I follow Hughes Leblanc and Robert Meyer [9] in their semantics for higher-order logic. In that paper Leblanc and Meyer give a semantics for the full simple theory of types. To understand their treatment of higher-order quantification, consider the following example. $\forall P(Pa \supset Pa)$ is true according to the substitutional theory, if and only if, for all formulae $A(x)$ with only x free, $A(a) \supset A(a)$. This biconditional, however, does not give an adequate explanation of the truth of $\forall P(Pa \supset Pa)$. The formula, $A(x)$, may contain further second order quantifications. For example, $\forall P(Pb \supset Fx)$ is an instance of $A(x)$. Thus, if the biconditional given above is supposed to explain or determine the truth or falsity of $\forall P(Pa \supset Pa)$, it is a failure. Any such explanation will be circular, since it implies that $\forall P(Pa \supset Pa)$ only if $\forall P(Pb \supset Fa) \supset \forall P(Pb \supset Fa)$ is true. We are trying to explain the truth of a statement with second order quantification by appeal to further statements that contain second-order quantification. There is no advance here in the understanding of the meaning of second-order quantifiers.

Instead, Leblanc and Meyer add predicate parameters to their language. I use "F" and "G" as predicate parameters. They formulate the truth condition for second-order quantification in terms of predicate parameters:

$\forall PA(P)$ is true iff $A(F)$ is true for all predicate parameters F of the same type as P.

There is no circularity here.

Taken on its own, however, the truth condition for second-order quantification is quite weak. The universal instantiation principle supported by this truth condition is the following:

$$\forall PA(P) \supset A(F),$$

where F is a predicate parameter of the same type as P. What we want is something quite a bit stronger, such as,

$$(\forall^2 E) \quad \forall PA(P) \supset A(B),$$

where B is an open formula of the same type as P and where P is free for B in A.[8] To derive $\forall^2 E$, we can add a *comprehension scheme* to the logic. Here is the general form of the sort of comprehension scheme that I have in mind:

$$(\text{CS}) \quad \exists P \forall x_1 \ldots \forall x_n (B \equiv P x_1 \ldots x_n),$$

where $x_1, \ldots, x_n$ is the complete list of first-order variables that are free in B. Adding CS entails the validity of some version of $\forall^2 E$. But further clarification is still needed. We need to know what class of formulae are represented by "B" in CS to tell us what class of formulae B are indicated in $\forall^2 E$.

At the very least, Prior's supervenience thesis requires that the metavariable B ranges over all the formulae of the original first-order language. Thus, in any second-order model, we require that for each first-order formula, B, there be a predicate parameter F such that

$$\forall x_1 \ldots \forall x_n (B \equiv F x_1 \ldots x_n).$$

The Prior models defined in section 6 below satisfy this requirement.

If predicate parameters are only to represent first-order formulae of the original language, then we can do away with predicate parameters altogether. We can define satisfaction for second-order formulae in terms of first-order formulae. But I am not sure that this is what Prior has in mind. Recall that Prior rejects a purely substitutional approach to first-order quantification because natural languages typically do not have names for all the entities over which they quantify. This could be true for second-order quantification as well. This issue, however, is quite complicated and Prior does not really discuss it. He holds that in a statement such as "Sally runs," the verb "runs" does not refer. The issue, then, is not whether we have verbs that refer to every property, but rather that we have verbs that pick out all the features of the world that in fact we talk about. Our language does increase sometimes by the addition of new nouns, but it also expands by the addition of new verbs. "Twerking" was not in our language three decades ago, but it may be that someone still twerked then. We could represent there being activities, and other features of things, for which there are no verbs in the language by the use of predicate parameters in a model for which there are no corresponding first-order formulae.

I call the situation in which there are more properties in the world (so-to-speak) than there are predicates to represent them, *overhang*. In section 8, I define overhang models and examine their use in a Priorian semantics.

6. Prior Models

Now that I have sketched the substitutional approach to the semantics of second-order logic, in this section I use this approach to extend Robinson models to treat second-order quantification. For obvious reasons, I call the extensions of Robinson

---

[8]For a very rigorous treatment of the notion of being "free for" see Church [3] for a good explanation of this idea.

models "Prior models." A Prior model combines a Robinson model with a map from open formulae of $\mathcal{L}_C$ to predicate parameters. The idea is that each formula is represented by a predicate parameter and that this, combined with the truth conditions for second-order quantification guarantees the validity of CS, when CS is restricted to treating formulae of $\mathcal{L}_C$.[9]

Definition 2 (Prior Model). A Prior model is a pair $\langle \mathfrak{M}, \iota \rangle$, where $\mathfrak{M}$ is a Robinson first-order model and ι is a *surjective* 1–1 function from open formulae of $\mathcal{L}_C$ with n-free variables to n-ary predicate parameters.

I call ι a *correlation* function. It helps to determine the truth or falsity of atomic formulae that contain predicate parameters:

$\langle \mathfrak{M}, \iota \rangle \vDash F c_1 \ldots c_n$ if and only if $\mathfrak{M} \vDash A(c_1, \ldots, c_n)$, where $\iota(A(x_1, \ldots, x_n)) = F$.

The choice of correlation function does not affect the truth or falsity of the first-order formulae of $\mathcal{L}_C$, hence I state the following proposition:

Proposition 3. *If A is a formula of $\mathcal{L}$, $\langle \mathfrak{M}, \iota \rangle$ is a Prior model, and $\langle \mathfrak{M}, \iota \rangle \vDash A$, then $\mathfrak{M} \vDash A$.*

The following theorem and corollary show that the choice of correlation function does not even affect the truth or falsity of second-order formulae that do not contain members of C or predicate parameters — i.e., that are formulae of $\mathcal{L}^2$.

Theorem 4. *If A is a formula of $\mathcal{L}^2_{C,Pr}$ and ι and ν are correlation functions, then $\langle \mathfrak{M}, \iota \rangle \vDash A$ if and only if $\langle \mathfrak{M}, \nu \rangle \vDash A'$, where A' results from the replacement of all propositional parameters F in A with G where $\iota^{-1}(F) = n^{-1}(G)$.*

Proof. By induction on the complexity of A.

Base case: A is $F a_1 \ldots a_n$. Suppose that $\iota^{-1}(F) = n^{-1}(G)$. Then, by proposition 3, $\langle \mathfrak{M}, \iota \rangle \vDash F a_1 \ldots a_n$ if and only if $\langle \mathfrak{M}, \iota \rangle \vDash B(a_1 \ldots a_n)$ if and only if $\langle \mathfrak{M}, \iota \rangle \vDash G a_1 \ldots a_n$ where $\iota(B(x_1 \ldots x_n)) = F$ and $\nu(B(x_1 \ldots x_n)) = G$.

The conjunction and negation cases follow by the inductive hypothesis.

First-Order Quantification case. A is of the form $\forall x B$. Suppose that $\langle \mathfrak{M}, \iota \rangle \vDash \forall x B$. Then $\langle \mathfrak{M}, \iota \rangle \vDash B[a/x]$ for all $a \in C$. By the inductive hypothesis, $\langle \mathfrak{M}, \nu \rangle \vDash B'[a/x]$ for all $a \in C$. Therefore, $\langle \mathfrak{M}, \iota \rangle \vDash \forall x B'$. The proof of the other direction is the same.

Second-Order Quantification case. A is of the form $\forall P B(P)$. $\langle \mathfrak{M}, \iota \rangle \vDash \forall P B(P)$ if and only if, for all H, $\langle \mathfrak{M}, \iota \rangle \vDash B(H)$, where H is a propositional parameter of the same type as P. By the inductive hypothesis, $\langle \mathfrak{M}, \iota \rangle \vDash B(H)$ if and only it $\langle \mathfrak{M}, \nu \rangle \vDash B(J)$ where $\iota^{-1}(H) = \nu^{-1}(J)$. Hence, $\langle \mathfrak{M}, \iota \rangle \vDash \forall P B(P)$ if and only if $\langle \mathfrak{M}, \iota \rangle \vDash B'(P)$. ◁

If A does not contain any predicate parameters, then A' is just A, so we can state the following corollary to theorem 4:

Corollary 5. *If A is a formula of $\mathcal{L}^2_C$ then for any two correlation functions ι and ν, $\langle \mathfrak{M}, \iota \rangle \vDash A$ if and only if $\langle \mathfrak{M}, \nu \rangle \vDash A$.*

---

[9]Note that I have assumed that there are at least as many predicate parameters in Pr as there are formulae of $\mathcal{L}_C$.

The truth of this corollary allows one to talk of the formulae of $\mathcal{L}_C^2$ that are true on a Robinson model, because the same set of such formulae are true on any Prior model based on a given Robinson model.

We can define different consequence operators on the class of Prior models. Consider a formula A and a set of formulae Γ of the full language, $\mathcal{L}_{C,Pr}^2$. We can say that $\Gamma \vDash^C A$ if and only if for every Prior model based on the set of constants C, if all the members of Γ are true on that model then so is A. We can also restrict that relation to create the relation $\vDash_\iota^C$, so that we consider only Prior models that contain the correlation function ι.

7. Supervenience and Prior Models

The use of Prior models makes the supervenience thesis quite straightforward. The following lemma and theorem show that if two Prior models differ in the second-order statements that they make true, they must differ in the first-order statements that they make true. Thus, the first-order statements are the supervenience basis for the second-order statements.

Lemma 6. *If A is a formula of $\mathcal{L}_C^2$ that is true on $\langle \mathfrak{M}, \iota \rangle$, then there is a set Γ of first-order formulae of $\mathcal{L}_C$ such that $\Gamma \vDash_\iota^C A$ and every member of Γ is true in $\langle \mathfrak{M}, \iota \rangle$.*

Proof. Suppose that A is true in $\langle \mathfrak{M}, \iota \rangle$. Let Γ be the set of $\mathcal{L}_C$ formulae true in $\langle \mathfrak{M}, \iota \rangle$.

Case 1. $A = Fc_1 \ldots c_n$. Then $\langle \mathfrak{M}, \iota \rangle \vDash B(c_1 \ldots c_n)$ where $\iota(B(x_1 \ldots x_n)) = F$. But $B(c_1 \ldots c_n) \in \Gamma$, so $\Gamma \vDash_\iota^C A$.

Case 2. $A = B \wedge C$. Follows by inductive hypothesis.

Case 3. $A = \neg B$. Suppose that $\Gamma \nvDash_\iota \neg B$, then there is some $\langle \mathfrak{M}', \iota \rangle \vDash B$, and $\langle \mathfrak{M}', \iota \rangle \vDash G$ for all $G \in \Gamma$. By the inductive hypothesis, there is some set of first-order formulae of $\mathcal{L}_C$, Γ', such that $\Gamma' \vDash_\iota^C B$. But Γ is bivalent, that is, for every first order formula of the non-extended language, G, either $G \in \Gamma$ or $\neg G \in \Gamma$. Since the set of formulae true in any model is negation consistent, $\Gamma' \subseteq \Gamma$. The consequence relation, $\vDash_\iota^C$, is monotonic, and so $\Gamma \vDash_\iota B$, but then $\langle \mathfrak{M}, \iota \rangle$ is inconsistent, and this can't be. Therefore, $\Gamma \vDash_\iota^C \neg B$.

Case 4. $A = \forall x B(x)$. If A is true on $\langle \mathfrak{M}, \iota \rangle$, then $\Gamma \vDash_\iota^C B(c)$ for all names c. Then, $\Gamma \vDash_\iota^C \forall x B(x)$.

Case 5. $A = \forall P B(P)$. $\langle \mathfrak{M}, \iota \rangle \vDash \forall P B(P)$ if and only if, for all parameters F of the same type as P, $\langle \mathfrak{M}, \iota \rangle \vDash B(F)$. By the inductive hypothesis, $\Gamma \vDash_\iota B(F)$ for all such Fs. Therefore, $\Gamma \vDash_\iota \forall P B(P)$. ◁

Theorem 7 (Supervenience). *For any two Prior models $\langle \mathfrak{M}, \iota \rangle$ and $\langle \mathfrak{M}', \iota \rangle$ both with set of constants C, for any formula A of $\mathcal{L}_C^2$, if $\langle \mathfrak{M}, \iota \rangle$ and $\langle \mathfrak{M}', \iota \rangle$ differ in the truth value attributed to A, then there is a first-order formula of $\mathcal{L}_C$, B, that differs in truth value on $\langle \mathfrak{M}, \iota \rangle$ and $\langle \mathfrak{M}', \iota \rangle$.*

Proof. Suppose that A is true in $\langle \mathfrak{M}, \iota \rangle$ but not in $\langle \mathfrak{M}', \iota \rangle$. By lemma 6, There is some set of first order formulae, Γ of $\mathcal{L}_C$ such that $\Gamma \vDash_\iota A$. Then if A fails to be true in $\langle \mathfrak{M}', \iota \rangle$, some member of Γ must be false in $\langle \mathfrak{M}', \iota \rangle$. ◁

Sticking with $\mathcal{L}_C$ as the supervenience basis might seem to make the supervenience relation rather superficial. In a purely substitutional framework in which every parameter expresses a first-order formula, it would seem, supervenience is too straightforward. But the current framework is not purely substitutional. There is a domain of individuals in every Robinson model (and, hence, in every Prior model). Suppose that there is a Prior model, $\langle\langle D', v', C'\rangle, v\rangle$ such that the cardinality of C' is the same as the cardinality of C. Let f be a 1–1 surjection from C' to C and let μ be a correlation function such that $v(A) = \mu(A')$, where A' is just like A except that every name c' that occurs in A is replaced with $f(c')$. Then it is clear that $\langle\langle D', v'', C\rangle, \mu\rangle$ is a Prior model, where v'' is an assignment function on $\mathcal{L}_C$, and $\langle\langle D', v', C'\rangle, v\rangle \vDash A$ if and only if $\langle\langle D', v'', C\rangle, \mu\rangle \vDash A'$. From this fact, we can derive a deeper sense of supervenience. Therefore, by theorem 7:

Theorem 8. *Let C and C' be sets of names of the same cardinality and let A be a formula of $\mathcal{L}_C^2$, such that A is true on some Prior model based on C. Then there are sets of formulae Γ and Γ' of $\mathcal{L}_C$ and $\mathcal{L}_{C'}$, respectively, such that for any formula B of $\mathcal{L}_C$, $B \in \Gamma$ iff $B' \in \Gamma'$ and $\Gamma \vDash^C A$ and $\Gamma' \vDash^{C'} A'$.*

The supervenience theorems of the present section are rather straightforward. In the next section, I look at ways in which the relationship between language and the world could make supervenience false or at least more complicated.

8. Supervaluations and Supervenience

In the definition of a Prior model I assume that the set of parameters, Pr, is identical to the range of the correlation function of a model, or rather, that we do not use the parameters that are not in that range. Now I discuss models in which there is *overhang* in the sense of section 5. That is to say, I now talk about models in which there are parameters that are not coextensive with open formulae of $\mathcal{L}_C$.

If our language use, or the world, or both the world and language use, determines that there is a determinate class of basic features of the world that are not expressed in our language, then we can represent this easily using a simple modification of Prior models. An overhang Prior model is triple $\langle \mathfrak{M}, V, \iota \rangle$ such that $\mathfrak{M}$ is a Robinson model, ι is a 1–1 map from open formulae of $\mathcal{L}_C$ **into** Pr and V is a function that takes n-place predicate parameters from $Pr - Rg(\iota)$ to subsets of D^n. A truth clause is then added for atomic formulae that contain predicate parameters not in the range of the correlation function:

$$\langle \mathfrak{M}, V, \iota \rangle \vDash F c_1 \ldots c_n \text{ if and only if } \langle v(c_1), \ldots, v(c_n) \rangle \in V(F)$$

Clearly, the theorems of section 7 hold for overhang models where $\mathcal{L}_C$ is replaced by $\mathcal{L}_{C,Pr}$.

A problem arises, however, if we think that our language does not necessarily exhaust the basic features of the world, but that the world and our language use together do not determine what exactly these basic features are. In other words, depending on how we extend our language, there could be different ways that the world is. If we take this line on the language/world interface, then I think we should use a form of supervaluations to represent it.

Definition 9 (Supervaluational Prior Model). A supervaluational Prior model $\mathcal{M}$ consists of a set of overhang Prior models $\langle \mathfrak{M}, V, \iota \rangle$ such that they all contain the same Robinson model $\mathfrak{M}$ and the same correlation function ι, but they may vary on their valuation V.

The theory of truth for supervaluational Prior models is quite simple. For any formula A of the full language, $\mathcal{M} \vDash A$ if and only if $\langle \mathfrak{M}, V, \iota \rangle \vDash A$ for every overhang Prior model $\langle \mathfrak{M}, V, \iota \rangle$ in $\mathcal{M}$.

Supervenience fails on the class of supervaluational Prior models. For suppose that Pr contains countably many binary predicates. Let's enumerate them $F_1, F_2, \ldots$. Consider a supervaluational Prior model that consists of countably many overhang Prior models, also enumerated, $M_1, M_2, \ldots$. Suppose that on each M_i, F_i is transitive, that is, $M_i \vDash \forall x \forall y \forall z ((F_i xy \wedge F_i yz) \supset F_i xz)$ and, for $i > 1$, there is a counterexample to the transitivity of F_{i-1}, that is, there are some c_1, c_2 and c_3 such that $M_i \vDash F_{i-1} c_1 c_2 \wedge F_{i-1} c_2 c_3 \wedge \neg F_{i-1} c_1 c_3$. Thus, we have $\mathcal{M} \vDash \exists P \forall x \forall y \forall z ((Pxy \wedge Pyz) \supset Pxz)$ but there is no particular parameter F such that $\mathcal{M} \vDash \forall x \forall y \forall z ((Fxy \wedge Fyz) \supset Fxz)$. We can see that there is a failure of supervenience. In this model, there is no set of "super-true" first-order formulae that entails this second order formula.

To maintain supervenience, one might remove all the parameters in Pr that are not in the range of the correlation function ι and replace each one of these parameters, F_i with countably many new parameters, $F_i^1, F_i^2, \ldots$. Then, set for all names $c_1, \ldots, c_n$, $\mathcal{M} \vDash F_i^j c_1 \ldots c_n$ if and only if $M_j \vDash F_i c_1 \ldots c_n$. In this way, we create a single overhang Prior model for second order logic. The acceptance of this sort of model implies that one takes every overhang model in $\mathcal{M}$ as expressing a legitimate way in which to represent the basic facts of the world and that these can simultaneously all be accepted as accurate.

What this discussion of the issue of overhangs and supervaluations shows us is that Prior's supervenience claim is very fragile. It depends heavily on the exact way in which a natural language together with the world are supposed to determine the set of second-order truths. The contribution of the world is really in question here. Prior's use of the doctrine of logical fictions, in my opinion, adds to the obscurity of this issue. A doctrine (like that of the *Tractatus*) that tells us that there are atomic facts at least tells us that the world determines what second-order truths there are even if we don't know what they are. One that refuses to say what the basic facts of the world are, or how they are determined, leaves obscure what we really mean by the first-order truths and whether and how they determine the second-order truths.

9. COMPACTNESS AND NON-COMPACTNESS

If the second-order truths do supervene on the first-order truths, then what is the point of using second-order logic? The use of quantifiers often allows us to express a lot of information in a compressed manner. We can often state facts about infinitely many things using quantifiers, such as, "Every point in space is arbitrarily close to another point." Second-order quantifiers allow the same sort of compression, e.g., "Nova has every property of a good dog," "Everything has some property," and so on.

One way in which supervenience is connected with information compression is through the failure of compactness. As we have seen, the sort of supervenience that second-order statements bear to first-order statements is one of entailment. If A is a true second-order statement, then there is a set Γ of true first-order statements such that $\Gamma \models^C A$. But even the first-order fragment of $\models^C$ is not compact. For let G be a unary predicate constant of $\mathcal{L}$, then, if C is infinite, $\{Gc: c \in C\} \models^C \forall x Gx$ but there is no finite subset of that premise set that entails $\forall x Gx$. Where A is not a valid formula and Γ is consistent, if Γ entails A, then A captures some of the information carried by the statements in Γ. In cases where Γ is irreducibly infinite, A captures information that is represented by infinitely many first-order formulae. A lot more study is needed to make precise the way in which second-order statements compress information of infinitely many first-order statements.

Historical Note. Dunn and Belnap [6] were perhaps the first to point out that compactness typically fails for substitutional interpretations of the quantifiers. Although the property was given the name "compactness" by Tarski [5, p.25] in 1950, the name was not in general use in philosophical logic until the 1970s, nor was the notion of compactness (under any name) widely discussed by philosophical logicians in the 1960s. So, it was of some interest that in 1968 Dunn and Belnap pointed out its failure in the substitutional semantics.

10. SIMPLE OR RAMIFIED TYPE THEORY?

The form of second-order logic that I have attributed to Prior is what Alonzo Church calls the *ramified functional-calculus of second-order* in [3, §58].[10] The salient contrast here is between the second-order fragment of the *simple* theory of types and the second-order fragment of the ramified theory of types. A ramified theory of types distinguishes between the types of predicate expressions both in terms of the arguments they take and in terms of sorts of quantifiers in them. A simple theory of types, in contrast, only distinguishes between the types of predicate expressions in terms of the arguments that they take.

If I am right about the way in which Prior views supervenience, then every predicate parameter represents a first-order predicate expression. Thus, the second-order quantifiers just range over the first-order predicates. Hence, the form of universal instantiation that Prior models validate is

$$(\forall^2 E) \quad \forall P A(P) \supset A(B)$$

where B is a first-order predicate expression of the same type as the variable, P. This is the same axiom scheme that Church uses to characterise ramified second-order logic in [3, p. 349].

Prior, however, rejects ramified type theory both in *Formal Logic* [10, pp. 285–287] and in *Objects of Thought*. In both books, his argument is the same. Ramified type theory is too complicated, it requires the axiom of reducibility, and the simple theory of types is adequate to solve the paradoxes. Here is a passage from *Objects of Thought* that presents this argument.

---

[10][3] may still be the only textbook that treats the second-order fragment of ramified type theory. I think, however, that looking at this fragment is perhaps the best way to introduce students to the ramified hierarchy. Church gives a very clear formulation of full ramified type theory in [4].

[In the ramified theory of types] propositions and predicates of higher 'order' are not absolutely ruled out ... but they are treated as not being propositions and predicates in he same sense as of those of lower 'orders.' The resulting restrictions on substitution turned out to exclude certain quite important forms of mathematical reasoning, and to save these Russell and Whitehead introduced a rather implausible 'axiom of reducibility,' the details of which need not concern us. In order to eliminate the necessity for this axiom, various logicians in the 1920s suggested 'simplifying' the theory of types by removing, for purposes of instantiating and substitution in theorems, the discrimination made between propositions and predicates of different 'order,' and dealing in other ways with the problems for which this discrimination was originally made. [11, pp. 41–42]

The axiom of reducibility is a comprehension axiom, like CS given in section 5 above. In terms of the logic that I attribute to Prior, what it says is that for every second order predicate, there is a first-order predicate that is coextensive with it. I.e.,

$$\exists P \forall x_1 \ldots x_n (A(x_1,\ldots,x_n) \equiv P x_1 \ldots x_n)$$

for every formula A of $\mathcal{L}^2$. In Prior's framework, this is indeed an implausible axiom.

The rejection of the axiom of reducibility by itself, however, does not turn the ramified theory of types into simple type theory. Instead, it turns it into *predicative type theory*. This is a rather weak theory, from the point of view of mathematical logic, since in it one can no longer prove the principle of induction or the the reliability of the method of Dedekind cuts used to construct the real numbers. Note that although there is no *need* to have an axiom of reducibility in simple type theory, it is not because simple type theory rejects reducibility. Rather, like ramified type theory with reducibility, it claims that all predicates of individuals (say) determine properties of individuals. Simple type theory does this in a more direct and less complicated manner, this is true, but ramified type theory and simple type theory from a mathematical point of view are very similar.

Prior suggests, however, that in addition to rejecting the axiom of reducibility, he is rejecting the distinction between predicates or propositions of different orders. I think he is not doing this at all. His view of higher-order quantification and his argument against Quine require that there be this distinction. In the passage against Quine quoted in section 5 Prior says that the truth of a statement of the form $\forall \varphi \varphi a$ requires only the truth of φa, ψa, and so on. If it also requires the truth of statements such as $\exists \chi \exists x \chi x a$, then Prior's statement does little to explain away the commitments of $\forall \varphi \varphi a$. His view becomes hopelessly circular and some way out is required. It leaves the Quinean with the option of saying that treating second-order variables as referring to sets is the best way out of this circle.

What Prior should have done, in my opinion, was explicitly accept predicative ramified type theory and reject logicism. This is the doctrine that mathematics (or some important part of it) is reducible to logic. Whitehead and Russell's argument for logicism in *Principia* required reducibility. He can say that the logic that is needed to understand natural language is not powerful enough to derive the fundamental theorems of mathematical theories and non-logical axioms are required to do so.

Acknowledgments. I'd like to thank Max Cresswell and Adriane Rini, who are my fellow investigators on Marsden Grant, The Logic of Natural Language (of which this essay is a part), Jack Copeland and Diane Proudfoot who invited me to give a talk at the 2018 Arthur Prior Day at the University of Canterbury which was a very early version of this paper, Nino Cocchiarella for email correspondence about Prior, Rob Goldblatt for discussions about Robinson and compactness, and Allen Hazen and Bernie Linsky and the rest of the Edmonton Logic Reading Group for discussions about the history of logic and about ramified type theory in particular. And I am grateful to Katalin Bimbó for editing this fine volume. The research for this paper was funded by a grant from the Marsden Fund of the Royal Society of New Zealand.

REFERENCES

[1] Armstrong, D. M. (1978). *Universals and Scientific Realism*, Cambridge University Press, Cambridge, UK.
[2] Bricker, P. (2016). Ontological Commitment, *in* E. N. Zalta (ed.), *The Stanford Encyclopedia of Philosophy*, Winter 2016 edn, Metaphysics Research Lab, Stanford University.
[3] Church, A. (1956). *Introduction to Mathematical Logic*, Princeton University Press, Princeton, NJ.
[4] Church, A. (1976). Comparison of Russell's resolution of the semantic antinomies with that of Tarski, *The Journal of Symbolic Logic* **41**: 747–760.
[5] Dawson, J. W. (1993). The compactness of first-order logic: From Gödel to Lindström, *History and Philosophy of Logic* **14**: 15–37.
[6] Dunn, J. M. and Belnap, N. D. (1968). The substitutional interpretation of the quantifiers, *Noûs* **2**: 177–185.
[7] Grover, D. (1992). *A Prosentential Theory of Truth*, Princeton University Press, Princeton.
[8] Grover, D., Camp, J. and Belnap, N. D. (1975). A prosentential theory of truth, *Philosophical Studies* **27**: 75–124.
[9] Leblanc, H. and Meyer, R. K. (1980). Truth value semantics for the theory of types, *in* K. Lambert (ed.), *Philosophical Problems in Logic: Some Recent Developments*, D. Reidel, Dordrecht.
[10] Prior, A. N. (1955). *Formal Logic*, Oxford University Press, Oxford, UK.
[11] Prior, A. N. (1971). *Objects of Thought*, Oxford University Press, Oxford, UK.
[12] Quine, W. V. O. (1957). The scope and language of science, *The British Journal for the Philosophy of Science* **8**: 1–17.
[13] Quine, W. V. O. (1971). *Set Theory and its Logic*, Harvard University Press, Cambridge, MA.
[14] Quine, W. V. O. (1989). Reflections on models and logical truth. Manuscript in the W. V. Quine Papers held in the Huntington Library, Harvard. ITEM Identifier: MS Am 2587, (2649).
[15] Robinson, A. (1947). *On the Metamathematics of Algebra*, PhD thesis, Birkbeck College, University of London, London, UK.
[16] Russell, B. (1912). *The Problems of Philosophy*, Oxford University Press, Oxford, UK.
[17] Steinhardt, L. D. (1940). *The Variable and its Relation to Semantic Problems*, PhD thesis, Harvard University, Cambridge, MA.
[18] Wisdom, J. (1931). Logical constructions, *Mind* **40**: 188–216.

PHILOSOPHY, VICTORIA UNIVERSITY OF WELLINGTON, WELLINGTON, NEW ZEALAND
Email: edwin.mares@vuw.ac.nz

A Completeness Result for Inequational Reasoning in a Full Higher-Order Setting

Lawrence S. Moss and Thomas F. Icard

Dedicated to the memory of J. Michael Dunn

ABSTRACT. This paper obtains a completeness result for inequational reasoning with applicative terms without variables in a setting where the intended semantic models are the full structures, the full type hierarchies over preorders for the base types. The syntax allows for the specification that a given constant be interpreted as a monotone function, or an antitone function, or both. There is a natural set of five rules for this inequational reasoning. One can add variables and also add a substitution rule, but we observe that this logic would be incomplete for full structures. This is why the completeness result in this paper pertains to terms without variables. Since the completeness is already known for the class of general (Henkin) structures, we are interested in full structures. We obtain the first result on this topic. Our result is not optimal because we restrict to base preorders which have a weak completeness property: every pair of elements has an upper bound and a lower bound. To compensate we add several rules to the logic. We also present extensions and variations of our completeness result.

Keywords. Completeness, Full models, Inequational reasoning, Monotone & antitone functions

1. Introduction

Tonoids recast. In his work on very general algebraic semantics of non-classical logics, Dunn [4] introduces the notion of a *tonoid*. This is a structure of the form $(A, \leq, OP)$, where $\mathbb{A} = (A, \leq)$ is a poset, and OP is a set of finite-arity *function symbols*, each with a *tonic type* $(s_1, \ldots, s_n)$, where each s_i is either $+$ or $-$. A familiar example done this way takes $\mathbb{A}$ to be $2 = \{0, 1\}$ with $0 < 1$, and $OP = \{\to\}$, where $\to$ is taken as an operation with tonicity type $(-, +)$. The formal requirement is that if $f \in OP$ is of arity n, then $f : A^n \to A$ is either isotone or antitone in the ith argument, depending on whether s_i is $+$ or $-$. To spell out the requirement in more detail, recall that a function $g : A \to A$ is *isotone* (here called *monotone*) if $a \leq b$ implies $g(a) \leq g(b)$; and $g : A \to A$ is *antione* if $a \leq b$ implies $g(b) \leq g(a)$. Suppose that f is of arity 3 and its tonic type is $(+, -, +)$. Then our requirement is:

(1.1)   whenever $a_1 \leq a_2$, $b_2 \leq b_1$, and $c_1 \leq c_2$, $f(a_1, b_1, c_1) \leq f(a_2, b_2, c_2)$.

2020 *Mathematics Subject Classification.* Primary: 03B16, Secondary: 03B38, 03B40.

Bimbó, Katalin, (ed.), *Relevance Logics and other Tools for Reasoning. Essays in Honor of J. Michael Dunn*, (Tributes, vol. 46), College Publications, London, UK, 2022, pp. 282–308.

The idea is to abstract a feature of material implication: it is antitone in its first argument and monotone in its second. Here are two equivalent ways to state the general requirement (1.1). The first uses the concept of the *opposite* poset $\mathbb{A}^{op}$; this is $\mathbb{A}$ with the converse order. Our requirement (1.1) now would say that

(1.2) $$f : \mathbb{A} \times \mathbb{A}^{op} \times \mathbb{A} \to \mathbb{A}.$$

In this, $\times$ denotes the product operation on posets, and the arrow $\to$ means "monotone function." This formulation (1.2) can be recast by currying, replacing a function of arity 3 by a higher-order function of the following form:

(1.3) $$f : \mathbb{A} \to \mathbb{A}^{op} \to \mathbb{A} \to \mathbb{A}.$$

So $\mathbb{A} \to \mathbb{A}$ is the set of monotone functions, taken as a poset $\mathbb{P}$ with the pointwise order. Then $\mathbb{A}^{op} \to \mathbb{A} \to \mathbb{A}$ is the set of monotone functions from $\mathbb{A}^{op}$ to $\mathbb{P}$, again taken as a poset which we call $\mathbb{Q}$. Finally, $\mathbb{A} \to \mathbb{A}^{op} \to \mathbb{A} \to \mathbb{A}$ is the set of monotone functions from $\mathbb{A}$ to $\mathbb{Q}$. Going one step further from (1.3), our requirement may be rephrased once again.

(1.4) $$f : \mathbb{A} \xrightarrow{+} \mathbb{A} \xrightarrow{-} \mathbb{A} \xrightarrow{+} \mathbb{A}.$$

In (1.4), the operative notation is that $\mathbb{P} \xrightarrow{+} \mathbb{Q}$ denotes the set of monotone functions from $\mathbb{P}$ to $\mathbb{Q}$, and $\mathbb{P} \xrightarrow{-} \mathbb{Q}$ denotes the set of antitone functions from $\mathbb{P}$ to $\mathbb{Q}$. In both cases, the order is pointwise.

Up until now, all we have done is to rephrase the definition of a tonoid in terms of higher-order functions in the realm of posets, something that Dunn did not need to do. We are indeed interested exactly in higher-order reasoning about ordered structures. Instead our result is aimed at settings where reasoning about monotone/antitone functions plays a central role. One such setting is the area of programming language semantics where the order represents subtyping. Another is natural language inference where higher-order functions are commonplace, following the tradition in Montague grammar and type-logical grammar. Concerning inference, van Benthem [11] pointed out the usefulness of monotonicity in connection with the higher-order semantics of determiners and saw that this topic would be a central part of logical studies connected to natural language. The connection to higher-order preorders in this area was first made in [9], and that paper is also the source of the observations behind the moves from (1.1) to (1.4).

Friedman's Theorem on the STLC. The results that we are after in this paper are modeled on the completeness result established by Friedman [5] for the simply typed lambda calculus (STLC). To explain our contribution, let us review part of Friedman's contribution. We change the notation and presentation of [5] to set the stage for our work.

The STLC begins with a set B of base types β. The full set $\mathcal{T}$ of types is the closure of B under the following rule: if σ and τ are types, so is $\sigma \to \tau$. Then one forms the set of typed terms $t : \sigma$ of the STLC using application of one term to another, variables, and abstraction. The main assertions in the STLC are identities $t = u$ between terms of the same type. The semantics is of interest here. The primary models are *full* (or *standard*) type structures. Beginning with sets X_β for $\beta \in B$, one constructs sets X_σ for all types $\sigma \in \mathcal{T}$ by recursion: $X_{\sigma \to \tau}$ is the set $(X_\tau)^{X_\sigma}$ of *all* functions from X_σ to

X_τ. Then one interprets each typed term $t : \sigma$ by an element $[\![t]\!] \in X_\sigma$. Naturally, one is interested in the relation on terms $\vDash t = u$ defined by:

(1.5) $\qquad\qquad \vDash t = u$ iff $[\![t]\!] = [\![u]\!]$ in all full structures

The main completeness result from [5] is that $\vDash t = u$ iff the statement $t = u$ can be proved in a certain logical system with very natural rules. The rules of the system are the reflexive, symmetric, and transitive rules of identity, the congruence rule for application, and the α, β, and η rules of the STLC. So the completeness of the system tells us that an identity assertion holds in all full structures iff it is provable from α, β, and η on top of the expected rules of identity.

What we are doing. Here is how things change in this paper. We would like the main assertions in our system to be *inequalities* $t \leq u$ instead of identites. Thus, we want our semantic spaces to be *preorders* rather than unstructured sets. Beginning with an assignment of preorders $(\mathbb{P}_\beta)_{\beta \in B}$ to the base types, we take preorders for function types $\mathbb{P}_{\sigma \to \tau}$ to be the set of all functions, as above, but endowed with the pointwise order. This is what we mean by *full* models in our title and throughout the paper. Moreover, we allow our type system to insist that a given function symbol be interpreted by a monotone function, or an antitone one (or both).

The logical systems in this paper are formulated *without variables*: the only terms are those which can be constructed from the typed *constants* using application. This might seem to be a severe limitation, so let us motivate it from several angles. First of all, monotonicity calculi without variables are useful in several settings (see Icard and Moss [6]). Second, the completeness results of interest in this paper are not available if one has variables (see Section 2.4). This is a parallel to the matter of equational reasoning with second-order terms (even without abstraction): the natural logical system would add substitution to the rules mentioned above. This system is not complete for full models. Finally, the authors and William Tune have formulated "order-aware" versions of the lambda calculus (see [7; 8; 10]). The type system expands that of the usual simply typed lambda calculus by permitting the formation of several additional kinds of function types: monotone functions $\sigma \xrightarrow{+} \tau$, antitone functions $\sigma \xrightarrow{-} \tau$, and others. Tune [10] is a variation on this which incorporates something like the "op" operation on preorders which we have seen above in (1.2). All work in this area expands the syntax of terms using variables and abstraction operations. Finally, the basic assertions in the language include inequalities between objects of the same type. What is more, it includes some inequalities between objects of different (but related) types. The formulation of the syntax is non-trivial, and the same goes for the proof rules. In any case, as we already mentioned, the logical systems in the area cannot be complete for full models. They are complete for wider classes of "Henkin" models. (The analogous structures for the STLC in [5] are called *pre-structures*, and sometimes they are called *applicative structures*.) But this is rather an expected result, since one can build a model canonically from the proof system. This is not what we are doing in this paper. We are building full models, and we are studying applicative order terms without variables or abstraction. Our work is thus drastically simpler on the syntactic side, and more complex on the semantic side: we call on and develop results specific to preorders (see Sections 1.1, 3.2, and 3.3).

The main logical system in this paper is presented in Sections 2.1 (the syntax and semantics) and 2.2 (the proof system). Briefly, the syntax allows us to declare that a given function symbol f be interpreted as a monotone function by writing f^+. We also might declare that f be interpreted as an antitone function by writing f^-. The basic assertions in the system are inequalities $t \leq u$ between terms of the same type. The main semantic objects are full structures in the setting of preorders. The consequence relation $\Gamma \vDash t^* \leq u^*$ is defined much as in (1.5), except that we use an order relation in the obvious way, and that we permit a set Γ of extra hypotheses.

The main completeness result ought to be a completeness result for a logical system. We would like to have $\Gamma \vdash t^* \leq u^*$ iff $\Gamma \vDash t^* \leq u^*$. We have a sound logical system; the rules are in Figure 1. We did not obtain a completeness theorem for this system, though we believe it to hold. But we do have a related completeness result, Theorem 42. The formulation restricts the full models to full models whose base preorders are *weakly complete* (every pair of elements has an upper bound and a lower bound) and the logic adds a few rules to compensate. Curiously, there is an echo here from *distributoids* that were introduced by Dunn in [3]. In distributoids there is a requirement that the underlying poset $\mathbb{A}$ be a bounded distributive lattice and that the operation symbols either respect 0 or 1. Every lattice is trivially a weakly complete preorder.

Related work. We have already mentioned papers on monotonicity calculi. This paper is the first in the area to present a completeness results for full structures, the intended semantic models.

The original completeness theorem of Friedman which we mentioned above has been extended in a few directions. Dougherty and Subrahmanyam [2] extend the STLC by adding product and coproduct types and a terminal type, and they obtain the completeness theorem for full structures. As far as we know, this is the only extension that obtains completeness for full models on sets. Several papers move from sets to other categories in order to obtain completeness results, and the completeness here is the strong completeness theorem $\Gamma \vDash t = u$ iff $\Gamma \vdash t = u$ which is not available in sets. For more on this topic, see Awodey [1].

1.1. Background: Preorders and Polarized Preorders.

Definition 1. A *preorder* is a pair $\mathbb{P} = (P, \leq)$, where P is a set, and $\leq$ is a reflexive and transitive relation on P. Although we technically should use P for the universe of the preorder, we sometimes write $p \in \mathbb{P}$ when we mean $p \in P$. If $p \leq q$ and $q \leq p$, then we write $p \equiv q$. It is possible that $p \equiv q$ without having $p = q$.

Let $\mathbb{P}$ and $\mathbb{Q}$ be preorders, and consider a function $f : P \to Q$.

1. f is *monotone* if whenever $p \leq q$ in $\mathbb{P}$, $f(p) \leq f(q)$ in $\mathbb{Q}$. We also say that f is *order-preserving* in this case. We write $f^+ : \mathbb{P} \to \mathbb{Q}$.
2. f is *antitone* if whenever $p \leq q$ in $\mathbb{P}$, $f(q) \leq f(p)$ in $\mathbb{Q}$. We write $f^- : \mathbb{P} \to \mathbb{Q}$.
3. f is *order-reflecting* whenever $f(p) \leq f(q)$ in $\mathbb{Q}$, $p \leq q$ in $\mathbb{P}$.
4. f is an *order embedding* if f is one-to-one, and preserves and reflects the order.

The logical systems in this paper are about monotone and antitone functions. But some of the proofs also use the concepts of order-reflecting functions and order embeddings.

Example 2. Here are some examples of the kinds of facts of interest in this paper.
1. If $f^+ \leq g^- : \mathbb{P} \to \mathbb{Q}$, and $a \leq c \geq b$, then $f(a) \leq g(b)$. This is because $f^+(a) \leq f^+(c) \leq g^-(c) \leq g^-(b)$.
2. If $f^- \leq g^+ : \mathbb{P} \to \mathbb{Q}$, and $a \geq c \leq b$, then again $f(a) \leq g(b)$. This is similar: $f^-(a) \leq f^-(c) \leq g^+(c) \leq g^+(b)$.
3. On the other hand, here is an example where $f^- \leq g^+$, $a \leq c \geq b$, but $f(a) \not\leq g(b)$. Let $\mathbb{P}$ be the poset $\{a,b,c\}$ with $a < c > b$, and let $\mathbb{Q}$ be $\{0,1\}$ with $0 < 1$. Let $f(a) = 1$, $f(b) = 0$, and $f(c) = 0$. Let $g(a) = 1$, $g(b) = 0$, and $g(c) = 1$.
4. It is possible for a function to be both monotone and antitone. In our notation, it is possible that $f^+ : \mathbb{P} \to \mathbb{Q}$ and also $f^- : \mathbb{P} \to \mathbb{Q}$. One way for this to happen is when f is a constant function. Another way is when $\mathbb{P}$ is the flat preorder (also called the discrete preorder) on some set S: $p_1 \leq p_2$ iff $p_1 = p_2$.

Some additional definitions and constructions concerning preorders appear later in this paper, closer to where they are used. At this point, we introduce *polarized preorders*, a type of structure that extends preorders.

Definition 3. A *polarized preorder* is a tuple $\mathbb{F} = (F, \leq, +, -)$, where F is a set, $\leq$ is a pre-order on F, and $+$ and $-$ are subsets of F.

Example 4. For any preorders $\mathbb{P}$ and $\mathbb{Q}$, we have a polarized preorder $\mathbb{Q}^\mathbb{P}$ defined as follows. The set of points of $\mathbb{Q}^\mathbb{P}$ is the set Q^P of all functions from P to Q. The order is the pointwise order. We take $+$ to be the set of monotone $f : \mathbb{P} \to \mathbb{Q}$, and $-$ to be the set of antitone $f : \mathbb{P} \to \mathbb{Q}$.

We also have abstract examples. In such polarized preorders, we think of the sets $+$ and $-$ as providing a specification for what we want them to be in an interpretation. Thus we think of them as "tagged" $+$ or $-$ (or possibly neither, or both). To say that f is tagged $+$ just means that $f \in +$; similarly for $-$. (A given function symbol might thus be tagged with neither $+$ or $-$, and it might also be tagged with both symbols.) We use f^+ to range over elements $f \in F$ which are tagged $+$, and we also use f^- to range over elements $f \in F$ which are tagged $-$. (And when we write f without $+$ or $-$, we mean an arbitrary element of F.)

Definition 5. Let $\mathbb{F}$ be a polarized preorder, and let $\mathbb{P}$ and $\mathbb{Q}$ be preorders. An *interpretation of $\mathbb{F}$ in $\mathbb{P}$ and $\mathbb{Q}$* is a function $\langle\!\langle\ \rangle\!\rangle : \mathbb{F} \to \mathbb{Q}^\mathbb{P}$ which is monotone and preserves polarities. That is, if f^+, then $\langle\!\langle f \rangle\!\rangle$ is monotone, and if f^-, then $\langle\!\langle f \rangle\!\rangle$ is antitone.

This definition will not be used much in this paper, but is shows where things are going. We think of $\mathbb{F}$ as "syntax" and $\mathbb{Q}^\mathbb{P}$ as the "semantic space," and $\langle\!\langle\ \rangle\!\rangle$ as the interpretation of the syntax in that space.

2. SYNTAX AND SEMANTICS

This section sets the stage for the rest of the paper by presenting the syntax and semantics of our system.

2.1. Syntax, and Semantics in Full Structures.

We begin with a set B of *base types*. We use the letter β for these. We make no assumption on the set B, and we also do not vary it in what follows. Henceforth we leave B out of our notation.

The full set $\mathcal{T}$ of types is the smallest set such that every base type β belongs to $\mathcal{T}$, and if σ and τ belong to $\mathcal{T}$, then so does $\sigma \to \tau$. The types which are not base types are called *function types*.

Definition 6. An *(ordered) signature* is a family $(\mathbb{F}_\beta)_\beta$ of preorders, one for each base type, and a family $(\mathbb{F}_\sigma)_\sigma$ of polarized preorders, one for each function type $\sigma \in \mathcal{T}$. We form *typed terms* $t : \sigma$ by the following recursion:

1. If $f \in \mathbb{F}_\sigma$, then $f : \sigma$ is a typed term.
2. If $t : \sigma \to \tau$ and $u : \sigma$ are typed terms, then $tu : \tau$ is a typed term.

When we need notation for a signature, we usually write $\mathbb{F} = (\mathbb{F}_\sigma)_\sigma$ and think of these as polarized, except for the base types.

We use notation like $t : \sigma$, $u : \tau$, etc., for typed terms. Usually we drop the types for readability. Indeed, we only supply the types to make a point about them. For example, in (2.2) below, the second equation exhibits the types. If we were to write $[\![tu]\!] = [\![t]\!]([\![u]\!])$ without the types, it could cause a confusion on first reading. We could use parentheses as well, but these will not be necessary. When we speak of terms, we usually do not mention the underlying signature.

The assertions in the language are inequalities $t : \sigma \leq u : \sigma$ between terms of the same type. Again, we usually drop the types and just write $t \leq u$.

Example 7. For all relations R, we write $R^\star$ for the reflexive and transitive closure of R. So $R^\star$ is the smallest preorder including R.

Let β be a base type, let τ be any type, so that $\beta \to \tau$ is a function type, and let $\mathbb{F}$ be the signature given by

$$\begin{aligned}
\mathbb{F}_\beta &= (\{a,b,c\}, \emptyset^\star) \\
\mathbb{F}_{\beta \to \tau} = (F_{\beta \to \tau}, \leq, +, -) &= (\{f,g\}, \{(f,g)\}^\star, \{f\}, \{g\}) \\
\mathbb{F}_{\tau \to (\beta \to \tau)} = (F_{\tau \to (\beta \to \tau)}, \leq, +, -) &= (\{\varphi\}, \emptyset^\star, \emptyset, \{\varphi\})
\end{aligned}$$

For other function types μ, we take $F_\mu = (\emptyset, \emptyset, \emptyset, \emptyset)$. In this signature, we are taking a, b, and c to be symbols of type β. There is no order relation among these, but our signature does have the reflexivity assertions $a \leq a$, $b \leq b$, $c \leq c$. Since β is a base type, there is no polarization assertion for these symbols. As for $\beta \to \tau$, we have two symbols f and g. Our signature records $f \leq g$ and that $f \in +$ and $g \in -$. When working with this signature, we usually will keep the polarization assertions in mind by repeatedly tagging the symbols. So we would summarize the polarized preorder $\mathbb{F}_{\beta \to \tau}$ by simply writing $f^+ \leq g^-$.

Typed terms in our signature include $\varphi^-(f^+(a)) : \beta \to \tau$. We could omit the parentheses without risking confusion and also the type; we then would just write $\varphi^- f^+ a$. An example term of type τ is $(\varphi^- f^+ a)b$.

Semantics: full structures. Fix a family of preorders $(\mathbb{P}_\beta)_\beta$, one for each base type β. The family $(\mathbb{P}_\beta)_\beta$ induces a family of preorders $(\mathbb{P}_\sigma)_\sigma$ by

$$\mathbb{P}_{\sigma \to \tau} = ((\mathbb{P}_\tau)^{\mathbb{P}_\sigma}, \leq) \tag{2.1}$$

where $\leq$ is the *pointwise order* on the function set $(P_\tau)^{P_\sigma}$. For function types $\sigma \to \tau$ we use the polarized preorder structure mentioned in Example 4: for $f : P_\sigma \to P_\tau$, we tag f^+ if f is monotone, and we tag f^- if f is antitone. If f is neither monotone nor antitone, it would be tagged with neither polarity. If f were both monotone and antitone (see Example 7(4)), it would be tagged both $+$ and $-$. What we have built is called the *full preorder type structure* over $(\mathbb{P}_\beta)_\beta$.

Definition 8. Fix a signature $\mathbb{F}$. A *full $\mathbb{F}$-structure* is a family of preorders $\mathcal{M} = ((\mathbb{P}_\sigma)_\sigma, [\![\,]\!])$, where $(\mathbb{P}_\sigma)_\sigma$ is the full preorder type structure over $(\mathbb{P}_\beta)_\beta$ together with a function $[\![\,]\!]$ defined on the typed terms over $\mathbb{F}$ with the following properties:

1. If f in $\mathbb{F}_\sigma$, then $[\![f]\!] \in P_\sigma$.
2. For function types σ, $[\![\,]\!]$ restricts to a map $[\![\,]\!]_\sigma : \mathbb{F}_\sigma \to \mathbb{P}_\sigma$ which is monotone and preserves polarity.

In other words: if $f \leq g$ in $\mathbb{F}_\sigma$, then $[\![f]\!] \leq [\![g]\!]$ in $\mathbb{P}_\sigma$; if $f^+ : \sigma \to \tau$, then $[\![f]\!] : \mathbb{P}_\sigma \to \mathbb{P}_\tau$ is monotone; and if $f^- : \sigma \to \tau$, then $[\![f]\!] : \mathbb{P}_\sigma \to \mathbb{P}_\tau$ is antitone.

Let us emphasize that in a full structure, (2.1) holds. Thus, in a full $\mathbb{F}$-structure, each function type $\sigma \to \tau$ gives us an interpretation of $\mathbb{F}_{\sigma \to \tau}$ in $\mathbb{P}_\sigma$ and $\mathbb{P}_\tau$ in the sense of Definition 5. Indeed, a full $\mathbb{F}$-structure amounts to a family of such interpretations together with maps $\mathbb{F}_\beta \to \mathbb{P}_\beta$ for the base types which preserve the order.

Interpreting typed terms in full structures. Fix a full $\mathbb{F}$-structure $\mathcal{M}$. By recursion on typed terms $t : \sigma$, we define $[\![t : \sigma]\!]$:

$$
\begin{aligned}
[\![f : \sigma]\!] &\quad \text{is given in } \mathcal{M}, \text{ when } f \in \mathbb{F}_\sigma \\
[\![tu : \tau]\!] &= [\![t : \sigma \to \tau]\!]([\![u : \sigma]\!])
\end{aligned}
\tag{2.2}
$$

We are using (2.1) when we see that $[\![t : \sigma \to \tau]\!]$ is a function and hence may apply it to $[\![u : \sigma]\!]$. An easy induction shows that when $t : \sigma$, $[\![t : \sigma]\!] \in P_\sigma$. As mentioned before, we usually omit the types. This holds when we use the $[\![\,]\!]$ notation.

Semantic assertions. Let t, u be terms of the same type σ. We say that $\mathcal{M} \vDash t \leq u$ if $[\![t]\!] \leq [\![u]\!]$. Let Γ be a set of inequalities $t \leq u$, and let $\mathcal{M}$ be a full structure. We say that $\mathcal{M} \vDash \Gamma$ if $\mathcal{M} \vDash t \leq u$ whenever Γ contains $t \leq u$. We then speak of a *full model* of Γ. Let $\Gamma \cup \{t^* \leq u^*\}$ be a set of inequalities in our language, omitting the types. We write $\Gamma \vDash t^* \leq u^*$ if every full $\mathbb{F}$-structure which satisfies Γ also satisfies $t^* \leq u^*$. (Incidentally, there is no real reason why we use the $*$ notation on the conclusion $t^* \leq u^*$. It just permits us to use letters t and u in the rest of an argument, and it also focuses our attention on one particular assertion of interest.)

In addition, we will need variations on this definition of $\Gamma \vDash t^* \leq u^*$. For example, we will be contracting the class of preorders to *weakly complete* preorders (see Section 3). We will change our notation slightly to clarify the meaning of semantic assertions. For example we write $\Gamma \vDash_{\mathrm{WC}} A$ if every *weakly complete* model of Γ is a model of A.

An important point is that our language is built on an ordered signature $\mathbb{F}$, and we do not display $\mathbb{F}$ in our notation $\Gamma \vDash t^* \leq u^*$. But this is something to keep in mind.

Example 9. Let $\mathbb{F}$ be as in Example 7. Example 2 shows that $f^+ \leq g^-, a \leq c, b \leq c \vDash fa \leq gb$.

Remark 10. This is perhaps a good place to mention a way in which our overall framework is more permissive than we need it to be. We allow our signatures to have order assertions $f \leq g$, but all such assertions could be absorbed into a given set Γ. So we could have just taken signatures to be families of polarized *sets* rather than polarized preorders.

2.2. Proof System. The proof system for the basic logic (without the rules which we shall introduce in Figure 2) is shown in Figure 1. One point to highlight is that in the (MONO) and (ANTI) rules, we have assumptions f^+ and f^- that are part of the underlying signature $\mathbb{F}$. There are two ways that we could take these assumptions. First, we could take them to be *side conditions* on the rules. Doing things that way would mean that we would not show those assumptions in examples. The second way would be to take the polarity assumptions to be "first class." This would mean that our proof trees would not consist solely of inequalities: they could also have assertions from the signature. This second alternative is the one we adopt. (However, very little would change if we went the other way.) In Section 4.2, we further extend the proof system in order to infer polarity statements about terms; up until then, all polarity assertions in proof trees occur at the leaves. With this forward view, we are led to the formulation which we chose.

Definition 11. We write $\Gamma \vdash s^* \leq t^*$, where $\Gamma \cup \{s^* \leq t^*\}$ is a set of assertions in our language, if there is a tree labeled by assertions in the language whose root is labeled with $s^* \leq t^*$, whose leaves are labeled with elements of Γ or with assertions from the underlying signature $\mathbb{F}$, and such that every non-leaf-node is justified by one of the rules in Figure 1.

Example 12. This is a version of Example 2, but done in our proof system. Let $\mathbb{F}$ be an ordered signature, and assume that for some type $\sigma \to \tau$, $\mathbb{F}_{\sigma \to \tau}$ contains symbols f and g, and that $f^+, g^- : \sigma \to \tau$.

Let t, u, and v be terms of the same type σ. Then $f^+ \leq g^-, t \leq v \geq u \vdash ft \leq gu$ via the following derivation:

$$\cfrac{\cfrac{f^+ \quad t \leq v}{ft \leq fv} \text{MONO} \quad \cfrac{\cfrac{f \leq g}{fv \leq gv} \text{POINT}}{ft \leq gv} \text{TRANS} \quad \cfrac{g^- \quad u \leq v}{gv \leq gu} \text{ANTI}}{ft \leq gu} \text{TRANS}$$

Observe that the leaves of the tree are assertions in $\mathbb{F}$.

Similarly, if $f^-, g^+ : \sigma \to \tau$, then we have $f^- \leq g^+, t \geq v \leq u \vdash ft \leq gu$. This assertion is more naturally made on top of a different ordered signature. (However, our framework allows the symbols f and g to be declared as both $+$ and $-$ in a given signature.)

2.3. The Syntactic Preorder of a Set Γ, and a Construction Lemma. In this section, we fix a signature $\mathbb{F}$ and a set Γ of inequalities over it.

Definition 13. For each type σ, $\mathbb{P}_\sigma^{syn}$ is the set of all terms of type σ (not just the constants, the symbols in F), the order is provability from Γ, and $+$ and $-$ are the constant symbols with the relevant tagging:

$$\frac{}{t \leq t} \text{ REFL} \qquad \frac{s \leq t \quad t \leq u}{s \leq u} \text{ TRANS} \qquad \frac{s \leq t}{su \leq tu} \text{ POINT}$$

$$\frac{f^+ \quad t \leq u}{ft \leq fu} \text{ MONO} \qquad \frac{f^- \quad t \leq u}{fu \leq ft} \text{ ANTI}$$

FIGURE 1. Basic rules of the logic for interpretations in full structures. See Figure 2 for additional rules sound for weakly complete preorder structures.

$$\begin{array}{lll} \mathbb{P}^{syn}_\sigma & = & \{t : t \text{ is an } \mathbb{F}\text{-term of type } \sigma\} \\ t \leq u & \text{iff} & \Gamma \vdash t \leq u \\ t^+ & \text{iff} & t \in \mathbb{F}_\sigma, \text{ and } t^+ \text{ in } \mathbb{F} \\ t^- & \text{iff} & t \in \mathbb{F}_\sigma, \text{ and } t^- \text{ in } \mathbb{F} \end{array}$$

We call $\mathbb{P}^{syn}_\sigma$ the *canonical polarized preorder of type σ*.

Doing this for all σ gives a family $(\mathbb{P}^{syn}_\sigma)_\sigma$ of polarized preorders. Please note that the family $(\mathbb{P}^{syn}_\sigma)_\sigma$ is *not* a full hierarchy over the base preorders.

Definition 14. Let $(\mathbb{Q}_\sigma)_\sigma$ be a full hierarchy over the base preorders. An *applicative family of interpretations (of Γ)* is a family $\mathcal{N} = (\langle\!\langle\ \rangle\!\rangle_\sigma)_\sigma$ of functions indexed by the types

$$(2.3) \qquad \langle\!\langle\ \rangle\!\rangle_\sigma : \mathbb{P}^{syn}_\sigma \to \mathbb{Q}_\sigma$$

such that each $\langle\!\langle\ \rangle\!\rangle_\sigma$ is monotone and preserves polarities on the function types, and with the following property: for all $t \in \mathbb{P}^{syn}_{\sigma \to \tau}$ and $u \in \mathbb{P}^{syn}_\sigma$,

$$(2.4) \qquad \langle\!\langle t \rangle\!\rangle_{\sigma \to \tau}(\langle\!\langle u \rangle\!\rangle_\sigma) = \langle\!\langle tu \rangle\!\rangle_\tau.$$

On the left we have function application in the usual sense, and on the right tu is an application on the level of terms. Please note that an applicative family $\mathcal{N}$ depends on a full hierarchy $(\mathbb{Q}_\sigma)_\sigma$, and as with everything in this section it depends on Γ (and thus ultimately on $\mathbb{F}$).

Lemma 15. *Let $(\mathbb{Q}_\sigma)_\sigma$ be a full hierarchy over the base preorders. Let $\mathcal{N}$ be an applicative family of interpretations of Γ as in Definition 14, so that (2.4) holds. Then there is a full structure $\mathcal{M} = ((\mathbb{Q}_\sigma)_\sigma, (\llbracket f \rrbracket)_{f \in \mathbb{F}})$ using the same preorders at each type, such that the following hold:*

1. *For all t, u of the same type σ, $\mathcal{M} \models t \leq u$ iff in $\mathcal{N}$, $\langle\!\langle t \rangle\!\rangle_\sigma \leq \langle\!\langle u \rangle\!\rangle_\sigma$.*
2. *If each function $\langle\!\langle\ \rangle\!\rangle_\sigma$ preserves the order, then $\mathcal{M} \models \Gamma$.*
3. *If $t^*, u^* : \sigma$, and $\langle\!\langle\ \rangle\!\rangle_\sigma$ reflects the order and $\mathcal{M} \models t^* \leq u^*$, we have $\Gamma \vdash t^* \leq u^*$.*

Proof. We define $\llbracket\ \rrbracket$ by recursion on typed terms (see Definition 6), starting with the case of elements $f \in \mathbb{F}_\sigma$ $\llbracket f \rrbracket = \langle\!\langle f \rangle\!\rangle_\sigma$. Then we extend to all typed terms by $\llbracket tu : \tau \rrbracket = \llbracket t : \sigma \to \tau \rrbracket (\llbracket u : \sigma \rrbracket)$. The difference between $\llbracket\ \rrbracket$ and $(\langle\!\langle\ \rangle\!\rangle_\sigma)_\sigma$ is that the former is a single function defined on all typed terms by recursion on those terms, while $(\langle\!\langle\ \rangle\!\rangle_\sigma)_\sigma$ is a family of functions. The content of our claim just below is that the two definitions agree.

Claim 16. *For all* $t \in \mathbb{P}_\sigma^{syn}$, $[\![t]\!] = \langle\!\langle t \rangle\!\rangle_\sigma$.

Proof. By induction on the typed term $t : \sigma$. The fact that $[\![f]\!] = \langle\!\langle f \rangle\!\rangle_\sigma$ is immediate for $f \in \mathbb{F}_\sigma$. Assuming our claim for $t : \sigma \to \tau$ and $u : \sigma$, we see that

$$[\![tu]\!] = [\![t]\!]([\![u]\!]) = \langle\!\langle t \rangle\!\rangle_{\sigma \to \tau}(\langle\!\langle u \rangle\!\rangle_\sigma) = \langle\!\langle tu \rangle\!\rangle_\tau$$

We used (2.4) at the end. ◁

This claim easily implies part (1): $\mathcal{M} \models t \leq u$ iff in $\mathcal{N}$, $\langle\!\langle t \rangle\!\rangle_\sigma \leq \langle\!\langle u \rangle\!\rangle_\sigma$.

For (2), suppose that Γ contains an assertion $t \leq u$. Let σ be the type of these terms. Then $t \leq u$ in $\mathbb{P}_\sigma^{syn}$. Since $\langle\!\langle \ \rangle\!\rangle_\sigma$ preserves the order, $\langle\!\langle t \rangle\!\rangle_\sigma \leq \langle\!\langle u \rangle\!\rangle_\sigma$. By part (1), $\mathcal{M} \models t \leq u$.

For (3), suppose that in $\mathcal{M}$, $[\![t^*]\!] \leq [\![u^*]\!]$. Then by Claim 16, $\langle\!\langle t^* \rangle\!\rangle_\sigma \leq \langle\!\langle u^* \rangle\!\rangle_\sigma$. Since $\langle\!\langle \ \rangle\!\rangle_\sigma$ reflects the order, $t^* \leq u^*$ in $\mathbb{P}_\sigma^{syn}$. By the definition of $\mathbb{P}_\sigma^{syn}$, $\Gamma \vdash t^* \leq u^*$. ◁

The reason that we will be using Lemma 15 in our main result, Theorem 42, is that it will be more natural for us to define the functions $\langle\!\langle \ \rangle\!\rangle_\sigma$ by recursion on σ than to define $[\![\ ]\!]$ by recursion on the typed terms t.

2.4. Digression: Incompleteness of the Logic with Variables on Full Structures.

Now that we have the semantics of our inequational typed lambda calculus and also the proof system, we can explain why this paper is about a logic without variables. The idea behind our construction comes from an example in Awodey [1] concerning the usual typed lambda calculus: when formulated with variables, it cannot have set theoretic full models. Take a base type β and function symbols $i : (\beta \to \beta) \to \beta$ and $r : \beta$ and the equation $r(i(x)) = x$. Any full model will interpret the base type β by a singleton set. This leads easily to an incompleteness result for the full semantics in sets. Although our language does not have the identity symbol $=$, we still get the same result.

In this section, we allow variables and also the rule of substitution: from $t \leq u$, infer $t[s] \leq u[s]$, where s is any substitution. (That is, any map s which maps variables to terms, respecting the types.) Let us write $\Gamma \vdash t \leq u$ for the proof relation which extends the main proof relation in this paper with this additional rule.

We take one base type, β, and symbols ψ^-, φ^+, c, and d with the types shown below:

$$c, d : \beta \underset{\varphi^+}{\overset{\psi^-}{\rightleftarrows}} \beta \to \beta$$

For Γ we take three inequalities:

$$\varphi(\psi(y)) \leq y \qquad y \leq \varphi(\psi(y)) \qquad \psi(\varphi(x)) \leq x$$

Here we are using a variable x of type $\beta \to \beta$ and a variable y of type β.

Proposition 17. $\Gamma \models c \leq d$, *but* $\Gamma \nvdash c \leq d$ *in the logic using our rules, including substitution.*

Proof. Let $\mathcal{M}$ be a full model of Γ. We first observe that for $p, q \in P_\beta$, if $p \leq q$, then $q \leq p$. To see this, write k for $[\![\varphi]\!] \circ [\![\psi]\!]$. So k is antitone, since it is the composition of a monotone and an antitone function. Notice that $k(r) \equiv r$ for all $r \in P_\beta$, by our first two assertions in Γ. Hence if $p \leq q$ then also $q \equiv k(q) \leq k(p) \equiv p$.

Next, we claim that for all elements $a, b \in \mathbb{P}_\beta$, $a \equiv b$ in $\mathbb{P}_\beta$. If not, let $a \not\equiv b$. Define $f : P_\beta \to P_\beta$ by
$$f(x) = \begin{cases} a & \text{if } [\![\psi]\!](x)(x) \equiv b \\ b & \text{otherwise} \end{cases}$$
Since $P_{\beta \to \beta}$ is the full function space, f belongs to it. Let $x^* = [\![\varphi]\!](f)$. By our last assertion in Γ, $[\![\psi]\!](x^*) \leq f$. Then
$$[\![\psi]\!](x^*)(x^*) \leq f(x^*) = \begin{cases} a & \text{if } [\![\psi]\!](x^*)(x^*) \equiv b \\ b & \text{otherwise} \end{cases}$$
If $[\![\psi]\!](x^*)(x^*) \equiv b$, then we would also have $[\![\psi]\!](x^*)(x^*) \leq a$. But by our first paragraph, we then would have $b \equiv [\![\psi]\!](x^*)(x^*) \equiv a$, and this is a contradiction to our choice of a and b. So we have $[\![\psi]\!](x^*)(x^*) \not\equiv b$, and thus $[\![\psi]\!](x^*)(x^*) \leq b$. Our first paragraph now shows that $[\![\psi]\!](x^*)(x^*) \equiv b$, giving a contradiction again.

It follows from this claim that in our model (hence in any model of Γ), $[\![c]\!] \leq [\![d]\!]$.

To complete the proof of our proposition, we make an observation about the particular set Γ that we have and also our rules, including substitution: if $\Gamma \vdash t \leq u$, and if either t or u contains some given variable of either type or constant symbol of base type β, then the other term contains it as well. (For example, we can prove $\psi(\varphi(x))(y) \leq x(y)$; both sides contain x and y. We can also prove $\psi(\varphi(\psi(d)))(y) \leq \psi(d)(y)$, and both sides contain d and y.) This observation is proved by an easy induction. Thus $\Gamma \nvdash c \leq d$, since $c \leq d$ has c on only one side. ◁

2.5. Lemmas on New Constants.

Definition 18. Let $\mathbb{F} = (\mathbb{F}_\sigma)_\sigma$ be a signature, and write each $\mathbb{F}_\sigma$ as $(F_\sigma, \leq, +, -)$. For each σ, let $\square_\sigma \notin F_\sigma$. Let $\mathbb{G}_\sigma = (F_\sigma \cup \{\square_\sigma\}, \leq^*_\sigma, +, -)$. We have added the new symbol $\square_\sigma$ to $\mathbb{F}_\sigma$. Notice that the reflexive-transitive closure $\leq^*_\sigma$ of $\leq_\sigma$ just adds to $\leq_\sigma$ the assertion $\square_\sigma \leq \square_\sigma$, and $+$ and $-$ are exactly the same as in $\mathbb{F}_\sigma$. Thus, we add no monotonicity information about the new symbols; for a type $\sigma \to \tau$, we do not add to Γ assertions like $\square^+_{\sigma \to \tau}$. $\mathbb{G}_\sigma$ does not have any ordering relation between any new constant and any other symbol.

This gives a new signature $\mathbb{G} = (\mathbb{G}_\sigma)_\sigma$. Note that we have an inclusion map $\iota_\sigma : \mathbb{F}_\sigma \to \mathbb{G}_\sigma$ which preserves the order and polarities. For a set Γ of inequalities over $\mathbb{F}$, we write $\Gamma_\square$ for the same set, but taking it to be a set of assertions over $\mathbb{G}$.

For a set Γ over $\mathbb{F}$, the syntactic and semantic consequence relations $\Gamma_\square \vdash t^* \leq u^*$ and $\Gamma_\square \models t^* \leq u^*$ are different from the ones involving Γ. The main point of the next results is that moving from Γ to $\Gamma_\square$ is a *conservative extension* in the relevant senses. First, a semantic fact.

Let $\mathcal{M} = ((\mathbb{P}_\beta)_\beta, ([\![\ ]\!]_\sigma)_\sigma : \mathbb{G}_\sigma \to \mathbb{P}_\sigma)$ be a full model over $\mathbb{G}$. Let $\mathcal{M}^0$ be the *reduct* to $\mathbb{F}$. This is $\mathcal{M}^0 = ((\mathbb{P}_\beta)_\beta, ([\![\ ]\!]^0_\sigma)_\sigma : \mathbb{F}_\sigma \to \mathbb{P}_\sigma)$, where $[\![\ ]\!]^0_\sigma : \mathbb{F}_\sigma \to \mathbb{P}_\sigma$, is $[\![\ ]\!]_\sigma \circ \iota_\sigma$.

Lemma 19. *For every inequality $t \leq u$ over $\mathbb{F}$, $\mathcal{M} \models t \leq u$ iff $\mathcal{M}^0 \models t \leq u$.*

Proof. An easy induction shows that for all terms t over $\mathbb{F}$, the interpretations of t in $\mathcal{M}$ and $\mathcal{M}^0$ are the same: $[\![t]\!] = [\![t]\!]^0$. ◁

Lemma 20. *If $\Gamma \vDash t^* \leq u^*$ and $\Gamma_\Box$ comes from extending the underlying signature with new constants, then $\Gamma_\Box \vDash t^* \leq u^*$.*

Proof. Let $\mathcal{M}$ be a $\mathbb{G}$-structure which satisfies $\Gamma_\Box$. By Lemma 19, $\mathcal{M}^0 \vDash \Gamma$. By hypothesis, $\mathcal{M}^0 \vDash t^* \leq u^*$. Then by Lemma 19 again, $\mathcal{M} \vDash t^* \leq u^*$. ◁

We now turn to some syntactic results that again point to a conservative extension.

Lemma 21. *If $\Gamma_\Box \vdash t \leq \Box$ or $\Gamma_\Box \vdash \Box \leq t$, then $t = \Box$.*

Proof. By induction on derivations. With a conclusion like $t \leq \Box$, the derivation can only use (REFL) or (TRANS). The inductive step for (TRANS) is trivial. ◁

Lemma 22. *If $t : \sigma \to \tau$ and $v : \sigma$ and $\Gamma_\Box \vdash t \leq v\Box$, then there is some $u : \sigma \to \tau$ so that $t = u\Box$, and $\Gamma_\Box \vdash u \leq v$.*
Similarly, if $\Gamma_\Box \vdash v\Box \leq t$, then there is some u so that $t = u\Box$, and $\Gamma_\Box \vdash v \leq u$.

Proof. Each part is proved by induction on the derivation. The step for (TRANS) is easy. If the root uses (MONO), or (ANTI), then $t\Box$ is $v\Box$ by Lemma 21; in this case, $t = v$. If it uses (POINT), then we directly have that $t \leq v$. ◁

Lemma 23. *If $\Gamma_\Box \vdash t\Box \leq u\Box$, then $\Gamma_\Box \vdash t \leq u$.*

Proof. By induction on the derivation. If the root uses (POINT), $t \leq u$. If it uses (REFL), (MONO) or (ANTI), t is u. Suppose that the root uses (TRANS), say

$$\frac{t\Box \leq v \quad v \leq u\Box}{t\Box \leq u\Box} \text{ TRANS}$$

The previous lemma applies to both subproofs. There is some $t \leq x$ so that $v = x\Box$. There is also some $w \leq u$ so that $v = w\Box$. So $x = w$. And then $t \leq w \leq u$ tells us that $t \leq u$. ◁

Lemma 24. *If $\Gamma_\Box \vdash t^* \leq u^*$ with none of the new symbols $\Box_\sigma$ occurring in t^* or u^*, then $\Gamma \vdash t^* \leq u^*$.*

Proof. Call a type σ *inhabited* (in a given signature) if there is a term of type σ other than $\Box_\sigma$. For each inhabited type, pick a term t_σ of that type. Consider the following substitution:

$$s(\Box_\sigma) = \begin{cases} t_\sigma & \text{if } \sigma \text{ is inhabited} \\ \Box_\sigma & \text{otherwise} \end{cases}$$

We claim that if we take any proof tree $\mathcal{T}$ over $\Gamma_\Box$ and apply this substitution to all terms, the result $\mathcal{T}[s]$ is a valid proof tree over $\Gamma_\Box$. The proof of this is by induction.

We next claim that in $\mathcal{T}[s]$ every assertion $t \leq u$ has the property that some $\Box_\sigma$ occurs in t iff it occurs in u. The proof is by induction, and the main interesting steps are for (TRANS).

We now fix a proof tree $\mathcal{T}$ showing that $\Gamma_\Box \vdash t^* \leq u^*$. Now none of the $\Box$ symbols occur in the root $t^* \leq u^*$, or in any of the leaves of the tree. So the leaves and root of $\mathcal{T}[s]$ are the same as those of $\mathcal{T}$. We claim that in $\mathcal{T}[s]$, every σ which occurs is inhabited. For this we argue by contradiction; suppose it is false. Since the root has no $\Box$ occurrences, there must be a node in the proof tree which does have a $\Box$-occurrence

$$\frac{f^+ : \sigma \to \tau \quad g^- : \sigma \to \tau \quad f \leq g}{ft \leq gu} \text{ WC}_1 \qquad \frac{f^- : \sigma \to \tau \quad g^+ : \sigma \to \tau \quad f \leq g}{ft \leq gu} \text{ WC}_2$$

$$\frac{f^-, g^+, h^+, k^- : \sigma \to \tau \quad g \leq f \quad k \leq h \quad ft \leq ku}{g \leq h} \text{ WC}_3$$

FIGURE 2. Additional rules of the logic which are sound for weakly complete preorders. (WC$_3$) stands in for four rules; we could also have the following arrangements at the front: (a) f^+, g^-, h^+, k^-; (b) f^-, g^+, h^-, k^+; (c) f^+, g^-, h^-, k^+.

but whose child (downward) in the tree has no $\Box$-occurrences. The only way this can happen is at the transitivity step:

$$\frac{t \leq u \quad u \leq v}{t \leq v}$$

But the observation above applies (twice) and tells us that both t and v have a $\Box$-subterm; hence $t \leq v$ has at least two of them — a contradiction! Therefore, every type in $\mathcal{T}[s]$ is inhabited. And then in passing from $\mathcal{T}$ to $\mathcal{T}[s]$, we removed $\Box_\sigma$ in favor of a term t_σ. We conclude that $\mathcal{T}[s]$ has no $\Box$-terms. Thus, $\mathcal{T}[s]$ is a proof tree over Γ. And as we have seen, its leaves and root are the same as those of $\mathcal{T}$. ◁

3. COMPLETENESS FOR FULL WEAKLY COMPLETE STRUCTURES IN THE EXTENDED LOGIC

The work in the previous section suggests that we should prove a completeness theorem for reasoning in full structures $\Gamma \vdash t^* \leq u^*$ iff $\Gamma \models t^* \leq u^*$, where the proof system is the one in Figure 1 and the semantic notion is based on the full structures which we have introduced in Definition 8. We have not been able to obtain this result. On the other hand, we have related results. First, we might well relax the condition of fullness to the natural weaker condition associated with Henkin-like models of the typed lambda calculus. Doing this leads to a completeness result fairly easily, not just for the logic of this paper but for much more expressive formalisms that have a richer type system, variables, abstraction, and arbitrary sets of hypotheses. This is not the topic of this paper, but for work in this area, see [7; 8; 10]. (We should mention that [8] has an error that will be fixed in a follow-up publication.)

Definition 25. A preorder is *weakly complete* if every x and y have some upper bound z and also some lower bound w. The bounds required need not be least upper bounds or greatest lower bounds. A full structure is called *weakly complete* if every base preorder $\mathbb{P}_\beta$ is weakly complete. (It follows that each $\mathbb{P}_\sigma$ is weakly complete.)

As the name suggests, weak completeness is a fairly weak property. Every lattice has this property, for example. Every preorder with a greatest and a least element is weakly complete. On the other hand, a flat preorder containing two or more points is not weakly complete. A disjoint union of two non-empty preorders is also not weakly complete.

The logic relevant to weakly complete full structures is given in Figure 2, taken in addition to the rules which we saw in Figure 1.

Suppose that f and g are function symbols of the same type, say σ. We write $f \leq^{+-} g$ to mean that either $f^+ \leq g^-$ or else that $f^- \leq g^+$. With this notation, the six (WC) rules maybe written as two:

$$\frac{f \leq^{+-} g \quad a \leq b}{fa \leq gb} \text{WC}_{1,2} \qquad \frac{g \leq^{+-} f \quad k \leq^{+-} h \quad ft \leq ku}{g \leq h} \text{WC}_3$$

We write $\Gamma \vdash_{\text{WC}} s^* \leq t^*$ if there is a derivation (a proof tree) that also allows the weak completeness rules in Figure 2. And $\Gamma \vDash_{\text{WC}} s^* \leq t^*$ means that every weakly complete full model of Γ is also a model of $s^* \leq t^*$.

Proposition 26. *If $\Gamma \vdash_{\text{WC}} s^* \leq t^*$, then $\Gamma \vDash_{\text{WC}} s^* \leq t^*$.*

Proof. By induction on proofs in the system. We only consider the (WC) rules. For (WC$_1$), fix a weakly complete full structure $\mathcal{M}$. We know that $[\![f]\!] : \mathbb{P}_\sigma \to \mathbb{P}_\tau$ is a monotone function, $[\![g]\!] : \mathbb{P}_\sigma \to \mathbb{P}_\tau$ is an antitone function, and also $[\![t]\!], [\![u]\!] \in \mathbb{P}_\sigma$. By weak completeness of $\mathbb{P}_\sigma$, let $x \in \mathbb{P}_\sigma$ be such that $[\![t]\!], [\![u]\!] \leq x$. Then by Example 2, $[\![f]\!]([\![t]\!]) \leq [\![g]\!]([\![u]\!])$. Thus $[\![ft]\!] \leq [\![gu]\!]$.

The soundness of (WC$_2$) is similar, and it uses the fact that every pair of elements of $\mathbb{P}_\sigma$ have some lower bound.

Next, let us consider (WC$_3$) with the same notation as just above. The important thing is that the premises do not include $f \leq k$, just the much weaker assertion that for particular terms t and u, $ft \leq ku$. But this is enough: take any $x \in \mathbb{P}_\sigma$ and observe

$$\begin{aligned}[]
[\![g]\!](x) &\leq [\![f]\!]([\![t]\!]) && \text{by Example 2} \\
&= [\![ft]\!] && \text{by the recursive clauses in the semantics} \\
&\leq [\![ku]\!] && \text{by the overall induction hypothesis, and } ft \leq ku \\
&= [\![k]\!]([\![u]\!]) \\
&\leq [\![h]\!](x)
\end{aligned}$$

Since x was arbitrary, we have shown that $[\![g]\!] \leq [\![h]\!]$ pointwise. ◁

The calculation just above makes it clear that the last two premises could be changed. For example, we could have g^-, h^-, f^+, and k^+. The only thing that matters is that the arrow directions g and f have to be opposite, and the same goes for h and k. So there are four (WC$_3$) rules.

3.1. Additional Lemmas on New Constants.

We proved results in Section 2.5 that showed how adding fresh constants to a signature gives a conservative extension both for the semantics and the proof theory. At this point, we need to re-work that section in light of the new (WC) rules. Definition 18 mentioned notation having to do with new constants. This needs no change. The semantic results in Lemma 19 and 20 do not change: the reduct of a weakly complete model is weakly complete. No change is needed in Lemma 21, since none of the (WC) rules allow us to conclude an inequality whose left- or right-hand side is a new symbol $\square_\sigma$ by itself. Lemma 22 does need to change.

Lemma 27. *If $\Gamma_\square \vdash_{\text{WC}} t \leq v\square$, then one of the following holds:*

1. There is some $u \leq v$ so that $t = u\Box$.
2. There is a term $s : \sigma$, and constants $f, g : \sigma \to \tau$ such that $\Gamma_\Box \vdash_{\text{WC}} t \leq fs$, and $f \leq^{+-} g \leq v$.

Proof. By induction on the number of (TRANS) steps in the derivation. We cannot have a derivation where the root is $t \leq u\Box$ justified by (WC$_3$). Applications of (WC$_3$) conclude an inequation between function symbols which have a declared + or − marking. ◁

We also have a parallel result for the situation $\Gamma_\Box \vdash u\Box \leq t$.

Lemma 28. *If $\Gamma_\Box \vdash_{\text{WC}} t\Box \leq u\Box$, then $\Gamma_\Box \vdash_{\text{WC}} t \leq u$.*

Proof. By induction on the the height of the derivation. If the root is (REFL), (MONO) or (ANTI), t is u. If the root is (POINT), we see that $t \leq u$. If the root is (WC$_1$) or (WC$_2$), then we have $t \leq^{+-} u$. In particular, $t \leq y$. As in Lemma 27, we cannot have a derivation where the root is (WC$_3$) and where the assertion at the root is $t\Box \leq u\Box$.

The main work is when the root is (TRANS), say

(3.1)
$$\frac{t\Box \leq v \quad v \leq u\Box}{t\Box \leq u\Box} \text{ TRANS}$$

The first case is when we have two instances of the first option in Lemma 27. The proof works as in Lemma 22.

Suppose first that we have the first option in Lemma 27 on the left premise of (3.1), say with v being $w\Box$ and $t \leq w$. Then the right premise above is $w\Box \leq u\Box$. By induction hypothesis $\Gamma_\Box \vdash w \leq u$. But then using (TRANS) we have $t \leq u$, as desired.

The same reasoning applies if the first option in Lemma 27 applied to the right premise of (3.1).

The most interesting case is when both premises of (3.1) give instances of the second option in Lemma 27. From the left premise $t\Box \leq v$, we get x, y, and z such that $xy \leq v$, and $t \leq z \leq^{+-} x$. From the right premise $v \leq u\Box$ we get f, d, and e such that $v \leq fd$, and $f \leq^{+-} e \leq u$. Then $xy \leq v \leq fd$, and also $z \leq^{+-} x$ and $f \leq^{+-} e$. From (WC$_3$), we get $z \leq e$. By this fact together with $t \leq z$ and $e \leq u$, we have $t \leq u$. ◁

Lemma 29. *If $\Gamma_\Box \vdash_{\text{WC}} t^* \leq u^*$ with none of the new symbols $\Box_\sigma$ occurring in t^* or u^*, then $\Gamma \vdash_{\text{WC}} t^* \leq u^*$.*

Proof. The proof of this result elaborates the proof of Lemma 24. We begin again with the observation that if we take any proof tree $\mathcal{T}$ over $\Gamma_\Box$ in this system and replace, for every inhabited type σ, every occurrence of $\Box_\sigma$ by a fixed term $t : \sigma$ which is not $\Box_\sigma$, the result is a valid proof tree $\mathcal{T}[s]$ over $\Gamma_\Box$.

We also claim that in $\mathcal{T}[s]$ every assertion $t \leq u$ has the following property: for all types σ, $\Box_\sigma$ occurs in t iff it occurs in u. In the induction this time, we do not have to worry about (WC$_3$), since conclusions of (WC$_3$) cannot involve a new symbol. But we do need to think about (WC$_1$) and (WC$_2$). It allows us to conclude an inequality $fx \leq gy$ where x and y are possibly new (not in the subterm above) terms of the same type σ. Indeed, x and y might possibly be $\Box$. If both or neither is $\Box$, then we are done. And we cannot have one being $\Box$ and the other not, since this would imply that σ is inhabited and that the $\Box$-occurrence would have been replaced.

The end of the proof expands on that of Lemma 24. In the other proof, we took a proof tree $\mathcal{T}$ showing that $\Gamma_{\hat{\Box}} \vdash_{WC} t^* \leq u^*$ and directly showed that $\mathcal{T}[s]$ could have no $\Box$-occurrences. This time we might have applications of (WC$_3$) that get rid of two $\Box$-occurrences as we go from top to bottom on the left below:

$$\dfrac{\vdots \quad \vdots \quad \vdots}{\dfrac{g \leq^{+-} f \quad k \leq^{+-} h \quad f\Box_\sigma \leq k\Box_\sigma}{g \leq h}} \text{WC}_3 \qquad \dfrac{\dfrac{\vdots \quad \vdots}{g \leq f \quad f \leq k}}{\dfrac{g \leq k}{g \leq h}} \text{TRANS} \quad \dfrac{\vdots}{k \leq h} \text{TRANS}$$

However, in view of Lemma 28, the third premise implies that $\Gamma_{\hat{\Box}} \vdash f \leq k$. We can thus replace the entire application of (WC$_3$) above by two applications of (TRANS) in order to conclude that $\Gamma_{\hat{\Box}} \vdash g \leq h$, as on the right above. We do this replacement for every application of (WC$_3$) that dropped two $\Box$-occurrences. After that, the same proof by induction as in Lemma 24 shows that $\mathcal{T}[s]$ has no $\Box$-occurrences. This completes the proof. ◁

3.2. Complete Preorders. Our completeness theorem is for full structures which use weakly complete preorders for every type. But the proof uses the stronger notion of a complete preorder. The high-level reason is that in building a model of some set Γ of assumptions, it is very useful to define functions using joins of sets of elements. The kind of definition we have in mind would not work out in general on weakly complete preorders. The pleasant fact is that every preorder has an order embedding into some complete preorder. Our eventual proof strategy will involve taking the syntactic preorders for the base types $\mathbb{P}^{syn}_\beta$ determined by Γ, choosing completions for them, and then building the full hierarchy over the completions.

Definition 30. A *complete preorder* is a preorder $\mathbb{P}$ with the property that every subset $S \subseteq P$ has a least upper bound. This is an element $\bigvee S \in P$ with the property that for all $x \in S$, $x \leq \bigvee S$; and if y is such that for all $x \in S$, $x \leq y$, then $\bigvee S \leq y$.

The least upper bound of a set S is not in general unique, but any two least upper bounds x and y have the property that $x \equiv y$.

In a complete preorder we can fix an operation $\bigvee$ on subsets which gives the least upper bound. This uses the Axiom of Choice. Our definition does not build in $\bigvee$ as part of the structure of a complete preorder. That is, we did not take a complete preorder to be a structure $(P, \leq, \bigvee)$. But nothing much would change if we had done so.

Notice that if $\mathbb{P}$ is a complete preorder then $\bigvee \emptyset \leq x$ for all x, and $x \leq \bigvee P$. So $\bigvee P$ is a "top." Similarly $\bigvee \emptyset$ is a "bottom." In particular, every complete preorder is weakly complete.

Proposition 31. *If X is a set and $\mathbb{L} = (L, \leq)$ is a complete preorder, then for all sets X, the function set L^X is a complete preorder under the pointwise $\leq$ relation. To see this, fix an $\bigvee$ operation for the subsets of L. For $S \subseteq L^X$, we define*

$$(\bigvee S)x = \bigvee(\{f(x) : f \in S\})$$

Then it is easy to see that $\bigvee$ turns L^X into a complete preorder.

Remark 32. Here are two facts worth keeping in mind.

1. For sets $A_1,\ldots,A_k$ and $B_1,\ldots,B_\ell$ of subsets of M,

$$\bigvee(A_1 \cup A_2 \cup \cdots A_k) \leq \bigvee(B_1 \cup B_2 \cup \cdots \cup B_\ell)$$

provided that every A_i is a subset of some B_j. (This sufficient condition is not necessary, but it is sufficient and useful.)
2. Thus, for sets $A, B \subseteq M$, $\bigvee A \leq \bigvee B$ provided that every $a \in A$ is $\leq$ some $b \in B$.

Proposition 33. *Let $\mathbb{P} = (P, \leq)$ be a preorder. Then there is a complete preorder $\mathbb{P}^* = (P^*, \leq)$ and an order embedding $i : \mathbb{P} \to \mathbb{P}^*$.*

Remark 34. Before we turn to the proof, let us make two comments. First, we are not claiming any uniqueness of $\mathbb{P}^*$ of i in Proposition 33. There are in fact many ways to take a preorder and complete it in our sense.

Second, for $\mathbb{P}$ a poset (that is, a preorder additionally satisfying antisymmetry), we may use the usual construction of a complete lattice extending $\mathbb{P}$ by taking down-closed sets. However, we need a construction in which distinct elements $p, q \in P$ which are equivalent ($p \leq q \leq p$) are *not* identified by i. So the construction using down-closed sets will not work. However, it will be close. We are going to take the product of the complete lattice of down-closed subsets of P by the indiscrete preorder on the set P.

Proof. We define the preorder $(P^*, \leq)$ and the map i by

$$P^* = \{(A,p) : A \subseteq P \text{ is down-closed in } \leq \text{ and } p \in P\} \cup \{\bot\}$$
$$(A,p) \leq (B,q) \text{ iff } A \subseteq B$$
$$\bot \leq x \quad \text{for all } x \in P^*$$
$$i(p) = (\{q \in P : q \leq p\}, p)$$

The symbol $\bot$ in P^* is just intended to be some object which is fresh: it should not be a down-closed subset of P. In the definition of $(A,p) \leq (B,q)$, p and q play no role. To prove that every subset has a least upper bound, we need some extra machinery and a piece of notation. Fix a choice function $\varepsilon : \mathcal{P}(P) \setminus \{\emptyset\} \to P$ such that $\varepsilon(W) \in W$ for all nonempty subsets $W \subseteq P$. For a set $S \subseteq P^*$, define $W = W_S$ by $W = \{q \in P : \text{ for some } (A,p) \in S, q \in A\}$. Then for each $S \subseteq P^*$ define

$$\bigvee S = \begin{cases} (W, \varepsilon(W)) & \text{if } W \neq \emptyset \\ \bot & \text{if } W = \emptyset \end{cases}$$

The reason that we need ε is that we could take $\bigvee S$ to be (W, p_0) whenever W is non-empty and $p_0 \in W$. All such elements (W, p_0) will be equivalent in $\mathbb{P}^*$.

It is easy to check that $\mathbb{P}^*$ is a preorder, and that for all $S \subseteq P^*$, $\bigvee S$ is a least upper bound of S. Here is the verification of the required properties of i. First, if $i(p) = i(q)$, then by considering the second components of $i(p)$ and $i(q)$, we see that $p = q$. Continuing, if $p \leq q$, then every $r \leq p$ is also $\leq q$, and so

$$i(p) = (\{r : r \leq p\}, p) \leq (\{r : r \leq q\}, q) = i(q).$$

Conversely, if $i(p) \leq i(q)$, then since p belongs to the first component of $i(p)$, we see that $p \leq q$. ◁

3.3. The Extension Lemma.
We are going to use a technical lemma which allows us to take preorders $\mathbb{M}$ and $\mathbb{L}$ and to define a map $\mathbb{F} \to \mathbb{M}^{\mathbb{L}}$ from a map $\mathbb{F} \to \mathbb{M}^{\mathbb{S}}$, where $\mathbb{S}$ is a "sub-preorder" of $\mathbb{L}$. The work in this section will surely seem unmotivated at first glance. In fact, it will play a key role in our proof of the completeness theorem for the (WC)-deductive system. The reason for separating out this lemma and presenting it here is that it will be used infinitely many times as part of an inductive construction (see Lemma 43). The reader may wish to omit the proof of Lemma 35 on first reading.

Lemma 35 (Extension Lemma). *Let $\mathbb{F}$ be a polarized preorder. Let $\mathbb{L}$, $\mathbb{M}$, and $\mathbb{S}$ be preorders with $\mathbb{M}$ complete. Let $j : \mathbb{S} \to \mathbb{L}$ be an order embedding. Let $p : \mathbb{F} \to \mathbb{M}^{\mathbb{S}}$ preserve the order and polarity, and write p_f for $p(f) : \mathbb{S} \to \mathbb{M}$. Assume the following weak-completeness-like property:*

(3.2) $\qquad$ *whenever $f \leq^{+-} g$ in $\mathbb{F}$, and $x, y \in S$, then $p_f(x) \leq p_g(y)$.*

Then p has an extension $q : \mathbb{F} \to \mathbb{M}^{\mathbb{L}}$: q preserves the order and polarity, and for all $f \in \mathbb{F}$, $q_f \circ j = p_f$:

$$\begin{array}{ccc} \mathbb{S} & \xrightarrow{p_f} & \mathbb{M} \\ {\scriptstyle j}\downarrow & \nearrow_{q_f} & \\ \mathbb{L} & & \end{array}$$

Proof. For each $f \in \mathbb{F}$ and $x \in L$, define the following four subsets of M:

$$\begin{aligned} A(f,x) &= \{p_{h^+}(s) : h^+ \leq f, j(s) \leq x, \text{ and } s \in S\} \\ B(f,x) &= \{p_{h^-}(s) : h^- \leq f, x \leq j(s), \text{ and } s \in S\} \\ C(f) &= \{p_{h^-}(s) : (\exists k^+ \leq f)(h^- \leq k^+), \text{ and } s \in S\} \\ D(f) &= \{p_{h^+}(s) : (\exists k^- \leq f)(h^+ \leq k^-), \text{ and } s \in S\} \end{aligned}$$

For each $f \in \mathbb{F}$ and $x \in L$, we then define $q_f(x) \in M$ by

(3.3) $\quad q_f(x) = \begin{cases} p_f(s), & \text{if for some (unique) } s \in S, x = j(s); \\ \bigvee\Big(A(f,x) \cup B(f,x) \cup C(f) \cup D(f)\Big), & \text{if } x \notin j[S]. \end{cases}$

Here and also below, we use the fact that if $x = j(s)$, then s is unique. This is because j is an order-embedding, hence, it is one-to-one by definition. The join in (3.3) exists because $\mathbb{M}$ is a complete preorder.

Claim 36. *If $x = j(s)$, then every element of $A(f,x) \cup B(f,x) \cup C(f) \cup D(f)$ is $\leq p_f(s)$.*

Proof. Take an element of $A(f, j(s))$, say $p_{h^+}(t)$ where $j(t) \leq j(s)$. Since j reflects order, $t \leq s$. Then $p_{h^+}(t) \leq p_{h^+}(s) \leq p_f(s)$. At the end we used the assumption that p preserves order and polarity: since $h^+ \leq f$, $p_{h^+} \leq p_f$ in $\mathbb{M}^{\mathbb{S}}$ and p_{h^+} is monotone.

This time, take an element of $B(f, j(s))$, say $p_{h^-}(t)$ where $j(s) \leq j(t)$. Since j reflects order, $s \leq t$. Then $p_{h^-}(t) \leq p_{h^-}(s) \leq p_f(s)$.

We turn to $C(f)$. Let $h^- \leq k^+ \leq f$ in $\mathbb{F}$ and $t \in S$. We have $p_{h^-}(t) \leq p_{k^+}(s) \leq p_f(s)$.

Finally, for $D(f)$, let $h^+ \leq k^- \leq f$ and $t \in S$. Then $p_{h^+}(t) \leq p_{k^-}(s) \leq p_f(s)$.

Please note that the points about $C(f)$ and $D(f)$ used the weak-completeness-like property (3.2). $\triangleleft$

Claim 37. *Suppose that $f \leq g$ in $\mathbb{F}$. Then for all $x \in \mathbb{L}$, $A(f,x) \subseteq A(g,x)$, $B(f,x) \subseteq B(g,x)$, $C(f) \subseteq C(g)$, and $D(f) \subseteq D(g)$.*

Proof. All parts of this claim are consequences of the transitivity of $\leq$ in $\mathbb{F}$.   ◁

Claim 38. *Suppose that $x \leq y$ in L. Then $A(f,x) \subseteq A(f,y)$, and $B(f,y) \subseteq B(f,x)$.*

Proof. These are consequences of the transitivity of $\leq$ in $\mathbb{L}$.   ◁

In the next few claims, we show that $q_f(x) \leq q_g(y)$ by showing that every set involved in the definition of $q_f(x)$ in (3.3) is a subset of some set involved in the definition of $q_g(y)$. This comes from Remark 32.

Claim 39. *If $f \leq g$, then $q_f(x) \leq q_g(x)$ for all $x \in L$. Thus, $q : \mathbb{F} \to \mathbb{M}^{\mathbb{L}}$ is monotone.*

Proof. If $x \in j[S]$, say $x = j(s)$, then $q_f(x) = p_f(s) \leq p_g(s) = q_g(x)$. If $x \notin S$, we see from Claim 37 that each of the sets involved in $q_f(x)$ is a subset of the corresponding set involved in $q_g(x)$. So $q_f(x) \leq q_g(x)$.   ◁

Claim 40. *If f^+, then q_{f^+} is monotone.*

Proof. Let $x \leq y$. We show that $q_{f^+}(x) \leq q_{f^+}(y)$. If $x \in j[S]$, say $x = j(s)$, then $q_{f^+}(x) = p_{f^+}(s) \in A(f^+,y)$. So $p_{f^+}(s) \leq \bigvee A(f^+,y) \leq q_{f^+}(y)$.

If $x \notin j[S]$, we show that $B(f^+,x) \subseteq C(f^+)$. For then, by Claims 37 and 38, we would have the desired inequality $q_{f^+}(x) \leq q_{f^+}(y)$. In more detail, we would have $A(f^+,x) \subseteq A(f^+,y)$, $B(f^+,x) \subseteq C(f^+)$, and obviously $C(f^+) \subseteq C(f^+)$ and $D(f^+) \subseteq D(f^+)$. Let $p_{h^-}(s) \in B(f^+,x)$, where $h^- \leq f^+$ in $\mathbb{F}$ and $s \in S$. (We also have $x \leq j(s)$, but this is not used.) Then $p_{h^-}(s) \in C(f^+)$: take $k^+ = f^+$ in the definition of $C(f^+)$.   ◁

Claim 41. *If f^-, then q_{f^-} is antitone.*

Proof. Let $x \leq y$. We show that $q_{f^-}(y) \leq q_{f^-}(x)$. If $y \in j[S]$, say $y = j(s)$, then $q_{f^-}(y) = p_{f^-}(s) \in B(f^-,x)$. So $p_{f^-}(s) \leq \bigvee B(f^-,x) \leq q_{f^-}(x)$.

If $y \notin j[S]$, we show that $A(f^-,y) \subseteq D(f^-)$. For then, by Claims 37 and 38, we would have the desired inequality $q_{f^-}(y) \leq q_{f^-}(x)$. Let $p_{h^+(s)} \in A(f^-,y)$, where $h^+ \leq f^-$ in $\mathbb{F}$, $s \in S$, and $j(s) \leq y$. Then $p_{h^+}(s) \in D(f^-)$: take $k^- = f^-$ in the definition of $D(f^-)$.   ◁

We complete the proof of Lemma 35. We began with $p : \mathbb{F} \to \mathbb{M}^S$ and defined $q : \mathbb{F} \to \mathbb{M}^{\mathbb{L}}$. The verifications that q is monotone and preserves polarity come from Claims 39–41. For all $f \in F$, (3.3) tells us that $q_f \circ j = p_f$. This completes the proof.   ◁

3.4. Completeness Theorem.

Theorem 42 (Completeness). *If $\Gamma \vDash_{WC} s^* \leq t^*$, then $\Gamma \vdash_{WC} s^* \leq t^*$.*

Proof. Fix a set Γ of inequalities over some signature $\mathbb{F}$. Let $\mathbb{G}$ come from $\mathbb{F}$ by adding a fresh constant $\square_\sigma$ of every type σ. Let $\Gamma_\square$ be Γ, taken as a set of inequalities over $\mathbb{G}$. Let $\mathbb{P}_\sigma^{syn}$ be as in Definition 13, using $\mathbb{G}$ and $\Gamma_\square$. For each base type β, use

Proposition 33 to choose a complete preorder $\mathbb{Q}_\beta$ and an order embedding $i_\beta : \mathbb{P}_\beta^{syn} \to \mathbb{Q}_\beta$. Let the preorders $\mathbb{P}_\sigma^{sem}$ be as defined below:

$$(3.4) \quad \begin{aligned} \mathbb{P}_\beta^{sem} &= \mathbb{Q}_\beta \text{ from just above} \\ \mathbb{P}_{\sigma \to \tau}^{sem} &= \text{the full function set } (\mathbb{P}_\tau^{sem})^{\mathbb{P}_\sigma^{sem}}, \text{ ordered pointwise} \end{aligned}$$

On the function types σ, we construe $\mathbb{P}_\sigma$ as a polarized preorder in the obvious way. By Proposition 31, each preorder $\mathbb{P}_\sigma^{sem}$ is complete. The family $(\mathbb{P}_\sigma^{sem})_\sigma$ is a full hierarchy.

In the lemma below, recall the notion of an applicative family of interpretations. We construct such a family using our signature $\mathbb{F}$ and the full hierarchy $(\mathbb{P}_\sigma^{sem})_\sigma$.

Lemma 43. *There is an applicative family of interpretations* $\mathcal{N} = (\langle\!\langle \; \rangle\!\rangle_\sigma)_\sigma$, *where*

$$\langle\!\langle \; \rangle\!\rangle_\sigma : \mathbb{P}_\sigma^{syn} \to \mathbb{P}_\sigma^{sem},$$

such that for base types β, $\langle\!\langle \; \rangle\!\rangle_\beta = i_\beta$, *and for all* σ, $\langle\!\langle \; \rangle\!\rangle_\sigma$ *is an order embedding.*

Proof. We define $\mathbb{P}_\sigma$ and $\langle\!\langle \; \rangle\!\rangle_\sigma$ by recursion on the type σ. We verify that $\langle\!\langle \; \rangle\!\rangle_\sigma$ is an order embedding and also for function types that the relevant applicative family property (2.3) holds.

The recursion begins with base types. The order embedding fact is stated in Proposition 33, and there is nothing to check concerning the applicative family property.

In the induction step, we assume that $\langle\!\langle \; \rangle\!\rangle_\sigma$ and $\langle\!\langle \; \rangle\!\rangle_\tau$ are order embeddings. We shall define $\langle\!\langle \; \rangle\!\rangle_{\sigma \to \tau}$ using Lemma 35. The role of $\mathbb{F}$ in the lemma will be played by the polarized preorder $\mathbb{P}_{\sigma \to \tau}^{syn}$; please note that we are not using the preorder given by the original signature but by its closure under the logic. We further take $\mathbb{L} = \mathbb{P}_\sigma^{sem}$, $\mathbb{M} = \mathbb{P}_\tau^{sem}$, $\mathbb{S} = \mathbb{P}_\sigma^{syn}$, $j : \mathbb{S} \to \mathbb{L}$ to be $\langle\!\langle \; \rangle\!\rangle_\sigma$, and $p : \mathbb{F} \to \mathbb{M}^\mathbb{S}$ to be given by $p_t(u) = \langle\!\langle tu \rangle\!\rangle_\tau$. In pictures, here is what is going on. For each term $t : \sigma \to \tau$, we obtain $\langle\!\langle t \rangle\!\rangle_{\sigma \to \tau}$ as shown below:

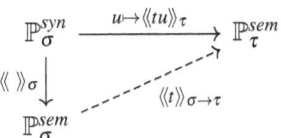

The rules of the logic translate to properties which we need p to have in order to apply Lemma 35: (POINT) implies that p preserves the order, while (MONO) and (ANTI) ensure that p preserves polarities. The induction hypothesis on σ includes the statement that j is an order embedding.

We also must check the weak-completeness-like property (3.2) which is a hypothesis of Lemma 35. Suppose that we have f and g in $\mathbb{P}_{\sigma \to \tau}^{syn}$ with $f^+ \leq g^-$. The only tagged symbols in that preorder are those in $\mathbb{G}_{\sigma \to \tau}$, so f and g are symbols in $\mathbb{G}_{\sigma \to \tau}$; indeed they come from the original signature. Let $t, u : \sigma$. Using the rule (WC$_1$), $\Gamma_\Box \vdash_{WC} f^+ t : \tau \leq g^- u : \tau$. That is, $f^+ t \leq g^- u$ in $\mathbb{S} = \mathbb{P}_\tau^{syn}$. Since $\langle\!\langle \; \rangle\!\rangle_\tau$ preserves the order, $\langle\!\langle ft \rangle\!\rangle_\tau \leq \langle\!\langle gu \rangle\!\rangle_\tau$ in $\mathbb{P}_\tau^{sem}$. This means that $p_f(t) \leq p_g(u)$, as required. We also verify (3.2) when $f^- \leq g^+$. The work is the same, using (WC$_2$) instead of (WC$_1$).

Lemma 35 tells us that p extends to $q : \mathbb{F} \to \mathbb{M}^\mathbb{L}$. We define $\langle\!\langle \; \rangle\!\rangle_{\sigma \to \tau} : \mathbb{P}_{\sigma \to \tau}^{syn} \to \mathbb{P}_{\sigma \to \tau}^{sem}$ by $\langle\!\langle t \rangle\!\rangle_{\sigma \to \tau} = q_t$. For each term $t : \sigma \to \tau$, q_t is an element of $\mathbb{M}^\mathbb{L}$ and hence a function of the right type. The fact that q preserves polarities and the order implies the same properties of $\langle\!\langle\ \rangle\!\rangle_{\sigma\to\tau}$. We have several further verifications.

The applicative family property (2.3). Let $t \in \mathbb{P}^{syn}_{\sigma\to\tau}$ and let $u \in \mathbb{P}^{syn}_{\sigma}$. Using the fact from Lemma 35 that $p_t = q_t \circ j$,

$$\langle\!\langle t \rangle\!\rangle_{\sigma\to\tau}(\langle\!\langle u \rangle\!\rangle_\sigma) = q_t(\langle\!\langle u \rangle\!\rangle_\sigma) = q_t(j(u)) = p_t(u) = \langle\!\langle tu \rangle\!\rangle_\tau.$$

$\langle\!\langle\ \rangle\!\rangle_{\sigma\to\tau}$ **reflects the order.** Suppose that in $\mathbb{P}^{sem}_{\sigma\to\tau}$, $\langle\!\langle t \rangle\!\rangle_{\sigma\to\tau} \leq \langle\!\langle u \rangle\!\rangle_{\sigma\to\tau}$. Let $x = \langle\!\langle \Box_\sigma \rangle\!\rangle_\sigma$. Then using the applicative family property which we just showed,

$$(3.5) \qquad \langle\!\langle t\Box_\sigma \rangle\!\rangle_\tau = \langle\!\langle t \rangle\!\rangle_{\sigma\to\tau}(x) \leq \langle\!\langle u \rangle\!\rangle_{\sigma\to\tau}(x) = \langle\!\langle u\Box_\sigma \rangle\!\rangle_\tau.$$

Since $\langle\!\langle\ \rangle\!\rangle_\tau$ reflects order, in $\mathbb{P}^{syn}_\tau$, $t\Box_\sigma \leq u\Box_\sigma$. Thus, $\Gamma_\Box \vdash t\Box_\sigma \leq u\Box_\sigma$. By Lemma 27, $\Gamma_\Box \vdash t \leq u$. This tells us that $t \leq u$ in $\mathbb{P}^{syn}_{\sigma\to\tau}$.

$\langle\!\langle\ \rangle\!\rangle_{\sigma\to\tau}$ **is one-to-one.** Suppose that in $\mathbb{P}^{sem}_{\sigma\to\tau}$, $\langle\!\langle t \rangle\!\rangle_{\sigma\to\tau} = \langle\!\langle u \rangle\!\rangle_{\sigma\to\tau}$. As in (3.5) above, we have $\langle\!\langle t\Box_\sigma \rangle\!\rangle_\tau = \langle\!\langle u\Box_\sigma \rangle\!\rangle_\tau$. Since $\langle\!\langle\ \rangle\!\rangle_\tau$ is one-to-one, $t\Box_\sigma = u\Box_\sigma$. Thus $t = u$.

This concludes the proof of Lemma 43. ◁

Let us complete the proof of Theorem 42. Suppose that $\Gamma \vDash_{WC} t^* \leq u^*$. By our remarks at the beginning of Section 3.1, this assertion holds when we add new symbols to the underlying signature. Let $\mathcal{N} = (\langle\!\langle\ \rangle\!\rangle_\sigma)_\sigma$ be the applicative family provided by Lemma 43. Let $\mathcal{M}$ be the full structure associated to $\mathcal{N}$ using Lemma 15. Each $\mathcal{M}_\sigma$ is (weakly) complete, since $\mathcal{M}_\sigma$ is the same preorder as $\mathcal{N}_\sigma$. Thus, $\mathcal{M} \vDash \Gamma$. Since the maps $\langle\!\langle\ \rangle\!\rangle_\sigma$ are monotone, $\mathcal{M} \vDash \Gamma_\Box$. By the assumption in our theorem, $\mathcal{M} \vDash t^* \leq u^*$. Since all of the maps $\langle\!\langle\ \rangle\!\rangle_\sigma$ reflect the order, Lemma 15 tells us that $\Gamma_\Box \vdash_{WC} t^* \leq u^*$. By Lemma 29, $\Gamma \vdash_{WC} t^* \leq u^*$. ◁

4. Variations and Extensions

Our next section contains results that build on what we saw in the previous section.

4.1. The Logic of Full Poset Structures.
A structure is a *poset structure* if each preorder $\mathbb{P}_\sigma$ is a partially ordered set: if $p \leq q$ and $q \leq p$, then $p = q$. For such structures, the following rule is sound:

$$\frac{s \leq t \quad t \leq s}{fs \leq ft} \text{ POS}$$

In this rule, $f \in \mathbb{F}_{\sigma\to\tau}$ is arbitrary; it need not be tagged $+$ or $-$. (When f is tagged either way, (POS) is obviously derivable.) In fact, we have a complete logic of weakly complete poset structures: take the rules in Figures 1 and 2 and add the (POS) rule. Here are the reasons: Every preorder $\mathbb{Q}$ has an associated poset $\mathbb{Q}^*$ obtained by taking the quotient $\mathbb{P}/\equiv$, where $p \equiv q$ iff $p \leq q \leq p$. The syntactic preorders $\mathbb{P}^{syn}_\sigma$ determined by a set Γ in the logic with (POS) may be taken to be a poset; we take the associated poset $(\mathbb{P}^{syn}_\sigma)^*$. To interpret function symbols on $\mathbb{P}^{syn}_\sigma$, we need a short well-definedness argument using (POS). We also tag an equivalence class $[f]$ with $+$ if some $g \equiv f$ is tagged $+$.

Continuing, the constructions of weakly complete preorders which we saw in Propositions 31 and 33 go through when we replace "preorder" by "poset" in the hypothesis and the conclusion. (In fact, Proposition 33 is a little easier in the poset setting, and it is rather well-known.)

In the proof of Theorem 42, we need to check that some functions are well-defined on $(\mathbb{P}^{syn}_\sigma)^*$. Each p_f is well-defined in Lemma 43; this comes from (POS). And in the Extension Lemma 35, we observe that if $f \equiv g$, then $q_f \equiv q_g$; this implies that each q_f is well-defined as a function on $(\mathbb{P}^{syn}_{\sigma \to \tau})^*$.

Identities. Another way to deal with poset structures would be to expand the basic assertions in the language to include identity statements $t = u$ with the obvious semantics. (This is also possible even with preordered structures, so we could have made this move early on.) Doing this, we would have the evident rules

$$\frac{p = q}{q = p} \text{ SYMM} \qquad \frac{p = q}{p \leq q} \text{ WEAK} \qquad \frac{p \leq q \quad q \leq p}{p = q} \text{ POS}'$$

Here is how the first two rules above are used. We need these rules in order to build the syntactic preorders in the first place. Their elements are equivalence classes $[t]$ of terms t under the $=$ equivalence relation. Using (WEAK) and (POS'), we can derive the reflexivity and transitivity rules for $=$. We also need them to define the order structure on these classes in such a way that $[t] \leq [u]$ iff $t \leq u$. This is needed at the very end of the proof of Theorem 42: our previous proof would go from $\langle\!\langle t \rangle\!\rangle \leq \langle\!\langle u \rangle\!\rangle$ to $[t] \leq [u]$. We need this extra step to know that $\Gamma \vdash t \leq u$ (rather than knowing that $\Gamma \vdash t' \leq u'$ for some $t' \equiv t$ and $u' \equiv u$.) The rule (POS') implies (POS). This rule (POS') would also be used at the very end of the proof of Theorem 42. We show that if $\Gamma \vDash_{WC} t^* = u^*$, then $\Gamma \vdash_{WC} t^* = u^*$. Our hypothesis easily implies that $\Gamma \vDash_{WC} t^* \leq u^*$ and that $\Gamma \vDash_{WC} u^* \leq t^*$. By the argument which have seen just above, $\Gamma \vdash_{WC} t^* \leq u^*$ and $\Gamma \vdash_{WC} u^* \leq t^*$. Hence using (POS'), $\Gamma \vdash_{WC} t^* = u^*$.

4.2. Arrow Assertions as Conclusions. Up until now, the main assertions in our language have been inequalities between terms of the same type. The polarity assertions f^+ and f^- were not "first-class" (despite what we said at the beginning of Section 2.2): our proof system contained no rules that allowed us to conclude a polarity assertion. To do this, we need to specify the semantics in full structures and to see what must be added to the proof system. For the semantics, suppose we are given a full structure $\mathcal{M}$ and a symbol $f : \sigma$ of a function type. Then we say

$$\mathcal{M} \vDash f^+ \text{ iff } [\![f]\!] \text{ is a monotone function.}$$

The proof theory adds two rules:

$$\frac{f^+ \quad f \leq g \quad g \leq f}{g^+} \text{ POL}^+ \qquad \frac{f^- \quad f \leq g \quad g \leq f}{g^-} \text{ POL}^-$$

Here, f and g are symbols from the underlying signature $\mathbb{F}$, and they should be of function type. The soundness of this rule appears in Theorem 45 below. When we write $\vdash_{WC}$ in the rest of this section, we mean provability with the rules in Figures 1 and 2, together with the rules (POL$^+$) and (POL$^-$).

The completeness proof adds to what we have seen in several ways. To begin, we need an analog of the construction where we add new symbols $\square_\sigma$. This time, we add *two* fresh constants. To ease our notation, we shall elide the type symbols and simply write these symbols as $\Box_1$ and $\Box_2$. Given a set Γ, we write Δ for the set of assertions that adds $\Box_1 \leq \Box_2$ for all types.

We need results on adding these new constants in this way, building on what we saw in Lemmas 22, 23, 27, and 28. In Lemma 44 below, note that some of the assertions appear to be weaker than one would want. Specifically, point (44) implies that "If $\Delta \vdash_{WC} t\Box_1 \leq u\Box_2$, then $\Delta \vdash_{WC} t \leq u$." The reason why we prefer the more involved statement is that this is what will be used in Theorem 45 below. (An additional support from our formulation is that the converses of all parts of Lemma 44 are true as well.)

Lemma 44. *Let Δ be defined from Γ as above.*

1. *If $\Delta \vdash_{WC} t \leq \Box_1$, then $t = \Box_1$. If $\Delta \vdash_{WC} \Box_2 \leq t$, then $t = \Box_2$. If $\Delta \vdash_{WC} t \leq \Box_2$, then either $t = \Box_1$ or $t = \Box_2$. If $\Delta \vdash_{WC} \Box_1 \leq t$, then either $t = \Box_1$ or $t = \Box_2$.*
2. *If $\Delta \vdash_{WC} t \leq v\Box_1$, then one of the following holds:*
   - (a) *there is some u such that $t = u\Box_1$ and $\Delta \vdash_{WC} u \leq v$, or else*
   - (b) *there are u and g^- such that $t = u\Box_2$ and $\Delta \vdash_{WC} u \leq g^- \leq v$; or*
   - (c) *there is a term $s : \sigma$, and constants $f, g : \sigma \to \tau$ such that $\Delta \vdash_{WC} t \leq fs$, and $f \leq^{+-} g \leq v$.*

   There are also similar facts when $\Delta \vdash_{WC} v\Box_1 \leq t$, $\Delta \vdash_{WC} t \leq v\Box_2$, and $\Delta \vdash_{WC} v\Box_2 \leq t$.
3. *If $\Delta \vdash_{WC} t\Box_1 \leq u\Box_1$, then $\Delta \vdash_{WC} t \leq u$. If $\Delta \vdash_{WC} t\Box_2 \leq u\Box_2$, then $\Delta \vdash_{WC} t \leq u$.*
   If $\Delta \vdash_{WC} t\Box_1 \leq u\Box_2$, then there is a function symbol f^+ such that $\Delta \vdash_{WC} t \leq f^+ \leq u$.
   If $\Delta \vdash_{WC} t\Box_2 \leq u\Box_1$, then there is a function symbol f^- such that $\Delta \vdash_{WC} t \leq f^- \leq u$.
4. *If $\Delta \vdash_{WC} f^+$, then $\Gamma \vdash_{WC} f^+$; similarly for $-$.*

Proof. Each assertion in part (1) is a straightforward induction.

Part (2) also is proved by four straightforward inductions. For one step, suppose that $\Delta \vdash_{WC} t \leq v\Box_1$ with a proof that ends with (ANTI) using $\Box_1 \leq \Box_2$. Then there is an antitone function symbol from the signature, say g^-, such that $v\Box_1 = g^- \Box_1$ and $t = g^- \Box_2$. So in this case, we have $u = g^- = v$.

Part (3) is proved by simultaneous induction on the number of transitivity steps in derivations. Here is the transitivity step in the first assertion. Suppose that the root uses (TRANS), say

$$\frac{t\Box_1 \leq v \quad v \leq u\Box_1}{t\Box_1 \leq u\Box_1} \text{ TRANS}$$

The previous lemma applies to both subproofs, and thus we have $3 \times 3 = 9$ cases. Let us suppose first that above the right subproof we have (a). There is some w such that v is $w\Box_1$, and $\Delta \vdash_{WC} w \leq u$. The left subproof ends $t\Box_1 \leq w\Box_1$, so by induction hypothesis, $\Delta \vdash_{WC} t \leq w$. And thus $\Delta \vdash_{WC} t \leq u$ as well.

Suppose next that above the right subproof we have (b). We thus have w and h^- such that $v = w\Box_2$ and $\Delta \vdash_{WC} w \leq h^- \leq u$. Thus, the second subproof concludes $t\Box_1 \leq w\Box_2$. By our induction hypothesis, there is some g^+ such that $\Delta \vdash_{WC} t \leq g^+ \leq w$. Hence $\Delta \vdash_{WC} t \leq u$, as desired.

The other assertions in part (3) are similar to what we have seen, either above or in Lemmas 22 and 23.

For part (4). We first show that $\Delta \nvdash_{WC} \square_i^+$ and $\Delta \nvdash_{WC} \square_i^-$. The proof is an easy induction on derivations, and it also uses part (1) of this result. We next show something stronger than the assertion in part (4): if φ is any assertion in this language which has no new $\square$ symbols and $\Delta \vdash_{WC} \varphi$, then $\Gamma \vdash_{WC} \varphi$. The proof is basically the same as that of Lemma 29: we observe that the rules (POL$^+$) and (POL$^-$) cannot eliminate the new $\square$ symbols: f in these rules cannot be $\square_i^+$ since $\Delta \nvdash_{WC} \square_i^+$; and if g were $\square_1$ or $\square_2$, then since one of the premises is $f \leq g$, we would have $f = \square_j^+$ for some j by part (44). This again contradicts $\Delta \nvdash_{WC} \square_j^+$. ◁

We turn to our main result on the system. We state Theorem 45 only mentioning assertions of the form f^+, but it also holds for inequality assertions $t^* \leq u^*$, with basically the same statement and proof as in Theorem 42.

Theorem 45. $\Gamma \vDash_{WC} f^{*+}$ iff $\Gamma \vdash_{WC} f^{*+}$, and similarly for $-$.

Proof. Here is the soundness half. Let $\mathcal{M}$ be a full hierarchy, and assume the hypotheses of (POL$^+$). Let the type involved be the (function) type σ. Since f^+, we know that $[\![f]\!]$ is monotone. Also, $[\![f]\!] \leq [\![g]\!] \leq [\![f]\!]$. Thus $[\![g]\!]$ is also monotone, as desired. The argument for (POL$^-$) is similar.

We turn to the completeness of the logic. Suppose that $\Gamma \vDash_{WC} f^+$. Starting from Γ, we form a theory Δ as mentioned earlier: for each type σ, we add two fresh constants $\square_1$ and $\square_2$ to the signature, and the assertion $\square_1 \leq \square_2$ to the theory. We need to know that $\Delta \vDash_{WC} f^+$, and this is straightforward by considering reducts: every full model of Δ is (after throwing away the interpretations of the new symbols) a model of Γ, and so the interpretation of f will be monotone.

At this point we are going to replay the proof of Theorem 42 and dwell only on the changes that are to be made. Form $\mathbb{P}_\sigma^{syn}$ and $\mathbb{P}_\sigma^{sem}$ as before, except that now we regard them as *polarized* preorders in the evident way: in $\mathbb{P}_\sigma^{syn}$ we use provability from Γ to determine the polarities, and in $\mathbb{P}_\sigma^{sem}$ we use the monotonicity/antitonicity of actual functions.

In Lemma 43 we amend the statement to also say that for a function type σ, $\langle\!\langle \ \rangle\!\rangle_\sigma$ reflects polarities. (This function *preserves* polarities, since this is part of the definition of an applicative family of interpretations.) We therefore must check that if $\langle\!\langle g \rangle\!\rangle$ is monotone, then $\Delta \vdash_{WC} g^+$. Since $\langle\!\langle \ \rangle\!\rangle_\sigma$ is monotone (by induction hypothesis), $\langle\!\langle \square_1 \rangle\!\rangle \leq \langle\!\langle \square_2 \rangle\!\rangle$. By monotonicity, $\langle\!\langle g \rangle\!\rangle(\langle\!\langle \square_1 \rangle\!\rangle) \leq \langle\!\langle g \rangle\!\rangle(\langle\!\langle \square_2 \rangle\!\rangle)$ in $\mathbb{P}_\tau^{sem}$. Since $\langle\!\langle \ \rangle\!\rangle_\tau$ reflects order, we get that $\Delta \vdash_{WC} g\square_1 \leq g\square_2$. By Lemma 44(3) with $t = g = u$, there is a symbol h in the underlying signature which is tagged $+$ such that $\Delta \vdash_{WC} g \leq h^+ \leq g$. By (POL$^+$), $\Delta \vdash_{WC} g^+$. This concludes the changes in Lemma 43.

To resume and complete the proof of our theorem, suppose that f^* is a symbol of function type and $\Gamma \vDash_{WC} f^{*+}$. Consider the full model $\mathcal{M}$ whose preorders are $\mathbb{P}_\sigma^{sem}$ with interpretations given by Lemma 43. Since those interpretations are monotone, $\mathcal{M} \vDash_{WC} \Delta$. Thus, $[\![f^*]\!] = \langle\!\langle f^* \rangle\!\rangle$ is monotone. Since $\langle\!\langle \ \rangle\!\rangle$ reflects polarities, $\Delta \vdash_{WC} f^{*+}$. In view of Lemma 44(4), $\Gamma \vdash_{WC} f^{*+}$. ◁

The result in this section may be recast as a "Lyndon-type" theorem. Statements like this may be found in [7] and [10]. But in both of these cases, the hypotheses are different, the languages include variables and abstraction but no polarity assertions, and the class of models includes more general models rather than just the full structures. But all of these are of the form "semantically monotone implies $+$; semantically antitone implies $-$."

Corollary 46. *Fix a set Γ. Let t be a term of function type, and assume that $[\![t]\!]$ is monotone in all (full) models of Γ; and also that for some function symbol f from the underlying signature, $\Gamma \vdash_{WC} f \leq t \leq f$. Then there is a symbol f with this property such that $\Gamma \vdash_{WC} f^+$.*

4.3. The Logic of Higher-order Applicative Terms and Equality.
For our last variation, we consider higher-order applicative terms and equality. In other words, we abandon the order structure entirely and consider the simply typed lambda calculus without variables or abstraction. The statements of interest are identities between terms of the same type, and the semantic notion is given by (1.5). For the logic, we take the reflexive, symmetric, and transitive laws for $=$, and also the congruence rule for application

$$\frac{t=t' \quad u=u'}{tu=t'u'} \text{ CONG}$$

This logic is complete, and we sketch the proof.

First, we need lemmas on constants in both the semantics and the proof theory. Let Γ be a set of identity assertions between terms, and let $\Gamma_{\square}^{\vphantom{}}$ add fresh constants of every type. In the syntax, the lemma would say that if $\Gamma_{\square} \vdash t\square = u\square$, then $\Gamma \vdash t = u$. In the semantics, we would want to know that for all assertions $t^* = u^*$ in the language of Γ, if $\Gamma \vDash t^* = u^*$, then also $\Gamma_{\square} \vDash t^* = u^*$.

Suppose that $\Gamma \vDash t^* = u^*$. As we have argued, we have $\Gamma_{\square} \vDash t^* = u^*$. For each type σ, let X_σ^{syn} be the set of terms of type σ in the expanded signature, modulo the equivalence relation $R(t, u) \leftrightarrow \Gamma_{\square} \vdash t = u$. So the elements X_σ^{syn} are equivalence classes $[t]$ of terms.

We build a full hierarchy of sets (X_σ^{sem}) in the evident way, by taking $X_\beta^{sem} = X_\beta^{syn}$ for base types β, and for other types, $X_{\sigma \to \tau}^{sem} = (X_\tau^{sem})^{X_\sigma^{sem}}$.

We now prove that there is a family of injective maps $\langle\!\langle \ \rangle\!\rangle_\sigma : X_\sigma^{syn} \to X_\sigma^{sem}$ with the property that $\langle\!\langle [tu] \rangle\!\rangle_\tau = \langle\!\langle [t] \rangle\!\rangle_{\sigma \to \tau}(\langle\!\langle [u] \rangle\!\rangle_\sigma)$. When σ is a base type, we take $\langle\!\langle \ \rangle\!\rangle_\sigma$ to be the identity. Suppose we are given $\langle\!\langle \ \rangle\!\rangle_\sigma$ and $\langle\!\langle \ \rangle\!\rangle_\tau$ with the desired properties, and we wish to define $\langle\!\langle \ \rangle\!\rangle_{\sigma \to \tau}$. The definition is

$$\langle\!\langle [t] \rangle\!\rangle_{\sigma \to \tau}(x) = \begin{cases} \langle\!\langle [tu] \rangle\!\rangle_\tau & \text{if for some (unique) } u : \sigma, x = \langle\!\langle [u] \rangle\!\rangle_\sigma \\ \langle\!\langle [\square_\tau] \rangle\!\rangle_\tau & \text{if there is no such term } u : \sigma \end{cases}$$

where $t : \sigma \to \tau$ is a term and $x \in \mathbb{P}_\sigma^{sem}$. In the bottom line, $\langle\!\langle [\square_\tau] \rangle\!\rangle_\tau$ is the only element of X_τ^{sem} that is sure to exist; no features of it are important. Here is the verification of the uniqueness of x in the top line: if $\langle\!\langle [u] \rangle\!\rangle_\sigma = x = \langle\!\langle [u'] \rangle\!\rangle_\sigma$, then since $\langle\!\langle \ \rangle\!\rangle_\sigma$ is injective (by our inductive assumption), $[u] = [u']$. We also check that the top line of this definition is independent of the choice of representatives of the classes $[t]$ and $[u]$. For if $\Gamma \vdash t = t'$ and also $\Gamma \vdash u = u'$, then also $\Gamma \vdash tu = t'u'$ by (CONG). Hence $[tu] = [t'u']$. It remains to check that $\langle\!\langle \ \rangle\!\rangle_{\sigma \to \tau}$ is injective. Suppose that $\langle\!\langle [t] \rangle\!\rangle_{\sigma \to \tau} = \langle\!\langle [t'] \rangle\!\rangle_{\sigma \to \tau}$. Then $\langle\!\langle t\square_\sigma \rangle\!\rangle_\tau = \langle\!\langle [t] \rangle\!\rangle_{\sigma \to \tau}(\langle\!\langle [\square_\sigma] \rangle\!\rangle_\sigma) = \langle\!\langle [t'] \rangle\!\rangle_{\sigma \to \tau}(\langle\!\langle [\square_\sigma] \rangle\!\rangle_\sigma) = \langle\!\langle t'\square_\sigma \rangle\!\rangle_\tau$. So by injectivity of $\langle\!\langle \ \rangle\!\rangle_\tau$, $\Gamma_{\square} \vdash t\square = t'\square$. Thus $\Gamma_{\square} \vdash t = t'$, and in other words $[t] = [t']$.

This completes the inductive step of the lemma. We conclude with a proof of the overall completeness theorem. Suppose that $\Gamma \vDash t^* = u^*$. Then also $\Gamma_\square \vDash t^* = u^*$. Let $\mathcal{M}$ be the full type hierarchy $(X^{sem}_\sigma)_\sigma$. We have defined maps $\langle\!\langle\ \rangle\!\rangle_\sigma : X^{syn}_\sigma \to X^{sem}_\sigma$. From these, we interpret the symbols in the original signature by taking $[\![t]\!] = \langle\!\langle [t] \rangle\!\rangle_\sigma$ for the unique σ such that $t : \sigma$. As in Lemma 15, for all terms $t : \sigma$, $[\![t]\!] = \langle\!\langle [t] \rangle\!\rangle_\sigma$. It follows that $\mathcal{M} \vDash \Gamma$. By our assumption that $\Gamma \vDash t^* = u^*$, we see that $[\![t^*]\!] = [\![u^*]\!]$. Let σ be the type of t^*. Then $\langle\!\langle [t^*] \rangle\!\rangle_\sigma = \langle\!\langle [u^*] \rangle\!\rangle_\sigma$. Since $\langle\!\langle\ \rangle\!\rangle$ is injective, $\Gamma_\square \vdash t^* = u^*$. By one our our points above, this tells us that $\Gamma \vdash t^* = u^*$, as desired.

5. CONCLUSION

The main results in this paper were the completeness theorems, Theorems 42 and 45, and also Corollary 46. The theorems suggest that the logical systems in the paper are the "right" ones: they are complete for the most natural semantics of higher-order applicative terms using a semantics where one can declare symbols to be interpreted in a monotone or antitone way, and also assert inequalities between terms. Corollary 46 does something similar, but not for entailment so much as for the expressive features of the system.

There are two ways in which it would be important to go beyond what we did here.

First, we return to the very start of this paper, the presentation of tonoids as operations defined by types as in (1.4). As the reader may have noticed, the type system in this paper was not sufficient to deal with (1.4). All of our types were "simpler arrows" $\to$ rather than $\xrightarrow{+}$ or $\xrightarrow{-}$. So we cannot type a function as in (1.4). It is thus of interest to extend our results to the system where we incorporate monotonicity/antitonicity information into the type system in a wholehearted manner, at all higher types. It is possible to formulate a syntax, semantics, and logical system that can handle this extension. The details are not so simple, and so we shall not enter in to them. Those details may be found in our paper [6, Section 5]. We expect that the methods of this paper show that the logical system there is complete for full models, at least when one works over weakly complete preorders.

Second, we have not been able to prove the completeness theorem that we are after in this subject, where one considers full preorder hierarchies built over arbitrary preorders, without assuming that the base preorders $\mathbb{P}_\beta$ are weakly complete. This would mean using the most natural logic for higher-order terms in our setting, the rules in Figure 1. In order to motivate the problem, let us review where in our work the assumption of weak completeness actually was used. Assuming weak completeness gives the additional (WC) rules stated in Figure 2. Those rules are not sound for all preorders, as shown in Example 2, part (2). Yet, they played a key role in Lemma 43. Specifically, Lemma 43 called on Lemma 35, and in order to apply Lemma 35, the logic needed to have the (WC) rules.

Here is a related point: our overall work made critical use of the passage from a preorder $\mathbb{P}$ to a completion $\mathbb{P}^*$, and it also made critical use of the Extension Lemma 43. To follow the general proof strategy of this paper, we seem to require a weaker type of completeness (one that adds fewer points), and a stronger Extension Lemma (one that works for the original logic). Getting all of this to work out is a challenge.

Acknowledgments. We are grateful to an anonymous referee for useful comments and corrections. All remaining errors are our own. We also thank Katalin Bimbó for all her work on this volume and other projects which keep alive the memory of Mike Dunn.

This work was partially supported by a grant from the Simons Foundation (#245591 to Lawrence Moss).

REFERENCES

[1] Awodey, S. (2000). Topological representation of the lambda-calculus, *Mathematical Structures in Computer Science* **10**(1): 81–96.

[2] Dougherty, D. J. and Subrahmanyam, R. (2000). Equality between functionals in the presence of coproducts, *Information and Computation* **157**(1-2): 52–83.

[3] Dunn, J. M. (1991). Gaggle theory: An abstraction of Galois connections and residuation, with applications to negation, implication, and various logical operators, *in* J. van Eijck (ed.), *Logics in AI: European Workshop JELIA '90*, number 478 in *Lecture Notes in Computer Science*, Springer, Berlin, pp. 31–51.

[4] Dunn, J. M. (1993). Partial gaggles applied to logics with restricted structural rules, *in* K. Došen and P. Schroeder-Heister (eds.), *Substructural logics*, Vol. 2 of *Studies in Logic and Computation*, Oxford University Press, New York, pp. 63–108.

[5] Friedman, H. (1975). Equality between functionals, *in* R. Parikh (ed.), *Proceedings of Logic Colloquium '73*, Vol. 53 of *Lecture Notes in Mathematics*, pp. 22–37.

[6] Icard, T. F. and Moss, L. S. (2014). Recent progress on monotonicity, *Linguistic Issues in Language Technology* **9**(7): 167–194.

[7] Icard, T. F. and Moss, L. S. (2021). Reasoning about monotonicity in a higher-order setting. unpublished ms.

[8] Icard, T. F., Moss, L. S. and Tune, W. (2017). A monotonicity calculus and its completeness, *Proceedings of the 15th Meeting on the Mathematics of Language*, Association for Computational Linguistics, pp. 75–87.

[9] Moss, L. S. (2012). The soundness of internalized polarity marking, *Studia Logica* **100**: 683–704.

[10] Tune, W. (2016). *A Lambda Calculus for Monotonicity Reasoning*, PhD thesis, Indiana University.

[11] van Benthem, J. (1986). *Essays in Logical Semantics*, Vol. 29 of *Studies in Linguistics and Philosophy*, D. Reidel Publishing Co., Dordrecht.

DEPARTMENT OF MATHEMATICS, INDIANA UNIVERSITY, BLOOMINGTON, IN 47401, U.S.A.
Email: lmoss@indiana.edu
DEPARTMENT OF PHILOSOPHY, STANFORD UNIVERSITY, STANFORD, CA, U.S.A.
Email: icard@stanford.edu

VARIETIES OF NEGATION AND CONTRA-CLASSICALITY IN VIEW OF DUNN SEMANTICS

Hitoshi Omori and Heinrich Wansing

ABSTRACT. In this paper, we discuss J. Michael Dunn's foundational work on the semantics for First Degree Entailment logic (**FDE**), also known as Belnap–Dunn logic (or Sanjaya–Belnap–Smiley–Dunn Four-valued Logic, as suggested by Dunn himself). More specifically, by building on the framework due to Dunn, we sketch a broad picture towards a systematic understanding of *contra-classicality*. Our focus will be on a simple propositional language with negation, conjunction, and disjunction, and we will systematically explore variants of **FDE**, **K3**, and **LP** by tweaking the falsity condition for negation.

Keywords. Bi-lateral natural deduction, Contraposition, Contra-classicality, Dunn semantics, Negation, Uni-lateral natural deduction, Variable sharing property

1. INTRODUCTION

Let us begin with a brief explanation of the three key notions included in the title of our paper, namely, *Dunn semantics*, *contra-classicality* and *negation*.

Dunn semantics. The logic of first-degree entailment **FDE**, also known as Belnap–Dunn logic (or Sanjaya–Belnap–Smiley–Dunn Four-valued Logic, as suggested by Dunn himself in [17, p. 95]), is a basic paraconsistent and paracomplete logic that has found many applications in philosophy and different areas of computer science, including the semantics of logic programs and inconsistency-tolerant description logics. The seminal papers [12; 4; 5] on **FDE** from the 1970s have been re-printed in [33], together with some recent essays devoted to Belnap–Dunn Logic.

The system **FDE** has various equivalent semantical presentations, cf. [31]. There exists a four-valued semantics, a so-called "star" semantics, an algebraic semantics, and a two-valued relational semantics due to Dunn [12]. (Note that the results published in [12] were already established and included in [11].) This semantics not only justifies the intuitive reading of the four truth values in the four-valued semantics but also enables a tweaking of the falsity condition of negation so as to obtain certain variants of **FDE**, the paracomplete three-valued strong Kleene logic **K3**, and the paraconsistent three-valued logic of paradox, **LP**. The four-valued semantics and the relational Dunn semantics are very closely related, and there exists a mechanical procedure to turn the many-valued truth tables into pairs of truth and falsity conditions, and vice versa, see [30].

2020 *Mathematics Subject Classification.* Primary: 03-03, Secondary: 03B53, 03B60.

Bimbó, Katalin, (ed.), *Relevance Logics and other Tools for Reasoning. Essays in Honor of J. Michael Dunn*, (Tributes, vol. 46), College Publications, London, UK, 2022, pp. 309–337.

Contra-classicality. The notion of a contra-classical logic has been coined by Lloyd Humberstone [26]. The most prominent non-classical logics such as, for example, minimal logic, intuitionistic logic, and the relevance logics **E** and **R** are subclassical. If they are presented in the vocabulary of classical logic, their consequence relations are subsets of the consequence relation of classical logic. In contrast to this, a contra-classical logic validates consequences that are not valid in classical logic. Various contra-classical logics have been studied in the literature. Examples include Abelian logic (cf. [29], [34]), systems with demi-negation (cf. [25; 26; 35]), certain systems of connexive logic (cf. [47], [49]), and the second-order Logic of Paradox (cf. [23]).

Some of the known contra-classical logic are contra-classical in a way that radically differs from logical orthodoxy insofar as they are non-trivial but negation inconsistent. These logics contain provable contradictions, i.e., they contain formulas A such that both A and the negation $\sim A$ of A are theorems. Whilst **FDE**, **K3**, and **LP** are subclassical logics, we will see that a tweaking of the falsity condition for negation in these logics can give rise to contra-classical systems. Some of the contra-classical variants of **FDE**, **K3**, and **LP** turn out to be negation inconsistent and some are negation incomplete.

Negation. There exists an extensive literature on the notion of negation and on which properties a genuine negation connective minimally ought to possess, see, for example, [22; 24; 50; 46; 7; 8; 10]. Although Michael Dunn has made substantial contributions to the study of negation as a modal operator of impossibility or "unnecessity" [13; 14; 15; 18], he clearly had a broader understanding of the concept of negation and even voiced the conviction that negation flip-flops between truth and falsity. Here is a quote from [15, p. 49] (notation adjusted):

> Tim Smiley once good-naturedly accused me of being a kind of lawyer for various non-classical logics. He flattered me with his suggestion that I could make a case for anyone of them, and in particular provide it with a semantics, no matter what the merits of the case [...] But I must say that my own favourite is the 4-valued semantics. I am persuaded that '$\sim A$ is true iff A is false', and that '$\sim A$ is false iff A is true'. And now to paraphrase Pontius Pilate, we need to know more about 'What are truth and falsity?'. It is of course the common view that they divide up the states into two exclusive kingdoms. But there are lots of reasons, motivated by applications, for thinking that this is too simple-minded.

In the present paper, we will study variants of logics in which negation flip-flops between truth and falsity, namely, variants of **FDE**, **K3**, and **LP**. A very weak requirement imposed on a unary connective in a logical system to deserve the classification as a negation connective is that for some formulas A and B, neither $A \vdash \sim A$ nor $\sim B \vdash B$, cf. [2; 28]. We will consider one-place connectives that not only satisfy this weak condition but also share the above truth condition for negation: $\sim A$ is true (under a given interpretation) iff A is false (under that interpretation). *Classically* falsity means untruth, so that the truth condition already fixes the falsity condition, but this is not the case in general, and in particular, it is not the case in **FDE**, **K3**, and **LP**, where truth and falsity are two primitive concepts that are on a par. A discussion of semantical opposition understood as an opposition between on the one hand truth and falsity, and on the other hand between truth and untruth can be found in [32], where it is observed that in the four-valued setting of **FDE**, the above truth condition for $\sim A$ together with the understanding of falsity of $\sim A$ as untruth of A results in the "demi-negation" of the system **CP** from [27].

According to Arnon Avron, the requirement that $\sim A$ is true iff A is false represents "the idea of falsehood within the language" [1, p. 160]. We shall keep this truth condition for negated formulas but abandon the classical understanding of falsity as untruth and instead treat truth and falsity as two separate primitive semantical notions of equal importance. There is thus a clear sense in which the unary connectives in this paper written as $\sim$, sometimes with a subscript, can be seen as negations. However, there is now room for tweaking the falsity condition for negation. We will consider all combinations that are possible for **FDE**, **K3**, and **LP** in a classical metatheory. This gives us sixteen variants of **FDE**, four variants of **K3**, and four variants of **LP**. By considering these logics, we are applying what Luis Estrada-González [19; 20] has called "the Bochum Plan."[1]

The themes dealt with in the present paper are among the topics addressed in nine questions we had posed to Prof. J. Michael Dunn in March 2021 together with Grigory Olkhovikov (the notion of negation, the tweaking of falsity conditions, negation inconsistency, bilateralism, contraposition), see [51]. Unfortunately, Mike was no longer able to answer these questions. He passed away on 5 April 2021, a few weeks after he informed us that he is willing to answer our questions.

Before moving further, let us recall some well known results related to **FDE**, **K3**, and **LP**. The language $\mathcal{L}$ consists of a set $\{\sim, \wedge, \vee\}$ of propositional connectives and a countable set Prop of propositional variables which we denote by $p, q, \ldots$. Furthermore, we denote by Form the set of formulas defined as usual in $\mathcal{L}$. We denote a formula of $\mathcal{L}$ by $A, B, C, \ldots$ and a set of formulas of $\mathcal{L}$ by $\Gamma, \Delta, \Sigma, \ldots$.

We begin with the many-valued representations of **FDE**, **K3** and **LP**.

Definition 1. A four-valued **FDE**-interpretation of $\mathcal{L}$ is a function $v_4 \colon \text{Prop} \longrightarrow \{\mathbf{t}, \mathbf{b}, \mathbf{n}, \mathbf{f}\}$. Given a four-valued interpretation v_4, this is extended to a function I_4 that assigns every formula a truth value by truth functions depicted in the form of truth tables as follows:

	$\sim$		$\wedge$	t	b	n	f		$\vee$	t	b	n	f
t	f		t	t	b	n	f		t	t	t	t	t
b	b		b	b	b	f	f		b	t	b	t	b
n	n		n	n	f	n	f		n	t	t	n	n
f	t		f	f	f	f	f		f	t	b	n	f

Then, the semantic consequence relation for **FDE** ($\vDash_{\mathbf{FDE}}$) is defined as follows.

Definition 2. For all $\Gamma \cup \{A\} \subseteq \text{Form}$, $\Gamma \vDash_{\mathbf{FDE}} A$ iff for all four-valued **FDE**-interpretations v_4, $I_4(A) \in \mathcal{D}$ if $I_4(B) \in \mathcal{D}$ for all $B \in \Gamma$, where $\mathcal{D} = \{\mathbf{t}, \mathbf{b}\}$.

Now, if we eliminate the value $\mathbf{b}$ from the semantics for **FDE**, then we obtain the three-valued semantics for **K3**, as follows.

Definition 3. A three-valued **K3**-interpretation of $\mathcal{L}$ is a function $v_3 \colon \text{Prop} \longrightarrow \{\mathbf{t}, \mathbf{n}, \mathbf{f}\}$. Given a three-valued interpretation v_3, this is extended to a function I_3 that

---

[1] Note that the Bochum Plan in general does not privilege truth over falsity, so that we could also keep the standard falsity condition for negation and systematically tweak the truth condition.

assigns every formula a truth value by truth functions depicted in the form of truth tables as follows:

~		∧	t	b	f		∨	t	b	f
t	f	t	t	n	f		t	t	t	t
n	n	n	n	n	f		n	t	n	n
f	t	f	f	f	f		f	t	n	f

Then, the semantic consequence relation for **K3** ($\vDash_{K3}$) is defined as follows.

Definition 4. For all $\Gamma \cup \{A\} \subseteq$ Form, $\Gamma \vDash_{K3} A$ iff for all three-valued interpretations v_3, $I_3(A) \in \mathcal{D}$ if $I_3(B) \in \mathcal{D}$ for all $B \in \Gamma$, where $\mathcal{D} = \{\mathbf{t}\}$.

Moreover, if we eliminate the value **n** from the semantics for **FDE**, then we obtain the three-valued semantics for **LP**, as follows.

Definition 5. A three-valued **LP**-interpretation of $\mathcal{L}$ is a function v_3: Prop ⟶ $\{\mathbf{t},\mathbf{b},\mathbf{f}\}$. Given a three-valued interpretation v_3, this is extended to a function I_3 that assigns every formula a truth value by truth functions depicted in the form of truth tables as follows:

~		∧	t	b	f		∨	t	b	f
t	f	t	t	b	f		t	t	t	t
b	b	b	b	b	f		b	t	b	b
f	t	f	f	f	f		f	t	b	f

Then, the semantic consequence relation for **LP** ($\vDash_{LP}$) is defined as follows.

Definition 6. For all $\Gamma \cup \{A\} \subseteq$ Form, $\Gamma \vDash_{LP} A$ iff for all three-valued interpretations v_3, $I_3(A) \in \mathcal{D}$ if $I_3(B) \in \mathcal{D}$ for all $B \in \Gamma$, where $\mathcal{D} = \{\mathbf{t},\mathbf{b}\}$.

Finally, let us recall the Dunn semantics for **FDE**.

Definition 7. A *Dunn-interpretation* of $\mathcal{L}$ is a relation, r, between propositional variables and the values 1 and 0, namely, $r \subseteq$ Prop $\times \{1,0\}$. Given an interpretation, r, this is extended to a relation between all formulas and truth values by the following clauses:

~$Ar1$ iff $Ar0$, $A \wedge Br1$ iff $Ar1$ and $Br1$, $A \vee Br1$ iff $Ar1$ or $Br1$,
~$Ar0$ iff $Ar1$, $A \wedge Br0$ iff $Ar0$ or $Br0$, $A \vee Br0$ iff $Ar0$ and $Br0$.

Definition 8. A formula A is a *two-valued semantic consequence* of Γ ($\Gamma \vDash_2 A$) iff for all Dunn-interpretations r, if $Br1$ for all $B \in \Gamma$ then $Ar1$.

Remark 9. We obtain the Dunn semantics for **K3** and **LP** by adding the following constraints, respectively, to r: (no-gap) for no p, $pr1$ and $pr0$; (no-glut) for all p, $pr1$ or $pr0$. Of course, if we add both constraints, then we obtain the semantics for classical logic.

Given our assumption concerning negation, we will systematically consider the variants of **FDE**, **K3** and **LP** by changing the falsity condition for negation, and explore their basic properties.[2]

---

[2]Note that in a recent article [21], Estrada-González considers the Bochum plan and suggests systematic changes in the evaluation conditions not only for negation, but also for other connectives. By doing so, he emphasized the tweaking of the evaluation clauses as a source of contra-classicality.

2. SEMANTICS

Let us now present the semantics for the variations of **FDE**, **K3** and **LP** we consider in the rest of paper. We begin with the variations of **FDE**.

By simple combinatorial considerations, the following sixteen operations exhaust the space of possible connectives that share the truth condition for negation.

A	$\sim_1 A$	$\sim_2 A$	$\sim_3 A$	$\sim_4 A$	$\sim_5 A$	$\sim_6 A$	$\sim_7 A$	$\sim_8 A$
t	f	f	f	f	f	f	f	f
b	b	b	b	b	t	t	t	t
n	n	n	f	f	n	n	f	f
f	t	b	t	b	t	b	t	b

A	$\sim_9 A$	$\sim_{10} A$	$\sim_{11} A$	$\sim_{12} A$	$\sim_{13} A$	$\sim_{14} A$	$\sim_{15} A$	$\sim_{16} A$
t	n	n	n	n	n	n	n	n
b	b	b	b	b	t	t	t	t
n	n	n	f	f	n	n	f	f
f	t	b	t	b	t	b	t	b

In view of the mechanical procedure described in [30, §2], we obtain falsity conditions for the above operators. We leave the details to interested readers as an easy exercise (the same applies to the variants of **K3** and **LP**, introduced below). Then, we define the semantic consequence relations for the variants with $\sim_i$ instead of $\sim$ (notation: $\vDash^i_{\mathbf{FDE}}$) as in Definition 1.

Remark 10. As one may easily observe, $\sim_1$ is the original negation included in **FDE**. Moreover, $\sim_{16}$ is the connective we discussed in [32]. The other fourteen operations are, to the best of our knowledge, not discussed in the literature.[3] Note that only three of the fourteen operations are *subclassical*. Further details of the operations will be explored in §5.

We now turn to variations of **K3**. By another simple combinatorial consideration, or by eliminating some cases starting from the above considerations for **FDE**, the following four operations exhaust the space of possible connectives that share the truth condition for negation.

A	$\sim_1 A$	$\sim_2 A$	$\sim_3 A$	$\sim_4 A$
t	f	f	n	n
n	n	f	n	f
f	t	t	t	t

Note here that $\sim_2$ is the connective discussed in [41]. Then, we define the semantic consequence relations for the variants with $\sim_i$ instead of $\sim$ (notation: $\vDash^i_{\mathbf{K3}}$) as in Definition 3.

Finally, we consider the variations of **LP**. By another simple combinatorial consideration, or again by eliminating some cases starting from the above considerations for **FDE**, the following four operations exhaust the space of possible connectives that share the truth condition for negation.

---

[3] A referee directed our attention to [36] as a reference that covers the connectives that we are discussing in this paper. This, however, is not the case. Note also that there is a crucial difference between [36] and the present paper insofar as we are *not* expanding the language of **FDE**, but only changing the interpretation of negation.

A	$\sim_1 A$	$\sim_2 A$	$\sim_3 A$	$\sim_4 A$
t	f	f	f	f
b	b	b	t	t
f	t	b	t	b

Note here that $\sim_2$ is the connective discussed in [40]. We define the semantic consequence relations for the variants with $\sim_i$ instead of $\sim$ (notation: $\vDash_{\mathbf{LP}}^i$) as in Definition 5.

3. Proof Systems

3.1. Unilateral Natural Deduction.
Let us first recall the natural deduction system for **FDE**, **K3** and **LP**. Our presentation below follows the one due to Dag Prawitz in [37, Appendix B], where he considers a certain expansion of **FDE** suggested by David Nelson, namely a logic that can be seen as an expansion of intuitionistic logic by a "strong" negation.[4]

Definition 11. The natural deduction rules $\mathcal{R}_{\mathbf{FDE}}$ for **FDE** are all the following rules:

$$\frac{[\sim A] \quad [\sim B]}{\sim(A \wedge B)} \frac{\vdots \quad \vdots}{C} \frac{C \quad C}{C} \qquad \frac{A \quad B}{A \wedge B} \quad \frac{A \wedge B}{A} \quad \frac{A \wedge B}{B} \quad \frac{\sim\sim A}{A}(\sim\sim 1) \quad \frac{A}{\sim\sim A}(\sim\sim 2) \quad \frac{A}{A \vee B} \quad \frac{B}{A \vee B} \quad \frac{A \vee B \quad \overset{[A]}{\vdots} \quad \overset{[B]}{\vdots}}{C}$$

$$\frac{\sim B}{\sim(A \wedge B)} \quad \frac{\sim A}{\sim(A \wedge B)} \quad \frac{\sim(A \vee B)}{\sim B} \quad \frac{\sim(A \vee B)}{\sim A} \quad \frac{\sim A \quad \sim B}{\sim(A \vee B)}$$

Moreover, for the natural deduction rules $\mathcal{R}_{\mathbf{K3}}$ and $\mathcal{R}_{\mathbf{LP}}$ for **K3** and **LP**, respectively, we add the ECQ and the Law of the Excluded Middle, respectively:

$$\frac{A \quad \sim A}{B} \text{ (ECQ)} \qquad \frac{}{A \vee \sim A} \text{ (LEM)}.$$

Then, given any set $\Sigma \cup \{A\}$ of formulas, $\Sigma \vdash_{\mathbf{FDE}} A$ iff for some finite $\Sigma' \subseteq \Sigma$, there is a derivation of A from Σ' in the calculus whose rule set is $\mathcal{R}_{\mathbf{FDE}}$. In the same way, we define $\vdash_{\mathbf{K3}}$ and $\vdash_{\mathbf{LP}}$.

We now turn to introduce the natural deduction systems for the variants of our basic systems.

Definition 12. The natural deduction rules $\mathcal{R}_{\mathbf{FDE}}^i$ for $\mathbf{FDE}^i$ are all the rules $\mathcal{R}_{\mathbf{FDE}}$ for **FDE** except that we replace $(\sim\sim 1)$ and $(\sim\sim 2)$ by the following rules.

$$\frac{\sim_1 \sim_1 A}{A}(\sim_1 \sim_1 1) \qquad \frac{A}{\sim_1 \sim_1 A}(\sim_1 \sim_1 2)$$

$$\frac{\sim_2 \sim_2 A}{A \vee \sim_2 A}(\sim_2 \sim_2 1) \qquad \frac{A}{\sim_2 \sim_2 A}(\sim_2 \sim_2 2) \qquad \frac{\sim_2 A}{\sim_2 \sim_2 A}(\sim_2 \sim_2 3)$$

$$\frac{\sim_3 \sim_3 A \quad \sim_3 A}{A}(\sim_3 \sim_3 1) \qquad \frac{A}{\sim_3 \sim_3 A}(\sim_3 \sim_3 2) \qquad \frac{}{\sim_3 A \vee \sim_3 \sim_3 A}(\sim_3 \sim_3 3)$$

$$\frac{}{\sim_4 \sim_4 A}(\sim_4 \sim_4)$$

---

[4]One can also present the system as in [38, p. 304] using two-way rules with double lines. However, for the purpose of making the connection more smooth to bilateral natural deduction systems, we will adopt the presentation by Prawitz.

$$\frac{\sim_5 A \quad \sim_5\sim_5 A}{B} \; (\sim_5\sim_5 1) \quad \frac{\sim_5\sim_5 A}{A} \; (\sim_5\sim_5 2) \quad \frac{A}{\sim_5 A \vee \sim_5\sim_5 A} \; (\sim_5\sim_5 3)$$

$$\frac{A \quad \sim_6 A \quad \sim_6\sim_6 A}{B} \; (\sim_6\sim_6 1) \quad \frac{\sim_6\sim_6 A}{A \vee \sim_6 A} \; (\sim_6\sim_6 2) \quad \frac{A}{\sim_6 A \vee \sim_6\sim_6 A} \; (\sim_6\sim_6 3)$$

$$\frac{\sim_6 A}{A \vee \sim_6\sim_6 A} \; (\sim_6\sim_6 4) \quad \frac{\sim_7 A \quad \sim_7\sim_7 A}{B} \; (\sim_7\sim_7 1) \quad \frac{}{\sim_7 A \vee \sim_7\sim_7 A} \; (\sim_7\sim_7 2)$$

$$\frac{A \quad \sim_8 A \quad \sim_8\sim_8 A}{B} \; (\sim_8\sim_8 1) \quad \frac{\sim_8\sim_8 A}{A \vee \sim_8\sim_8 A} \; (\sim_8\sim_8 2) \quad \frac{}{\sim_8 A \vee \sim_8\sim_8 A} \; (\sim_8\sim_8 3)$$

$$\frac{\sim_9\sim_9 A}{A} \; (\sim_9\sim_9 1) \quad \frac{\sim_9\sim_9 A}{\sim_9 A} \; (\sim_9\sim_9 2) \quad \frac{A \quad \sim_9 A}{\sim_9\sim_9 A} \; (\sim_9\sim_9 3)$$

$$\frac{\sim_{10}\sim_{10} A}{\sim_{10} A} \; (\sim_{10}\sim_{10} 1) \quad \frac{\sim_{10} A}{\sim_{10}\sim_{10} A} \; (\sim_{10}\sim_{10} 2)$$

$$\frac{A \quad \sim_{11}\sim_{11} A}{\sim_{11} A} \; (\sim_{11}\sim_{11} 1) \quad \frac{\sim_{11} A \quad \sim_{11}\sim_{11} A}{A} \; (\sim_{11}\sim_{11} 2)$$

$$\frac{}{A \vee \sim_{11} A \vee \sim_{11}\sim_{11} A} \; (\sim_{11}\sim_{11} 3) \quad \frac{A \quad \sim_{11} A}{\sim_{11}\sim_{11} A} \; (\sim_{11}\sim_{11} 4)$$

$$\frac{A \quad \sim_{12}\sim_{12} A}{\sim_{12} A} \; (\sim_{12}\sim_{12} 1) \quad \frac{}{A \vee \sim_{12}\sim_{12} A} \; (\sim_{12}\sim_{12} 2) \quad \frac{\sim_{12} A}{\sim_{12}\sim_{12} A} \; (\sim_{12}\sim_{12} 3)$$

$$\frac{\sim_{13}\sim_{13} A}{B} \; (\sim_{13}\sim_{13})$$

$$\frac{A \quad \sim_{14}\sim_{14} A}{B} \; (\sim_{14}\sim_{14} 1) \quad \frac{\sim_{14}\sim_{14} A}{\sim_{14} A} \; (\sim_{14}\sim_{14} 2) \quad \frac{\sim_{14} A}{A \vee \sim_{14}\sim_{14} A} \; (\sim_{14}\sim_{14} 3)$$

$$\frac{A \quad \sim_{15}\sim_{15} A}{B} \; (\sim_{15}\sim_{15} 1) \quad \frac{\sim_{15} A \quad \sim_{15}\sim_{15} A}{B} \; (\sim_{15}\sim_{15} 2)$$

$$\frac{}{A \vee \sim_{15} A \vee \sim_{15}\sim_{15} A} \; (\sim_{15}\sim_{15} 3) \quad \frac{A \quad \sim_{16}\sim_{16} A}{B} \; (\sim_{16}\sim_{16} 1) \quad \frac{}{A \vee \sim_{16}\sim_{16} A} \; (\sim_{16}\sim_{16} 2)$$

Based on these, given any set $\Sigma \cup \{A\}$ of formulas, $\Sigma \vdash^i_{\mathbf{FDE}} A$ iff for some finite $\Sigma' \subseteq \Sigma$, there is a derivation of A from Σ' in the calculus whose rule set is $\mathcal{R}_{\mathbf{FDE}^i}$.

Definition 13. The natural deduction rules $\mathcal{R}^i_{\mathbf{K3}}$ for $\mathbf{K3}^i$ are all the rules $\mathcal{R}_{\mathbf{K3}}$ for $\mathbf{K3}$ but replacing $(\sim\sim 1)$ and $(\sim\sim 2)$ by the following rules.

$$\frac{\sim_1\sim_1 A}{A} \; (\sim_1\sim_1 1) \quad \frac{A}{\sim_1\sim_1 A} \; (\sim_1\sim_1 2) \quad \frac{}{\sim_2 A \vee \sim_2\sim_2 A} \; (\sim_2\sim_2)$$

$$\frac{\sim_3\sim_3 A}{B} \; (\sim_3\sim_3) \quad \frac{A \quad \sim_4\sim_4 A}{B} \; (\sim_4\sim_4 1) \quad \frac{}{A \vee \sim_4 A \vee \sim_4\sim_4 A} \; (\sim_4\sim_4 2)$$

Based on these, given any set $\Sigma \cup \{A\}$ of formulas, $\Sigma \vdash^i_{\mathbf{K3}} A$ iff for some finite $\Sigma' \subseteq \Sigma$, there is a derivation of A from Σ' in the calculus whose rule set is $\mathcal{R}_{\mathbf{K3}^i}$.

Definition 14. The natural deduction rules $\mathcal{R}^i_{\mathbf{LP}}$ for $\mathbf{LP}^i$ are all the rules $\mathcal{R}_{\mathbf{LP}}$ for $\mathbf{LP}$ but replacing $(\sim\sim 1)$ and $(\sim\sim 2)$ by the following rules.

$$\frac{\sim_1\sim_1 A}{A} \; (\sim_1\sim_1 1) \quad \frac{A}{\sim_1\sim_1 A} \; (\sim_1\sim_1 2) \quad \frac{}{\sim_2\sim_2 A} \; (\sim_2\sim_2) \quad \frac{\sim_3 A \quad \sim_3\sim_3 A}{B} \; (\sim_3\sim_3)$$

$$\frac{A \quad \sim_4 A \quad \sim_4\sim_4 A}{B} \; (\sim_4\sim_4 1) \quad \frac{}{A \vee \sim_4\sim_4 A} \; (\sim_4\sim_4 2)$$

Based on these, given any set $\Sigma \cup \{A\}$ of formulas, $\Sigma \vdash^i_{\mathbf{LP}} A$ iff for some finite $\Sigma' \subseteq \Sigma$, there is a derivation of A from Σ' in the calculus whose rule set is $\mathcal{R}_{\mathbf{LP}^i}$.

3.2. Bilateral Natural Deduction.

We will present *bilateral* natural deduction systems for the consequence relations $\vDash^i_{\mathbf{FDE}}$ ($i \in \{1,\ldots,16\}$), $\vDash^i_{\mathbf{K3}}$ ($i \in \{1,\ldots,4\}$), and $\vDash^i_{\mathbf{LP}}$ ($i \in \{1,\ldots,4\}$) along the lines of [48]. These calculi make use of pure (separated) introduction and elimination rules, i.e., rules that introduce into the conclusion or eliminate from the premises only a single connective as the main connective of a compound formula. The systems are, therefore, interesting from the point of view of proof-theoretic semantics, because their rules can be seen as laying down the meaning of the connectives inferentially. We will present the bilateral rules in the style of the natural deduction rules from §3.1, but now with a distinction drawn between proofs and disproofs (refutations) from assumptions that are taken to be true and counterassumptions that are taken to be definitely false. We use single lines in the notation for proofs and double lines in the notation for refutations. Thus, in this section, double lines indicate disproofs. In particular, we write $\overline{A}$ to denote a proof of A from A as an assumption, and $\overline{\overline{A}}$ to denote a refutation of A from A as a counterassumption. This gives the inductive base for a definition of the set of proofs and refutations in any of the systems we will consider. A permitted discharge of assumptions is indicated by square brackets, [ ], and a permitted discharge of counterassumptions is indicated by double square brackets, [[ ]]. We will simplify the notation by writing $[A]$ instead of $[\overline{A}]$ and $[[A]]$ instead of $[[\overline{\overline{A}}]]$. Moreover, if Σ is a set of formulas, then Σ^+ is defined as the set $\{\overline{A}: A \in \Sigma\}$ and Σ^- as $\{\overline{\overline{A}}: A \in \Sigma\}$.

The introduction and elimination rules for conjunctions and disjunctions from §3.1 then take the following form:[5]

$$\frac{\overline{A} \quad \overline{B}}{A \wedge B} \qquad \frac{\overline{A \wedge B}}{A} \qquad \frac{\overline{A \wedge B}}{B} \qquad \frac{\overline{A}}{A \vee B} \qquad \frac{\overline{B}}{A \vee B} \qquad \frac{\overline{A \vee B} \quad \overset{[A]}{\vdots} C \quad \overset{[B]}{\vdots} C}{C}$$

In the present setup, the dotted lines indicate derivations that may be built up from *both* refutations and proofs. Instead of rules for introducing and eliminating negated conjunctions, disjunctions, and negations into and from proofs, we have rules for introducing and removing disjunctions, conjunctions, and negations into and from disproofs.

Definition 15. The set of natural deduction rules $\mathfrak{R}_{\mathbf{FDE}}$ for **FDE** consists of the above rules for $\wedge$ and $\vee$ together with:

$$\frac{\overline{\overline{A}} \quad \overline{\overline{B}}}{A \vee B} \qquad \frac{\overline{\overline{A \vee B}}}{A} \qquad \frac{\overline{\overline{A \vee B}}}{B} \qquad \frac{\overline{\overline{A}}}{A \wedge B} \qquad \frac{\overline{\overline{B}}}{A \wedge B} \qquad \frac{\overline{\overline{A \wedge B}} \quad \overset{[[A]]}{\vdots} C \quad \overset{[[B]]}{\vdots} C}{C}$$

and the following rules for introducing and eliminating negations into and from proofs and refutations:

$$\frac{\overline{\overline{A}}}{\sim A} \qquad \frac{\overline{\sim A}}{A} \qquad \frac{\overline{A}}{\sim A} \, (\sim\sim 1) \qquad \frac{\overline{\overline{\sim A}}}{A} \, (\sim\sim 2)$$

---

[5]This is the way how these rules are presented in [44], though without abbreviating $[\overline{A}]$ as $[A]$.

Moreover, for the sets of natural deduction rules $\mathfrak{R}_{\mathbf{K3}}$ and $\mathfrak{R}_{\mathbf{LP}}$ for **K3** and **LP**, respectively, we add the rule ECQ and the dilemma rule DIL, respectively, which express a certain interaction between proofs and disproofs:

$$\frac{\overline{A} \quad \overline{\overline{A}}}{B} \text{ (ECQ)} \qquad \frac{\begin{array}{cc}[A] & [\![A]\!]\\ \vdots & \vdots\\ B & \overline{\overline{B}}\end{array}}{B} \text{ (DIL)}$$

Let $\Sigma \cup \Gamma \cup \{A\}$ be a set of formulas. Then $\Sigma^+ \cup \Gamma^- \vdash^+_{\mathfrak{R}_{\mathbf{FDE}}} A$ ($\Sigma^+ \cup \Gamma^- \vdash^-_{\mathfrak{R}_{\mathbf{FDE}}} A$) iff for some finite $\Sigma' \subseteq \Sigma$ and finite $\Gamma' \subseteq \Gamma$, there is a proof (disproof) of A from $\Sigma'^+ \cup \Gamma'^-$ in the calculus whose rule set is $\mathfrak{R}_{\mathbf{FDE}}$. In the same way, we define the relations $\vdash^+_{\mathfrak{R}_{\mathbf{K3}}}$, $\vdash^-_{\mathfrak{R}_{\mathbf{K3}}}$, $\vdash^+_{\mathfrak{R}_{\mathbf{LP}}}$, and $\vdash^-_{\mathfrak{R}_{\mathbf{LP}}}$.

Definition 16. The set of rules $\mathfrak{R}^i_{\mathbf{FDE}}$ for **FDE**^{i}, with $i \in \{1, \ldots, 16\}$, consists of all the rules of $\mathfrak{R}_{\mathbf{FDE}}$ for **FDE**, but the rules for $\sim$ are replaced by the following introduction and elimination rules:

$$\frac{\overline{\overline{A}}}{\sim_i A} \qquad \frac{\sim_i A}{A} \qquad \frac{\overline{A}}{\sim_1 A} \qquad \frac{\sim_1 A}{A}$$

$$\frac{\sim_2 A \quad \begin{array}{cc}[A] & [\![A]\!]\\ \vdots & \vdots\\ B & B\end{array}}{B} \qquad \frac{\overline{A}}{\sim_2 A} \qquad \frac{\overline{\overline{A}}}{\sim_2 A} \qquad \frac{\sim_3 A \quad \overline{\overline{A}}}{A} \qquad \frac{\overline{A}}{\sim_3 A} \qquad \frac{\begin{array}{cc}[A] & [\![\sim_3 A]\!]\\ \vdots & \vdots\\ B & B\end{array}}{B}$$

$$\frac{}{\sim_4 A} \qquad \frac{\overline{A} \quad \overline{\sim_5 A}}{B} \qquad \frac{\sim_5 A \quad \overline{A}}{A} \qquad \frac{\begin{array}{cc}[A] & [\![\sim_5 A]\!]\\ \vdots & \vdots\\ B & B\end{array}}{B}$$

$$\frac{\overline{A} \quad \overline{\overline{A}} \quad \overline{\overline{\sim_6 A}}}{B} \qquad \frac{\sim_6 A \quad \begin{array}{cc}[A] & [\![A]\!]\\ \vdots & \vdots\\ B & B\end{array}}{B} \qquad \frac{\overline{A} \quad \begin{array}{cc}[A] & [\![\sim_6 A]\!]\\ \vdots & \vdots\\ B & B\end{array}}{B} \qquad \frac{\overline{\overline{A}} \quad \begin{array}{cc}[A] & [\![\sim_6 A]\!]\\ \vdots & \vdots\\ B & B\end{array}}{B}$$

$$\frac{\overline{A} \quad \sim_7 A}{B} \qquad \frac{\begin{array}{cc}[\![A]\!] & [\![\sim_7 A]\!]\\ \vdots & \vdots\\ B & B\end{array}}{B} \qquad \frac{\overline{A} \quad \overline{\overline{A}} \quad \overline{\overline{\sim_8 A}}}{B} \qquad \frac{\begin{array}{cc}[A] & [\![\sim_8 A]\!]\\ \vdots & \vdots\\ B & B\end{array}}{B} \qquad \frac{\begin{array}{cc}[\![A]\!] & [\![\sim_8 A]\!]\\ \vdots & \vdots\\ B & B\end{array}}{B}$$

$$\frac{\overline{\sim_9 A}}{A} \qquad \frac{\sim_9 A}{A} \qquad \frac{\overline{A} \quad \overline{\overline{A}}}{\sim_9 A} \qquad \frac{\overline{\overline{A}}}{\sim_{10} A} \qquad \frac{\sim_{10} A}{A}$$

$$\frac{\overline{A} \quad \sim_{11} A}{A} \qquad \frac{\overline{\overline{A}} \quad \overline{\sim_{11} A}}{A} \qquad \frac{\begin{array}{ccc}[A] & [\![A]\!] & [\![\sim_{11} A]\!]\\ \vdots & \vdots & \vdots\\ B & B & B\end{array}}{B} \qquad \frac{\overline{A} \quad \overline{\overline{A}}}{\sim_{11} A}$$

$$\frac{\overline{A} \quad \sim_{12} A}{A} \qquad \frac{\begin{array}{cc}[A] & [\![\sim_{12} A]\!]\\ \vdots & \vdots\\ B & B\end{array}}{B} \qquad \frac{\overline{\overline{A}}}{\sim_{12} A} \qquad \frac{\sim_{13} A}{B}$$

$$\frac{\overline{A} \quad \overline{\overline{\sim_{14}A}}}{B} \qquad \frac{\overline{\overline{\sim_{14}A}}}{A} \qquad \frac{\overline{A} \quad \overline{B} \quad [A] \quad [\![\sim_{14}A]\!]}{B} \quad \vdots \quad \vdots$$

$$\frac{\overline{A} \quad \overline{\overline{\sim_{15}A}}}{B} \qquad \frac{\overline{A} \quad \overline{\overline{\sim_{15}A}}}{B} \qquad \frac{[A] \quad [A] \quad [\![\sim_{15}A]\!]}{B} \qquad \frac{\overline{A} \quad \overline{\sim_{16}A}}{B} \qquad \frac{[A] \quad [\![\sim_{16}A]\!]}{B}$$

Let $\Sigma \cup \Gamma \cup \{A\}$ be a set of formulas. Then $\Sigma^+ \cup \Gamma^- \vdash^{i+}_{\mathfrak{R}\text{FDE}} A$ ($\Sigma^+ \cup \Gamma^- \vdash^{i-}_{\mathfrak{R}\text{FDE}} A$) iff for some finite $\Sigma' \subseteq \Sigma$ and finite $\Gamma' \subseteq \Gamma$, there is a proof (disproof) of A from $\Sigma'^+ \cup \Gamma'^-$ in the calculus whose rule set is $\mathfrak{R}^i_{\text{FDE}}$.

Definition 17. For $i \in \{1,2,3,4\}$, the set of natural deduction rules $\mathfrak{R}^i_{\text{K3}}$ for **K3**^{i} consists all the rules $\mathfrak{R}_{\text{K3}}$ for **K3**, but the rules for $\sim$ are replaced by the following rules:

$$\frac{\overline{\overline{A}}}{\sim_i A} \quad \frac{\overline{\sim_i A}}{A} \qquad \frac{\overline{A} \quad \overline{\sim_1 A}}{\sim_1 A} \quad \frac{\overline{\sim_1 A}}{A} \qquad \frac{[A] \quad [\![\sim_2 A]\!]}{B} \qquad \frac{[A] \quad [A] \quad [\![\sim_4 A]\!]}{B}$$

$$\frac{\overline{\sim_3 A}}{B} \qquad \frac{\overline{A} \quad \overline{\sim_4 A}}{B} \qquad \frac{\overline{B} \quad \overline{B}}{B}$$

Let $\Sigma \cup \Gamma \cup \{A\}$ be a set of formulas. Then $\Sigma^+ \cup \Gamma^- \vdash^{i+}_{\mathfrak{R}\text{K3}} A$ ($\Sigma^+ \cup \Gamma^- \vdash^{i-}_{\mathfrak{R}\text{K3}} A$) iff for some finite $\Sigma' \subseteq \Sigma$ and finite $\Gamma' \subseteq \Gamma$, there is a proof (disproof) of A from $\Sigma'^+ \cup \Gamma'^-$ in the calculus whose rule set is $\mathfrak{R}^i_{\text{K3}}$.

Definition 18. For $i \in \{1,2,3,4\}$, the set of rules $\mathfrak{R}^i_{\text{LP}}$ for **LP**^{i} comprises all the rules $\mathfrak{R}_{\text{LP}}$ for **LP**, but the rules for $\sim$ are replaced by the following rules:

$$\frac{\overline{\overline{A}}}{\sim_i A} \quad \frac{\overline{\overline{\sim_i A}}}{A} \qquad \frac{\overline{A} \quad \overline{\sim_1 A}}{\sim_1 A} \quad \frac{\overline{\overline{\sim_1 A}}}{A} \qquad \frac{}{\sim_2 A} \qquad \frac{[A] \quad [\![\sim_4 A]\!]}{B}$$

$$\frac{\overline{\overline{A}} \quad \overline{\overline{\sim_3 A}}}{B} \qquad \frac{\overline{A} \quad \overline{\overline{A}} \quad \overline{\sim_4 A}}{B}$$

Let $\Sigma \cup \Gamma \cup \{A\}$ be a set of formulas. Then $\Sigma^+ \cup \Gamma^- \vdash^{i+}_{\mathfrak{R}\text{LP}} A$ ($\Sigma^+ \cup \Gamma^- \vdash^{i-}_{\mathfrak{R}\text{LP}} A$) iff for some finite $\Sigma' \subseteq \Sigma$ and finite $\Gamma' \subseteq \Gamma$, there is a proof (disproof) of A from $\Sigma'^+ \cup \Gamma'^-$ in the calculus whose rule set is $\mathfrak{R}^i_{\text{LP}}$.

We show the bilateral systems to be equivalent with their unilateral counterparts.

Theorem 19. *Let* $\tau(\Delta^-) = \{\sim A : \overline{\overline{A}} \in \Delta^-\}$. *Then (1)* $\Sigma^+ \cup \Gamma^- \vdash^{i+}_{\mathfrak{R}\text{FDE}} A$ *iff* $\Sigma \cup \tau(\Gamma^-) \vdash^i_{\mathcal{R}\text{FDE}} A$ *and (2)* $\Sigma^+ \cup \Gamma^- \vdash^{i-}_{\mathfrak{R}\text{FDE}} A$ *iff* $\Sigma \cup \tau(\Gamma^-) \vdash^i_{\mathcal{R}\text{FDE}} \sim A$.

Proof. By induction on derivations in $\mathfrak{R}^i_{\text{FDE}}$ and $\mathcal{R}^i_{\text{FDE}}$. The cases of the rules for introducing and eliminating conjunctions and disjunctions into and from proofs are obvious. We present some of the remaining cases. Direction from left to right, claim (1). By applying the definition of derivations, the induction hypothesis, and rules of $\mathcal{R}^i_{\text{FDE}}$, for the derivations on the left we obtain the derivations on the right:

$$\dfrac{\overline{\overline{A}}\quad A}{\sim_9 A}\qquad \dfrac{\overline{\overline{\sim_2 A}}\quad \dfrac{[A]}{\vdots}\ \dfrac{[A]}{\vdots}}{B}\qquad \dfrac{\dfrac{\sim_2\sim_2 A}{A\vee\sim_2 A}\ \dfrac{[A]}{\vdots}\ \dfrac{[\sim_2 A]}{\vdots}}{B}\qquad \dfrac{\dfrac{[A]}{\vdots}\ \dfrac{[A]}{\vdots}\ \dfrac{[\sim_{11}A]}{\vdots}}{B}$$

$$\dfrac{[\sim_{11}A]}{\vdots}\ \dfrac{[\sim_{11}\sim_{11}A]}{\vdots}$$

$$\dfrac{\sim_9\sim_9 A}{A}\qquad \dfrac{A\vee(\sim_{11}A\vee\sim_{11}\sim_{11}A)\ \dfrac{[A]}{\vdots}\ \dfrac{[\sim_{11}A\vee\sim_{11}\sim_{11}A]}{\vdots}\ \dfrac{\overline{B}}{B}\ \dfrac{\overline{B}}{B}}{B}$$

Direction from left to right, claim (2). By applying the definition of derivations, the induction hypothesis, and rules of $\mathcal{R}^i_{\mathbf{FDE}}$, for the first derivations we obtain the second derivations with the same subscript:

$$\dfrac{\overline{\overline{A}}}{\sim A}\qquad \dfrac{\overline{A}\quad\overline{\overline{A}}}{\sim_9 A}\quad \dfrac{A\quad\sim_9 A}{\sim_9\sim_9 A}\qquad \dfrac{A\wedge B\ \dfrac{[A]}{\vdots}\ \dfrac{[B]}{\vdots}}{C}\qquad \dfrac{\dfrac{\sim_i(A\wedge B)}{\sim_i A\vee\sim_i B}\ \dfrac{[\sim_i A]}{\vdots}\ \dfrac{[\sim_i B]}{\vdots}}{\sim_i C}$$

Direction from right to left, claim (1). By applying the definition of derivations, the induction hypothesis, and rules of $\mathfrak{R}^i_{\mathbf{FDE}}$, for the first derivations we obtain the second derivations with the same subscript:

$$A\quad \overline{A}\qquad \dfrac{\sim_9\sim_9 A}{A}\qquad \dfrac{\overline{\sim_9 A}}{A}\qquad \dfrac{\sim_{13}\sim_{13}A}{B}\qquad \dfrac{\overline{\sim_{13}A}}{B}$$

Direction from right to left, claim (2). By applying the definition of derivations, the induction hypothesis, and rules of $\mathfrak{R}^i_{\mathbf{FDE}}$, for the first derivations we obtain the second derivations with the same subscript:

$$\sim A\quad \overline{\overline{A}}\qquad \dfrac{\sim_i A\wedge\sim_i B}{\sim_i(A\vee B)}\qquad \dfrac{\dfrac{\sim_i A\wedge\sim_i B}{\sim_i A}\ \dfrac{\sim_i A\wedge\sim_i B}{\sim_i B}}{A\vee B}\qquad \dfrac{A\quad\sim_9 A}{\sim_9\sim_9 A}\qquad \dfrac{\overline{A}\quad\overline{\overline{A}}}{\sim_9 A}\quad \triangleleft$$

Theorem 20. Let $\tau(\Delta^-) = \{\sim A\colon \overline{\overline{A}} \in \Delta^-\}$. Then (1) $\Sigma^+\cup\Gamma^-\vdash^{i+}_{\mathfrak{R}K3} A$ iff $\Sigma\cup\tau(\Gamma^-)\vdash^i_{\mathcal{R}K3} A$ and (2) $\Sigma^+\cup\Gamma^-\vdash^{i-}_{\mathfrak{R}K3} A$ iff $\Sigma\cup\tau(\Gamma^-)\vdash^i_{\mathcal{R}K3}\sim A$.

Proof. By induction on derivations in $\mathfrak{R}^i_{\mathbf{K3}}$ and $\mathcal{R}^i_{\mathbf{K3}}$. We present only one more interesting case for the direction from right to left, claim (1). By applying the definition of derivations, the induction hypothesis, and a rule of $\mathfrak{R}^4_{\mathbf{K3}}$, for the first derivations we obtain the second derivations with the same subscript:

$$\dfrac{}{A\vee(\sim_4 A\vee\sim_4\sim_4 A)}\qquad \dfrac{[A]}{A\vee(\sim_4 A\vee\sim_4\sim_4 A)}\quad \dfrac{\dfrac{[A]}{\sim_4 A}}{\dfrac{\sim_4 A\vee\sim_4\sim_4 A}{A\vee(\sim_4 A\vee\sim_4\sim_4 A)}}\quad \dfrac{\dfrac{[\sim_4 A]}{\sim_4\sim_4 A}}{\dfrac{\sim_4 A\vee\sim_4\sim_4 A}{A\vee(\sim_4 A\vee\sim_4\sim_4 A)}}$$

$$A\vee(\sim_4 A\vee\sim_4\sim_4 A)\qquad \triangleleft$$

Theorem 21. Let $\tau(\Delta^-) = \{\sim A : \overline{A} \in \Delta^-\}$. Then (1) $\Sigma^+ \cup \Gamma^- \vdash^{i+}_{\mathfrak{R}LP} A$ iff $\Sigma \cup \tau(\Gamma^-) \vdash^i_{\mathcal{R}LP} A$ and (2) $\Sigma^+ \cup \Gamma^- \vdash^{i-}_{\mathfrak{R}LP} A$ iff $\Sigma \cup \tau(\Gamma^-) \vdash^i_{\mathcal{R}LP} \sim A$.

Proof. By induction on derivations in $\mathfrak{R}^i_{LP}$ and $\mathcal{R}^i_{LP}$. We present only the more interesting cases. Direction from left to right, claim (1). By applying the induction hypothesis and the rule for eliminating disjunctions from proofs in $\mathcal{R}^i_{LP}$, for the first derivations we obtain the second derivations with the same subscript:

$$\frac{\begin{array}{cc} [A] & [A] \\ \vdots & \vdots \\ B & B \end{array}}{B} \text{ (DIL)} \qquad \frac{A \vee \sim_i A \quad \begin{array}{c}[A] \\ \vdots \\ B\end{array} \quad \begin{array}{c}[\sim_i A] \\ \vdots \\ B\end{array}}{B}$$

Direction from right to left, claim (1). By applying rules from $\mathfrak{R}^i_{LP}$, for the first derivations we obtain the second derivations with the same subscript:

$$\overline{A \vee \sim_i A} \qquad \frac{\begin{array}{cc} [A] & \begin{array}{c}[\![A]\!] \\ \sim_i A \\ A \vee \sim_i A\end{array} \\ A \vee \sim_i A & \end{array}}{A \vee \sim_i A} \text{ (DIL)}$$

4. Soundness and Completeness

Theorem 22 (Soundness). *For all $\Gamma \cup \{A\} \subseteq$ Form, (1) $\Gamma \vdash^i_{FDE} A$ only if $\Gamma \vDash^i_{FDE} A$, (2) $\Gamma \vdash^i_{K3} A$ only if $\Gamma \vDash^i_{K3} A$, and (3) $\Gamma \vdash^i_{LP} A$ only if $\Gamma \vDash^i_{LP} A$.*

Proof. Tedious, but standard. ◁

For the completeness direction, we prepare some well known notions and lemmas.

Definition 23. Let Σ be a set of formulas. Then, Σ is a *theory* iff $\Sigma \vdash A$ implies $A \in \Sigma$, and Σ is *prime* iff $A \vee B \in \Sigma$ implies $A \in \Sigma$ or $B \in \Sigma$.

Lemma 24 (Lindenbaum). *If $\Sigma \nvdash A$, then there is $\Sigma' \supseteq \Sigma$ such that $\Sigma' \nvdash A$ and Σ' is a prime theory.*

We now define the canonical valuation in the usual manner.

Definition 25. For any $\Sigma \subseteq$ Form, let v^i_Σ from Prop to $\{\mathbf{t}, \mathbf{b}, \mathbf{n}, \mathbf{f}\}$ be defined as follows:

$$v^i_\Sigma(p) := \begin{cases} \mathbf{t} \text{ iff } \Sigma \vdash^i_{FDE} p \text{ and } \Sigma \nvdash^i_{FDE} \sim p; \\ \mathbf{b} \text{ iff } \Sigma \vdash^i_{FDE} p \text{ and } \Sigma \vdash^i_{FDE} \sim p; \\ \mathbf{n} \text{ iff } \Sigma \nvdash^i_{FDE} p \text{ and } \Sigma \nvdash^i_{FDE} \sim p; \\ \mathbf{f} \text{ iff } \Sigma \nvdash^i_{FDE} p \text{ and } \Sigma \vdash^i_{FDE} \sim p. \end{cases}$$

The following lemma is the key for the completeness result.

Lemma 26. *If Σ is a prime theory, then the following hold for all $B \in$ Form:*

$$v^i_\Sigma(B) = \begin{cases} \mathbf{t} \text{ iff } \Sigma \vdash^i_{FDE} B \text{ and } \Sigma \nvdash^i_{FDE} \sim B; \\ \mathbf{b} \text{ iff } \Sigma \vdash^i_{FDE} B \text{ and } \Sigma \vdash^i_{FDE} \sim B; \\ \mathbf{n} \text{ iff } \Sigma \nvdash^i_{FDE} B \text{ and } \Sigma \nvdash^i_{FDE} \sim B; \\ \mathbf{f} \text{ iff } \Sigma \nvdash^i_{FDE} B \text{ and } \Sigma \vdash^i_{FDE} \sim B. \end{cases}$$

Proof. Note first that it is obvious that v_Σ well defined. Then the desired result is proved by induction on the construction of B. The base case, for atomic formulas, is obvious by the definition. For the induction step, the cases are split based on the connectives. We will here only deal with the case for negation of **FDE**^{16}.

$v_\Sigma^{16}(\sim_{16} B) = \mathbf{t}$ iff $v_\Sigma^{16}(B) = \mathbf{b}$ (by the definition of v_Σ^{16}) iff $\Sigma \vDash_{\mathbf{FDE}}^{16} B$ and $\Sigma \vDash_{\mathbf{FDE}}^{16} \sim_{16} B$ (by IH) iff $\Sigma \vDash_{\mathbf{FDE}}^{16} \sim_{16} B$ and $\Sigma \nvDash_{\mathbf{FDE}}^{16} \sim_{16}\sim_{16} B$ (by ($\sim_{15}\sim_{15} 1$)) for the left-to-right direction and ($\sim_{16}\sim_{16} 2$) for the other direction).

$v_\Sigma^{16}(\sim_{16} B) = \mathbf{b}$ iff $v_\Sigma^{16}(B) = \mathbf{f}$ (by the definition of v_Σ^{16}) iff $\Sigma \nvDash_{\mathbf{FDE}}^{16} B$ and $\Sigma \vDash_{\mathbf{FDE}}^{16} \sim_{16} B$ (by IH) iff $\Sigma \vDash_{\mathbf{FDE}}^{16} \sim_{16} B$ and $\Sigma \vDash_{\mathbf{FDE}}^{16} \sim_{16}\sim_{16} B$ (by ($\sim_{15}\sim_{15} 2$)) for the left-to-right direction and ($\sim_{16}\sim_{16} 1$) for the other direction).

$v_\Sigma^{16}(\sim_{16} B) = \mathbf{n}$ iff $v_\Sigma^{16}(B) = \mathbf{t}$ (by the definition of v_Σ^{16}) iff $\Sigma \vDash_{\mathbf{FDE}}^{16} B$ and $\Sigma \nvDash_{\mathbf{FDE}}^{16} \sim_{16} B$ (by IH) iff $\Sigma \nvDash_{\mathbf{FDE}}^{16} \sim_{16} B$ and $\Sigma \nvDash_{\mathbf{FDE}}^{16} \sim_{16}\sim_{16} B$ (by ($\sim_{16}\sim_{16} 1$)) for the left-to-right direction and ($\sim_{16}\sim_{16} 2$) for the other direction).

$v_\Sigma^{16}(\sim_{16} B) = \mathbf{f}$ iff $v_\Sigma^{16}(B) = \mathbf{n}$ (by the definition of v_Σ^{16}) iff $\Sigma \nvDash_{\mathbf{FDE}}^{16} B$ and $\Sigma \nvDash_{\mathbf{FDE}}^{16} \sim_{16} B$ (by IH) iff $\Sigma \nvDash_{\mathbf{FDE}}^{16} \sim_{16} B$ and $\Sigma \vDash_{\mathbf{FDE}}^{16} \sim_{16}\sim_{16} B$ (by ($\sim_{16}\sim_{16} 2$)) for the left-to-right direction and ($\sim_{16}\sim_{16} 1$) for the other direction).

The other cases are left to the interested readers to be written out in detail. ◁

For the variations of **K3** and **LP**, we need to eliminate the values **b** and **n**, respectively.

We are now ready to prove the completeness result.

Theorem 27 (Completeness). *For all $\Gamma \cup \{A\} \subseteq$ Form, (1) $\Gamma \vDash_{\mathbf{FDE}}^i A$ only if $\Gamma \vdash_{\mathbf{FDE}}^i A$, (2) $\Gamma \vDash_{\mathbf{K3}}^i A$ only if $\Gamma \vdash_{\mathbf{K3}}^i A$, and (3) $\Gamma \vDash_{\mathbf{LP}}^i A$ only if $\Gamma \vdash_{\mathbf{LP}}^i A$.*

Proof. We only deal with the case for **FDE**^{i} since other cases can be established in the same manner. Assume $\Gamma \nvdash_{\mathbf{FDE}}^i A$. Then, by Lemma 24, there is a $\Sigma \supseteq \Gamma$ such that Σ is a prime theory and $A \notin \Sigma$, and by Lemma 26, a four-valued valuation v_Σ can be defined with $I_\Sigma(B) \in \mathcal{D}$ for every $B \in \Gamma$ and $I_\Sigma(A) \notin \mathcal{D}$. Thus it follows that $\Gamma \nvDash_{\mathbf{FDE}}^i A$, as desired. ◁

5. BASIC RESULTS

5.1. Negation Inconsistency and Negation Incompleteness.
As one may easily observe, all variants of **FDE** are both paraconsistent and paracomplete, all variants of **K3** are paracomplete, but not paraconsistent, and all variants of **LP** are paraconsistent, but not paracomplete. However, for some of the variants, some stronger properties than paraconsistency and paracompleteness hold. The stronger properties we have in mind are the following.

Definition 28. A logic **L** is *negation inconsistent* if for some A, we have both $B \vDash_\mathbf{L} A$ and $B \vDash_\mathbf{L} \sim A$ for all B. Moreover, a logic **L** is *negation incomplete* if for some A, both $A \vDash_\mathbf{L} B$ and $\sim A \vDash_\mathbf{L} B$ for all B.

Then, we obtain the following results.

Theorem 29. *$\mathbf{LP}^2$, $\mathbf{LP}^4$, $\mathbf{FDE}^4$, $\mathbf{FDE}^8$, $\mathbf{FDE}^{12}$ and $\mathbf{FDE}^{16}$ are negation inconsistent.*

Proof. We prove the result by showing the specific instances of inconsistency. For $\mathbf{LP}^2$, we have $B \vDash_{\mathbf{LP}}^2 \sim_2\sim_2 A$ and $B \vDash_{\mathbf{LP}}^2 \sim_2\sim_2\sim_2 A$. For $\mathbf{LP}^4$, we have $B \vDash_{\mathbf{LP}}^4 \sim_4 (A \wedge$

$\sim_4 A \wedge \sim_4 \sim_4 A)$ and $B \vDash^4_{\mathbf{LP}} \sim_4 \sim_4 (A \wedge \sim_4 A \wedge \sim_4 \sim_4 A)$. For $\mathbf{FDE}^4$, we have $B \vDash^4_{\mathbf{FDE}} \sim_4 \sim_4 A$ and $B \vDash^4_{\mathbf{FDE}} \sim_4 \sim_4 \sim_4 A$. For $\mathbf{FDE}^8$, we have $B \vDash^8_{\mathbf{FDE}} \sim_8 (A \wedge \sim_8 A \wedge \sim_8 \sim_8 A)$ and $B \vDash^8_{\mathbf{FDE}} \sim_8 \sim_8 (A \wedge \sim_8 A \wedge \sim_8 \sim_8 A)$. For $\mathbf{FDE}^{12}$, we have $B \vDash^{12}_{\mathbf{FDE}} \sim_{12} \sim_{12} \sim_{12} A$ and $B \vDash^{12}_{\mathbf{FDE}} \sim_{12} \sim_{12} \sim_{12} \sim_{12} A$. Finally, for $\mathbf{FDE}^{16}$, we have $B \vDash^{16}_{\mathbf{FDE}} \sim_{16}(A \wedge \sim_{16} \sim_{16} A)$ and $B \vDash^{16}_{\mathbf{FDE}} \sim_{16} \sim_{16}(A \wedge \sim_{16} \sim_{16} A)$. ◁

Remark 30. The other variants of **FDE**, **K3** and **LP** are not negation inconsistent. Indeed, for the variants of **K3**, this is obvious since they are all explosive. For the other variants of **FDE** and **LP**, note that negation inconsistency implies that there is a formula that receives the value **b** for all interpretations. But it is easy to see that this cannot be the case with these variants. For example, the subclassical variants have the set $\{\mathbf{t}, \mathbf{f}\}$ being closed under all three connectives. Similar arguments by looking at sets $\{\mathbf{t}, \mathbf{n}, \mathbf{f}\}$ or $\{\mathbf{n}\}$ will establish the desired results.

Theorem 31. $K3^3$, $K3^4$, FDE^{13}, FDE^{14}, FDE^{15} and FDE^{16} are negation incomplete.

Proof. We prove the result by showing the specific instances of incompleteness. For $\mathbf{K3}^3$, we have $\sim_3 \sim_3 A \vDash^3_{\mathbf{K3}} B$ and $\sim_3 \sim_3 \sim_3 A \vDash^3_{\mathbf{K3}} B$. For $\mathbf{K3}^4$, we have $\sim_4 \sim_4 (A \wedge \sim_4 A \wedge \sim_4 \sim_4 A) \vDash^4_{\mathbf{K3}} B$ and $\sim_4 \sim_4 \sim_4 (A \wedge \sim_4 A \wedge \sim_4 \sim_4 A) \vDash^4_{\mathbf{K3}} B$. For $\mathbf{FDE}^{13}$, we have $\sim_{13} \sim_{13} A \vDash^{13}_{\mathbf{FDE}} B$ and $\sim_{13} \sim_{13} \sim_{13} A \vDash^{13}_{\mathbf{FDE}} B$. For $\mathbf{FDE}^{14}$, we have $\sim_{14} \sim_{14} \sim_{14} A \vDash^{14}_{\mathbf{FDE}} B$ and $\sim_{14} \sim_{14} \sim_{14} \sim_{14} A \vDash^{14}_{\mathbf{FDE}} B$. For $\mathbf{FDE}^{15}$, we have $\sim_{15} \sim_{15}(A \wedge \sim_{15} A \wedge \sim_{15} \sim_{15} A) \vDash^{15}_{\mathbf{FDE}} B$ and $\sim_{15} \sim_{15} \sim_{15}(A \wedge \sim_{15} A \wedge \sim_{15} \sim_{15} A) \vDash^{15}_{\mathbf{FDE}} B$. Finally, for $\mathbf{FDE}^{16}$, we have $\sim_{16}(A \vee \sim_{16} \sim_{16} A) \vDash^{16}_{\mathbf{FDE}} B$ and $\sim_{16} \sim_{16}(A \vee \sim_{16} \sim_{16} A) \vDash^{16}_{\mathbf{FDE}} B$. ◁

Remark 32. The other variants of **FDE**, **K3** and **LP** are not negation incomplete. Indeed, for the variants of **LP**, this is obvious since they all have (LEM). For the other variants of **FDE** and **K3**, note that negation incompleteness implies that there is a formula that receives the value **n** for all interpretations. But it is easy so see that this cannot be the case by similar considerations we sketched above for the cases with negation inconsistency.

5.2. Functional Completeness. We now turn to show that the matrices that characterize some of the contra-classical variants of **FDE**, **K3** and **LP** are functionally complete as a corollary of a general characterization of functional completeness. To this end, we first introduce some related notions.

Definition 33 (Functional completeness). An algebra $\mathfrak{A} = \langle A, f_1, \ldots, f_n \rangle$, is said to be *functionally complete* provided that every finitary function $f \colon A^m \to A$ is definable by compositions of the functions $f_1, \ldots, f_n$ alone. A matrix $\langle \mathfrak{A}, \mathcal{D} \rangle$ is *functionally complete* if $\mathfrak{A}$ is functionally complete.

Definition 34 (Definitional completeness). A logic **L** is *definitionally complete* if there exists a functionally complete matrix that is strongly adequate for L.

For the characterization of the functional completeness, the following theorem of Jerzy Słupecki is elegant and useful. In order to state the result, we need the following definition.

Definition 35. Let $\mathfrak{A} = \langle A, f_1, \ldots, f_n \rangle$ be an algebra, and f be a binary operation defined in $\mathfrak{A}$. Then, f is *unary reducible* iff for some unary operation g definable in

$\mathfrak{A}$, $f(x,y) = g(x)$ for all $x,y \in A$ or $f(x,y) = g(y)$ for all $x,y \in A$. And f is *essentially binary* if f is not unary reducible.

Theorem 36 (Słupecki, [42]). $\mathfrak{A} = \langle\langle \mathcal{V}, f_1, \ldots, f_n\rangle, \mathcal{D}\rangle$ ($|\mathcal{V}| \geq 3$) *is functionally complete iff in* $\langle \mathcal{V}, f_1, \ldots, f_n\rangle$ *(1) all unary functions on* $\mathcal{V}$ *are definable, and (2) at least one surjective and essentially binary function on* $\mathcal{V}$ *is definable.*

Based on this elegant characterization by Słupecki, the desired result is obtained as follows. In case of expansions of the algebra related to **FDE**, we can simplify even further, as we observed in [32, Theorem 4.8].

Theorem 37. *Given any expansion* $\mathcal{F}$ *of the algebra* $\langle\{\mathbf{t},\mathbf{b},\mathbf{n},\mathbf{f}\},\wedge,\vee\rangle$ *the following (1) and (2) are equivalent: (1)* $\mathcal{F}$ *is functionally complete; (2) all of the* δ_a*'s as well as* C_a*'s* ($a \in \{\mathbf{t},\mathbf{b},\mathbf{n},\mathbf{f}\}$) *are definable, where* $\delta_a(b) = \mathbf{t}$*, if* $a = b$*, otherwise* $\delta_a(b) = \mathbf{f}$*; and* $C_a(b) = a$*, for all* $a,b \in \mathcal{V}$*.*

Similarly, we obtain the next result for the three-element cases, where $\mathbf{i} \in \{\mathbf{b},\mathbf{n}\}$.

Theorem 38. *Given any expansion* $\mathcal{F}$ *of the algebra* $\langle\{\mathbf{t},\mathbf{i},\mathbf{f}\},\wedge,\vee\rangle$ *the following (1) and (2) are equivalent: (1)* $\mathcal{F}$ *is functionally complete; (2) all of the* δ_a*'s as well as* C_a*'s* ($a \in \{\mathbf{t},\mathbf{i},\mathbf{f}\}$) *are definable, where* δ_a *and* C_a *are defined as in Theorem 37.*

Building on these results, we obtain the following.

Theorem 39. $\mathbf{FDE}^{16}$, $\mathbf{K3}^4$ *and* $\mathbf{LP}^4$ *are definitionally complete.*

Proof. For $\mathbf{FDE}^{16}$, in view of the above theorem, it suffices to prove that all of the δ_as as well as C_a's ($a \in \{\mathbf{t},\mathbf{b},\mathbf{n},\mathbf{f}\}$) are definable in $\langle\{\mathbf{t},\mathbf{b},\mathbf{n},\mathbf{f}\},\sim_{16},\wedge,\vee\rangle$, and this can be done as follows: $\delta_\mathbf{t}(x) := \neg(\sim_{16}x \vee \sim_{16}\sim_{16}x)$, $\delta_\mathbf{b}(x) := \neg(\neg\sim_{16}x \vee \sim_{16}\sim_{16}x)$, $\delta_\mathbf{n}(x) := \neg(\neg\sim_{16}\sim_{16}x \vee \sim_{16}x)$, $\delta_\mathbf{f}(x) := \neg\neg(\sim_{16}x \wedge \sim_{16}\sim_{16}x)$, $C_\mathbf{t}(x) := x \vee \sim_{16}\sim_{16}x$, $C_\mathbf{b}(x) := \sim_{16}(x \wedge \sim_{16}\sim_{16}x)$, $C_\mathbf{n}(x) := \sim_{16}(x \vee \sim_{16}\sim_{16}x)$, and $C_\mathbf{f}(x) := x \wedge \sim_{16}\sim_{16}x$, where $\neg x := \sim_{16}(\sim_{16}\sim_{16}((x \wedge \sim_{16}x) \wedge \sim_{16}(x \wedge \sim_{16}x)) \wedge ((x \wedge \sim_{16}x) \vee \sim_{16}(x \wedge \sim_{16}x)))$.

For $\mathbf{K3}^4$, in view of the above theorem, it suffices to prove that all of the δ_a's as well as C_a's ($a \in \{\mathbf{t},\mathbf{n},\mathbf{f}\}$) are definable in $\langle\{\mathbf{t},\mathbf{n},\mathbf{f}\},\sim_4,\wedge,\vee\rangle$, and this can be done as follows: $\delta_\mathbf{t}(x) := x \wedge \sim_4(x \wedge \sim_4\sim_4 x)$, $\delta_\mathbf{n}(x) := \sim_4\sim_4(x \vee \sim_4 x)$, $\delta_\mathbf{f}(x) := \sim_4\sim_4(x \vee \sim_4 x)$, $C_\mathbf{t}(x) := \sim_4\sim_4(x \wedge \sim_4 x \wedge \sim_4 \sim_4 x)$, $C_\mathbf{n}(x) := \sim_4 \sim_4(x \wedge \sim_4 x \wedge \sim_4 \sim_4 x)$, and $C_\mathbf{f}(x) := x \wedge \sim_4 x \wedge \sim_4 \sim_4 x$.

Similarly, for $\mathbf{LP}^4$, in view of the above theorem, it suffices to prove that all of the δ_a's as well as C_a's ($a \in \{\mathbf{t},\mathbf{b},\mathbf{f}\}$) are definable in $\langle\{\mathbf{t},\mathbf{b},\mathbf{f}\},\sim_4,\wedge,\vee\rangle$, and this can be done as follows: $\delta_\mathbf{t}(x) := x \wedge \sim_4 \sim_4(x \wedge \sim_4 x)$, $\delta_\mathbf{b}(x) := \sim_4(x \vee \sim_4 \sim_4 x)$, $\delta_\mathbf{f}(x) := \sim_4(x \vee \sim_4 x)$, $C_\mathbf{t}(x) := \sim_4 \sim_4(x \wedge \sim_4 x \wedge \sim_4 \sim_4 x)$, $C_\mathbf{b}(x) := \sim_4(x \wedge \sim_4 x \wedge \sim_4 \sim_4 x)$, and $C_\mathbf{f}(x) := x \wedge \sim_4 x \wedge \sim_4 \sim_4 x$. This completes the proof. ◁

Remark 40. Note that it is not difficult to see that other variants are *not* functionally complete.

Finally, we add a brief remark on the Post completeness.

Definition 41. The logic **L** is *Post complete* iff for every formula A such that $\nvdash A$, the extension of **L** by A becomes trivial, i.e., $\vdash_{\mathbf{L} \cup \{A\}} B$ for any B.

Theorem 42 (Tokarz, [45]). *Definitionally complete logics are Post complete.*

In view of Theorems 39 and 42, we obtain the following result.

Corollary 43. FDE^{16}, $K3^4$ *and* LP^4 *are Post complete.*

Remark 44. Note that the converse of Theorem 42 does not hold, i.e., there are logics that are Post complete without being definitionally complete, such as the negation-free fragment of classical propositional logic. Therefore, one may ask if other variants of **FDE**, **LP** and **K3** are Post complete. The answer is that in our case, none of the variants other than $\mathbf{FDE}^{16}$, $\mathbf{K3}^4$ and $\mathbf{LP}^4$ are Post complete, as observed in the following proposition.

Proposition 45. *None of the variants other than* $\mathbf{FDE}^{16}$, $\mathbf{K3}^4$ *and* $\mathbf{LP}^4$ *are Post complete.*

Proof. The results hold by considering extensions by (ECQ) or (LEM). ◁

5.3. Variable Sharing Property and Admissibility of Contraposition.
Let us now turn our attention to two more properties that **FDE** is well known for enjoying, namely, the variable sharing property and the admissibility of the rule of contraposition. We will first deal with the variable sharing property, by recalling the definition.

Definition 46. A logic **L** satisfies the *variable sharing property* iff for all $A, B \in$ Form, $A \vdash_\mathbf{L} B$ implies that A and B share at least one propositional variable.

Remark 47. Usually, the variable sharing property is stated with respect to the conditional included in the object language, but since we do not have conditionals in the language, we will consider the version above.[6]

Then, we obtain the following result.

Theorem 48. $\mathbf{FDE}^1$, $\mathbf{FDE}^2$, $\mathbf{FDE}^9$ *and* $\mathbf{FDE}^{10}$ *satisfy the variable sharing property. The other systems, including the variants of* $\mathbf{K3}$ *and* $\mathbf{LP}$, *do not satisfy the variable sharing property.*

Proof. Suppose $A \vdash^i_{\mathbf{FDE}} B$ ($i \in \{1, 2, 9, 10\}$), but that A and B do not share any propositional variables. Then, if we consider a valuation v that assigns the value **b** to all the variables in A and the value **n** to all the variables in B, then we obtain $v(A) = \mathbf{b}$ and $v(B) = \mathbf{n}$. Indeed, by a simple inductive proof, we may observe that both values **b** and **n** are closed under the set of operations $\{\sim_i, \wedge, \vee\}$, where $i \in \{1, 2, 9, 10\}$. Then the above valuation is a counter-model for $A \vdash^i_{\mathbf{FDE}} B$, an absurdity in view of our assumption.

For the latter half, it easy to check that one of the rules of the unilateral natural deduction system serves as a counterexample of the variable sharing property. ◁

Remark 49. Our result shows that there are no sub-classical variants of **FDE** with the variable sharing property (VSP), but there are three other systems if we widen our scope beyond sub-classicality.

---

[6]As a referee pointed out, there is a system with the variable sharing property in the original form, but not in the above form. Such examples include the logic determined by the matrix M_0 presented by Nuel Belnap in [6].

Let us now turn to the status of the rule of contraposition, which is seen as crucial for the understanding of negation by, for example, the advocates of the so-called *Australian plan* for negation.[7] We first clarify the form of contraposition we have in mind.

Definition 50. A logic **L** admits the *rule of contraposition* iff for all $A,B \in$ Form,

(Contra) $\qquad A \vdash_\mathbf{L} B$ implies $\sim B \vdash_\mathbf{L} \sim A$.

Then, as is well known, **FDE** admits the rule of contraposition. The easiest way to see this is from the perspective of the star semantics, defined as follows.

Definition 51. A *Routley interpretation* is a structure $\langle W, *, v \rangle$, where W is a set of worlds, $*: W \longrightarrow W$ is a function with $w^{**} = w$, and $v: W \times$ Prop $\longrightarrow \{0,1\}$. The function v is extended to an assignment I of truth values for all pairs of worlds and formulas by the conditions:

(1) $I(w,p) = v(w,p)$,
(2) $I(w, \sim A) = 1$ iff $I(w^*, A) \neq 1$,
(3) $I(w, A \wedge B) = 1$ iff $I(w,A) = 1$ and $I(w,B) = 1$,
(4) $I(w, A \vee B) = 1$ iff $I(w,A) = 1$ or $I(w,B) = 1$.

Definition 52. A formula A is a *star semantic consequence* of Γ ($\Gamma \vDash^*_{\mathbf{FDE}} A$) iff for all Routley interpretations $\langle W, *, v \rangle$ and for all $w \in W$, if $I(w,B) = 1$ for all $B \in \Gamma$ then $I(w,A) = 1$.

Then, the following result is well known, due to Richard Routley and Valerie Routley (cf. [39]).[8]

Theorem 53 (Routley & Routley). *For all* $\Gamma \cup \{A\} \subseteq$ Form, $\Gamma \vdash_{\mathbf{FDE}} A$ *iff* $\Gamma \vDash^*_{\mathbf{FDE}} A$.

As a corollary, we obtain that **FDE** satisfies the rule of contraposition. Now, the question is that if there are other systems within the variations we are considering that satisfy the rule of contraposition. The answer is yes, since **FDE**[7] also admits (Contra), and for the purpose of establishing this result, we introduce a variation of Routley interpretations as follows.

Definition 54. Let *one-step Routley interpretation* be a structure $\langle W, *, v \rangle$ as in Routley interpretation, except that $w^{**} = w$ is replaced by $w^* = w^{**}$.

Remark 55. One-step here means that it starts to "loop" after one application of the star operator. We can also consider n-step Routley interpretations in general, but we will not consider them in this paper.

Definition 56. A formula A is a *one step star semantic consequence* of Γ ($\Gamma \vDash^{*1}_{\mathbf{FDE}} A$) iff for all one step Routley interpretations $\langle W, *, v \rangle$ and for all $w \in W$, if $I(w,B) = 1$ for all $B \in \Gamma$ then $I(w,A) = 1$.

Then, we obtain the following result.

Theorem 57. *For all* $\Gamma \cup \{A\} \subseteq$ Form, $\Gamma \vdash^7_{\mathbf{FDE}} A$ *iff* $\Gamma \vDash^{*1}_{\mathbf{FDE}} A$.

---

[7]For one of the most recent discussions on this topic, see [7; 8]. Note also that the Dunn semantics offers the key insight for the so-called *American plan* for negation.

[8]Or, more precisely as Dunn writes in [16, p. 440], the star semantics was "actually mathematically in 1957 anticipated by A. Białynicki-Birula and H. Rasiowa, and shown equivalent by Dunn in 1966."

Proof. For the soundness direction, we establish the result by a straightforward verification that each rule is truth-preserving. We only deal with $(\sim_7\sim_7 1)$ and $(\sim_7\sim_7 2)$. To this end, it suffices to observe that $I(w, \sim_7 A) = 1$ iff $I(w^*, A) \neq 1$ and that $I(w, \sim_7\sim_7 A) = 1$ iff $I(w^*, \sim_7 A) \neq 1$ iff $I(w^{**}, A) = 1$ iff $I(w^*, A) = 1$ (by $w^* = w^{**}$), and thus for all $A \in$ Form and for all $w \in W$, we obtain $I(w, \sim_7 A \lor \sim_7\sim_7 A) = 1$ and $I(w, \sim_7 A \land \sim_7\sim_7 A) \neq 1$.

For the completeness direction, we make use of the completeness result with respect to the four-valued semantics (cf. Theorem 27), and establish that if $\Gamma \not\vDash^7_{\mathbf{FDE}} A$ then $\Gamma \not\vdash^{*1}_{\mathbf{FDE}} A$. So, assume that $\Gamma \not\vDash^7_{\mathbf{FDE}} A$. Then, there is a four-valued **FDE**^{7} interpretation v_0 such that $v_0(B) \in \mathcal{D}$ for all $B \in \Gamma$, and $v_0(A) \notin \mathcal{D}$. Given v_0, we define a one-step Routley interpretation as follows: $W := \{a, b\}$, $a^* = b$ and $b^* = b$, and

$$v(a, p) = 1 \text{ iff } v_0(p) \in \{\mathbf{t}, \mathbf{b}\}, \qquad v(b, p) = 1 \text{ iff } v_0(p) \in \{\mathbf{t}, \mathbf{n}\}.$$

Then, once we show that the above condition holds for all $A \in$ Form, we obtain the desired result. That the above condition holds for all $A \in$ Form can be proved by induction on the construction of A. The base case, for atomic formulas, is obvious by the definition. For the induction step, the cases are split based on the connectives. Since the cases for conjunction and disjunction can be done in exactly the same way as we do for **FDE**, we will focus on the case for negation, namely, the case when A is of the form $\sim_7 B$. Then,

(1) $v(a, \sim_7 B) = 1$ iff $v(a^*, B) \neq 1$ iff $v(b, B) \neq 1$ iff $v_0(B) \notin \{\mathbf{t}, \mathbf{n}\}$ (by IH) iff $v_0(\sim_7 B) = \mathbf{t}$ (by the truth table) iff $v_0(\sim_7 B) \in \{\mathbf{t}, \mathbf{b}\}$ (since $v_0(\sim_7 B)$ is never $\mathbf{b}$ by the truth table).

(2) $v(b, \sim_7 B) = 1$ iff $v(b^*, B) \neq 1$ iff $v(b, B) \neq 1$ iff $v_0(B) \notin \{\mathbf{t}, \mathbf{n}\}$ (by IH) iff $v_0(\sim_7 B) = \mathbf{t}$ (by the truth table) iff $v_0(\sim_7 B) \in \{\mathbf{t}, \mathbf{n}\}$ (since $v_0(\sim_7 B)$ is never $\mathbf{n}$ by the truth table).

This completes the proof. ◁

As an immediate corollary, we obtain the following.

Corollary 58. ***FDE***^{7} *admits (Contra).*

Remark 59. Note that (Contra) is not admissible for the other systems. First, for **K3** and **LP**, this is immediate since (ECQ) and (Contra) will establish (LEM), and (LEM) and (Contra) will establish (ECQ). Second, for the variants of **FDE** that are negation inconsistent or negation incomplete, it is not difficult to prove the desired results. Indeed, assume (Contra) and take B to be an instance of the negation inconsistent formula. Then, we obtain $p \vdash B$ holds, and thus by (Contra), we obtain $\sim B \vdash \sim p$. But, since B is an instance of the negation inconsistent formula, we obtain $\vdash \sim p$, but this is absurd. The proof is similar for the negation incomplete case. Finally, for the rest of systems, we show specific counterexamples.

1. For **FDE**^{2}: $p \vdash^2_{\mathbf{FDE}} \sim_2\sim_2 p$ but $\sim_2\sim_2\sim_2 p \not\vdash^2_{\mathbf{FDE}} \sim_2 p$.
2. For **FDE**^{3}: $q \vdash^3_{\mathbf{FDE}} \sim_3 p \lor \sim_3\sim_3 p$ but $\sim_3(\sim_3 p \lor \sim_3\sim_3 p) \not\vdash^3_{\mathbf{FDE}} \sim_3 q$.
3. For **FDE**^{5}: $q \vdash^5_{\mathbf{FDE}} \sim_5(p \land \sim_5 p \land \sim_5\sim_5 p)$ but $\sim_5\sim_5(p \land \sim_5 p \land \sim_5\sim_5 p) \not\vdash^5_{\mathbf{FDE}} \sim_5 q$.
4. For **FDE**^{6}: $p \land \sim_6 p \land \sim_6\sim_6 p \vdash^6_{\mathbf{FDE}} q$ but $\sim_6 q \not\vdash^5_{\mathbf{FDE}} \sim_6(p \land \sim_6 p \land \sim_6\sim_6 p)$.

5. For **FDE**^{9}: $\sim_9 \sim_9 p \vdash^9_{\mathbf{FDE}} p$ but $\sim_9 p \nvdash^9_{\mathbf{FDE}} \sim_9 \sim_9 \sim_9 p$.
6. For **FDE**^{10}: $\sim_{10} p \vee \sim_{10} q \vdash^{10}_{\mathbf{FDE}} \sim_{10}(p \wedge q)$ but $\sim_{10} \sim_{10}(p \wedge q) \nvdash^{10}_{\mathbf{FDE}} \sim_{10}(\sim_{10} p \vee \sim_{10} q)$.
7. For **FDE**^{11}: $q \vdash^{11}_{\mathbf{FDE}} p \vee \sim_{11} p \vee \sim_{11} \sim_{11} p$ but $\sim_{11}(p \vee \sim_{11} p \vee \sim_{11} \sim_{11} p) \nvdash^{11}_{\mathbf{FDE}} \sim_{11} q$.

Remark 60. In view of the above result, none of the *contra-classical* variants can be captured by the Australian plan with the local consequence relation. In other words, if one is in deep favor of (Contra), then the contra-classical variants cannot be captured. However, one may still work with the Australian plan, but take pointed models and define the semantic consequence relation in terms of truth preservation at the distinguished point. Whether this way will allow the Australian plan advocates to capture any of the contra-classical variants or not is an interesting question that we will leave to interested readers.

6. Reflections: Too Many Varieties?

Given all the variants, one may conclude that there are far too many options, and wonder about the implications of all this. This, of course, is a natural and even a pressing question. For the purpose of addressing the question, at least partially, we will make use of non-deterministic semantics. More specifically, we will consider some family of negations under certain classification, put them together along the framework of non-deterministic semantics, and explore the shared property for those negations. Let us first recall the basic definition of non-deterministic semantics (cf. [3] for an overview).

Definition 61. A *non-deterministic matrix* (Nmatrix for short) for $\mathcal{L}$ is a tuple $M = \langle \mathcal{V}, \mathcal{D}, \mathcal{O} \rangle$, where $\mathcal{V}$ is a non-empty set of truth values, $\mathcal{D}$ is a non-empty proper subset of $\mathcal{V}$, and for every n-ary connective $*$ of $\mathcal{L}$, $\mathcal{O}$ includes a corresponding n-ary function $\tilde{*}$ from $\mathcal{V}^n$ to $2^{\mathcal{V}} \setminus \{\emptyset\}$. We say that M is (in)finite if so is $\mathcal{V}$. A *legal valuation* in an Nmatrix M is a function $v : \text{Form} \to \mathcal{V}$ that satisfies the following condition for every n-ary connective $*$ of $\mathcal{L}$ and $A_1, \ldots, A_n \in \text{Form}$:

(gHom) $\qquad v(*(A_1, \ldots, A_n)) \in \tilde{*}(v(A_1), \ldots, v(A_n))$.

The condition (gHom) can be interpreted as a generalized homomorphism condition.

Let us now consider four kinds of non-deterministic matrices. The first one is obtained by putting together the truth table for subclassical negations, with a motivation to explore the common core of subclassical variants of **FDE**.

Definition 62. A *four-valued subclassical **FDE**-Nmatrix* for $\mathcal{L}$ is a tuple $M = \langle \mathcal{V}, \mathcal{D}, \mathcal{O} \rangle$, where $\mathcal{V} = \{\mathbf{t}, \mathbf{b}, \mathbf{n}, \mathbf{f}\}$, $\mathcal{D} = \{\mathbf{t}, \mathbf{b}\}$, and for every n-ary connective $*$ of $\mathcal{L}$, $\mathcal{O}$ includes a corresponding n-ary function $\tilde{*}$ from $\mathcal{V}^n$ to $2^{\mathcal{V}} \setminus \{\emptyset\}$ as follows (we omit the braces for sets):

A	$\tilde{\sim} A$		$A \tilde{\wedge} B$	$\mathbf{t}$	$\mathbf{b}$	$\mathbf{n}$	$\mathbf{f}$		$A \tilde{\vee} B$	$\mathbf{t}$	$\mathbf{b}$	$\mathbf{n}$	$\mathbf{f}$
$\mathbf{t}$	$\mathbf{f}$		$\mathbf{t}$	$\mathbf{t}$	$\mathbf{b}$	$\mathbf{n}$	$\mathbf{f}$		$\mathbf{t}$	$\mathbf{t}$	$\mathbf{t}$	$\mathbf{t}$	$\mathbf{t}$
$\mathbf{b}$	$\mathbf{t},\mathbf{b}$		$\mathbf{b}$	$\mathbf{b}$	$\mathbf{b}$	$\mathbf{f}$	$\mathbf{f}$		$\mathbf{b}$	$\mathbf{t}$	$\mathbf{b}$	$\mathbf{t}$	$\mathbf{b}$
$\mathbf{n}$	$\mathbf{n},\mathbf{f}$		$\mathbf{n}$	$\mathbf{n}$	$\mathbf{f}$	$\mathbf{n}$	$\mathbf{f}$		$\mathbf{n}$	$\mathbf{t}$	$\mathbf{t}$	$\mathbf{n}$	$\mathbf{n}$
$\mathbf{f}$	$\mathbf{t}$		$\mathbf{f}$	$\mathbf{f}$	$\mathbf{f}$	$\mathbf{f}$	$\mathbf{f}$		$\mathbf{f}$	$\mathbf{t}$	$\mathbf{b}$	$\mathbf{n}$	$\mathbf{f}$

A *four-valued subclassical* **FDE**-*valuation* in a four-valued subclassical **FDE**-Nmatrix M is a function v: Form $\to \mathcal{V}$ that satisfies (gHom). Finally, A is a *four-valued subclassical* **FDE**-*consequence* of Γ ($\Gamma \vDash_{4s} A$) iff for every four-valued subclassical **FDE**-valuation v, if $v(B) \in \mathcal{D}$ for every $B \in \Gamma$ then $v(A) \in \mathcal{D}$.

The corresponding (unilateral) natural deduction system is introduced as follows.

Definition 63. The natural deduction rules $\mathcal{R}^{sub}_{\textbf{FDE}}$ for **sub-FDE** are all the rules $\mathcal{R}_{\textbf{FDE}}$ for **FDE** but replacing $(\sim\sim 1)$ and $(\sim\sim 2)$ by the following rules.

$$\dfrac{\sim A \quad \sim\sim A}{A} \; (\sim\sim 1) \qquad \dfrac{A}{\sim A \vee \sim\sim A} \; (\sim\sim 2)$$

Based on these, given any set $\Sigma \cup \{A\}$ of formulas, $\Sigma \vdash^{sub}_{\textbf{FDE}} A$ iff for some finite $\Sigma' \subseteq \Sigma$, there is a derivation of A from Σ' in the calculus whose rule set is $\mathcal{R}^{sub}_{\textbf{FDE}}$.

Remark 64. One can also devise a bilateral natural deduction for **sub-FDE** replacing $(\sim\sim 1)$ and $(\sim\sim 2)$ by the following rules to the bilateral presentation of **FDE**.

$$\dfrac{\overline{A} \quad \overline{\sim A}}{\overline{A}} \qquad \dfrac{A \quad \begin{array}{c}[A]\\ \vdots \\ B\end{array} \quad \begin{array}{c}[\sim A]\\ \vdots \\ B\end{array}}{B}$$

The further details are left for the interested readers.

Then, we may establish soundness and completeness results. The soundness is again tedious but not difficult.

Theorem 65. *For all $\Gamma \cup \{A\} \subseteq$ Form, if $\Gamma \vdash^{sub}_{\textbf{FDE}} A$ then $\Gamma \vDash_{4s} A$.*

Proof. It can be shown by a straightforward verification that each rule preserves designated values. Here we only spell out the details for the validity of $(\sim\sim 1)$ and $(\sim\sim 2)$.

Ad $(\sim\sim 1)$: Suppose, for reductio, that there is a four-valued subclassical **FDE**-valuation v_0 such that $v_0(\sim A) \in \mathcal{D}$, $v_0(\sim\sim A) \in \mathcal{D}$, but $v_0(A) \notin \mathcal{D}$. Then, the first and the third assumption together with the Nmatrices imply that $v_0(A) = \textbf{f}$, and thus $v_0(\sim\sim A) = \textbf{f}$. But, this is absurd in view of the second assumption.

Ad $(\sim\sim 2)$: Suppose, for reductio, that there is a four-valued subclassical **FDE**-valuation v_0 such that $v_0(A) \in \mathcal{D}$, but $v_0(\sim A \vee \sim\sim A) \notin \mathcal{D}$. Then, the second assumption together with the Nmatrices imply that $v_0(\sim A) \notin \mathcal{D}$ and $v_0(\sim\sim A) \notin \mathcal{D}$. By $v_0(A) \in \mathcal{D}$ and $v_0(\sim A) \notin \mathcal{D}$, we obtain that $v_0(A) = \textbf{t}$, and thus $v_0(\sim\sim A) = \textbf{t}$. But, this is absurd in view of $v_0(\sim\sim A) \notin \mathcal{D}$. ◁

For completeness, we prepare a definition and a lemma.

Definition 66. For any $\Sigma \subseteq$ Form, let v^{sub}_{Σ} from Form to $\{\textbf{t}, \textbf{b}, \textbf{n}, \textbf{f}\}$ be defined as follows:

$$v^{sub}_{\Sigma}(A) := \begin{cases} \textbf{t} \text{ iff } \Sigma \vdash^{sub}_{\textbf{FDE}} A \text{ and } \Sigma \nvdash^{sub}_{\textbf{FDE}} \sim A; \\ \textbf{b} \text{ iff } \Sigma \vdash^{sub}_{\textbf{FDE}} A \text{ and } \Sigma \vdash^{sub}_{\textbf{FDE}} \sim A; \\ \textbf{n} \text{ iff } \Sigma \nvdash^{sub}_{\textbf{FDE}} A \text{ and } \Sigma \nvdash^{sub}_{\textbf{FDE}} \sim A; \\ \textbf{f} \text{ iff } \Sigma \nvdash^{sub}_{\textbf{FDE}} A \text{ and } \Sigma \vdash^{sub}_{\textbf{FDE}} \sim A. \end{cases}$$

Note that we are defining the canonical valuation in a different manner compared to Definition 25, reflecting the difference of how deterministic and non-deterministic semantics are introduced.

Lemma 67. *If Σ is a prime theory, then v_Σ^{sub} is a well-defined four-valued subclassical FDE-valuation.*

Proof. Note first that the well-definedness of v_Σ^{sub} is obvious. Then the desired result is proved by induction on the number n of connectives. Base case: For atomic formulas, it is obvious by the definition. Induction step: We split the cases based on the connectives. Here we only deal with $\sim$. If $A = \sim B$, then we have the following cases.

Cases	$v_\Sigma(B)$	condition for B	$v_\Sigma(A)$	condition for A i.e., $\sim B$
(i)	t	$\Sigma \vdash_{FDE}^{sub} B$ and $\Sigma \not\vdash_{FDE}^{sub} \sim B$	f	$\Sigma \not\vdash_{FDE}^{sub} \sim B$ and $\Sigma \vdash_{FDE}^{sub} \sim\sim B$
(ii)	b	$\Sigma \vdash_{FDE}^{sub} B$ and $\Sigma \vdash_{FDE}^{sub} \sim B$	t,b	$\Sigma \vdash_{FDE}^{sub} \sim B$
(iii)	n	$\Sigma \not\vdash_{FDE}^{sub} B$ and $\Sigma \not\vdash_{FDE}^{sub} \sim B$	n,f	$\Sigma \not\vdash_{FDE}^{sub} \sim B$
(iv)	f	$\Sigma \not\vdash_{FDE}^{sub} B$ and $\Sigma \vdash_{FDE}^{sub} \sim B$	t	$\Sigma \vdash_{FDE}^{sub} \sim B$ and $\Sigma \not\vdash_{FDE}^{sub} \sim\sim B$

By induction hypothesis, we have the conditions for B, for cases (ii) and (iii), it is easy to see that the conditions for A i.e., $\sim B$ are provable. For (i) and (iv), we can use ($\sim\sim 2$) and ($\sim\sim 1$), respectively. ◁

We are now ready to establish the completeness result.

Theorem 68. *For all $\Gamma \cup \{A\} \subseteq$ Form, if $\Gamma \vDash_{4s} A$ then $\Gamma \vdash_{FDE}^{sub} A$.*

Proof. Assume $\Gamma \not\vdash_{FDE}^{sub} A$. Then, by Lemma 24, there is a $\Sigma \supseteq \Gamma$ such that Σ is a prime theory and $A \notin \Sigma$, and by Lemma 67, a four-valued subclassical valuation v_Σ^{sub} can be defined with $v_\Sigma^{sub}(B) \in \mathcal{D}$ for every $B \in \Gamma$ and $v_\Sigma^{sub}(A) \notin \mathcal{D}$. Thus it follows that $\Gamma \not\vDash_{FDE}^{sub} A$, as desired. ◁

Let us now turn to the second kind of non-deterministic matrices, which is obtained by combining the negations that produce negation inconsistency.

Definition 69. A *four-valued negation inconsistent FDE-Nmatrix* for $\mathcal{L}$ is a tuple $M = \langle \mathcal{V}, \mathcal{D}, \mathcal{O} \rangle$, where $\mathcal{V} = \{t, b, n, f\}$, $\mathcal{D} = \{t, b\}$, and for every n-ary connective $*$ of $\mathcal{L}$, $\mathcal{O}$ includes a corresponding n-ary function $\tilde{*}$ from $\mathcal{V}^n$ to $2^\mathcal{V} \setminus \{\emptyset\}$. Definition 62 gives $\tilde{\wedge}$ and $\tilde{\vee}$; $\tilde{\sim}$ is

A	t	b	n	f
$\tilde{\sim}A$	n,f	t,b	f	b

A *four-valued negation inconsistent FDE-valuation* in a four-valued negation inconsistent FDE-Nmatrix M is a function v: Form $\to \mathcal{V}$ that satisfies (gHom). Finally, A is a *four-valued negation inconsistent FDE-consequence* of Γ ($\Gamma \vDash_{4b} A$) iff for every four-valued negation inconsistent FDE-valuation v, if $v(B) \in \mathcal{D}$ for every $B \in \Gamma$ then $v(A) \in \mathcal{D}$.

The corresponding (unilateral) natural deduction system is introduced as follows.

Definition 70. The natural deduction rules $\mathcal{R}_{FDE}^b$ for **b-FDE** are all the rules $\mathcal{R}_{FDE}$ for **FDE** but replacing ($\sim\sim 1$) and ($\sim\sim 2$) by the following rule.

$$\frac{}{A \vee \sim\sim A} \, (\sim\sim)$$

Based on these, given any set $\Sigma \cup \{A\}$ of formulas, $\Sigma \vdash^b_{FDE} A$ iff for some finite $\Sigma' \subseteq \Sigma$, there is a derivation of A from Σ' in the calculus whose rule set is $\mathcal{R}^b_{FDE}$.

Remark 71. One can also devise a bilateral natural deduction for **b-FDE** by replacing $(\sim\sim 1)$ and $(\sim\sim 2)$ by the following rule to the bilateral presentation of **FDE**.

$$\frac{\begin{matrix}[A] & [\![\sim A]\!]\\ \vdots & \vdots \\ \overline{B} & \overline{B}\end{matrix}}{B}$$

The further details are left for the interested readers.

Then, we may establish the following result.

Theorem 72. *For all* $\Gamma \cup \{A\} \subseteq$ Form, $\Gamma \vdash^b_{FDE} A$ *iff* $\Gamma \models_{4b} A$.

Proof. For the soundness direction, we establish the result by a straightforward verification that each rule preserves designated values. Here we only spell out the details for the validity of $(\sim\sim)$.

Suppose, for reductio, that there is a four-valued negation inconsistent **FDE**-valuation v_0 such that $v_0(A \vee \sim\sim A) \notin \mathcal{D}$. Then, together with the Nmatrices, the assumption implies that $v_0(A) \notin \mathcal{D}$ and $v_0(\sim\sim A) \notin \mathcal{D}$. By $v_0(A) \notin \mathcal{D}$, there are two cases. If $v_0(A) = \mathbf{n}$, then $v_0(\sim\sim A) = \mathbf{b}$, which is absurd in view of $v_0(\sim\sim A) \notin \mathcal{D}$. If $v_0(A) = \mathbf{f}$, then $v_0(\sim\sim A) \in \mathcal{D}$ which is absurd in view of $v_0(\sim\sim A) \notin \mathcal{D}$.

For the completeness direction, we need to define v^b_Σ as in Definition 66 with an obvious modification, and establish the analogue of Lemma 67. In particular, we need to check the following.

Cases	$v_\Sigma(B)$	condition for B	$v_\Sigma(A)$	condition for A i.e., $\sim B$
(i)	t	$\Sigma \vdash^b_{FDE} B$ and $\Sigma \nvdash^b_{FDE} \sim B$	n, f	$\Sigma \nvdash^b_{FDE} \sim B$
(ii)	b	$\Sigma \vdash^b_{FDE} B$ and $\Sigma \vdash^b_{FDE} \sim B$	t, b	$\Sigma \vdash^b_{FDE} \sim B$
(iii)	n	$\Sigma \nvdash^b_{FDE} B$ and $\Sigma \nvdash^b_{FDE} \sim B$	f	$\Sigma \nvdash^b_{FDE} \sim B$ and $\Sigma \vdash^b_{FDE} \sim\sim B$
(iv)	f	$\Sigma \nvdash^b_{FDE} B$ and $\Sigma \vdash^b_{FDE} \sim B$	b	$\Sigma \vdash^b_{FDE} \sim B$ and $\Sigma \vdash^b_{FDE} \sim\sim B$

By induction hypothesis, we have the conditions for B, for cases (i) and (ii), it is easy to see that the conditions for A i.e., $\sim B$ are provable. For (iii) and (iv), we can use $(\sim\sim)$. ◁

Remark 73. Note that although we combined the negations that produce negation inconsistency, and thus named the Nmatrix including the phrase "negation inconsistent," it is not clear to us at the time of writing if the resulting system **b-FDE** is negation inconsistent or not.

The third one now is obtained by combining the negations that produce negation incompleteness.

Definition 74. A *four-valued negation incomplete* **FDE**-*Nmatrix* for $\mathcal{L}$ is a tuple $M = \langle \mathcal{V}, \mathcal{D}, \mathcal{O} \rangle$, where $\mathcal{V} = \{\mathbf{t}, \mathbf{b}, \mathbf{n}, \mathbf{f}\}$, $\mathcal{D} = \{\mathbf{t}, \mathbf{b}\}$, and for every n-ary connective $*$ of $\mathcal{L}$, $\mathcal{O}$ includes a corresponding n-ary function $\tilde{*}$ from $\mathcal{V}^n$ to $2^\mathcal{V} \setminus \{\emptyset\}$. Definition 62 gives $\tilde{\wedge}$ and $\tilde{\vee}$; $\tilde{\sim}$ is

A	t	b	n	f
$\sim A$	n	t	n,f	t,b

A *four-valued negation incomplete **FDE**-valuation* in a four-valued negation incomplete **FDE**-Nmatrix M is a function v: Form $\to \mathcal{V}$ that satisfies (gHom). Finally, A is a *four-valued negation incomplete **FDE**-consequence* of Γ ($\Gamma \vDash_{4n} A$) iff for every four-valued negation inconsistent **FDE**-valuation v, if $v(B) \in \mathcal{D}$ for every $B \in \Gamma$ then $v(A) \in \mathcal{D}$.

The corresponding (unilateral) natural deduction system is as follows.

Definition 75. The natural deduction rules $\mathcal{R}^n_{\mathbf{FDE}}$ for **n-FDE** are all the rules $\mathcal{R}_{\mathbf{FDE}}$ for **FDE** but replacing $(\sim\sim 1)$ and $(\sim\sim 2)$ by the following rule.

$$\frac{A \quad \sim\sim A}{B} \; (\sim\sim)$$

Based on these, given any set $\Sigma \cup \{A\}$ of formulas, $\Sigma \vdash^n_{\mathbf{FDE}} A$ iff for some finite $\Sigma' \subseteq \Sigma$, there is a derivation of A from Σ' in the calculus whose rule set is $\mathcal{R}^n_{\mathbf{FDE}}$.

Remark 76. One can also devise a bilateral natural deduction for **n-FDE** by replacing $(\sim\sim 1)$ and $(\sim\sim 2)$ by the following rules to the bilateral presentation of **FDE**.

$$\frac{\overline{A} \quad \overline{\sim A}}{B}$$

The further details are left for the interested readers.

Then, we may establish the following result.

Theorem 77. *For all $\Gamma \cup \{A\} \subseteq$ Form, $\Gamma \vdash^n_{\mathbf{FDE}} A$ iff $\Gamma \vDash_{4n} A$.*

Proof. For the soundness direction, we establish the result by a straightforward verification that each rule preserves designated values. Here we only spell out the details for the validity of $(\sim\sim)$.

Suppose, for reductio, that there is a four-valued negation inconsistent **FDE**-valuation v_0 such that $v_0(A) \in \mathcal{D}$ and $v_0(\sim\sim A) \in \mathcal{D}$, but $v_0(B) \notin \mathcal{D}$. Then, the first assumption together with the Nmatrices imply that $v_0(\sim\sim A) \notin \mathcal{D}$. But this is absurd in view of the second assumption.

For the completeness direction, we again need to define v^n_Σ as in Definition 66 with an obvious modification, and establish the analogue of Lemma 67. In particular, we need to check the following.

Cases	$v_\Sigma(B)$	condition for B	$v_\Sigma(A)$	condition for A i.e., $\sim B$
(i)	t	$\Sigma \vdash^n_{\mathbf{FDE}} B$ and $\Sigma \nvdash^n_{\mathbf{FDE}} \sim B$	n	$\Sigma \nvdash^n_{\mathbf{FDE}} \sim B$ and $\Sigma \nvdash^n_{\mathbf{FDE}} \sim\sim B$
(ii)	b	$\Sigma \vdash^n_{\mathbf{FDE}} B$ and $\Sigma \vdash^n_{\mathbf{FDE}} \sim B$	t	$\Sigma \vdash^n_{\mathbf{FDE}} \sim B$ and $\Sigma \nvdash^n_{\mathbf{FDE}} \sim\sim B$
(iii)	n	$\Sigma \nvdash^n_{\mathbf{FDE}} B$ and $\Sigma \nvdash^n_{\mathbf{FDE}} \sim B$	n,f	$\Sigma \nvdash^n_{\mathbf{FDE}} \sim B$
(iv)	f	$\Sigma \nvdash^n_{\mathbf{FDE}} B$ and $\Sigma \vdash^n_{\mathbf{FDE}} \sim B$	t,b	$\Sigma \vdash^n_{\mathbf{FDE}} \sim B$

By induction hypothesis, we have the conditions for B, for cases (iii) and (iv), it is easy to see that the conditions for A i.e., $\sim B$ are provable. For (i) and (ii), we can use $(\sim\sim)$. ◁

Remark 78. Similarly to the case with **b-FDE**, although we combined the negations that produce negation incompleteness, and thus named the Nmatrix including the phrase "negation incomplete," it is not clear to us at the time of writing if the resulting system **n-FDE** is negation incomplete or not.

Finally, let us consider the *fully* contra-classical kind, by combining all the contra-classical negations.

Definition 79. A *four-valued contra-classical* **FDE**-*Nmatrix* for $\mathcal{L}$ is a tuple $M = \langle \mathcal{V}, \mathcal{D}, \mathcal{O} \rangle$, where $\mathcal{V} = \{\mathbf{t}, \mathbf{b}, \mathbf{n}, \mathbf{f}\}$, $\mathcal{D} = \{\mathbf{t}, \mathbf{b}\}$, and for every n-ary connective $*$ of $\mathcal{L}$, $\mathcal{O}$ includes a corresponding n-ary function $\tilde{*}$ from $\mathcal{V}^n$ to $2^{\mathcal{V}} \setminus \{\emptyset\}$. Definition 62 gives $\tilde{\wedge}$ and $\tilde{\vee}$; $\tilde{\sim}$ is

A	t	b	n	f
$\tilde{\sim}A$	n,f	t,b	n,f	t,b

A *four-valued contra-classical* **FDE**-*valuation* in a four-valued contra-classical **FDE**-Nmatrix M is a function $v \colon \text{Form} \to \mathcal{V}$ that satisfies (gHom). Finally, A is a *four-valued contra-classical* **FDE**-*consequence* of Γ ($\Gamma \vDash_{4c} A$) iff for every four-valued contra-classical **FDE**-valuation v, if $v(B) \in \mathcal{D}$ for every $B \in \Gamma$ then $v(A) \in \mathcal{D}$.

The corresponding (unilateral) natural deduction system is introduced as follows.

Definition 80. The natural deduction rules $\mathcal{R}_{\mathbf{FDE}}^{con}$ for **con-FDE** are all the rules $\mathcal{R}_{\mathbf{FDE}}$ for **FDE** but eliminating the rules $(\sim\sim 1)$ and $(\sim\sim 2)$. Based on this, given any set $\Sigma \cup \{A\}$ of formulas, $\Sigma \vdash_{\mathbf{FDE}}^{con} A$ iff for some finite $\Sigma' \subseteq \Sigma$, there is a derivation of A from Σ' in the calculus whose rule set is $\mathcal{R}_{\mathbf{FDE}}^{con}$.

Remark 81. One can also devise a bilateral natural deduction for **con-FDE** by eliminating $(\sim\sim 1)$ and $(\sim\sim 2)$. The further details are left for the interested readers.

Then, we may establish the following result.

Theorem 82. *For all* $\Gamma \cup \{A\} \subseteq \text{Form}$, $\Gamma \vdash_{\mathbf{FDE}}^{con} A$ *iff* $\Gamma \vDash_{4c} A$.

Proof. For the soundness direction, we having nothing specific to do for rules solely involving negation since we do not have any after eliminating the double negation introduction/elimination rules.

For the completeness, we again need to define v_{Σ}^{con} as in Definition 66 with an obvious modification, and establish the analogue of Lemma 67. In particular, we need to check the following.

Cases	$v_{\Sigma}(B)$	condition for B	$v_{\Sigma}(A)$	condition for A i.e., $\sim B$
(i)	t	$\Sigma \vdash_{\mathbf{FDE}}^{con} B$ and $\Sigma \nvdash_{\mathbf{FDE}}^{con} \sim B$	n,f	$\Sigma \nvdash_{\mathbf{FDE}}^{con} \sim B$
(ii)	b	$\Sigma \vdash_{\mathbf{FDE}}^{con} B$ and $\Sigma \vdash_{\mathbf{FDE}}^{con} \sim B$	t,b	$\Sigma \vdash_{\mathbf{FDE}}^{con} \sim B$
(iii)	n	$\Sigma \nvdash_{\mathbf{FDE}}^{con} B$ and $\Sigma \nvdash_{\mathbf{FDE}}^{con} \sim B$	n,f	$\Sigma \nvdash_{\mathbf{FDE}}^{con} \sim B$
(iv)	f	$\Sigma \nvdash_{\mathbf{FDE}}^{con} B$ and $\Sigma \vdash_{\mathbf{FDE}}^{con} \sim B$	t,b	$\Sigma \vdash_{\mathbf{FDE}}^{con} \sim B$

By induction hypothesis, we have the conditions for B, for all the cases, and it is easy to see that the conditions for A i.e., $\sim B$ are provable without any additional rules. ◁

Remark 83. Somewhat surprisingly, the contra-classicality vanishes in the resulting system that is obtained by combining, with the help of non-deterministic semantics, all the contra-classical variants of **FDE** with respect to negation. In particular, we end up in a subsystem of **FDE**, which is obtained by removing the falsity condition for negation. This is in sharp contrast with the case in which we combined the sub-classical variants of **FDE**. Of course, if we take the combination of all variants of **FDE**, both sub-classical and contra-classical, then the result will be the same as with the case of focusing on contra-classical variants.

Remark 84. Given that the corresponding Dunn semantics will be to simply leave the falsity condition for negation unspecified, this system can be also seen as reflecting the position that there is nothing more to negation than expressing falsity. A similar consideration for *classical negation* in the context of expansions of **FDE**, in which there are again 16 candidates as explored in [9], can be found in [43].

7. Concluding Remarks

By building on the framework of Dunn semantics, we explored variants of **FDE**, **K3**, and **LP** by fixing the truth condition for negation, but making changes in the falsity condition. We also offered proof systems in the style of natural deduction, both in the unilateral and in the bilateral manner, and established soundness and completeness results for all systems. This was followed by an investigation into the basic properties of the given variants. Our results, for the variants of **FDE**, are summarized in the following table.

	FDE^1	**FDE**^{2}	**FDE**^{3}	**FDE**^{4}	**FDE**^{5}	**FDE**^{6}	**FDE**^{7}	**FDE**^{8}
Subclassical	✓	×	✓	×	✓	×	✓	×
Contra-classical	×	✓	×	✓	×	✓	×	✓
Neg. inconsistent	×	×	×	✓	×	×	×	✓
Neg. incomplete	×	×	×	×	×	×	×	×
Func. complete	×	×	×	×	×	×	×	×
Post complete	×	×	×	×	×	×	×	×
Adm. of (Contra)	✓	×	×	×	×	×	✓	×
VSP	✓	✓	×	×	×	×	×	×

	FDE^9	**FDE**^{10}	**FDE**^{11}	**FDE**^{12}	**FDE**^{13}	**FDE**^{14}	**FDE**^{15}	**FDE**^{16}
Subclassical	×	×	×	×	×	×	×	×
Contra-classical	✓	✓	✓	✓	✓	✓	✓	✓
Neg. inconsistent	×	×	×	✓	×	×	×	✓
Neg. incomplete	×	×	×	×	✓	✓	✓	✓
Func. complete	×	×	×	×	×	×	×	✓
Post complete	×	×	×	×	×	×	×	✓
Adm. of (Contra)	×	×	×	×	×	×	×	×
VSP	✓	✓	×	×	×	×	×	×

This may seem to be too many variations. With that possible objection in mind, we also explored four combinations of systems, by putting together (i) sub-classical systems, (ii) negation inconsistent systems, (iii) negation incomplete systems, and (iv) contra-classical systems. The resulting systems are semantically described in terms of non-deterministic semantics, and we also offered unilateral and bilateral proof systems that are sound and complete.

Moreover, our results, for the variants of **K3** and **LP**, are summarized in the following table.

	$K3^1$	$K3^2$	$K3^3$	$K3^4$	LP^1	LP^2	LP^3	LP^4
Subclassical	✓	✓	×	×	✓	×	✓	×
Contra-classical	×	×	✓	✓	×	✓	×	✓
Negation inconsistent	×	×	×	×	×	✓	×	✓
Negation incomplete	×	×	✓	✓	×	×	×	×
Functionally complete	×	×	×	✓	×	×	×	✓
Post complete	×	×	×	✓	×	×	×	✓
Admissibility of (Contra)	×	×	×	×	×	×	×	×
Variable sharing property	×	×	×	×	×	×	×	×

Unsurprisingly, the variants of **K3** and **LP** do not enjoy the variable sharing property and thus fail to be relevance logics. The tweaking of the falsity condition of negation in **K3** may lead to negation incomplete systems, whereas the tweaking of the falsity condition of negation in **LP** may give one a negation inconsistent logic.

There are a number of different directions to pursue for further investigation. Beside those already mentioned in passing, we will note a few more questions. First, let us briefly note that if one emphasizes the symmetry of truth and falsity, and make that carry over for various properties, then among the contra-classical variations, the one with both negation inconsistency and negation incompleteness might be seen as the most favorable, not only satisfying one of them, and that will single out **FDE**[16] as the plausible variant of **FDE**. Given that **FDE**[16] also enjoys the functional completeness, the system, at least from a purely technical perspective, seems worth investigating further.

Second, a related direction to the previous one, is to explore if we can specify further properties, beside the very basic ones we discussed in this paper, so that each of the variants can be singled out by different desiderata. A full answer to this problem seems to contribute substantially to our systematic understanding of both subclassicality and contra-classicality.

Third, given the origin of **FDE** as the first-degree entailment of relevance logics **R** and **E**, we may ask, especially with those having the variable sharing property, if there are variants of relevance logics that will have our variants as their first degree entailment.

Fourth, our variations mainly focused on *deterministic* ones, and only explored four *non-deterministic* ones. However, for the case with **FDE**, from a purely combinatorial perspective assuming the framework of non-deterministic semantics, there are 81 possibilities, and we have only covered 20 of them (16 deterministic and 4 non-deterministic cases). What can be learnt from the other 41 cases is also a problem that seems to be worth addressing.

Finally, but not the least, we focused on a simple propositional language in this paper, but there are a lot of motivations to expand the language both with further propositional connectives (conditionals, modalities, etc.) as well as quantifiers. What kind of insight we gain in these various expansions is yet another direction that is natural and important.

Acknowledgments. We would like to thank the editor for the invitation to the volume, and an anonymous referee for some helpful comments. The research by Hitoshi Omori has been supported by a Sofja Kovalevskaja Award of the Alexander von Humboldt-Foundation, funded by the German Ministry for Education and Research. The research by Heinrich Wansing has been supported by the European Research Council (ERC) under the European Union's Horizon 2020 research and innovation programme, grant agreement ERC-2020-ADG, 101018280, ConLog.

REFERENCES

[1] Avron, A. (2005). A non-deterministic view on non-classical negations, *Studia Logica* **80**(2–3): 159–194.
[2] Avron, A., Arieli, O. and Zamansky, A. (2018). *Theory of Effective Propositional Paraconsistent Logics*, College Publications, London, UK.
[3] Avron, A. and Zamansky, A. (2011). Non-deterministic semantics for logical systems, *in* D. Gabbay and F. Guenthner (eds.), *Handbook of Philosophical Logic*, Vol. 16, Springer, pp. 227–304.
[4] Belnap, N. (1976). How a computer should think, *in* G. Ryle (ed.), *Contemporary Aspects of Philosophy*, Oriel Press, Stocksfield, pp. 30–55.
[5] Belnap, N. (1977). A useful four-valued logic, *in* J. M. Dunn and G. Epstein (eds.), *Modern Uses of Multiple-valued Logic*, Springer, Dordrecht, pp. 5–37.
[6] Belnap, N. D. (1960). Entailment and relevance, *Journal of Symbolic Logic* **25**: 144–146.
[7] Berto, F. (2015). A modality called 'negation', *Mind* **124**(495): 761–793.
[8] Berto, F. and Restall, G. (2019). Negation on the Australian plan, *Journal of Philosophical Logic* **48**(6): 1119–1144.
[9] De, M. and Omori, H. (2015). Classical negation and expansions of Belnap–Dunn logic, *Studia Logica* **103**: 825–851.
[10] De, M. and Omori, H. (2018). There is more to negation than modality, *Journal of Philosophical Logic* **47**(2): 281–299.
[11] Dunn, J. M. (1966). *The Algebra of Intensional Logics*, Doctoral dissertation, University of Pittsburgh, Pittsburgh, PA. Published as Vol. 2 in the *Logic PhDs* series by College Publications, London, UK, 2019.
[12] Dunn, J. M. (1976). Intuitive semantics for first-degree entailment and 'coupled trees', *Philosophical Studies* **29**(3): 149–168.
[13] Dunn, J. M. (1993). Star and perp: Two treatments of negation, *Philosophical Perspectives* **7**: 331–357. (Language and Logic, 1993, J. E. Tomberlin (ed.)).
[14] Dunn, J. M. (1996). Generalized ortho negation, *in* H. Wansing (ed.), *Negation: A Notion in Focus*, W. de Gruyter, Berlin, pp. 3–26.
[15] Dunn, J. M. (1999). A comparative study of various model-theoretic treatments of negation: A history of formal negation, *in* D. M. Gabbay and H. Wansing (eds.), *What is Negation?*, Kluwer Academic Publishers, Dordrecht, pp. 23–51.
[16] Dunn, J. M. (2001). The concept of information and the development of modern logic, *in* W. Stelzner and M. Stoeckler (eds.), *Zwischen traditioneller und moderner Logik: Nichtklassiche Ansatze*, Mentis-Verlag, Paderborn, pp. 423–447.
[17] Dunn, J. M. (2019). Two, three, four, infinity: The path to the four-valued logic and beyond, *in* H. Omori and H. Wansing (eds.), *New Essays on Belnap–Dunn Logic*, Springer, Switzerland, pp. 77–97.
[18] Dunn, J. M. and Zhou, C. (2005). Negation in the context of gaggle theory, *Studia Logica* **80**: 235–264.
[19] Estrada-González, L. (2019). The Bochum Plan and the foundations of contra-classical logics, *CLE e-Prints* **19**: 1–21.

[20] Estrada-González, L. (2021). An easy road to multi-contra-classicality, *Erkenntnis* DOI: 10.1007/s10670-021-00468-9.
[21] Estrada-González, L. (2022). Dunn semantics for contra-classical logics, *EPTCS* **358**: 298–309.
[22] Gabbay, D. M. and Wansing, H. (eds.) (1999). *What is Negation?*, Kluwer Academic Publishers, Dordrecht.
[23] Hazen, A. and Pelletier, J. (2018). Second-order logic of paradox, *Notre Dame Journal Formal Logic* **59**: 547–558.
[24] Horn, L. R. and Wansing, H. (2015). Negation, *in* E. N. Zalta (ed.), *The Stanford Encyclopedia of Philosophy*, Spring 2015 edn, URL: plato.stanford.edu/entries/negation/.
[25] Humberstone, L. (1995). Negation by iteration, *Theoria* **61**(1): 1–24.
[26] Humberstone, L. (2000). Contra-classical logics, *Australasian Journal of Philosophy* **78**(4): 438–474.
[27] Kamide, N. (2017). Paraconsistent double negations as classical and intuitionistic negations, *Studia Logica* **105**(6): 1167–1191.
[28] Marcos, J. (2005). On negation: Pure local rules, *Journal of Applied Logic* **3**: 185–219.
[29] Meyer, R. K. and Slaney, J. K. (1989). Abelian logic (from A to Z), *in* R. Routley, G. Priest and J. Norman (eds.), *Paraconsistent Logic: Essays on the Inconsistent*, Philosophia Verlag, Munich, pp. 245–288.
[30] Omori, H. and Sano, K. (2015). Generalizing functional completeness in Belnap–Dunn logic, *Studia Logica* **103**(5): 883–917.
[31] Omori, H. and Wansing, H. (2017). 40 years of FDE: An introductory overview, *Studia Logica* **105**: 1021–1049.
[32] Omori, H. and Wansing, H. (2018). On contra-classical variants of Nelson logic N4 and its classical extension, *Review of Symbolic Logic* **11**(4): 805–820.
[33] Omori, H. and Wansing, H. (eds.) (2019). *New Essays on Belnap–Dunn Logic*, Springer.
[34] Paoli, F. (2001). Logic and groups, *Logic and Logical Philosophy* **9**: 109–128.
[35] Paoli, F. (2019). Bilattice logics and demi-negation, *in* H. Omori and H. Wansing (eds.), *New Essays on Belnap–Dunn Logic*, Synthese Library, Springer, pp. 233–253.
[36] Petrukhin, Y. and Shangin, V. (2020). Correspondence analysis and automated proof-searching for first degree entailment, *European Journal of Mathematics* **6**(4): 1452–1495.
[37] Prawitz, D. (2006). *Natural deduction: A proof-theoretical study*, Dover Publications.
[38] Priest, G. (2002). Paraconsistent logic, *Handbook of philosophical logic*, Springer, pp. 287–393.
[39] Routley, R. and Routley, V. (1972). The semantics of first-degree entailment, *Noûs* **6**(4): 335–390.
[40] Sette, A. (1973). On the propositional calculus P^1, *Mathematica Japonicae* **16**: 173–180.
[41] Sette, A. M. and Carnielli, W. A. (1995). Maximal weakly-intuitionistic logics, *Studia Logica* **55**(1): 181–203.
[42] Słupecki, J. (1972). A criterion of fullness of many-valued systems of propositional logic, *Studia Logica* **30**: 153–157.
[43] Szmuc, D. and Omori, H. (forthcoming). Liberating classical negation from falsity conditions, *2022 IEEE 52th International Symposium on Multiple-Valued Logic*, Vol. 52, IEEE.
[44] Tennant, N. (1978). *Natural Logic*, Edinburgh University Press, Edinburgh.
[45] Tokarz, M. (1973). Connections between some notions of completeness of structural propositional calculi, *Studia Logica* **32**(1): 77–89.
[46] Wansing, H. (2001). Negation, *in* L. Goble (ed.), *The Blackwell Guide to Philosophical Logic*, Basil Blackwell Publishers, Cambridge, MA, pp. 415–436.

[47] Wansing, H. (2005). Connexive modal logic, *in* R. Schmidt, I. Pratt-Hartmann, M. Reynolds and H. Wansing (eds.), *Advances in Modal Logic*, Vol. 5, King's College Publications, pp. 367–383.
[48] Wansing, H. (2016). Falsification, natural deduction and bi-intuitionistic logic, *Journal of Logic and Computation* **26**(1): 425–450.
[49] Wansing, H. (2021). Connexive logic, *in* E. N. Zalta (ed.), *The Stanford Encyclopedia of Philosophy*, Spring 2021 edn, URL: plato.stanford.edu/archives/spr2021/entries/logic-connexive/.
[50] Wansing, H. (ed.) (1996). *Negation: A Notion in Focus*, W. de Gruyter, Berlin.
[51] Wansing, H., Olkhovikov, G. and Omori, H. (2021). Questions to Michael Dunn, *Logical Investigations* **27**: 9–19.

RUHR UNIVERSITY, BOCHUM, *Email:* heinrich.wansing@rub.de
RUHR UNIVERSITY, BOCHUM, *Email:* hitoshi.omori@rub.de

A Class of 4-valued Implicative Expansions of First-degree Entailment Logic (FDE) with the Variable-sharing Property

Gemma Robles and José M. Méndez

ABSTRACT. Anderson and Belnap consider the variable-sharing property (VSP) a necessary property a relevance logic has to fulfill. A logic L has the VSP if in all L-theorems of implication form antecedent and consequent share at least a propositional variable. If a propositional logic has the VSP then it is free from "paradoxes of relevance." The aim of this paper is to define a class of implicative expansions with the VSP of the well-known *first-degree entailment logic*, FDE. The properties the elements in this class enjoy make them important logics, not mere artificial constructs.

Keywords. First degree entailment logic, 4-valued relevance logics, Relevance logics, Two-valued Belnap–Dunn semantics, Variable-sharing property

1. Introduction

Anderson and Belnap consider the variable-sharing property (VSP) a necessary property a relevance logic has to fulfill. A logic L has the VSP if in all L-theorems of implication form antecedent and consequent share at least a propositional variable (cf. Anderson and Belnap [1] and Anderson, Belnap and Dunn [2]). Given that in a propositional logic the non-logical semantical content is conveyed by propositional variables, if L is a propositional logic with the VSP, then it is free from "paradoxes of relevance" in the sense that L does not contain implicational formulas where the semantical content of antecedent and consequent is disjoint.

As regards standard relevance logics, *first degree entailment logic* (FDE) is the minimal member in Anderson and Belnap's De Morgan family of relevance logics (cf. [1; 2]). FDE is also known as Belnap and Dunn's 4-valued logic BD4 (cf. Omori and Wansing [22]). BD4 (our label) can be viewed as a 4-valued logic in which formulas can be both true and false, neither true nor false, in addition to being true or false (cf. Belnap [7; 6], Dunn [12; 13; 14]).

The question of expanding FDE with a full implicative connective emerges, since as the name of the logic suggests, formulas of the form $A \to B$ are not considered in FDE if either A or B contains $\to$ (cf. [1, p. 158]; cf. Definition 1 on the logical language used in the paper). And, according to [22], there is still a lot of investigation to be done in the topic (cf. also Omori and Wansing [23]).

2020 *Mathematics Subject Classification.* Primary: 03B47, Secondary: 03B50, 03B53.

Bimbó, Katalin, (ed.), *Relevance Logics and other Tools for Reasoning. Essays in Honor of J. Michael Dunn*, (Tributes, vol. 46), College Publications, London, UK, 2022, pp. 338–364.

Some full implicative expansions of FDE have been given in the literature (cf. Brady [8], Hazen and Pelletier [17], López [18], Robles and Méndez [19; 20; 28], [22], Petrukhin and Shangin [24] and references in the last two items). Among these, Brady's 4-valued logic BN4 (cf. [8]) seems to be widely regarded as the adequate implicative 4-valued logic by specialists on relevance logic. In this respect, Meyer et al. note: "BN4 is the correct logic for the 4-valued situation where extra values are to be interpreted in the *both* and *neither* senses" (cf. [21, p. 25]). On his part, Slaney thinks that BN4 has the truth-functional implication most naturally associated with FDE (cf. [32, p. 289]; but cf. also the recent paper [10] by Brady). Nevertheless, BN4 lacks the VSP as it is the case with the 4-valued logic of entailment E4 introduced in [28] and in fact, with all implicative expansions of FDE proposed so far in the literature, to the best of our knowledge; for example, the formula (wff) $\sim(A \to A) \to (B \to B)$, is provable in BN4 and E4.

The aim of this paper is to define a class of interesting implicative expansions of FDE with the VSP, that is, a class of interesting implicative expansions of FDE free from paradoxes of relevance as this notion has been understood and explained above. This class is dubbed MI4^{VSP} (i.e., "implicative expansions of $\mathcal{FOUR}$ with the VSP" — cf. §2).

As it is shown in the concluding remarks to the paper, it is very easy to design a broad class of binary expansions of FDE with the sole purpose that the elements in this class comply with the VSP regardless of other properties required for a formal translation of an acceptable notion of a conditional or an implication. However, the elements in MI4^{VSP} share a number of properties that we think do not support their consideration as mere artificial constructs.

Without trying to be exhaustive, all logics in MI4^{VSP} have the ensuing properties:

(1) They have "natural conditionals" in the sense of Robles and Méndez [29] akin to that of Tomova [33] where the notion was originally defined (cf. Definition 7 in §2).

(2) They fulfill all conditions required of implicative logics in the classical Polish logical tradition, except, of course, that of complying with the rule VEQ, $A \Rightarrow B \to A$ (cf. Definition 25 in §5).

(3) They all are 4-valued extensions of a strong restriction of Brady's important weak relevance logic DJ^d (cf. Remark 17 in §4) instead of being 4-valued extensions of strong relevance logics, as it happens with BN4 and E4, the former extending contractionless relevance logic R, while the latter extends reductioless logic of entailment E (cf. [8; 28]).

(4) They have considerable expressive power (cf. the concluding remarks to the paper). For instance, Gödel-type and dual Gödel-type negations are definable. Also, necessity and possibility operators similar to these definable in BN4 and E4 (cf. [20; 28]).

(5) They preserve, of course, the paraconsistency and paracompleteness of FDE.

(6) Some expansions in MI4^{VSP} have the "Ackermann property" or the "Converse Ackermann property" in addition to the VSP (cf. §7).

And last but not least:

(7) They all are interpretable with the clear and important two-valued Belnap–Dunn semantics (BD-semantics), which is grounded in the truth values T and F (truth and falsity, respectively).

Before explaining the structure of the paper, let us point out an observation. It is known that there are infinitely many logics with the VSP (cf. Dziobiak [15]). Furthermore, some many-valued logics with the VSP have been studied in the literature. For example, the logic characterized by Belnap's eight-element matrix M_0 (cf. Belnap [5], axiomatized in Brady [11]); or the logic determined by Meyer's six-element crystal lattice CL, also axiomatized in [11]). But it does not seem possible to interpret the meaning of the logical values in these matrices in an intuitively clear way. However, the meaning of the four truth-values in FDE (or BD4) and its expansions is crystalline.

The paper is organized as follows. In §2, the class of implicative expansions of $\mathcal{FOUR}$ with the VSP, MI4^{VSP}, is defined. MI4^{VSP} is attained by restricting a broad class of implicative expansions of $\mathcal{FOUR}$ with some desirable basic properties such as being C-extending matrices satisfying the self-identity axiom, as well as the rules *modus ponens* and *transitivity*. In §3, a two-valued Belnap–Dunn semantics equivalent to the matrix semantics based upon each one of the 24 matrices in MI4^{VSP} is defined. By L_i we refer to the logic determined by M_i ($1 \leq i \leq 24$). Then we sketch the soundness and completeness proofs for the L_i-logics w.r.t. both the matrix semantics based on the M_i-matrices and the BD-semantics equivalent to them. We follow the method in [8], as applied in, e.g., [19; 20] and [28]. We will only sketch soundness and completeness proofs here; the details can be filled in as in the aforementioned papers. In §4, the basic logics b_0, b_1 and b_2 are defined. In §5, the L_i-logics are built upon the basic logics and some of their proof-theoretical properties are proved. In §6, completeness of the L_i-logics is proved by using a canonical model construction. In §7, it is proved that each L_i-logic has the VSP. Also, that some of them have the "Ackermann property" or the "Converse Ackermann property." In §8, we note some concluding remarks on the results obtained as well as some suggestions on future work that could be done in the topic. The paper is ended with an appendix displaying a part of the proof of the generation of MI4^{VSP} in §2.

As pointed out above, the logics determined by the M_i-matrices in MI4^{VSP} are given a Hilbert-style formulation using a two-valued BD-semantics following the method developed in [8] as applied in our papers quoted above. Of course, we could have used the methods in Avron et al. [3; 4] (resp., those in [24]) in order to define a Gentzen-type system, (resp. a natural deduction system) equivalent to each one of the L_i-logics formulated as Hilbert-style systems. But let us stress that the aim of this paper is *not* to axiomatize the M_i-matrices, but to highlight them and the properties the L_i-logics they determine enjoy when defined from the point of view of a Hilbert-style calculus. No doubt, other properties of the L_i-logics can be emphasized when defined as Gentzen-type systems or natural deduction ones (concerning the relative merits of these three methods just mentioned, cf. Robles [26, §6] and [24, §8]).

A last remark. As indicated above, there are 24 L_i-logics determined by each one of the M_i-matrices in MI4^{VSP}. We think that, contrarily to what the market rules state, this abundance does not devaluate the product: maybe there is a really outstanding logic between the 24 candidates, or else, they are in some sense the same logic formulated with different choices of the set of primitive connectives (cf. the concluding remarks to the paper).

2. A CLASS OF IMPLICATIVE EXPANSIONS OF $\mathcal{FOUR}$ WITH THE VSP

In this section we define a class of implicative expansions of Belnap and Dunn's matrix $\mathcal{FOUR}$ characterizing Anderson and Belnap's *first-degree entailment logic*, FDE (cf. [1, §15.2], [2; 7; 6; 12; 13; 14]). This class is dubbed MI4^{VSP} (i.e., "implicative expansions of $\mathcal{FOUR}$ with the VSP"). Firstly, some preliminary notions are noted. Then, the matrix $\mathcal{FOUR}$ is recalled.

Definition 1 (Preliminary notions). The propositional language consists of a denumerable set of propositional variables $p_0, p_1, \ldots, p_n, \ldots$, and the following connectives: $\to$ (conditional), $\wedge$ (conjunction), $\vee$ (disjunction) and $\sim$ (negation). The biconditional and the set of wffs is defined in the customary way. A, B, C, etc. are metalinguistic variables. Then logics are formulated as Hilbert-type axiomatic systems, the notions of "theorem" and "proof from a set of premises" being the usual ones, while the following notions are understood in a fairly standard sense (cf., e.g., [19; 20; 28]): extension and expansion of a given logic, logical matrix M and M-interpretation, M-consequence, M-validity and, finally, M-determined logic.

Definition 2 (Belnap and Dunn's matrix $\mathcal{FOUR}$). The propositional language consists of the connectives $\wedge, \vee$ and $\sim$. Belnap and Dunn's matrix $\mathcal{FOUR}$ is the structure $(\mathcal{V}, D, F)$ where (1) $\mathcal{V}$ is $\{0, 1, 2, 3\}$ and is partially ordered as shown in the following lattice (it is also displayed with the subsets of $\{T, F\}$):

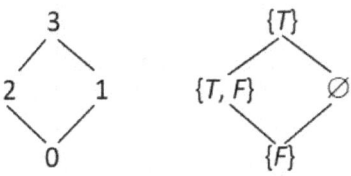

(2) $D = \{2, 3\}$; $F = \{f_\wedge, f_\vee, f_\sim\}$ where $f_\wedge$ and $f_\vee$ are defined as the glb (or lattice meet) and the lub (or lattice joint), respectively. Finally, $f_\sim$ is an involution with $f_\sim(0) = 3, f_\sim(3) = 0, f_\sim(1) = 1, f_\sim(2) = 2$ (cf. [7; 6; 12; 13; 14]). We display the tables for $\wedge, \vee$ and $\sim$:

$\wedge$	0	1	2	3
0	0	0	0	0
1	0	1	0	1
*2	0	0	2	2
*3	0	1	2	3

$\vee$	0	1	2	3
0	0	1	2	3
1	1	1	3	3
2	2	3	2	3
3	3	3	3	3

	$\sim$
0	3
1	1
2	2
3	0

Remark 3 (On the symbols for referring to the four truth-values). It is customary to use f, n, b and t instead of $0, 1, 2$ and 3, respectively (cf., e.g., [22]). The former stand for false only, neither true or false, both true and false and true only, respectively. The

latter have been chosen in order to use the tester in González [16], in case one is needed and to put in connection the results in the present paper with previous work by us.

Next, we proceed to define the class MI4^{VSP}. The $f_\rightarrow$-functions expanding $\mathcal{FOUR}$ we are interested in need to have at least the ensuing properties: (a) they are C-extending $f_\rightarrow$-functions; (b) they satisfy *modus ponens*; (c) they verify the self-identity axiom $A \rightarrow A$; (d) they are such that $f_\rightarrow(2,2) = 2$ and $f_\rightarrow(2,3) \in \{0,1\}$ (together with (b), this condition guarantees the fulfillment of the VSP); (e) they satisfy the rule contraposition (Con), i.e., $A \rightarrow B \Rightarrow \sim B \rightarrow \sim A$. (An $f_\rightarrow$-function is C-extending if it coincides with (the $f_\rightarrow$-function for) the classical conditional when restricted to the "classical" values 0 and 3.) But in order to define a member in MI4^{VSP} an $f_\rightarrow$-function has to fulfill the following conditions, in addition to properties (a) through (e): (f) it verifies the contraposition axiom (i.e., $(A \rightarrow B) \rightarrow (\sim B \rightarrow \sim A)$); (g) it satisfies the rules prefixing (Pref) and suffixing (Suf) (Pref is $B \rightarrow C \Rightarrow (A \rightarrow B) \rightarrow (A \rightarrow C)$; Suf is $A \rightarrow B \Rightarrow (B \rightarrow C) \rightarrow (A \rightarrow C)$).

It will be shown that the requirements just demanded are such that the logics determined by the members in MI4^{VSP} are not mere artificial constructs with the VSP. But let us advance to describe the implicative expansions of $\mathcal{FOUR}$ in MI4^{VSP}. The class MI4^{VSP} is defined by following 4 steps the 3 last of which consist in successively restricting a broad class of implicative expansions of $\mathcal{FOUR}$ with the VSP built up in step 1. The 4 steps are the following.

(1) Implicative expansions of $\mathcal{FOUR}$ with properties (a)–(d).
(2) Implicative expansions of $\mathcal{FOUR}$ with properties (a)–(d) which satisfy the rule Con (i.e., implicative expansions of $\mathcal{FOUR}$ with properties (a)–(e)).
(3) Implicative expansions of $\mathcal{FOUR}$ with properties (a)–(d) which verify the contraposition axiom (i.e., implicative expansions of $\mathcal{FOUR}$ with properties (a)–(f)).
(4) Implicative expansions of $\mathcal{FOUR}$ with properties (a)–(d) which verify the contraposition axiom and satisfy the rules Pref and Suf (i.e., implicative expansions of $\mathcal{FOUR}$ with properties (a)–(g)).

(1) $f_\rightarrow$-functions fulfilling properties (a)–(d). An implicative truth-table describing an $f_\rightarrow$-function fulfilling (a)–(d) has to present the structure displayed in the general table TI recorded below (blank spaces can be filled with no matter which truth-values in $\mathcal{FOUR}$)

TI

$\rightarrow$	0	1	2	3
0	3			3
1		a		
2	b_1	b_2	2	b_3
3	0	b_4		3

where $a \in \{2,3\}$ and b_i $(1 \leq i \leq 4) \in \{0,1\}$.

Notice that implicative expansions of $\mathcal{FOUR}$ the $f_\rightarrow$-function of which is one of the 2^{17} $f_\rightarrow$-functions described in TI determine logics with the VSP. For let M be any such expansion and suppose that A and B do not share propositional variables in $A \rightarrow B$. Define then an M-interpretation I such that $I(p) = 2$ (resp., $I(p) = 0$) for all propositional variables in A (resp., B). It follows that $I(A) = 2$ and $I(B) \in \{0,3\}$ as

{2} and {0,3} are closed under $\to$, $\wedge$, $\vee$ and $\sim$. Consequently, $I(A \to B) \in \{0,1\}$, that is, $A \to B$ is not M-valid (cf. Proposition 41 below).

(2) $f_\to$-functions fulfilling properties (a)–(e). We restrict the general table TI in order to obtain $f_\to$-functions satisfying the rule Con. The fact to be noted is that Con is not satisfied if any of $f_\to(0,2), f_\to(1,0), f_\to(1,2)$ and $f_\to(3,2)$ equals 2 or 3. Then we are left with the $f_\to$-functions described by the general truth tables TII and TIII recorded below.

TII

$\to$	0	1	2	3
0	3	a_1	b_1	3
1	b_2	c	b_3	a_2
2	b_4	b_5	2	b_6
3	0	b_7	b_8	3

TIII

$\to$	0	1	2	3
0	3	d_1	b_1	3
1	b_2	c	b_3	d_2
2	b_4	b_5	2	b_6
3	0	b_7	b_8	3

where $a_i (1 \leq i \leq 2) \in \{2,3\}$, $b_i (1 \leq i \leq 8) \in \{0,1\}$, $c \in \{2,3\}$ and $d_1, d_2 \in \{0,1\}$.

We have:

Proposition 4 (TII and TIII satisfy Con). *Let M be an implicative expansion of $\mathcal{FOUR}$ built up by adding any of the 2^{11} (resp., 2^{11}) $f_\to$-functions in the general table TII (resp., TIII). Then M satisfies Con.*

Proof. (1) TII: Let M be built upon TII, and I be an M-interpretation. The cases of interest are (a) $I(A) = 0 \& I(B) = 1$; (b) $I(A) = 0 \& I(B) = 3$; (c) $I(A) = 1 \& I(B) = 3$. But it is clear that in each one of these cases I assigns a designated value to $\sim B \to \sim A$. (2) TIII: the proof is similar to that of case (1). ◁

(3) $f_\to$-functions fulfilling properties (a)–(f). We have the following fact.

Proposition 5 (Table TIV verifying the contraposition axiom). *The general table TIV recorded below contains the set of all truth-tables in TII and TIII verifying the contraposition axiom:*

TIV

$\to$	0	1	2	3
0	3	a	b	3
1	c	f	d	a
2	e	d	2	b
3	0	c	e	3

where $a \in \{0,1,2,3\}$, $b,c,d,e \in \{0,1\}$ and $f \in \{2,3\}$.

Proof. Given the diagonal in tables TII and TIII, the verification of the contraposition axiom requires, in addition, the following conditions: (a) $f_\to(0,1) = f_\to(1,3)$; (b) $f_\to(0,2) = f_\to(2,3)$; (c) $f_\to(1,0) = f_\to(3,1)$; (d) $f_\to(1,2) = f_\to(2,1)$ and (e) $f_\to(2,0) = f_\to(3,2)$. The accomplishment of these conditions gives us the general table TIV containing 128 functions verifying the contraposition axiom. ◁

(4) $f_\to$-functions fulfilling properties (α)–(η). The last step is to select tables in TIV satisfying both Pref and Suf. In this regard, we note that if one of the conditions (α)–(η) (resp., θ) below obtains, then the rule Suf (resp., Pref) is not satisfied as it is summarily shown in Diagram 1: (α) $f_\to(0,1) = 2$ (equivalently, $f_\to(1,3) = 2$); (β) $f_\to(1,1) = 2$ if $f_\to(0,1) = 3$; (γ) $f_\to(1,2) = 1$ and $f_\to(0,2) = 0$ if $f_\to(0,1) = 3$; (δ) $f_\to(3,1) = 1$ and $f_\to(0,1) = 0$; (ε) $f_\to(3,1) = 0$ and $f_\to(0,1) = 1$; (ζ) $f_\to(3,2) = 0$

and $f_\to(0,2) = 1$ if $f_\to(0,1) \neq 3$; (η) $f_\to(3,2) = 1$ and $f_\to(0,2) = 0$; (θ) $f_\to(2,0) = 1$ and $f_\to(2,1) = 0$ if $f_\to(0,1) = 3$.

	p	$\to$	q	$\Rightarrow$	$(q$	$\to$	$r)$	$\to$	$(p$	$\to$	$r)$
α	0	2	1		1	2	3	b	0	3	3
β	0	3	1		1	2	1	b	0	3	1
γ	0	3	1		1	1	2	c	0	0	2
δ	0	3	3		3	1	1	c	0	0	1
ε	0	3	3		3	0	1	1	0	1	1
ζ	0	3	3		3	0	2	b	0	1	2
η	0	3	3		3	1	2	c	0	0	2
	q	$\to$	r	$\Rightarrow$	$(p$	$\to$	$q)$	$\to$	$(p$	$\to$	$r)$
θ	0	3	1		2	1	0	c	2	0	1

Diagram 1.

(p, q and r are distinct propositional variables.)

Thus, there are 24 $f_\to$-functions in TIV satisfying the rules Pref and Suf, in addition to the contraposition axiom. These 24 $f_\to$-functions can be described as shown in the general tables recorded below (b, c, d, e and f are read as in table TIV).

$\to$	0	1	2	3
0	3	3	b	3
1	c	3	0	3
2	0	0	2	b
3	0	c	0	3

$\to$	0	1	2	3
0	3	3	1	3
1	c	3	1	3
2	e	1	2	1
3	0	c	e	3

$\to$	0	1	2	3
0	3	0	0	3
1	0	f	d	0
2	0	d	2	0
3	0	0	0	3

$\to$	0	1	2	3
0	3	0	1	3
1	0	f	d	0
2	1	d	2	1
3	0	0	1	3

$\to$	0	1	2	3
0	3	1	0	3
1	1	f	d	1
2	0	d	2	0
3	0	1	0	3

$\to$	0	1	2	3
0	3	1	1	3
1	1	f	d	1
2	1	d	2	1
3	0	1	1	3

In particular, the 24 $f_\to$-functions are described by the following truth-tables t1–t24. With regard to the remaining tables in TIV not satisfying either Pref, Suf or both rules, they are generally displayed in the Appendix.

Definition 6 (The class MI4^{VSP}). The class MI4^{VSP} consists of 24 implicative expansions of $\mathcal{FOUR}$, $M_1, M_2, \ldots, M_{24}$. Each M_i ($1 \leq i \leq 24$) is the structure $(\mathcal{V}, D, F)$ where $\mathcal{V}, D, f_\wedge, f_\vee$ and $f_\sim$ are defined exactly as in $\mathcal{FOUR}$ (Definition 2) and $f_\to$ is defined according to table t_i. Tables $t_1, t_2, \ldots, t_{24}$ are displayed below.

t1	$\to$	0	1	2	3
	0	3	3	0	3
	1	0	3	0	3
	2	0	0	2	0
	3	0	0	0	3

t2	$\to$	0	1	2	3
	0	3	3	0	3
	1	1	3	0	3
	2	0	0	2	0
	3	0	1	0	3

t3	$\to$	0	1	2	3
	0	3	3	1	3
	1	0	3	0	3
	2	0	0	2	1
	3	0	0	0	3

t4:

→	0	1	2	3
0	3	3	1	3
1	1	3	0	3
2	0	0	2	1
3	0	1	0	3

t5:

→	0	1	2	3
0	3	3	1	3
1	0	3	1	3
2	1	1	2	1
3	0	0	1	3

t6:

→	0	1	2	3
0	3	3	1	3
1	1	3	1	3
2	1	1	2	1
3	0	1	1	3

t7:

→	0	1	2	3
0	3	3	1	3
1	0	3	1	3
2	0	1	2	1
3	0	0	0	3

t8:

→	0	1	2	3
0	3	3	1	3
1	1	3	1	3
2	0	1	2	1
3	0	1	0	3

t9:

→	0	1	2	3
0	3	0	0	3
1	0	2	0	0
2	0	0	2	0
3	0	0	0	3

t10:

→	0	1	2	3
0	3	0	0	3
1	0	3	0	0
2	0	0	2	0
3	0	0	0	3

t11:

→	0	1	2	3
0	3	0	0	3
1	0	2	1	0
2	0	1	2	0
3	0	0	0	3

t12:

→	0	1	2	3
0	3	0	0	3
1	0	3	1	0
2	0	1	2	0
3	0	0	0	3

t13:

→	0	1	2	3
0	3	0	1	3
1	0	2	0	0
2	1	0	2	1
3	0	0	1	3

t14:

→	0	1	2	3
0	3	0	1	3
1	0	3	0	0
2	1	0	2	1
3	0	0	1	3

t15:

→	0	1	2	3
0	3	0	1	3
1	0	2	1	0
2	1	1	2	1
3	0	0	1	3

t16:

→	0	1	2	3
0	3	0	1	3
1	0	3	1	0
2	1	1	2	1
3	0	0	1	3

t17:

→	0	1	2	3
0	3	1	0	3
1	1	2	0	1
2	0	0	2	0
3	0	1	0	3

t18:

→	0	1	2	3
0	3	1	0	3
1	1	3	0	1
2	0	0	2	0
3	0	1	0	3

t19:

→	0	1	2	3
0	3	1	0	3
1	1	2	1	1
2	0	1	2	0
3	0	1	0	3

t20:

→	0	1	2	3
0	3	1	0	3
1	1	3	1	1
2	0	1	2	0
3	0	1	0	3

t21:

→	0	1	2	3
0	3	1	1	3
1	1	2	0	1
2	1	0	2	1
3	0	1	1	3

t22:

→	0	1	2	3
0	3	1	1	3
1	1	3	0	1
2	1	0	2	1
3	0	1	1	3

t23:

→	0	1	2	3
0	3	1	1	3
1	1	2	1	1
2	1	1	2	1
3	0	1	1	3

t24:

→	0	1	2	3
0	3	1	1	3
1	1	3	1	1
2	1	1	2	1
3	0	1	1	3

We remark that the conditional defined by M_i ($1 \leq i \leq 24$) is a natural conditional in accordance with the following definition (cf. Definition 2.5 in [29]).

Definition 7 (Natural conditionals)**.** Let $\mathcal{V}$ and D be defined as in Definition 2. Then, an $f_\rightarrow$-function on $\mathcal{V}$ defines a natural conditional if the following conditions are satisfied:

(1) $f_\rightarrow$ coincides with (the $f_\rightarrow$-function for) the classical conditional when restricted to the subset $\{0,3\}$ of $\mathcal{V}$.

(2) $f_\to$ satisfies modus ponens, that is, for any $a,b \in \mathcal{V}$, if $a \to b \in D$ and $a \in D$, then $b \in D$.
(3) For any $a,b \in \mathcal{V}$, $a \to b \in D$ if $a = b$.

Remark 8 (Natural conditionals in Tomova's original paper). We note that natural conditionals are defined in [33] exactly as in Definition 7 except for condition (3), which reads there as follows: For any $a,b \in \mathcal{V}$, $a \to b \in D$ if $a \leq b$.

The section is ended by noting that each one of the implicative expansions in $MI4^{VSP}$ has considerable expressive power. As a way of an example, we show some unary functions definable in each of them. (For simplicity, we use the wffs defined by the functions in question instead of the functions themselves.)

Gödel-type negation ($\overset{\bullet}{\neg}$) and dual Gödel-type negation ($\overset{\circ}{\neg}$) can be defined as follows for any wff A: $\overset{\bullet}{\neg}A = (A \to \sim A) \wedge \sim A$ and $\overset{\circ}{\neg}A = \sim \overset{\bullet}{\neg} \sim A$ in expansions where $f_\to(1,1) = 2$; $\overset{\bullet}{\neg}A = \sim(\sim A \to A)$ and $\overset{\circ}{\neg}A = A \to \sim A$ in expansions where $f_\to(1,1) = 3$. Also, necessity ($\Box$), possibility ($\Diamond$), truth guaranteeing connective ($\overset{\bullet}{t}$) and falsity ensuring connective ($\overset{\bullet}{f}$) are defined in all expansions in $MI4^{VSP}$ similarly as in [20] and [28]. For any wff A, $\Box A = \sim \overset{\bullet}{\neg} A$, $\Diamond A = \sim \overset{\circ}{\neg} A$, $\overset{\bullet}{t}A = A \vee \overset{\bullet}{\neg}A$ and $\overset{\bullet}{f}A = A \wedge \overset{\circ}{\neg}A$. (Recall that the truth-value 2 represents both truth and falsity.)

3. Belnap–Dunn Semantics for the L_i-logics

As it is well-known, Belnap–Dunn two-valued semantics (BD-semantics) is characterized by the possibility of assigning T, F, both T and F or neither T nor F to the formulas of a given language (cf. [7; 6; 12; 13; 14]; T represents truth and F represents falsity).

Given M an implicative expansion of $\mathcal{FOUR}$ (cf. Definition 2), the idea for defining a BD-semantics, M', equivalent to the matrix semantics based upon M is simple: a wff A is assigned both T and F in M' iff it is assigned 2 in M; A is assigned neither T nor F in M' iff it is assigned 1 in M; finally, A is assigned T but not F (resp., F but not T) in M' iff it is assigned 3 (resp., 0) in M.

The BD-semantics for each one of the L_i-logics, equivalent to the matrix semantics based upon M_i ($1 \leq i \leq 24$) (cf. Definition 6) to be defined below has been built by following the simple intuitive ideas just exposed.

In the sequel, the notion of an L_i-model and the accompanying notions of L_i-consequence and L_i-validity are defined for each i ($1 \leq i \leq 24$). L_i-models and annexed notions constitute a BD-semantics for each one of the L_i-logics (an L_i-semantics) equivalent to the one based upon the matrix M_i ($1 \leq i \leq 24$) in the sense explained above. It will be proved that for each i ($1 \leq i \leq 24$), the logic L_i is sound and complete w.r.t. L_i-semantics.

We will define two types of L_i-models: Eb_1-models, for extensions of b_1 and Eb_2-models, for extensions of b_2. (The logics b_1 and b_2 are defined in §4.)

Definition 9 (Eb_1-models). An L_i-model ($1 \leq i \leq 8$) is a structure (K,I) where (i) $K = \{\{T\},\{F\},\{T,F\},\emptyset\}$, and (ii) I is an L_i-interpretation from the set of all wffs to K, this notion being defined according to the following conditions ("clauses") for each propositional variable p and wffs A,B:

1. $I(p) \in K$
2a. $T \in I(\neg A)$ iff $F \in I(A)$
2b. $F \in I(\neg A)$ iff $T \in I(A)$
3a. $T \in I(A \wedge B)$ iff $T \in I(A) \,\&\, T \in I(B)$
3b. $F \in I(A \wedge B)$ iff $F \in I(A)$ or $F \in I(B)$
4a. $T \in I(A \vee B)$ iff $T \in I(A)$ or $T \in I(B)$
4b. $F \in I(A \vee B)$ iff $F \in I(A) \,\&\, F \in I(B)$
5a. $T \in I(A \to B)$ iff $[T \notin I(A) \,\&\, F \notin I(B)]$ or
$[T \notin I(A) \,\&\, F \in I(A) \,\&\, T \notin I(B)]$ or
$[F \notin I(A) \,\&\, T \in I(B) \,\&\, F \notin I(B)]$ or
$[T \in I(A) \,\&\, F \in I(A) \,\&\, T \in I(B) \,\&\, F \in I(B)]$

Clause 5b for assigning F to conditionals is different for each Eb_1-model. Thus, we have the following 8 conditions.

$F \in I(A \to B)$ iff:

(5b1) $[T \in I(A) \,\&\, T \notin I(B)]$ or $[F \notin I(A) \,\&\, F \in I(B)]$ or $[T \in I(A) \,\&\, F \in I(A)]$ or $[T \in I(B) \,\&\, F \in I(B)]$.
(5b2) $[T \in I(A) \,\&\, F \in I(B)]$ or $[T \in I(B) \,\&\, F \in I(B)]$ or $[T \in I(A) \,\&\, F \in I(A)]$.
(5b3) $[F \notin I(A) \,\&\, F \in I(B)]$ or $[T \in I(A) \,\&\, T \notin I(B)]$ or $[T \in I(A) \,\&\, F \in I(B)]$.
(5b4) $[T \in I(A) \,\&\, F \in I(B)]$ or $[F \notin I(A) \,\&\, T \in I(B) \,\&\, F \in I(B)]$ or $[T \in I(A) \,\&\, F \in I(A) \,\&\, T \notin I(B)]$.
(5b5) $[F \notin I(A) \,\&\, T \notin I(B) \,\&\, F \in I(B)]$ or $[T \in I(A) \,\&\, F \notin I(A) \,\&\, T \notin I(B)]$ or $[T \in I(A) \,\&\, F \in I(A) \,\&\, T \in I(B) \,\&\, F \in I(B)]$.
(5b6) $[T \in I(A) \,\&\, F \in I(A) \,\&\, T \in I(B) \,\&\, F \in I(B)]$ or $[T \in I(A) \,\&\, F \notin I(A) \,\&\, T \notin I(B) \,\&\, F \in I(B)]$.
(5b7) $[T \in I(A) \,\&\, F \in I(B)]$ or $[F \notin I(A) \,\&\, T \notin I(B) \,\&\, F \in I(B)]$ or $[T \in I(A) \,\&\, F \notin I(A) \,\&\, T \notin I(B)]$.
(5b8) $T \in I(A) \,\&\, F \in I(B)$.

Definition 10 (Eb_2-models). An L_i-model ($9 \leq i \leq 24$), is a structure (K, I) where K and I are defined similarly as in Eb_1-models, save for clauses (5a) and (5b), which are now as follows:

(5a) $T \in I(A \to B)$ iff $[T \notin I(A) \,\&\, F \in I(A) \,\&\, T \notin I(B) \,\&\, F \in I(B)]$ or $[T \notin I(A) \,\&\, F \notin I(A) \,\&\, T \notin I(B) \,\&\, F \notin I(B)]$ or $[T \in I(A) \,\&\, F \in I(A) \,\&\, T \in I(B) \,\&\, F \in I(B)]$ or $[T \in I(A) \,\&\, F \notin I(A) \,\&\, T \in I(B) \,\&\, F \notin I(B)]$ or $[T \notin I(A) \,\&\, F \in I(A) \,\&\, T \in I(B) \,\&\, F \notin I(B)]$.

Regarding clause 5b for assigning F to conditionals, as in the case of Eb_1-models, it is different for each Eb_2-model. So, we have the following 16 conditions.

$F \in I(A \to B)$ iff:

(5b9) $[T \in I(A) \,\&\, F \in I(B)]$ or $[T \notin I(A) \,\&\, F \notin I(A)]$ or $[T \in I(A) \,\&\, F \in I(A)]$ or $[T \notin I(B) \,\&\, F \notin I(B)]$ or $[T \in I(B) \,\&\, F \in I(B)]$.
(5b10) $[F \notin I(A) \,\&\, F \in I(B)]$ or $[T \in I(A) \,\&\, T \notin I(B)]$ or $[T \in I(A) \,\&\, F \in I(A)]$ or $[T \in I(B) \,\&\, F \in I(B)]$ or $[F \in I(A) \,\&\, T \notin I(B) \,\&\, F \notin I(B)]$ or $[T \notin I(A) \,\&\, F \notin I(A) \,\&\, T \in I(B) \,\&\, F \notin I(B)]$.

(5b11) $[T \in I(A) \& F \in I(B)]$ or $[F \notin I(A) \& T \notin I(B)]$ or $[T \notin I(A) \& T \notin I(B) \& F \notin I(B)]$ or $[F \in I(A) \& T \in I(B) \& F \in I(B)]$ or $[T \notin I(A) \& F \notin I(A) \& F \notin I(B)]$ or $[T \in I(A) \& F \in I(A) \& T \in I(B) \& F \notin I(B)]$.

(5b12) $[T \in I(A) \& F \in I(B)]$ or $[F \in I(A) \& T \in I(B) \& F \in I(B)]$ or $[F \notin I(A) \& T \notin I(B) \& F \in I(B)]$ or $[T \in I(A) \& F \in I(A) \& T \in I(B)]$ or $[T \in I(A) \& F \notin I(A) \& T \notin I(B)]$ or $[T \notin I(A) \& F \notin I(A) \& T \in I(B) \& F \notin I(B)]$ or $[T \notin I(A) \& F \in I(A) \& T \notin I(B) \& F \notin I(B)]$.

(5b13) $[T \notin I(A) \& F \notin I(A)]$ or $[T \notin I(B) \& F \notin I(B)]$ or $[F \notin I(A) \& T \notin I(B)]$ or $[T \in I(A) \& F \in I(A) \& T \in I(B) \& F \in I(B)]$.

(5b14) $[T \notin I(A) \& F \notin I(A) \& T \in I(B)]$ or $[F \notin I(A) \& T \notin I(B) \& F \in I(B)]$ or $[T \in I(A) \& F \notin I(A) \& T \notin I(B)]$ or $[T \in I(A) \& T \notin I(B) \& F \notin I(B)]$ or $[T \notin I(A) \& F \in I(A) \& T \notin I(B) \& F \notin I(B)]$ or $[T \in I(A) \& F \in I(A) \& T \in I(B) \& F \in I(B)]$.

(5b15) $[F \notin I(A) \& T \notin I(B)]$ or $[T \notin I(A) \& F \notin I(A) \& F \notin I(B)]$ or $[T \notin I(A) \& T \notin I(B) \& F \notin I(B)]$ or $[T \in I(A) \& F \in I(A) \& T \in I(B) \& F \in I(B)]$.

(5b16) $[F \notin I(A) \& T \notin I(B) \& F \in I(B)]$ or $[T \in I(A) \& F \notin I(A) \& T \notin I(B)]$ or $[T \notin I(A) \& F \notin I(A) \& T \in I(B) \& F \notin I(B)]$ or $[T \in I(A) \& F \in I(A) \& T \in I(B) \& F \in I(B)]$ or $[T \notin I(A) \& F \in I(A) \& T \notin I(B) \& F \notin I(B)]$.

(5b17) $[T \in I(A) \& F \in I(B)]$ or $[T \in I(A) \& F \in I(A)]$ or $[T \in I(B) \& F \in I(B)]$ or $[T \notin I(A) \& F \notin I(A) \& T \notin I(B) \& F \notin I(B)]$.

(5b18) $[T \in I(A) \& F \in I(B)]$ or $[T \in I(A) \& F \in I(A)]$ or $[T \in I(B) \& F \in I(B)]$.

(5b19) $[T \in I(A) \& F \in I(B)]$ or $[F \in I(A) \& T \in I(B) \& F \in I(B)]$ or $[T \in I(A) \& F \in I(A) \& T \in I(B)]$ or $[T \notin I(A) \& F \notin I(A) \& T \notin I(B) \& F \notin I(B)]$.

(5b20) $[T \in I(A) \& F \in I(B)]$ or $[F \in I(A) \& T \in I(B) \& F \in I(B)]$ or $[T \in I(A) \& F \in I(A) \& T \in I(B)]$.

(5b21) $[T \notin I(A) \& F \notin I(A) \& T \notin I(B) \& F \notin I(B)]$ or $[T \notin I(A) \& F \notin I(A) \& T \in I(B) \& F \in I(B)]$ or $[T \in I(A) \& F \in I(A) \& T \notin I(B) \& F \notin I(B)]$ or $[T \in I(A) \& F \in I(A) \& T \in I(B) \& F \in I(B)]$ or $[T \in I(A) \& F \notin I(A) \& T \notin I(B) \& F \in I(B)]$.

(5b22) $[T \notin I(A) \& F \notin (A) \& T \in I(B) \& F \in I(B)]$ or $[T \in I(A) \& F \in I(A) \& T \notin I(B) \& F \notin I(B)]$ or $[T \in I(A) \& F \in I(A) \& T \in I(B) \& F \in I(B)]$ or $[T \in I(A) \& F \notin I(A) \& T \notin I(B) \& F \in I(B)]$.

(5b23) $[T \notin I(A) \& F \notin I(A) \& T \notin I(B) \& F \notin I(B)]$ or $[T \in I(A) \& F \in I(A) \& T \in I(B) \& F \in I(B)]$ or $[T \in I(A) \& F \notin I(A) \& T \notin I(B) \& F \in I(B)]$.

(5b24) $[T \in I(A) \& F \in I(A) \& T \in I(B) \& F \in I(B)]$ or $[T \in I(A) \& F \notin I(A) \& T \notin I(B) \& F \in I(B)]$.

Definition 11 (L_i-consequence, L_i-validity). Let M be an L_i-model ($1 \leq i \leq 24$). For any set of wffs Γ and a wff A:

(1) $\Gamma \vDash_M A$ (A is a consequence of Γ in M) iff $T \in I(A)$ whenever $T \in I(\Gamma)$. ($T \in I(\Gamma)$ iff $\forall A \in \Gamma (T \in I(A))$; $F \in I(\Gamma)$ iff $\exists A \in \Gamma (F \in I(A))$.)
(2) $\Gamma \vDash_{L_i} A$ (A is a consequence of Γ in L_i-semantics) iff $\Gamma \vDash_M A$ for each L_i-model M.
(3) In particular, $\vDash_{L_i} A$ (A is valid in L_i-semantics) iff $\vDash_M A$ for each L_i-model M (i.e., iff $T \in I(A)$ for each L_i-model M).

By $\vDash_{L_i}$ we shall refer to the relation just defined.

Now, given Definition 6 together with the adjoined notions of M_i-interpretation and M_i-validity ($1 \leq i \leq 24$) (cf. Definition 1) and Definitions 9, 10 and 11, we easily prove (by $\vDash_{M_i}$ we refer to the consequence relation definable in M_i ($1 \leq i \leq 24$) — cf. Definitions 1 and 6):

Proposition 12 (Coextensiveness of $\vDash_{M_i}$ and $\vDash_{L_i}$). *For each i ($1 \leq i \leq 24$), a set of wffs Γ and a wff A, $\Gamma \vDash_{M_i} A$ iff $\Gamma \vDash_{L_i} A$. In particular, $\vDash_{M_i} A$ iff $\vDash_{L_i} A$.*

Proof. See the proof of Theorem 8 in [8] or Proposition 4.4 in [19] where the simple proof procedure is exemplified in the cases of the logics BN4 and Sm4.   ◁

Proposition 12 simply formalizes the intuitive translation (explained above) of the matrix semantics based upon M_i into Belnap and Dunn's two-valued type L_i-semantics ($1 \leq i \leq 24$). Nevertheless, Proposition 12 is a most useful proposition: it gives us the possibility of easily proving soundness of L_i w.r.t. $\vDash_{M_i}$ while proving completeness w.r.t. $\vDash_{L_i}$ by using a canonical model construction.

Suppose that the L_i-logics have been defined (cf. Definitions 20 and 21). Then soundness is proved as follows.

Theorem 13 (Soundness of the L_i-logics). *For any i ($1 \leq i \leq 24$), a set of wffs Γ and a wff A, if $\Gamma \vdash_{L_i} A$ then (1) $\Gamma \vDash_{M_i} A$ and (2) $\Gamma \vDash_{L_i} A$.*

Proof. The 24 L_i-logics are axiomatized in Definitions 20 and 21. Then let I be an M_i-interpretation (defined in the M_i-model M). (1) It is easy to check the following facts. (i) Let r be an inference rule of L_i. Then I assigns a designated valued to the conclusion of r if it assigns a designated value to the premise(s) of r; (ii) all axioms of L_i are assigned 2 or 3; (iii) regarding the metarule MR, suppose $I(D \vee A) = I(D \vee B) = 2$ or 3 but $I(D \vee C) = 0$ or 1 for some wffs A, B and C. Then, it is clear that C is not a consequence of A, B (i.e., $A, B \Rightarrow C$ is falsified). (2) It is immediate from (1) and Proposition 12.[1]   ◁

As remarked above, completeness is proved by a canonical model construction similarly as in e.g., [19; 20] or [28]. Let us see how this proof proceeds.

Consider, for example, L5-models. Let $\mathcal{T}$ be a prime L5-theory (cf. Definition 26). A canonical L5-model is a structure $(K, I_\mathcal{T})$ where K is defined as in Definition 9 and $I_\mathcal{T}$ is a function from the set of all wffs to K defined as follows: for each wff A, $T \in I(A)$ iff $A \in \mathcal{T}$, and $F \in I(A)$ iff $\sim A \in \mathcal{T}$. It is shown that $(K, I_\mathcal{T})$ is a canonical L5-model by proving that $I_\mathcal{T}$ fulfills clauses (2a), (2b), (3a), (3b), (4a), (4b), (5a) and (5b5) in Definition 9, by induction on the structure of A. (It is immediate that $I_\mathcal{T}$ complies with clause (1).) That is, we have to prove: (i) $B \wedge C \in \mathcal{T}$ iff $B \in \mathcal{T} \,\&\, C \in \mathcal{T}$; (ii) $\sim(B \wedge C) \in \mathcal{T}$ iff $\sim B \in \mathcal{T}$ or $\sim C \in \mathcal{T}$; (iii) $B \vee C \in \mathcal{T}$ iff $B \in \mathcal{T}$ or $C \in \mathcal{T}$; (iv) $\sim(B \vee C) \in \mathcal{T}$ iff $\sim B \in \mathcal{T} \,\&\, \sim C \in \mathcal{T}$; (v) $\sim\sim B \in \mathcal{T}$ iff $B \in \mathcal{T}$; (vi) $B \to C \in \mathcal{T}$ iff $[B \notin \mathcal{T} \,\&\, \sim C \notin \mathcal{T}]$ or $[B \notin \mathcal{T} \,\&\, \sim B \in \mathcal{T} \,\&\, C \notin \mathcal{T}]$ or $[\sim B \notin \mathcal{T} \,\&\, C \in \mathcal{T} \,\&\, \sim C \notin \mathcal{T}]$ or $[B \in \mathcal{T} \,\&\, \sim B \in \mathcal{T} \,\&\, C \in \mathcal{T} \,\&\, \sim C \in \mathcal{T}]$; (vii) $\sim(B \to C) \in \mathcal{T}$ iff $[\sim B \notin \mathcal{T} \,\&\, C \notin \mathcal{T} \,\&\, \sim C \in \mathcal{T}]$ or $[B \in \mathcal{T} \,\&\, \sim B \notin \mathcal{T} \,\&\, C \notin \mathcal{T}]$ or $[B \in \mathcal{T} \,\&\, \sim B \in \mathcal{T} \,\&\, C \in \mathcal{T} \,\&\, \sim C \in \mathcal{T}]$.

Once canonical L5-models are shown L5-models, completeness is proved as follows. Suppose $\Gamma \nvdash_{L5} A$, i.e., that A is not included in the set of consequences derivable

---

[1] In case a tester is needed, the one in [16] can be used.

in L5 from Γ (in symbols, $A \notin Cn\Gamma[L5]$). Then $Cn\Gamma[L5]$ is extended to a prime L5-theory $\mathcal{T}$ such that $A \notin \mathcal{T}$. Next, the canonical model $\mathbf{M} = (K, I_\mathcal{T})$ based upon $\mathcal{T}$ is defined, and we have $\Gamma \nvDash_\mathbf{M} A$ since $\mathcal{T} \in I_\mathcal{T}(\Gamma)$ (as $\mathcal{T} \in I_\mathcal{T}(Cn\Gamma[L5])$) but $\mathcal{T} \notin I_\mathcal{T}(A)$ whence $\Gamma \nvDash_{L5} A$ (by Definitions 9 and 11), as it was to be proved.

Completeness of the rest of the Li-logics is proved in a similar way. In §6, we prove the two facts that are required in the completeness proofs, as shown above. For each i ($1 \leq i \leq 24$):

(1) An L$_i$-theory without a given wff can be extended to a prime L$_i$-theory without the same wff.
(2) Let $\mathcal{T}$ be a prime L$_i$-theory. Then, the canonical translations of clauses (1), (2a), (2b), (3a), (3b), (4a), (4b) and those of the corresponding clauses for the conditional are provable in $\mathcal{T}$.

As pointed out in the introduction, we follow the strategy set up in [8] as applied in, e.g., [19; 20] or [28]. Thus, it is possible to be reasonably general about the details and most of the proofs will be referred to the papers quoted above.

4. THE BASIC LOGIC B_0 AND ITS EXTENSIONS B_1 AND B_2

The main relevance logics are Routley and Meyer's basic logic B, T (*Ticket Entailment*), E (*Entailment*) and R (*Relevance*, (cf. [1; 2], Routley et al. [31], [11]), although some relevantists have argued that weaker relevance logics may be preferable (cf. e.g., [31, Chapter 3]). Now, BN4 and its companion E4 are based upon strong relevance logics. In particular, Brady's relevance logic BN4 (cf. [8]) can intuitively be described as a 4-valued extension of contractionless relevance logic R, whereas the logic of entailment E4 (cf. [28]) can be viewed as a 4-valued extension of reductioless entailment logic E. However, all L$_i$-logics are 4-valued extensions of a basic logic we label b_0, which is related to weak relevance logics in the vicinity of Brady's important logic DJ (cf. [9] and references therein), as it will be shown in this section.

In this section, the basic logics b_0, b_1 and b_2 are defined. All L$_i$-logics are extensions of one of the two basic logics b_1 and b_2 introduced below. Especially, L1 through L8 are extensions of b_1, while L9 through L24 extend b_2. Both b_1 and b_2 (which are independent from each other) are built upon the more basic logic commented upon above.

Definition 14 (The logic b_0). The logic b_0 can be formulated with the following axioms, rules of inference and metarule ($A_1, \ldots, A_n \Rightarrow B$ means "if $A_1, \ldots, A_n$, then B"):

Axioms:
  A1. $A \to A$
  A2. $(A \wedge B) \to (B \wedge A)$
  A3. $[A \wedge (B \wedge C)] \to [(A \wedge B) \wedge C]$
  A4. $[A \wedge (B \vee C)] \leftrightarrow [(A \wedge B) \vee (A \wedge C)]$
  A5. $\sim(A \vee B) \leftrightarrow (\sim A \wedge \sim B)$
  A6. $(A \to \sim B) \to (B \to \sim A)$
  A7. $\sim\sim A \to A$

Rules of inference:
  R1. $A, B \Rightarrow A \wedge B$ (Adj)

R2. $A \to B, A \Rightarrow B$ (MP)
R3. $A \wedge B \Rightarrow A, B$ (E$\wedge$)
R4. $A \Rightarrow A \vee B, B \vee A$ (I$\vee$)
R5. $A \to B, A \to C \Rightarrow A \to (B \wedge C)$ (CI$\wedge$)
R6. $A \leftrightarrow B \Rightarrow (A \wedge C) \leftrightarrow (B \wedge C)$ (Fac$\leftrightarrow$)
R7. $B \to C \Rightarrow (A \to B) \to (A \to C)$ (Pref)

Metarule:
MR. If $A, B \Rightarrow C$, then $D \vee A, D \vee B \Rightarrow D \vee C$.

Remark 15 (On the axiomatization of b_0). CI$\wedge$, Fac$\leftrightarrow$ and Pref abbreviate "conditioned introduction of conjunction," "factor w.r.t. $\leftrightarrow$" and "prefixing," respectively. The metarule MR can be dropped if a "disjunctive version" of each rule is added (the disjunctive version of, e.g., MP is the following rule: $C \vee (A \to B), C \vee A \Rightarrow C \vee B$). On the role of disjunctive rules in certain logics, cf. [31], [11] and references therein.

In what follows, we prove some proof-theoretical properties of b_0.

Proposition 16 (Some theorems and rules of b_0). *We note some wffs and rules provable in b_0. A proof sketch is recorded to the right of each item. References to the transitivity rules T3 and T4 are generally omitted.*

T1.	$A \leftrightarrow A$		A1
T2.	$(A \leftrightarrow B) \to (B \leftrightarrow A)$		A2
T3.	$A \to B, B \to C \Rightarrow A \to C$	(Trans)	Pref
T4.	$A \leftrightarrow B, B \leftrightarrow C \Rightarrow A \leftrightarrow C$	(Trans$\leftrightarrow$)	Trans (T3)
T5.	$A \to {\sim}{\sim}A$		A1, A6
T6.	$A \leftrightarrow {\sim}{\sim}A$		A7, T5
T7.	$A \leftrightarrow B \Rightarrow (C \to A) \leftrightarrow (C \to B)$	(Pref$\leftrightarrow$)	Pref
T8.	$(A \to B) \to ({\sim}B \to {\sim}A)$		A6, T6, Pref$\leftrightarrow$
T9.	$({\sim}A \to {\sim}B) \to (B \to A)$		A6, T6, Pref$\leftrightarrow$
T10.	$(A \to B) \leftrightarrow ({\sim}B \to {\sim}A)$		T8, T9
T11.	$({\sim}A \to B) \to ({\sim}B \to A)$		T6, T8, Pref$\leftrightarrow$
T12.	$A \to B \Rightarrow {\sim}B \to {\sim}A$	(Con)	T8
T13.	$A \leftrightarrow B \Rightarrow {\sim}B \leftrightarrow {\sim}A$	(Con$\leftrightarrow$)	Con
T14.	$A \to B, {\sim}B \Rightarrow {\sim}A$	(Modus Tollens — MT)	Con, MP
T15.	$A \to B \Rightarrow (B \to C) \to (A \to C)$	(Suf)	Pref, Con, T10
T16.	$A \leftrightarrow B \Rightarrow (B \to C) \leftrightarrow (A \to C)$	(Suf$\leftrightarrow$)	Suf
T17.	$(A \vee B) \leftrightarrow {\sim}({\sim}A \wedge {\sim}B)$		A5, Con$\leftrightarrow$, T6
T18.	$A \to C, B \to C \Rightarrow (A \vee B) \to C$	(E$\vee$)	Con, CI$\wedge$, T11, T17
T19.	$A \leftrightarrow B \Rightarrow (A \vee C) \leftrightarrow (B \vee C)$	(Sum$\leftrightarrow$)	Con$\leftrightarrow$, Fac$\leftrightarrow$, T17
T20.	$(A \wedge B) \leftrightarrow (B \wedge A)$		A2
T21.	$[(A \wedge B) \wedge C] \to [A \wedge (B \wedge C)]$		A2, A3
T22.	$[A \wedge (B \wedge C)] \leftrightarrow [(A \wedge B) \wedge C]$		A3, T21
T23.	$(A \vee B) \leftrightarrow (B \vee A)$		T20, Con$\leftrightarrow$, T17
T24.	$(A \vee A) \to A$		A1, E$\vee$
T25.	$A \to (A \wedge A)$		A1, CI$\wedge$
T26.	$A \leftrightarrow B \Rightarrow (C \wedge A) \leftrightarrow (C \wedge B)$	(Fac'$\leftrightarrow$)	Fac$\leftrightarrow$, T20
T27.	$A \leftrightarrow B \Rightarrow (C \vee A) \leftrightarrow (C \vee B)$	(Sum'$\leftrightarrow$)	Sum$\leftrightarrow$, T23

T28. $\sim(A \wedge B) \leftrightarrow (\sim A \vee \sim B)$	T17, T6, Fac↔, Fac'↔, Con↔
T29. $(A \wedge B) \leftrightarrow \sim(\sim A \vee \sim B)$	T28, Con↔, T6
T30. $[A \vee (B \vee C)] \leftrightarrow [(A \vee B) \vee C]$	T22, Con↔, T29, T6
T31. $[A \vee (B \wedge C)] \leftrightarrow [(A \vee B) \wedge (A \vee C)]$	A4, Con↔, A5, T29, T6

Remark 17 (On b_0 and Brady's logic DJ^d). Routley and Meyer's basic logic B is axiomatized as follows (cf. [31, Chapter 4]): (a1) $A \to A$; (a2) $(A \wedge B) \to A$, $(A \wedge B) \to B$; (a3) $A \to (A \vee B)$, $B \to (A \vee B)$; (a4) $[(A \to B) \wedge (A \to C)] \to [A \to (B \wedge C)]$; (a5) $[(A \to C) \wedge (B \to C)] \to [(A \vee B) \to C]$; (a6) $[A \wedge (B \vee C)] \to [(A \wedge B) \vee (A \wedge C)]$; (a7) $A \to \sim\sim A$; (a8) $\sim\sim A \to A$. Rules of inference: Adj, MP, Pref, Suf (T15) and Con (T12). Then the logic DW is a useful weak relevant logic extending B, which is axiomatized when deleting a7 and Con, while adding the axiom (a7') $(A \to \sim B) \to (B \to \sim A)$. Finally, Brady's important logic DJ is formulated by adding to DW the axiom (a9) $[(A \to B) \wedge (B \to C)] \to (A \to C)$ (cf. [9], [31, Chapter 4]). In addition, the logics B^d and DW^d and DJ^d (including — but not included in — B, DW and DJ, respectively) are the result of adding the metarule MR to B, DW and DJ, respectively. Well then, the logic b_0 can intuitively be viewed as the result of restricting in DJ^d, a2, a3, a4, a5 and a9 to their respective rule form.

Definition 18 (The logics b_1 and b_2). The logics b_1 and b_2 are axiomatized when adding the following axioms and rules to b_0.

b_1:
A8. $(A \vee \sim B) \vee (A \to B)$
A9. $[(A \wedge \sim A) \wedge (B \wedge \sim B)] \to (A \to B)$
R8. $\sim A \Rightarrow (A \vee B) \vee (A \to B)$
R9. $A \to B, A \wedge \sim A \Rightarrow \sim B$
R10. $A \to B, B \wedge \sim B \Rightarrow \sim A$

b_2: R9 of b_1 plus:
A10. $[(A \vee \sim A) \vee (B \vee \sim B)] \vee (A \to B)$
R11. $A \wedge B \Rightarrow (\sim A \vee \sim B) \vee (A \to B)$
R12. $\sim A \wedge B \Rightarrow (A \vee \sim B) \vee (A \to B)$
R13. $(A \wedge \sim A) \wedge (B \wedge \sim B) \Rightarrow (A \to B)$
R14. $A \to B, \sim A \Rightarrow B \vee \sim B$

Proposition 19 (Some rules provable in b_1 and b_2). *By using T6 and T10 the following alternative versions of the characteristic rules of b_1 and b_2 are provable:*

R8'. $B \Rightarrow (\sim A \vee \sim B) \vee (A \to B)$
R9'. $A \to B, B \wedge \sim B \Rightarrow A$
R10'. $A \to B, A \wedge \sim A \Rightarrow B$
R11'. $\sim A \wedge \sim B \Rightarrow (A \vee B) \vee (A \to B)$
R14'. $A \to B, B \Rightarrow A \vee \sim A$

All the above rules except R14' are provable in b_1, while R9', R11' and R14' are rules of b_2.

5. The 4-valued Logics Determined by M1–M24

In this section, the L_i-logics, the logics determined by the M_i-matrices ($1 \leq i \leq 24$) are defined and some of their proof-theoretical properties are proved. As pointed out in the preceding section, each L_i-logic is a 4-valued extension of the basic logic b_0. Especially, L1–L8 are extensions of b_1, and L9–L24 are extensions of b_2.

The L_i-logics are axiomatized with some subset of the following set of axioms and rules. Most of the items are accompanied by an alternative version. Both versions are equivalent by T6 and T10.

A11. $(\sim A \vee B) \vee \sim(A \rightarrow B)$
A12. $(A \vee \sim A) \vee \sim(A \rightarrow B)/(B \vee \sim B) \vee \sim(A \rightarrow B)$
A13. $[(A \vee \sim A) \vee \sim B] \vee \sim(A \rightarrow B)/[(B \vee \sim B) \vee A] \vee \sim(A \rightarrow B)$
A14. $[(A \vee \sim A) \vee (B \vee \sim B)] \vee \sim(A \rightarrow B)$
R15. $A \wedge \sim B \Rightarrow \sim(A \rightarrow B)$
R16. $A \wedge \sim A \Rightarrow \sim(A \rightarrow B)/B \wedge \sim B \Rightarrow \sim(A \rightarrow B)$
R17. $A \wedge \sim B \Rightarrow (\sim A \vee B) \vee \sim(A \rightarrow B)$
R18. $A \wedge \sim A \Rightarrow B \vee \sim(A \rightarrow B)/B \wedge \sim B \Rightarrow \sim A \vee \sim(A \rightarrow B)$
R19. $A \wedge \sim A \Rightarrow (B \vee \sim B) \vee \sim(A \rightarrow B)/B \wedge \sim B \Rightarrow (A \vee \sim A) \vee \sim(A \rightarrow B)$
R20. $(A \wedge \sim A) \wedge B \Rightarrow \sim(A \rightarrow B)/(B \wedge \sim B) \wedge \sim A \Rightarrow \sim(A \rightarrow B)$
R21. $(A \wedge \sim A) \wedge (B \wedge \sim B) \Rightarrow \sim(A \rightarrow B)$
R22. $A \Rightarrow B \vee \sim(A \rightarrow B)/\sim B \Rightarrow \sim A \vee \sim(A \rightarrow B)$
R23. $A \Rightarrow (\sim A \vee B) \vee \sim(A \rightarrow B)/\sim B \Rightarrow (\sim A \vee B) \vee \sim(A \rightarrow B)$
R24. $\sim A \Rightarrow (B \vee \sim B) \vee \sim(A \rightarrow B)/B \Rightarrow (A \vee \sim A) \vee \sim(A \rightarrow B)$
R25. $A \Rightarrow (B \vee \sim B) \vee \sim(A \rightarrow B)/\sim B \Rightarrow (A \vee \sim A) \vee \sim(A \rightarrow B)$
R26. $\sim A \Rightarrow [(B \vee \sim B) \vee A] \vee \sim(A \rightarrow B)/B \Rightarrow [(A \vee \sim A) \vee \sim B] \vee \sim(A \rightarrow B)$
R27. $\sim(A \rightarrow B) \Rightarrow \sim A \vee \sim B/\sim(A \rightarrow B) \Rightarrow A \vee B$
R28. $\sim(A \rightarrow B) \Rightarrow A \vee \sim B$
R29. $\sim(A \rightarrow B) \Rightarrow A \wedge \sim B$
R30. $\sim(A \rightarrow B) \Rightarrow A \vee \sim A/\sim(A \rightarrow B) \Rightarrow B \vee \sim B$
R31. $\sim(A \rightarrow B) \Rightarrow (A \vee \sim A) \vee (B \vee \sim B)$
R32. $\sim(A \rightarrow B) \wedge \sim A \Rightarrow A/\sim(A \rightarrow B) \wedge B \Rightarrow \sim B$
R33. $\sim(A \rightarrow B) \wedge A \Rightarrow \sim B/\sim(A \rightarrow B) \wedge \sim B \Rightarrow A$
R34. $\sim(A \rightarrow B) \wedge \sim A \Rightarrow A \wedge \sim B/\sim(A \rightarrow B) \wedge B \Rightarrow A \wedge \sim B$
R35. $\sim(A \rightarrow B) \wedge \sim A \Rightarrow (B \wedge \sim B) \wedge A/\sim(A \rightarrow B) \wedge B \Rightarrow (A \wedge \sim A) \wedge \sim B$
R36. $\sim(A \rightarrow B) \wedge A \Rightarrow \sim A \vee \sim B/\sim(A \rightarrow B) \wedge \sim B \Rightarrow A \vee B$
R37. $\sim(A \rightarrow B) \wedge \sim A \Rightarrow A \vee B/\sim(A \rightarrow B) \wedge B \Rightarrow \sim A \vee \sim B$
R38. $\sim(A \rightarrow B) \wedge \sim A \Rightarrow A \vee \sim B/\sim(A \rightarrow B) \wedge B \Rightarrow A \vee \sim B$
R39. $\sim(A \rightarrow B) \wedge (A \vee \sim A) \Rightarrow B \vee \sim B/\sim(A \rightarrow B) \wedge (B \vee \sim B) \Rightarrow A \vee \sim A$
R40. $\sim(A \rightarrow B) \wedge (\sim A \wedge \sim B) \Rightarrow A \vee B/\sim(A \rightarrow B) \wedge (A \wedge B) \Rightarrow \sim A \vee \sim B$
R41. $\sim(A \rightarrow B) \wedge (\sim A \wedge B) \Rightarrow A \vee \sim B$
R42. $\sim(A \rightarrow B) \wedge (A \wedge \sim A) \Rightarrow B \vee \sim B/\sim(A \rightarrow B) \wedge (B \wedge \sim B) \Rightarrow A \vee \sim A$
R43. $\sim(A \rightarrow B) \wedge (\sim A \wedge \sim B) \Rightarrow A \wedge B/\sim(A \rightarrow B) \wedge (A \wedge B) \Rightarrow \sim A \wedge \sim B$
R44. $\sim(A \rightarrow B) \wedge (\sim A \wedge B) \Rightarrow A \wedge \sim B$
R45. $\sim(A \rightarrow B) \wedge (A \wedge \sim A) \Rightarrow B \wedge \sim B/\sim(A \rightarrow B) \wedge (B \wedge \sim B) \Rightarrow A \wedge \sim A$

In particular, we have:

Definition 20 (Extensions of b_1). The following logics are axiomatized by adding to b_1 the following axioms or rules:

L1: R16, R22, R27, R28.

L2: R15, R16, R27, R28.
L3: R15, R22, R28, R32.
L4: R15, R18, R27, R28, R32.
L5: R21, R23, R28, R35.
L6: R17, R21, R29, R35.
L7: R15, R23, R28, R34.
L8: R15, R29.

Definition 21 (Extensions of b_2). The following logics are axiomatized by adding to b_2 the following axioms or rules:

L9: A12, R15, R16, R40, R41.
L10: R16, R22, R24, R26, R31, R40, R41.
L11: A11, A13, R15, R20, R40, R41, R42.
L12: R15, R20, R23, R26, R31, R40, R41, R42.
L13: A11, A12, R21, R43, R44.
L14: R21, R23, R25, R26, R31, R43, R44.
L15: A11, A13, R21, R43, R44, R45.
L16: R21, R23, R26, R31, R43, R44, R45.
L17: A14, R15, R16, R36, R37, R38.
L18: R15, R16, R27, R28.
L19: A14, R15, R20, R36, R37, R38, R39.
L20: R15, R20, R27, R28, R30.
L21: A14, R17, R19, R21, R32, R36, R43.
L22: R17, R19, R21, R27, R28, R32, R43.
L23: A14, R17, R21, R33, R35.
L24: R17, R21, R29. R35.

Next, a couple of proof-theoretical properties of the L_i-logics are proved. Besides being significant in themselves, these properties, are instrumental in the completeness proofs to be developed in the following section.

Proposition 22 (Replacement). *For any wffs A, B, $A \leftrightarrow B \Rightarrow C[A] \leftrightarrow C[A/B]$ where $C[A]$ is a wff in which A appears and $C[A/B]$ is the result of replacing A by B in $C[A]$ in one or more places where A occurs.*

Proof. By induction on the structure of $C[A]$ using Fac$\leftrightarrow$ (R6), Trans$\leftrightarrow$ (T4), Con$\leftrightarrow$ (T13), Suf$\leftrightarrow$ (T16), Sum$\leftrightarrow$ (T19), Fac$'\leftrightarrow$ (T26) and Sum$'\leftrightarrow$ (T27). ◁

Proposition 23 (Arrangement in conjunctive and disjunctive wffs). *Let A be a wff of the form $B_1 \wedge \cdots \wedge B_n$ (resp., $B_1 \vee \cdots \vee B_n$) where the n wffs are arranged in a given way. And let A' be the result of associating $B_1, \ldots, B_n$ in any way whichever. Then for any i $(1 \leq i \leq 24)$ $\vdash_{L_i} A \leftrightarrow A'$.*

Proof. By Replacement and the commutative and associative properties of $\wedge$ and $\vee$ (T20, T22, T23 and T30). ◁

Proposition 24 (Summation w.r.t. $\vdash_L$ — Sum $\vdash_L$). *Let us refer by ρ to the set of inference rules formed by the metarule MR and the rules R1 through R45. And let L* be an extension of b_0 whose primitive rules of inference are in the set ρ. Then, for any set of wffs $A, C, B_1, \ldots, B_n$, if $B_1, \ldots, B_n \vdash_L A$, then $C \vee (B_1 \wedge \cdots \wedge B_n) \vdash_L C \vee A$.

Proof. By induction on the structure of the proof $B_1, \ldots, B_n \vdash_L A$. If A is some B_i, the proof follows by E$\wedge$ and Propositions 22 and 23; and if A is an axiom, then the proof is immediate by I$\vee$. Next, if A has been obtained by application of one of the rules of inference or by a disjunctive rule as a result of an application of the metarule MR, then the case is proved by the rule in question and MR. Let us consider an example. Suppose A has been derived by CI$\wedge$ (R5). Then A is of the form $D \to (E \wedge F)$. By hypothesis, $C \vee (B_1 \wedge \cdots \wedge B_n) \vdash_L C \vee (D \to E), C \vee (D \to F)$, whence $C \vee (B_1 \wedge \cdots \wedge B_n) \vdash_L C \vee [D \to (E \wedge F)]$ follows by CI$\wedge$ and MR. ◁

In §2, we have seen that the M_i-matrices determine "natural conditionals" in a sense akin to that defined in [33]. We conclude the section by noting that the L_i-logics comply with the requirements imposed on "implicative logics" in the classical Polish logical tradition, except, of course, VEQ (cf. Rasiowa [25, pp. 179–180] or Wójcicki [34, p. 228]). Consider the following definition.

Definition 25 (Implicative logics). A logic L is implicative if the following properties (C1)–(C5) are predicable of L:

C1. $A \to A$	Reflexivity
C2. $A \to B, A \Rightarrow B$	Modus Ponens
C3. $A \Rightarrow B \to A$	VEQ
C4. $A \to B, B \to C \Rightarrow A \to C$	Transitivity
C5. $A \leftrightarrow B \Rightarrow C[A] \leftrightarrow C[A/B]$	Replacement

(VEQ abbreviates "verum e quodlibet" — "Any true proposition follows from no matter which proposition".)

Now, each L_i-logic has properties C1, C2, C4 and C5 (cf. Definition 14, Proposition 16 and Proposition 22; none of them, however, satisfies VEQ).

6. EXTENSION AND PRIMENESS LEMMAS. CANONICAL TRANSLATION OF THE VALUATION CLAUSES. COMPLETENESS

In this section, we proceed to prove the facts (1) and (2), instrumental in the completeness proofs, as discussed at the end of section 3. Then we prove the completeness for the L_i-logics.

(1) An L_i-theory without a given wff can be extended to a prime theory without the same wff.
(2) Let $\mathcal{T}$ be a prime L-theory. Then the canonical translations of the valuation clauses are provable in $\mathcal{T}$.

Definition 26 (Preliminary concepts). Let us refer by Eb to the family of extensions of the basic logics b_o, b_1 and b_2 (cf. Definitions 14 and 18) and the L_i-logics ($1 \leq i \leq 24$) (cf. Definitions 20 and 21). (In general, by EL we mean an extension of the logic L — cf. Definition 1) And let L be an Eb-logic. An L-theory is a set of wffs containing all L-theorems and closed under all rules of inference of L and under the metarule MR. For example, an L18-theory is a set of wffs containing all L18-theorems and closed under the metarule MR, all the rules of b_2 and R15, R16, R27 and R28. Then, an L-theory t is prime if for any wffs A, B, if $A \vee B \in t$, then $A \in t$ or $B \in t$.

Definition 27 (Disjunctive derivability). Let L be an Eb-logic. For any sets of wffs Γ, Θ, Θ is disjunctively derivable from Γ in L (in symbols, $\Gamma \vdash_L^d \Theta$) iff $A_1 \wedge \cdots \wedge A_n \vdash_L B_1 \vee \cdots \vee B_m$ for some wffs $A_1, \ldots, A_n \in \Gamma$ and $B_1, \ldots, B_m \in \Theta$.

Definition 28 (Maximal sets). Let L be an Eb-logic. Γ is an L-maximal set of wffs if $\Gamma \nvdash_L^d \overline{\Gamma}$. ($\overline{\Gamma}$ is the complement of Γ.)

Lemma 29 (Extension to maximal sets). *Let us refer by ρ to the set of inference rules formed by the metarule MR and rules R1 through R45, as in Proposition 24 and let L be an Eb-logic whose rules of inference are in the set ρ. Furthermore, let Γ, Θ be sets of wffs such that $\Gamma \nvdash_L^d \Theta$. Then there are sets of wffs Γ', Θ' such that $\Gamma \subseteq \Gamma'$, $\Theta \subseteq \Theta'$, $\Theta' = \overline{\Gamma}'$ and $\Gamma' \nvdash_L^d \Theta'$ (that is, Γ' is an L-maximal set such that $\Gamma' \nvdash_L^d \Theta'$).*

Proof. Thanks to Proposition 24 in section 5, the proof can proceed similarly as in Lemma 3.11 in [20]. ◁

Lemma 30 (Primeness). *Let L be an Eb-logic whose rules of inference are in the set ρ (cf. Lemma 29). If Γ is an L-maximal set, then it is a prime L-theory.*

Proof. Similar to that of Lemma 3.12 in [20]. ◁

The fundamental fact (1) is proved. Next, we advance to the proof of the fundamental fact (2).

Proposition 31 (Conj., disj. and neg. in prime Eb-theories). *Le L be an Eb-logic and t be a prime L-theory. Then (1) $A \wedge B \in t$ iff $A \in t$ and $B \in t$; (2) $\sim(A \wedge B) \in t$ iff $\sim A \in t$ or $\sim B \in t$; (3) $A \vee B \in t$ iff $A \in t$ or $B \in t$; (4) $\sim(A \vee B) \in t$ iff $\sim A \in t$ and $\sim B \in t$; (5) $A \in t$ iff $\sim\sim A \in t$.*

Proof. (1): Adj and E∧; (2): T28, I∨ and primeness; (3): I∨ and primeness, (4). A5, Adj, E∧; (5): T6. ◁

Concerning the conditional, we have Propositions 32 through 35.

Proposition 32 (The conditional in prime Eb_1-theories). *Let L be an Eb_1-logic and t be a prime L-theory. Then $A \to B \in t$ iff $[A \notin t \,\&\, \sim B \notin t]$ or $[A \notin t \,\&\, \sim A \in t \,\&\, B \notin t]$ or $[\sim A \notin t \,\&\, B \in t \,\&\, \sim B \notin t]$ or $[A \in t \,\&\, \sim A \in t \,\&\, B \in t \,\&\, \sim B \in t]$.*

Proof. ($\Rightarrow$) Suppose (1) $A \to B \in t$ and for reductio (2) $[A \in t$ or $\sim B \in t] \,\&\, [A \in t$ or $\sim A \notin t$ or $B \in t] \,\&\, [\sim A \in t$ or $B \notin t$ or $\sim B \in t] \,\&\, [A \notin t$ or $\sim A \notin t$ or $B \notin t$ or $\sim B \notin t]$. There are 72 possibilities to consider but each one of them is impossible by using one of the following rules: MP, MT(T14), R9, R10 and R9' (cf. Proposition 19). Consider, for example, the case where $\sim B \in t, B \in t, \sim A \notin t$. This situation is impossible by R10.

($\Leftarrow$) $A \to B \in t$ follows by using A8, A9, R8 and R8'. ◁

Proposition 33 (The conditional in prime Eb_2-theories). *Let L be an Eb_2-logic and t be a prime L-theory. Then $A \to B \in t$ iff $[A \notin t \,\&\, \sim A \in t \,\&\, B \notin t \,\&\, \sim B \in t]$ or $[A \notin t \,\&\, \sim A \notin t \,\&\, B \notin t \,\&\, \sim B \notin t]$ or $[A \in t \,\&\, \sim A \in t \,\&\, B \in t \,\&\, \sim B \in t]$ or $[A \in t \,\&\, \sim A \notin t \,\&\, B \in t \,\&\, \sim B \notin t]$ or $[A \notin t \,\&\, \sim A \in t \,\&\, B \in t \,\&\, \sim B \notin t]$.*

Proof. Similar to that of Proposition 32. ($\Rightarrow$) We use MP, MT(T14), R9, R9′, R14 and R14′. ($\Leftarrow$) It follows by R11′, A10, R13, R11 and R12. ◁

Proposition 34 (Negated conditionals in Eb$_1$-theories). *Let L be an EL$_i$-logic where L$_i$ will refer in each case to one of the extensions of b_1 displayed in Definition 20, and let t be a prime L-theory. We have that* $\sim(A \to B) \in t$ *iff:*

L1: $[A \in t \& B \notin t]$ or $[\sim A \notin t \& \sim B \in t]$ or $[A \in t \& \sim A \in t]$ or $[B \in t \& \sim B \in t]$.
L2: $[A \in t \& \sim B \in t]$ or $[B \in t \& \sim B \in t]$ or $[A \in t \& \sim A \in t]$.
L3: $[\sim A \notin t \& \sim B \in t]$ or $[A \in t \& B \notin t]$ or $[A \in t \& \sim B \in t]$.
L4: $[A \in t \& \sim B \in t]$ or $[\sim A \notin t \& B \in t \& \sim B \in t]$ or $[A \in t \& \sim A \in t \& B \notin t]$.
L5: $[\sim A \notin t \& B \notin t \& \sim B \in t]$ or $[A \in t \& \sim A \notin t \& B \notin t]$ or $[A \in t \& \sim A \in t \& B \in t \& \sim B \in t]$.
L6: $[A \in t \& \sim A \in t \& B \in t \& \sim B \in t]$ or $[A \in t \& \sim A \notin t \& B \notin t \& \sim B \in t]$.
L7: $[A \in t \& \sim B \in t]$ or $[\sim A \notin t \& B \notin t \& \sim B \in t]$ or $[A \in t \& \sim A \notin t \& B \notin t]$.
L8: $A \in t \& \sim B \in t$.

Proof. Similar to that of Propositions 32 and 33 by using now the characteristic axioms of each L$_i$-logic ($1 \leq i \leq 8$), as displayed in Definition 20 (cf. the proof of Proposition 35). ◁

Proposition 35 (Negated conditional is Eb$_2$-theories). *Let L be an EL$_i$-logic where L$_i$ will refer in each case to one of the extensions of b_2 displayed if Definition 21, and let t be a prime L-theory. We have that* $\sim(A \to B) \in t$ *iff:*

L9: $[A \in t \& \sim B \in t]$ or $[A \notin t \& \sim A \notin t]$ or $[A \in t \& \sim A \in t]$ or $[B \notin t \& \sim B \notin t]$ or $[B \in t \& \sim B \in t]$.
L10: $[\sim A \notin t \& \sim B \in t]$ or $[A \in t \& B \notin t]$ or $[A \in t \& \sim A \in t]$ or $[B \in t \& \sim B \in t]$ or $[\sim A \in t \& B \notin t \& \sim B \notin t]$ or $[A \notin t \& \sim A \notin t \& B \in t \& \sim B \notin t]$.
L11: $[A \in t \& \sim B \in t]$ or $[\sim A \notin t \& B \notin t]$ or $[A \notin t \& B \in t \& \sim B \notin t]$ or $[\sim A \in t \& B \in t \& \sim B \in t]$ or $[A \notin t \& \sim A \notin t \& \sim B \notin t]$ or $[A \in t \& \sim A \in t \& B \in t \& \sim B \notin t]$.
L12: $[A \in t \& \sim B \in t]$ or $[\sim A \in t \& B \in t \& \sim B \in t]$ or $[\sim A \notin t \& B \notin t \& \sim B \in t]$ or $[A \in t \& \sim A \in t \& B \in t]$ or $[A \notin t \& \sim A \notin t \& B \in t \& \sim B \notin t]$ or $[A \notin t \& \sim A \in t \& B \notin t \& \sim B \notin t]$.
L13: $[A \notin t \& \sim A \notin t]$ or $[B \notin t \& \sim B \notin t]$ or $[\sim A \notin t \& B \notin t]$ or $[A \in t \& \sim A \in t \& B \in t \& \sim B \in t]$.
L14: $[A \notin t \& \sim A \notin t \& B \in t]$ or $[\sim A \notin t \& B \notin t \& \sim B \in t]$ or $[A \in t \& \sim A \notin t \& B \notin t]$ or $[A \in t \& B \notin t \& \sim B \notin t]$ or $[A \notin t \& \sim A \in t \& B \notin t \& \sim B \notin t]$ or $[A \in t \& \sim A \in t \& B \in t \& \sim B \in t]$.
L15: $[\sim A \notin t \& B \notin t]$ or $[A \notin t \& \sim A \notin t \& \sim B \notin t]$ or $[A \notin t \& B \notin t \& \sim B \notin t]$ or $[A \in t \& \sim A \in t \& B \in t \& \sim B \in t]$.
L16: $[\sim A \notin t \& B \notin t \& \sim B \in t]$ or $[A \in t \& \sim A \notin t \& B \notin t]$ or $[A \in t \& \sim A \notin t \& B \in t \& \sim B \notin t]$ or $[A \in t \& \sim A \in t \& B \in t \& \sim B \in t]$ or $[A \notin t \& \sim A \in t \& B \notin t \& \sim B \notin t]$.
L17: $[A \in t \& \sim B \in t]$ or $[A \in t \& \sim A \in t]$ or $[B \in t \& \sim B \in t]$ or $[A \notin t \& \sim A \notin t \& B \notin t \& \sim B \notin t]$.
L18: $[A \in t \& \sim B \in t]$ or $[A \in t \& \sim A \in t]$ or $[B \in t \& \sim B \in t]$.
L19: $[A \in t \& \sim B \in t]$ or $[\sim A \in t \& B \in t \& \sim B \in t]$ or $[A \in t \& \sim A \in t \& B \in t]$ or $[A \notin t \& \sim A \notin t \& B \notin t \& \sim B \notin t]$.

L20: $[A \in t \,\&\, \sim B \in t]$ or $[\sim A \in t \,\&\, B \in t \,\&\, \sim B \in t]$ or $[A \in t \,\&\, \sim A \in t \,\&\, B \in t]$.

L21: $[A \notin t \,\&\, \sim A \notin t \,\&\, B \notin t \,\&\, \sim B \notin t]$ or $[A \notin t \,\&\, \sim A \notin t \,\&\, B \in t \,\&\, \sim B \in t]$ or $[A \in t \,\&\, \sim A \in t \,\&\, B \notin t \,\&\, \sim B \notin t]$ or $[A \in t \,\&\, \sim A \in t \,\&\, B \in t \,\&\, \sim B \in t]$ or $[A \in t \,\&\, \sim A \notin t \,\&\, B \notin t \,\&\, \sim B \in t]$.

L22: $[A \notin t \,\&\, \sim A \notin t \,\&\, B \in t \,\&\, \sim B \in t]$ or $[A \in t \,\&\, \sim A \in t \,\&\, B \notin t \,\&\, \sim B \notin t]$ or $[A \in t \,\&\, \sim A \in t \,\&\, B \in t \,\&\, \sim B \in t]$ or $[A \in t \,\&\, \sim A \notin t \,\&\, B \notin t \,\&\, \sim B \in t]$.

L23: $[A \notin t \,\&\, \sim A \notin t \,\&\, B \notin t \,\&\, \sim B \notin t]$ or $[A \in t \,\&\, \sim A \in t \,\&\, B \in t \,\&\, \sim B \in t]$ or $[A \in t \,\&\, \sim A \notin t \,\&\, B \notin t \,\&\, \sim B \in t]$.

L24: $[A \in t \,\&\, \sim A \in t \,\&\, B \in t \,\&\, \sim B \in t]$ or $[A \in t \,\&\, \sim A \notin t \,\&\, B \notin t \,\&\, \sim B \in t]$.

Proof. Similar to that of the preceding Propositions 32–34. Let us consider L12 as a way of an example. ($\Rightarrow$) Suppose $\sim(A \to B) \in t$ and, for reductio, $[A \notin t$ or $\sim B \notin t] \,\&\, [\sim A \notin t$ or $B \notin t$ or $\sim B \notin t] \,\&\, [\sim A \in t$ or $B \in t$ or $\sim B \notin t] \,\&\, [A \notin t$ or $\sim A \notin t$ or $B \notin t] \,\&\, [A \notin t$ or $\sim A \in t$ or $B \in t] \,\&\, [A \in t$ or $\sim A \in t$ or $B \notin t$ or $\sim B \in t] \,\&\, [A \in t$ or $\sim A \notin t$ or $B \in t$ or $\sim B \in t]$. There are 2,592 possibilities to consider but each one of them is either a contradictory statement or it contains one of the following items (1)–(6): (1) $A \in t, B \in t, \sim A \notin t, \sim B \notin t$; (2) $\sim A \in t, \sim B \in t, A \notin t, B \notin t$; (3) $\sim A \in t, B \in t, A \notin t, \sim B \notin t$; (4) $B \in t, \sim B \in t, A \notin t, \sim A \notin t$; (5) $A \in t, \sim A \in t, B \notin t, \sim B \notin t$; (6) $A \notin t, \sim A \notin t, B \notin t, \sim B \notin t$. But (1)–(6) are impossible by using R40, R40, R41, R42, R42 and R31, respectively. ($\Leftarrow$) $\sim(A \to B) \in t$ follows by using R15, R20, R23, R20, R26, R23 and R26, respectively. ◁

Remark 36 (On the proofs of Propositions 32–35). Notice that, as shown above, the characteristic axioms of b_1 (resp., b_2) suffice to prove Proposition 32 (resp., 33) and consequently, the canonical validity of clause (5a). Regarding the clauses for assigning F to conditionals, in Propositions 34 and 35, we have seen that these are proved to hold canonically by using the characteristic axioms or rules added to the basic logics b_1 and b_2 in order to define the particular L_i-logics (cf. Definitions 20 and 21).

Remark 37 (On consistency of L_i-theories). Notice that Propositions 31–35 have been proved without the L_i-theories in question needing to be consistent in any sense of the term.

On the basis of the discussion developed so far in this section, we consider proved the fundamental facts (1) and (2); then on the basis of the argumentation developed at the end of section 3 together with facts (1) and (2), we think that we are entitled to state the following theorem.

Theorem 38 (Completeness of the L_i-logics). *For any i ($1 \leq i \leq 24$), a set of wffs Γ and a wff A, (1) if $\Gamma \vDash_{M_i} A$ then $\Gamma \vdash_{L_i} A$; (2) if $\Gamma \vDash_{L_i} A$ then $\Gamma \vdash_{L_i} A$.*

7. Variable-sharing Property. Ackermann Property

In this section, it is shown that each L_i-logic has the *variable-sharing property* (VSP). Also, that 5 of the 24 L_i-logics have the *Ackermann property*. These properties read as follows.

Definition 39 (Variable-sharing property — VSP). A logic L has the variable-sharing property (VSP) if in all L-theorems of the form $A \to B$, A and B share at least a propositional variable.

Definition 40 (Ackermann property — AP). A logic L has the Ackermann property (AP) if in all L-theorems of the form $A \to (B \to C)$, A contains at least an implication connective ($\to$).

Intuitively, the VSP amounts to rule out implications in which antecedent and consequent do not share some (minimal) semantical content; the AP, on its part, would intuitively exclude implications in which pure non-necessitive wffs entail necessitive ones (cf. [1]; A is *necessitive* if A is equivalent to a wff of the form $\Box B$). According to Anderson and Belnap (cf. [1]), the VSP is a necessary property of any *relevance logic*, but, in addition to the VSP, a logic L has to comply with the AP in order to be considered an *entailment logic*. Thus, for example, T (*Ticket Entailment*) and E (*Entailment*) are entailment logics as both have the VSP and the AP, but R (*Relevance Logic*) and Lewis' S3 are not entailment logics, given that the former lacks the AP and the latter the VSP, although the AP is predicable of S3 and the VSP is a property of R.

Next, it is shown that the VSP is a property of each L_i-logic and that five of them in addition have the AP.

Proposition 41 (Each L_i-logic has the VSP). *Let $A \to B$ be an L_i-theorem ($1 \leq i \leq 24$). Then A and B share at least a propositional variable.*

Proof. Suppose that A and B do not share propositional variables in $A \to B$. It is proved that $A \to B$ is not M_i-valid, M_i being the matrix determining L_i. Let I be an M_i-interpretation assigning 2 (resp., 0) to each propositional variable in A (resp., B). Then $I(A) = 2$ and $I(B) \in \{0,3\}$ since $\{2\}$ and $\{0,3\}$ are closed under $\to, \wedge, \vee$ and $\sim$. Consequently, $I(A \to B) \in \{0,1\}$, i.e., $A \to B$ is not M_i-valid. Now, it follows from the soundness theorem (Theorem 13) that $A \to B$ is not an L_i-theorem. ◁

Proposition 42 (L_i-logics with the AP). *The logics L9, L10, L12, L14 and L16 have the AP.*

Proof. As in the case of Proposition 41, we lean upon the soundness theorem. Suppose then that no implication connective ($\to$) appears in A. It is proved that $A \to (B \to C)$ is not M_i-valid ($i \in \{9, 10, 12, 14, 16\}$).

(1) M9, M10: Let I be an M_i-interpretation ($i \in \{9, 10\}$) assigning 1 to each propositional variable in A, B and C. Then it is clear that $I(A) = 1$ and $I(B \to C) \in \{0, 2, 3\}$. Thus, $I(A \to (B \to C)) = 0$, that is $A \to (B \to C)$ is neither M9-valid nor M10-valid.

(2) M12, M14, M16: As in case (1), let I be an M_i-interpretation ($i \in \{12, 14, 16\}$) assigning 1 to each variable in A, B and C. Then, clearly $I(A) = 1$ and $I(B \to C) \in \{0, 3\}$ since in order to have $I(B \to C) = 1$ we need either $I(B) = 2$ or $I(C) = 2$, but $\{2\}$ is closed under $\to, \wedge, \vee$ and $\sim$. So $I(A \to (B \to C)) \in \{0, 1\}$, i.e., $A \to (B \to C)$ is not M_i-valid ($i \in \{12, 14, 16\}$).

Consequently, if $\to$ does not appear in A, then $A \to (B \to C)$ is not an L_i-theorem ($i \in \{9, 10, 12, 14, 16\}$). ◁

However, the rest of the L_i-logics do not have the AP.

Proposition 43 (L_i-logics not having the AP). *Let L be an L_i-logic other than L9, L10, L12, L14 and L16. Then L does not have the AP.*

Proof. Consider the ensuing one-variable wffs: (a) $p \to (p \to p)$; (b) $p \to [p \to (p \to p)]$; (c) $p \to [p \to [p \to (p \to p)]]$; (d) $p \to [(p \to p) \to [(p \to p) \to p]]$ (we note that (a) is an instance of the *mingle axiom*, $A \to (A \to A)$). We have : (a) is valid in M1 through M8; (b) is valid in M11, M15, M18–M20, M22–M24; (c) is valid in M17 and M21; and finally, (d) is M13-valid. Therefore, we use Theorem 38 and conclude that all L_i-logics except L9, L10, L12, L14 and L16 lack the AP. ◁

The section is ended with some remarks on the *Converse Ackermann property* (CAP). The CAP is defined as follows.

Definition 44 (Converse Ackermann property — CAP). *A logic L has the Converse Ackermann property (CAP) if in all L-theorems of the form $(A \to B) \to C$, C contains at least an implication connective $(\to)$.*

The CAP is introduced in [1, §8.12] (cf. Robles Vázquez [30] and references therein for detailed account on the property and the systems having it). We note that L9, L10, L12 and L14 and L16 have the CAP in addition to the VSP and AP (the fact can be proved by using the same M_i-interpretations defined in Proposition 42). The rest of the L_i-logics do not have the CAP. The proof is left to the reader. (The wff $[[(p \to p) \to p] \to p] \to (p \vee \sim p)$ can be used in order to prove that L17 and L21 lack the CAP; the wff $[(p \to p) \to p] \to p$ serves the same purpose in the case of the rest of the L_i-logics, save, of course, the ones with the CAP remarked above.)

8. Concluding Remarks

The aim of this paper was to define useful implicative expansions of $\mathcal{FOUR}$ with the VSP. That is, important implicative expansions of $\mathcal{FOUR}$ free from paradoxes of relevance. The present paper generalizes the results in Robles [27] where the matrix M8 and its corresponding logic L8 are investigated.

It is trivial to build up binary expansions of $\mathcal{FOUR}$ with the VSP. Consider the ensuing general truth-table T_* (a_i ($1 \leq i \leq 12$) $\in \{0,1,2,3\}$; b_i ($1 \leq i \leq 3$) $\in \{0,1\}$).

T_*	$*$	0	1	2	3
	0	a_1	a_2	a_3	a_4
	1	a_5	a_6	a_7	a_8
	2	b_1	b_2	2	b_3
	3	a_9	a_{10}	a_{11}	a_{12}

Now, it is clear that some of the functions in T_* could be considered adequate interpretations of $\to$. Suppose now that A and B do not share propositional variables in $A * B$. Let M be any expansion of $\mathcal{FOUR}$ built by adding any of the $*$-functions described in T_* and let I be an M-interpretation such that $I(p) = 2$ (resp., $I(p) = c$) for each propositional variable p in A (resp., in B) (c is a truth-value other than 2). Then $I(A * B) \in \{0,1\}$. Nevertheless, the 24 implicative functions contained in T_* we have selected are important, as we have tried to prove in the present paper, due to the properties predicable of them. These properties include: "natural conditionals" (in some sense of the term — cf. §2); C-extensionality; self-extensionality (i.e., "replacement"); satisfiability of the *self-identity* axiom and the rules *modus ponens* and *transitivity*; considerable expressive power; considerable syntactical strength, as they extend a strong restriction of Brady's weak relevance logic DJ^d and finally, their being interpretable in the clear and important two-valued Belnap–Dunn semantics.

As regards future work, there is a number of ways in which the investigation carried out in this paper could be pursued. We remark a couple of them.

(1) Let TI′, TII′, TIII′ and TIV′ be the result of deleting t1–t24 in TI, TII, TIII and TIV, respectively (cf. §2). Surely it will be worthwhile to investigate if it is possible to build up interesting implicative expansions of $\mathcal{FOUR}$ based upon the tables contained in TI′, TII′, TIII′ and TIV′. The expansions in TI′, TII′, TIII′ and TIV′ can be treated similarly as those in MI4^{VSP} have been treated in the present paper. For instance, the basic logic b″$_0$, for expansions in TIV′, is the result of modifying the basic logic b$_0$ as follows: delete Pref (R7) and add Pref↔ (T7) and Trans (T3). Then the basic logic b′$_0$ adequate for expansions in TII′ and TIII′, is formulated by adding to b″$_0$ T5, Sum↔ (T19), Suf↔ (T16) and Con (T12) while deleting A6. Consider for example the following truth-table t25 in TIV′.

t25 →	0	1	2	3
0	3	3	0	3
1	1	3	1	3
2	0	1	2	0
3	0	1	0	3

The particular expansion of $\mathcal{FOUR}$ L25 built upon the $f_\to$-function described by t25 can be axiomatized by adding A8, A9, R8, R9, R10, R15, R20, R27, R28 and R30 to b″$_0$. (We note that all theorems and rules of b$_0$, except Suf and Pref — cf. Proposition 16 — are provable in L25.)

(2) It will be interesting to investigate the functional relations the 24 L$_i$-logics maintain to each other. Are they actually the same logic but for a different choice of the set of primitive connectives? In this sense, we note that, for instance, the M3-conditional function is definable from M4 by using $(\overset{\bullet}{\neg}B \to \overset{\circ}{\neg}A) \wedge (\overset{\circ}{\neg}B \to \overset{\circ}{\neg}A)$ where → is the M4-conditional function and $\overset{\bullet}{\neg}$, $\overset{\circ}{\neg}$, the negation functions defined in §2. But on the other hand, is it maybe one of the L$_i$-logics, say L, preferable to the rest of them on the basis of some property or other none of them possess but L?

A THE 104 TABLES NOT SATISFYING AT LEAST ONE OF PREF AND SUF

The $f_\to$-functions described in the general tables displayed in (α)–(η) below fail to satisfy the rule Suf; those in the table in (θ) do not satisfy the rule Pref (cf. point (4) in §2. The symbols b, c, d, e and f are read as in table TIV in §2).

(α) $f_\to$-functions such that $f_\to(0,1) = 2$ (equivalently, $f_\to(1,2) = 2$)

$\to$	0	1	2	3
0	3	2	b	3
1	c	f	d	2
2	e	d	2	b
3	0	c	e	3

(β) $f_\to$-functions such that $f_\to(0,1) = 3$ and $f_\to(1,1) = 2$

$\to$	0	1	2	3
0	3	3	b	3
1	c	2	d	3
2	e	d	2	b
3	0	c	e	3

(γ) $f_\to$-functions such that $f_\to(1,2) = 1$, $f_\to(0,2) = 0$ and $f_\to(0,1) = 3$

$\to$	0	1	2	3
0	3	3	0	3
1	c	3	1	3
2	e	1	2	0
3	0	c	e	3

(δ) $f_\to$-functions such that $f_\to(3,1) = 1$ and $f_\to(0,1) = 0$

$\to$	0	1	2	3
0	3	0	b	3
1	1	f	d	0
2	e	d	2	b
3	0	1	e	3

(ε) $f_\to$-functions such that $f_\to(3,1) = 0$ and $f_\to(0,1) = 1$

$\to$	0	1	2	3
0	3	1	b	3
1	0	f	d	1
2	e	d	2	b
3	0	0	e	3

(ζ) $f_\to$-functions such that $f_\to(3,2) = 0$, $f_\to(0,2) = 1$ and $f_\to(0,1) = 3$

$\to$	0	1	2	3
0	3	0	1	3
1	0	f	d	0
2	0	d	2	1
3	0	0	0	3

$\to$	0	1	2	3
0	3	1	1	3
1	1	f	d	1
2	0	d	2	1
3	0	1	0	3

(η) $f_\to$-functions such that $f_\to(3,2) = 1$ and $f_\to(0,2) = 0$

$\to$	0	1	2	3
0	3	0	0	3
1	0	f	d	0
2	1	d	2	0
3	0	0	1	3

$\to$	0	1	2	3
0	3	1	0	3
1	1	f	d	1
2	1	d	2	0
3	0	1	1	3

$\to$	0	1	2	3
0	3	3	0	3
1	c	3	0	3
2	1	0	2	0
3	0	c	1	3

(θ) $f_\to$-functions such that $f_\to(2,1) = 0$ and $f_\to(2,1) = 0$ if $f_\to(0,1) = 3$

$\to$	0	1	2	3
0	3	3	1	3
1	c	3	0	3
2	1	0	2	1
3	0	c	0	3

Acknowledgments. We sincerely thank the editor of the volume and an anonymous referee for their comments and suggestions on a previous version of this paper. This work is supported by MCIN/AEI/ 10.13039/501100011033 [Grant PID2020-116502GB-I00].

REFERENCES

[1] Anderson, A. R. and Belnap, N. D. (1975). *Entailment. The Logic of Relevance and Necessity*, Vol. I, Princeton University Press, Princeton, NJ.

[2] Anderson, A. R., Belnap, N. D. and Dunn, J. M. (1992). *Entailment, The Logic of Relevance and Necessity*, Vol. II, Princeton University Press, Princeton, NJ.

[3] Avron, A., Ben-Naim, J. and Konikowska, B. (2007). Cut-free ordinary sequent calculi for logics having generalized finite-valued semantics, *Logica Universalis* **1**(1): 41–70.

[4] Avron, A., Konikowska, B. and Zamansky, A. (2013). Cut-free sequent calculi for C-systems with generalized finite-valued semantics, *Journal of Logic and Computation* **23**(3): 517–540.

[5] Belnap, N. D. (1960). Entailment and relevance, *Journal of Symbolic Logic* **25**(2): 144–146.

[6] Belnap, N. D. (1977a). How a computer should think, in G. Ryle (ed.), *Contemporary Aspects of Philosophy*, Oriel Press Ltd., Stocksfield, pp. 30–55.

[7] Belnap, N. D. (1977b). A useful four-valued logic, in G. Epstein and J. M. Dunn (eds.), *Modern Uses of Multiple-Valued Logic*, D. Reidel Publishing Co., Dordrecht, pp. 8–37.

[8] Brady, R. T. (1982). Completeness proofs for the systems RM3 and BN4, *Logique et Analyse* **25**(97): 9–32.

[9] Brady, R. T. (2006). *Universal Logic*, CSLI Publications, Stanford, CA.

[10] Brady, R. T. (2019). The number of logical values, in C. Başkent and T. M. Ferguson (eds.), *Graham Priest on Dialetheism and Paraconsistency*, Vol. 18 of *Outstanding Contributions to Logic*, Springer International Publishing, Cham, pp. 21–37.

[11] Brady, R. T. (ed.) (2003). *Relevant Logics and Their Rivals*, Vol. II, Ashgate, Aldershot.

[12] Dunn, J. M. (1966). *The Algebra of Intensional Logics*, PhD thesis, University of Pittsburgh, UMI, Ann Arbor, MI. (Published as Vol. 2 in the *Logic PhDs* series by College Publications, London, UK, 2019).

[13] Dunn, J. M. (1976). Intuitive semantics for first-degree entailments and 'coupled trees', *Philosophical Studies* **29**: 149–168.

[14] Dunn, J. M. (2000). Partiality and its dual, *Studia Logica* **66**(1): 5–40.

[15] Dziobiak, W. (1983). There are $2^{\aleph_0}$ logics with the relevance principle between R and RM, *Studia Logica* **42**(1): 49–61.

[16] González, C. (2011). *MaTest*. URL: sites.google.com/site/sefusmendez/matest.

[17] Hazen, A. P. and Pelletier, F. J. (2019). K3, Ł3, LP, RM3, A3, FDE, M: How to make many-valued logics work for you, in H. Omori and H. Wansing (eds.), *New Essays on Belnap–Dunn Logic*, Vol. 418 of *Synthese Library*, Springer International Publishing, Cham, pp. 155–190.

[18] López, S. M. (2021). Belnap–Dunn semantics for the variants of BN4 and E4 which contain Routley and Meyer's logic B, *Logic and Logical Philosophy* **30**: 1–28.
[19] Méndez, J. M. and Robles, G. (2016a). The logic determined by Smiley's matrix for Anderson and Belnap's first-degree entailment logic, *Journal of Applied Non-Classical Logics* **26**(1): 47–68.
[20] Méndez, J. M. and Robles, G. (2016b). Strengthening Brady's paraconsistent 4-valued logic BN4 with truth-functional modal operators, *Journal of Logic, Language and Information* **25**(2): 163–189.
[21] Meyer, R. K., Giambrone, S. and Brady, R. T. (1984). Where gamma fails, *Studia Logica* **43**: 247–256.
[22] Omori, H. and Wansing, H. (2017). 40 years of FDE: An introductory overview, *Studia Logica* **105**(6): 1021–1049.
[23] Omori, H. and Wansing, H. (eds.) (2019). *New Essays on Belnap–Dunn Logic*, Vol. 418 of *Synthese Library (Studies in Epistemology, Logic, Methodology, and Philosophy of Science)*, Springer, Cham.
[24] Petrukhin, Y. and Shangin, V. (2019). Correspondence analysis and automated proof-searching for first degree entailment, *European Journal of Mathematics* **6**: 1452–1495.
[25] Rasiowa, H. (1974). *An Algebraic Approach to Non-classical Logics*, Vol. 78, North-Holland Publishing Company, Amsterdam.
[26] Robles, G. (2021a). The class of all 3-valued implicative expansions of Kleene's strong logic containing Anderson and Belnap's first degree entailment logic, *Journal of Applied Logics – IFCoLog Journal of Logics and their Applications* **8**(7): 2035–2071.
[27] Robles, G. (2021b). A variant with the variable-sharing property of Brady's 4-valued implicative expansion BN4 of Anderson and Belnap's logic FDE, *in* P. Baroni, C. Benzmüller and Y. N. Wáng (eds.), *Logic and Argumentation. CLAR 2021*, number 13040 in *Lecture Notes in Computer Science*, Springer, Cham, pp. 362–376.
[28] Robles, G. and Méndez, J. M. (2016). A companion to Brady's 4-valued relevant logic BN4: The 4-valued logic of entailment E4, *Logic Journal of the IGPL* **24**(5): 838–858.
[29] Robles, G. and Méndez, J. M. (2020). The class of all natural implicative expansions of Kleene's strong logic functionally equivalent to Łukasiewicz's 3-valued logic Ł3, *Journal of Logic, Language and Information* **29**(3): 349–374.
[30] Robles Vázquez, G. (2006). *Negaciones subintuicionistas para lógicas con la Conversa de la Propiedad Ackermann*, PhD thesis, Universidad de Salamanca, Salamanca, Spain (Published in Colección Vítor, 179, Ediciones Universidad de Salamanca, ISBN 84-7800-468-8).
[31] Routley, R., Meyer, R. K., Plumwood, V. and Brady, R. T. (1982). *Relevant Logics and their Rivals*, Vol. 1, Ridgeview Publishing Co., Atascadero, CA.
[32] Slaney, J. (2005). Relevant logic and paraconsistency, *in* L. Bertossi, A. Hunter and T. Schaub (eds.), *Inconsistency Tolerance*, number 3300 in *Lecture Notes in Computer Science*, Springer, pp. 270–293.
[33] Tomova, N. (2012). A lattice of implicative extensions of regular Kleene's logics, *Reports on Mathematical logic* **47**: 173–182.
[34] Wójcicki, R. (1988). *Theory of Logical Calculi: Basic Theory of Consequence Operations*, number vol. 199 in *Synthese Library*, Kluwer, Dordrecht.

DPTO. DE PSICOLOGÍA, SOCIOLOGÍA Y FILOSOFÍA, UNIVERSIDAD DE LEÓN. CAMPUS DE VEGAZANA, S/N, 24071, LEÓN, SPAIN, *Email:* gemma.robles@unileon.es

UNIVERSIDAD DE SALAMANCA, EDIFICIO FES, CAMPUS UNAMUNO, 37007, SALAMANCA, SPAIN, *Email:* sefus@usal.es

THE DIAMOND OF MINGLE LOGICS:
A FOUR-FOLD INFINITE WAY TO BE SAFE FROM PARADOX

Yaroslav Shramko

ABSTRACT. System **R**-Mingle (**RM**) was invented by J. Michael Dunn in the middle of the 1960s. This system got its name due to the characteristic logical principle called "Mingle." Although this principle allows for certain irrelevant inferences, it can protect us from (the worst effects of) the paradoxes of relevance. Furthermore, separating the first-degree entailment fragment of a mingle logic allows one to concentrate on the characteristic principle of that fragment, known as "Safety." Based on a purely Tarskian formulation of first-degree entailment systems, four types of Safety can be distinguished and corresponding proof systems can be constructed, forming a diamond-shaped lattice with infinitely many systems between its vertices. The corner systems of the diamond can be supplied with uniform and rather natural semantics, which reaffirms the rightful place of the mingle logics in the family of the first-degree entailment systems.

Keywords. Binary consequence system, First-degree entailment, Generalized truth-value functions, Paradoxes of relevance, R-Mingle, Variable-sharing property

1. INTRODUCTION

The first-degree entailment fragment of Dunn's logic **R**-Mingle (**RM**) has occasionally appeared in the literature under various names and characterizations. For one, Makinson in [43, p. 38] presents it as a *system of Kalman implication*, reflecting the fact that the algebraic counterpart of its characteristic axiom $x \cap -x \leq y \cup -y$ was first considered by Kalman [42] for defining a "normal" lattice with involution (*i-lattice* for short, nowadays standardly called *De Morgan lattice*). Dunn in [32, p. 43] also pays tribute to Kalman by calling *Kalman consequence system* essentially the same system, but formulated with a turnstile instead of an arrow. Dunn also notes that it is in fact the first-degree entailment fragment of the relevance logic **RM**. In Ermolaeva and Muchnik [37] still the same system is presented as a "fragment of Łukasiewicz's logic," and Dunn in [34, p. 15] observes as well that the system, which comprises "the first-degree entailments of Dunn and McCall's '**R**-Mingle' [...] is also the first-degree entailment fragment of Łukasiewicz's 3-valued logic."

There is also a tradition of naming Kalman's normal i-lattice *Kleene algebra*. Apparently, this tradition was initiated by Brignole and Monteiro, see [22, p. 4, especially Definition 2.4], and then continued by Cignoli [23], Balbes and Dwinger [9], Blyth

---

2020 *Mathematics Subject Classification.* Primary: 03B47, Secondary: 03B50.

Bimbó, Katalin, (ed.), *Relevance Logics and other Tools for Reasoning. Essays in Honor of J. Michael Dunn*, (Tributes, vol. 46), College Publications, London, UK, 2022, pp. 365–393.

and Varlet [21], and others. However, this word usage can be misleading in two ways. First, as Dunn remarks [35, p. 444], it should not be confused with Kleene algebras "which arise in the study of relation algebras and regular expressions" (see, e.g., Bimbó and Dunn [18] and also [19, Ch. 7]). Second, confusion may arise by extrapolating this terminology to a logical level, which happens in Font [38, p. 26], where it was proposed to define "Kleene's three-valued logic" through $\varphi \wedge \sim \varphi \vdash \psi \vee \sim \psi$ as the characteristic consequence. This proposal is unfortunate indeed, taking into account the fact that for Kleene's logic a more general principle $\varphi \wedge \sim \varphi \vdash \psi$ is usually considered to be characteristic.

Rivieccio in [54, p. 325] tries to reconcile the above mentioned algebraic tradition with logical usage by introducing the name "Kleene's logic of order" for "the logic that corresponds to the lattice order of Kleene lattices" in the sense of Brignole and Monteiro. (He also corrects a mistake in Font's deductive formalization of this logic based on his "Hilbert-style system.") It is unclear how much this new term will help to clear up the confusion. In Section 7, I will explain why it is better to associate the system in question with a specific fragment of **R**-Mingle or Łukasiewicz's logic rather than with that of Kleene's logic.

Anyway, one can only agree with the assessment of Albuquerque, Přenosil and Rivieccio [1, p. 1025] that this logic "has received considerably less attention in the literature" than some of its cousins, such as Kleene's strong three-valued logic and Priest's Logic of Paradox. This lack of attention appears to be unjustified given certain important features of the mingle principle, particularly, its ability to neutralize so-called *paradoxes of relevance*.

In this paper, I will explain, in which sense this principle can secure us against paradoxes even if they appear in our logic, which thus turns out to be "semi-relevant" (Section 2). Moreover, since the separation of the first-degree entailment fragment of a mingle logic makes it possible to leave out further irrelevant properties (such as the "Chain Property"), I will focus on the characteristic principle of that fragment, known as "Safety" (Section 3). Based on a "purely Tarskian" formulation of the first-degree entailment systems (Section 4), I will differentiate between four types of Safety and construct the corresponding proof systems (Section 5). It turns out that these systems form a diamond-shaped lattice with an infinite number of systems connecting its vertices. Furthermore, the diamond's corner systems can be supplied with uniform and rather natural semantics (Section 6), reaffirming the mingle logics' rightful place in the family of first-degree entailment systems.

An important caveat. When in this paper I speak of "first-degree entailment" I always mean a relation between *single formulas*, as it is originally conceived by Belnap, see [11]. Occasionally, I also involve the entailment relation between *sets* of formulas and *formulas*, but only for the sake of comparison.

2. Preliminaries. R-Mingle and Variable Sharing Property

J. Michael Dunn's contributions to the evolution of modern logic are significant and varied. His achievements in investigating and eliminating so-called *paradoxes of relevance*, in particular, are widely acknowledged. These usually refer to certain properties of material (and strict) implication, which may be true even if its antecedent and consequent have nothing to do with each other (are mutually irrelevant). Central among these paradoxes are:

(Positive Paradox) $\quad\quad\quad\quad\quad\quad \varphi \to (\psi \to \varphi),$

which says that a true proposition is implied by any proposition whatsoever, and

(Negative Paradox) $\quad\quad\quad\quad\quad\quad \sim\varphi \to (\varphi \to \psi),$

according to which a false proposition is implied by any proposition whatever it might be.[1]

Objections to these paradoxes have given rise to a whole branch of logical investigation, *relevance logic*, initiated by pioneering work of Wilhelm Ackermann, Alan Ross Anderson and Nuel Belnap, see [3].[2] People who have chosen relevance logic as a field of their scientific interest — Mike Dunn among them — are often called *relevance logicians*.

One can describe a relevance logician as a person who "seeks an entailment connective $\to$ which is such that $A \to B$ holds only if B is relevant to A" Copeland [24, p. 325]. In search of such a connective Belnap [12] has proposed a certain criterion that is now considered a necessary condition for any logic ($\mathfrak{L}$) to be relevant, the so-called *variable sharing property* (VSP):

$\varphi \to \psi$ is a theorem of $\mathfrak{L}$ only if φ and ψ share a sentential variable.

As is well known, all of the major relevance logic systems, such as **B**, **T**, **R** and **E**, have this property. However, in the vicinity of relevance logics, there is a remarkable system that lacks VSP. Dunn invented this system, known as **R**-Mingle (**RM**), in the mid-1960s. The system gets its name from a corresponding logical principle, that is characteristic for it. The principle in question has its origin in a paper by Ohnishi and Matsumoto [48], where the term "mingle" was used for the following rule:

(Mingle rule) $\quad\quad\quad\quad \dfrac{\Gamma \vdash \Theta \quad \Sigma \vdash \Pi}{\Gamma, \Sigma \vdash \Theta, \Pi}$

Anderson and Belnap in [3, p. 97] consider a version of mingle in the context of an intuitionistic sequent system, dealing with sequents having at most one formula in the succedents:

(A–B Mingle) $\quad\quad\quad\quad \dfrac{\Gamma \vdash \varphi \quad \Sigma \vdash \varphi}{\Gamma, \Sigma \vdash \varphi}.$

---

[1] Strictly speaking, the formulas just presented are most commonly called "paradoxes of material implication." I will use a more general name, however, considering these formulas as paradigmatic representatives of a wider group of implicative statements (and also consequence expressions), in which "the antecedents and consequents (or premises and conclusions) are on completely different topics" (Mares [44]), or *irrelevant* to each other. Note also, that the terms "Positive Paradox" and "Negative Paradox" are sometimes used differently by different authors. In this paper, I will use these terms in the sense just described, namely, to specify the situations in which truth (maybe necessary, or logical) is implied (or entailed) by any proposition, and falsehood (maybe necessary, or logical) implies (or entails) any proposition.

[2] Another powerful branch of modern logic, which arose from a consideration of these paradoxes is, of course, modal logic.

They observe, that adding this rule to a system with the following "Arrow on the right" rule:

$(\vdash \to)$
$$\frac{\Gamma, \varphi \vdash \psi}{\Gamma \vdash \varphi \to \psi},$$

yields the following derived rule (even in the absence of Weakening, and provided that $\vdash \varphi \to \psi$ and $\varphi \vdash \psi$ are interderivable, see Dunn [33, p. 149]):

(O–M Mingle)
$$\frac{\varphi \to \gamma \quad \psi \to \gamma}{\varphi \to (\psi \to \gamma)}.$$

The latter leads to the following axiom with the same name (in the presence of $\varphi \to \varphi$):

(Mingle axiom) $\qquad \varphi \to (\varphi \to \varphi).$[3]

Dunn in [33, p. 146] tells an interesting story about inventing **R**-Mingle, formulated by him in a Storrs McCall's graduate seminar at the University of Pittsburgh (cf. [3, p. 94]). By modifying a suggestion of McCall, Dunn simply added Mingle to the set of axioms of relevance logic **R**, and formulated thus the system **RM**, the first appearance of which in print seems to be Dunn [27]. In that paper Dunn shows that, based on Meyer's completeness result for **RM**, every proper normal extension of **RM** has a finite characteristic matrix, despite the fact that **RM** lacks such a matrix.[4] As a result, **RM** is pretabular (or has the so-called *Scroggs's property*, see [33, p. 147]). Furthermore, Dunn demonstrates the strong completeness of **RM** with respect to Sugihara matrices.

Despite the perception (perhaps not quite unjustified) that "**RM** deserves more respect than it has gotten" [33, p. 142], it should be noted that many important aspects of **R**-Mingle and some of its fragments were discussed in detail by different authors, Avron [5; 6; 7; 8], Blok and Raftery [20], Meyer [46], Metcalfe [45], Parks [50], Robles, Méndez and Salto [55] among them. Moreover, it is sometimes argued that the "Dunn–McCall logic **RM** is by far the best understood and the most well-behaved logic in the family of logics developed by the school of Anderson and Belnap" [8, p. 15].

As to the variable sharing property, although **RM** does not possess it in full generality, see [3, p. 417], it still has the following *weak variable sharing property* (WVSP):

> If $\varphi \to \psi$ is a theorem of **RM**, then either (i) φ and ψ share a sentential variable, or (ii) both $\sim \varphi$ and ψ are theorems of **RM**. (Cf. [3, p. 417] and [33, p. 142].)

As Meyer has shown (see [27, p. 4]), the following formula is provable in **RM**.

(ii) $\qquad \sim(\varphi \to \varphi) \to (\psi \to \psi)$

This theorem is denoted by (ii) here because it exemplifies exactly item (ii) in the definition of WVSP by representing an implication in which the negation of a theorem implies (even if irrelevantly) a theorem. It is worthy of note that Mingle and (ii) are

---

[3]My presentation of various "mingles" (either in the form of a rule or an axiom) employs a unified symbolism. "A–B Mingle" and "O–M Mingle" mark the contributions of Anderson and Belnap, and Ohnishi and Matsumoto, respectively. The latter label is taken from [33, p. 152], where it stands for a corresponding axiom. In what follows when I say "Mingle," I will always mean the Mingle *axiom*.

[4]An extension of **RM** is called *normal* iff it is closed under substitution and the **RM**-rules of inference.

interderivable within **RM**, see Schechter [57], so **R** + (ii) yields **RM**. Furthermore, there is another remarkable property of **RM**, that appears to be related to the paradoxes of relevance and states, in effect, that a false proposition implies *any* true proposition. Dunn calls this principle *Ex Falso Verum*:

(EFV) $\qquad \sim \varphi \to (\psi \to (\varphi \to \psi))$.

Mingle is provable in **R** + EFV (see [33, pp. 154–155]), and thus, all three principles — Mingle, (ii) and EFV — turn out to be equivalent within **RM**. However, unlike Mingle, principles (ii) and EFV allow to identify explicitly a subtle but rather important distinction between (1) being *free* from the paradoxes of relevance, and (2) being *safe* from them. This distinction between freedom and safety deserves special consideration. The first, of course, implies the second, but there may well be the second without the first. That is to say, there may well be irrelevant inferences which, nevertheless, are guaranteed to do no harm, and occurrence of which does not cause damage to our knowledge.

Imagine we are developing some theory, and we do so in a standard way by establishing certain axioms (which are by definition true), and then continue to prove theorems, step by step. We want to avoid at least two bad situations. *First*, we want to rule out the possibility of inferring irrelevant conclusions from *true* premises, i.e., we want our arguments for what follows from the axioms to be always on point. *Second*, even if a false statement is inadvertently introduced into our theory, we want to avoid the multiplication of falsity, i.e., we want falsehood to remain isolated and, at the very least, not reproduce itself (till we will be able to find and remove it). As soon as these minimal conditions are met, we feel safe and unaffected by contradictions, even if some of them manage to penetrate our theory.

It turns out that **RM** is quite capable to guarantee the fulfillment of those two conditions. Indeed, according to condition (ii) of WVSP and its proof-theoretic counterpart, even if our inference comes to be irrelevant, the worst thing that can happen, is that we stay with our truths. While the variable sharing property is meant to ensure freedom from paradoxes of relevance, its weak version (WVSP) is well suited to prevent their possible destructive effect.

It is exactly because of WVSP that **RM** is often regarded as a "semi-relevant" system, see, e.g., [3, p. 375], [65, p. 768]. Recently Avron [8] elaborated a more precise notion of "semi-relevance" which encompasses logics in which (1) for every two sets of formulas Γ, Δ, and any formula ψ we have $\Gamma \vdash \psi$ whenever $\Gamma \cup \Delta \vdash \psi$ and Δ has no atomic formulas in common with $\Gamma \cup \{\psi\}$; and which (2) does not have a finite weakly characteristic matrix. Avron proved that **RM** is semi-relevant in this sense.

Dunn in [33, p. 157] specifies some useful properties of **RM**, which make it desirable for use as a "logical tool." Namely, it is decidable, has a low complexity, and can be equipped with a simple, easy-to-understand Kripke-style semantics with a binary accessibility relation, which can be extended to obtain a constant domain semantics for quantifiers. Although **RM** does not have the variable sharing property, it is semi-relevant in the sense of Avron, and it is paraconsistent "in the sense that contradiction does not imply every sentence whatsoever ('Explosion')" [33, p. 160]. Dunn observes that whereas **RM** does have some irrelevant implications like (ii), they are "safe in that, unlike Explosion lead to nothing new" [33, p. 161].[5]

In this context another principle derivable in **RM** might be even more telling, expressing *explicitly* the property of being safe from the paradoxes of relevance as discussed above; namely,

(Safety) $\qquad (\varphi \wedge \sim \varphi) \to (\psi \vee \sim \psi).$

Safety essentially says the same thing as (ii), but with a single implication as the main connective, which opens up the possibility of explaining the idea of protection against paradoxes of relevance on the level of *first-degree entailment* (by replacing implication with a consequence relation). It is called "Safety" in [34, p. 14], because, as Dunn observes, with this principle we can always feel safe: "even if a theory has a contradiction as a theorem, all that can be derived from it are tautologies" [35, p. 443]. Note, that although an informal content of Safety is similar to (ii), their deductive strength is not the same. The former is still weaker than the latter, since neither Mingle, nor (ii) is derivable in **R** + Safety, whereas Safety is derivable not only in **RM**, but also in **R** + (ii), see [57].

The significance of Safety may become clearer if we consider another **RM** property located between Mingle and Safety, which is rather counter-intuitive from a "relevance standpoint" — the so-called *Chain Property* expressed by the formula

(CP) $\qquad (\varphi \to \psi) \vee (\psi \to \varphi).$

As Dunn explains, "[i]t says that given two possibly very distinct sentences, say $p =$ 'The moon is made of green cheese,' and $q =$ 'The cat is on the mat,' one of the two will imply the other" [33, p. 161]. It is hardly possible to find for CP any kind of intuitive justification similar to the one available for (ii) and Safety, and thus, from a relevantist perspective, the validity of the Chain Property may be taken as a "serious weakness for **RM**" [ibid.].

The claim that CP is located "between" (ii) and Safety has an exact sense, since these three principles constitute a non-reversible "chain of derivabilities." (1) CP is derivable in **R** + (ii), but (ii) is not derivable in **R** + CP; and (2) Safety is derivable in **R** + CP, but CP is not derivable in **R** + Safety (see [57, p. 120], taking also into account that Mingle and (ii) are interderivable as mentioned above). As a result, if we want to remove the aforementioned weakness from **RM**, focusing on Safety as "the most safe" result of adding Mingle to **R** may be promising. To that end, one can begin by separating a specific fragment of **RM** that retains some of its most useful features, but lacks the Chain Property. This is the *first-degree entailment* fragment of **RM**.

3. First-degree Entailment and Safety from the Paradoxes

First-degree entailment is a rather remarkable field of study, which has arisen within relevance logic research program, and to which Dunn's contribution is not only

---

[5]Dunn provides a rather witty illustration of how an irrelevant inference can be harmless within **RM**, which I quote here in full: "I have been trying to think of an analogy and the best I have been able to come up with goes something like this. Suppose I am building an electrical circuit and I want to protect against faults. Normally, a small fault will turn all the switches on. But what if I somehow insert a clever circuit that allows a switch to be turned on only if it is already on?" [33, p. 158].

significant, but indeed fundamental. The idea of first-degree entailment has been introduced by Belnap already in his doctoral dissertation [10] and then put into print in [11], who considered an expression of the form $\varphi \to \psi$ a first-degree entailment iff "φ and ψ are both written solely in terms of propositional variables, $\wedge$, $\vee$, and $\sim$ (other truth-functional connectives being treated as defined by these)" [11, notation adjusted]. Thus, by considering first-degree entailments $\varphi \to \psi$, "where φ and ψ can be truth functions of any degree but cannot contain any arrows," one "ignores the possibility and problems of *nested entailments*" [3, p. 150, italics mine].

To grasp this idea, keep in mind that Belnap originally conceived of first-degree entailment as a fragment of the relevance logic system **E** (of entailment). The latter, being the favorite system of Anderson and Belnap, has been designed to provide a "formal analysis of the notion of logical implication, variously referred to also as 'entailment,' [...] expressed in such logical locutions as 'if... then–,' 'implies,' 'entails,' etc., and answering to such conclusion-signaling logical phrases as 'therefore,' 'it follows that,' 'hence,' 'consequently,' and the like" [3, p. 5]. According to this interpretation, the system **E** represents an object-language theory that explains the properties of the entailment relation by identifying it (within this theory) with an object-language implicational connective. In this way, nested implications express statements about entailments between entailments; for instance, $\varphi \to (\psi \to \chi)$ says that φ entails that ψ entails χ.

However, if we want to treat entailment as a meta-language relation in its own right (i.e., "as signifying a metalinguistic relation of logical consequence" [3, p. 150]), it may be appropriate to separate "logical implication" from the rest of the propositional connectives, and consider expressions, in which statements formed only with these remaining connectives are *consequences* of the others. To make this separation more explicit, one can use turnstile ($\vdash$) instead of arrow ($\to$), and obtain in this way *consequence expressions*. Among these expressions one can distinguish the so-called *binary consequence expressions* (cf. [32, p. 24]), or expressions from the FMLA-FMLA logical framework, see [39, p. 198], which represent consequences between *single* formulas. (Dunn and Hardegree consider in this respect "binary implicational systems" [36, p. 194].) A *binary consequence system* is then a proof system which manipulates consequence expressions as formal objects. Anderson and Belnap's concept of first-degree entailment corresponds to the concept of a binary consequence system. Let me summarize (and in a way generalize) this by means of precise definitions.

If $\{\circ_1, \ldots, \circ_m\}$ is a set of binary, and $\{\diamond_1, \ldots, \diamond_n\}$ — a set of of unary propositional connectives, then propositional language $\mathcal{L}_{\{\circ_1, \ldots, \circ_m, \diamond_1, \ldots, \diamond_n\}}$ can be defined as usual:

$$\varphi ::= p \mid \varphi \circ_1 \varphi \mid \ldots \mid \varphi \circ_m \varphi \mid \diamond_1 \varphi \mid \ldots \mid \diamond_n \varphi.$$

Definition 1. A *binary consequence relation* $\vdash$ over $\mathcal{L}_{\{\circ_1, \ldots, \circ_m, \diamond_1, \ldots, \diamond_n\}}$ is a subset of the set $\mathcal{L}_{\{\circ_1, \ldots, \circ_m, \diamond_1, \ldots, \diamond_n\}} \times \mathcal{L}_{\{\circ_1, \ldots, \circ_m, \diamond_1, \ldots, \diamond_n\}}$.

Definition 2. A *binary consequence expression* (or simply a consequence) of $\mathcal{L}_{\{\circ_1, \ldots, \circ_m, \diamond_1, \ldots, \diamond_n\}}$ is a pair $(\varphi, \psi) \in \vdash$, usually written as $\varphi \vdash \psi$ (to be read as "φ has ψ as a consequence" Dunn [31, p. 302]), where $\varphi, \psi \in \mathcal{L}_{\{\circ_1, \ldots, \circ_m, \diamond_1, \ldots, \diamond_n\}}$.

Definition 3. A *logic* in language $\mathcal{L}_{\{\circ_1, \ldots, \circ_m, \diamond_1, \ldots, \diamond_n\}}$ is a binary consequence relation over $\mathcal{L}_{\{\circ_1, \ldots, \circ_m, \diamond_1, \ldots, \diamond_n\}}$, closed at least under the usual Tarskian conditions of *reflexivity* and *transitivity*:

(ref) $\varphi \vdash \varphi$, (tr) $\varphi \vdash \psi, \psi \vdash \chi \Rightarrow \varphi \vdash \chi.$[6]

Definition 4. A *binary consequence system* $(\mathcal{L}_{\{\circ_1,\ldots,\circ_m,\diamond_1,\ldots,\diamond_n\}}, \vdash)$ in the language $\mathcal{L}_{\{\circ_1,\ldots,\circ_m,\diamond_1,\ldots,\diamond_n\}}$ is a proof system, which manipulates binary consequences of $\mathcal{L}_{\{\circ_1,\ldots,\circ_m,\diamond_1,\ldots,\diamond_n\}}$ as formal objects.

Definition 5. Let $\mathfrak{L}$ be a logical system formulated in language $\mathcal{L}_{\{\circ_1,\ldots,\circ_m,\to,\diamond_1,\ldots,\diamond_n\}}$ with an implication $\to$ among its binary connectives. Then binary consequence system $\mathfrak{L}_{\text{fde}} = (\mathcal{L}_{\{\circ_1,\ldots,\circ_m,\diamond_1,\ldots,\diamond_n\}}, \vdash)$ is the *first-degree entailment fragment* of $\mathfrak{L}$ iff for any $\varphi, \psi \in \mathcal{L}_{\{\circ_1,\ldots,\circ_m,\diamond_1,\ldots,\diamond_n\}}$, $\vdash_{\mathfrak{L}} \varphi \to \psi \Leftrightarrow \varphi \vdash_{\mathfrak{L}_{\text{fde}}} \psi$.

In Section 4, I will present a more precise definition of the notion of a binary consequence system (see Definition 12). For now, Definition 4 will suffice for our purposes.

I now return to system **E** and Belnap's elaboration of the idea of first-degree entailment in the context of this system. **E** is usually formulated in the language $\mathcal{L}_{\{\to,\wedge,\vee,\sim\}}$. In his doctoral dissertation [10] (see also [13]) Belnap introduced a proof system in the same language, which was then presented in [3, §5.2] under the label $\mathbf{E}_{\text{fde}}$. A distinctive feature of that system is that one of the connectives, namely, implication ($\to$) could have only one occurrence in a formula of the system as the main operator. I reproduce here this system under the same label as a binary consequence system formulated in language $\mathcal{L}_{\{\wedge,\vee,\sim\}}$.

System $\mathbf{E}_{\text{fde}}$

(ce_1) $\varphi \wedge \psi \vdash \varphi$ $\qquad\qquad$ (ce_2) $\varphi \wedge \psi \vdash \psi$
(di_1) $\varphi \vdash \varphi \vee \psi$ $\qquad\qquad$ (di_2) $\psi \vdash \varphi \vee \psi$
(ni) $\varphi \vdash \sim\sim\varphi$ $\qquad\qquad$ (ne) $\sim\sim\varphi \vdash \varphi$
(dis_1) $\varphi \wedge (\psi \vee \chi) \vdash (\varphi \wedge \psi) \vee \chi$ $\qquad$ (tr) $\varphi \vdash \psi, \psi \vdash \chi / \varphi \vdash \chi$
(ci) $\varphi \vdash \psi, \varphi \vdash \chi / \varphi \vdash \psi \wedge \chi$ $\qquad$ (de) $\varphi \vdash \chi, \psi \vdash \chi / \varphi \vee \psi \vdash \chi$
(con) $\varphi \vdash \psi / \sim\psi \vdash \sim\varphi$

The following theorem shows that $\mathbf{E}_{\text{fde}}$ is indeed the first-degree entailment fragment of system **E**.

Theorem 6. *For any $\varphi, \psi \in \mathcal{L}_{\{\wedge,\vee,\sim\}}$, $\varphi \to \psi$ is provable in **E** iff $\varphi \vdash \psi$ is provable in $\mathbf{E}_{\text{fde}}$.*

Proof. Following Belnap [11], define *primitive entailment* as a consequence $\chi_1 \wedge \cdots \wedge \chi_k \vdash \xi_1 \vee \cdots \vee \xi_l$, in which every χ_i and ξ_j is an *atom* (i.e., a propositional variable or its negation). A primitive entailment is *explicitly tautological* iff some atom χ_i is the same as some ξ_j. Now, a consequence $\varphi \vdash \psi$ represents a *tautological entailment* iff it is reducible by replacements through commutativity, associativity, distributivity, De Morgan and double negation rules to a consequence $\varphi_1 \vee \cdots \vee \varphi_m \vdash \psi_1 \wedge \cdots \wedge \psi_n$, where every $\varphi_i \vdash \psi_j$ is an explicitly tautological entailment.

It can be shown that a formula $\varphi \to \psi$ (where $\varphi, \psi \in \mathcal{L}_{\{\wedge,\vee,\sim\}}$) is provable in **E** iff $\varphi \vdash \psi$ represents a tautological entailment, see [2, p. 14]. Moreover, it is

---
[6]Normally, Tarskian conditions for a consequence relation include monotonicity ($\varphi \vdash \psi \Rightarrow \Gamma, \varphi \vdash \psi$), but it is inexpressible in the FMLA-FMLA framework.

known that $\varphi \vdash \psi$ represents a tautological entailment iff it is provable in $\mathbf{E}_{\text{fde}}$, see [3, pp. 159–161]. ◁

As Dunn observes, already in his doctoral dissertation, this proof may be easily adapted to demonstrate that $\mathbf{E}_{\text{fde}}$ is also the first-degree entailment fragment of system $\mathbf{R}$, "which shows that $\mathbf{E}$ and $\mathbf{R}$ agree in their first degree entailment fragments" [25, p. 115]. To highlight this fact, Dunn in [30, p. 146] calls this system $\mathbf{R}_{\text{fde}}$. In [34] a somewhat different formulation of the system labeled as $\mathbf{R}_{\text{fde}}$ is presented, in which the contraposition rule (con) is replaced by four De Morgan laws, which are derivable in $\mathbf{E}_{\text{fde}}$.

(dm_1) $\sim(\varphi \vee \psi) \vdash \sim\varphi \wedge \sim\psi$ $\qquad$ (dm_2) $\sim\varphi \wedge \sim\psi \vdash \sim(\varphi \vee \psi)$
(dm_3) $\sim(\varphi \wedge \psi) \vdash \sim\varphi \vee \sim\psi$ $\qquad$ (dm_4) $\sim\varphi \vee \sim\psi \vdash \sim(\varphi \wedge \psi)$

It is not difficult to demonstrate that (con) remains admissible (although not derivable) in the system so formulated, see [34, Proposition 11]. In what follows, I will mark by $\mathbf{R}_{\text{fde}}$ the system with De Morgan laws taken as axioms instead of contraposition rule (con), while retaining the label $\mathbf{E}_{\text{fde}}$ for the original formulation from [3] with the contraposition rule. In view of the admissibility of (con) in $\mathbf{R}_{\text{fde}}$ and derivability of (dm_1)–(dm_4) in $\mathbf{E}_{\text{fde}}$, both formulations are deductively equivalent in the sense that they determine the same set of provable consequences.

We now turn to Dunn's fundamental contribution to the metatheory of first-degree entailment. In his doctoral dissertation [25] (see also the seminal paper [28]), Dunn provides $\mathbf{E}_{\text{fde}}$ with an intuitively appealing semantics, the main point of which is to allow for underdetermined and overdetermined valuations, allowing a sentence to be *rationally* considered to be *neither* true nor false, as well as *both* true and false. This has given rise to a highly innovative research program in modeling entailment, which is sometimes called "the American Plan," see Routley [56], (cf. also Shramko [59]). Belnap in [14; 15] has developed Dunn's idea further by introducing specific truth values for such non-standard valuations. These new truth values allow for a quite natural informational interpretation, namely, as information that has been communicated, say, to a computer.

Let $\{t, f\}$ be the set of classical truth values. Define a *generalized truth value function* v^4 as a map from the set of propositional variables into the *subsets* of $\{t, f\}$. These subsets can then be considered *generalized truth values* understood as "mere truth" ($T = \{t\}$), "mere falsehood" ($F = \{f\}$), "neither truth nor falsehood" ($N = \{\}$), and "both truth and falsehood" ($B = \{t, f\}$). Function v^4 is then determined on the set $\{T, F, N, B\}$, and can be extended to compound formulas by the following definition.

Definition 7.

(1) $t \in v^4(\varphi \wedge \psi) \Leftrightarrow t \in v^4(\varphi)$ and $t \in v^4(\psi)$,
  $f \in v^4(\varphi \wedge \psi) \Leftrightarrow f \in v^4(\varphi)$ or $f \in v^4(\psi)$;
(2) $t \in v^4(\varphi \vee \psi) \Leftrightarrow t \in v^4(\varphi)$ or $t \in v^4(\psi)$,
  $f \in v^4(\varphi \vee \psi) \Leftrightarrow f \in v^4(\varphi)$ and $f \in v^4(\psi)$;
(3) $t \in v^4(\sim\varphi) \Leftrightarrow f \in v^4(\varphi)$,
  $f \in v^4(\sim\varphi) \Leftrightarrow t \in v^4(\varphi)$.

Generalized truth values can be explicated as the outcomes of applying the function v^4 to propositions of our language, being thus the elements from the power-set of the set of classical truth values, $\mathcal{P}(\{t,f\})$:

$v^4(\varphi) = T \Leftrightarrow t \in v^4(\varphi)$ and $f \notin v^4(\varphi)$ (φ is true only);
$v^4(\varphi) = F \Leftrightarrow t \notin v^4(\varphi)$ and $f \in v^4(\varphi)$ (φ is false only);
$v^4(\varphi) = B \Leftrightarrow t \in v^4(\varphi)$ and $f \in v^4(\varphi)$ (φ is both true and false);
$v^4(\varphi) = N \Leftrightarrow t \notin v^4(\varphi)$ and $f \notin v^4(\varphi)$ (φ is neither true nor false).

Note the difference between expressions $v^4(\varphi) = T$ and $t \in v^4(\varphi)$. Whereas the former expression says that φ is *only* true (i.e., true and not false), the latter means that φ is *at least* true (which does not exclude that it can be false as well). And similarly for the expressions $v^4(\varphi) = F$ and $f \in v^4(\varphi)$.

We have the following definition of entailment as a relation between single formulas (of the FMLA-FMLA type).

Definition 8. $\varphi \vDash_{\mathsf{fde}} \psi =_{df} \forall v^4 (t \in v^4(\varphi) \Rightarrow t \in v^4(\psi))$.

Entailment relation so defined (call it *FDE-entailment*) is faithful to the consequence relation of $\mathbf{E}_{\mathsf{fde}}$ ($\mathbf{R}_{\mathsf{fde}}$), that is, $\mathbf{E}_{\mathsf{fde}}$ ($\mathbf{R}_{\mathsf{fde}}$) is sound and complete with respect to Definition 8.

Theorem 9. *For any* $\varphi, \psi \in \mathcal{L}_{\{\wedge, \vee, \sim\}}$, $\varphi \vdash \psi$ *is provable in* $\mathbf{E}_{\mathsf{fde}}$ ($\mathbf{R}_{\mathsf{fde}}$) *iff* $\varphi \vDash_{\mathsf{fde}} \psi$.

Proof. See, e.g., Dunn [34, Theorem 7]. ◁

Now, consider the following consequences, which are not derivable in $\mathbf{E}_{\mathsf{fde}}$ (and, of course, neither in $\mathbf{R}_{\mathsf{fde}}$).

(*veq*) $\qquad\qquad\qquad\qquad \varphi \vdash \psi \vee \sim \psi$

(*efq*) $\qquad\qquad\qquad\qquad \varphi \wedge \sim \varphi \vdash \psi$

(*saf*) $\qquad\qquad\qquad\qquad \varphi \wedge \sim \varphi \vdash \psi \vee \sim \psi$

Principles *verum ex quodlibet* (*veq*), *ex falso quodlibet* (*efq*), and (*saf*) are the consequence (first-degree entailment) analogues of the Positive Paradox, Negative Paradox and Safety, respectively (in the sense outlined in footnote 1). Extending $\mathbf{E}_{\mathsf{fde}}$ or $\mathbf{R}_{\mathsf{fde}}$ with these consequences as axioms yields consequence systems for some well-known logics, including the first-degree entailment fragment of **RM**. The latter consequence system is obtained by adding (*saf*) either to $\mathbf{E}_{\mathsf{fde}}$ or $\mathbf{R}_{\mathsf{fde}}$, and it is labeled by $\mathbf{RM}_{\mathsf{fde}}$ in [34]. Thus, (*saf*) is the characteristic consequence for the first-degree entailment fragment of **R-Mingle**, and in this sense one can consider Safety to be a representative of Mingle *on the first-degree entailment level*.

Notably, if $(x) \in \{(veq), (efq), (saf)\}$, then adding (x) to $\mathbf{E}_{\mathsf{fde}}$ is not always deductively equivalent to $\mathbf{R}_{\mathsf{fde}} + (x)$. For example, $\mathbf{E}_{\mathsf{fde}} + (veq)$ is deductively equivalent to $\mathbf{E}_{\mathsf{fde}} + (efq)$, and yields a binary consequence system of classical entailment that includes all valid consequences between formulas of classical logic, cf. Shramko [58, pp. 255–256]. In contrast, $\mathbf{R}_{\mathsf{fde}} + (veq)$ is not deductively equivalent to $\mathbf{R}_{\mathsf{fde}} + (efq)$, and neither of these extensions alone produces classical consequence. To obtain the classical consequence based of $\mathbf{R}_{\mathsf{fde}}$, one must add *both* (*veq*) and (*efq*). On the other hand, $\mathbf{E}_{\mathsf{fde}} + (saf)$ is deductively equivalent to $\mathbf{R}_{\mathsf{fde}} + (saf)$. We thus have the following observation.

Observation 10. *Let the equality sign* $(=)$ *mean the deductive equivalence between consequence systems. Then,* $\mathbf{E}_{fde} + (efq) \neq \mathbf{R}_{fde} + (efq)$, $\mathbf{E}_{fde} + (veq) \neq \mathbf{R}_{fde} + (veq)$ *but* $\mathbf{E}_{fde} + (saf) = \mathbf{R}_{fde} + (saf)$.

This observation shows that systems $\mathbf{E}_{fde}$ and $\mathbf{R}_{fde}$, even being deductively equivalent, are not of equal strength in terms of their possible extensions. Whereas $\mathbf{E}_{fde}$ allows *only two* nontrivial extensions, namely, the first-degree entailment fragment of $\mathbf{RM} = \mathbf{E}_{fde} + (saf)$, see, e.g., [28, p. 157 and note 7], and a system for classical consequence $\mathbf{E}_{fde} + (veq)$ (or equivalently $\mathbf{E}_{fde} + (efq)$), system $\mathbf{R}_{fde}$, even being just another formalization of first-degree entailment, nevertheless, allows *two more* nontrivial extensions. These two additional systems (which are indistinguishable from each other and from classical logic within the deductive framework of $\mathbf{E}_{fde}$) are the consequence system for Kleene's strong three-valued logic $\mathbf{R}_{fde} + (efq)$, and Priest's Logic of Paradox $\mathbf{R}_{fde} + (veq)$, see [34, Theorem 12].

We thus should differentiate between extending a consequence *system* and extending a *logic*. To extend a logic it is enough to add some consequence to the corresponding set of consequences, and to ensure that the resulting set is closed under reflexivity and transitivity, see Definition 3. In contrast, extending a consequence system assumes adding a consequence not derivable in this system to its axioms. Clearly, the resulting logic will be automatically closed under all of this system's primitive inference rules.

Taking into account that one and the same logic can be generated (determined) by different consequence systems, we can generalize Observation 10 by the following proposition.

Proposition 11. *Let S_1 and S_2 be two different consequence systems, which formalize one and the same logic (are deductively equivalent), and let C be a binary consequence, such that neither $S_1 + C$, nor $S_2 + C$ is trivial. Then logics generated by $S_1 + C$ and by $S_2 + C$ may, but do not need be the same.*

In the context of this Proposition, and given the fact that $\mathbf{E}_{fde} + (saf) = \mathbf{R}_{fde} + (saf) = \mathbf{RM}_{fde}$, the following questions arise.

1. Is there another consequence system (**X**) formalizing the logic of first-degree entailment, such that $\mathbf{X} + (saf) \neq \mathbf{RM}_{fde}$?
2. If such a formalization exists, which is the consequence C, such that $\mathbf{X} + C = \mathbf{RM}_{fde}$?
3. Provided there are such **X** and C, are there other systems between $\mathbf{X} + (saf)$ and $\mathbf{X} + C$, which embody the idea of safety from the paradoxes of relevance, and if yes, how many are they?

In the following sections, I will address these questions, by introducing the notion of a purely Tarskian consequence system and applying it to a particular construction of first-degree entailment logic.

4. A Purely Tarskian Deductive Formalization of First-degree Entailment

Consider the following precise definition of what is a binary consequence system, obtained through the notion of a binary consequence rule.

Definition 12. A *binary consequence rule* is a construction of the form
$$\frac{C_1,\ldots,C_n}{C},$$
where $C_1,\ldots,C_n,C$ are binary consequence expressions. If $n=0$, the rule is an *axiom scheme*. A binary consequence rule is *proper* iff $n \geq 1$. A *binary consequence system* is a nonempty set of binary consequence rules, of which at least one is an axiom. A binary consequence system is *proper* iff it has at least one proper binary consequence rule.

In what follows I will keep saying simply "rule" (or "inference rule") instead of "binary consequence rule," and also write rules in the form $C_1,\ldots,C_n/C$. Having a consequence system, it is important to differentiate between derivable and admissible rules of this system.

Definition 13. Rule $C_1,\ldots,C_n/C$ is *derivable* in the binary consequence system S iff there is a derivation of C in S with $C_1,\ldots,C_n$ as the premisses of this derivation.

Definition 14. Rule $C_1,\ldots,C_n/C$ is *admissible* in the binary consequence system S iff whenever all of $C_1,\ldots,C_n$ are derivable in S, then so is C.

All the primitive rules of a binary consequence system are derivable by definition. Clearly, every derivable rule is admissible, but not *vice versa*. Because adding an admissible rule to a consequence system does not change the set of derivable consequences, a binary consequence system is closed under all of its admissible rules. However, while any extension of a consequence system is closed under all derivable rules of the initial system, this is not true for admissible rules. It is possible that a rule that is admissible in a consequence system will no longer be admissible in some of its extensions. Thus, the fewer derivable rules a consequence system has, the more extensions it may allow.

To ensure that a consequence system generates a logic, one should show, in particular, that (tr) is admissible in this system. The easiest way to guarantee this is to take (tr) as a primitive rule. In fact, having transitivity as a derivable rule and reflexivity as an axiom is all that is needed for a consequence system to determine a logic.

Let me call a consequence system *purely Tarskian* iff $\varphi \vdash \varphi$ is derivable for any φ, and (tr) is *the only* primitive inference rule of this system. It may well be that a certain logic cannot be generated by a purely Tarskian consequence system. Remarkably, the logic of first-degree entailment *can* be formalized by a system of this kind. Such a system has been introduced in Shramko [61] as a *genuinely structural* binary consequence system (see also Shramko [60, pp. 1234–1235]).

System **FDE**$_S$

(di_1) $\quad \varphi \vdash \varphi \vee \psi$ $\qquad$ (dco) $\quad \varphi \vee \psi \vdash \psi \vee \varphi$ $\qquad$ (did) $\quad \varphi \vee \varphi \vdash \varphi$

(ce_1) $\quad \varphi \wedge \psi \vdash \varphi$ $\qquad$ (cco) $\quad \varphi \wedge \psi \vdash \psi \wedge \varphi$ $\qquad$ (cid) $\quad \varphi \vdash \varphi \wedge \varphi$

$(das^\vee)$ $\quad (\varphi \vee (\psi \vee \chi)) \vee \xi \vdash ((\varphi \vee \psi) \vee \chi) \vee \xi$

$(cas^\wedge)$ $\quad ((\varphi \wedge \psi) \wedge \chi) \wedge \xi \vdash (\varphi \wedge (\psi \wedge \chi)) \wedge \xi$

$(dis_2^{\vee\wedge})$ $\quad ((\varphi \vee (\psi \wedge \chi)) \vee \xi) \wedge \tau \vdash (((\varphi \vee \psi) \wedge (\varphi \vee \chi)) \vee \xi) \wedge \tau$

$(dis_3^{\vee\wedge})$ $\quad (((\varphi \vee \psi) \wedge (\varphi \vee \chi)) \vee \xi) \wedge \tau \vdash ((\varphi \vee (\psi \wedge \chi)) \vee \xi) \wedge \tau$

$(dis_4^{\vee\wedge})$ $(((\varphi \wedge \psi) \vee (\varphi \wedge \chi)) \vee \xi) \wedge \tau \vdash ((\varphi \wedge (\psi \vee \chi)) \vee \xi) \wedge \tau$
$(dis_5^{\vee\wedge})$ $((\varphi \wedge (\psi \vee \chi)) \vee \xi) \wedge \tau \vdash (((\varphi \wedge \psi) \vee (\varphi \wedge \chi)) \vee \xi) \wedge \tau$
$(ni^{\vee\wedge})$ $(\varphi \vee \psi) \wedge \chi \vdash (\sim\sim\varphi \vee \psi) \wedge \chi$
$(ne^{\vee\wedge})$ $(\sim\sim\varphi \vee \psi) \wedge \chi \vdash (\varphi \vee \psi) \wedge \chi$
$(dm_1^{\vee\wedge})$ $(\sim(\varphi \vee \psi) \vee \chi) \wedge \xi \vdash ((\sim\varphi \wedge \sim\psi) \vee \chi) \wedge \xi$
$(dm_2^{\vee\wedge})$ $((\sim\varphi \wedge \sim\psi) \vee \chi) \wedge \xi \vdash (\sim(\varphi \vee \psi) \vee \chi) \wedge \xi$
$(dm_3^{\vee\wedge})$ $(\sim(\varphi \wedge \psi) \vee \chi) \wedge \xi \vdash ((\sim\varphi \vee \sim\psi) \vee \chi) \wedge \xi$
$(dm_4^{\vee\wedge})$ $((\sim\varphi \vee \sim\psi) \vee \chi) \wedge \xi \vdash (\sim(\varphi \wedge \psi) \vee \chi) \wedge \xi$
(tr) $\varphi \vdash \psi, \psi \vdash \chi \,/\, \varphi \vdash \chi$

In comparison to $\mathbf{E}_{\mathsf{fde}}$ and $\mathbf{R}_{\mathsf{fde}}$, this system has only one primitive inference rule (tr). To compensate for the removal of the rules (ci) and (de), and to ensure their admissibility, axioms for the mutual distributivity of disjunction and conjunction, double negation introduction and elimination, and De Morgan laws are provided with a combined disjunctive-conjunctive context of the form $(\cdots \vee \chi) \wedge \xi$. Moreover, the associativity axioms for conjunction and disjunction are four-termed (and not three-termed, as usual).

A notable feature of a purely Tarskian consequence system, and particularly system $\mathbf{FDE}_\mathsf{S}$, is the ability to be transformed directly into a (purely inferential) Hilbert-style system. Indeed, a binary consequence expression is nothing more than a (one-premise) Hilbertian inference rule, and (tr) can be viewed as a tool for connecting such rules in a logical derivation process. Thus, a binary consequence system with (tr) as the only binary consequence rule can be easily reshaped in a form of a Hilbert system. Such Hilbert-style systems for the first-degree entailment $\mathbf{FDE}_\mathsf{H}$ and a family of its extensions have been elaborated in detail in Shramko [62].

I will now reproduce a number of lemmas and theorems, proofs of which can be found in [61]. In particular, the following lemma makes derivations in $\mathbf{FDE}_\mathsf{S}$ more manageable, by ridding them of the disjunctive/conjunctive context if redundant, and thus, securing an unrestricted implementation of all the usual properties of first-degree entailment.

Lemma 15. *For axioms $(dis_2^{\vee\wedge})$, $(dis_3^{\vee\wedge})$, $(dis_4^{\vee\wedge})$, $(dis_5^{\vee\wedge})$, $(ni^{\vee\wedge})$, $(ne^{\vee\wedge})$, $(dm_1^{\vee\wedge})$, $(dm_2^{\vee\wedge})$, $(dm_3^{\vee\wedge})$, $(dm_4^{\vee\wedge})$ of the form $(\alpha \vee \chi) \wedge \xi \vdash (\beta \vee \chi) \wedge \xi$,*
  (1) *the respective consequences (dis_2)–(dm_4) of the form $\alpha \vdash \beta$ are derivable;*
  (2) *the respective dual consequences $(dis_2^{\wedge\vee})$–$(dm_4^{\wedge\vee})$ of the form $(\alpha \wedge \chi) \vee \xi \vdash (\beta \wedge \chi) \vee \xi$ are derivable;*
  (3) *the respective consequences $(dis_2^{\vee})$–$(dm_4^{\vee})$ of the form $\alpha \vee \chi \vdash \beta \vee \chi$, and $(dis_2^{\wedge})$–$(dm_4^{\wedge})$ of the form $\alpha \wedge \chi \vdash \beta \wedge \chi$ are derivable.*

Moreover, standard formulations of associativity for disjunction (das) $\varphi \vee (\psi \vee \chi) \vdash (\varphi \vee \psi) \vee \chi$, and conjunction (cas) $(\varphi \wedge \psi) \wedge \chi \vdash \varphi \wedge (\psi \wedge \chi)$ are derivable as well.

Lemma 16. *All the inference rules of $\mathbf{E}_{\mathsf{fde}}$ are admissible in $\mathbf{FDE}_\mathsf{S}$.*

Lemma 17. *System $\mathbf{FDE}_\mathsf{S}$ is deductively equivalent to systems $\mathbf{E}_{\mathsf{fde}}$ and $\mathbf{R}_{\mathsf{fde}}$ in the sense that they determine the same set of provable consequences.*

And system $\mathbf{FDE}_\mathsf{S}$ is sound and complete with respect to Definition 8.

Theorem 18. *For any φ, ψ, $\varphi \vdash_{\mathsf{fde}} \psi \Leftrightarrow \varphi \vDash_{\mathsf{fde}} \psi$.*

It is observed in [60, p. 1237] that system **FDE**$_S$ provides the most suitable basis for the family of all its possible extensions. Among these extensions, there is a noteworthy subfamily that represents logics that implement a "minglish idea" to varying degrees on the first-degree entailment level. This subfamily will be discussed further in the following section.

5. Variations on Safety and the Corresponding Consequence Systems

System **RM**$_{fde}$, obtained from **E**$_{fde}$ or **R**$_{fde}$ by adding (saf) as an axiom, is indeed the first-degree entailment fragment of **RM**.

Lemma 19. *For any $\varphi, \psi \in \mathcal{L}_{\{\wedge,\vee,\sim\}}$, $\varphi \to \psi$ is provable in* **RM** *iff* $\varphi \vdash \psi$ *is provable in* **E**$_{fde}$ + (saf), *or equivalently in* **R**$_{fde}$ + (saf).

The proof of this lemma will be given in Section 7. It might appear that adding (saf) to any (deductively equivalent) formalization of the first-degree entailment will produce the same result. However, this would be a mistake, as Rivieccio has shown in [54, p. 328] by the case study of certain extensions of the SET-FMLA system $\vdash_H$ introduced in [38]. Namely, consequence $(\varphi \wedge \sim \varphi) \vee \chi \vdash (\psi \vee \sim \psi) \vee \chi$ turns out not to be derivable in $\vdash_H + (saf)$, although it is valid in **RM**$_{fde}$. This is because the disjunction elimination rule (i.e., (de)) of **E**$_{fde}$, which is admissible in $\vdash_H$, is not admissible in $\vdash_H + (saf)$. In fact, it is proved in [1, p. 1066] that adding $(\varphi \wedge \sim \varphi) \vee \chi \vdash (\psi \vee \sim \psi) \vee \chi$ to $\vdash_H$ provides a correct deductive formalization of the SET-FMLA analogue of **RM**$_{fde}$.[7]

In the case of the first-degree entailment constructed in the FMLA-FMLA framework, the situation can be even more intricate. **FDE**$_S$ is closed under both *disjunction elimination* (de) and *conjunction introduction* (ci) (see Lemma 16), but its various extensions may no longer be so. This allows us to distinguish *four* different versions of Safety, which correspond to the situations, where (1) neither of the two closures holds, (2)–(3) one of them holds, but the other does not, and (4) both of them hold. To express these situations by means of different deductive systems we will need the following four consequences, neither of which is derivable in **FDE**$_S$.

(saf) $\quad \varphi \wedge \sim \varphi \vdash \psi \vee \sim \psi$
$(saf^\wedge)$ $\quad (\varphi \wedge \sim \varphi) \wedge \chi \vdash (\psi \vee \sim \psi) \wedge \chi$
$(saf^\vee)$ $\quad (\varphi \wedge \sim \varphi) \vee \chi \vdash (\psi \vee \sim \psi) \vee \chi$
$(saf^{\vee\wedge})$ $\quad ((\varphi \wedge \sim \varphi) \vee \chi) \wedge \xi \vdash ((\psi \vee \sim \psi) \vee \chi) \wedge \xi$

That is, side by side with the standard principle of Safety (saf), we can consider other three versions of it, augmented with a conjunctive context, a disjunctive context, and a combined disjunctive-conjunctive context.[8] This opens the way for the corresponding binary consequence systems, which formalize different variations of Safety on the basis of **FDE**$_S$. Let me call these systems "FDE-based mingle logics."[9] We thus have:

---

[7] Cf. also Přenosil [52, p. 12], where it is observed that instead of $(\varphi \wedge \sim \varphi) \vee \chi \vdash (\psi \vee \sim \psi) \vee \chi$ one can likewise take a consequence in two variables, namely, $(\varphi \wedge \sim \varphi) \vee \psi \vdash \psi \vee \sim \psi$.

[8] I take this opportunity to correct an erroneous formulation of $(saf^{\vee\wedge})$ in [60, p. 1238] and [62].

[9] If the term "mingle logic" is taken literally, it means that the corresponding system includes Mingle as one of its axioms. In view of this, it might be more appropriate to call these systems "safety logics." It should be noted, however, that we are dealing with the first-degree entailment framework, and Safety is

$$\mathbf{SM_S} = \mathbf{FDE_S} + (saf) \qquad \mathbf{RM_S^\wedge} = \mathbf{FDE_S} + (saf^\wedge)$$
$$\mathbf{RM_S^\vee} = \mathbf{FDE_S} + (saf^\vee) \qquad \mathbf{RM_S} = \mathbf{FDE_S} + (saf^{\vee\wedge})$$

$\mathbf{SM_S}$ stands for "sub-mingle," and it is the weakest FDE-based logic that validates Safety, the characteristic principle of a mingle logic on the first-degree entailment level. It is not, however, the first-degree entailment fragment of $\mathbf{RM}$, because some consequences, whose implicational counterparts are derivable in $\mathbf{RM}$, are not derivable in $\mathbf{SM_S}$. In particular, $(saf^\wedge)$, $(saf^\vee)$ and $(saf^{\vee\wedge})$ do not hold in $\mathbf{SM_S}$, although $(\varphi \wedge \sim\varphi) \wedge \chi \to (\psi \vee \sim\psi) \wedge \chi$, $(\varphi \wedge \sim\varphi) \vee \chi \to (\psi \vee \sim\psi) \vee \chi$ and $((\varphi \wedge \sim\varphi) \vee \chi) \wedge \xi \to ((\psi \vee \sim\psi) \vee \chi) \wedge \xi$ are theorems of $\mathbf{RM}$. This enables further unrolling of Safety in two different directions by means of $\mathbf{RM_S^\wedge}$ and $\mathbf{RM_S^\vee}$. The first of these two systems turns out to be an intermediate stage on the road to Pietz/Kapsner and Rivieccio's "Exactly True Logic" [51], whereas the second one leads to "Non-Falsity Logic" introduced in Shramko et al. [63], see in more detail in [62]. The union of $\mathbf{RM_S^\wedge}$ and $\mathbf{RM_S^\vee}$ results in $\mathbf{RM_S}$, which is exactly the first-degree entailment fragment of $\mathbf{RM}$ (the proof of this fact will be given in Section 7 along with the proof of Lemma 19).

It is not difficult to extend (1)–(3) from Lemma 15 to the case with $(saf^{\vee\wedge})$, and thus to show that (saf) is derivable in all FDE-based mingle logics, whereas both $(saf^\vee)$ and $(saf^\wedge)$ are derivable in $\mathbf{RM_S}$. As an example, consider the derivation of (saf) in $\mathbf{RM_S^\vee}$.

1. $\varphi \wedge \sim\varphi \vdash (\varphi \wedge \sim\varphi) \vee (\psi \vee \sim\psi)$ (di_1)
2. $(\varphi \wedge \sim\varphi) \vee (\psi \vee \sim\psi) \vdash (\psi \vee \sim\psi) \vee (\psi \vee \sim\psi)$ $(saf^\vee)$
3. $(\psi \vee \sim\psi) \vee (\psi \vee \sim\psi) \vdash \psi \vee \sim\psi$ (did)
4. $\varphi \wedge \sim\varphi \vdash \psi \vee \sim\psi$ (1–3; (tr), twice)

Relations between the FDE-based mingle systems are such that they constitute a four-element lattice presented in Figure 1 (together with the base system $\mathbf{FDE_S}$). In this Hasse diagram, the order is the subset relation between the sets of provable consequences of the systems. It is also a sublattice of the lattice of FDE family constructed in [60, Figure 1] (cf. Figure 4 in [62]).

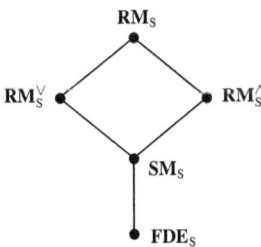

FIGURE 1. Diamond of FDE-based mingle logics

Remarkably, the four system defined above are not the only FDE-based mingle logics. In fact, one can show that there are *infinitely many* of such logics between $\mathbf{SM_S}$ and $\mathbf{RM_S^\wedge}$, as well as between $\mathbf{RM_S^\wedge}$ and $\mathbf{RM_S}$ on the one side, and also between

---

the characteristic principle of the first-degree entailment fragment of the basic mingle logic $\mathbf{RM}$. Therefore, I employ the label which reflects the interconnection between Mingle and Safety on the first-degree entailment level.

$\mathbf{SM}_S$ and $\mathbf{RM}_S^\vee$, as well as between $\mathbf{RM}_S^\vee$ and $\mathbf{RM}_S$ on the other side. To show this, consider the following lemma.

Lemma 20. *Rules (ci) and (de) of the system $\mathbf{E}_{\mathrm{fde}}$ are not admissible in $\mathbf{SM}_S$; (ci) is admissible in $\mathbf{RM}_S^\wedge$, but not in $\mathbf{RM}_S^\vee$; (de) is admissible in $\mathbf{RM}_S^\vee$, but not in $\mathbf{RM}_S^\wedge$; and both (ci) and (de) are admissible in $\mathbf{RM}_S$.*

Proof. To see that neither (ci), nor (de) is admissible in $\mathbf{SM}_S$, note that the following consequences are not derivable in this system

$(saf_{\wedge 1})$ $\qquad (\varphi \wedge \sim \varphi) \vdash (\psi \vee \sim \psi) \wedge (\chi \vee \sim \chi),$

$(saf_{\vee 1})$ $\qquad (\varphi \wedge \sim \varphi) \vee (\psi \wedge \sim \psi) \vdash (\chi \vee \sim \chi),$

although $\varphi \wedge \sim \varphi \vdash \chi \vee \sim \chi$, $\psi \wedge \sim \psi \vdash \chi \vee \sim \chi$, $\varphi \wedge \sim \varphi \vdash \psi \vee \sim \psi$ are derivable in it. The mentioned non-derivability can be established semantically by using the completeness result from Theorem 23 in the next section. I postpone this task till then.

Moreover, (ci) is not admissible in $\mathbf{RM}_S^\vee$, because $(saf_{\wedge 1})$ is not $\mathbf{RM}_S^\vee$-derivable, and (de) is not admissible in $\mathbf{RM}_S^\wedge$, because $(saf_{\vee 1})$ is not $\mathbf{RM}_S^\wedge$-derivable.

To see that both (ci) and (de) are admissible in $\mathbf{RM}_S$, it is enough to show that the following consequences are derivable in it (cf. [62, Lemma 4.2]).

$$(((\varphi \wedge \sim \varphi) \vee \chi) \wedge \xi) \wedge \upsilon \vdash (((\psi \vee \sim \psi) \vee \chi) \wedge \xi) \wedge \upsilon,$$
$$(((\varphi \wedge \sim \varphi) \vee \chi) \wedge \xi) \vee \upsilon \vdash (((\psi \vee \sim \psi) \vee \chi) \wedge \xi) \vee \upsilon,$$

which is not a difficult exercise.

Analogously, to see that (ci) is admissible in $\mathbf{RM}_S^\wedge$ it is enough to observe that $((\varphi \wedge \sim \varphi) \wedge \chi) \wedge \xi \vdash ((\psi \vee \sim \psi) \wedge \chi) \wedge \xi$ is derivable in it, whereas for the admissibility of (de) in $\mathbf{RM}_S^\vee$, one has to show the derivability of $((\varphi \wedge \sim \varphi) \vee \chi) \vee \xi \vdash ((\psi \vee \sim \psi) \vee \chi) \vee \xi$. ◁

In view of Lemma 20, we can apply the methodology from [54], and first observe, that $\mathbf{SM}_S$ turns out to be stronger than the system $\mathbf{SM}_{S\wedge 1} = \mathbf{SM}_S + (saf_{\wedge 1})$, in the sense that the set of consequences derivable in the first system is a proper subset of the set of consequences derivable in the second system, i.e., $\mathbf{SM}_S \subset \mathbf{SM}_{\wedge 1S}$. It is possible to generalize this observation by considering the consequences

$(saf_{\wedge n})$ $\qquad \varphi \wedge \sim \varphi \vdash (\psi_0 \vee \sim \psi_0) \wedge \cdots \wedge (\psi_n \vee \sim \psi_n)$

for each $n \geq 1$. The corresponding systems are then defined as $\mathbf{SM}_{S\wedge n} = \mathbf{SM}_S + (saf_{\wedge n})$. We obtain then a denumerable chain of extensions of $\mathbf{SM}_S$:

$$\mathbf{SM}_S \subset \mathbf{SM}_{S\wedge 1} \subset \cdots \subset \mathbf{SM}_{S\wedge n} \subset \cdots \subset \mathbf{SM}_{S\wedge \infty} \subset \mathbf{RM}_S^\wedge,$$

such that $\mathbf{SM}_{S\wedge n} \subset \mathbf{SM}_{S\wedge n+1}$ for any $n \geq 1$, and $\mathbf{SM}_{S\wedge \infty}$ is the union of all the elements of the chain, except of $\mathbf{RM}_S^\wedge$ (cf. [54, p. 330]).

Furthermore, since $\mathbf{RM}_S^\wedge$ is not closed under (de), it turns out that, e.g.,

$(saf_{\vee 1}^\wedge)$ $\qquad ((\varphi_0 \wedge \sim \varphi_0) \vee (\varphi_1 \wedge \sim \varphi_1)) \wedge \chi \vdash (\psi \vee \sim \psi) \wedge \chi$

is not derivable in $\mathbf{RM}_S^\wedge$, although both $(\varphi_0 \wedge \sim \varphi_0) \wedge \chi \vdash (\psi \vee \sim \psi) \wedge \chi$ and $(\varphi_1 \wedge \sim \varphi_1) \wedge \chi \vdash (\psi \vee \sim \psi) \wedge \chi$ are derivable in it. Generalizing this observation, consider the consequences

$(saf_{\vee n}^\wedge)$ $\qquad ((\varphi_0 \wedge \sim \varphi_0) \vee \cdots \vee (\varphi_n \wedge \sim \varphi_n)) \wedge \chi \vdash (\psi \vee \sim \psi) \wedge \chi$

for each $n \geq 1$, and define the corresponding systems $\mathbf{RM}^\wedge_{S \vee n} = \mathbf{RM}^\wedge_S + (\mathit{saf}^\wedge_{\vee n})$. We then again obtain a denumerable chain of extensions of $\mathbf{RM}^\wedge_S$:

$$\mathbf{RM}^\wedge_S \subset \mathbf{RM}^\wedge_{S \vee 1} \subset \cdots \subset \mathbf{RM}^\wedge_{S \vee n} \subset \cdots \subset \mathbf{RM}^\wedge_{S \vee \infty} \subset \mathbf{RM}_S,$$

such that $\mathbf{RM}^\wedge_{S \vee n} \subset \mathbf{RM}^\wedge_{S \vee n+1}$ for any $n \geq 1$.

Dually, for each $n \geq 1$, we can first consider the consequence

$$(\mathit{saf}_{\vee n}) \qquad (\varphi_0 \wedge \sim \varphi_0) \vee \cdots \vee (\varphi_n \wedge \sim \varphi_n) \vdash \psi \vee \sim \psi,$$

and the corresponding system $\mathbf{SM}_{S \vee n} = \mathbf{SM}_S + (\mathit{saf}_{\vee n})$. A denumerable chain of extensions of $\mathbf{SM}_S$ in other direction looks then as follows:

$$\mathbf{SM}_S \subset \mathbf{SM}_{S \vee 1} \subset \cdots \subset \mathbf{SM}_{S \vee n} \subset \cdots \subset \mathbf{SM}_{S \vee \infty} \subset \mathbf{RM}^\vee_S,$$

such that $\mathbf{SM}_{S \vee n} \subset \mathbf{SM}_{S \vee n+1}$ for any $n \geq 1$. Moving on this path further towards $\mathbf{RM}_S$, we can consider the consequences:

$$(\mathit{saf}^\vee_{\wedge n}) \qquad (\varphi \wedge \sim \varphi) \vee \chi \vdash ((\psi_0 \vee \sim \psi_0) \wedge \cdots \wedge (\psi_n \vee \sim \psi_n)) \vee \chi$$

for each $n \geq 1$, and define the corresponding systems $\mathbf{RM}^\vee_{S \wedge n} = \mathbf{RM}^\vee_S + (\mathit{saf}^\vee_{\wedge n})$. We then obtain a denumerable chain of extensions of $\mathbf{RM}^\vee_S$:

$$\mathbf{RM}^\vee_S \subset \mathbf{RM}^\vee_{S \wedge 1} \subset \cdots \subset \mathbf{RM}^\vee_{S \wedge n} \subset \cdots \subset \mathbf{RM}^\vee_{S \wedge \infty} \subset \mathbf{RM}_S,$$

such that $\mathbf{RM}^\vee_{S \wedge n} \subset \mathbf{RM}^\vee_{S \wedge n+1}$ for any $n \geq 1$.

Thus, each side of the diamond in Figure 1 contains infinitely many FDE-based mingle systems. Algebraic properties of these chains of systems deserve special consideration.

6. Semantics for Safety

Belnap, when explaining his useful four-valued logic of how a computer should think, argues, in particular, that a computer should "say that the inference from A to B is valid, or that A entails B, if the inference never leads us from the True to the absence of the True (preserves Truth), *and also* never leads us from the absence of the False to the False (preserves non-Falsity)", see in [49, p. 44, emphasis in original]. He immediately adds: "Dunn, 1976, has shown that it suffices to mention truth-preservation, since if some inference form fails to always preserve non-Falsity, then it can be shown by a technical argument that it also fails to preserve Truth" [ibid.]. The following lemma confirms the latter remark with respect to the Dunn–Belnap intuitive semantics for first-degree entailment.

Lemma 21. *For any* $\varphi, \psi \in \mathcal{L}_{\{\wedge, \vee, \sim\}}$, $\varphi \vDash_{\mathsf{fde}} \psi \Leftrightarrow \forall v (f \in v^4(\psi) \Rightarrow f \in v^4(\varphi))$.

Proof. See, e.g., Dunn [34, Proposition 4]. ◁

Belnap interprets this result of Dunn in the sense that "the False really is on all fours with the True, so that it is profoundly natural to state our account of 'valid' or 'acceptable' inference in a way which is neutral with respect to the two" [ibid.]. Now, it can also make sense to focus on the emphasis in the initial quotation. Why not interpret it as meaning that the conditions of truth preservation *and* non-falsity preservation are not interchangeable, but rather mutually complementary (and thus jointly necessary)? To be specific, one can assume, that, at least *in some cases*, it is not sufficient for a valid inference to preserve truth *or* (equivalently) non-falsity, as is the case with FDE-entailment, but that *both* preservations must be used as essential components of an entailment framework.

Quite notably, if we will get literal with Belnap's initial suggestion that a valid inference should *explicitly* preserve *both* truth and non-falsity, we can obtain semantics for various FDE-based mingle systems. Dunn in [28, Note 7] explains that we "can capture semantically the first-degree implications of the system **RM**" (for which, he says, (*saf*) is "kind of characteristic") when, together with the usual requirement of truth preservation, we "bring into the definition of 'validity' the *additional* requirement that whenever the conclusion is false so is the premiss" (italics mine). Moreover, it turns out that if by these requirements we will appropriately incorporate two types of truth and falsity — to be *only* true (false), and to be *at least* true (false), then we can develop a general semantic framework which covers all the four "corner systems" of our Diamond. Whereas truth and falsity conditions for $\mathbf{RM_S}$ can be defined in terms of "at least values" (see [34, p. 15]), and truth and falsity conditions for $\mathbf{SM_S}$ employ "only values" (see [62, p. 19]), the semantics for $\mathbf{RM_S^\wedge}$ and $\mathbf{RM_S^\vee}$ can be constructed as certain combinations thereof.

Recall generalized truth-value function v^4, which has been defined in Section 3 as a map from the set of propositional variables ($\mathcal{V}ar$) into Dunn–Belnap's set of (four) generalized truth values $\{T,F,B,N\}$. Define a three-valued truth-value function on a subset of the set of four truth values, namely, $v^3: \mathcal{V}ar \mapsto \{T,F,N\}$, and extend it to compound formulas as in Definition 7, *mutatis mutandis*. We have then the following definition of entailment relations for the four main FDE-based mingle logics.

Definition 22.
1. $\varphi \vDash_{sm} \psi =_{df} \forall v^4(v^4(\varphi) = T \Rightarrow v^4(\psi) = T) \,\&\, \forall v^4(v^4(\psi) = F \Rightarrow v^4(\varphi) = F)$;
2. $\varphi \vDash_{rm\wedge} \psi =_{df} \forall v^4(v^4(\varphi) = T \Rightarrow v^4(\psi) = T) \,\&\, \forall v^3(f \in v^3(\psi) \Rightarrow f \in v^3(\varphi))$;
3. $\varphi \vDash_{rm\vee} \psi =_{df} \forall v^3(t \in v^3(\varphi) \Rightarrow t \in v^3(\psi)) \,\&\, \forall v^4(v^4(\psi) = F \Rightarrow v^4(\varphi) = F)$;
4. $\varphi \vDash_{rm} \psi =_{df} \forall v^3(t \in v^3(\varphi) \Rightarrow t \in v^3(\psi)) \,\&\, \forall v^3(f \in v^3(\psi) \Rightarrow f \in v^3(\varphi))$.

The entailment relation of $\mathbf{SM_S}$ ensures the preservation of value T from a premise to a conclusion, *and also* the preservation of value F in the backward direction (from conclusion to premise) in the Dunn–Belnap four-valued framework. Thus, T plays here the role of the *designated* truth value, whereas F can be considered the *antidesignated* one, cf. [64, p. 492]. The definition of entailment relation in $\mathbf{RM_S}$ retains the same designated and antidesignated truth values, but now in a three-valued setting $\{T,F,N\}$.[10] It is also noteworthy that if in the definition of $\vDash_{sm}$ we keep just the first part, which deals with the value T, we obtain semantics for the Exactly True Logic from [51], and if we focus only on the second part with the value F, the result will be semantics for the Non-Falsity Logic from [63]. Quite remarkably, semantics for Kleene's strong three-valued logic and Priest's Logic of Paradox are obtained in the same way from the definition of $\vDash_{rm}$, see [62, Definition 5.1]. Therefore, relations of $\mathbf{SM_S}$ to Exactly True Logic and Non-Falsity Logic are exactly the same as the relations of $\mathbf{RM_S}$ to Kleene's logic and Priest's logic.

---

[10] Observe, that it could equivalently be defined by means of another truth-value function $v^{3'}: \mathcal{V}ar \mapsto \{T,F,B\}$, cf. [34, Theorem 12 (*iii*)].

Now, since **RM**$_S^\wedge$ is the intersection of **RM**$_S$ and Exactly True Logic, the definition of $\vDash_{rm^\wedge}$ is in fact the combination of definitions of their entailment relations. Note, that in the definition of $\vDash_{rm^\wedge}$ we use simultaneously two truth-value functions, v^4 and v^3. The definition of $\vDash_{rm^\vee}$ is obtained dually.

In what follows, I will denote by $\varphi \vdash_{sm} \psi$, $\varphi \vdash_{rm^\wedge} \psi$, $\varphi \vdash_{rm^\vee} \psi$ and $\varphi \vdash_{rm} \psi$ the facts that the consequence $\varphi \vdash \psi$ is provable in the system **SM**$_S$, **RM**$_S^\wedge$, **RM**$_S^\vee$ and **RM**$_S$ respectively. We have then the following soundness theorem.

Theorem 23. *Let s be sm, rm$^\wedge$, rm$^\vee$ or rm. Then $\varphi \vdash_s \psi \Rightarrow \varphi \vDash_s \psi$.*

Proof. Let us check only the characteristic consequences of each system.

1. Consider (saf) and $\vDash_{sm}$. Assume, that $\varphi \wedge \sim\varphi \nvDash_{sm} \psi \vee \sim\psi$. Then, (a) there is v^4, such that $[v^4(\varphi \wedge \sim\varphi) = T$, and $v^4(\psi \vee \sim\psi) \neq T]$; or (b) there is v^4, such that $[v^4(\psi \vee \sim\psi) = F$, and $v^4(\varphi \wedge \sim\varphi) \neq F]$. In the case of (a) we have, in particular, $v^4(\varphi) = T$, and $v^4(\varphi) = F$, which is impossible. In the case of (b), we have that $v^4(\psi) = F$ and $v^4(\psi) = T$, which is again impossible.

2. Consider $(saf^\wedge)$ and $\vDash_{rm^\wedge}$. Assume $(\varphi \wedge \sim\varphi) \wedge \chi \nvDash_{rm^\wedge} (\psi \vee \sim\psi) \wedge \chi$. Then at least one of the following two cases should be possible:
   (a) $\exists v^4 [v^4((\varphi \wedge \sim\varphi) \wedge \chi) = T$ and $v^4((\psi \vee \sim\psi) \wedge \chi) \neq T]$.
   (b) $\exists v^3 [f \in v^3((\psi \vee \sim\psi) \wedge \chi)$ and $f \notin v^3((\varphi \wedge \sim\varphi) \wedge \chi)]$.
   Take (a). We immediately get $v^4(\varphi) = T$, and $v^4(\varphi) = F$, which is impossible.
   Take (b). Then $[(f \in v^3(\psi)$ and $t \in v^3(\psi))$, or $f \in v^3(\chi)]$, and $f \notin v^3(\varphi), t \notin v^3(\varphi), f \notin v^3(\chi)$. In the first case we have $f \in v^3(\psi), t \in v^3(\psi)$, which is impossible for v^3. In the second case we obtain $f \in v^3(\chi)$ and $f \notin v^3(\chi)$, a contradiction.

3. Consider $(saf^\vee)$ and $\vDash_{rm^\vee}$. Assume $(\varphi \wedge \sim\varphi) \vee \chi \nvDash_{rm^\vee} (\psi \vee \sim\psi) \vee \chi$. Then at least one of the following two cases should be possible:
   (a) $\exists v^3 [t \in v^3((\varphi \wedge \sim\varphi) \vee \chi)$ and $t \notin v^3((\psi \vee \sim\psi) \vee \chi)]$.
   (b) $\exists v^4 [v^4((\psi \vee \sim\psi) \vee \chi) = F$ and $v^4((\varphi \wedge \sim\varphi) \vee \chi) \neq F]$.
   Take (a). Then $[(t \in v^3(\varphi)$ and $f \in v^3(\varphi))$, or $t \in v^3(\chi)]$, and $t \notin v^3(\psi), f \notin v^3(\psi), t \notin v^3(\chi)$. In the first case, we get $t \in v^3(\varphi)$, and $f \in v^3(\varphi)$, which is impossible for v^3. In the second case, we obtain $t \in v^3(\chi)$, and $t \notin v^3(\chi)$, a contradiction.
   Take (b). We immediately get $v^4(\psi) = F$, and $v^4(\psi) = T$, which is impossible.

4. Consider $(saf^{\vee\wedge})$ and $\vDash_{rm}$. Assume $((\varphi \wedge \sim\varphi) \vee \chi) \wedge \xi \nvDash_{rm} ((\psi \vee \sim\psi) \vee \chi) \wedge \xi$. Then at least one of the following two cases should be possible:
   (a) $\exists v^3 [t \in v^3(((\varphi \wedge \sim\varphi) \vee \chi) \wedge \xi)$, and $t \notin v^3(((\psi \vee \sim\psi) \vee \chi) \wedge \xi)]$.
   (b) $\exists v^3 [f \in v^3(((\psi \vee \sim\psi) \vee \chi) \wedge \xi)$, and $f \notin v^3(((\varphi \wedge \sim\varphi) \vee \chi) \wedge \xi)]$.
   Take (a). Then, first, $t \in v^3((\varphi \wedge \sim\varphi) \vee \chi)$, and $t \in v^3(\xi)$, whereas $t \notin v^3((\psi \vee \sim\psi) \vee \chi)$, or $t \notin v^3(\xi)$. We rule out the case $t \in v^3(\xi)$, and $t \notin v^3(\xi)$ (a contradiction), and what remains is: $[(t \in v^3(\varphi)$, and $f \in v^3(\varphi))$, or $t \in v^3(\chi)]$, and $t \notin v^3(\psi), f \notin v^3(\psi), t \notin v^3(\chi)$. In the first case, we get $t \in v^3(\varphi)$, and $f \in v^3(\varphi)$, which is impossible for v^3, and in the second case, we obtain $t \in v^3(\chi)$, and $t \notin v^3(\chi)$, a contradiction.
   Case (b) is covered analogously. ◁

Having Theorem 23, one can show, as promised, that, e.g., $(saf_{\wedge 1})$ from the proof of Lemma 20 is not derivable in **SM**$_S$. Indeed, let, e.g., $v^4(\varphi) = B$, $v^4(\psi) = B$ and $v^4(\chi) = N$. In this case $v^4((\psi \vee \sim\psi) \wedge (\chi \vee \sim\chi)) = F$, but $v^4(\varphi \wedge \sim\varphi) = B$. Hence, $(\varphi \wedge \sim\varphi) \nvDash_{sm} (\psi \vee \sim\psi) \wedge (\chi \vee \sim\chi)$.

For a completeness proof, we employ the canonical model constructions in terms of theories. For any system S we as usual define *S-theory* $\mathcal{T}$ as the set of formulas closed under $\vdash_s$ and conjunction introduction, that is, $\varphi \in \mathcal{T}, \varphi \vdash_s \psi \Rightarrow \psi \in \mathcal{T}$ and $\varphi \in \mathcal{T}, \psi \in \mathcal{T} \Rightarrow \varphi \wedge \psi \in \mathcal{T}$. A theory is *prime* iff it has the disjunction property, that is, $\varphi \vee \psi \in \mathcal{T} \Rightarrow \varphi \in \mathcal{T}$ or $\psi \in \mathcal{T}$. A theory $\mathcal{T}$ is *consistent* iff there is no φ, such that both $\varphi \in \mathcal{T}$ and $\sim \varphi \in \mathcal{T}$. $\mathcal{T}$ is *decisive* iff for each φ, $\varphi \in \mathcal{T}$ or $\sim \varphi \in \mathcal{T}$.

We first prove the completeness of **RM**$_\mathbf{s}$ by considering **RM**-theories. The following variation of Lindenbaum's lemma holds:

Lemma 24. *If $\varphi \nvdash_{\mathsf{rm}} \psi$, then there is a consistent prime **RM**-theory $\mathcal{T}$, such that $\varphi \in \mathcal{T}$ and $\psi \notin \mathcal{T}$, or there is a consistent prime **RM**-theory $\mathcal{T}$, such that $\sim \varphi \notin \mathcal{T}$ and $\sim \psi \in \mathcal{T}$.*

Proof. As usual, one starts from the set of formulas $\mathcal{T}_0 = \{\psi' : \varphi \vdash_{\mathsf{rm}} \psi'\}$. It is easy to see that $\mathcal{T}_0$ is an **RM**-theory (since (tr) and (ci) are admissible in **RM**$_\mathbf{s}$), and moreover, $\varphi \in \mathcal{T}_0$, and $\psi \notin \mathcal{T}_0$. Enumerate all the sentences of our language $\chi_0, \chi_1, \chi_2, \ldots$ (where χ_0 is φ), and consider a series of **RM**-theories $\mathcal{T}_0, \mathcal{T}_1, \mathcal{T}_2, \ldots$, defining for any $\mathcal{T}_n$ theory $\mathcal{T}_{n+1}$ as follows: (1) if $\psi \notin \mathcal{T}_n + \varphi_n$, then $\mathcal{T}_{n+1} = \mathcal{T}_n + \varphi_n$; (2) $\mathcal{T}_{n+1} = \mathcal{T}_n$, otherwise. Consider the union of all $\mathcal{T}_n$. It is easy to see that $\mathcal{T}$ is an **RM**-theory, such that $\varphi \in \mathcal{T}$ and $\psi \notin \mathcal{T}$. By using the closure of **RM**-theories under (de) and distributivity rules one can show that $\mathcal{T}$ is also prime, cf. e.g., Dunn [34, Lemma 8].

Now, if this theory is consistent, we are through. If it is inconsistent, then by (saf), which is derivable in **RM**$_\mathbf{s}$, it is decisive. Define theory $\mathcal{T}^*$ as follows (for any χ): (1) $\chi \in \mathcal{T}^* \Leftrightarrow \sim \chi \notin \mathcal{T}$; (2) $\sim \chi \in \mathcal{T}^* \Leftrightarrow \chi \notin \mathcal{T}$. Using the closure of **RM**$_\mathbf{s}$ under (con), it is not difficult to show that $\mathcal{T}^*$ is indeed a prime **RM**-theory, which is consistent. We also have $\sim \varphi \notin \mathcal{T}^*$ and $\sim \psi \in \mathcal{T}^*$. ◁

We now have the following valuation lemma.

Lemma 25. *Let $\mathcal{T}$ be a prime **RM**-theory, and define a canonical valuation v_τ so that $t \in v_\tau(p)$ iff $p \in \mathcal{T}$, and $f \in v_\tau(p)$ iff $\sim p \in \mathcal{T}$. Then truth and falsity conditions of compound formulas (Definition 7) hold for the canonical valuation so defined.*

Proof. A simple check. ◁

Theorem 26. *For any $\varphi, \psi \in \mathcal{L}_{\{\wedge, \vee, \sim\}}$, $\varphi \vDash_{\mathsf{rm}} \psi \Rightarrow \varphi \vdash_{\mathsf{rm}} \psi$.*

Proof. Let $\varphi \nvdash_{\mathsf{rm}} \psi$. By Lemma 24 there is a consistent prime **RM**-theory $\mathcal{T}$, such that $\varphi \in \mathcal{T}$ and $\psi \notin \mathcal{T}$, or there is a consistent prime **RM**-theory $\mathcal{T}$, such that $\sim \varphi \notin \mathcal{T}$ and $\sim \psi \in \mathcal{T}$. Thus, there is a canonical valuation v_τ, such that $t \in v_\tau(\varphi)$, and $t \notin v_\tau(\psi)$, or there is a canonical valuation v_τ, such that $f \notin v_\tau(\varphi)$, and $f \in v_\tau(\psi)$. Consistency of $\mathcal{T}$ ensures that v_τ is v^3. Hence, $\varphi \nvDash_{\mathsf{rm}} \psi$. ◁

For the completeness proof of **SM**$_\mathbf{s}$, **RM**$_\mathbf{s}^\wedge$, and **RM**$_\mathbf{s}^\vee$, we will deal with FDE-theories to obtain the corresponding variations of the Lindenbaum lemma.

Lemma 27. *Let $\varphi \nvdash_{\mathsf{sm}} \psi$. Then there is a prime FDE-theory $\mathcal{T}$, such that $\varphi \in \mathcal{T}$, $\sim \varphi \notin \mathcal{T}$, and $\psi \notin \mathcal{T}$, or there is a prime FDE-theory $\mathcal{T}$, such that $\sim \varphi \notin \mathcal{T}$, $\sim \psi \in \mathcal{T}$, and $\psi \notin \mathcal{T}$.*

Proof. First, take as the starting theory $\mathcal{T}_0 = \{\psi' : \varphi \vdash_{\mathsf{fde}} \psi'\}$. We have $\varphi \in \mathcal{T}_0$, and since $\varphi \vdash_{\mathsf{fde}} \psi \Rightarrow \varphi \vdash_{\mathsf{sm}} \psi$, also $\psi \notin \mathcal{T}_0$. Either $\sim\varphi \in \mathcal{T}_0$ or $\sim\varphi \notin \mathcal{T}_0$. If the latter is the case, then enumerate all the sentences of the language: $\chi_1, \chi_2, \ldots$, and build up a series of theories by defining for every $\mathcal{T}_n$ the next theory $\mathcal{T}_{n+1}$ as follows: (1) if $\sim\varphi \vee \psi \notin \mathcal{T}_n + \varphi_n$, then $\mathcal{T}_{n+1} = \mathcal{T}_n + \varphi_n$; (2) $\mathcal{T}_{n+1} = \mathcal{T}_n$, otherwise. The required theory $\mathcal{T}$ is then defined as the union of all the $\mathcal{T}_n$'s. $\mathcal{T}$ is a theory containing φ that is maximal with respect to the property of not containing $\sim\varphi \vee \psi$. By the usual argument one can show that $\mathcal{T}$ is prime. By using (di_1) and (dco) we also get that $\sim\varphi \notin \mathcal{T}$ and $\psi \notin \mathcal{T}$.

Assume $\sim\varphi \in \mathcal{T}_0$; then $\varphi \vdash_{\mathsf{fde}} \sim\varphi$. Since $\varphi \vdash_{\mathsf{fde}} \varphi$, by (ci) we have $\varphi \vdash_{\mathsf{fde}} \varphi \wedge \sim\varphi$. Hence, $\varphi \vdash_{\mathsf{sm}} \varphi \wedge \sim\varphi$, and by (saf) it turns out that φ is such that $\varphi \vdash_{\mathsf{sm}} \psi \vee \sim\psi$. In this case, consider theory $\mathcal{T}_0' = \{\sim\varphi' : \varphi' \vdash_{\mathsf{fde}} \psi\}$. We have $\sim\psi \in \mathcal{T}_0'$, and $\sim\varphi \notin \mathcal{T}_0'$. Moreover, $\psi \notin \mathcal{T}_0'$. Indeed, assume $\psi \in \mathcal{T}_0'$; then $\sim\psi \vdash_{\mathsf{fde}} \psi$. Using $\psi \vdash_{\mathsf{fde}} \psi$, by (de) we get $\psi \vee \sim\psi \vdash_{\mathsf{fde}} \psi$, and hence $\psi \vee \sim\psi \vdash_{\mathsf{sm}} \psi$. By transitivity of $\vdash_{\mathsf{sm}}$ we get to $\varphi \vdash_{\mathsf{sm}} \psi$, contrary to the assumption of the lemma. The required theory $\mathcal{T}$ can be constructed as above. ◁

Lemma 28. *Let $\varphi \not\vdash_{\mathsf{rm}^\wedge} \psi$. Then there is a prime FDE-theory $\mathcal{T}$, such that $\varphi \in \mathcal{T}$, $\sim\varphi \notin \mathcal{T}$ and $\psi \notin \mathcal{T}$, or there is a consistent prime FDE-theory $\mathcal{T}$, such that $\sim\varphi \notin \mathcal{T}$ and $\sim\psi \in \mathcal{T}$.*

Proof. Again, take as the starting theory $\mathcal{T}_0 = \{\psi' : \varphi \vdash_{\mathsf{fde}} \psi'\}$. We have $\varphi \in \mathcal{T}_0$, and since $\varphi \vdash_{\mathsf{fde}} \psi \Rightarrow \varphi \vdash_{\mathsf{rm}^\wedge} \psi$, also $\psi \notin \mathcal{T}_0$. Either $\sim\varphi \in \mathcal{T}_0$ or $\sim\varphi \notin \mathcal{T}_0$. If the latter is the case, then we continue as in Lemma 27 and are through.

Assume $\sim\varphi \in \mathcal{T}_0$; then $\varphi \vdash_{\mathsf{fde}} \sim\varphi$. Since $\varphi \vdash_{\mathsf{fde}} \varphi$, then by (ci) we have $\varphi \vdash_{\mathsf{fde}} \varphi \wedge \sim\varphi$. Now, consider theory $\mathcal{T}_0' = \{\sim\varphi' : \varphi' \vdash_{\mathsf{fde}} \psi\}$. We have $\sim\psi \in \mathcal{T}_0'$, and $\sim\varphi \notin \mathcal{T}_0'$. Starting from $\mathcal{T}_0'$ we build up a series of theories by defining for every $\mathcal{T}_n$ the next theory $\mathcal{T}_{n+1}$ as follows: (1) if $\sim\varphi \notin \mathcal{T}_n + \varphi_n$, then $\mathcal{T}_{n+1} = \mathcal{T}_n + \varphi_n$; (2) $\mathcal{T}_{n+1} = \mathcal{T}_n$, otherwise. The required theory $\mathcal{T}$ can be defined as the union of all the $\mathcal{T}_n$, which is a maximal theory containing $\sim\psi$ with respect to the property of not having $\sim\varphi$. By the usual argument one can show that $\mathcal{T}$ is prime. Moreover, $\mathcal{T}$ is consistent. Indeed, assume it is not. Then, there is a formula χ, such that $\chi \in \mathcal{T}$, and $\sim\chi \in \mathcal{T}$. Hence, $\chi \vdash_{\mathsf{fde}} \psi$, and $\sim\chi \vdash_{\mathsf{fde}} \psi$. By (de) we obtain $\chi \vee \sim\chi \vdash_{\mathsf{fde}} \psi$. Recall that $\varphi \vdash_{\mathsf{fde}} \varphi \wedge \sim\varphi$. By (saf) we get to $\varphi \vdash_{\mathsf{fde}} \psi$. Then, $\varphi \vdash_{\mathsf{rm}^\wedge} \psi$, contrary to the assumption of the lemma. ◁

Lemma 29. *Let $\varphi \not\vdash_{\mathsf{rm}^\vee} \psi$. Then there is a consistent prime FDE-theory $\mathcal{T}$, such that $\varphi \in \mathcal{T}$ and $\psi \notin \mathcal{T}$, or there is a prime FDE-theory $\mathcal{T}$, such that $\sim\varphi \notin \mathcal{T}$, $\sim\psi \in \mathcal{T}$ and $\psi \notin \mathcal{T}$.*

Proof. The lemma is proved dually to Lemma 28. ◁

And finally, we have the completeness theorem.

Theorem 30. *Let s be sm, $\mathsf{rm}^\wedge$ or $\mathsf{rm}^\vee$. Then for any $\varphi, \psi \in \mathcal{L}_{\{\wedge, \vee, \sim\}}$, $\varphi \vDash_s \psi \Rightarrow \varphi \vdash_s \psi$.*

Proof. 1. Let $\varphi \not\vdash_{sm} \psi$. By Lemma 27, there is a prime FDE-theory $\mathcal{T}$, such that $\varphi \in \mathcal{T}$, $\sim\varphi \notin \mathcal{T}$, and $\psi \notin \mathcal{T}$, or there is a prime FDE-theory $\mathcal{T}$, such that $\sim\varphi \notin \mathcal{T}$, $\sim\psi \in \mathcal{T}$, and $\psi \notin \mathcal{T}$. Consider the canonical valuation defined in Lemma 25. We have $t \in v_\tau(\varphi)$, $f \notin v_\tau(\varphi)$, and $t \notin v_\tau(\psi)$; or $f \in v_\tau(\psi)$, $t \notin v_\tau(\psi)$, and $f \notin v_\tau(\varphi)$. That is, $v_\tau(\varphi) = T$ and $v_\tau(\psi) \neq T$, or $v_\tau(\psi) = F$ and $v_\tau(\varphi) \neq F$. Hence, $\varphi \not\vDash_{sm} \psi$.

2. Let $\varphi \not\vdash_{rm\wedge} \psi$. Then, by Lemma 28, there is a prime FDE-theory $\mathcal{T}$, such that $\varphi \in \mathcal{T}$, $\sim\varphi \notin \mathcal{T}$ and $\psi \notin \mathcal{T}$, or there is a consistent prime FDE-theory $\mathcal{T}$, such that $\sim\varphi \notin \mathcal{T}$ and $\sim\psi \in \mathcal{T}$. Using the canonical valuation v_τ, we get $v_\tau(\varphi) = T$ and $v_\tau(\psi) \neq T$, or $f \in v_\tau(\psi)$ and $f \notin v_\tau(\varphi)$. Consistency of the theory in the second case ensures that v_τ is in fact v^3. Hence, $\varphi \not\vDash_{rm\wedge} \psi$.

3. The proof for $\varphi \not\vdash_{rm\vee} \psi$ is analogous. ◁

7. Safety, R-Mingle and Łukasiewicz's Logic

In this section, I will consider the issue of a proper characterization of $\mathbf{RM_S}$, and in particular, a disputability of its association with some kind of "Kleene's logic," as, e.g., in [54; 1; 52]. As observed in Section 1, the latter connection arose from consideration of certain algebraic structure which is sometimes named after Kleene. In what follows, I will argue in favor of other logics as more suitable companions for $\mathbf{RM_S}$, and also cast some doubt on the supposed connection of the algebraic structure in question with Kleene.

To this effect, I will give a proof of the fact mentioned above, namely, that $\mathbf{RM_S}$ (and hence $\mathbf{RM_{fde}}$) is indeed the first-degree entailment fragment of both R-Mingle (RM) and Łukasiewicz's three-valued logic Ł3. The proof will be essentially semantical. For axiomatic formulations of RM and Ł3 in the language $\mathcal{L}_{\{\rightarrow,\wedge,\vee,\sim\}}$, consider the following list of axioms and rules of inference:

A1. $\varphi \rightarrow \varphi$
A2. $(\varphi \rightarrow \psi) \rightarrow ((\psi \rightarrow \chi) \rightarrow (\varphi \rightarrow \chi))$
A3. $\varphi \rightarrow ((\varphi \rightarrow \psi) \rightarrow \psi)$
A4. $(\varphi \rightarrow (\varphi \rightarrow \psi)) \rightarrow (\varphi \rightarrow \psi)$
A5. $\varphi \rightarrow (\varphi \rightarrow \varphi)$
A6. $\varphi \rightarrow (\psi \rightarrow \varphi)$
A7. $(\varphi \rightarrow (\psi \rightarrow \chi)) \rightarrow ((\varphi \rightarrow \psi) \rightarrow (\varphi \rightarrow \chi))$
A8. $((\varphi \rightarrow \chi) \rightarrow \psi) \rightarrow (((\psi \rightarrow \varphi) \rightarrow \chi) \rightarrow \chi)$
A9. $(\varphi \wedge \psi) \rightarrow \varphi$ A10. $(\varphi \wedge \psi) \rightarrow \psi$
A11. $((\varphi \rightarrow \psi) \wedge (\varphi \rightarrow \chi)) \rightarrow (\varphi \rightarrow (\psi \wedge \chi))$
A12. $\varphi \rightarrow (\varphi \vee \psi)$ A13. $\psi \rightarrow (\varphi \vee \psi)$
A14. $((\varphi \rightarrow \chi) \wedge (\psi \rightarrow \chi)) \rightarrow ((\varphi \vee \psi) \rightarrow \chi)$
A15. $(\varphi \wedge (\psi \vee \chi)) \rightarrow ((\varphi \wedge \psi) \vee \chi)$
A16. $(\varphi \rightarrow \sim\psi) \rightarrow (\psi \rightarrow \sim\varphi)$
A17. $\sim\sim\varphi \rightarrow \varphi$ A18. $\varphi \rightarrow \sim\sim\varphi$
A19. $(\varphi \wedge \sim\varphi) \rightarrow (\psi \vee \sim\psi)$
R1. $\varphi \rightarrow \psi, \varphi \Rightarrow \psi$
R2. $\varphi, \psi \Rightarrow \varphi \wedge \psi$
R3. $\varphi \rightarrow \psi \Rightarrow \sim\psi \rightarrow \sim\varphi$

System **RM** is determined by A1–A5, A9–A17, R1, R2, see, e.g., [3, p. 341] and [4, §R], and system **Ł3** is determined by A6–A14, A17–A19, R1, R3, see Iturrioz [40, p. 618].

First consider semantics for **RM**. Dunn in [29] constructs a Kripke-style semantics for **R**-Mingle using a binary accessibility relation. Dunn's construction is explicitly based on the valuation $v^{3'}$ from footnote 10, but for the sake of uniformity with the approach adopted in the present paper, I will modify here that semantics with respect to valuation v^3.

Namely, define an **RM**-*model* as a triple $\langle W, R, v^3 \rangle$, where W is a set, R is a reflexive, anti-symmetric, transitive, connected relation on W, and valuation v^3 is now relativized with respect to W, being thus defined as a map from $Var \times W$ into $\{T, F, N\}$, subject to the following *hereditary condition* (for any $p \in Var$, and $\alpha, \beta \in W$).

$$R\alpha\beta \Rightarrow v^3(p,\alpha) \subseteq v^3(p,\beta).$$

The valuation v^3 is extended to compound formulas as follows:

Definition 31.
1. $t \in v^3(\varphi \wedge \psi, \alpha) \Leftrightarrow t \in v^3(\varphi, \alpha)$ and $t \in v^3(\psi, \alpha)$,
   $f \in v^3(\varphi \wedge \psi, \alpha) \Leftrightarrow f \in v^3(\varphi, \alpha)$ or $f \in v^3(\psi, \alpha)$;
2. $t \in v^3(\varphi \vee \psi, \alpha) \Leftrightarrow t \in v^3(\varphi, \alpha)$ or $t \in v^3(\psi, \alpha)$,
   $f \in v^3(\varphi \vee \psi, \alpha) \Leftrightarrow f \in v^3(\varphi, \alpha)$ and $f \in v^3(\psi, \alpha)$;
3. $t \in v^3(\varphi \to \psi, \alpha) \Leftrightarrow \forall \beta [R\alpha\beta \Rightarrow [(t \in v^3(\varphi, \beta) \Rightarrow t \in v^3(\psi, \beta))$ and $(f \in v^3(\psi, \beta) \Rightarrow f \in v^3(\varphi, \beta))]]$,
   $f \in v^3(\varphi \to \psi, \alpha) \Leftrightarrow \exists \beta [R\alpha\beta$ and $[(t \in v^3(\varphi, \beta)$ and $f \in v^3(\psi, \beta))$ or $(f \in v^3(\psi, \beta)$ and $t \in v^3(\varphi, \beta))]]$;
4. $t \in v^3(\sim\varphi, \alpha) \Leftrightarrow f \in v^3(\varphi, \alpha)$, $f \in v^3(\sim\varphi, \alpha) \Leftrightarrow t \in v^3(\varphi, \alpha)$.

Let $\vDash_{RM} \varphi$ mean that the formula φ is valid in the logic **RM**, and let $\vdash_{RM} \varphi$ mean that φ is derivable in **RM**. One can define the notion of **RM**-*validity* as follows.

Definition 32. $\vDash_{RM} \varphi$ iff for any valuation v^3 in every RM-model $\langle W, R, v^3 \rangle$, we have that $v^3(\varphi, \alpha) = T$.

And we have the following soundness and completeness theorem, the proof of which can be extracted from [29] *mutatis mutandis*.

Theorem 33. *For any* $\varphi \in \mathcal{L}_{\{\to, \wedge, \vee, \sim\}}$, $\vdash_{RM} \varphi \Leftrightarrow \vDash_{RM} \varphi$.

Remark 34. The completeness proof of **RM** is given in [29] by a canonical model construction, where an **RM**-model is defined on the base of prime **RM**-theories. The canonical valuation v^c is then relativized with respect to the theories in a canonical model, so that for a propositional variable p, and a theory $\mathcal{T}$, $t \in v^c(p, \mathcal{T}) \Leftrightarrow p \in \mathcal{T}$, and $f \in v^c(p, \mathcal{T}) \Leftrightarrow \sim p \in \mathcal{T}$. It can be shown that the canonical valuation so defined can be extended to compound formulas by Definition 31.

To obtain semantics for **Ł3** define a matrix $\mathbf{Ł}_3 = \langle \{T, N, F\}, T, v^3 \rangle$, where $\{T, N, F\}$ is as in Section 6, T is the designated element, and v^3 is again a map $Var \mapsto \{T, F, N\}$ extended to compound formulas by means of the following definitions for each connective from the language $\mathcal{L}_{\{\to, \wedge, \vee, \sim\}}$:

$v^3_\to$	T	N	F		$v^3_\wedge$	T	N	F		$v^3_\vee$	T	N	F			$v^3_\sim$
T	T	N	F		T	T	N	F		T	T	T	T		T	F
N	T	T	N		N	N	N	F		N	T	N	N		N	N
F	T	T	T		F	F	F	F		F	T	N	F		F	T

Ł3-validity ($\vDash_{Ł3}$) can be defined as usual.

Definition 35. $\vDash_{Ł3} \varphi$ iff for any v^3 in valuation system $Ł_3$, $v^3(\varphi) = T$.

Ł3 is sound and complete with respect to this semantics:

Theorem 36. *For any* $\varphi \in \mathcal{L}_{\{\to,\wedge,\vee,\sim\}}$, $\vdash_{Ł3} \varphi \Leftrightarrow \vDash_{Ł3} \varphi$.

We are now in a position to establish the relationships between **RM**$_S$ on the one hand, and **RM** and **Ł3** on the other hand (cf. [34, Theorem 12]).

Lemma 37. *For any* $\varphi, \psi \in \mathcal{L}_{\{\wedge,\vee,\sim\}}$,
(1) $\varphi \to \psi$ *is provable in* **RM** *iff* $\varphi \vdash \psi$ *is provable in* **RM**$_S$;
(2) $\varphi \to \psi$ *is provable in* **Ł3** *iff* $\varphi \vdash \psi$ *is provable in* **RM**$_S$.

Proof. The direction from right to left is easy to establish by demonstrating that for any rule of **RM**$_S$ of the form $\varphi \vdash \psi$ the corresponding implication $\varphi \to \psi$ is provable both in **RM** and in **Ł3**, which is a routine exercise.

Moving in the opposite direction, assume that $\varphi \vdash \psi$ is *not* derivable in **RM**$_S$. Then:

(1) By Lemma 24, there is a consistent prime **RM**-theory $\mathcal{T}'$, such that $\varphi \in \mathcal{T}'$ and $\psi \notin \mathcal{T}'$, or there is a consistent prime **RM**-theory $\mathcal{T}''$, such that $\sim\varphi \notin \mathcal{T}''$ and $\sim\psi \in \mathcal{T}''$. Consider a canonical **RM**-model $\langle W^c, R^c, v^c \rangle$, where W^c is a set of consistent prime **RM**-theories, such that $\mathcal{T}', \mathcal{T}'' \in W^c$; $R^c \alpha \beta \Leftrightarrow \alpha \subseteq \beta$; and v^c is defined as in the remark above. In this **RM**-model, we thus have $\exists \alpha \in W^c (t \in v^c(\varphi, \alpha)$ and $t \notin v^c(\psi, \alpha))$, or $\exists \beta \in W^c (f \in v^c(\psi, \beta)$ and $f \notin v^c(\varphi, \beta))$. In both cases, we have $\nvDash_{RM} \varphi \to \psi$, and, since **RM** is sound, $\varphi \to \psi$ is not provable in **RM**.

(2) By Theorem 18, $\varphi \nvDash_{rm} \psi$. Hence, $\exists v^3 (t \in v^3(\varphi)$ and $t \notin v^3(\psi))$, or $\exists v^3 (f \in v^3(\psi)$ and $f \notin v^3(\varphi))$. A simple inspection of the definition of $v^3_\to$ in the matrix $Ł_3$ shows that in both cases $t \notin v^3(\varphi \to \psi)$. Thus, $\nvDash_{Ł3} \varphi \to \psi$, and so $\nvdash_{Ł3} \varphi \to \psi$. ◁

Taking into account the deductive equivalence between **RM**$_S$ and **RM**$_{fde}$, this proof provides also the proof of Lemma 19.

Now, returning to the problem of a suitable characterization of the logic with Safety as a distinctive principle, consider algebraic structures usually employed on this issue. The basic structure here is De Morgan algebra.[11] It can be defined as a structure $\langle A, \cap, \cup, -, 1 \rangle$, where $\langle A, \cap, \cup, 1 \rangle$ is a distributive lattice with greatest element 1, and $-$ is a unary operation on A satisfying the following conditions:

(1) $$--x = x,$$
(2) $$-(x \cup y) = -x \cap -y.$$

---

[11]Béziau in [16, p. 280] explains: "The idea of De Morgan algebra can be traced back to Moisil's paper [47]. They were later on studied in Poland and called *quasi-boolean algebras* by Rasiowa [17]. They also have been called *distributive i-lattices* by Kalman [42]."

Brignole and Monteiro in [22] came up with a new name "Kleene algebra" for a structure obtained by equipping a De Morgan algebra with an additional condition, which is an algebraic counterpart of (saf):[12]

(3) $$x \cap -x \leq y \cup -y.$$

They seem to justify this name by an observation that (saf) is verified in Kleene's (strong) three-valued logic, see [22, p. 4]. Kaarli and Pixley propose in [41, p. 296] another definition of Kleene algebra by taking the equation

(4) $$(x \cap -x) \cup (y \cup -y) = y \cup -y$$

instead of (3). Clearly, (3) and (4) are interderivable. It is observed that the variety of Kleene algebras so defined is generated by a special structure $K_3 = \langle \{0, a, 1\}, \cap, \cup, -, 1 \rangle$ (notation adjusted), with $0 < a < 1$, and $-a = a$. The label K_3 apparently suggests the association of this structure with Kleene's logic. It is, however, noteworthy that the structure in question is not uniquely Kleenean, and may well serve as an algebraic background for other three-valued logics.

Most importantly, neither (3) nor (4) is characteristic for Kleene's logic. Indeed, although (saf) is a valid consequence of Kleene's logic, it is so only as a substitutional case of a more general principle (efq) provable there. Moreover, the implicational version of Safety is *not* a theorem of Kleene's logic, since the latter has no theorems at all. Therefore, Brignole and Monteiro's justification of the name "Kleene algebra" seems not very persuasive.

At the same time, an easy check shows that implicational version of Safety is valid in Łukasiewicz's three-valued logic, and (saf) is valid there by itself, without (veq) or (efq) being valid. Lemma 37 states essentially that (saf) is indeed characteristic for the non-implicational fragments of both **R**-Mingle and Łukasiewicz's three-valued logic. Thus, it could be more appropriate to associate the algebraic structure in question with the name of Łukasiewicz rather than Kleene.

Itturioz in [40] defines a three-valued Łukasiewicz algebra as a structure $\langle A, \cap, \cup, \Rightarrow, -, 1 \rangle$, where $\langle A, \cap, \cup, \Rightarrow, 1 \rangle$ is a relatively pseudo-complemented lattice with the following additional condition for $\Rightarrow$:[13]

(5) $$((x \Rightarrow z) \Rightarrow y) \Rightarrow (((y \Rightarrow x) \Rightarrow y) \Rightarrow y) = 1,$$

and the unary operation $-$ is subject to the conditions (1) and (2) above, as well as the following additional condition:

(6) $$(x \cap -x) \cap (y \cup -y) = x \cap -x.$$

Of course, (3), (4) and (6) are all equivalent. Furthermore, a Łukasiewicz algebra so defined is explicitly formulated in the signature $\Omega = \langle \cap, \cup, \Rightarrow, - \rangle$. Now, if one removes from this signature the operation of relative pseudo-complement together with the corresponding conditions, one obtains exactly the structure, which Brignole and

---

[12] As already said in Section 1, Kalman called such structure a "normal i-lattice."

[13] A relatively pseudo-complemented lattice is a structure $\langle A, \cap, \cup, \Rightarrow, 1 \rangle$, where $\langle A, \cap, \cup \rangle$ is a lattice, and the following condition holds: $x \cap y \leq z$ iff $x \leq y \Rightarrow z$. The element $x \Rightarrow y$ is called the pseudo-complement of x relative to y, see [53, pp. 52–53].

Monteiro called "Kleene algebra," but which might be better called "Kalman algebra" or "quasi-Łukasiewicz algebra."

The corresponding consequence system is the first-degree entailment fragment of Łukasiewicz's three-valued logic, which is coincident with the first-degree entailment fragment of the logic **R**-Mingle, and which is a *subsystem* of Kleene's strong three-valued logic.

8. Concluding Remarks

Remarkably, the first individual full-fledged paper [27] published by J. Michael Dunn (if not to take into account a short note [26]) was devoted to **R**-Mingle. The development of this logical system at the start of Dunn's rich and fruitful scientific career, which spanned more than half a century, attested to his exceptional talent and abilities in the field of philosophical logic. It is also rather symbolic that the last (individual) paper Dunn apparently was working on, which has been posthumously published in [33] was also dealing with this logical system. This clearly reaffirms the importance of the mingle logics and the mingle principle in modern logical investigations.

In this paper, I argued for the applicability of the mingle principle as a kind of safety-lock that helps to avoid the most disastrous consequences of the paradoxes of relevance even in the presence of *some* irrelevant inferences. Furthermore, I have concentrated on a specific implementation of the mingle principle on the first-degree entailment level, dubbed "Safety." As it turns out, the first-degree entailment framework allows for a more subtle distinction between four main versions of this principle, which form a four-element diamond-shaped lattice of what can be called "FDE-based mingle logics," with infinitely many intermediate systems in between. The corner systems of that lattice have a very natural and uniform semantics in terms of the forward truth preservation and backward falsity preservation. It would be interesting to extend the proposed semantic framework to the whole infinity of systems from our diamond.

Moreover, it was also observed that the first-degree fragment of **RM** is devoid of a rather problematic irrelevant property (CP), which links together any propositions of our language. In the first-degree entailment context this property is inexpressible on the level of the object language; moreover, its meta-language formulation does not hold in **RM**$_S$ either. That is, there are formulas φ and ψ in the language $\mathcal{L}_{\{\wedge,\vee,\sim\}}$, such that neither $\varphi \vdash \psi$ nor $\psi \vdash \varphi$ is provable in **RM**$_S$. This observation suggests a promising direction of future work — to consider the system **RS** = **R** + Safety.

Acknowledgments. I would like to thank Katalin Bimbó for valuable comments and suggestions that helped me to improve the paper.

References

[1] Albuquerque, H., Přenosil, A. and Rivieccio, U. (2017). An algebraic view of super-Belnap logics, *Studia Logica* **105**: 1051–1086.

[2] Anderson, A. R. and Belnap, N. D. (1962). Tautological entailments, *Philosophical Studies* **13**: 9–24.

[3] Anderson, A. R. and Belnap, N. D. (1975). *Entailment. The Logic of Relevance and Necessity*, Vol. I, Princeton University Press, Princeton, NJ.

[4] Anderson, A. R., Belnap, N. D. and Dunn, J. M. (1992). *Entailment: The Logic of Relevance and Necessity*, Vol. II, Princeton University Press, Princeton, NJ.
[5] Avron, A. (1984). Relevant entailment—semantics and formal systems, *Journal of Symbolic Logic* **49**: 334–342.
[6] Avron, A. (1986). On an implication connective of RM, *Notre Dame Journal of Formal Logic* **27**: 201–209.
[7] Avron, A. (2014). What is relevance logic?, *Annals of Pure and Applied Logic* **165**: 26–48.
[8] Avron, A. (2016). R-mingle and its nice properties, *in* K. Bimbó (ed.), *J. Michael Dunn on Information Based Logics (Outstanding Contributions to Logic, vol. 8)*, Springer, Switzerland, pp. 15–44.
[9] Balbes, R. and Dwinger, P. (1974). *Distributive lattices*, University of Missouri Press, Columbia, MO.
[10] Belnap, N. D. (1959a). *The Formalization of Entailment*, PhD thesis, Yale University, New Haven, CT.
[11] Belnap, N. D. (1959b). Tautological entailments, (abstract), *Journal of Symbolic Logic* **24**: 316.
[12] Belnap, N. D. (1960a). Entailment and relevance, *Journal of Symbolic Logic* **25**: 144–146.
[13] Belnap, N. D. (1960b). A formal analysis of entailment, *Technical Report 7*, U.S. Office of Naval Research.
[14] Belnap, N. D. (1977a). A useful four-valued logic, *in* J. M. Dunn and G. Epstein (eds.), *Modern Uses of Multiple-Valued Logic*, D. Reidel, Dordrecht, pp. 8–37.
[15] Belnap, N. D. (1977b). How a computer should think, *in* G. Ryle (ed.), *Contemporary Aspects of Philosophy*, Oriel Press Ltd., Stocksfield, pp. 30–55.
[16] Béziau, J. (2012). A history of truth-values, *in* D. M. Gabbay, F. Pelletier and J. Woods (eds.), *Handbook of the History of Logic*, Vol. 11, Logic: a History of its Central Concepts, North Holland, Amsterdam, pp. 235–307.
[17] Białynicki-Birula, A. and Rasiowa, H. (1957). On the representation of quasi-boolean algebra, *Bulletin de l'Académie Polonaise des Sciences, Classe III* **5**: 259–261.
[18] Bimbó, K. and Dunn, J. M. (2005). Relational semantics for Kleene logic and action logic, *Notre Dame Journal of Formal Logic* **46**: 461–490.
[19] Bimbó, K. and Dunn, J. M. (2008). *Generalized Galois Logics: Relational Semantics of Nonclassical Logical Calculi*, Vol. 188 of *CSLI Lecture Notes*, CSLI Publications, Stanford, CA.
[20] Blok, W. J. and Raftery, J. G. (2004). Fragments of R-mingle, *Studia Logica* **78**: 59–106.
[21] Blyth, T.S., V. J. (1994). *Ockham Algebras*, Oxford University Press, New York.
[22] Brignole, D. and Monteiro, A. (1964). Caractérisation des algèbres de Nelson par des égalités, *Notas de Lógica Matemática 20*, Instituto de Matemática Universidad del sur Bahía Blanca.
[23] Cignoli, R. (1975). Injective De Morgan and Kleene algebras, *Proceedings of the American Mathematical Society*, Vol. 47, pp. 269–278.
[24] Copeland, B. J. (1980). The trouble Anderson and Belnap have with relevance, *Philosophical Studies* **37**: 325–334.
[25] Dunn, J. M. (1966). *The Algebra of Intensional Logics*, PhD thesis, University of Pittsburgh, University Microfilms, Ann Arbor, MI. (Published as J. Michael Dunn, *The Algebra of Intensional Logics*, with an Introductory Essay by Katalin Bimbó, *Logic PhDs*, vol. 2, College Publications, London, UK, 2019.).
[26] Dunn, J. M. (1967). Drange's paradox lost, *Philosophical Studies* **18**: 94–95.
[27] Dunn, J. M. (1970). Algebraic completeness results for R-mingle and its extensions, *Journal of Symbolic Logic* **35**: 1–13.

[28] Dunn, J. M. (1976a). Intuitive semantics for first-degree entailment and 'coupled trees', *Philosophical Studies* **29**: 149–168.
[29] Dunn, J. M. (1976b). A Kripke-style semantics for R-mingle using a binary accessibility relation, *Studia Logica* **35**: 163–172.
[30] Dunn, J. M. (1986). Relevance logic and entailment, *in* D. Gabbay and F. Guenthner (eds.), *Handbook of Philosophical Logic*, Vol. III, D. Reidel, Dordrecht, pp. 117–224.
[31] Dunn, J. M. (1995). Positive modal logic, *Studia Logica* **55**: 301–317.
[32] Dunn, J. M. (1999a). A comparative study of various model-theoretic treatmets of negation: A history of formal negation, *in* D. M. Gabbay and H. Wansing (eds.), *What is Negation?*, Kluwer Academic Publishers, Dordrecht, pp. 23–51.
[33] Dunn, J. M. (1999b). R-mingle is nice, and so is Arnon Avron, *in* O. Arieli and A. Zamansky (eds.), *Arnon Avron on Semantics and Proof Theory of Non-Classical Logics*, Vol. 21 of *Outstanding Contributions to Logic*, Springer, Switzerland, pp. 23–51.
[34] Dunn, J. M. (2000). Partiality and its dual, *Studia Logica* **66**: 5–40.
[35] Dunn, J. M. (2010). Contradictory information: Too much of a good thing, *Journal of Philosophical Logic* **39**: 425–452.
[36] Dunn, J. M. and Hardegree, G. M. (2001). *Algebraic Methods in Philosophical Logic*, Vol. 41 of *Oxford Logic Guides*, Clarendon Press, Oxford University Press, Oxford, UK.
[37] Ermolaeva, N. M. and Muchnik, A. A. (1974). Modalnye rasshireniya logicheskikh ischislenij tipa Hao Vana, *in* D. A. Bochvar (ed.), *Issledovaniya po Formalizovannym Yazykam i Neklassicheskim Logikam*, Nauka, Moscow, pp. 172–193. (Modal extensions of logical calculi of Hao Wang type, in: D. A. Bochvar (ed.) Investigations into Formalized Languages and Non-classical Logics, [in Russian]).
[38] Font, J. M. (1997). Belnap's four-valued logic and De Morgan lattices, *Logic Journal of the IGPL* **5**: 413–440.
[39] Humberstone, L. (2011). *The Connectives*, MIT Press, Cambridge, MA.
[40] Iturrioz, L. (1977). An axiom system for three-valued Łukasiewicz propositional calculus, *Notre Dame Journal of Formal Logic* **18**: 616–620.
[41] Kaarli, K. and Pixley, A. F. (2018). *Polynomial Completeness in Algebraic Systems*, CRC Press,Taylor & Francis Group, Boca Raton, FL.
[42] Kalman, J. A. (1958). Lattices with involution, *Transactions of the American Mathematical Society* **87**: 485–491.
[43] Makinson, D. C. (1973). *Topics in Modern Logic*, Methuen, London.
[44] Mares, E. (2020). Relevance logic, *in* E. N. Zalta (ed.), *The Stanford Encyclopedia of Philosophy*, Winter 2020 edn, Stanford, CA.
[45] Metcalfe, G. (2016). An Avron rule for fragments of R-mingle, *Journal of Logic and Computation* **26**: 381–393.
[46] Meyer, R. K. (1975). §29.3 Sugihara is a characteristic matrix for **RM**, in Anderson, A. R. and Belnap, N. D., *Entailment. The Logic of Relevance and Necessity*, Vol. I, Princeton University Press, Princeton, NJ, pp. 393–420.
[47] Moisil, G. C. (1935). Recherche sur l'algèbre, de la logique, *Annales Scientifiques de l'Université de Jassy* **22**: 1–117.
[48] Ohnishi, M. and Matsumoto, K. (1962). A system for strict implication, *Proceedings of the Symposium on the Foundations of Mathematics*, Vol. 1, Katada, Japan, pp. 98–108. (Reprinted in *Annals of the Japan Assoc. for Philosophy of Science* 2(4), 1964, 183–188.).
[49] Omori, H. and Wansing, H. (eds.) (2019). *New Essays on Belnap–Dunn Logic*, Vol. 418 of *Synthese Library, Studies in Epistemology, Logic, Methodology, and Philosophy of Science*, Springer, Switzerland.

[50] Parks, R. Z. (1972). A note on R-Mingle and Sobociński's three-valued logic, *Notre Dame Journal of Formal Logic* **13**: 227–228.
[51] Pietz, A. and Rivieccio, U. (2013). Nothing but the truth, *Journal of Philosophical Logic* **42**: 125–135.
[52] Přenosil, A. (n.d.). The lattice of super-Belnap logics, *Review of Symbolic Logic* .
[53] Rasiowa, H. (1974). *An Algebraic Approach to Non-Classical Logics*, PWN – Polish Scientific Publishers, Warszawa.
[54] Rivieccio, U. (2012). An infinity of super-Belnap logics, *Journal of Applied Non-Classical Logics* **22**: 319–335.
[55] Robles, G., Méndez, J. and Salto, F. (2010). A modal restriction of R-Mingle with the variable-sharing property, *Logic and Logical Philosophy* **19**: 341–351.
[56] Routley, R. (1984). The American plan completed: Alternative classical-style semantics, without stars, for relevant and paraconsistent logics, *Studia Logica* **43**: 131–158.
[57] Schechter, E. (2004). Equivalents of mingle and positive paradox, *Studia Logica* **77**: 117–128.
[58] Shramko, Y. (2016a). A modal translation for dual-intuitionistic logic, *The Review of Symbolic Logic* **9**: 251–265.
[59] Shramko, Y. (2016b). Truth, falsehood, information and beyond: the American plan generalized, *in* K. Bimbó (ed.), *J. Michael Dunn on Information Based Logics*, Vol. 8 of *Outstanding Contributions to Logic*, Springer, Switzerland, pp. 191–212.
[60] Shramko, Y. (2020). First-degree entailment and binary consequence systems, *Journal of Applied Logics – IFCoLoG Journal of Logics and their Applications* **7**: 1221–1240.
[61] Shramko, Y. (2021a). Between Hilbert and Gentzen: Four-valued consequence systems and structural reasoning, *Archive for Mathematical Logic* .
[62] Shramko, Y. (2021b). Hilbert-style axiomatization of first-degree entailment and a family of its extensions, *Annals of Pure and Applied Logic* **172**: 103011.
[63] Shramko, Y., Zaitsev, D. and Belikov, A. (2017). First-degree entailment and its relatives, *Studia Logica* **105**: 1291–1317.
[64] Wansing, H. and Shramko, Y. (2008). Harmonious many-valued propositional logics and the logic of computer networks, *in* C. Dégremont, L. Keiff and H. Rückert (eds.), *Dialogues, Logics and Other Strange Things. Essays in Honour of Shahid Rahman*, College Publications, London, pp. 491–516.
[65] Yang, E. (2013). R and relevance principle revisited, *Journal of Philosophical Logic* **42**: 767–782.

DEPARTMENT OF PHILOSOPHY, KRYVYI RIH STATE PEDAGOGICAL UNIVERSITY, KRYVYI RIH, 50086, UKRAINE, *Email:* `shramko@rocketmail.com`

COMPLETENESS VIA METACOMPLETENESS

Shawn Standefer

ABSTRACT. We show that all logics in a certain class of modal relevant logics are complete with respect to their reduced frames. The proof uses a combination of the canonical frame method and metacompleteness results.

Keywords. Completeness, Metavaluations, Modal relevant logics, Reduced frames

This paper is dedicated to the memory of Mike Dunn. I had the good fortune of taking a class at Pitt with Mike, on relevant logics, although he preferred the term "relevance logics." That class was the first time I felt like I understood completeness proofs for relevant logics with respect to ternary relational frames. Mike was a wonderful teacher, and his work on relevant logics has influenced my research greatly. I hope he would have enjoyed this paper for connecting a few dots and answering an open question.

1. INTRODUCTION

The study of relevant logics has been concerned, from its early days, with modal elements. (Dunn and Restall [15] and Bimbó [7] are excellent overviews of the field of relevant logic, and the interested reader can also consult Anderson and Belnap [1], Read [46], Anderson et al. [2], Routley et al. [56], Brady [9], and Mares [34].) The logic E of [1] is the logic of relevance and *necessity*, and Meyer [38] introduced an alethic modal extension of R, which was algebraized by Dunn [13]. The modal aspect of E has been further investigated by Mares and Standefer [37], Standefer [64], and Standefer and Brady [68]. Meyer conjectured that E and the alethic extension of R would coincide under translation, but this was refuted by Maksimova [28].

Early work on models for relevant logics was concerned with modality, such as Urquhart [70, ch. 5] Routley and Meyer [53], and Fine [18, 359*ff*.]. This concern was revitalized several years later, as evidenced by Fuhrmann [19], Mares and Meyer [35; 36], Mares [29; 32], Meyer and Mares [40], and others. Certain approaches to negation in the setting of frames for relevant logics take negation to be a modal notion, such as Restall [49], Berto [3], and Berto and Restall [4]. Modal relevant logics are not just restricted to alethic modal logics, as demonstrated by Goble [21; 23], Wansing [71], Lokhorst [26; 27], Bílková et al. [5], Sedlár [58; 59], Punčochář and Sedlár [45], Standefer [65; 66], and Savić and Studer [57], for example. Nor are modal relevant logics restricted to entirely propositional concerns, as demonstrated by Ferenz [16]

2020 *Mathematics Subject Classification.* Primary: 03B47.

and Tedder and Ferenz [69]. Modal relevant logics are an active area of ongoing research.

Another theme in the area of frames for relevant logics is an interest in reduced frames, frames whose set of regular points is a singleton.[1] This goes back to the first papers by Routley and Meyer [53; 54; 55], although it recurs in later works, such as [56] and Slaney [63].[2]

The two themes come together in the work of [19]. Furhmann notes the interest in reduced frames, just before proving a result showing the incompleteness of an S4-ish extension of R with respect to its reduced frames. The result is generalized by Standefer [67], extending the incompleteness to weaker base logics and more modal extensions.[3] This leads naturally to the question of whether any modal relevant logics are complete with respect to some class of reduced frames. In this paper, we will show that there are. En route to proving this result, we will highlight a slight simplification of Slaney's [63] completeness proof for relevant logics that lack the axiom (WI).

The plan of the paper is as follows. In §2, we will present an overview of the logics we are interested in and provide basic axiom systems. In §3, we will present an overview of the ternary relational frames for relevant logics and their modal extensions and we will define reduced frames. Then in §4, we will give an overview of the method of proving Completeness via the canonical model method, including the adjustments made for canonical reduced frames. Metavaluations are used in the latter construction as well as in the main result of this paper, so they are explained in §5. Finally, in §6, we bring the pieces together to prove that there are modal relevant logics that are complete with respect to their reduced frames.

2. Logics

There are many relevant logics, and there are different ways of distinguishing relevant and non-relevant logics. The logics of interest for this paper are the weaker relevant logics. The stronger relevant logics will not play a prominent role, since their modal extensions have been shown to be incomplete with respect to reduced modal frames.

We will work with a language $\mathcal{L}$ built from a countably infinite set of atoms and the connectives $\{\rightarrow, \wedge, \vee, \sim\}$ extended to include $\Box$.[4] We will use $\mathcal{L}$ to mean either the basic relevant language or the modal extension, leaving it to context to settle which is under discussion. The basic logic B is the smallest set of formulas containing the following axioms and closed under the following rules.

---

[1] See [66] for some discussion of the interest in reduced frames.

[2] Reduced frames are in some work on simplified semantics, such as Priest and Sylvan [44] and Restall [48].

[3] There are incompleteness results for modal relevant logics that do not focus on reduced frames. Goble [22] and Mares [33] both obtain incompleteness results for modal extensions of relevant logics without restricting to reduced frames.

[4] To reduce parentheses, I will adopt the convention that the arrow binds least tightly, followed by conjunction and disjunction, with negation and necessity binding most tightly.

(A1) $A \to A$
(A2) $A \wedge B \to A$, $A \wedge B \to B$
(A3) $(A \to B) \wedge (A \to C) \to (A \to B \wedge C)$
(A4) $A \to A \vee B$, $A \to B \vee A$
(A5) $(B \to A) \wedge (C \to A) \to (B \vee C \to A)$
(A6) $A \wedge (B \vee C) \to (A \wedge B) \vee (A \wedge C)$
(A7) $\sim\sim A \to A$
(R1) $A, A \to B \Rightarrow B$
(R2) $A, B \Rightarrow A \wedge B$
(R3) $A \to \sim B \Rightarrow B \to \sim A$
(R4) $A \to B \Rightarrow (B \to C) \to (A \to C)$
(R5) $A \to B \Rightarrow (C \to A) \to (C \to B)$

There are many different axioms one can add to obtain other relevant logics. Meyer and Routley [41], [56, ch. 4] and Brady [10] provide some examples of common axioms to add to base relevant logics. For the main results of this paper, the upper bound for the strength of the base logic is marked by the addition of the following axioms.[5]

(A9) $A \wedge (A \to B) \to B$ (WI)
(A10) $(A \to B) \to ((B \to C) \to (A \to C))$ (B')
(A11) $(A \to B) \to (\sim B \to \sim A)$ (Contra)

Adding these three axioms to B gives one the logic C of [56], who show this logic to be complete with respect to its reduced frames. C has the distinction of being the weakest logic to be shown complete with respect to reduced frames using the techniques of [56]. [63] showed how to obtain completeness results for weaker logics, notably those lacking (WI), with respect to their reduced frames.[6]

Given a base logic L, the minimal modal extension L.M is obtained by adding the following axiom and rule.

(Agg) $\Box A \wedge \Box B \to \Box (A \wedge B)$
(Mono) $A \to B \Rightarrow \Box A \to \Box B$

One gets further modal extensions by adding other modal axioms and rules. Some standard ones to be considered below are the following.

(T) $\Box A \to A$
(D) $\Box A \to \sim \Box \sim A$
(4) $\Box A \to \Box \Box A$
(Nec) $A \Rightarrow \Box A$

(B) $A \to \Box \sim \Box \sim A$
(5) $\sim \Box A \to \Box \sim \Box A$
(K) $\Box (A \to B) \to (\Box A \to \Box B)$

Adding a set X of the axioms and rule above to L.M will result in the logic L.MX. Below, "L" will at times be used in a way that is indifferent between a base relevant logic and a modal relevant logic, since many of the points do not depend on modal elements being absent. When a modal relevant logic is specifically under consideration, the "L.M" or "L.MX" notation will be used.

The last two items on the list deserve comment, since they are included in normal modal logics whose base logic is classical. The standard relational models for classically based modal logics ensure that (K) and (Nec) are valid. Despite the fact that

---

[5]This is not a common upper bound for logical strength, as it is properly weaker than T, perhaps the weakest of the well known strong relevant logics defended by Anderson and Belnap. To get T from C, one strengthens (A9) to $(A \to (A \to B)) \to (A \to B)$ (W).

[6]Giambrone [20] made a correction to Slaney's work, but the details do not matter for present purposes.

modal frames for relevant logics use a binary relation to interpret the necessity operator, as is done with relational models for normal modal logics, (K) and (Nec) are not valid.

We will say that a formula A is a theorem of the logic L just in case there is a proof using the axioms and rules of L ending in A. When this is the case, we write $\vdash_L A$. It will be useful, at times, to identify a logic with its set of theorems.

Let us turn to the frames for relevant logics.

3. FRAMES

We will use ternary relational frames to define validity.[7] For that, we need some definitions.

Definition 1. A ternary relational frame is a quadruple $\langle K, N, R, * \rangle$, where $K \neq \emptyset$, $N \subseteq K$, $R \subseteq K \times K \times K$, $* : K \mapsto K$, and which obeys the following conditions, where $a \leq b =_{\mathrm{Df}} \exists x \in N \, Rxab$:

(i) $\leq$ is a partial order (reflexive, transitive, and anti-symmetric);
(ii) if $a \in N$ and $a \leq b$, then $b \in N$;
(iii) if $a \leq b$, then $b^* \leq a^*$;
(iv) $a^{**} = a$; and
(v) if $Rabc$, $d \leq a$, $e \leq b$, and $c \leq f$, then $Rdef$.

The basic frames are for the logic B, defined in §2. Frames for stronger logics can be obtained by imposing frame conditions. We will return to these conditions later in this section.

The main result of the paper deals with modal extensions of relevant logics, so we will define modal frames.

Definition 2. A modal frame is a quintuple $\langle K, N, R, *, S \rangle$, where the first four components make up a ternary relational frame and $S \subseteq K \times K$ such that if Sbc and $a \leq b$, then Sac.

We have defined *modal frames* apart from *ternary relational frames* because there are a few points at which it will be useful to have the two notions separate. In particular, the completeness results with respect to reduced frames have mostly been proven for non-modal, reduced ternary relational frames. We will use "frame" indifferently for non-modal ternary relational frames and modal frames.

From frames, whether ternary relational or modal, we obtain models by adding a valuation.

Definition 3. A model M is a pair of a ternary relational frame F and a valuation V, where V is a function from $\mathrm{At} \times K$ to $\{0,1\}$ such that if $a \leq b$ and $V(p,a) = 1$ then $V(p,b) = 1$. The valuation is extended to the whole language as follows.

- $a \Vdash p$ iff $V(p,a) = 1$;
- $a \Vdash \sim B$ iff $a^* \nVdash B$;
- $a \Vdash B \wedge C$ iff $a \Vdash B$ and $a \Vdash C$;
- $a \Vdash B \vee C$ iff $a \Vdash B$ or $a \Vdash C$;

---
[7]For more on ternary relational frames, see, for example, Restall [50] or Bimbó and Dunn [8].

- $a \Vdash B \to C$ iff $\forall b, c \in K(Rabc \wedge b \Vdash B \Rightarrow c \Vdash C)$;
- $a \Vdash \Box B$ iff $\forall b \in K(Sab \Rightarrow b \Vdash B)$.

The heredity condition on $\leq$ is postulated only for atoms. One can show that in a given model, for all formulas A, if $a \Vdash A$ and $a \leq b$, then $b \Vdash A$. This heredity fact, while important for the overall development of the model theory, will be appealed to only implicitly in what follows.

With that background in place, we can define validity.

Definition 4. A formula A holds in a model M iff $\forall a \in N$, $a \Vdash A$.
A formula A is valid on a frame F iff A holds in every model M built on F.
A formula A is valid in a class of frames $\mathcal{C}$ iff A is valid on every frame F in $\mathcal{C}$.

As suggested by the definition of validity, we are interested in logics in the framework FMLA, that is, as sets of formulas.[8]

The goal of this paper is to demonstrate completeness with respect to reduced frames for a range of logics. So, we will define what it is for a frame to be reduced.

Definition 5. A frame F is *reduced* iff there is a unique $\leq$-minimal point $a \in N$ such that $N = \{b \in K : a \leq b\}$. We will denote the minimal element of N in a reduced frame by 0.

Where $\mathcal{C}$ is a class of frames, $\mathfrak{r}(\mathcal{C})$ is the class of reduced frames in $\mathcal{C}$.

We will say that a class of frames that does not satisfy the condition that all frames be reduced is *unreduced*.

In the definition of ternary relational frames, $\leq$ was required to be a partial order. This can be relaxed to be a pre-order, at the cost of adjusting the definition of being a reduced frame. If $\leq$ is a pre-order, then there may be multiple, $\leq$-equivalent minimal worlds, any one one of which could act as 0.[9]

In discussions of reduced frames, there is a different definition that is sometimes used, according to which N is a singleton, $\{0\}$.[10] In that context, a slightly different definition of heredity is used, one that involves only 0, namely, $a \preceq b$ iff $R0ab$. We can show that this definition of heredity agrees with the usual definition on reduced frames.

Lemma 6. *Suppose $\langle K, N, R, * \rangle$ is a reduced frame. Let $\leq$ be the usual heredity ordering. Define $a \preceq b$ as $R0ab$. Then, $a \leq b$ iff $a \preceq b$.*

The lemma extends to modal frames, with the additional frame condition on S following immediately. There is no difference between the definition of reduced frame used here in terms of N having a $\leq$-least element and the other definition in terms of N being a singleton as far as validity and holding in a model go. This follows from the next lemma.

Lemma 7. *Let $\langle K, N, R, *, S \rangle$ be a reduced frame. Then for any model on the frame, $0 \Vdash A$ iff for all $a \in N$, $a \Vdash A$.*

---
[8] See Humberstone [25, 103*ff.*] for more on logical frameworks.
[9] I would like to thank Greg Restall for discussion of this point.
[10] The definition adopted here is used by [50, 304*ff.*].

Proof. The right to left direction is immediate. For the converse, suppose $0 \Vdash A$. Let $a \in N$ be arbitrary. From the definition of 0, $0 \leq a$, so $a \Vdash A$. Therefore, for all $a \in N$, $a \Vdash A$, as desired. ◁

For many logics L, the additional axioms and rules added to B, or to B.M, to obtain L have corresponding frame conditions. For example, the frame condition for (Contra) is $Rabc \Rightarrow Rac^*b^*$, and the frame condition for (4) is that S is transitive.[11] The class of frames obeying the frame conditions for the axioms and rules of L will be $\mathcal{C}_L$. We will call these L-frames.

In $\mathcal{C}_L$, the axioms and rules of L are sound, i.e., if A is a theorem of L then A is valid in $\mathcal{C}_L$, and further, the logic is complete with respect to the class of frames obeying these conditions, that is, if A is valid in $\mathcal{C}_L$, then A is a theorem of L. Not every axiom and rule has a frame condition to which it corresponds in the present sense. For the present paper, we are, for the most part, focusing on axioms and rules that do have corresponding frame conditions.

Soundness with respect to an unreduced class of frames $\mathcal{C}$ implies soundness with respect to a class of reduced frames, $\mathfrak{r}(\mathcal{C})$. Completeness, however, is another matter. It is a surprising feature of many non-modal relevant logics that they are complete with respect to their reduced frames. The primary result of this paper is that completeness extends to some, but not all, modal extensions of relevant logics.

The logic B is sound and complete with respect to $\mathcal{C}$, the class of all ternary relational frames. Completeness extends to $\mathfrak{r}(\mathcal{C})$.[12] The logic B.M is sound and complete with respect to $\mathcal{M}$, the class of all modal frames. If we let X be KT45Nec, for example, then the logic B.MX is complete with respect to $\mathcal{M}_{\mathsf{B.MX}}$ and in fact it is complete with respect to $\mathfrak{r}(\mathcal{M}_{\mathsf{B.MX}})$, a fact that follows from the results of §6. In §4 we will look at the relevant details of the completeness proof.

4. CANONICAL MODEL

The proof of Completeness for relevant logics proceeds via a Henkin-style canonical model method. In this method, one uses the logic L to obtain a large set of appropriate L-theories. One then defines the set of regular points as the set of L-theories containing all the theorems of the logic. One then defines the relations R and S and the operation $*$ in terms of certain formulas being in, or not, certain L-theories. Let us look at the details.

First, let us define L-theories.

Definition 8. A set of formulas X is an L-theory iff (i) if $\vdash_L A \to B$ and $A \in X$ then $B \in X$, and (ii) if $A, B \in X$, then $A \wedge B \in X$.

An L-theory X is *prime* iff $A \vee B \in X$ only if $A \in X$ or $B \in X$.

When proving Completeness, with respect to unreduced frames, one usually proves a lemma showing that there are enough prime L-theories, where there being enough

---

[11] The particular frame conditions will not matter for the results of this paper, so they will be omitted. For detailed lists of frame conditions, see [41], [56, ch. 4], [50, ch. 11], or Goldblatt and Kane [24], for example.

[12] The simplified semantics of [44] uses reduced frames. A version of the Completeness result is proved there.

implies that for any non-theorem A, there is a regular, prime L-theory that does not contain A. We will not recapitulate those details here, since detailed Completeness proofs for the logics we are interested in can be found elsewhere.[13]

The canonical frame for L is defined as follows, where S is omitted if a non-modal L is under consideration, in which case the language $\mathcal{L}$ is understood not to contain formulas with $\Box$.

Definition 9. The canonical frame for L is $\langle K,N,R,^*,S \rangle$, where the components are defined as follows.

- K is the set of prime L-theories.
- $a \in N$ iff $\mathsf{L} \subseteq a$.
- $Rabc$ iff for all $B,C \in \mathcal{L}$, if $B \to C \in a$ and $B \in b$, then $C \in c$.
- $a^* = \{B \in \mathcal{L}: \sim B \notin a\}$.
- Sab iff $\{B \in \mathcal{L}: \Box B \in a\} \subseteq b$.

In the proof of Completeness, the canonical frame for L is shown to be in the class of L-frames, $\mathcal{C}_\mathsf{L}$. This holds for the logics whose axioms have been listed above, as well as other logics. The canonical valuation V is defined as $V(p,a) = 1$ iff $p \in a$, and one shows that for all formulas A, $V(A,a) = 1$ iff $A \in a$. This is often called the Truth Lemma, and it requires substantive proof, the details of which need not concern us here. One then uses the fact that there is a regular, prime L-theory a with $A \notin a$, for the target non-theorem A, to conclude that A is not valid in $\mathcal{C}_\mathsf{L}$. We then conclude that if A is valid in $\mathcal{C}_\mathsf{L}$, then A is a theorem of L.

To adapt the canonical model method to the case where reduced frames are being considered, using the method of Routley et al., one needs to consider some special L-theories. They use L-theories that are closed under T-implications, for some regular, prime L-theory T, where a theory X is closed under T-implication iff $A \to B \in T$ and $A \in X$ only if $B \in X$. L-theories closed under such a theory T are called T-theories. In the canonical frame, one defines K as the set of prime T-theories, with the remaining definitions unchanged.

The proofs demonstrating that the canonical model works, as developed by Routley et al, depend on the theoremhood of (WI), (B), and (Contra) in the target logic. This works for many of the logics stronger than their logic C, including some of the best known stronger relevant logics, such as Anderson and Belnap's R and T. This approach does not work for logics weaker than C, in particular for those lacking (WI), as discussed by Slaney. The question of completeness with respect to reduced frames was, then, left open for those weaker logics for many years. This was, perhaps, unfortunate, since those weaker logics have many virtues.

[63] showed how to prove completeness with respect to reduced frames for many weaker logics, in fact for most of the better known weaker logics.[14] Slaney's approach is somewhat different from that of Routley et al. Rather than require that the target

---

[13] [50, ch. 5] is a good reference with the relevant details.

[14] Some of the weaker logics are known not to be complete with respect to reduced frames. An example is the logic obtained by adding $A \lor \sim A$ to B. It may be worth noting that $A \lor \sim A$ was included in a logic called "B" by [41], although that axiom was later dropped from the now standard B.

logic L contain the axioms (WI), (B), and (Contra), he defines properties of L-theories that will work in the axioms' absence:
- T is *detached* iff $A \to B \in T$ and $A \in T$ only if $B \in T$;
- T is *affixed* iff whenever $A \to B \in T$, both $(C \to A) \to (C \to B) \in T$ and $(B \to C) \to (A \to C) \in T$; and
- T is *transpositive* iff whenever $A \to B \in T$, $\sim B \to \sim A \in T$.

We then take a prime, detached, affixed, transpositive, regular L-theory T. We can use this theory to define the set K in the canonical frame as the set of all prime T-theories. The other parts of the canonical frame are defined as before.

The remaining issue is showing that there are enough prime theories, in particular that there is an appropriate regular prime theory. The construction of a regular, prime L-theory T, excluding the target non-theorem and obeying the conditions above, via Lindenbaum's lemma runs into a problem: the end result of the construction may not be prime. Slaney's second major innovation was to use metavaluations, to be explained in §5, to obtain an appropriate prime subtheory U of the constructed theory T, building on an idea of Meyer. The resulting theory, U, is regular, prime, obeys the conditions above, and excludes the target non-theorem. The construction of the canonical frame can proceed using U as 0. We are, then, left to provide the details of metavaluations, to which we now turn.

5. Metacompleteness

In this section, we will define metavaluations. We will not present any of the details of the history of metavaluations, an excellent overview of which is provided by Brady [12]. We will begin with the metavaluations presented by [63]. Slaney distinguishes two classes of logics, M1 and M2, depending on the additional axioms and rules included. The axioms and rules for each class of logics are displayed in Table 1.[15]

(B1)	$(A \to B) \land (B \to C) \to (A \to C)$	(B6)	$A \to (B \to A)$
(B2)	$(A \to \sim B) \to (B \to \sim A)$	(B7)	$A \Rightarrow B \to A$
(B3)	$(A \to B) \to ((C \to A) \to (C \to B))$	(B8)	$A \to ((A \to B) \to B)$
(B4)	$(A \to B) \to ((B \to C) \to (A \to C))$	(B9)	$(A \to (B \to C)) \to (B \to (A \to C))$
(B5)	$A \to (A \to A)$	(B10)	$A \Rightarrow (A \to B) \to B$

Table 1. Axioms and rules for the M1 and M2 logics.

The M1 logics are obtained by adding zero or more of (B1)–(B7) to B, and the M2 logics are obtained by adding (B10) and zero or more of (B2)–(B9) to B.

Definition 10. A *metavaluation* for L is a pair of functions $\mathfrak{m}(\cdot)$ and $\mathfrak{m}^\star(\cdot)$ from $\mathcal{L} \mapsto \{0,1\}$ such that
- $\mathfrak{m}(p) = 0$, $\mathfrak{m}^\star(p) = 1$, for $p \in $ At;
- $\mathfrak{m}(t) = 1$, $\mathfrak{m}^\star(t) = 1$;
- $\mathfrak{m}(\sim A) = 1$ iff $\mathfrak{m}^\star(A) = 0$, $\mathfrak{m}^\star(\sim A) = 1$ iff $\mathfrak{m}(A) = 0$;

---

[15] The metavaluations used here are the metavaluations for the logic. A more general definition is available, which is appropriate for applying metavaluations to arbitrary theories. The extra generality, while interesting and sometimes useful, is not needed here, so it is omitted. I will also note that the tables omit an axiom that does not have a corresponding frame condition for the ternary relational models.

- $m(A \wedge B) = 1$ iff $m(A) = 1$ and $m(B) = 1$,
  $m^\star(A \wedge B) = 1$ iff $m^\star(A) = 1$ and $m^\star(B) = 1$;
- $m(A \vee B) = 1$ iff $m(A) = 1$ or $m(B) = 1$,
  $m^\star(A \vee B) = 1$ iff $m^\star(A) = 1$ or $m^\star(B) = 1$;
- $m(A \to B) = 1$ iff (i) $\vdash_L A \to B$, (ii) if $m(A) = 1$ then $m(B) = 1$, and (iii) if $m^\star(A) = 1$, then $m^\star(B) = 1$.
  For M1 logics, $m^\star(A \to B) = 1$.
  For M2 logics, $m^\star(A \to B) = 1$ iff $m(A) = 1$ only if $m^\star(B) = 1$.

As is clear from the definition, the clause for the conditional differs depending on whether the logic is an M1 logic or an M2 logic.

[63] proves a metacompleteness theorem for the M1 and M2 logics.[16]

Theorem 11 (Metacompleteness). *Let L be an M1 logic or an M2 logic. Then $\vdash_L A$ iff $m(A) = 1$.*

Seki [61] extends Slaney's metavaluations to modal vocabulary. We will only use Seki's extension for necessity, although he provides clauses for a primitive possibility operator.

In addition to the distinction between M1 and M2 logics, Seki needs to distinguish Ms and Mt logics, depending on the modal axioms and rules that are included. The Ms/Mt distinction is independent of the M1/M2 distinction, so there are four classes of logics one can consider. The basic modal axioms and rules included in Ms and Mt logics are in Table 2.[17]

(C1)	$\Box A \Rightarrow A$	(CoNec)	(C3)	$\Box(A \to B) \to (\Box A \to \Box B)$	(K)
(C2)	$A \Rightarrow \Box A$	(Nec)	(C4)	$\Box(A \to B) \to (\sim\Box\sim A \to \sim\Box\sim B)$	
			(C5)	$A \Rightarrow \sim\Box\sim A$	(Poss)

Table 2. Axioms and rules for the Ms and Mt logics.

The Ms logics are obtained by adding zero or more of (C1)–(C4) to L.M, where L is an M1 or M2 logic, and the Mt logics are obtained by adding (C5) to an Ms logic.

Definition 12. A *modal metavaluation* for a logic L.MX is a pair of functions $m(\cdot)$ and $m^\star(\cdot)$ from $\mathcal{L}_\Box$ to $\{0,1\}$ that satisfy the conditions to be M1 or M2 metavaluations and also satisfy the following additional conditions.

- $m(\Box A) = 1$ iff $\vdash_L \Box A$ and $m(A) = 1$.
  For Ms logics, $m^\star(\Box A) = 1$.
  For Mt logics, $m^\star(\Box A) = 1$ iff $m^\star(A) = 1$.

[61] proves a metacompleteness theorem for Ms and Mt logics. He goes on to show that several additional axioms can be added to the different logics, including various Sahlqvist axiom forms. Several of the more common modal axioms can be

---

[16]Slaney [62] proves metacompleteness for some particular M1 and M2 logics, but the later paper provides the general definition of the classes of logics.

[17]Seki includes a disjunctive meta-rule, if $A \Rightarrow B$, then $A \vee C \Rightarrow B \vee C$, as one of the optional additions to both the Ms and Mt classes. For more on the use of disjunctive meta-rules in reduced completeness proofs, see [9, pp. 7–9].

added to Ms and Mt logics while maintaining metacompleteness. Both Ms and Mt logics can be augmented with (4) ($\Box A \to \Box\Box A$) or (5) ($\sim\Box A \to \Box\sim\Box A$) and remain metacomplete. Mt logics can also be augmented with any of (D) ($\Box A \to \sim\Box\sim A$), (T) ($\Box A \to A$) and (B) ($A \to \Box\sim\Box\sim A$) and still be metacomplete.[18] These results hold regardless of whether the non-modal base logic is an M1 or M2 logic, so the modal element of the metavaluation has considerable freedom from the non-modal base logic.

Metacompleteness, whether for a base logic or for a modal logic, has many consequences. For our purposes, the primary consequence is that metacomplete logics are prime.

Corollary 13. *Suppose* L *is metacomplete. Then* $\vdash_L A \vee B$ *only if* $\vdash_L A$ *or* $\vdash_L B$.

This is the crucial fact that we will appeal to in the completeness results, to which we now turn.

6. COMPLETENESS

We are now almost in a position to prove that there are modal relevant logics that are complete with respect to their reduced frames.

Let us say that a frame condition is *reducible* iff it uses only $S, R, *, \leq$, terms for points, existential quantifiers, universal quantifiers, conjunction, disjunction, and the material conditional. In particular, a frame condition is not reducible if it uses a restricted quantification over N that is not used in an instance of $\leq$.[19] For example, the frame condition for (Contra), $Rabc \Rightarrow Rac^*b^*$, is reducible, as is the frame condition for (T), that S is reflexive. By contrast, the frame condition for (Nec), $a \in N \wedge Sab \Rightarrow b \in N$, is not, since it uses "N" outside of "$\leq$." Similarly, $\exists x \in N\, Raxa$ is not a reducible condition, since it has a restricted quantification on N occurring outside of "$\leq$."

Many of the frame conditions for common axioms for relevant logics are reducible. Adding these axioms, individually or in a group, to B, or to B.M results in a logic that is sound and complete with respect to the class of frames obeying the corresponding conditions. The proof of Completeness for L with respect to L-frames, using the canonical model method, demonstrates that the canonical frame for L belongs to the class of L-frames. For our main result, we will record two lemmas, noting a connection between L-frames and reducible frame conditions.

Lemma 14. *Suppose* $\langle K, N, R, *\rangle$ *is an* L*-frame, all the conditions on* L*-frames are reducible, and there is* $0 \in N$ *such that* $0 \leq a$, *for all* $a \in N$. *Then* $\langle K, N, R, *\rangle$ *is a reduced* L*-frame.*

Proof. All the conditions on B-frames are satisfied in virtue of being an L-frame. Further, the additional frame conditions hold as the frame is unchanged. Therefore, it is a reduced L-frame. ◁

---

[18]NB: The axiom (B), "B" for "Brouwersche," is a modal axiom, not to be confused with the relevant logic B, "B" for "Basic," or the names for the implicational axioms (B) and (B'), whose designations come from combinatory logic.

[19]It should be clear that a frame condition may be reducible while a logically equivalent condition is not. A more general definition treats a condition as reducible iff it is equivalent to one of the highlighted form.

The lemma still holds if the frame is a modal frame, which we will state separately.

Lemma 15. *Suppose $\langle K,N,R,^*,S \rangle$ is an L-frame, all the conditions on L-frames are reducible, and there is $0 \in N$ such that $0 \leq a$, for all $a \in N$. Then $\langle K,N,R,^*,S \rangle$ is a reduced L-frame.*

The proof is the same as the previous lemma, so we omit it. These lemmas tell us that if the canonical frame for L has a $\leq$-minimal world and is an L-frame, then the canonical frame is a reduced L-frame.

We now come to our main lemma.

Lemma 16. *Suppose L is prime. If the canonical frame for L is in a class $\mathcal{C}$ of frames, then the canonical frame is in $\mathfrak{r}(\mathcal{C})$.*

Proof. Suppose L is prime and that the canonical frame for L is in the class $\mathcal{C}$. If L is a modal logic, then the frame includes S as well, defined as above. Since L is prime, $L \in K$. In fact, $L \in N$, and for all $a \in N$, $L \subseteq a$, so $L \leq a$. Thus, the canonical frame for L is reduced, with L acting as 0, and so the canonical frame is in $\mathfrak{r}(\mathcal{C})$. ◁

The lemma's proof does not establish that there are any prime logics, but that is what the metavaluations do. In particular, the final bit of the argument we need is the corollary from the previous section, namely that metacomplete logics are prime.

Corollary 17. *If L is metacomplete and complete with respect to a class $\mathcal{C}$ of frames that includes the canonical frame for L, then L is complete with respect to $\mathfrak{r}(\mathcal{C})$.*

There are metacomplete relevant logics, so there are prime relevant logics. The axioms and rules for some of these metacomplete relevant logics have reducible frame conditions. Thus, the previous corollary applies to them and they are complete with respect to their reduced frames.

Theorem 18. *Let L be a metacomplete logic such that L is complete with respect to L-frames and its canonical frame is an L-frame. Then L is complete with respect to reduced L-frames.*

Proof. This follows from the previous corollary. ◁

The logic TW is close to C, as their axiomatizations differ only in the presence of (WI). That difference makes a difference, as TW is metacomplete, so many of the modal extensions of TW are complete with respect to their reduced frames. In contrast, modal extensions of C.M are not complete with respect to their reduced frames, as shown by [67].

It is worth dwelling on the proof above to note a few points. First, we do not use metavaluations on (non-logical) theories, which is to say theories that are not themselves logics, whereas Slaney does, in general, in his construction of a canonical reduced frame.[20] Second, a logic being prime implies that it is in the set K of its canonical frame, which means that the canonical frame is, in fact, reduced. Since the theory L contains no non-theorems, by definition, every non-theorem is refuted at L, considered as 0, in the canonical model for L. As long as the frame conditions

---

[20][50, ch. 5] also applies metavaluations to non-logical theories, as does Seki [60].

corresponding to the axioms for L are reducible, then the canonical frame for L is in fact a reduced L-frame. There is, then, no need to use a separate construction for the canonical reduced frame.[21] The canonical frame that is obtained from the more standard completeness proof for the unreduced frames does the job. This part of the argument applies equally to non-modal relevant logics as to modal relevant logics.

In the definition of the Ms and Mt logics, the (Nec) rule is one of the optional extras that can be added included in a metacomplete logic. While the frame condition for (Nec) is not reducible, an alternative condition, $0 \leq a \wedge Sab \Rightarrow 0 \leq b$, can be shown to work in the context of reduced models. This condition is obtained from the condition for unreduced frames by replacing $x \in N$ with $0 \leq x$. In fact, an equivalent, simplified condition can be used instead, $S0a \Rightarrow 0 \leq a$.[22]

The rule (CoNec), $\Box A \Rightarrow A$, is also an optional extra for the Ms and Mt logics. Unlike (Nec), it does not appear to have a corresponding frame condition that the canonical frame is guaranteed to satisfy. Some Ms and Mt logics may not be complete with respect to any class of reduced frames, because of the lack of a suitable frame condition for (CoNec).

Since the axioms (Contra) and (B') are both available in the M1 and M2 logics, the axiom (WI) presents a stark boundary for completeness with respect to reduced frames for modal relevant logics. There are non-modal relevant logics containing (WI) that are complete with respect to their reduced frames, so the issue, or at least *this issue*, with (WI) only emerges when one looks at modal relevant logics.[23]

The results above imply that there are many modal relevant logics that are complete with respect to their reduced frames, even S4-ish and S5-ish logics. [19] and [67] show that S4-ish and S5-ish extensions of R are incomplete with respect to their reduced frames. When the base logic is weakened to an M1 or M2 logic whose axioms have appropriate frame conditions, the S4-ish and S5-ish extensions are complete with respect to their reduced frames. Since many of the more common weaker logics are metacomplete, including B, DJ, TW, and RW, many of their modal extensions, including S4-ish and S5-ish extensions will be complete with respect to their reduced frames. Thus, the question left open by [67] has a positive answer. Completeness with respect to reduced frames can be had by modal relevant logics, if the logic is metacomplete. This covers a wide range of modal relevant logics, although it does not

---

[21] There is no need as far as completeness goes. [63, pp. 405–406] notes a reason to prefer one's reduced models satisfy some contingent truths.

[22] The simplified condition is consequence of the other in virtue of the fact that $0 \leq 0$, setting a in the condition to be 0 and b to be a. The equivalence is obtained by noting that $0 \leq a$ and Sab yields $S0b$, from a frame condition for modal frames, whence $0 \leq b$, by the simplified condition.

[23] (WI) is known to lead to other problems. For example, Meyer et al. [42] show how it leads to triviality in combination with naive set theory via a variation on Curry's paradox. Indeed, Brady [11] uses metavaluations to show that for many weaker relevant logics, the addition of the naive set theory axioms is consistent. It is known that (WI) is not unique for leading to triviality in combination with naive set theory axioms. Restall [47], Rogerson and Restall [52], Bimbó [6], Robles and Méndez [51], Øgaard [43], and Field et al. [17] all discuss different routes to triviality via contraction-like axioms, of which (WI) is a prominent instance.

include all, and generally does not include logics containing the modal confinement axiom, (MC) $\Box(A \lor B) \to \sim\Box\sim A \lor \Box B$.[24]

Finally, there are some multi-modal relevant logics in the literature. For example, logics of alethic necessity and actuality are suggested by [66]. Seki's metavaluations can be extended with clauses to cover the additional modalities, with the Ms and Mt distinction being duplicated for the new modalities. This opens the doorway for metacompleteness results for these multi-modal logics.[25] Some metacompleteness results will be straightforward consequences of the extensions. Whether logics with interaction axioms, such as $\mathbb{A}B \to \Box B$, are metacomplete will depend on the details of the metavaluations and the classes of logics. For logics that are metacomplete, however, completeness with respect to reduced frames will be available, provided the axioms added to B.M, or its adaptation to a multi-modal setting, have reducible frame conditions.

Acknowledgments. I would like to thank Greg Restall, Katalin Bimbó, and an anonymous referee for feedback on this paper.

REFERENCES

[1] Anderson, A. R. and Belnap, N. D. (1975). *Entailment: The Logic of Relevance and Necessity, Vol. I*, Princeton University Press, Princeton, NJ.

[2] Anderson, A. R., Belnap, N. D. and Dunn, J. M. (1992). *Entailment: The Logic of Relevance and Necessity, Vol. II*, Princeton University Press, Princeton, NJ.

[3] Berto, F. (2015). A modality called 'negation', *Mind* **124**(495): 761–793.

[4] Berto, F. and Restall, G. (2019). Negation on the Australian plan, *Journal of Philosophical Logic* **48**(6): 1119–1144.

[5] Bilková, M., Majer, O., Peliš, M. and Restall, G. (2010). Relevant agents, *in* L. Beklemishev, V. Goranko and V. Shehtman (eds.), *Advances in Modal Logic, 8*, College Publications, London, UK, pp. 22–38.

[6] Bimbó, K. (2006). Curry-type paradoxes, *Logique et Analyse* **49**(195): 227–240.

[7] Bimbó, K. (2007). Relevance logics, *in* D. Jacquette (ed.), *Philosophy of Logic*, Vol. 5 of *Handbook of the Philosophy of Science*, Elsevier, Amsterdam, pp. 723–789.

[8] Bimbó, K. and Dunn, J. M. (2008). *Generalized Galois Logics: Relational Semantics of Nonclassical Logical Calculi*, Vol. 188 of *CSLI Lecture Notes*, CSLI Publications, Stanford, CA.

[9] Brady, R. (ed.) (2003). *Relevant Logics and their Rivals, Volume II, A continuation of the work of Richard Sylvan, Robert Meyer, Val Plumwood and Ross Brady*, Ashgate, Burlington, VT.

[10] Brady, R. T. (1984). Natural deduction systems for some quantified relevant logics, *Logique et Analyse* **27**(8): 355–377.

[11] Brady, R. T. (2014). The simple consistency of naive set theory using metavaluations, *Journal of Philosophical Logic* **43**(2–3): 261–281.

[12] Brady, R. T. (2017). Metavaluations, *Bulletin of Symbolic Logic* **23**(3): 296–323.

---

[24]Logics containing (MC) have been studied by Dunn [14] and Mares [30, 31, 32], among others.

[25]It also opens the doorway for other sorts of results that use metacompleteness, such as γ-admissibility arguments, showing that if $\vdash_L A$ and $\vdash_L \sim A \lor B$, then $\vdash_L B$. For such arguments, see [60]. Although modal relevant logics are not considered by Meyer et al. [39], it would be interesting to see whether any distinctively modal principles result in failures of γ.

[13] Dunn, J. M. (1966). *The Algebra of Intensional Logics*, PhD thesis, University of Pittsburgh. Published as Vol. 2 in the Logic PhDs series by College Publications, London (UK), 2019.
[14] Dunn, J. M. (1995). Positive modal logic, *Studia Logica* **55**(2): 301–317.
[15] Dunn, J. M. and Restall, G. (2002). Relevance logic, *in* D. M. Gabbay and F. Guenthner (eds.), *Handbook of Philosophical Logic*, 2nd edn, Vol. 6, Kluwer, Amsterdam, pp. 1–136.
[16] Ferenz, N. (2021). Quantified modal relevant logics, *The Review of Symbolic Logic* pp. 1–31.
[17] Field, H., Lederman, H. and Øgaard, T. F. (2017). Prospects for a naive theory of classes, *Notre Dame Journal of Formal Logic* **58**(4): 461–506.
[18] Fine, K. (1974). Models for entailment, *Journal of Philosophical Logic* **3**(4): 347–372.
[19] Fuhrmann, A. (1990). Models for relevant modal logics, *Studia Logica* **49**(4): 501–514.
[20] Giambrone, S. (1992). Real reduced models for relevant logics without WI, *Notre Dame Journal of Formal Logic* **33**(3): 442–449.
[21] Goble, L. (1999). Deontic logic with relevance, *in* P. McNamara and H. Prakken (eds.), *Norms, Logics and Information Systems: New Studies on Deontic Logic and Computer Science*, IOS Press, pp. 331–345.
[22] Goble, L. (2000). An incomplete relevant modal logic, *Journal of Philosophical Logic* **29**(1): 103–119.
[23] Goble, L. (2001). The Andersonian reduction and relevant deontic logic, *in* B. Brown and J. Woods (eds.), *New Studies in Exact Philosophy: Logic, Mathematics and Science. Proceedings of the 1999 Conference of the Society of Exact Philosophy*, Hermes Science Publications, Paris, pp. 213–246.
[24] Goldblatt, R. and Kane, M. (2009). An admissible semantics for propositionally quantified relevant logics, *Journal of Philosophical Logic* **39**(1): 73–100.
[25] Humberstone, L. (2011). *The Connectives*, MIT Press, Cambridge, MA.
[26] Lokhorst, G. C. (2006). Andersonian deontic logic, propositional quantification, and Mally, *Notre Dame Journal of Formal Logic* **47**(3): 385–395.
[27] Lokhorst, G. C. (2008). Anderson's relevant deontic and eubouliatic systems, *Notre Dame Journal of Formal Logic* **49**(1): 65–73.
[28] Maksimova, L. (1973). A semantics for the system E of entailment, *Bulletin of the Section of Logic of the Polish Academy of Sciences* **2**: 18–21.
[29] Mares, E. D. (1992a). Andersonian deontic logic, *Theoria* **58**(1): 3–20.
[30] Mares, E. D. (1992b). The semantic completeness of RK, *Reports on Mathematical Logic* pp. 3–10.
[31] Mares, E. D. (1993). Classically complete modal relevant logics, *Mathematical Logic Quarterly* **39**(1): 165–177.
[32] Mares, E. D. (1994). Mostly Meyer modal models, *Logique et Analyse* **37**(146): 119–128.
[33] Mares, E. D. (2000). The incompleteness of RGL, *Studia Logica* **65**(3): 315–322.
[34] Mares, E. D. (2004). *Relevant Logic: A Philosophical Interpretation*, Cambridge University Press, Cambridge, UK.
[35] Mares, E. D. and Meyer, R. K. (1992). The admissibility of γ in R4, *Notre Dame Journal of Formal Logic* **33**(2): 197–206.
[36] Mares, E. D. and Meyer, R. K. (1993). The semantics of R4, *Journal of Philosophical Logic* **22**(1): 95–110.
[37] Mares, E. D. and Standefer, S. (2017). The relevant logic E and some close neighbours: A reinterpretation, *IFCoLog Journal of Logics and Their Applications* **4**(3): 695–730.
[38] Meyer, R. K. (1966). *Topics in Modal and Many-valued Logic*, PhD thesis, University of Pittsburgh, UMI, Ann Arbor, MI.

[39] Meyer, R. K., Giambrone, S. and Brady, R. T. (1984). Where gamma fails, *Studia Logica* **43**(3): 247–256.
[40] Meyer, R. K. and Mares, E. D. (1993). Semantics of entailment 0, *in* K. Došen and P. Schroeder-Heister (eds.), *Substructural Logics*, Oxford Science Publications, Oxford, UK, pp. 239–258.
[41] Meyer, R. K. and Routley, R. (1972). Algebraic analysis of entailment I, *Logique et Analyse* **15**: 407–428.
[42] Meyer, R. K., Routley, R. and Dunn, J. M. (1979). Curry's paradox, *Analysis* **39**(3): 124–128.
[43] Øgaard, T. (2016). Paths to triviality, *Journal of Philosophical Logic* **45**(3): 237–276.
[44] Priest, G. and Sylvan, R. (1992). Simplified semantics for basic relevant logics, *Journal of Philosophical Logic* **21**(2): 217–232.
[45] Punčochář, V. and Sedlár, I. (2017). Substructural logics for pooling information, *in* A. Baltag, J. Seligman and T. Yamada (eds.), *Logic, Rationality, and Interaction*, Springer, Berlin, pp. 407–421.
[46] Read, S. (1988). *Relevant Logic: A Philosophical Examination of Inference*, Blackwell, Oxford, UK.
[47] Restall, G. (1993a). How to be really contraction free, *Studia Logica* **52**(3): 381–391.
[48] Restall, G. (1993b). Simplified semantics for relevant logics (and some of their rivals), *Journal of Philosophical Logic* **22**(5): 481–511.
[49] Restall, G. (1999). Negation in relevant logics (how I stopped worrying and learned to love the Routley star), *in* D. M. Gabbay and H. Wansing (eds.), *What is Negation?*, Kluwer Academic Publishers, pp. 53–76.
[50] Restall, G. (2000). *An Introduction to Substructural Logics*, Routledge.
[51] Robles, G. and Méndez, J. M. (2014). Blocking the routes to triviality with depth relevance, *Journal of Logic, Language and Information* **23**(4): 493–526.
[52] Rogerson, S. and Restall, G. (2004). Routes to triviality, *Journal of Philosophical Logic* **33**(4): 421–436.
[53] Routley, R. and Meyer, R. K. (1972a). The semantics of entailment—II, *Journal of Philosophical Logic* **1**(1): 53–73.
[54] Routley, R. and Meyer, R. K. (1972b). The semantics of entailment—III, *Journal of Philosophical Logic* **1**(2): 192–208.
[55] Routley, R. and Meyer, R. K. (1973). The semantics of entailment, *in* H. Leblanc (ed.), *Truth, Syntax, and Modality: Proceedings of the Temple University Conference on Alternative Semantics*, Amsterdam: North-Holland Publishing Company, pp. 199–243.
[56] Routley, R., Plumwood, V., Meyer, R. K. and Brady, R. T. (1982). *Relevant Logics and Their Rivals*, Vol. 1, Ridgeview, Atascadero, CA.
[57] Savić, N. and Studer, T. (2019). Relevant justification logic, *Journal of Applied Logics* **6**(2): 395–410.
[58] Sedlár, I. (2013). Justifications, awareness and epistemic dynamics, *in* S. Artemov and A. Nerode (eds.), *Logical Foundations of Computer Science*, number 7734 in *Lecture Notes in Computer Science*, Springer, pp. 307–318.
[59] Sedlár, I. (2016). Epistemic extensions of modal distributive substructural logics, *Journal of Logic and Computation* **26**(6): 1787–1813.
[60] Seki, T. (2011). The γ–admissibility of relevant modal logics II — the method using metavaluations, *Studia Logica* **97**(3): 351–383.
[61] Seki, T. (2013). Some metacomplete relevant modal logics, *Studia Logica* **101**(5): 1115–1141.

[62] Slaney, J. K. (1984). A metacompleteness theorem for contraction-free relevant logics, *Studia Logica* **43**(1-2): 159–168.
[63] Slaney, J. K. (1987). Reduced models for relevant logics without WI, *Notre Dame Journal of Formal Logic* **28**(3): 395–407.
[64] Standefer, S. (2018). Trees for E, *Logic Journal of the IGPL* **26**(3): 300–315.
[65] Standefer, S. (2019). Tracking reasons with extensions of relevant logics, *Logic Journal of the IGPL* **27**(4): 543–569.
[66] Standefer, S. (2020). Actual issues for relevant logics, *Ergo* **7**(8): 241–276.
[67] Standefer, S. (2021). An incompleteness theorem for modal relevant logics, *Notre Dame Journal of Formal Logic* **62**(4): 669–681.
[68] Standefer, S. and Brady, R. T. (2019). Natural deduction systems for E, *Logique et Analyse* **61**(242): 163–182.
[69] Tedder, A. and Ferenz, N. (2021). Neighbourhood semantics for quantified relevant logics, *Journal of Philosophical Logic,* (forthcoming).
[70] Urquhart, A. (1972). *The Semantics of Entailment*, PhD thesis, University of Pittsburgh. UMI, Ann Arbor, MI.
[71] Wansing, H. (2002). Diamonds are a philosopher's best friends, *Journal of Philosophical Logic* **31**(6): 591–612.

DEPARTMENT OF PHILOSOPHY, NATIONAL TAIWAN UNIVERSITY, TAIPEI, TAIWAN
Email: standefer@ntu.edu.tw

SITUATIONS, PROPOSITIONS, AND INFORMATION STATES

Andrew Tedder

ABSTRACT. Two families of relational semantics for relevant logics, the ternary relation and the Fine-style, or operational-relational, semantics are compared on point of interpretation. Following Punčochář, it's noted that the former kind tend to be given *ontic* or *realist* styles of interpretation, whereas the latter tend to be given *epistemic* or *informational* styles. The equivalence between these semantic approaches means that we can have both in one setting (with one grounded in the other), but it's argued that, nonetheless, there are reasons to prefer a version which takes the realist interpretation as basic and the informational one as grounded in it. The resulting, layered, semantic picture is sketched, and an application to the Mares–Goldblatt interpretation of quantifiers is proposed.

Keywords. Philosophy of logic, Relational semantics, Relevant logic, Theory of meaning

1. INTRODUCTION

Relational semantics for relevant and substructural logics can be put into two camps: there is the *ternary relation* (TR) framework most famously studied by Sylvan (né Routley) and Meyer [39; 40] and the operational–relational, or *Fine-style* (F) framework most famously studied by Fine [20]. (For further details on the history of these developments, see Bimbó and Dunn [7]; Bimbó et al. [8].) Punčochář [33, §6], noting that these two frameworks are formally equivalent, suggests that the difference between them should be understood as having to do with the kind of explanation they tend to proffer for the meanings of the logical vocabulary. As he sees it, the TR framework tends to be understood *ontically*, as having to do with the real world and objects therein, while the F framework tends to be understood *epistemically*, or perhaps a better word is *informationally*, as having to do with the sorts of things grasped by agents, communicated by assertions, and which comprise theories.

While one should not put too much weight on the claim for TR semantics are read realistically and the F semantics otherwise (realist versions of F semantics have been given, for instance, by Jago [24], where the elements of an F model are taken to represent exact truthmakers), it does track a tendency. For example, Barwise and Perry [4], whose *situation theory* is often invoked as a philosophical story motivating TR semantics (for instance, in Restall [35]; Mares [30] and Tedder [42]), took the realism of their picture as one of its theoretical strengths. On the other hand, Logan

---

2020 *Mathematics Subject Classification.* Primary: 03B47, Secondary: 03B80.

Bimbó, Katalin, (ed.), *Relevance Logics and other Tools for Reasoning. Essays in Honor of J. Michael Dunn,* (Tributes, vol. 46), College Publications, London, UK, 2022, pp. 410–426.

[27; 28] has recently proposed a form of the F semantics, which takes the objects literally to be *theories* (i.e., deductively closed sets of sentences).

These two frameworks are formally equivalent in the sense that any model of one kind can be transformed into a model of the other kind satisfying all the same formulas as the original (this will be properly spelled out, and proved, in §4, building on [33]). Punčochář takes this fact to indicate that one is free to choose whichever framework one prefers for a particular application, and this is, of course, true. Indeed, there are mathematical reasons why one might prefer one to the other: the TR semantics is often simpler in a mechanical sense (there are fewer things in the frames and fewer constraints), whereas the F semantics is often simpler in a conceptual sense (binary operations and relations are more familiar as mathematical objects than are ternary relations).[1] So in a sense there is, and need be, no rivalry between these two approaches.

It has, however, been suggested for some time that the TR semantics of relevant logics are unmotivated, *ad hoc* formalisms that do not provide a true meaning theory for the logical vocabulary. Perhaps the most famous version of this line of criticism is due to Copeland [9; 10], and one gets the impression that the F semantics has usually been taken to be the more natural account, as providing a more natural interpretation of the central conditional connective.[2] So when the goal is to provide a philosophically significant theory of meaning of the relevant vocabulary, which is one of the things that one may take the relational semantics to be for, there is a serious question on the table, and a reason to want to take sides.

In this paper, I will harness the equivalence results to suggest that we approach this apparent distinction in a different way, by focusing on the fact that we can always construct one kind of model from another. The resulting picture is a *layered* semantics, where one kind of model is *grounded in* a model of the other kind. Rather than taking either an ontic or informational stand, *tout court*, I'll suggest that we can always have both, and that the question comes down to which we take to be *basic*. We can always capture the ontic flavor of the TR semantics and the informational flavor of the F semantics in one framework; the real question concerns the *direction of explanation*. Do we account the ontic properties of a TR model in terms of the informational properties of an underlying F model, or vice versa? I'll argue that, in general, an ontically-based presentation provides a more satisfying route for explanation of the facts to be accounted for by a semantics (namely, facts about *entailments*), and that therefore, if we take Punčochář's distinction seriously, we ought to prefer to take the TR semantics as basic. With this argument made, I'll briefly discuss an interesting upshot for the Mares and Goldblatt [32] semantics for quantifiers.

---

[1]It might be, cheekily, put that part of the miracle performed in Urquhart's [45] undecidability proof was simply in *finding* a ternary relation in the wild, in the form of co-linearity.

[2]It's been suggested to the author, in conversation with an interlocutor who shall remain nameless (you know who you are), that the TR truth condition is a *kludge*, trying, and failing, to capture the beauty and simplicity of the operational truth condition, due to Urquhart [44], and that the resulting framework is ugly and unmotivated.

2. THEORIES OF MODELS AND THEORIES OF ENTAILMENT

The formal structure of the options for orders of explanation — either accounting for ontic features in terms of informational ones or vice versa — closely mirrors a related dispute in the philosophical discussion surrounding possible worlds and their use in frame semantics for modal logics. In that literature a salient distinction is that between grounding propositions in possible worlds and grounding possible worlds in propositions. Versions of these approach are discussed in Loux [29].

The dispute between these two approaches, as discussed in Divers [12], can be seen as circling around the question of how best to account for the modal properties of propositions. An account of possible worlds, and the relationship they bear to propositions, should provide an account of when propositions are necessary or possible. Ideally, it should give us insight into questions about which particular propositions are necessary or possible, and why. The account which the realist line on possible worlds gives of these is familiar, namely, that a proposition is necessary when it is true in every possible world and it is possible when true in some. So when we are tasked with accounting for, say, what makes it the case that a particular proposition is possible, if we take the realist line of explanation our task is to provide reasons to believe that there is a possible world which makes the proposition true.

Theories of relational semantics for relevant logics are not aimed at providing explanations for why propositions are necessary or possible, but rather are aimed at providing explanations for why certain propositions *entail* others. As Anderson and Belnap [1, §1] stress, entailment is the heart of logic. So if we aim to give an account of the meanings of the logical connectives in terms of a theory of models, a major part of our project will be to do so in a way that accounts for why, in general, some propositions entail others.[3] Furthermore, the account should provide us with the means to answer questions about which particular propositions entail which others. So when it comes to deciding between rival semantic theories, it will be in terms of their accounts of entailments, and the kinds of explanation they proffer for particular entailment facts, that I propose we make our decisions. One reason to prefer a realist picture, as opposed to one like that of [28], which takes the basic elements to be theories, is that the realist approach seems to stand a better shot at providing satisfying answers to questions of the form "why does such-and-so particular implication claim express an entailment (or not)?" In the case of a negative answer to such a question, the realist line would have us describe a situation which supports an instance of the antecedent but not the appropriate instance of the consequent. With such a countermodel in mind, one will have provided a strong reason to reject the claim that the implication in question is an entailment.[4] Positive answers will concern the properties of situations, and how they go about satisfying propositions. The account taking theories as basic seems hard-pressed to provide a similarly satisfying, non-circular explanation of entailment facts.

---

[3]Throughout, I'll discuss "explanation" in a *metaphysical* sense. In this sense, a fact explains another one when it features in an account of why the latter holds.

[4]One example of doing such a thing can be found in [42], where it is argued that implications of the form $(\mathcal{A} \to \mathcal{B}) \to ((\mathcal{B} \to \mathcal{C}) \to (\mathcal{A} \to \mathcal{C}))$ do not express entailments, by means of describing a concrete situation which would falsify this implication formula.

Having said this, note that even theories of meaning for the logical vocabulary which are formally equivalent may yet differ with respect to the explanations they offer of these facts. (I venture to suggest that non-realist approaches will generally fare worse on this front.) Consider the case of possible worlds again. Whether we take propositions to be composed of possible worlds or take possible worlds to be composed of propositions, we may wind up with the same collection of propositions being necessary. There is, nonetheless, still a debate between these about which approach provides the better explanation of those modal facts. How can such debates proceed?

2.1. Data-Fit, Parsimony, and Explanatory Power. One way to cash out such debates concerns the extent to which different proposals satisfy different principles of theory choice. One such principle, the fit to the data, does not decide between these: being equivalent, both theories will fit the data to the same extent. So the choice comes down to other principles, and for my purposes there seem to be two, which are most salient:

(1) Ontological Parsimony: One should choose a theory which, *ceteris paribus*, involves commitment to fewer kinds of entity.
(2) Explanatory Power: One should choose a theory which, *ceteris paribus*, provides a more satisfying explanation of the phenomena underlying the data.

In the case of modal theories, we can take the salient data to concern the modal status of propositions, and choose between the candidate theories based on (1) and (2). Following Lewis [25], there seem to be good reasons to think that a realist account performs better on (2), but non-realist accounts seem to fare better on (1). The question then becomes when we're forced to choose one of (1), (2), which should we prefer?

I think that we should pretty much always prefer to gain explanatory power at the loss of parsimony than go the other way around, at least when all other things are, indeed, equal. Let me sketch a brief argument why.[5] What is the theoretical cost of having more ontological commitments? As far as I can tell, the main cost is that taking on commitments to more kinds of entities runs the risk of falsifying the theory. If we commit ourselves to the existence of something which turns out not to exist, we'll have made an error, and have a false theory on our hands. Such risks are, indeed, theoretical costs, as are any commitments we take on which might wind up false. However, they are *just as costly* as any other such risky commitments we take on by making assertions — they are not *more costly*. So when we decide which theories to adopt, and we weigh the costs of ontological commitments, which come along with the theory, we should weigh these the same way we do any potentially false claims the theory makes.

If this is correct, then when we are in a position to decide whether to adopt an ontologically profligate, but more explanatory theory or one which is more parsimonious and less explanatory, the question comes down to whether we should take on a greater risk of falsehood in the hope of having a more explanatory theory. I think the answer

---

[5]This argument is, of course, deeply indebted to Lewis [25], though I'll refrain from citing chapter and verse as I go into it. It's worth noting that it involves an appeal to *inference to the best explanation*, and this has been discussed in detail by Lipton, information concerning which can be found in [26]. The notion of a "satisfying explanation" to which I appeal is, perhaps, best understood as an appeal to an explanation being the "loveliest," in his terminology, but I'll leave this appeal somewhat vague here.

here should be a resounding "yes." In general, we are better off taking the liberal attitude of seeking truth than we are taking the more conservative approach of avoiding falsehood; from this perspective, if the only thing we have to lose by taking onboard commitments to further entities that provide us with better explanatory power is the risk of falsehood, we should do so.

3. Ontic TR Frames and Informational F Frames

Let's cash out the sense in which you might understand TR semantics as being more ontic and F semantics as being more informational. As mentioned, this distinction is not hard and fast. There are realist readings offered of some forms of F semantics (as in [24]), and there have certainly been informationally flavored readings of semantics in the TR framework, such as in Dunn's work on program interpretations [17; 14; 15].

Having said this, it does seem to be the case that, as Punčochář [33] notes, there is a tendency for proponents of the TR semantics to defend realist readings and those of the F semantics to defend informational readings. Perhaps, the most clear point of distinction between these approaches concerns the interpretation of disjunction; a bit of discussion of the history here is salient. Urquhart [44] first attempted to give an operational semantics for relevant logics employing a frame with points obeying the truth condition for disjunction common from Kripke semantics for modal and intuitionist logic. Taking $[\![\mathcal{A}]\!]^M$ as the collection of points of a frame satisfying a formula $\mathcal{A}$ in a model M, this truth condition is:

$$[\![\mathcal{A} \vee \mathcal{B}]\!]^M = [\![\mathcal{A}]\!]^M \cup [\![\mathcal{B}]\!]^M.$$

The problem, well known, is that if one attempts to interpret the conditional in terms of a binary operation $\otimes$, as

$$[\![\mathcal{A} \to \mathcal{B}]\!]^M = \{s \colon \forall t (t \in [\![\mathcal{A}]\!]^M \Rightarrow s \otimes t \in [\![\mathcal{B}]\!]^M)\},$$

then one winds up with models for which standard relevant logics are not complete (see Dunn and Restall [18], for more details). In order to resolve the problem, one must adopt a different truth condition either for disjunction or the conditional. The F framework takes the former route, and the TR framework takes the latter.

It's been noted many times over the years (e.g., [23; 13; 22; 24]) that the standard truth condition for disjunction is ill-suited to interpretations of points in the frame as informational, motivating the move made by proponents of the F framework. For instance, suppose we take frame elements to represent information states, such as those available to an agent in the course of a reasoning task. There's no good reason to suppose that whenever such an agent has information supporting a disjunction, they'll have information supporting either disjunct. For instance, Sherlock Holmes may have enough information to know "either Moriarty or Queen Victoria committed the murder" without having information adequate to pin down the identity of the killer. This is one way that an informational reading is especially well suited to the F style semantics.

The kind of situation-theoretic reading often offered for TR semantics can avoid this issue by taking situations themselves *not* to be the sorts of things which agents directly cognize. On this sort of picture, what an agent cognizes is not a situation but rather a proposition (or collection thereof), which *type* situations, but need not *be* situations themselves. Situations are, perhaps, well understood as *inexact truthmakers* which support the truth of propositions — in this case, it is a plausible claim that they support a disjunction just in case they support one of the disjuncts.[6]

Another way this tendency comes up is that the most standard interpretation of the ternary relation, using *channel theory* Barwise [3], as in [35; 42], has a realist flavour. It posits mind-independent links between situations to interpret the ternary relation. The operation of the F semantics, on the other hand, is usually read informationally, as concerning the result of applying an *epistemic* or *informational* action on bits of information, sentences, or theories [44; 41; 28]. I won't go into further detail, but hopefully this suffices to bolster Punčochář's case that there is a tendency for TR semantics to be read ontically and F semantics to be read epistemically/informationally.

Now let's turn to the equivalence of the frameworks.

4. A Sketch of Equivalence Between TR and F

This section is, as the title suggests, just a sketch — a more detailed investigation of these matters is certainly possible. I'll give basic details to provide the reader an indication of how the construction works, going in either direction, and how it naturally proceeds through the three layers on which I'll be focused later. A fuller presentation of a narrower result concerning the logic **R** can be found in [33].

The main aim is to show that the TR and F semantic frameworks are equivalent in the sense that from a model on a frame of one kind, we can construct a model on a frame of the other kind which satisfies just the same formulas. This goes to show that the frameworks capture, in a sense, the same *data*, leaving the question of the choice between them up to other theoretical considerations. In this section, I'll introduce a form of the TR semantics and a form of the F semantics and then show how to construct one from the other in a simple, uniform way.

There are a number of available variations on the theme of "TR semantics" and "F semantics," and the versions I sketch here are chosen in a way which is partially due to my own, perhaps idiosyncratic, preferences and partially in order to simplify the presentation. I'll deal here with basic forms of the TR and F semantics appropriate for the relevant logic **B** — the correspondence available between frame conditions and further axioms or rules which may be added to **B** to obtain further logics is well known, and we do not need to go into it here. I take the propositional language to be defined, as usual, from a set of atomic formulas $\mathbb{P}$, the logical constant **t**, and the connectives $\neg, \wedge, \vee, \rightarrow$ (of arities 1, 2, 2, 2, respectively). I'll use $\mathcal{L}$ to denote the language.

---

[6]For related discussion, into which I'll not go further here, see Deigan [11]. As a related point, note that both the TR and F semantic frameworks commonly employ the standard truth condition for conjunction in terms of set intersection. One upshot of this, in the case of the situation-theoretic picture, is that we obtain the validity of the distribution law immediately from the fact that a powerset algebra, with unions and intersections, is a distributive lattice. The justification of the distribution law has been discussed in relevant circles (e.g., in Belnap [5] and Restall [37]), and this raises a potential avenue of objection against reading the situation-theoretic line as realist. I won't go into this question further, but note it as a potential difficulty.

4.1. Ternary Relation Frames and Models.

Definition 1. A ternary relation (TR) frame $\mathfrak{F}$ is a tuple $\langle W, N, R, * \rangle$ where $\varnothing \neq N \subseteq W$, $R \subseteq W^3$, and $*: W \longrightarrow W$ are such that, given the following definitions:

$$\leq = \{\langle \alpha, \beta \rangle \in W^2 : \exists \gamma \in N(R\gamma\alpha\beta)\}$$
$$\mathcal{P}(W)^\uparrow = \{X \subseteq W : \forall \beta \in W(\exists \alpha \in X(\alpha \leq \beta) \Rightarrow \beta \in X)\},$$

the following constraints are satisfied:

- (tr1) $\langle W, \leq \rangle$ is a poset.
- (tr2) $N \in \mathcal{P}(W)^\uparrow$.
- (tr3) If $\alpha' \leq \alpha, \beta' \leq \beta, \gamma \leq \gamma'$ and $R\alpha\beta\gamma$, then $R\alpha'\beta'\gamma'$.
- (tr4) If $\alpha \leq \beta$ then $\beta^* \leq \alpha^*$, and furthermore, $\alpha^{**} = \alpha$.

Before defining models on TR frames, let's fix a couple other definitions. First,

Definition 2. Given a set $\Gamma \subseteq \mathcal{P}(W)^\uparrow$, we fix the following:

$$[\Gamma) := \{Y \in \mathcal{P}(W)^\uparrow : \exists X_1, \ldots, X_n \in \Gamma(\bigcap_{j \leq n} X_j \subseteq Y)\}$$

Briefly, $[\Gamma)$ is the least *filter*, on the distributive lattice $\langle \mathcal{P}(W)^\uparrow, \cap, \cup \rangle$, containing Γ.[7]

Definition 3. Given $X, Y \in \mathcal{P}(W)^\uparrow$, let

$$X \to Y = \{\alpha : \forall \gamma(\exists \beta \in X(R\alpha\beta\gamma) \Rightarrow \gamma \in Y)\}$$
$$\neg X = \{\alpha : \alpha^* \notin X\}$$

Definition 4. A model M on a TR frame $\mathfrak{F}$ is a function of type $\mathbb{P} \longrightarrow \mathcal{P}(W)^\uparrow$, extended to a valuation $[\![\cdot]\!]^M : \mathcal{L} \longrightarrow \mathcal{P}(W)^\uparrow$ as follows:

(1) $[\![p]\!]^M = M(p)$
(2) $[\![\mathbf{t}]\!]^M = N$
(3) $[\![\neg \mathcal{A}]\!]^M = \neg [\![\mathcal{A}]\!]^M$
(4) $[\![\mathcal{A} \wedge \mathcal{B}]\!]^M = [\![\mathcal{A}]\!]^M \cap [\![\mathcal{B}]\!]^M$
(5) $[\![\mathcal{A} \vee \mathcal{B}]\!]^M = [\![\mathcal{A}]\!]^M \cup [\![\mathcal{B}]\!]^M$
(6) $[\![\mathcal{A} \to \mathcal{B}]\!]^M = [\![\mathcal{A}]\!]^M \to [\![\mathcal{B}]\!]^M$

A formula $\mathcal{A}$ is satisfied by M on $\mathfrak{F}$ just in case $N \subseteq [\![\mathcal{A}]\!]^M$; it is satisfied by $\mathfrak{F}$ in case it is satisfied by any model on $\mathfrak{F}$; it is valid on a class $\mathcal{F}$ of TR frames just in case it is satisfied by each $\mathfrak{F} \in \mathcal{F}$.

4.2. Fine-Style Frames and Models.

The semantics in this section does not quite follow Fine's original presentation. The most salient point is that I explicitly include an operation, $\sqcap$, which interprets disjunction. Since much of what I say here concerns disjunction, I pull this out explicitly and state some conditions concerning it ((f1), (f5), and (f6)) in order to clarify its behavior. For instance, (f5) is the constraint, noted by Humberstone [23], which enforces distribution in a general setting. Fine does not include a detailed discussion of disjunction, but I render these constraints explicit for the purposes of comparison.

Definition 5. A Fine-style (F) frame $\mathfrak{G}$ is a tuple $\langle S, S_P, \sqsubseteq, \otimes, -, @ \rangle$, where $\varnothing \neq S_P \subseteq S$, $\sqsubseteq \subseteq S^2$, $\otimes: S^2 \longrightarrow S$, $-: S_P \longrightarrow S_P$, and $@ \in S$ so that the following constraints are satisfied:

---

[7] For further information on lattices and related topics, the reader may consult Dunn and Hardegree [16].

(f1) $\langle S, \sqsubseteq \rangle$ is a meet (written $\sqcap$) semi-lattice.
(f2) If $s \sqsubseteq s'$ and $t \sqsubseteq t'$ then $s \otimes t \sqsubseteq s' \otimes t'$. Also, $@ \otimes s = s$, for any $s \in S$.
(f3) $-$ is an order-inverting involution: so $s \sqsubseteq t \Rightarrow -t \sqsubseteq -s$ and $--s = s$.
(f4) If $x, y \in S$, $a \in S_P$, and $x \otimes y \sqsubseteq a$, then there are $x', y' \in S_P$ s.t. $x \sqsubseteq x'$, $y \sqsubseteq y'$, $x' \otimes y \sqsubseteq a$, and $x \otimes y' \sqsubseteq a$.
(f5) If $s \sqcap t \sqsubseteq u$, then there are $s', t' \in S_P$ s.t. $s \sqsubseteq s', t \sqsubseteq t'$, and $s' \sqcap t' \sqsubseteq u$.
(f6) If $s \sqcap t \not\sqsubseteq u$, then there are $s', t' \in S_P$ s.t. $s \sqsubseteq s', t \sqsubseteq t'$, and $s' \sqcap t' \not\sqsubseteq u$.

Definition 6. Given $X, Y \in \mathcal{P}(S)^\uparrow$, let

$$X \rightsquigarrow Y = \{s \in S : \forall t \in S(t \in X \Rightarrow s \otimes t \in Y)\}$$
$$X \sqcup Y = \{s \in S : \exists t, u \in S(t \sqcap u \sqsubseteq s \ \& \ t \in X \ \& \ u \in Y)\}$$
$$\sim X = \{s \in S : \forall t \in S_P(s \sqsubseteq t \Rightarrow -t \notin X)\}$$

Definition 7. A model on an F frame $\mathfrak{G}$ is a function L of type $\mathbb{P} \longrightarrow \mathcal{P}(S)^\uparrow$ required to satisfy the constraint, for any $p \in \mathbb{P}$,

$$\forall t \in S_P(s \sqsubseteq t \Rightarrow t \in L(p)) \Rightarrow s \in L(p)$$

L is extended to a full valuation $|\cdot|^L : \mathcal{L} \longrightarrow \mathcal{P}(S)^\uparrow$ required to satisfy the following clauses:

(1) $|p|^L = L(p)$
(2) $|\neg A|^L = \sim |A|^L$
(3) $|A \wedge B|^L = |A|^L \cap |B|^L$
(4) $|A \vee B|^L = |A|^L \sqcup |B|^L$
(5) $|A \rightarrow B|^L = |A|^L \rightsquigarrow |B|^L$
(6) $\forall t \in S_P(s \sqsubseteq t \Rightarrow t \in |A|^L) \Rightarrow s \in |A|^L$

A formula A is satisfied on L just in case $@ \in |A|^L$. Satisfaction on a frame and validity w.r.t. a class of frames are defined as for TR models.

4.3. From TR to F. Now let's show that from an arbitrary TR frame, we can construct an F frame such that for any model on the original frame, we can obtain a model on the new frame which satisfies the same formulas as the original model.

Definition 8. Given a TR frame $\mathfrak{F}$, let its F-mate $\mathfrak{F}^F = \langle S^F, S_P^F, \sqsubseteq^F, @^F, \otimes^F, -^F \rangle$ be defined as follows:

(i) $S^F = \{\Gamma \subseteq \mathcal{P}(W)^\uparrow : \forall Y \in \mathcal{P}(W)^\uparrow (\exists_{i \leq n} X_i \in \mathcal{P}(W)^\uparrow (\bigcap_{i \leq n} X_i \in \Gamma \Rightarrow Y \in \Gamma))\}$
(ii) $S_P^F = \{\Gamma \in S^F : \forall X, Y \in \mathcal{P}(W)^\uparrow (X \cup Y \in \Gamma \Rightarrow X \in \Gamma \text{ or } Y \in \Gamma)\}$
(iii) $\sqsubseteq^F = \subseteq$
(iv) $@^F = [\{N\}]$
(v) $\Gamma \otimes^F \Delta = \{Y \in \mathcal{P}(W)^\uparrow : \exists X \in \Delta(X \rightarrow Y \in \Gamma)\}$
(vi) $-^F \Gamma = \{X \in \mathcal{P}(W)^\uparrow : \neg X \notin \Gamma\}$, for $\Gamma \in S_P^F$.

The idea of the construction is that we take the set of filters, w.r.t. $\langle \mathcal{P}(W)^\uparrow, \cap, \cup \rangle$, S^F, the set of prime filters thereon, S_P^F, an order $\sqsubseteq^F$, a logical point $@^F$, and a pair of operations appropriate for the interpretation of implication and negation on Fine-style frames, $\otimes^F$ and $-^F$. So we have, essentially, constructed a set of *non-prime* points out of (up)sets of prime points and defined operations appropriate to interpret the logical vocabulary, all in accordance with the structure of F frames.

This construction proceeds by a two step process. We start from the TR frame, and then we consider the space of *propositions* thereon, given by $\mathcal{P}(W)^\uparrow$, and it is out of this space that we define our desired F frame. Intuitively speaking, we construct propositions out of elements of W, and from there we construct elements of the desired F frame. Let us verify that $\mathfrak{F}^F$ is, indeed, an F frame when $\mathfrak{F}$ is a TR frame.

Fact 9. *If $\mathfrak{F}$ is a TR frame, then $\mathfrak{F}^F$ is an F frame.*

Proof. It suffices to ensure that $\mathfrak{F}^F$ verifies conditions (f1)–(f6). For (f1), just note that $\sqsubseteq^F = \subseteq$ does have a meet, namely, $\cap$, and so $\langle S^F, \sqsubseteq^F \rangle$ is, indeed, a meet semi-lattice.

For (f2) we have two things to check. First, suppose $\Gamma \sqsubseteq^F \Delta$ and $\Sigma \sqsubseteq^F \Theta$ are the case, and furthermore, that $X \in \Gamma \otimes^F \Sigma$. Therefore, there is a $Y \in \Sigma$ such that $Y \to X \in \Gamma$, and so $Y \in \Theta$ and $Y \to X \in \Delta$, and so $X \in \Delta \otimes^F \Theta$. Since X was arbitrary, this suffices to prove that $\Gamma \otimes^F \Sigma \sqsubseteq^F \Delta \otimes^F \Theta$. Next, we want to show that for any $\Gamma \in S^F$, $@^F \otimes^F \Gamma = \Gamma$. First, if $X \in @^F \otimes^F \Gamma$, then there is a $Y \in \Gamma$ s.t. $Y \to X \in @^F$. But $Y \to X \in @^F$ holds iff $N \subseteq Y \to X$ holds in F and so $Y \subseteq X$ holds there, and so if $Y \in \Gamma$ then $X \in \Gamma$, since $\Gamma \in S^F$. For the converse, if $X \in \Gamma$ then, since $X \subseteq X$ always holds, we have $X \to X \in @^F$, and so $X \in @^F \otimes^F \Gamma$, as desired.

For (f3), we again have two things to prove. First, suppose that $\Gamma \subseteq \Delta$ and that $X \in -^F \Delta$. Thus, $\neg X \notin \Delta$ and so $\neg X \notin \Gamma$, and so $X \in -^F \Gamma$, as desired. Note, further, that $X \in -^F -^F \Gamma$ holds iff $\neg\neg X \in \Gamma$ iff $X \in \Gamma$.

For (f4), suppose that we have $\Gamma, \Delta \in S^F$ and $\Theta \in S_P^F$ s.t. $\Gamma \otimes^F \Delta \subseteq \Theta$. To obtain a $\Delta' \supseteq \Delta$ such that $\Gamma \otimes^F \Delta' \subseteq \Theta$, consider the pair

$$\langle \Delta, \Delta^- = \{X \in \mathcal{P}(W)^\uparrow : \exists Y \notin \Theta (X \to Y \in \Gamma)\} \rangle.$$

Now by definition Δ is a filter on $\langle \mathcal{P}(W)^\uparrow, \cap, \cup \rangle$, and it is fairly easy to verify that Δ^- is an ideal.[8] Furthermore, we can show that $\Delta \cap \Delta^- = \emptyset$. In fact, there are no $X_1, \ldots, X_m \in \Delta$ and $Y_1, \ldots, Y_n \in \Delta^-$ s.t. $\bigcap_{1 \leq i \leq m} X_i \subseteq \bigcup_{1 \leq j \leq n} Y_j$. With this fact, since $\langle \mathcal{P}(W)^\uparrow, \cap, \cup \rangle$ is a distributive lattice, we can employ [16, Corollary 13.4.6] to infer that there is a $\Delta' \in S_P^F$ s.t. $\Delta' \supseteq \Delta$ and $\Delta' \cap \Delta^- \neq \emptyset$.[9] From this, we can infer that $\Gamma \otimes^F \Delta' \subseteq \Theta$, as desired.

Let's show the key fact, assuming, for contradiction, that for each Y_j there is a $Z_j \notin \Theta$ such that $Y_j \to Z_j \in \Gamma$, and so $\bigcap_{1 \leq i \leq m} X_i \to \bigcup_{1 \leq j \leq n} Z_j \in \Gamma$ and so $\bigcup_{1 \leq j \leq n} Z_j \in \Theta$, contradicting the assumption that $\Theta \in S_P^F$.[10]

The argument needed to obtain a $\Gamma' \supseteq \Gamma$ s.t. $\Gamma' \in S_P$ and $\Gamma' \otimes \Delta \subseteq \Theta$ is similar, so elided. Also, the proofs of (f5) and (f6) are straightforward and, for reasons of space, are left to the reader. ◁

This suffices to show that the F-mate of a TR frame is an F frame, as desired. It remains to show how, given a model M on a TR frame $\mathfrak{F}$, to obtain a model on $\mathfrak{F}^F$ which will satisfy the same formulas. For this, we adapt the definition of a *canonical valuation*, given in Bimbó and Dunn [6, p. 23].

---

[8]The key facts are: $(X \to Y) \cap (Z \to U) \subseteq (X \cup Z) \to (Y \cup U)$, and if $X \subseteq Y$ then $Y \to Z \subseteq X \to Z$.

[9]Note, this step is analogous to the use of the Pair Extension lemma in completeness proofs for relevant logics w.r.t. their TR frame semantics, for instance in [2, §48.3].

[10]This relies on the fact that $(X \to Y) \cap (Z \to U) \subseteq (X \cap Z) \to (Y \cup U)$.

Definition 10. Given a model M on a TR frame $\mathfrak{F}$, we fix $|\cdot|^M : \mathcal{L} \longrightarrow \mathcal{P}(S^F)^\uparrow$ by setting $|\mathcal{A}|^M = \{\Gamma \in S^F : [\![\mathcal{A}]\!]^M \in \Gamma\}$.

Fact 11. *Given a model M on a TR frame $\mathfrak{F}$, the model $|\cdot|^M$ on $\mathfrak{F}^F$ has the following properties:*

(1) $|\mathcal{A}|^M \in \mathcal{P}(S^F)^\uparrow$, for every $\mathcal{A} \in \mathcal{L}$ (2) $|\mathbf{t}|^M = \{\Gamma \in S^F : @ \subseteq \Gamma\}$
(3) $|\mathcal{A} \wedge \mathcal{B}|^M = |\mathcal{A}|^M \cap |\mathcal{B}|^M$ (4) $|\mathcal{A} \vee \mathcal{B}|^M = |\mathcal{A}|^M \sqcup |\mathcal{B}|^M$
(5) $|\mathcal{A} \to \mathcal{B}|^M = |\mathcal{A}|^M \leadsto |\mathcal{B}|^M$ (6) $|\neg \mathcal{A}|^M = \sim |\mathcal{A}|^M$
(7) $\forall \Delta \in S^F (\forall \Gamma \in S^F_P (\Delta \subseteq \Gamma \Rightarrow \Gamma \in |\mathcal{A}|^M) \Rightarrow \Delta \in |\mathcal{A}|^M)$
(8) *For any* $\mathcal{A} \in \mathcal{L}, N \subseteq [\![\mathcal{A}]\!]^M$ *iff* $@^F \in |\mathcal{A}|^M$.

Proof. (1) is immediate from the definition. The others we can prove by induction on the complexity of formulas. For (2), the only atomic case, we can show:

$$|\mathbf{t}|^M = \{\Gamma \in S^F : [\![\mathbf{t}]\!]^M = N \in \Gamma\} = \{\Gamma \in S^F : [\{N\}] = @ \subseteq \Gamma\}$$

(3) is immediate, and left to the reader. For (4), note that the right-to-left direction is immediate from the fact that $[\![\mathcal{C}]\!]^M \subseteq [\![\mathcal{A} \vee \mathcal{B}]\!]^M$ holds for $\mathcal{C} \in \{\mathcal{A}, \mathcal{B}\}$, so let's consider the converse. Note that if $[\![\mathcal{A} \vee \mathcal{B}]\!]^M = [\![\mathcal{A}]\!]^M \cup [\![\mathcal{B}]\!]^M \in \Gamma$, then it's immediate that $[\![\mathcal{C}]\!]^M \in [\{[\![\mathcal{C}]\!]^M\}]$ holds for $\mathcal{C} \in \{\mathcal{A}, \mathcal{B}\}$ and $[\{[\![\mathcal{A}]\!]^M\}] \cap [\{[\![\mathcal{B}]\!]^M\}] \subseteq \Gamma$, which suffices to show that $\Gamma \in |\mathcal{A} \vee \mathcal{B}|^M$, as desired. For (5) and (6), the standard kind of arguments given in completeness proofs (for instance, those in Restall [36]) suffice, and verifying these are left to the reader.

For (7), we proceed by contraposition. Suppose that $\Delta \in S^F \cap \overline{|\mathcal{A}|^M}$, so that $[\![\mathcal{A}]\!]^M \notin \Delta$. We want to show that there is a $\Gamma \in S^F_P$ s.t. $\Delta \subseteq \Gamma$ and $[\![\mathcal{A}]\!]^M \notin \Gamma$. For this, however, it suffices to employ Dunn and Hardegree [16, Corollary 13.4.6], fixing $\{X \in \mathcal{P}(W)^\uparrow : X \subseteq [\![\mathcal{A}]\!]^M\}$, noting that this is an ideal which doesn't overlap Δ, and thus we can obtain a prime filter Γ on $\langle \mathcal{P}(W)^\uparrow, N, \cap, \cup, \to, \neg \rangle$ s.t. $\Gamma \supseteq \Delta$ and $\Gamma \cap \{X : X \subseteq [\![\mathcal{A}]\!]^M\} = \varnothing$, so that $[\![\mathcal{A}]\!]^M \notin \Gamma$ as desired.

For (8), it suffices to note that:

$$@ \in |\mathcal{A}|^M \iff [\![\mathcal{A}]\!]^M \in @ = [\{N\}] \iff N \subseteq [\![\mathcal{A}]\!]^M \qquad \triangleleft$$

Points (1)–(7) guarantee that $|\cdot|^M$ is well-defined, giving rise to a model on $\mathfrak{F}^F$. Point (8) gives the desired property, that the formulas satisfied by $|\cdot|^M$ on $\mathfrak{F}^F$ are just those satisfied by M on $\mathfrak{F}$. So we can state:

Theorem 12. *Given any TR frame $\mathfrak{F}$, and model M thereon, we can construct an F-frame $\mathfrak{F}^F$ and a model satisfying just those formulas satisfied by M.*

This gives us one half of our puzzle; that any formulas satisfiable on a TR frame are satisfiable on some F frame.

4.4. From F to TR. This direction is quite similar, and is, in any case, better understood. In Fine's original paper, especially, the part reproduced in Anderson et al. [2, §51.5], he considered in some detail the relationship between his frames and the TR frames presented by Sylvan, Meyer, and their collaborators Routley et al. [40]. The method I'll employ is a bit different from his, but shares some similarities.

Definition 13. Given the an F frame $\mathfrak{G}$, we construct $\mathfrak{G}^{TR} = \langle W^{TR}, N^{TR}, R^{TR}, *^{TR} \rangle$, $\mathfrak{G}$'s TR-mate, as follows:

$$W^{TR} = \{\Gamma \subseteq \mathcal{P}(S)^\uparrow : \forall Y \in \mathcal{P}(S)^\uparrow (\exists_{i \leq n} X_i \in \Gamma (\bigcap_{i \leq n} X_i \subseteq Y \Rightarrow Y \in \Gamma)) \&$$

$$\forall X, Y \in \mathcal{P}(S)^\uparrow (X \sqcup Y \in \Gamma \Rightarrow (X \in \Gamma \text{ or } Y \in \Gamma))\}$$

$$N^{TR} = \{\Gamma \in W^{TR} : @ \in \Gamma\}$$

$$R^{TR} = \{\langle \Gamma, \Delta, \Theta \rangle \in (W^{TR})^3 : \forall X, Y \in \mathcal{P}(S)^\uparrow ((X \rightsquigarrow Y \in \Gamma \& X \in \Delta) \Rightarrow Y \in \Theta)\}$$

$$\Gamma^{*^{TR}} = \{X \in \mathcal{P}(S)^\uparrow : {\sim} X \notin \Gamma\}$$

Fact 14. *If $\mathfrak{G}$ is an F frame then $\mathfrak{G}^{TR}$ is a TR frame.*

Proof. It suffices to prove that (tr1)–(tr4) hold of $\mathfrak{G}^{TR}$.

For (tr1), it suffices to show that the defined $\leq^{TR}$ is, in fact, just $\subseteq$, i.e., that $\exists \Gamma \in N^{TR}(R^{TR}\Gamma\Delta\Theta) \iff \Delta \subseteq \Theta$. For the left-to-right, suppose that $\exists \Gamma \in N^{TR}(R^{TR}\Gamma\Delta\Theta)$ and $X \in \Delta$. If $\Gamma \in N^{TR}$, then $@ \in \Gamma$ and since $@ \subseteq X \rightsquigarrow X$, we have that $X \rightsquigarrow X \in \Gamma$, and thus $X \in \Theta$. Since X was arbitrary, this suffices to show that $\Delta \subseteq \Theta$, as desired. For the converse, suppose that $\Delta \subseteq \Theta$; in fact, since for any $X \in \Delta$ and any $\Gamma \in N^{TR}$ we have $X \rightsquigarrow X \in \Gamma$, we have that $R^{TR}\Gamma\Delta\Theta$, which suffices to show the result (given that $N^{TR} \neq \varnothing$, verification of which fact we leave to the reader).

For the remainder, we'll take the order concerned just to be $\subseteq$ without further comment. For (tr2), we want to show that if $\Gamma \in N^{TR}$ and $\Gamma \subseteq \Delta$ then $\Delta \in N^{TR}$. This is immediate from the definition of N^{TR}.

The arguments needed for (tr3) and (tr4) are quite similar to arguments standardly given to show that the *canonical frame* of a logic is a TR frame, and the reader may consult [2, §48.3] or [40, Ch. 4] for details of this style of argument. ◁

Definition 15. Given the TR-mate $\mathfrak{G}^{TR}$ of an F frame $\mathfrak{G}$ and a model L on $\mathfrak{G}$, let $[\![\mathcal{A}]\!]^L = \{\Gamma \in W^{TR} : |\mathcal{A}|^L \in \Gamma\}$.

Now, once again, we just have to verify that the resulting model satisfies the required properties.

Fact 16. *Given a model L on a F frame $\mathfrak{G}$, the evaluation $[\![\cdot]\!]^L$ on $\mathfrak{G}^{TR}$ has the following properties:*

(1) *If $\Gamma \in [\![\mathcal{A}]\!]^L$ and $\Gamma \subseteq \Delta$, then $\Delta \in [\![\mathcal{A}]\!]^L$.*
(2) $[\![t]\!]^L = N^{TR}$
(3) $[\![\mathcal{A} \wedge \mathcal{B}]\!]^L = [\![\mathcal{A}]\!]^L \cap [\![\mathcal{B}]\!]^L$
(4) $[\![\mathcal{A} \vee \mathcal{B}]\!]^L = [\![\mathcal{A}]\!]^L \cup [\![\mathcal{B}]\!]^L$
(5) $[\![\mathcal{A} \to \mathcal{B}]\!]^L = \{\Gamma \in W^{TR} : \forall \Delta, \Theta((R^{TR}\Gamma\Delta\Theta \& \Delta \in [\![\mathcal{A}]\!]^L) \Rightarrow \Theta \in [\![\mathcal{B}]\!]^L)\}$
(6) $[\![\neg \mathcal{A}]\!]^L = \{\Gamma \in W^{TR} : \Gamma^{*^{TR}} \notin [\![\mathcal{A}]\!]^L\}$
(7) *For any $\mathcal{A} \in \mathcal{L}$, $N^{TR} \subseteq [\![\mathcal{A}]\!]^L$ iff $@ \in |\mathcal{A}|^L$.*

Proof. The reader is encouraged to check [6] for details of proving completeness for relational frames for distributive multi-gaggles. The details there suffice here, as can be noted by the fact that the complex algebra of an F frame will be a multi-gaggle. The only part of verifying this that is not standard involves verifying that distribution obtains, and the argument style, using (f5), can be found in [23]. ◁

From this we can infer the key fact, which is:

Theorem 17. *From any F frame $\mathfrak{G}$ we can obtain a TR frame $\mathfrak{G}^{TR}$ such that, for any model L on $\mathfrak{G}$ there is a model $[\![\cdot]\!]^L$ on $\mathfrak{G}^{TR}$ satisfying exactly the same formulas as L.*

I've only dealt here with frames appropriate for the basic logic **B**, but there is a well-known correspondence theory for accommodating stronger logics, and it seems likely that these results allow for the above to be generalized to frames appropriate for a wider range of logics (as can be done in the case of **R**, as shown in [33]). For my purposes, the basic form I've given here is enough to make my point, so I'll leave it at that and get back to the philosophical work.

5. Layered Semantics

As per §2.1, a realist account provides for a more explanatorily satisfying picture, and the equivalence results of §4 indicate how it is that, starting from this basis, we can recapture the working of the information-based semantics of the F approach in a more satisfying way using the TR semantics.[11] In any case, regardless of which way one proceeds to do the grounding, the equivalence provides a way of capturing both in one framework with some nice results.

The three-layer picture can be represented as follows — the arrows on the left side indicate explanatory priority (the arrows go from from the thing-grounded to the thing-grounding), and those on the right side order of the "defined in terms of" relation:

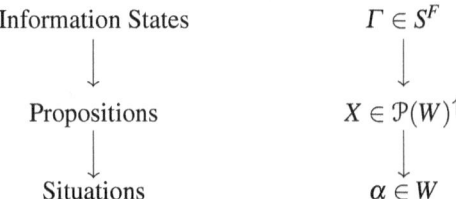

As indicated, situations provide the ground of the truth of propositions, and elements of S^F represent the states of information to which agents can find themselves having access. As before, there are good reasons that these should not be required to be prime, as they are not. One can have an information state which includes/supports a disjunctive proposition without supporting either disjunct. Situations, however, understood as inexact truthmakers are prime.

On the base level, we have objects, situations — particularly, something like the *abstract* situations of [4] — to which we have an existential commitment. We take them to be real things, and take propositions to be constructed out of these in systematic ways. Propositions, or representations thereof, are then the constituents of information states, to which agents have cognitive access. For instance, it is by taking in visual information that an agent learns what information the witnessed situation conveys, and they are then in a position to perform various cognitive tasks with that information. Part of the story here is that we don't directly perceive situations, nor do

---

[11] Assuming, of course, that the version of the F semantics involved is read in a *non-realist* and the TR semantics in a *realist* way.

we express situations directly by our various linguistic/cognitive actions. Rather, what we perceive/express/cognize are propositions and collections thereof into information states. On this line, when we open our eyes and perceive the world around us, what we perceive is not the world directly, but information carried by the world — what we perceive is a fact, not an object.

One place where the distinction becomes most salient is that many situations will be typed by a proposition. This captures the intuitive idea that our available information *underdetermines* the state of the world (the situation) we have information about. When I look at my office, and out the window, there is a great deal of information I get, but the actual world situation I, the office, and the window inhabit supports a great deal more information than that which I obtain by perception. For instance, I may see a drawer, and have a vague idea of what is in it, but may not have access to the more precise information supported by the situation of my office, which specifies precisely what is in the drawer. It is this underdetermination which explains why our information has certain imperfections, such as not being prime.

While in need of further precisification this story provides a skeleton for how a reasonably natural theory of meaning could be constructed on this sort of layered picture, and this in a way which accommodates the nice features of both the ontic and the epistemic/informational readings.

6. Mares–Goldblatt Quantifiers in Layered Semantics

One nice feature of the three-layered semantic picture is that we have three places where we can locate meanings. I've suggested that entailment facts should be understood to be grounded in the world. Having said that, however, we can locate the meanings of other expressions in other places, namely in the proposition or information state layer. One natural kind of expression which would seem to have its meaning most naturally in one of these higher layers may be certain modals which concern the interactions between agents and their available information.

The example I want to consider is the Mares–Goldblatt (MG) [32] interpretation of quantifiers, which I'll suggest most naturally lives at the propositional layer.[12] This provides an interesting contrast with the standard, Quinean, picture of the quantifiers wherein their meanings are to be found in the world, and the arrangements of properties over objects. The picture I'll sketch is similar to Mares' [31] proposed interpretation of the MG semantics, though it differs from his in some respects. Let me begin by recapping the basic elements of the MG semantics, building on the basis of the TR framework.

6.1. MG Quantifiers in TR Semantics.
First we extend the basic propositional language (implicit up until now) by a denumerable collection of variables $Var = \{x_n\}_{n \in \omega}$, and the quantifiers $\forall, \exists$. A language signature consists of a set of name constants Con and a collection $Pred$ of predicate letters of varying arities: the letter c will function as a metavariable over Con and P^n over $Pred$, having arity n.

---

[12]The original form of this semantics was given for quantified extensions of **R**, but it has recently been expanded to include a range of weaker logics in Ferenz [19]; Tedder and Ferenz [43].

Definition 18. An MG frame is a tuple $\langle W, N, R, *, Prop, D, PropFun \rangle$, where $F = \langle W, N, R, * \rangle$ is a TR frame, $Prop \subseteq \mathcal{P}(W)^\uparrow$, $D \neq \varnothing$, and $PropFun \subseteq \{\varphi \colon \varphi \colon D^\omega \longrightarrow Prop\}$. We stipulate a range of constraints on these things. To that end, given $f \in D^\omega$ (called a "variable assignment"), if $f' \in D^\omega$ is such that for any $m \neq n$, $fm = f'm$, then f' is an x_n-variant of f, written $f' \sim_{x_n} f$.

The constraints, taking the definitions of $\rightarrow, \neg$ as operations on $\mathcal{P}(W)^\uparrow$ from Definition 6, are:

(MG1) There is a $\varphi_N \in PropFun$ s.t. for all $f \in D^\omega$, $\varphi_N f = N$.
(MG2) If $\varphi \in PropFun$, then there is a $\neg \varphi \in PropFun$ s.t. $(\neg \varphi)f = \neg(\varphi f)$.
(MG3) If $\varphi, \psi \in PropFun$, then there is a $\varphi \otimes \psi \in PropFun$ s.t. $(\varphi \otimes \psi)f = \varphi f \otimes \psi f$ for each $\otimes \in \{\cap, \cup, \rightarrow\}$.
(MG4) If $\varphi \in PropFun, n \in \omega$, then there is a $\forall_n \varphi \in PropFun$ s.t.

$$(\forall_n \varphi)f = \bigcup \{X \in Prop \colon X \subseteq \bigcap_{f' \sim_{x_n} f} \varphi f'\}.$$

(MG5) If $\varphi \in PropFun, n \in \omega$, then there is a $\exists_n \varphi \in PropFun$ s.t.

$$(\exists_n \varphi)f = \bigcap \{X \in Prop \colon \bigcup_{f' \sim_{x_n} f} \varphi f' \subseteq X\}.$$

A model M on a MG frame is a multi-type function: it is of types $Con \longrightarrow D$ and $Pred^n \longrightarrow D^n$ (where $Pred^n \subseteq Pred$ is the set of n-ary predicate letters), and we define the combination of M with $f \in D^\omega$ as follows, for any $\tau \in Con \cup Var$:

$$M_f(\tau) = \begin{cases} fn & \text{if } \tau = x_n \in Var; \\ M(\tau) & \text{if } \tau \in Con. \end{cases}$$

We define $[\![\cdot]\!]^M$ assigning formulas to elements of $PropFun$ inductively as follows (note that $([\![\mathcal{A}]\!]^M)f$, often written $[\![\mathcal{A}]\!]^M_f$, takes a value in $Prop$):

(1) $[\![P^n(\tau_1, \ldots, \tau_n)]\!]^M_f = M(P^n)(M_f(\tau_1), \ldots, M_f(\tau_n))$
(2) $[\![\neg \mathcal{A}]\!]^M_f = \neg([\![\mathcal{A}]\!]^M_f)$
(3) $[\![\mathcal{A} \wedge \mathcal{B}]\!]^M_f = [\![\mathcal{A}]\!]^M_f \cap [\![\mathcal{B}]\!]^M_f$
(4) $[\![\mathcal{A} \vee \mathcal{B}]\!]^M_f = [\![\mathcal{A}]\!]^M_f \cup [\![\mathcal{B}]\!]^M_f$
(5) $[\![\mathcal{A} \rightarrow \mathcal{B}]\!]^M_f = [\![\mathcal{A}]\!]^M_f \rightarrow [\![\mathcal{B}]\!]^M_f$
(6) $[\![\forall x_n \mathcal{A}]\!]^M_f = (\forall_n [\![\mathcal{A}]\!]^M)f$
(7) $[\![\exists x_n \mathcal{A}]\!]^M_f = (\exists_n [\![\mathcal{A}]\!]^M)f$

A formula $\mathcal{A}$ is satisfied by the pair M, f just in case $N \subseteq [\![\mathcal{A}]\!]^M_f$. $\mathcal{A}$ is satisfied by M just in case it it satisfied by M, f for any $f \in D^\omega$. $\mathcal{A}$ is satisfied by an MG frame if satisfied by every model on the frame, and it is valid w.r.t. a class of MG frames if satisfied by every frame in the class.

The key innovation in this semantic framework concerns, naturally, the quantifiers. In particular, it is the introduction of the clauses (MG4) and (MG5). Note that, unlike in the standard, Tarskian framework, these are not interpreted just as generalized intersections/unions of "instances." Rather, these are mediated by elements of $Prop$ — we don't just consider, when evaluating a quantified claim at a world α, whether all/some instance of the quantified formula holds at α, or even at worlds related to α. Rather, we consider the state of information from α, that is, how α fits into the structure of propositions; in effect, we consider what the information supported by α commits one to. It's by working with *Prop* like this explicitly that Mares and Goldblatt are able to avoid the problems, discovered by Fine [21], with employing the standard Tarskian truth condition. So the interpretation of the quantifiers concerns not just a frame, but this *combined* with a particular complex algebra over that frame — that is to say, it is a form of *general frame semantics*. However this difference isn't just interesting for technical purposes, but also for philosophical purposes.

In particular, by working with this larger structure of information, the MG interpretation of the quantifiers seems to open itself up to readings of these objects other than the traditional reading made famous by Quine [34]. For example, the truth condition for the existential quantifier can be spelled out as

$$\alpha \in [\![\exists x_n \mathcal{A}]\!]_f^M \iff \forall X \in Prop(\alpha \in X \Rightarrow \forall \beta (\exists f' \sim_{x_n} f(\beta \in [\![\mathcal{A}]\!]_{f'}^M) \Rightarrow \beta \in X)).$$

That is, any proposition X which α supports contains any situation β which supports at least one instance of $\mathcal{A}$. That is, the information supported by α must be supported by a situation which supports at least one instance. We are concerned not with an existential commitment at the world of evaluation, but rather with a situation-independent *informational* commitment. In order to be so committed, one *does not need* to be committed to the existence of an $\mathcal{A}$ in any *particular* situation, but rather just be committed to infer the information supported by α in any situation, which does support the existence of an $\mathcal{A}$. To use the preferred terminology of Sylvan [38], we might call this a *particular* quantifier, which simply tracks the commitments associated with commitment to *a particular* one satisfying the formula.

The important thing for my purposes is that the three-layer semantic framework provides the grist both for a realist interpretation of the propositional vocabulary, and an informational interpretation of the quantifiers, in one setting. That this is a strength of the account is, of course, the kind of point Punčochář [33] noted, but it's an advantage we retain even when we are more picky about the grounding of the framework.

Acknowledgments. Many thanks to Katalin Bimbó, Nicholas Ferenz, Teresa Kouri Kissel, Shay Allen Logan, Franci Mangraviti, Eileen Nutting, Vít Punčochář, Heinrich Wansing, and an anonymous referee for discussion of these matters and comments on earlier versions. I gratefully acknowledge fellowship funding from the Humboldt Foundation.

REFERENCES

[1] Anderson, A. R. and Belnap, N. D. (1975). *Entailment: The Logic of Relevance and Necessity*, Vol. 1, Princeton University Press, Princeton, NJ.

[2] Anderson, A. R., Belnap, N. D. and Dunn, J. M. (1992). *Entailment: The Logic of Relevance and Necessity*, Vol. 2, Princeton University Press, Princeton, NJ.

[3] Barwise, J. (1993). Constraints, channels, and the flow of information, in P. Aczel, D. Israel, Y. Katagiri and S. Peters (eds.), *Situation Theory and Its Applications, 3*, Vol. 37 of *CSLI Lecture Notes*, CSLI Publications, Stanford, CA, pp. 3–27.

[4] Barwise, J. and Perry, J. (1983). *Situations and Attitudes*, MIT Press, Cambridge, MA.

[5] Belnap, N. D. (1994). Life in the undistributed middle, in K. Došen and P. Schroeder-Heister (eds.), *Substructural Logics*, Clarendon Press, Oxford UK, pp. 31–42.

[6] Bimbó, K. and Dunn, J. M. (2008). *Generalized Galois Logics: Relational Semantics for Non-Classical Logical Calculi*, Vol. 188 of *CSLI Lecture Notes*, CSLI Publications, Stanford, CA.

[7] Bimbó, K. and Dunn, J. M. (2016). The emergence of set-theoretical semantics for relevance logics around 1970, *IFCoLog Journal of Logics and their Applications* 4(3): 557–589. (Special issue: Bimbó, K. and Dunn, J. M., (eds.), *Proceedings of the Third Workshop, May 16–17, 2016, Edmonton, Canada*).

[8] Bimbó, K., Dunn, J. M. and Ferenz, N. (2018). Two manuscripts, one by Routley, one by Meyer: The origins of the Routley–Meyer semantics for relevance logics, *Australasian Journal of Logic* 15(2): 171–209.

[9] Copeland, B. J. (1979). On when a semantics is not a semantics: Some reasons for disliking the Routley–Meyer semantics for relevance logic, *Journal of Philosophical Logic* 8(1): 399–413.

[10] Copeland, B. J. (1983). Pure semantics and applied semantics: A response to Routley, Routley, Meyer, and Martin, *Topoi* 2: 197–204.

[11] Deigan, M. (2020). A plea for inexact truthmaking, *Linguistics and Philosophy* 43: 515–536.

[12] Divers, J. (2002). *Possible Worlds*, Routledge, London, UK.

[13] Došen, K. (1989). Sequent-systems and groupoid models. II, *Studia Logica* 48(1): 41–65.

[14] Dunn, J. M. (2001a). A representation of relation algebras using Routley–Meyer frames, in C. A. Anderson and M. Zelëny (eds.), *Logic, Meaning, and Computation: Essays in Memory of Alonzo Church*, Kluwer, Dordrecht, pp. 77–108.

[15] Dunn, J. M. (2001b). Ternary relational semantics and beyond: Programs as arguments (data) and programs as functions (programs), *Logical Studies* 7: 1–20.

[16] Dunn, J. M. and Hardegree, G. (2001). *Algebraic Methods in Philosophical Logic*, Vol. 41 of *Oxford Logic Guides*, Oxford University Press, Oxford, UK.

[17] Dunn, J. M. and Meyer, R. K. (1997). Combinators and structurally free logic, *Logic Journal of the IGPL* 5(4): 505–537.

[18] Dunn, J. M. and Restall, G. (2002). Relevance logic, in D. Gabbay and F. Guenthner (eds.), *Handbook of Philosophical Logic*, 2 edn, Vol. 6, Kluwer, Amsterdam, pp. 1–136.

[19] Ferenz, N. (Forthcoming). Quantified modal relevant logics, *Review of Symbolic Logic* .

[20] Fine, K. (1974). Models for entailment, *Journal of Philosophical Logic* 3: 347–371.

[21] Fine, K. (1989). Incompleteness for quantified relevance logics, in J. Norman and R. Sylvan (eds.), *Directions in Relevant Logic*, Kluwer, Dordrecht, pp. 205–225.

[22] Hartonas, C. (2016). Reasoning with incomplete information in generalized Galois logics without distribution: The case of negation and modal operators, in K. Bimbó (ed.), *J. Michael Dunn on Information Based Logics*, Springer, Switzerland, pp. 279–312.

[23] Humberstone, L. (1988). Operational semantics for positive **R**, *Notre Dame Journal of Formal Logic* 29(1): 61–80.

[24] Jago, M. (2020). Truthmaker semantics for relevant logic, *Journal of Philosophical Logic* 49: 681–701.

[25] Lewis, D. (1986). *On the Plurality of Worlds*, Blackwell, Cambridge, MA.

[26] Lipton, P. (2000). Inference to the best explanation, in W. H. Newton-Smith (ed.), *A Companion to the Philosophy of Science*, Blackwell, Malden, MA, pp. 184–193.

[27] Logan, S. A. (2020). Putting the stars in their places, *Thought* 9(3): 188–197.

[28] Logan, S. A. (Forthcoming). Deep fried logic, *Erkenntnis* .

[29] Loux, M. (1979). *The Possible and the Actual*, Cornell University Press, Ithaca, NY.

[30] Mares, E. (2004). *Relevant Logic: A Philosophical Interpretation*, Cambridge University Press, Cambridge, UK.

[31] Mares, E. D. (2009). General information in relevant logic, *Synthese* **167**: 343–362.
[32] Mares, E. D. and Goldblatt, R. (2006). An alternative semantics for quantified relevant logic, *Journal of Symbolic Logic* **71**(1): 163–187.
[33] Punčochář, V. (2020). A relevant logic of questions, *Journal of Philosophical Logic* **49**: 905–939.
[34] Quine, W. V. (1948). On what there is, *Review of Metaphysics* **2**(5): 21–38.
[35] Restall, G. (1995). Information flow and relevant logics, *in* J. Seligman and D. Westerståhl (eds.), *Logic, Language and Computation: The 1994 Moraga Proceedings*, CSLI Publications, Stanford, CA, pp. 463–477.
[36] Restall, G. (2000). *An Introduction to Substructural Logics*, Routledge, London, UK.
[37] Restall, G. (2007). Review of Ross T. Brady's *Universal Logic*, *Bulletin of Symbolic Logic* **13**(4): 544–546.
[38] Routley, R. (1980). *Exploring Meinong's Jungle and Beyond*, Philosophy Department, RSSS, Australian National University, Canberra. Departmental Monograph #3.
[39] Routley, R. and Meyer, R. K. (1973). The semantics of entailment, *in* H. Leblanc (ed.), *Truth, Syntax, and Modality. Proceedings of the Temple University Conference on Alternative Semantics*, North-Holland, Amsterdam, pp. 199–243.
[40] Routley, R., Plumwood, V., Meyer, R. K. and Brady, R. T. (1982). *Relevant Logics and Their Rivals: The Basic Philosophical and Semantical Theory*, Ridgeview, Atascadero.
[41] Sequoiah-Grayson, S. (2016). Epistemic relevance and epistemic actions, *in* K. Bimbó (ed.), *J. Michael Dunn on Information Based Logics*, Springer, Switzerland, pp. 133–146.
[42] Tedder, A. (2021). Information flow in logics in the vicinity of **BB**, *Australasian Journal of Logic* **18**(1): 1–24.
[43] Tedder, A. and Ferenz, N. (Forthcoming). Neighbourhood semantics for first-order relevant logics, *Journal of Philosophical Logic* .
[44] Urquhart, A. (1972). Semantics for relevant logics, *Journal of Symbolic Logic* **37**(1): 159–169.
[45] Urquhart, A. (1984). The undecidability of entailment and relevant implication, *Journal of Symbolic Logic* **49**(4): 1059–1073.

DEPARTMENT OF PHILOSOPHY I, RUHR UNIVERSITY BOCHUM, *Email:* `ajtedder.at@gmail.com`

WHO WAS SCHILLER JOE SCROGGS?

Alasdair Urquhart

ABSTRACT. Schiller Joe Scroggs was the author of one of the most renowned papers in modal logic. Several logicians have conjectured that he in fact never existed, being the invention of J. C. C. McKinsey. The main aim of this article is to refute this conjecture, and to explain who he was. The precise origins of Scroggs's paper remain somewhat mysterious.

Keywords. Finite characteristic matrix, J. C. C. McKinsey, Modal logic, S5, Schiller Joe Scroggs

1. THE SCROGGS PROBLEM

The title of our article is a historical puzzle that has exercised the ingenuity of a number of distinguished logicians. From time to time, I have discussed this question with (among others) Mike Dunn, Bob Meyer and Kit Fine — there are probably other logicians who have at least wondered about this intriguing question. Before listing earlier investigations of this problem, let me explain how the question arose.

1.1. The Famous Scroggs Article. The publication that gave rise to all this head-scratching is a 9-page article Scroggs [21] that is a classic of modal logic. The main theorem of the paper is that every logic extending **S5** closed under *modus ponens* and substitution has a finite characteristic matrix. This result has inspired a large number of similar theorems, such as the work of Kit Fine [8] on the logics extending **S4.3**, Dunn's theorem [4] on the extensions of the relevant logic **RM**, and the work of Dunn and Meyer [6] on the logic **LC** of Michael Dummett. Dunn's results on **RM** are described in §3.3.

A footnote to Scroggs's paper reads: "This paper was prepared under the direction of J. C. C. McKinsey as a Master's thesis at Oklahoma Agricultural and Mechanical College." However, the paper obviously bears the strong imprint of the style of McKinsey, who had been engaged in a research program since the early 1940s, investigating systems of modal logic using algebraic techniques. Scroggs's article cites three of McKinsey's papers [12; 13; 17], the last co-authored with Alfred Tarski.

Since the techniques and ideas of the Scroggs paper of 1951 are so closely related to those employed by McKinsey since the 1940s, it is understandable that some logicians have taken it to be a paper written by McKinsey himself, and even that "Schiller Joe Scroggs" was his own invention. As we shall see shortly, Scroggs was in fact a real

---

2020 *Mathematics Subject Classification.* Primary: 01A60, Secondary: 03B45, 03B47.

Bimbó, Katalin, (ed.), *Relevance Logics and other Tools for Reasoning. Essays in Honor of J. Michael Dunn*, (Tributes, vol. 46), College Publications, London, UK, 2022, pp. 427–434.

FIGURE 1. McKinsey in 1942

person, though the precise circumstances leading to the publication of [21] are not at all clear.

1.2. Some Earlier Investigations. As I mentioned initially, I have discussed the Scroggs problem with a number of logicians who were also interested in the question. Perhaps the most assiduous of these was Mike Dunn. I recall that he once told me that he had made a trip to Oklahoma State University (originally, "Oklahoma Agricultural and Mechanical College") to investigate the matter. However, this was quite some time ago, and I don't recall that his pilgrimage bore fruit.

What appeared to be a promising lead on the Scroggs problem came to light when Mike Dunn received a letter from Schiller Joe himself! In it, Scroggs claimed to have served in the Second World War, and expressed an interest in graduate work at Indiana University. Sad to say, however, this letter shortly proved to be a hoax perpetrated by the Maximum Leader of the Logicians Liberation League, Robert K. Meyer.

I attended a lattice theory and universal algebra conference at Asilomar in 1987, where Leon Henkin was in attendance. Since I was aware that Henkin knew McKinsey when the latter taught at Stanford University, I took the opportunity to ask him about Scroggs. Henkin was not able to answer my query, but he did tell me two interesting facts about McKinsey. First, he was a chain smoker; second, he was gay, but completely open about his sexual orientation at a time when most gay men were strictly in the closet.

2. J. C. C. MCKINSEY

2.1. Biography. John Charles Chenoweth McKinsey (1908–1953), known to his friends as "Chen," was a very productive and original mathematician, as well as a remarkably interesting person, who became Alfred Tarski's closest friend. A great deal of material about him is now available in the Fefermans' fascinating biography [7] of Tarski. His open homosexuality and the fears it induced in university administrators led him to move from position to position.

McKinsey received his B.S. and M.S. degrees from New York University and his Ph.D. degree from the University of California, where his advisor was Benjamin Abram Bernstein. He was a Blumenthal Research Fellow at New York University from 1936 to 1937 and a Guggenheim Fellow from 1942 to 1943. He also taught at Montana State College, and in Nevada, then Oklahoma. In 1947, he joined a research group, project RAND, at Douglas Aircraft Corporation. He was fired as a security risk in 1951, and then took up a position in the Stanford University Philosophy Department, where he collaborated with Patrick Suppes on the foundations of physics. In his book on the history of RAND Corporation, Alex Abella tells the story of McKinsey's dismissal:

> An open homosexual, McKinsey had been in a committed relationship for years when the FBI decided he was a security risk. When told that his sexual orientation could subject him to blackmail, McKinsey complained to Roberta Wohlstetter, "How can anyone threaten me with disclosure when everybody already knows?" A few years after his clearance was revoked and Frank Collbohm himself had fired him, McKinsey committed suicide. [1, Chapter 5]

McKinsey's tragically early death at the age of 45 was a major loss to logic and philosophy [3].

McKinsey had the misfortune of living through a period of extreme suspicion and paranoia in the United States. The early 1950s are remembered as a time when communists and people with left-wing political views were persecuted. However, it is often forgotten that there were equally violent attacks on gay men and lesbians. David K. Johnson has written an excellent history of this period entitled *The Lavender Scare*. In February 1950, Deputy Undersecretary John Peurifoy revealed to a congressional committee that a number of persons considered to be security risks had been forced out of the State Department, among them ninety-one homosexuals. This was the beginning of the "Lavender Scare" [9, p. 1].

In the context of that time, McKinsey's attitude as an openly gay man seems dangerously naïve. In fact, politicians at that time seem to have had trouble distinguishing between homosexuals and communists. Three of President Truman's top advisors wrote him a joint memorandum warning that "the country is more concerned about the charges of homosexuals in the Government that about communists" [9, p. 2]. Senator Kenneth Wherry, an enthusiastic persecutor of the gay community, remarked in 1950:

> You can't hardly separate homosexuals from subversives. Mind you, I don't say every homosexual is a subversive, and I don't say every subversive is a homosexual. But a man of low morality is a menace in the government, whatever he is, and they are all tied up together.
> [9, pp. 37–38]

In August 1954, after McKinsey was fired, the mathematician John Nash was arrested in a police sting operation and charged with indecent exposure. Subsequently, he also was fired from RAND [19, Chapter 25].

Game theory was one of the central concerns at RAND Corporation while McKinsey worked there, and he wrote one of the first textbooks on game theory, a monograph [14] that is still in print. In the late 1940s, while at RAND, McKinsey wrote to von

FIGURE 2. Schiller Joe Scroggs in 1949

Neumann about his idea for solving n-person games by using a computer. I have discussed this interesting correspondence and its connection to computational complexity in an article [22] in the *Bulletin of Symbolic Logic*.

2.2. **McKinsey's Work in Modal Logic.** In the 1940s, McKinsey published a series of important papers in the area of modal logic, some co-authored with Alfred Tarski. Among the latter papers is a groundbreaking article on the algebra of topology [15] in which a topological space is defined as a closure algebra, that is to say, a Boolean algebra with an algebraic operator satisfying the usual laws of topological closure. A closure algebra is an algebraic version of the modal logic **S4**; this connection with topology had already been exploited by McKinsey in an earlier article [12]. In that paper, he proved decidability for the Lewis systems **S2** and **S4** by showing that they have the finite model property. This result also yields decidability for the algebra of topology. McKinsey and Tarski extended this research in two subsequent papers to include results about Brouwerian algebras and Brouwerian logics [16] and [17], as well as the Lewis systems.

The techniques employed in all of the above papers by McKinsey, as well as in the joint work with Tarski all belong to algebraic logic. The usual apparatus of subalgebras, free algebras, homomorphisms, product algebras and so forth is deployed to prove some penetrating results about the algebras in question, that are then used to prove results about the logical systems from which the algebras were derived.

3. SCHILLER JOE SCROGGS

3.1. **Biography.** Schiller Joe Scroggs was born in Shawnee, Oklahoma on 17 January 1929. He entered Oklahoma Agricultural and Mechanical College in 1945, where his father served as Dean of Arts and Sciences, after he graduated from Stillwater High School. He earned his Bachelor of Science from the College in 1949 and his Master of Science in 1950.

In the 1951 and 1952 yearbooks of the University of California, Berkeley, Scroggs is listed as a member of Beta Theta Pi, and described as a "University Associate."

It seems likely (given his later employment) that he graduated from Berkeley with a two-year engineering degree. He died in Orlando, Florida on November 7, 1993, of throat cancer. He moved to Florida in 1983, where he was employed as an aerospace engineer and adjunct mathematics instructor at the University of Central Florida.

The Fourjay Foundations established a graduate award in memory of Schiller J. Scroggs after he passed away in 1993. The recipient of this award is a student who displays academic leadership and outstanding leadership ability.

3.2. The 1951 Scroggs Paper. The *JSL* article of 1951 [21] is a slightly altered version of Scroggs's Master's thesis of 1950. The ideas, techniques and style of the paper are very similar to those of McKinsey, as well as those of McKinsey and Tarski, in the series of papers on modal logic that we discussed in §2.2. The article fits seamlessly into this stream of research from the 1940s.

Scroggs proves the main result of the paper as follows. A *matrix* for a logic **S** is an algebra $\mathfrak{M} = \langle K, D, \times, -, ^* \rangle$ containing operators $\times$ and $-$ corresponding to the Boolean connectives of conjunction and negation, a modal operator *, and a subset D of *designated elements*. A logic **S** is a *quasi-normal* extension of **S5** if is closed under substitution and *modus ponens*. An **S**-matrix $\mathfrak{M}$ is a matrix such that any theorem of **S** always takes a designated value under any assignment to the variables. A matrix is *normal* [12] if the family D of designated elements is closed under detachment and adjunction and in addition, if $x \leftrightarrow y \in D$, then $x = y$.

Theorem 1 of [21] shows that any quasi-normal extension **S** of **S5** has the finite model property with respect to normal matrices, that is to say, any unprovable formula of **S** can be refuted in a finite normal **S**-matrix. Later, Scroggs strengthens Theorem 1 in the following way: any non-theorem of a quasi-normal extension of **S5** can be refuted in a finite Henle matrix $\mathfrak{M}$ with only one designated element, so that $D = \{1\}$, where 1 is the unit element of the Boolean algebra of $\mathfrak{M}$.

To explain the notion of a "Henle matrix," rather than follow Scroggs's paper at this point, it is easier to use the well-known duality theory for modal algebras expounded by Jónsson and Tarski [10; 11]. If R is an equivalence relation on a non-empty set S, then we can define a modal algebra $\mathfrak{M}(S,R)$ on the family of all subsets of S by defining for $X \subseteq S$, $\lozenge X = \{a \in S: \exists b \in X(aRb)\}$. If we take S to be the only designated value, then the resulting matrix validates all theorems of **S5**. If the relation R is the universal relation on S, then the matrix defined in this way is a Henle matrix in the sense of Definition 1 of Scroggs's paper. Furthermore, any finite Henle matrix with 1 the only designated value is isomorphic to one defined by this dual construction [10, §3].

Up to isomorphism there is exactly one finite Henle matrix with n atoms and $D = \{1\}$. Consequently, there is a sequence $\mathfrak{H}_1, \ldots, \mathfrak{H}_n, \ldots$ of finite Henle matrices, where every $\mathfrak{H}_k$ contains all earlier matrices in the sequence as submatrices. Hence, if **S** is a quasi- normal extension of **S5**, then either there is a k so that $\mathfrak{H}_k$ is the largest matrix in the sequence validating **S**, or no such k exists. In the first case, $\mathfrak{H}_k$ is a finite characteristic matrix for **S**, while in the second case, **S** is **S5**. This is the main theorem of Scroggs's paper.

3.3. Dunn's Continuation of Scroggs's Work.
Scroggs's paper of 1951 has served as a model for a number of later results, such as the work of J. M. Dunn [4] on the logic **R**-mingle (**RM**), the extension of the relevant logic **R** that results by adding the axiom schema $A \to (A \to A)$. For more details on **RM**, the reader can consult the treatise of Anderson and Belnap [2, §29], as well as Dunn's informative survey [5] of work on the system, including that of Arnon Avron. A logic **X** is an extension of **RM** if every theorem of **RM** is a theorem of **X**; it is a *proper extension* if it contains theorems not in **RM**. An extension is *normal* if it is closed under substitution, as well as *modus ponens* and adjunction.

As in the case of Scroggs's main theorem, Dunn defines a family of finite matrices, the *Sugihara matrices*, defined on a subset of the integers. These matrices come in two flavors. The matrices S_n are defined on the interval $[-n, \ldots, +n] \setminus \{0\}$, the integers from $-n$ to $+n$ with 0 omitted. The matrices $S_n + 0$ are defined on the interval $[-n, \ldots, +n]$ including 0. The negation of an element m in either case is defined as $-m$.

In both cases, the non-negative elements are designated. The matrices S_n are *normal* (there is no element m in S_n so that m and $-m$ are both designated), while the matrices $S_n + 0$ are not, since $-0 = 0$. Dunn uses the notation $S_n(+0)$ to refer ambiguously to these two structures. The Sugihara matrices $S_n(+0)$ play the same role in Dunn's result as the Henle matrices in Scroggs's paper. Robert K. Meyer [18] proved that **R**-mingle is complete with respect to the matrices S_n; consequently, the rule γ is admissible in **R**-mingle, since these matrices are all normal. Later, Dunn and Meyer generalized this result to logics such as **E** and **R**; I discuss this history in my paper on γ [23].

Dunn's main theorem [4, Theorem 9] follows the pattern of Scroggs's paper. Theorem 4 of [4] defines the sequence of matrices $S_0 + 0, S_1, S_1 + 0, S_2, \ldots$, and shows that if a sentence is valid in a matrix in the sequence, then it is valid in all earlier matrices. Then any proper normal extension **X** of **RM** has a finite characteristic matrix, by an argument paralleling that of Scroggs. If **X** is a normal extension of **RM**, then either there is a largest k so that the matrix $S_k(+0)$ in the sequence defined above validates **X**, or no such k exists. In the first case, $S_k(+0)$ is a finite characteristic matrix for **X**; in the second case, **X** is the full logic **RM**, by the result of Meyer [18].

The main theorem of [4] has a very pleasing corollary [4, Corollary 4] — a complete classification of those extensions of **RM** for which the rule γ is admissible. Ackermann's rule of γ is admissible for all of those consistent proper normal extensions of **RM** that have a characteristic matrix of the form S_n, and inadmissible for those with a characteristic matrix of the form $S_n + 0$.

3.4. The Origins of Scroggs's Master's Thesis.
The biographical information on McKinsey and Scroggs answers at least one basic question, since it establishes beyond doubt that Scroggs was a real person, contrary to speculations that we reported above. However, in another sense, it deepens the mystery of the origin of the Master's thesis. Scroggs submitted his thesis in April 1950 at a time when McKinsey was working at RAND Corporation. McKinsey must have left his job in Oklahoma in 1947, and it is hard to believe that he spent a lot of time there supervising Scroggs's thesis work — though he could of course have supervised it through correspondence.

The thesis itself is available online Scroggs [20], but unfortunately does not shed a great deal of further light on our problem. The typewritten thesis is essentially the same as the 1951 article in the *JSL*. It bears the signature of McKinsey as thesis adviser, together with those of a faculty representative, O. H. Hamilton, and Dean D. C. McIntosh of the graduate school. The first footnote reads: "This paper was prepared under the direction of J. C. C. McKinsey as a Master's thesis at Oklahoma Agricultural and Mechanical College and was submitted April 30, 1950." The last page bears the inscription "Typist: Mrs. J. P. Gardner."

In terms of ideas, techniques and style, the thesis resembles closely the work of McKinsey. Furthermore, the 1951 *JSL* paper is the one and only mathematical publication by Scroggs. However, we should perhaps not jump too readily to the conclusion that Scroggs contributed nothing at all to the work. Scroggs seems to have been a competent mathematician, judging by his later employment as an aerospace engineer and adjunct professor of mathematics in Florida.

A reasonably plausible story about the origins of the thesis is as follows. McKinsey may have taught Schiller Joe as an undergraduate and could have become friendly with him and his family (presumably it was Schiller Joe's father, Dean Schiller Scroggs, who hired him). It seems conceivable that McKinsey remained in touch with the Scroggs family and may have explained the basic ideas of the paper to Schiller Joe, who then wrote it up as his Master's thesis.

McKinsey was noted for his generosity to students. In their biography of Tarski, the Fefermans quote reminiscences of Ruth Barcan Marcus, who was McKinsey's student at NYU in 1940:

> He took me under his wing and invited me to do a tutorial in logic. We met two or three times a week in Bickford's cafeteria near Washington Square and we reviewed my work on the exercises he had given me from his own translation of Hilbert and Bernays. He urged me to go to graduate school but not to Harvard where, he said, Quine would clip my wings. I subsequently realized how much I took his mentorship and generosity for granted; I thought that was what all professors did. [7, p. 141]

Schiller Joe Scroggs may well have been a student whose talent McKinsey recognized. Given McKinsey's generous nature, he may have explained the outline of the 1951 paper to Scroggs and allowed him to publish the results as a Master's thesis. However, in the absence of further evidence, this is all just speculation, so we have achieved only a partial solution to the Scroggs problem.

I would like to thank Katalin Bimbó for very helpful correspondence about my research on Schiller Joe Scroggs, as well as for her unearthing of Bob Meyer's hoax letter at Indiana University.

REFERENCES

[1] Abella, A. (2009). *Soldiers of Reason: The RAND Corporation and the Rise of the American Empire*, Mariner Books.

[2] Anderson, A. R. and Belnap, N. D. (1975). *Entailment*, Vol. 1, Princeton University Press, Princeton, NJ.

[3] Davidson, D., Goheen, J. and Suppes, P. (1953–1954). J. C. C. McKinsey, *Proceedings and Addresses of the American Philosophical Association* **27**: 103–104.

[4] Dunn, J. (1970). Algebraic completeness results for R-Mingle and its extensions, *Journal of Symbolic Logic* **35**: 1–13.

[5] Dunn, J. M. (2021). R-mingle is nice and so is Arnon Avron, *in* O. Arieli and A. Zamansky (eds.), *Arnon Avron on Semantics and Proof Theory of Non-classical Logics*, Springer, pp. 141–165. Outstanding Contributions to Logic, Volume 21.

[6] Dunn, J. M. and Meyer, R. K. (1971). Algebraic completeness results for Dummett's LC and its extensions, *Zeitschrift für mathematische Logik und Grundlagen der Mathematik* **17**: 225–230.

[7] Feferman, A. B. and Feferman, S. (2004). *Alfred Tarski: Life and Logic*, Cambridge University Press, Cambridge, UK.

[8] Fine, K. (1973). The logics containing S4.3, *Zeitschrift für mathematische Logik und Grundlagen der Mathematik* **3**: 347–372.

[9] Johnson, D. K. (2004). *The Lavender Scare: The Cold War Persecution of Gays and Lesbians in the Federal Government*, The University of Chicago Press, Chicago, IL.

[10] Jónsson, B. and Tarski, A. (1951). Boolean algebras with operators. Part I., *American Journal of Mathematics* **73**: 891–939.

[11] Jónsson, B. and Tarski, A. (1952). Boolean algebras with operators. Part II., *American Journal of Mathematics* **74**: 127–162.

[12] McKinsey, J. (1941). A solution of the decision problem for the Lewis systems S2 and S4, with an application to topology, *Journal of Symbolic Logic* **6**: 117–134.

[13] McKinsey, J. (1944). On the number of complete extensions of the Lewis systems of sentential calculus, *Journal of Symbolic Logic* **9**: 42–45.

[14] McKinsey, J. (1952). *Introduction to the Theory of Games*, McGraw-Hill, New York. The RAND Series: reprinted by Dover Publications.

[15] McKinsey, J. and Tarski, A. (1944). The algebra of topology, *Annals of Mathematics* **45**: 141–191.

[16] McKinsey, J. and Tarski, A. (1946a). On closed elements in closure algebras, *Annals of Mathematics* **47**: 122–162.

[17] McKinsey, J. and Tarski, A. (1946b). Some theorems about the sentential calculi of Lewis and Heyting, *Journal of Symbolic Logic* **13**: 1–15.

[18] Meyer, R. (1971). R-mingle and relevant disjunction, (abstract), *Journal of Symbolic Logic* **36**: 366.

[19] Nasar, S. (1998). *A Beautiful Mind: A Biography of John Forbes Nash, Jr., Winner of the Nobel Prize in Economics, 1994*, Simon and Schuster, New York, NY.

[20] Scroggs, S. J. (1950). *Extensions of S5*, Master's thesis, Oklahoma Agricultural and Mechanical College. Available at the institutional repository SHAREOK.

[21] Scroggs, S. J. (1951). Extensions of the Lewis System S5, *Journal of Symbolic Logic* **16**(2): 112–120.

[22] Urquhart, A. (2010). Von Neumann, Gödel and complexity theory, *Bulletin of Symbolic Logic* **16**: 516–530.

[23] Urquhart, A. (2016). The story of γ, *in* K. Bimbó (ed.), *J. Michael Dunn on Information Based Logics*, Springer Nature, Switzerland, pp. 93–105. Outstanding Contributions to Logic 8.

DEPARTMENTS OF PHILOSOPHY AND COMPUTER SCIENCE, UNIVERSITY OF TORONTO, TORONTO, CANADA, *Email:* urquhart@cs.toronto.edu

REVISITING CONSTRUCTIVE MINGLE: ALGEBRAIC AND OPERATIONAL SEMANTICS

Yale Weiss

ABSTRACT. Among Dunn's many important contributions to relevance logic was his work on the system **RM** (R-mingle). Although **RM** is an interesting system in its own right, it is widely considered to be too strong. In this paper, I revisit a closely related system, **RM0** (sometimes known as "constructive mingle"), which includes the mingle axiom while not degenerating in the way that **RM** itself does. My main interest will be in examining this logic from two related semantical perspectives. First, I give a purely operational bisemilattice semantics for it by adapting previous work of Humberstone. Second, I examine a more conventional algebraic semantics for it and discuss how this relates to the operational semantics. A novel operational semantics for **J** (intuitionistic logic) as well as its conventional Heyting algebraic semantics emerge as special cases of the corresponding semantics for **RM0**. The results of this paper suggest that **RM0** is a more interesting logic than has been appreciated and that Humberstone's operational semantic framework similarly deserves more attention than it has received.

Keywords. Bisemilattices, Intuitionistic logic, Mingle, Operational semantics, Relevance logic

1. INTRODUCTION

Among Mike Dunn's many important contributions to relevance logic was his work on the system **RM** (R-mingle) [11; 15; 12]. Indeed, with Storrs McCall, Dunn is one of the system's "parents" (some of the history is recounted in Dunn [14, §7.3]). **RM**, which results by adding $\varphi \to (\varphi \to \varphi)$ to **R**, is one of the best behaved systems in the broader family of (quasi-)relevance logics and, not unrelatedly, also rather a disappointment (that **RM** is disappointing is, as far as I am aware, the consensus view, though Dunn has suggested—pace Meyer in Anderson and Belnap [1, §29.3, pp. 393–394]—that **RM** is superior to **R** "when all things are considered" [14, p. 143]). On the one hand, it is semantically natural, possessing both elegant binary relational and algebraic semantics, is decidable, and prima facie looks like an eminently reasonable axiomatic extension of **R**. On the other hand, it is just way too strong, producing such unsavory theorems as $(\varphi \to \psi) \lor (\psi \to \varphi)$ (sometimes called the "chain theorem") and ultimately tilting into the abyss of irrelevance.

2020 *Mathematics Subject Classification.* Primary: 03B47, Secondary: 03B20, 03G10.

Bimbó, Katalin, (ed.), *Relevance Logics and other Tools for Reasoning. Essays in Honor of J. Michael Dunn*, (Tributes, vol. 46), College Publications, London, UK, 2022, pp. 435–455.

The cognoscenti have long appreciated that the original sin of **RM** has less to do with the innocuous seeming mingle axiom than to do with the negation postulates of **R**:

> But the breakdowns that afflicted **RM** rested on **R**-style negation, which [...] is not as transparent as the other truth-functional connectives. Accordingly, further pursuit of the original Dunn-McCall insights, dropping the **R**-style negation [...] appears an interesting present alternative. (Meyer, in [1, §29.3, p. 394].)

This seems to me—and has seemed to others—to be an eminently reasonable suggestion.[1] The result of adding the mingle axiom to the pure implicational fragment of **R** yields a system, **RM0**$_\rightarrow$, which does *not* in fact coincide with the pure implicational fragment of **RM** but which is, on any reasonable understanding, relevant.[2] Anderson and Belnap call this "constructive mingle," as it is a subsystem of the implicational fragment of **J** (intuitionistic logic) [1, §8.15, pp. 98–99]. I will extend this name to all of **RM0**, which I take to be **RM0**$_\rightarrow$ extended with conjunction, disjunction, and the constant $\bot$—all governed by their usual axioms—and potentially some further connectives, though *not* the negation of **R** (Section 2).

This paper is primarily devoted to a study of **RM0** from two semantical perspectives. In Section 3, I give a purely operational *bisemilattice semantics* (cf. Urquhart's semilattice semantics of [38]) for **RM0** by adapting previous work of Humberstone from [20]. An operational semantics for **J** then emerges as the special case in which the bisemilattices—which here play the role of frames—are lattices. In Section 4, I examine a more conventional algebraic semantics for **RM0** and relate it to the previously developed operational semantics; here, the familiar Heyting algebraic semantics for **J** emerges as the special case.

Let me emphasize that my main interest in this paper is not so much novelty (though there will be some novelty) as it is in reframing existing ideas and situating them in a more abstract, broadly lattice-theoretic context. I will point out a number of connections and conceptual links which do not appear to have been adequately appreciated and also highlight certain ways in which Humberstone's ideas, properly situated, have anticipated subsequent developments (e.g., in inquisitive semantics). Some concluding remarks on such morals and outstanding problems are offered in Section 5.

2. Axiomatics

In this section, I present an axiom (Hilbert) system for **RM0** as well as certain extensions thereof. In what follows, the basic propositional language contains a countable set of propositional variables Π, the propositional constant $\bot$, and the binary

---

[1]For example, in [25], Méndez discusses how various sorts of alternative negations might be added to the standard axiomatic (not actual) positive fragment of **RM** (the article also provides ternary relational—though not algebraic or operational—semantics for some of these variations on **RM**).

An alternative idea is pursued by Avron (see, e.g., [2; 3]), who considers and advocates for an implication-negation system—the standard axiomatic (not actual) fragment of **RM** in that language—in which intensional versions of conjunction and disjunction can be defined. This project certainly has its interest, though it is quite different from the project which I shall pursue here.

[2] In particular, **RM0**$_\rightarrow$ (as well as its extension with the usual axioms for disjunction and conjunction) satisfies the *variable sharing property* (i.e., $\varphi \rightarrow \psi$ is never a theorem when φ and ψ do not share propositional variables) [25, p. 286].

connectives $\{\to, \land, \lor\}$. Formulae, etc., are defined as usual. I will use $p, q, \ldots$ for arbitrary propositional variables and $\varphi, \psi, \ldots$ for arbitrary formulae. I denote the set of all formulae in this language by Φ.

The axioms for **RM0** are just those of positive **R** (see, e.g., Dunn and Restall [16, §1.3]), together with the mingle axiom M and $\bot$.[3]

Definition 1. The system **RM0** is the smallest set of formulae containing all instances of the following axiom schemata and closed under the following rules:

- (I) $\varphi \to \varphi$
- (B) $(\varphi \to \psi) \to ((\chi \to \varphi) \to (\chi \to \psi))$
- (C) $(\varphi \to (\psi \to \chi)) \to (\psi \to (\varphi \to \chi))$
- (W) $(\varphi \to (\varphi \to \psi)) \to (\varphi \to \psi)$
- (M) $\varphi \to (\varphi \to \varphi)$
- ($\land$E1) $(\varphi \land \psi) \to \varphi$
- ($\land$E2) $(\varphi \land \psi) \to \psi$
- ($\land$I) $((\varphi \to \psi) \land (\varphi \to \chi)) \to (\varphi \to (\psi \land \chi))$
- ($\lor$I1) $\varphi \to (\varphi \lor \psi)$
- ($\lor$I2) $\psi \to (\varphi \lor \psi)$
- ($\lor$E) $((\varphi \to \chi) \land (\psi \to \chi)) \to ((\varphi \lor \psi) \to \chi)$
- (DIS) $(\varphi \land (\psi \lor \chi)) \to ((\varphi \land \psi) \lor \chi)$
- ($\bot$) $\bot \to \varphi$
- (ADJ) $\dfrac{\varphi, \psi}{\varphi \land \psi}$
- (MP) $\dfrac{\varphi, \varphi \to \psi}{\psi}$

Theoremhood ($\vdash_{\mathbf{RM0}}$) is defined as usual.[4] This axiomatization of **RM0** contains some redundancy (e.g., I easily follows from M and W by MP), but it has the benefit of making clear the relationship between **RM0** and **R**. Also, note that $\top$ is definable as $\bot \to \bot$ and, so defined, it is clear that $\vdash_{\mathbf{RM0}} \varphi \to \top$.

For certain purposes, I will be interested in extensions of **RM0** with the propositional constant t as well as the binary connective $\circ$ (for intensional conjunction or fusion). If I need to refer to the set of formulae formulated in a language containing either or both of these additional connectives, I will refer to it by Φ'. Where these are included in the language, the corresponding axioms for them are as follows, where $\varphi \leftrightarrow \psi$ abbreviates $(\varphi \to \psi) \land (\psi \to \varphi)$:

- (t) $\varphi \leftrightarrow (t \to \varphi)$

---

[3] It bears emphasis that this is *not* the fragment of **RM** in this language. The easiest way to see this is to note that **RM0**, so formulated, is a subsystem of **J**, whereas **RM**, which contains the chain theorem [1, §29.3.1, p. 397], clearly is not.

[4] One could of course also define a suitable consequence relation, holding between sets of formulae and formulae, though I will not pursue this here.

(∘) $(\varphi \to (\psi \to \chi)) \leftrightarrow ((\varphi \circ \psi) \to \chi)$

In Subsection 3.4, I will have occasion to make special use of **RM0** extended by t. For emphasis, I will sometimes designate this system by **RM0**^{t}.

It is clear that **J**, intuitionistic logic, is axiomatized by extending **RM0** with the weakening axiom schema:

(K) $\varphi \to (\psi \to \varphi)$

Of course, this system has a number of redundancies, but that is alright. One could also add to **J**, formulated in the appropriate language, axioms t and $\circ$, but the result would be that t and $\circ$ are equivalent (in the obvious sense) to $\top$ and $\wedge$, respectively, so there is little point (though see Lemma 29).

Finally, note that constructive negation ($\neg$) is definable in both **RM0** and **J** in the usual way: $\neg\varphi$ abbreviates $\varphi \to \bot$.[5] Incidentally, it may be complained that **RM0** is not really a relevance logic as, for example, $(\varphi \wedge \neg\varphi) \to \psi$ will come out a theorem. Without wishing to digress for too long on what makes a logic relevant, let me nevertheless state that I do not view this as a serious objection to the relevant credentials of **RM0**. In any case, the reader should note that the positive fragment of **RM0** *does* satisfy the variable sharing property (see footnote 2), standard relevance logics like **R** are themselves not infrequently presented with constants including $\bot$, and **RM0** does not have as theorems "bad guys" like the chain theorem or K.[6]

3. OPERATIONAL SEMANTICS

In this section, I present a purely operational bisemilattice semantics for **RM0** as well as **J**. All of the essential features of this semantics were already isolated in [20], however, Humberstone's focus was on different systems and my own presentation will reframe the material by placing it in a broadly lattice-theoretic context, the benefits of which will become clear shortly.

In Subsection 3.1, I review some essential concepts from lattice theory and the theory of bisemilattices. In Subsection 3.2, I present the formal semantics and discuss its relationship to some other frameworks, including inquisitive semantics. I sketch the proofs of soundness and completeness in Subsection 3.3. Finally, in Subsection 3.4, I illustrate an application of this semantics and results concerning it by giving an embedding of **J** in **RM0**^{t}.

3.1. **Lattice-Theoretic Preliminaries.** I begin by briefly reviewing some familiar and less familiar algebraic structures and definitions. The lattice-theoretic material is standard (consult, for example, Davey and Priestley [7] and Grätzer [18]). The material on bisemilattices should also be fairly standard, though I will only be concerned with elementary results concerning them (for additional background and some more advanced results, the reader might consult Balbes [4], Romanowska [36] and Ledda [24], for example).

---

[5]For a recent study of various logics with intuitionistic-type negations from a broadly relevant perspective (i.e., using ternary relational semantics), consult Robles and Méndez [35].

[6]Omission of this last is how Bimbó characterizes relevance logics [5, p. 723].

Definition 2 (Semilattice). A *semilattice* is a structure $\langle S, \bullet \rangle$ where S is a set and $\bullet : S \times S \to S$ satisfies the following equations:

(AS) $(x \bullet y) \bullet z = x \bullet (y \bullet z)$;
(CO) $x \bullet y = y \bullet x$;
(ID) $x \bullet x = x$.

A semilattice $\langle S, \bullet \rangle$ can be used to define a partial order in two ways. In a *meet-semilattice*, the semilattice will generally be written as $\langle S, \wedge \rangle$ and the partial order $\langle S, \leq_\wedge \rangle$ is defined by putting $x \leq_\wedge y$ if and only if $x \wedge y = x$. Dually, in a *join-semilattice*, the semilattice will generally be written as $\langle S, \vee \rangle$ and the partial order $\langle S, \leq_\vee \rangle$ is defined by putting $x \leq_\vee y$ if and only if $x \vee y = y$.

There are notions of distributivity for both kinds of semilattice. So as not to overburden a limited terminology, however, I will follow Humberstone in describing these semilattice-distribution properties as decomposition properties [20, p. 67]. A join-semilattice $\langle S, \vee \rangle$ is said to be *join-decomposable* if $z \leq_\vee x \vee y$ implies $\exists x', y'$ such that $x' \leq_\vee x$, $y' \leq_\vee y$, and $z = x' \vee y'$. Dually, a meet-semilattice $\langle S, \wedge \rangle$ is said to be *meet-decomposable* if $x \wedge y \leq_\wedge z$ implies $\exists x', y'$ such that $x \leq_\wedge x'$, $y \leq_\wedge y'$, and $z = x' \wedge y'$. How decomposability relates to distribution will be discussed below.

There are also notions of bounds for both semilattices. A join-semilattice $\langle S, 0, \vee \rangle$ has a *least element (bottom)* 0 if for any x, $x \vee 0 = x$. A meet-semilattice $\langle S, 1, \wedge \rangle$ has a *greatest element (top)* 1 if for any x, $x \wedge 1 = x$.

Definition 3 (Bisemilattice). A *bisemilattice* is a structure $\langle S, \vee, \wedge \rangle$ where $\langle S, \vee \rangle$ and $\langle S, \wedge \rangle$ are semilattices.

A bisemilattice will be called join-decomposable (meet-decomposable) just when the underlying join-semilattice (meet-semilattice) is. It will simply be called *decomposable* if it is both join-decomposable and meet-decomposable. A *bounded bisemilattice* is a bisemilattice $\langle S, 0, 1, \vee, \wedge \rangle$ with both least and greatest elements. Let it be emphasized that "least" and "greatest" are relative to the orders $\leq_\vee$ and $\leq_\wedge$, respectively; what is greatest (least) in one order need not be greatest (least) in the other. A bounded bisemilattice in which $x \vee 1 = 1$ holds will be called *top respecting* and a bounded bisemilattice in which $x \wedge 0 = 0$ holds will be called *bottom respecting*.

A bisemilattice $\langle S, \vee, \wedge \rangle$ is *meet-distributive* if its operations satisfy the equation $x \wedge (y \vee z) = (x \wedge y) \vee (x \wedge z)$ and *join-distributive* if they satisfy the equation $x \vee (y \wedge z) = (x \vee y) \wedge (x \vee z)$. If a bisemilattice is both meet-distributive and join-distributive, it will be called *distributive*.

If $\langle S, \vee, \wedge \rangle$ is a bisemilattice, a set $\emptyset \neq T \subseteq S$ is called a *filter* if $x, y \in T$ if and only if $x \wedge y \in T$. Thinking in terms of the induced partial order, a filter is a nonempty set which is upwards-closed under $\leq_\wedge$ and closed under meet. I will call a filter T *join-closed* if whenever $x, y \in T$, $x \vee y \in T$. The following result will frequently be used (mostly implicitly) in the sequel:

Lemma 4. *If $\langle S, \vee, \wedge \rangle$ is either a meet-distributive or join-distributive bisemilattice and T is a filter in it, T is join-closed.*

Proof. Suppose that $\langle S, \vee, \wedge \rangle$ is meet-distributive. Then $(x \wedge y) \wedge (x \vee y) = ((x \wedge y) \wedge x) \vee ((x \wedge y) \wedge y) = (x \wedge y) \vee (x \wedge y) = x \wedge y$, so $x \wedge y \leq_\wedge x \vee y$. Clearly, then, if $x, y \in T$,

$x \vee y$ is as well by upwards-closure and the fact that $x \wedge y \in T$. Alternatively, suppose that $\langle S, \vee, \wedge \rangle$ is join-distributive. Then $(x \wedge y) = (x \wedge y) \vee (x \wedge y) = ((x \wedge y) \vee x) \wedge ((x \wedge y) \vee y) = ((x \vee x) \wedge (x \vee y)) \wedge ((x \vee y) \wedge (y \vee y)) = (x \wedge y) \wedge (x \vee y)$, that is, $x \wedge y \leq_\wedge x \vee y$, which suffices by parallel reasoning. ◁

If $\mathfrak{B}$ is a bisemilattice, I write $\mathcal{F}(\mathfrak{B})$ for the set of all filters in $\mathfrak{B}$ and I write $\uparrow x$ for the *principal filter* generated by x, i.e., $\{y : x \leq_\wedge y\}$. Ideals, meet-closed ideals, and principal ideals are defined dually, though I will have little use for them in this paper.

Definition 5 (Lattice). A *lattice* is a bisemilattice $\langle S, \vee, \wedge \rangle$ in which $\vee$ and $\wedge$ satisfy the absorption equations:

(A1) $x \vee (x \wedge y) = x$;
(A2) $x \wedge (x \vee y) = x$.

In any lattice $\langle S, \vee, \wedge \rangle$, the partial orders $\langle S, \leq_\wedge \rangle$ and $\langle S, \leq_\vee \rangle$ coincide. Consequently, where $\langle S, \vee, \wedge \rangle$ is a lattice, the unambiguous induced partial order will generally be written as $\langle S, \leq \rangle$. Over bisemilattices, all of join-decomposability, meet-decomposability, join-distributivity, and meet-distributivity are independent.[7] On the other hand—and this illustrates how strong the absorption laws really are—all of these properties are equivalent over lattices (consult, e.g., [18, p. 167]). Any filter T in a lattice, regardless of whether it is distributive, is join-closed (indeed, satisfies the stronger property that if $x \in T$, $x \vee y \in T$ for any y). Finally, any bounded lattice is both top and bottom respecting.

Remark 6. What separates lattices from bisemilattices are the absorption postulates (A1) and (A2). A weakening of the absorption postulates, that $x \vee (x \wedge y) = x \wedge (x \vee y)$, is sometimes known as *Birkhoff's equation*, and bisemilattices which satisfy this are known as *Birkhoff systems* (see, e.g., Harding and Romanowska [19, p. 46]). It is obvious that any join-distributive or meet-distributive bisemilattice is a Birkhoff system.[8]

Before rounding out this subsection by giving some examples of various of the foregoing algebraic structures, I will note two more facts concerning bisemilattices and lattices which will turn out to play an important role in semantically distinguishing (and relating) **RM0** and **J**.

Lemma 7. *If $\langle S, 0, 1, \vee, \wedge \rangle$ is a bounded join-distributive bisemilattice, it is a lattice if and only if it is bottom respecting.*[9]

Proof. For the easy direction, if $\langle S, 0, 1, \vee, \wedge \rangle$ is a lattice, then by (A2), $0 \wedge x = 0 \wedge (0 \vee x) = 0$. Conversely, suppose that $\langle S, 0, 1, \vee, \wedge \rangle$ is bottom respecting. It must be shown that the absorption equations from Definition 5 are satisfied. Ad (A2): $x = x \vee 0 = x \vee (y \wedge 0) = ((x \vee y) \wedge (x \vee 0)) = x \wedge (x \vee y)$. Ad (A1): $x \vee (x \wedge y) = ((x \vee x) \wedge (x \vee y)) = x \wedge (x \vee y) = x$, by (A2). ◁

---
[7] I am not sure if this exact fact is stated anywhere in the literature, but various parts of this independence result can be found (e.g., in [36, p. 37]) and the rest can be shown without too much difficulty.

[8] I am grateful to H. P. Sankappanavar for suggesting that Birkhoff systems may be relevant to the subject of this paper.

[9] Cf. Płonka [30, p. 195, Theorem 2].

Lemma 8. *If $\langle S, 0, 1, \vee, \wedge \rangle$ is a bounded join-distributive bisemilattice, $\langle \uparrow 0, 0, 1, \vee, \wedge \rangle$ is a bounded distributive lattice (where these operations are restricted to $\uparrow 0$).*

Proof. In view of Lemma 7, it suffices to show that $\langle \uparrow 0, 0, 1, \vee, \wedge \rangle$ is bottom respecting (which is obvious, since if $x \in \uparrow 0$, $0 \leq_\wedge x$, i.e., $x \wedge 0 = 0$) and closed under the relevant operations (and so, a sub-bisemilattice of $\langle S, 0, 1, \vee, \wedge \rangle$). The only case that requires thought involves $\vee$: if $x, y \in \uparrow 0$, by the assumption that $\langle S, 0, 1, \vee, \wedge \rangle$ is join-distributive, $x \vee y \in \uparrow 0$, by Lemma 4. ◁

I now briefly give some examples. The first two, reducts of the strong and weak Kleene algebras [22, §64, p. 334], are among the best-known lattices and bisemilattices in logic. The third, which I believe is original to this paper, combines them; this last structure turns out to be a (non-degenerate) frame for **RM0**.

Example 9 (Strong Kleene). Consider the structure $\langle \{0, .5, 1\}, 0, 1, \vee, \wedge \rangle$ where the operations $\wedge$ and $\vee$ are defined by the following strong Kleene tables:

$\wedge$	0	.5	1
0	0	0	0
.5	0	.5	.5
1	0	.5	1

$\vee$	0	.5	1
0	0	.5	1
.5	.5	.5	1
1	1	1	1

It is of course well-known that these tables determine a bounded distributive lattice.

Example 10 (Weak Kleene). Consider the structure $\langle \{0, .5, 1\}, 0, 1, \vee, \wedge \rangle$ where the operations $\wedge$ and $\vee$ are defined by the following weak Kleene tables:

$\wedge$	0	.5	1
0	0	.5	0
.5	.5	.5	.5
1	0	.5	1

$\vee$	0	.5	1
0	0	.5	1
.5	.5	.5	.5
1	1	.5	1

This is easily shown to be a bounded join-distributive meet-decomposable bisemilattice, but it is *not* a lattice: $0 \wedge (0 \vee .5) = .5$, contradicting (A2). It is also neither top respecting ($.5 \vee 1 = .5$) nor bottom respecting ($.5 \wedge 0 = .5$).

Example 11 (Moderate Kleene). Consider the structure $\langle \{0, .5, 1\}, 0, 1, \vee, \wedge \rangle$ where the operations $\wedge$ and $\vee$ are defined by the weak and strong Kleene tables, respectively:

$\wedge$	0	.5	1
0	0	.5	0
.5	.5	.5	.5
1	0	.5	1

$\vee$	0	.5	1
0	0	.5	1
.5	.5	.5	1
1	1	1	1

This is another example of a bounded join-distributive meet-decomposable bisemilattice that's not a lattice and is not bottom respecting. However, this one *is* top respecting.

3.2. Bisemilattice Models. In this subsection, I present bisemilattice frames and models for **RM0** and **J** and prove some basic results about the semantics which will be required in later parts of the paper. I also discuss connections between this semantics and Humberstone's semantics in [20] as well as Punčochář's semantics in [32].

As I have already indicated, the semantics to be presented here is directly inspired by, and largely follows, [20]. Nevertheless, there are important differences. Humberstone's focus is on positive **R** and the frames he proposes for it are structures of the form $\langle S, 1, 0, \cdot, + \rangle$ where $\langle S, 1, \cdot \rangle$ is an Abelian (commutative) monoid, $\langle S, 0, + \rangle$ is a join-decomposable join-semilattice, $\cdot$ distributes over $+$, $0 \cdot x = 0$, and $\cdot$ and $+$ satisfy "pseudo-idempotence," i.e., $x \cdot (x+1) = x \cdot x = x^2$ [20, pp. 66–67].

The condition of pseudo-idempotence is particularly aesthetically and otherwise unfortunate (which Humberstone actually concedes [20, p. 67]), but Humberstone also considers, if only briefly, what occurs if you adopt the real thing: you get bisemilattice frames which suffice to characterize **RM0** [20, pp. 75–76].[10] I will use the following bisemilattices to furnish a semantics for **RM0**:

Definition 12 (Mingle Frame). A *mingle frame* is a bounded, top respecting, join-distributive, meet-decomposable bisemilattice $\mathfrak{F} = \langle S, 0, 1, \vee, \wedge \rangle$.

It must be emphasized that the bisemilattice frames described by Definition 12 are still not exactly the same as those which Humberstone considered for **RM0**. The central distinction is that, in my proposal, everything is, as it were, flipped (thus, I have meet-decomposability where Humberstone has join-decomposability, etc.). The motivation for this is narrowly technical and has to do with the naturalness of certain constructions yet to come.

Concrete instances of mingle frames are given in Examples 9 and 11, though the first is degenerate in the sense that it is a lattice.[11] It turns out that the class of lattice mingle frames characterizes intuitionistic logic.

Definition 13 (Intuitionistic Frame). An *intuitionistic frame* is a structure $\mathfrak{F} = \langle S, 0, 1, \vee, \wedge \rangle$ where $\mathfrak{F}$ is a mingle frame which is a lattice (equivalently, in view of Lemma 7, which is bottom respecting). More succinctly, an intuitionistic frame is just a bounded distributive lattice.

Definition 13 marks a considerable departure from the frames used to characterize **J** in Humberstone's own semantics. For Humberstone, frames for (positive) **J** are just frames for positive **R** (as described above) which satisfy the added condition that $x + 1 = 1$ [20, p. 66]. Flipping, this amounts to the condition that I have called bottom respect. But, over the relevant class of bisemilattice structures, this turns out to be equivalent to being a lattice, per Lemma 7.

It is here, in the formal apparatus for **J**, that the real conceptual clarity afforded by the bisemilattice semantics shines. It allows us to mark the difference between relevant (**RM0**) and irrelevant (**J**) logics by those properties which distinguish bisemilattices from lattices: the absorption laws. As my principal interest in this paper is not philosophical, I will not dwell long on this, but allow me to point out that, of all the laws defining distributive lattices, these are the only *non-regular identities* (i.e., identities in

---

[10]Humberstone does not actually use the word "bisemilattice" or talk about **RM0** by that name, but this is effectively what he describes in [20, pp. 75–76].

[11]It is worth remarking that, while combining weak Kleene conjunction with strong Kleene disjunction yields a mingle frame, it would not do to combine weak Kleene disjunction with strong Kleene conjunction. The resulting structure would be bottom respecting, but not top respecting.

which the variables on the sides of = are mismatched)—a strong whiff of irrelevance, indeed.[12]

Definition 14 (Model). A *mingle (intuitionistic) model* is a structure $\mathfrak{M} = \langle \mathfrak{F}, V \rangle$ where $\mathfrak{F} = \langle S, 0, 1, \vee, \wedge \rangle$ is a mingle (intuitionistic) frame and $V : \Pi \to \mathcal{F}(\mathfrak{F})$.

Thus, a model is obtained by assigning filters to propositional variables in the underlying frame; note that, by Lemma 4, all such filters must be join-closed. As would be expected from what has been said so far, in Humberstone's own semantics, one gets a model by assigning ideals to variables (Humberstone proposes something a bit more convoluted in [20, p. 68], but this is what it would come to in a bisemilattice framework).

Turning now to the truth conditions, which are essentially those of [20, pp. 63–65, 72] (cf. [38, §§2, 4]) modulo "flipping," with respect to a mingle model $\mathfrak{M} = \langle S, 0, 1, \vee, \wedge, V \rangle$ where $x \in S$, the relation $\vDash_x^{\mathfrak{M}}$ is defined as follows:[13]

(1) $\vDash_x^{\mathfrak{M}} p$ if and only if $x \in V(p)$;
(2) $\vDash_x^{\mathfrak{M}} \bot$ if and only if $x = 1$;
(3) $\vDash_x^{\mathfrak{M}} t$ if and only if $0 \leq_\wedge x$;
(4) $\vDash_x^{\mathfrak{M}} \varphi \wedge \psi$ if and only if $\vDash_x^{\mathfrak{M}} \varphi$ and $\vDash_x^{\mathfrak{M}} \psi$;
(5) $\vDash_x^{\mathfrak{M}} \varphi \vee \psi$ if and only if $\exists y, z \in S$ such that $x = y \wedge z$, $\vDash_y^{\mathfrak{M}} \varphi$, and $\vDash_z^{\mathfrak{M}} \psi$;
(6) $\vDash_x^{\mathfrak{M}} \varphi \to \psi$ if and only if for all $y \in S$, $\nvDash_y^{\mathfrak{M}} \varphi$ or $\vDash_{x \vee y}^{\mathfrak{M}} \psi$;
(7) $\vDash_x^{\mathfrak{M}} \varphi \circ \psi$ if and only if $\exists y, z \in S$ such that $y \vee z \leq_\wedge x$, $\vDash_y^{\mathfrak{M}} \varphi$, and $\vDash_z^{\mathfrak{M}} \psi$.

With reference to a given model $\mathfrak{M} = \langle S, 0, 1, \vee, \wedge, V \rangle$ and formula φ, define $[\varphi]^{\mathfrak{M}} = \{x \in S : \vDash_x^{\mathfrak{M}} \varphi\}$. $[\varphi]^{\mathfrak{M}}$ may intuitively be thought of as the *proposition* expressed by φ in $\mathfrak{M}$.

The following two results (Lemma 15 and Corollary 16) are versions of Humberstone's Plus and Zero lemmata [20, pp. 68–69] though, in the present framework, the second is a mere corollary of the first.

Lemma 15 (Propositional Filters). *For any formula φ and any mingle model $\mathfrak{M} = \langle \mathfrak{F}, V \rangle$, $[\varphi]^{\mathfrak{M}} \in \mathcal{F}(\mathfrak{F})$.*

Proof. The result holds by Definition 14 for propositional variables. Since $\uparrow 1 = \{1\}$ is obviously a filter (indeed, the smallest one), $[\bot]^{\mathfrak{M}} \in \mathcal{F}(\mathfrak{F})$. It is also obvious that $\uparrow 0 = [t]^{\mathfrak{M}}$ is a filter. The other cases follow by induction. ◁

I have been rather brief with Lemma 15 because I will effectively cover some of the primary inductive cases as part of a more general and related result concerning the algebra of propositions below (Lemma 34).

Corollary 16. *For any formula φ and any mingle model $\mathfrak{M} = \langle \mathfrak{F}, V \rangle$, $1 \in [\varphi]^{\mathfrak{M}}$.*

Proof. Immediate from Lemma 15, noting that 1 is an element of any filter. ◁

---

[12]For more on regular identities and their importance, consult Padmanabhan [29].

[13]Note that all of the truth conditions are in fact purely operational. In particular, $\leq_\wedge$ is a *defined* relation. Therefore, the truth condition for t, for example, could instead have been given as $\vDash_x^{\mathfrak{M}} t$ if and only if $0 \wedge x = 0$. This feature of the semantic framework distinguishes it from Fine's hybrid partial order-operational framework in [17], which postulates a primitive relation $\leq$.

Definition 17 (Validity). Where $\mathfrak{M} = \langle S, 0, 1, \vee, \wedge, V \rangle$ is a mingle model, φ is valid in $\mathfrak{M}$ ($\vDash^{\mathfrak{M}} \varphi$) if $0 \in [\varphi]^{\mathfrak{M}}$. Where $\mathfrak{F} = \langle S, 0, 1, \vee, \wedge \rangle$ is a mingle frame, φ is valid in $\mathfrak{F}$ ($\vDash^{\mathfrak{F}} \varphi$) if $\vDash^{\mathfrak{M}} \varphi$ for every model $\mathfrak{M}$ over $\mathfrak{F}$. φ is *valid* in **RM0** ($\vDash_{\textbf{RM0}} \varphi$) if $\vDash^{\mathfrak{F}} \varphi$ for every mingle frame $\mathfrak{F}$ and valid in **J** ($\vDash_{\textbf{J}} \varphi$) if $\vDash^{\mathfrak{F}} \varphi$ for every intuitionistic frame $\mathfrak{F}$.

Before concluding this subsection, I wish to touch upon the relation of this semantics to inquisitive semantics or, in any case, the sort of "generalization" of inquisitive semantics developed for **J** by Punčochář in [32]. Punčochář shows (among other things) that **J** is characterized by all *distributive information models*, where a distributive information frame (algebra) is a join-decomposable join-semilattice with a least element and a model is obtained by assigning to each propositional variable an ideal in the algebra.

The truth conditions proposed by Punčochář in [32, p. 1648] for $\bot$, $\wedge$, and $\vee$ are identical to Humberstone's from [20], that is to say, to flipped versions of the conditions presented above. The condition for $\to$ offered by [32, p. 1648] is superficially different. Taking the liberty to flip things as appropriate, it amounts to the following:

(6′) $\vDash^{\mathfrak{M}}_x \varphi \to \psi$ if and only if for all $x \leq_\vee y$, $\nvDash^{\mathfrak{M}}_y \varphi$ or $\vDash^{\mathfrak{M}}_y \psi$.

In fact, though, this condition is just equivalent to (6) over the lattice frames given for **J** above. For suppose condition (6) obtains and $x \leq y$ (subscripts may be ignored in a lattice frame as there is only one unambiguous partial order) and $\vDash^{\mathfrak{M}}_y \varphi$; then $\vDash^{\mathfrak{M}}_{x \vee y} \psi$, that is, $\vDash^{\mathfrak{M}}_y \psi$, given that $y = x \vee y$, as required for (6′). Conversely, suppose condition (6′) obtains and $\vDash^{\mathfrak{M}}_y \varphi$; then as $x \leq x \vee y$ and $\vDash^{\mathfrak{M}}_{x \vee y} \varphi$—since $y \in [\varphi]^{\mathfrak{M}}$ and $[\varphi]^{\mathfrak{M}}$ is upwards closed—it follows that $\vDash^{\mathfrak{M}}_{x \vee y} \psi$, as required for condition (6).

It is clear, then, that there is significant overlap between the inquisitive semantic approach to **J** developed in [32], as well as related work by other inquisitive semanticists, and the decades-earlier but unfortunately not well-known work of [20] and my own presentation of that material here. Since the work of Punčochář and other inquisitive semanticists is, however, quite independent as far as I can tell,[14] the recurrence of these ideas should be taken to speak to their quality.

3.3. Soundness and Completeness. In this subsection, I prove that **RM0** and **J** (Section 2) are sound and complete with respect to their operational semantics from Subsection 3.2. The arguments straightforwardly adapt results from [20], but are worth including in some detail to make this paper self-contained.

Theorem 18 (Soundness). *If $\vdash_{\textbf{RM0}} \varphi$, then $\vDash_{\textbf{RM0}} \varphi$.*

Proof. I survey just a couple representative cases. Suppose that the mingle axiom M fails, i.e., that $\nvDash_{\textbf{RM0}} \psi \to (\psi \to \psi)$; then there is a mingle model $\mathfrak{M} = \langle S, 0, 1, \vee, \wedge, V \rangle$ and some $x, y \in S$ such that $x, y \in [\psi]^{\mathfrak{M}}$ and $x \vee y \notin [\psi]^{\mathfrak{M}}$. But $[\psi]^{\mathfrak{M}}$ is a join-closed filter by Lemmata 4 and 15, so $x \vee y \in [\psi]^{\mathfrak{M}}$, a contradiction. Suppose for contradiction that axiom $\bot$ fails, i.e., that $\nvDash_{\textbf{RM0}} \bot \to \psi$; then there is a mingle model $\mathfrak{M} = \langle S, 0, 1, \vee, \wedge, V \rangle$ and an $x \in S$ such that $x \in [\bot]^{\mathfrak{M}}$ and $x \notin [\psi]^{\mathfrak{M}}$. But then $x = 1$, so by Corollary 16, $x \in [\psi]^{\mathfrak{M}}$, a contradiction. ◁

---

[14] In fact, in a recent article, Punčochář and Tedder *do* note the connection to Humberstone's condition for $\vee$ in any case [33, p. 357]. In another fairly recent article, Humberstone himself discusses various accounts of disjunction including his own from [20] as well as inquisitive views [21].

Theorem 19 (Soundness). *If $\vdash_\mathbf{J} \varphi$, then $\vDash_\mathbf{J} \varphi$.*

Proof. There is only one further case to consider. To show the validity of axiom K, suppose for contradiction that $\nvDash_\mathbf{J} \psi \to (\theta \to \psi)$. Then this fails in some intuitionistic model $\mathfrak{M} = \langle S, 0, 1, \vee, \wedge, V \rangle$ which must be a lattice. So there are $x, y \in S$ such that $x \in [\psi]^\mathfrak{M}$ and $y \in [\theta]^\mathfrak{M}$ and $x \vee y \notin [\psi]^\mathfrak{M}$. But $[\psi]^\mathfrak{M}$ is a filter in a *lattice*, whence $x \in [\psi]^\mathfrak{M}$ implies $x \vee y \in [\psi]^\mathfrak{M}$, which gives the desired contradiction. ◁

To prove completeness, I construct a canonical model for **L** (I will use **L** to refer ambiguously to **RM0** or **J** in what follows, and disambiguate where it becomes relevant). A set of formulae Γ is a **L** *theory* if the following conditions are satisfied:

1. $\varphi \in \Gamma$ and $\psi \in \Gamma$ imply $\varphi \wedge \psi \in \Gamma$;
2. $\varphi \in \Gamma$ and $\vdash_\mathbf{L} \varphi \to \psi$ imply $\psi \in \Gamma$.

I write $\text{Th}(\Gamma)$ for the smallest theory containing the set of formulae Γ, or just $\text{Th}(\varphi)$ if $\Gamma = \{\varphi\}$.[15] By $\mathbb{TH}$, I denote the set of all theories; $\mathbb{TH} \setminus \{\emptyset\}$ is, then, obviously the set of all nonempty theories. Define $\Gamma \cdot \Delta = \{\psi \colon \exists \varphi \in \Delta (\varphi \to \psi \in \Gamma)\}$ (cf. [17, p. 353]).

Definition 20. The *canonical model* for **L** is the structure $\mathfrak{M}^c = \langle \mathbb{TH} \setminus \{\emptyset\}, \mathbf{L}, \Phi, \cdot, \cap, V^c \rangle$ where $V^c(p) = \{\Gamma \in \mathbb{TH} \setminus \{\emptyset\} \colon p \in \Gamma\}$.[16]

Remark 21. One reason for my preference for the flipped, filter semantics rather than Humberstone's ideal semantics is that the canonical model construction is more natural. In Humberstone's construction, $\cap$ counterintuitively plays the role of join with Φ as semilattice bottom [20, pp. 70–71].

Lemma 22. *The structure $\mathfrak{M}^c = \langle \mathbb{TH} \setminus \{\emptyset\}, \mathbf{RM0}, \Phi, \cdot, \cap, V^c \rangle$ is a mingle model.*

Proof. The argument is essentially that given by [20, pp. 70–72] (cf. [17, §3]). For the flavor, I show that $\cdot$ is idempotent, sketch the main ideas required for proving meet-decomposability and join-distributivity, and verify that V^c meets the condition required by Definition 14, i.e., that each $V^c(p)$ is a filter.

To show that $\cdot$ is idempotent, suppose that $\varphi \in x \cdot x$; then $\exists \psi \in x$ such that $\psi \to \varphi \in x$. Since x is closed under ADJ, $\psi \wedge (\psi \to \varphi) \in x$ whence $\varphi \in x$ by the fact that $\vdash_\mathbf{RM0} (\psi \wedge (\psi \to \varphi)) \to \varphi$ (note that the proof of this makes indispensable use of W). Conversely, suppose that $\varphi \in x$; then since $\vdash_\mathbf{RM0} \varphi \to (\varphi \to \varphi)$ by M, $\varphi \to \varphi \in x$, which suffices to show $\varphi \in x \cdot x$. Therefore, $x = x \cdot x$, as required by idempotence. To show that $\langle \mathbb{TH} \setminus \{\emptyset\}, \Phi, \cap \rangle$ is meet-decomposable, on the supposition that $x \cap y \subseteq z$, put $x' = \text{Th}(x \cup z)$ and $y' = \text{Th}(y \cup z)$. This immediately delivers everything that is needed except for the property that $x' \cap y' \subseteq z$, which follows making use of DIS. Ad join-distributivity, the difficult direction is showing that $(x \cdot y) \cap (x \cdot z) \subseteq x \cdot (y \cap z)$. Suppose $\varphi \in (x \cdot y) \cap (x \cdot z)$; then $\exists \psi \in y$ such that $\psi \to \varphi \in x$ and $\exists \theta \in z$ such that $\theta \to \varphi \in x$. By ADJ and $\vee$E, $(\psi \vee \theta) \to \varphi \in x$, and by $\vee$I1 and $\vee$I2, $\psi \vee \theta \in y \cap z$. Hence, $\varphi \in x \cdot (y \cap z)$, as required. Finally, to show that $V^c(p)$ is a filter, note that it

---

[15]In the interest of rigor, I really ought to write something like $\text{Th}_\mathbf{L}(\Gamma)$ for the smallest **L** theory containing Γ, but I will generally suppress what system **L** I am talking about when talking about theories.

[16]Technically, depending on the language, Φ' should be used instead of Φ. For the purposes of this subsection, I just intend by Φ the set of all formulae of whatever the language is. Incidentally, nothing in the basic completeness proof requires the use of the constants or ∘.

must be nonempty since $\Phi \in V^c(p)$ and $x, y \in V^c(p)$ if and only if $p \in x, y$ if and only if $p \in x \cap y$ if and only if $x \cap y \in V^c(p)$. ◁

Lemma 23. *The structure* $\mathfrak{M}^c = \langle \mathbb{TH} \setminus \{\emptyset\}, \mathbf{J}, \Phi, \cdot, \cap, V^c \rangle$ *is an intuitionistic model.*

Proof. The proof is identical to that of Lemma 22, except it also has to be shown that $\mathfrak{M}^c$ is a lattice. By Lemma 7, it suffices to show that $\mathfrak{M}^c$ is bottom respecting. Obviously, $\mathbf{J} \cap x \subseteq \mathbf{J}$, so, for the converse, suppose that $\varphi \in \mathbf{J}$; then, as there is some $\psi \in x$ and $\vdash_\mathbf{J} \psi \to \varphi$ (by K), $\varphi \in x$, which suffices to show $\mathbf{J} \subseteq \mathbf{J} \cap x$, as desired. ◁

Lemma 24 (Truth Lemma). *If* $\mathfrak{M}^c = \langle \mathbb{TH} \setminus \{\emptyset\}, \mathbf{L}, \Phi, \cdot, \cap, V^c \rangle$ *is the canonical model for* **L**, *then for any* $x \in \mathbb{TH} \setminus \{\emptyset\}$, $x \in [\varphi]^{\mathfrak{M}^c}$ *if and only if* $\varphi \in x$.

Proof. By induction on the complexity of φ. The result holds by definition when φ is a propositional variable and is obvious when φ is t, $\bot$, or of the form $\psi \wedge \theta$. I will just consider the cases in which φ is either of the form $\psi \to \theta$ or $\psi \vee \theta$, supposing the result holds for ψ and θ. (The arguments for $\to$ and $\vee$ are essentially the same as those found in [17, p. 355] and [20, p. 72], respectively.)

Suppose $\psi \to \theta \in x$ and $y \in [\psi]^{\mathfrak{M}^c}$; by the induction hypothesis, $\psi \in y$, therefore, $\theta \in x \cdot y$, i.e., $x \cdot y \in [\theta]^{\mathfrak{M}^c}$, which suffices to show $x \in [\psi \to \theta]^{\mathfrak{M}^c}$. Conversely, suppose that $\psi \to \theta \notin x$ and put $y = \text{Th}(\psi)$. Then $\theta \notin x \cdot y$; for otherwise, there would be a formula χ such that $\vdash_\mathbf{L} \psi \to \chi$ and $\chi \to \theta \in x$, which would imply that $\psi \to \theta \in x$ (by suffixing), a contradiction. Thus, by the induction hypothesis, $y \in [\psi]^{\mathfrak{M}^c}$ and $x \cdot y \notin [\theta]^{\mathfrak{M}^c}$, which suffices.

Suppose $\psi \vee \theta \in x$ and put $y = \text{Th}(\psi)$ and $z = \text{Th}(\theta)$. Then $y \cap z \subseteq x$, for if $\chi \in y \cap z$, then $\vdash_\mathbf{L} \psi \to \chi$ and $\vdash_\mathbf{L} \theta \to \chi$, whence $\vdash_\mathbf{L} (\psi \vee \theta) \to \chi$ by ∨E, so $\chi \in x$. By meet-decomposability, there are $y \subseteq y' \in \mathbb{TH} \setminus \{\emptyset\}$ and $z \subseteq z' \in \mathbb{TH} \setminus \{\emptyset\}$ such that $x = y' \cap z'$. By the induction hypothesis, $\psi \in y \subseteq y' \in [\psi]^{\mathfrak{M}^c}$ and $\theta \in z \subseteq z' \in [\theta]^{\mathfrak{M}^c}$, which yields the result. Conversely, suppose $x \in [\psi \vee \theta]^{\mathfrak{M}^c}$; then there are y, z such that $x = y \cap z$, $y \in [\psi]^{\mathfrak{M}^c}$, and $z \in [\theta]^{\mathfrak{M}^c}$. By the induction hypothesis, $\psi \in y$ and $\theta \in z$, whence it follows that $\psi \vee \theta \in y \cap z = x$ by ∨I1 and ∨I2. ◁

Theorem 25 (Completeness). *If* $\vDash_\mathbf{RM0} \varphi$, *then* $\vdash_\mathbf{RM0} \varphi$.

Proof. Suppose that $\nvdash_\mathbf{RM0} \varphi$; then $\varphi \notin \mathbf{RM0}$ and so, by Lemma 24, $\mathbf{RM0} \notin [\varphi]^{\mathfrak{M}^c}$, i.e., $\nvDash^{\mathfrak{M}^c} \varphi$. Moreover, by Lemma 22, $\mathfrak{M}^c$ is a mingle model, so $\nvDash_\mathbf{RM0} \varphi$, which suffices. ◁

Theorem 26 (Completeness). *If* $\vDash_\mathbf{J} \varphi$, *then* $\vdash_\mathbf{J} \varphi$.

Proof. The proof is essentially that for Theorem 25, except the role of Lemma 22 is played by Lemma 23. ◁

3.4. An Embedding of J in RM0′.

Using a well-known translation scheme (see, e.g., Meyer [26, pp. 198ff.] and Dunn and Meyer [15, pp. 229–230]), I shall now give an embedding of **J** into **RM0′**. The result (if I may say so) gives a nice illustration of an application of the foregoing semantics and some of the results concerning it.

Definition 27 (Translation). Define the function $\tau : \Phi \to \Phi'$ as follows:

1. $\tau(p) = p$;

2. $\tau(\bot) = \bot$;
3. $\tau(\varphi \wedge \psi) = \tau(\varphi) \wedge \tau(\psi)$;
4. $\tau(\varphi \vee \psi) = \tau(\varphi) \vee \tau(\psi)$;
5. $\tau(\varphi \to \psi) = (\tau(\varphi) \wedge t) \to \tau(\psi)$.

Lemma 28. *For any $\varphi \in \Phi$, if $\vdash_\mathbf{J} \varphi$, then $\vdash_{\mathbf{RM0}^t} \tau(\varphi)$.*

Proof. Suppose $\nvdash_{\mathbf{RM0}^t} \tau(\varphi)$. By Theorem 25, there is a mingle model $\mathfrak{M} = \langle S, 0, 1, \vee, \wedge, V \rangle$ such that $\nvDash_0^\mathfrak{M} \tau(\varphi)$. Define $\mathfrak{M}' = \langle \uparrow 0, 0, 1, \vee, \wedge, V' \rangle$, where $V'(p) = V(p) \cap \uparrow 0$ and the operations are likewise restricted. $\langle \uparrow 0, 0, 1, \vee, \wedge \rangle$ is an intuitionistic frame by Lemma 8 and, as intersections of filters are filters, $V'(p)$ is a filter for every p. Thus, $\mathfrak{M}'$ is an intuitionistic model.

It is to be shown by induction that, for all formulae $\psi \in \Phi$ and $x \in \uparrow 0$, $\vDash_x^{\mathfrak{M}'} \psi$ if and only if $\vDash_x^\mathfrak{M} \tau(\psi)$. The basis cases are immediate, so suppose the result holds for θ and χ. I examine just the cases concerning $\vee$ and $\to$.

Suppose $\vDash_x^\mathfrak{M} \tau(\theta \vee \chi)$, i.e., $\vDash_x^\mathfrak{M} \tau(\theta) \vee \tau(\chi)$. Then $\exists y, z \in S$ such that $x = y \wedge z$, $\vDash_y^\mathfrak{M} \tau(\theta)$, and $\vDash_z^\mathfrak{M} \tau(\chi)$. By the induction hypothesis and the fact that $y, z \in \uparrow 0$ since $y \wedge z = x \in \uparrow 0$, $\vDash_y^{\mathfrak{M}'} \theta$ and $\vDash_z^{\mathfrak{M}'} \chi$, i.e., $\vDash_x^{\mathfrak{M}'} \theta \vee \chi$. Conversely, if $\vDash_x^{\mathfrak{M}'} \theta \vee \chi$, then $\exists y, z \in \uparrow 0$ such that $x = y \wedge z$, $\vDash_y^{\mathfrak{M}'} \theta$, and $\vDash_z^{\mathfrak{M}'} \chi$, which immediately yields the result by the induction hypothesis.

Suppose $\vDash_x^\mathfrak{M} \tau(\theta \to \chi)$, i.e., $\vDash_x^\mathfrak{M} (\tau(\theta) \wedge t) \to \tau(\chi)$, and suppose $\vDash_y^{\mathfrak{M}'} \theta$. By the induction hypothesis and the fact that $0 \leq_\wedge y$, $\vDash_y^\mathfrak{M} \tau(\theta) \wedge t$, whence $\vDash_{x \vee y}^\mathfrak{M} \tau(\chi)$. $x, y \in \uparrow 0$ implies $x \vee y \in \uparrow 0$ (Lemma 4), so by the induction hypothesis, $\vDash_{x \vee y}^{\mathfrak{M}'} \chi$, which suffices to show $\vDash_x^{\mathfrak{M}'} \theta \to \chi$. Conversely, suppose $\nvDash_x^\mathfrak{M} \tau(\theta \to \chi)$, i.e., $\nvDash_x^\mathfrak{M} (\tau(\theta) \wedge t) \to \tau(\chi)$. Then $\exists y \in S$ such that $\vDash_y^\mathfrak{M} \tau(\theta) \wedge t$ and $\nvDash_{x \vee y}^\mathfrak{M} \tau(\chi)$. Then $0 \leq_\wedge y$ so, by the induction hypothesis, $\vDash_y^{\mathfrak{M}'} \theta$ and $\nvDash_{x \vee y}^{\mathfrak{M}'} \chi$, that is, $\nvDash_x^{\mathfrak{M}'} \theta \to \chi$.

Then $\nvDash_0^{\mathfrak{M}'} \varphi$ follows from $\nvDash_0^\mathfrak{M} \tau(\varphi)$. Therefore, by Theorem 19, $\nvdash_\mathbf{J} \varphi$, which was to be proved. ◁

Lemma 29. *For any $\varphi \in \Phi$, if $\vdash_{\mathbf{RM0}^t} \tau(\varphi)$, then $\vdash_\mathbf{J} \varphi$.*

Proof. Let $\mathbf{J}'$ be $\mathbf{J}$ formulated in the language with t and the corresponding axiom t. Then it is clear that $\mathbf{RM0}^t$ is a subsystem of $\mathbf{J}'$, so if $\vdash_{\mathbf{RM0}^t} \tau(\varphi)$ (ex hypothesi), we have $\vdash_{\mathbf{J}'} \tau(\varphi)$. By induction, $\tau(\varphi)$ and φ are provably equivalent in $\mathbf{J}'$, thus $\vdash_{\mathbf{J}'} \varphi$. Lastly, it must be shown that $\mathbf{J}'$ is a conservative extension of $\mathbf{J}$, i.e., that for any $\psi \in \Phi$, $\vdash_{\mathbf{J}'} \psi$ only if $\vdash_\mathbf{J} \psi$. But this clearly holds since in any proof in $\mathbf{J}'$ of such a ψ, t can be replaced with any theorem of $\mathbf{J}$ (e.g., $p \to p$) thereby yielding a proof of ψ in $\mathbf{J}$. Thus, $\vdash_\mathbf{J} \varphi$, as desired. ◁

Theorem 30. *For any $\varphi \in \Phi$, $\vdash_\mathbf{J} \varphi$ if and only if $\vdash_{\mathbf{RM0}^t} \tau(\varphi)$.*

Proof. Immediate from Lemmata 28 and 29. ◁

4. Algebraic Semantics

In this section, I present an algebraic semantics for **RM0**. The kind of algebraic structure used for modeling **RM0** is the obvious extension of what Meyer (in [27, p. 39], cf. [28, p. 408]) calls a *Dunn monoid*, in honor of Dunn's pioneering work in [10] (published as [13]).[17] Whereas Dunn monoids furnish an algebraic semantics for positive **R**, what I will call *Dunn semilattices* furnish an algebraic semantics for **RM0**.[18] The name is, in a sense, unfortunate, since Dunn semilattices are also bisemilattices and, indeed, lattices (under different operations). However, I hope the reader will indulge my penchant for semilattice nomenclature, if only because the name highlights that the pertinent (commutative) monoids are now required to be fully idempotent.

Definition 31 (Dunn Semilattice). A *Dunn semilattice* is a structure $\mathbf{D} = \langle D, \mathbf{1}, \mathbf{0}, \bullet, \Rightarrow, \sqcup, \sqcap \rangle$, where $\mathbf{0}, \mathbf{1} \in D$ and the binary operations $\bullet, \Rightarrow, \sqcup$, and $\sqcap$ satisfy the properties that:

1. $\langle D, \mathbf{0}, \sqcup, \sqcap \rangle$ is a distributive lattice with least element $\mathbf{0}$;[19]
2. $\langle D, \mathbf{1}, \bullet \rangle$ is a meet-semilattice with greatest element $\mathbf{1}$;
3. $a \bullet \mathbf{0} = \mathbf{0}$;
4. $a \bullet (b \sqcup c) = (a \bullet b) \sqcup (a \bullet c)$;
5. $a \bullet b \sqsubseteq c$ if and only if $a \sqsubseteq b \Rightarrow c$.

It is clear that a *Heyting algebra* (consult, e.g., Rasiowa and Sikorski [34]) is the special case of a Dunn semilattice in which $\bullet$ and $\sqcap$ are the same operation; for this reason, where $\mathbf{D} = \langle D, \mathbf{1}, \mathbf{0}, \bullet, \Rightarrow, \sqcup, \sqcap \rangle$ is a Heyting algebra, I will often omit $\bullet$. (Not every Dunn semilattice is a Heyting algebra; consult Example 36 below.)

A few elementary results concerning Dunn semilattices, some of which I will have occasion to appeal to in the sequel, are summarized without proof in Fact 32.

Fact 32. *In any Dunn semilattice* $\mathbf{D} = \langle D, \mathbf{1}, \mathbf{0}, \bullet, \Rightarrow, \sqcup, \sqcap \rangle$, *the following obtain:*

1. $a \sqsubseteq b$ *implies* $a \bullet c \sqsubseteq b \bullet c$;
2. $a \sqcap b \sqsubseteq a \bullet b \sqsubseteq a \sqcup b$;
3. $a \bullet (b \sqcap c) \sqsubseteq (a \bullet b) \sqcap (a \bullet c)$;
4. $(a \sqcap b) \bullet (c \sqcap d) \sqsubseteq (a \bullet c) \sqcap (b \bullet d)$.

There is an obvious way to generate a Dunn semilattice or Heyting algebra from a given mingle frame (Definition 12) or intuitionistic frame (Definition 13).

Definition 33 (Filter Algebra). Given a mingle frame $\mathfrak{F} = \langle S, 0, 1, \vee, \wedge \rangle$, the *filter algebra* over $\mathfrak{F}$, $\mathbb{A}(\mathfrak{F}) = \langle D, \mathbf{1}, \mathbf{0}, \bullet, \Rightarrow, \sqcup, \sqcap \rangle$, is defined as follows:

1. $D = \mathcal{F}(\mathfrak{F})$;
2. $\mathbf{1} = \uparrow 0$;
3. $\mathbf{0} = \uparrow 1$;
4. $I \bullet J = \{k \in S : \exists i \in I, \exists j \in J (i \vee j \leq_\wedge k)\}$;
5. $I \Rightarrow J = \bigcup \{K \in \mathcal{F}(\mathfrak{F}) : K \bullet I \subseteq J\}$;
6. $I \sqcup J = \{i \wedge j : i \in I, j \in J\}$;
7. $I \sqcap J = I \cap J$.

---

[17] Of course, much of the *mathematics* behind Dunn monoids is older; see, e.g., Ward and Dilworth [40].

[18] In the interest of completeness, I should note that Meyer and Routley discuss algebraic models for mingle-extended relevance logics en passant in [28, pp. 419–420].

[19] Any Dunn semilattice will also have a greatest element (with respect to $\sqsubseteq$), viz., $\mathbf{0} \Rightarrow \mathbf{0}$, which will not in general be identical to $\mathbf{1}$.

Lemma 34. *Any filter algebra* $\mathbb{A}(\mathfrak{F}) = \langle D, \mathbf{1}, \mathbf{0}, \bullet, \Rightarrow, \sqcup, \sqcap \rangle$ *over a mingle frame* $\mathfrak{F} = \langle S, 0, 1, \vee, \wedge \rangle$ *is a Dunn semilattice.*

Proof. First, it must be verified that the operations, so defined, actually are operations on $\mathcal{F}(\mathfrak{F})$, i.e., that given filters, they yield filters. I examine just the cases of $\bullet$ and $\Rightarrow$.

It is clear that $I \bullet J$ is nonempty if I and J are. So suppose that $x, y \in I \bullet J$; then $\exists i, i' \in I$ and $\exists j, j' \in J$ such that $i \vee j \leq_\wedge x$ and $i' \vee j' \leq_\wedge y$. By join-distributivity and the facts that $(i \vee j) \wedge (i' \vee j) \leq_\wedge i \vee j \leq_\wedge x$ and $(i \vee j') \wedge (i' \vee j') \leq_\wedge i' \vee j' \leq_\wedge y$, $(i \wedge i') \vee (j \wedge j') = ((i \vee j) \wedge (i' \vee j)) \wedge ((i \vee j') \wedge (i' \vee j')) \leq_\wedge x \wedge y$, where $i \wedge i' \in I$ and $j \wedge j' \in J$. Thus, $x \wedge y \in I \bullet J$, as desired. Conversely, if $x \wedge y \in I \bullet J$, $\exists i \in I$ and $\exists j \in J$ such that $i \vee j \leq_\wedge x \wedge y$. The result then follows immediately from the facts that $x \wedge y \leq_\wedge x$ and $x \wedge y \leq_\wedge y$.

For any filters I and J, since $I \bullet \uparrow 1 \subseteq J$, clearly $I \Rightarrow J \neq \emptyset$. Suppose that $x, y \in I \Rightarrow J$; then $\exists X, Y \in \mathcal{F}(\mathfrak{F})$ such that $x \in X$ and $y \in Y$ with $X \bullet I \subseteq J$ and $Y \bullet I \subseteq J$. Consider the filter $X \sqcup Y$; we wish to show $(X \sqcup Y) \bullet I \subseteq J$. Suppose $z \in (X \sqcup Y) \bullet I$. Then $\exists i \in I$, $x' \in X$, and $y' \in Y$ such that $(x' \wedge y') \vee i = (x' \vee i) \wedge (y' \vee i) \leq_\wedge z$. But $X \bullet I \subseteq J$ implies that $x' \vee i \in J$ and $Y \bullet I \subseteq J$ implies that $y' \vee i \in J$, so $(x' \vee i) \wedge (y' \vee i) \in J$ (as J is meet-closed) and $z \in J$ (as J is upwards closed). This suffices to show $x \wedge y \in I \Rightarrow J$, since $x \wedge y \in X \sqcup Y$. Conversely, suppose $x \wedge y \in I \Rightarrow J$; then $\exists K \in \mathcal{F}(\mathfrak{F})$ such that $x \wedge y \in K$ and $K \bullet I \subseteq J$. By upwards closure, $x, y \in K$, which suffices.

I omit the arguments that $\langle D, \mathbf{0}, \sqcup, \sqcap \rangle$ is a distributive lattice with bottom $\mathbf{0}$, that $\langle D, \mathbf{1}, \bullet \rangle$ is a meet-semilattice with top $\mathbf{1}$, and that $I \bullet \uparrow 1 = \uparrow 1$; these are fairly routine. It remains to verify the last two requirements from Definition 31. To show that $I \bullet (J \sqcup K) = (I \bullet J) \sqcup (I \bullet K)$, suppose that $x \in I \bullet (J \sqcup K)$; then for some $i \in I$, $j \in J$, and $k \in K$, $i \vee (j \wedge k) = (i \vee j) \wedge (i \vee k) \leq_\wedge x$. Clearly, $i \vee j \in I \bullet J$ and $i \vee k \in I \bullet K$, so $(i \vee j) \wedge (i \vee k) \in (I \bullet J) \sqcup (I \bullet K)$, from which the result follows by upwards closure. Conversely, suppose $x \in (I \bullet J) \sqcup (I \bullet K)$. Then $x = y \wedge z$ for some $i, i' \in I$, $j \in J$, and $k \in K$ such that $i \vee j \leq_\wedge y$ and $i' \vee k \leq_\wedge z$, and therefore, $(i \vee j) \wedge (i' \vee k) \leq_\wedge y \wedge z$. Then $j \wedge k \in J \sqcup K$ and $i \wedge i' \in I$, so $(i \wedge i') \vee (j \wedge k) \in I \bullet (J \sqcup K)$; but $(i \wedge i') \vee (j \wedge k) = ((i \vee j) \wedge (i' \vee j)) \wedge ((i \vee k) \wedge (i' \vee k)) \leq_\wedge (i \vee j) \wedge (i' \vee k) \leq_\wedge y \wedge z = x$, so $x \in I \bullet (J \sqcup K)$. Finally, it has to be verified that $I \bullet J \subseteq K$ if and only if $I \subseteq J \Rightarrow K$. From left to right, this is essentially immediate from the definition of $J \Rightarrow K$. Conversely, it suffices to show that $(J \Rightarrow K) \bullet J \subseteq K$.[20] Suppose $x \in (J \Rightarrow K) \bullet J$; then there is some y in some filter Y such that $Y \bullet J \subseteq K$ and some $z \in J$ such that $y \vee z \leq_\wedge x$. But then $y \vee z \in K$, so $x \in K$ by upwards closure, as desired. ◁

Lemma 35. *Any filter algebra* $\mathbb{A}(\mathfrak{F}) = \langle D, \mathbf{1}, \mathbf{0}, \bullet, \Rightarrow, \sqcup, \sqcap \rangle$ *over an intuitionistic frame* $\mathfrak{F} = \langle S, 0, 1, \vee, \wedge \rangle$ *is a Heyting algebra.*

Proof. The argument is the same as that for Lemma 34, except we have to check that $I \bullet J = I \sqcap J$ for all filters I, J. From right to left, if $x \in I \sqcap J = I \cap J$, then $x \in I, J$, so $x \in I \bullet J$ as $x \vee x \leq x$. Conversely, if $x \in I \bullet J$, then there are $i \in I$ and $j \in J$ such that $i \vee j \leq x$; but $i \leq i \vee j \leq x$ and $j \leq i \vee j \leq x$ imply that $x \in I \cap J$, as required. (Obviously this argument depends on the fact that $\leq$ is unambiguous in an intuitionistic frame.) ◁

---

[20]This follows from the general fact that $I \subseteq J$ and $J \bullet K \subseteq L$ imply $I \bullet K \subseteq L$. For if $x \in I \bullet K$, $i \vee k \leq_\wedge x$ for some $i \in I$ and $k \in K$. But $i \in I \subseteq J$, so $x \in L$ as $J \bullet K \subseteq L$.

Example 36 (RM3). Recall the moderate Kleene bisemilattice from Example 11. I will presently show that the filter algebra over this frame is a reduct of the characteristic algebra for the logic **RM3**.[21] In particular, our algebra is $\mathbb{A} = \langle \{-\mathbf{1}, \mathbf{0}, \mathbf{1}\}, \mathbf{0}, -\mathbf{1}, \bullet, \Rightarrow, \sqcup, \sqcap \rangle$ where $-\mathbf{1} = \{1\}$, $\mathbf{0} = \{0, 1\}$, and $\mathbf{1} = \{0, .5, 1\}$—these are all the filters in this bisemilattice—and the connectives, defined by Definition 33, are displayed table-wise for convenience:[22]

$\bullet$	$\{1\}$	$\{0,1\}$	$\{0,.5,1\}$
$\{1\}$	$\{1\}$	$\{1\}$	$\{1\}$
$\{0,1\}$	$\{1\}$	$\{0,1\}$	$\{0,.5,1\}$
$\{0,.5,1\}$	$\{1\}$	$\{0,.5,1\}$	$\{0,.5,1\}$

$\sqcap$	$\{1\}$	$\{0,1\}$	$\{0,.5,1\}$
$\{1\}$	$\{1\}$	$\{1\}$	$\{1\}$
$\{0,1\}$	$\{1\}$	$\{0,1\}$	$\{0,1\}$
$\{0,.5,1\}$	$\{1\}$	$\{0,1\}$	$\{0,.5,1\}$

$\Rightarrow$	$\{1\}$	$\{0,1\}$	$\{0,.5,1\}$
$\{1\}$	$\{0,.5,1\}$	$\{0,.5,1\}$	$\{0,.5,1\}$
$\{0,1\}$	$\{1\}$	$\{0,1\}$	$\{0,.5,1\}$
$\{0,.5,1\}$	$\{1\}$	$\{1\}$	$\{0,.5,1\}$

$\sqcup$	$\{1\}$	$\{0,1\}$	$\{0,.5,1\}$
$\{1\}$	$\{1\}$	$\{0,1\}$	$\{0,.5,1\}$
$\{0,1\}$	$\{0,1\}$	$\{0,1\}$	$\{0,.5,1\}$
$\{0,.5,1\}$	$\{0,.5,1\}$	$\{0,.5,1\}$	$\{0,.5,1\}$

Observe that **RM3** is not a Heyting algebra as, for example, $\mathbf{0} \bullet \mathbf{1} \neq \mathbf{0} \sqcap \mathbf{1}$. On the other hand, the filter algebra over strong Kleene (which is of course an intuitionistic frame, per Definition 13) does yield a Heyting algebra—indeed, the smallest Heyting algebra which is not a Boolean algebra.

I have examined how to obtain an algebraic structure from an operational frame; it is time to examine the converse. While there are several ways to get a mingle frame from a Dunn semilattice (cf. [32, §5]), I will just consider the one which I find most natural. The reader will observe that the construction mirrors, algebraically, the canonical model construction in Definition 20 from Subsection 3.3.[23]

Definition 37 (Filter Frame). Given a Dunn semilattice $\mathbf{D} = \langle D, \mathbf{1}, \mathbf{0}, \bullet, \Rightarrow, \sqcup, \sqcap \rangle$, the *filter frame* over $\mathbf{D}$, $\mathfrak{F}(\mathbf{D}) = \langle S, 0, 1, \vee, \wedge \rangle$, is defined as follows:

1. $S = \mathcal{F}(\mathbf{D})$;[24]
2. $0 = \uparrow \mathbf{1}$;
3. $1 = \uparrow \mathbf{0} = D$;
4. $I \vee J = \{k \in S : \exists i \in I, \exists j \in J (i \bullet j \sqsubseteq k)\}$;

---

[21] Consult, for example, Anderson and Belnap [1, §29.12, p. 470], Brady [6, p. 9], or Priest [31, §7.4, pp. 124–125]. Note that I am omitting the negation table for **RM3**.

[22] I have named the values of the algebra specifically to call to mind the fact that **RM3** is one member of the infinite class of so-called *Sugihara matrices* (named after the author of [37]); these play an important role in the algebraic theory of **RM** [11]. Here I should also note an interesting anticipation of my work by Meyer, who in [1, §29.3.2, p. 400] very nearly presents Sugihara matrices as bisemilattices, discussing extensional and intensional orders of the pertinent sets of integers. Of course, an important difference is that neither $\sqsubseteq$ nor $\leq_\bullet$ in a Dunn semilattice need be a chain.

[23] Here I should note that in the canonical model construction, where $\circ$ is included in the language, $\Gamma \cdot \Delta$ could have been equivalently defined as $\{\theta : \exists \varphi \in \Gamma, \exists \psi \in \Delta (\vdash_L (\varphi \circ \psi) \to \theta)\}$, which makes the connection even sharper.

[24] Just to be clear, $\mathcal{F}(\mathbf{D})$ is taken to be the set of $\sqcap$-filters in $\mathbf{D}$.

5. $I \wedge J = I \cap J$.

Lemma 38. *Any filter frame* $\mathfrak{F}(\mathbf{D}) = \langle S, 0, 1, \vee, \wedge \rangle$ *over a Dunn semilattice* $\mathbf{D} = \langle D, \mathbf{1}, \mathbf{0}, \bullet, \Rightarrow, \sqcup, \sqcap \rangle$ *is a mingle frame.*

Proof. The argument mirrors the proof of Lemma 22, so I will not belabor it for too long. It should, however, briefly be verified that when I and J are filters, $I \vee J$ is as well, since this is not entirely obvious. Suppose $a, b \in I \vee J$, so as to show that $a \sqcap b \in I \vee J$. Then $\exists i, i' \in I$ and $j, j' \in J$ such that $i \bullet j \sqsubseteq a$ and $i' \bullet j' \sqsubseteq b$; clearly, $(i \bullet j) \sqcap (i' \bullet j') \sqsubseteq a \sqcap b$. I and J are filters, so $i \sqcap i' \in I$ and $j \sqcap j' \in J$, whence $a \sqcap b \in I \vee J$ since $(i \sqcap i') \bullet (j \sqcap j') \sqsubseteq (i \bullet j) \sqcap (i' \bullet j') \sqsubseteq a \sqcap b$ by the assumptions, definition of $\vee$, and Fact 32. Conversely, if $a \sqcap b \in I \vee J$, that $a, b \in I \vee J$ is immediate from the facts that $a \sqcap b \sqsubseteq a$ and $a \sqcap b \sqsubseteq b$. Finally, it is obvious that $I \vee J$ is nonempty, since (ex hypothesi) I and J are. ◁

Lemma 39. *Any filter frame* $\mathfrak{F}(\mathbf{D}) = \langle S, 0, 1, \vee, \wedge \rangle$ *over a Heyting algebra* $\mathbf{D} = \langle D, \mathbf{1}, \mathbf{0}, \Rightarrow, \sqcup, \sqcap \rangle$ *is an intuitionistic frame.*

Proof. The result follows from Lemma 38 and the observation that $0 = {\uparrow}\mathbf{1} \subseteq I$ for any filter I because in a Heyting algebra, $\mathbf{1}$ is the top element in the $\sqsubseteq$ order and therefore is contained in any filter. ◁

Given a Dunn semilattice, an algebraic model is obtained by assigning elements of the algebra to propositional variables.[25]

Definition 40 (Model). A *Dunn semilattice model* is a structure $\mathfrak{M}^a = \langle \mathbf{D}, v \rangle$ where $\mathbf{D} = \langle D, \mathbf{1}, \mathbf{0}, \bullet, \Rightarrow, \sqcup, \sqcap \rangle$ is a Dunn semilattice and $v : \Pi \to D$ is extended to the full language in the obvious way:

1. $v(\bot) = \mathbf{0}$;
2. $v(t) = \mathbf{1}$;
3. $v(\varphi \wedge \psi) = v(\varphi) \sqcap v(\psi)$;
4. $v(\varphi \vee \psi) = v(\varphi) \sqcup v(\psi)$;
5. $v(\varphi \circ \psi) = v(\varphi) \bullet v(\psi)$;
6. $v(\varphi \to \psi) = v(\varphi) \Rightarrow v(\psi)$.

A *Heyting algebraic model* is defined in essentially the same way, with Heyting algebras playing the role of Dunn semilattices and the irrelevant connectives and clauses being omitted.

Definition 41 (Validity). Where $\mathfrak{M}^a = \langle \mathbf{D}, v \rangle$ is a Dunn semilattice model, φ is valid in $\mathfrak{M}^a$ ($\vDash^{\mathfrak{M}^a} \varphi$) if $\mathbf{1} \sqsubseteq v(\varphi)$. φ is Dunn semilattice *valid* ($\vDash^a_{\mathbf{RM0}} \varphi$) if $\vDash^{\mathfrak{M}^a} \varphi$ for every Dunn semilattice model $\mathfrak{M}^a = \langle \mathbf{D}, v \rangle$. Heyting validity ($\vDash^a_{\mathbf{J}} \varphi$) is defined analogously.

Lemma 42. *If* $\vDash^a_{\mathbf{RM0}} \varphi$ ($\vDash^a_{\mathbf{J}} \varphi$), *then* $\vDash_{\mathbf{RM0}} \varphi$ ($\vDash_{\mathbf{J}} \varphi$).

Proof. For the case of **RM0**, suppose $\nvDash_{\mathbf{RM0}} \varphi$; then there is some mingle model $\mathfrak{M} = \langle \mathfrak{F}, V \rangle$ such that $\nvDash^{\mathfrak{M}}_0 \varphi$. Let $\mathbb{A}(\mathfrak{F}) = \langle D, \mathbf{1}, \mathbf{0}, \bullet, \Rightarrow, \sqcup, \sqcap \rangle$ be the filter algebra over $\mathfrak{F}$; by Lemma 34, this is a Dunn semilattice. The Dunn semilattice countermodel is

---

[25]For the purposes of algebraic semantics, it is natural to assume **RM0** is formulated in the full language.

defined to be $\mathfrak{M}^a = \langle \mathbb{A}(\mathfrak{F}), v \rangle$ where $v(p) = V(p)$. By an induction that is essentially trivial in virtue of Lemmata 15 and 34, $v(\psi) = [\psi]^{\mathfrak{M}}$ for all ψ. But then, clearly, $\uparrow 0 = \mathbf{1} \not\sqsubseteq v(\varphi) = [\varphi]^{\mathfrak{M}}$, as $0 \notin [\varphi]^{\mathfrak{M}}$ ex hypothesi. So, $\not\models^{\mathfrak{M}^a} \varphi$, which suffices. The case of **J** is essentially the same, but Lemma 35 fulfills the role of Lemma 34. ◁

Lemma 43. *If $\models_{\mathbf{RM0}} \varphi$ ($\models_{\mathbf{J}} \varphi$), then $\models^a_{\mathbf{RM0}} \varphi$ ($\models^a_{\mathbf{J}} \varphi$).*

Proof. For the case of **RM0**, suppose $\not\models^a_{\mathbf{RM0}} \varphi$. Then there is a Dunn semilattice model $\mathfrak{M}^a = \langle \mathbf{D}, v \rangle$ where $\mathbf{D} = \langle D, \mathbf{1}, \mathbf{0}, \bullet, \Rightarrow, \sqcup, \sqcap \rangle$ is a Dunn semilattice and $\mathbf{1} \not\sqsubseteq v(\varphi)$. Let $\mathfrak{F}(\mathbf{D}) = \langle S, 0, 1, \vee, \wedge \rangle$ be the filter frame over $\mathbf{D}$; by Lemma 38, this is a mingle frame. The mingle countermodel is defined to be $\mathfrak{M} = \langle \mathfrak{F}(\mathbf{D}), V \rangle$, where, for all p, $V(p) = \{I \in S \colon v(p) \in I\}$. Clearly, each $V(p)$ is a filter in $\mathfrak{F}(\mathbf{D})$ since $I, J \in V(p)$ if and only if $v(p) \in I, J$ if and only if $v(p) \in I \cap J$ if and only if $I \cap J \in V(p)$ and every $V(p)$ is nonempty (containing, e.g., 1). Thus, $\mathfrak{M}$ is a mingle model.

It must be shown that for all ψ and all filters I, $\models^{\mathfrak{M}}_I \psi$ if and only if $v(\psi) \in I$. The argument for this result is entirely analogous to that for Lemma 24, so I will just briefly examine the case of $\to$. Suppose $\models^{\mathfrak{M}}_I \theta$ and $v(\theta \to \chi) = v(\theta) \Rightarrow v(\chi) \in I$. By the induction hypothesis, $v(\theta) \in J$, so as $(v(\theta) \Rightarrow v(\chi)) \bullet v(\theta) \sqsubseteq v(\chi)$, $v(\chi) \in I \vee J$, which suffices by the induction hypothesis. Conversely, suppose that $v(\theta \to \chi) = v(\theta) \Rightarrow v(\chi) \notin I$ and consider $I \vee \uparrow v(\theta)$. If it were the case that $v(\chi) \in I \vee \uparrow v(\theta)$, then $i \bullet k \sqsubseteq v(\chi)$ for some $i \in I$ and $v(\theta) \sqsubseteq k$. By Fact 32, $v(\theta) \sqsubseteq k$ implies $i \bullet v(\theta) \sqsubseteq i \bullet k \sqsubseteq v(\chi)$, whence $i \sqsubseteq v(\theta) \Rightarrow v(\chi)$ and $v(\theta) \Rightarrow v(\chi) \in I$, which is impossible. So $v(\theta) \in \uparrow v(\theta)$ and $v(\chi) \notin I \vee \uparrow v(\theta)$ imply $\models^{\mathfrak{M}}_{\uparrow v(\theta)} \theta$ and $\not\models^{\mathfrak{M}}_{I \vee \uparrow v(\theta)} \chi$ by the induction hypothesis, which yields the result.

Now, since $\mathbf{1} \not\sqsubseteq v(\varphi)$, $v(\varphi) \notin \mathbf{0} = \uparrow \mathbf{1}$, whence $\not\models^{\mathfrak{M}}_{\mathbf{0}} \varphi$ by the immediately preceding induction. Therefore, $\not\models_{\mathbf{RM0}} \varphi$, as desired. The case involving **J** is essentially the same, but Lemma 39 plays the role of Lemma 38. ◁

Theorem 44 (Algebraic Soundness and Completeness). *$\vdash_{\mathbf{RM0}} \varphi$ ($\vdash_{\mathbf{J}} \varphi$) if and only if $\models^a_{\mathbf{RM0}} \varphi$ ($\models^a_{\mathbf{J}} \varphi$).*

Proof. Immediate from Theorems 18, 19, 25, and 26 and Lemmata 42 and 43. ◁

Theorem 44 could of course have been proved much more directly, using a routine Lindenbaum construction for the algebraic completeness component; but the proof I have given sheds considerably more light on the relationship between the algebraic and operational semantics presented in this paper.

5. Concluding Remarks

In this paper, I examined operational and algebraic semantics for **RM0** and **J**. Adapting work of Humberstone from [20], I showed that **RM0** is determined by a certain class of bisemilattices, taken as frames, whereas **J** is determined by the subclass of those frames which are lattices. I also examined algebraic semantics for both **RM0** and **J** and showed how to transform operational models into equivalent algebraic models and vice versa.

One clear takeaway from this paper is that **RM0** and **J** are very closely related. This is not only apparent semantically, in the fact that intuitionistic frames and Heyting algebras are natural special cases of mingle frames and Dunn semilattices respectively, but in the fact that **J** can be straightforwardly exactly translated into **RM0**^{t} per Theorem 30. In [41], I presented extensions of Urquhart's semilattice relevance logic **S** which might be thought of as (quasi-)relevant companions of **J** and **KC** (Jankov's logic). Such logics, in my view, could hold appeal to relevantists of a constructivist bent (or constructivists of a relevantist bent). In view of the results of this paper, I think that **RM0** is another system that could hold appeal to such logicians.

Another clear takeaway is that the operational semantics of [20] deserves more attention than it has received. As I showed, Humberstone's semantics importantly anticipated more recent developments in inquisitive semantics (as illustrated in the work of, for example, [32]). In fact, though, this paper only scratches the surface of what can be done by extending or modifying the Humberstone framework. In unpublished work, I have shown how the operational semantics of this paper can be used to characterize a variety of intuitionistic and relevant modal logics, with embedding results forthcoming for intuitionistic modal systems and their relevant companions; without doubt, the algebra of such logics will also prove a rich vein for future study.

This paper leaves open a number of interesting problems, both philosophical and technical. I have not attempted to articulate a philosophical account of the operational semantics developed here for either **RM0** or **J** (in this respect, **RM0** would appear to be on worse footing than the systems surveyed in [41], which have clear philosophical motivation). This is emphatically *not* because I do not think the semantics can be well-motivated, but rather because this is, by design, a technical piece. I leave to future work, my own or others', the project of interpreting this semantics.[26]

On the technical side, much more work could still be done even just on the model theory of **RM0** and **J**. One example: while I have examined operational and algebraic models for both of these systems and shown how to move between them, both of these logics already have relational modelings (ternary in the case of **RM0**, binary in the case of **J** [25; 23]) which I have not discussed. It would be valuable to examine the relation of those semantics to the semantics presented here.[27]

Dedication. I dedicate this paper to the memory of J. Michael Dunn, a great logician and generous human being.

Acknowledgments. This paper is adapted from part of a presentation ("Bisemilattice Semantics for Intuitionistic and Relevant Modal Logics") that I gave to the Logic and Metaphysics Workshop in New York City on October 4, 2021. I thank the attendees

---

[26]It could just as well have been left to *past* and future work, in view of the fact that Humberstone (not to mention the inquisitive semanticists) has some informal things to say about how to interpret his semantics in [20]. But I confess that my own interpretive views, germinal though they are, do not entirely align with his.

[27]*Added in proof*: I regret that this paper neglected to discuss certain relevant work of Došen [8; 9]. Došen's semilattice-ordered groupoid semantics (cf. the monoid semantics from Wansing [39]), apparently developed independently of and roughly concurrent with Humberstone's operational semantics, bears various connections to the operational semantics presented here (though there are also differences, e.g., in the formulation of the truth condition for disjunction). I leave to future work a thorough comparison of these semantic approaches.

for feedback. Furthermore, I am grateful to an anonymous referee, Katalin Bimbó, Graham Priest, H. P. Sankappanavar, and Shawn Standefer for written comments and suggestions.

REFERENCES

[1] Anderson, A. R. and Belnap, N. D. (1975). *Entailment: The Logic of Relevance and Necessity*, Vol. I, Princeton University Press, Princeton, NJ.
[2] Avron, A. (1984). Relevant entailment—semantics and formal systems, *Journal of Symbolic Logic* **49**(2): 334–342.
[3] Avron, A. (1992). Whither relevance logic?, *Journal of Philosophical Logic* **21**(3): 243–281.
[4] Balbes, R. (1970). A representation theorem for distributive quasi-lattices, *Fundamenta Mathematicae* **68**(2): 207–214.
[5] Bimbó, K. (2007). Relevance logics, in D. Jacquette (ed.), *Philosophy of Logic*, Vol. 5 of *Handbook of the Philosophy of Science* (D. Gabbay, P. Thagard and J. Woods, eds.), Elsevier (North-Holland), Amsterdam, pp. 723–789.
[6] Brady, R. T. (1982). Completeness proofs for the systems RM3 and BN4, *Logique et Analyse* **25**(97): 9–32.
[7] Davey, B. A. and Priestley, H. A. (1990). *Introduction to Lattices and Order*, Cambridge University Press, New York, NY.
[8] Došen, K. (1988). Sequent-systems and groupoid models I, *Studia Logica* **47**(4): 353–385.
[9] Došen, K. (1989). Sequent-systems and groupoid models II, *Studia Logica* **48**(1): 41–65.
[10] Dunn, J. M. (1966). *The Algebra of Intensional Logics*, PhD thesis, University of Pittsburgh, PA.
[11] Dunn, J. M. (1970). Algebraic completeness results for R-mingle and its extensions, *Journal of Symbolic Logic* **35**(1): 1–13.
[12] Dunn, J. M. (1976). A Kripke-style semantics for R-Mingle using a binary accessibility relation, *Studia Logica* **35**(2): 163–172.
[13] Dunn, J. M. (2019). *The Algebra of Intensional Logics*, Vol. 2 of *Logic PhDs*, College Publications, London, UK.
[14] Dunn, J. M. (2021). R-Mingle is nice, and so is Arnon Avron, in O. Arieli and A. Zamansky (eds.), *Arnon Avron on Semantics and Proof Theory of Non-Classical Logics*, Vol. 21 of *Outstanding Contributions to Logic*, Springer, Cham, pp. 141–165.
[15] Dunn, J. M. and Meyer, R. K. (1971). Algebraic completeness results for Dummett's LC and its extensions, *Zeitschrift für mathematische Logik und Grundlagen der Mathematik* **17**(1): 225–230.
[16] Dunn, J. M. and Restall, G. (2002). Relevance logic, in D. Gabbay and F. Guenthner (eds.), *Handbook of Philosophical Logic*, Vol. 6, Kluwer Academic Publishers, Dordrecht, pp. 1–128.
[17] Fine, K. (1974). Models for entailment, *Journal of Philosophical Logic* **3**(4): 347–372.
[18] Grätzer, G. (2011). *Lattice Theory: Foundation*, Birkhäuser, Basel.
[19] Harding, J. and Romanowska, A. B. (2017). Varieties of Birkhoff systems part I, *Order* **34**(1): 45–68.
[20] Humberstone, I. L. (1988). Operational semantics for positive R, *Notre Dame Journal of Formal Logic* **29**(1): 61–80.
[21] Humberstone, L. (2019). Supervenience, dependence, disjunction, *Logic and Logical Philosophy* **28**(1): 3–135.
[22] Kleene, S. C. (1952). *Introduction to Metamathematics*, D. van Nostrand Company, Inc., New York, NY.

[23] Kripke, S. A. (1965). Semantical analysis of intuitionistic logic I, *in* J. N. Crossley and M. A. E. Dummett (eds.), *Formal Systems and Recursive Functions: Proceedings of the Eighth Logic Colloquium, Oxford, July 1963*, Vol. 40 of *Studies in Logic and the Foundations of Mathematics*, North-Holland Publishing Company, Amsterdam, pp. 92–130.
[24] Ledda, A. (2018). Stone-type representations and dualities for varieties of bisemilattices, *Studia Logica* **106**(2): 417–448.
[25] Méndez, J. M. (1988). The compatibility of relevance and mingle, *Journal of Philosophical Logic* **17**(3): 279–297.
[26] Meyer, R. K. (1966). *Topics in Modal and Many-Valued Logic*, PhD thesis, University of Pittsburgh, PA.
[27] Meyer, R. K. (1972). Conservative extension in relevant implication, *Studia Logica* **31**: 39–48.
[28] Meyer, R. K. and Routley, R. (1972). Algebraic analysis of entailment I, *Logique et Analyse* **15**(59/60): 407–428.
[29] Padmanabhan, R. (1971). Regular identities in lattices, *Transactions of the American Mathematical Society* **158**(1): 179–188.
[30] Płonka, J. (1967). On distributive quasi-lattices, *Fundamenta Mathematicae* **60**(2): 191–200.
[31] Priest, G. (2008). *An Introduction to Non-Classical Logic*, 2 edn, Cambridge University Press, Cambridge, UK.
[32] Punčochář, V. (2017). Algebras of information states, *Journal of Logic and Computation* **27**(5): 1643–1675.
[33] Punčochář, V. and Tedder, A. (2021). Disjunction and negation in information based semantics, *in* A. Silva, R. Wassermann and R. de Queiroz (eds.), *Logic, Language, Information, and Computation*, Lecture Notes in Computer Science, Springer, Cham, pp. 355–371.
[34] Rasiowa, H. and Sikorski, R. (1963). *The Mathematics of Metamathematics*, PWN, Warsaw.
[35] Robles, G. and Méndez, J. M. (2018). *Routley–Meyer Ternary Relational Semantics for Intuitionistic-type Negations*, Academic Press, London.
[36] Romanowska, A. (1980). On bisemilattices with one distributive law, *Algebra Universalis* **10**(1): 36–47.
[37] Sugihara, T. (1955). Strict implication free from implicational paradoxes, *Memoirs of the Faculty of Liberal Arts*, number 4 in *I*, Fukui University, pp. 55–59.
[38] Urquhart, A. (1972). Semantics for relevant logics, *Journal of Symbolic Logic* **37**(1): 159–169.
[39] Wansing, H. (1993). Informational interpretation of substructural propositional logics, *Journal of Logic, Language, and Information* **2**(4): 285–308.
[40] Ward, M. and Dilworth, R. P. (1939). Residuated lattices, *Transactions of the American Mathematical Society* **45**(3): 335–354.
[41] Weiss, Y. (2021). A reinterpretation of the semilattice semantics with applications, *Logica Universalis* **15**(2): 171–191.

THE SAUL KRIPKE CENTER, THE GRADUATE CENTER, CUNY, ROOM 7118, 365 FIFTH AVE., NEW YORK, NY 10016, U.S.A., *Email:* yweiss@gradcenter.cuny.edu

J. Michael Dunn's Publications
(Books and Articles)

Books

1. Dunn, J. M., Epstein, G., Cocchiarella, N. and Shapiro, S. (eds.) (1975). *Proceedings of the 1975 International Symposium on Multiple-Valued Logic*, (Indiana University, Bloomington, IN, May 13–16, 1975), IEEE Computer Society, Long Beach, CA.

2. Dunn, J. M. and Epstein, G. (eds.) (1977). *Modern Uses of Multiple-Valued Logic*, D. Reidel, Dordrecht.

3. Dunn, J. M. and Gupta, A. (eds.) (1990). *Truth or Consequences: Essays in Honor of Nuel Belnap*, Kluwer, Amsterdam.

4. Anderson, A. R., Belnap, N. D. and Dunn, J. M. (1992). *Entailment: The Logic of Relevance and Necessity*, Vol. II, Princeton University Press, Princeton, NJ.

5. Dunn, J. M. and Hardegree, G. M. (2001). *Algebraic Methods in Philosophical Logic*, Vol. 41 of *Oxford Logic Guides*, Oxford University Press, Oxford, UK.

6. Bimbó, K. and Dunn, J. M. (2008). *Generalized Galois Logics: Relational Semantics of Nonclassical Logical Calculi*, Vol. 188 of *CSLI Lecture Notes*, CSLI Publications, Stanford, CA.

7. Chen, M., Dunn, J. M., Golan, A. and Ullah, A. (eds.) (2021). *Advances in Info-Metrics: Information and Information Processing across Disciplines*, Oxford University Press, New York, NY.

Articles in journals and books

1. Dunn, J. M. (1967). Drange's paradox lost, *Philosophical Studies* **18**: 94–95.

2. Dunn, J. M. and Belnap, N. D. (1968a). Homomorphisms of intensionally complemented distributive lattices, *Mathematische Annalen* **176**: 28–38.

3. Dunn, J. M. and Belnap, N. D. (1968b). The substitution interpretation of the quantifiers, *Noûs* **2**: 177–185.

4. Meyer, R. K. and Dunn, J. M. (1969). E, R and γ, *Journal of Symbolic Logic* **34**: 460–474. Reprinted in Anderson, A. R. and Belnap, N. D., *Entailment: The Logic of Relevance and Necessity*, Vol. I, Princeton University Press, Princeton, NJ, 1975, §25.2, pp. 300–314.

Bimbó, Katalin, (ed.), *Relevance Logics and other Tools for Reasoning. Essays in Honor of J. Michael Dunn*, (Tributes, vol. 46), College Publications, London, UK, 2022, pp. 456–463.

5. Dunn, J. M. (1970). Algebraic completeness results for R-mingle and its extensions, *Journal of Symbolic Logic* **35**: 1–13.

6. Dunn, J. M. and Meyer, R. K. (1971). Algebraic completeness results for Dummett's LC and its extensions, *Zeitschrift für Mathematische Logik und Grundlagen der Mathematik* **17**: 225–230.

7. Dunn, J. M. (1972). A modification of Parry's analytic implication, *Notre Dame Journal of Formal Logic* **13**: 195–205.

8. Dunn, J. M. (1973). A truth value semantics for modal logic, *in* H. Leblanc (ed.), *Truth, Syntax and Modality. Proceedings of the Temple University Conference on Alternative Semantics*, North-Holland, Amsterdam, pp. 87–100.

9. Meyer, R. K., Dunn, J. M. and Leblanc, H. (1974). Completeness of relevant quantification theories, *Notre Dame Journal of Formal Logic* **15**: 97–121.

10. Dunn, J. M. (1975a). Axiomatizing Belnap's conditional assertion, *Journal of Philosophical Logic* **4**: 383–397.

11. Dunn, J. M. (1975b). The algebra of **R**, §28.2, *in* Anderson, A. R. and Belnap, N. D., *Entailment: The Logic of Relevance and Necessity*, Vol. I, Princeton University Press, Princeton, NJ, pp. 352–371.

12. Dunn, J. M. (1975c). Consecution formulation of positive **R** with co-tenability and **t**, §28.5, *in* Anderson, A. R. and Belnap, N. D., *Entailment: The Logic of Relevance and Necessity*, Vol. I, Princeton University Press, Princeton, NJ, pp. 381–391.

13. Dunn, J. M. (1975d). Extensions of **RM**, §29.3, *in* Anderson, A. R. and Belnap, N. D., *Entailment: The Logic of Relevance and Necessity*, Vol. I, Princeton University Press, Princeton, NJ, pp. 420–429.

14. Dunn, J. M. (1975e). Intensional algebras, §18, *in* Anderson, A. R. and Belnap, N. D., *Entailment: The Logic of Relevance and Necessity*, Vol. I, Princeton University Press, Princeton, NJ, pp. 180–206.

15. Dunn, J. M. (1976a). Intuitive semantics for first-degree entailments and 'coupled trees', *Philosophical Studies* **29**: 149–168. Reprinted in Omori, H. and Wansing, H. (eds.) (2019). *New Essays on Belnap–Dunn Logic*, Springer Nature, Switzerland, pp. 21–34.

16. Dunn, J. M. (1976b). A Kripke-style semantics for R-mingle using a binary accessibility relation, *Studia Logica* **35**: 163–172.

17. Dunn, J. M. (1976c). Quantification and **RM**, *Studia Logica* **35**: 315–322.

18. Dunn, J. M. (1976d). A variation on the binary semantics for R-Mingle, *Relevance Logic Newsletter* **1**: 56–67.

19. Dunn, J. M. (1979a). R-mingle and beneath. Extensions of the Routley–Meyer semantics for **R**, *Notre Dame Journal of Formal Logic* **20**: 369–376.

20. Dunn, J. M. (1979b). Relevant Robinson's arithmetic, *Studia Logica* **38**: 407–418.

21. Dunn, J. M. (1979c). A theorem in 3-valued model theory with connections to number theory, type theory, and relevant logic, *Studia Logica* **38**: 149–169.

22. Meyer, R. K., Routley, R. and Dunn, J. M. (1979). Curry's paradox, *Analysis (n.s.)* **39**: 124–128.

23. Dunn, J. M. (1980). A sieve for entailments, *Journal of Philosophical Logic* **9**: 41–57.

24. Belnap, N. D., Gupta, A. and Dunn, J. M. (1980). A consecutive calculus for positive relevant implication with necessity, *Journal of Philosophical Logic* **9**: 343–362.

25. Belnap, N. D. and Dunn, J. M. (1981). Entailment and the disjunctive syllogism, *in* G. Fløistad and G. H. von Wright (eds.), *Contemporary Philosophy: A New Survey*, Vol. I, Philosophy of Language/Philosophical Logic, Martinus Nijhoff, The Hague, pp. 337–366.

26. Dunn, J. M. (1981). Quantum mathematics, *in* P. D. Asquith and R. N. Giere (eds.), *PSA 1980: Proceedings of the 1980 Biennial Meeting of the Philosophy of Science Association*, Vol. 2, Philosophy of Science Association, East Lansing, MI, pp. 512–531.

27. Dunn, J. M. (1982a). Anderson and Belnap, and Lewy on entailment, *in* L. J. Cohen, J. Łoś, H. Pfeiffer and K.-P. Podewski (eds.), *Logic, Methodology and Philosophy of Science, VI. Proceedings of the Sixth International Congress of Logic, Methodology and Philosophy of Science, Hannover 1979*, North-Holland and PWN – Polish Scientific Publishers, Amsterdam and Warszawa, pp. 291–297.

28. Dunn, J. M. (1982b). A relational representation of quasi-Boolean algebras, *Notre Dame Journal of Formal Logic* **23**: 353–357.

29. Dunn, J. M. (1986). Relevance logic and entailment, *in* D. Gabbay and F. Guenthner (eds.), *Handbook of Philosophical Logic*, 1st edn, Vol. 3, D. Reidel, Dordrecht, pp. 117–224.

30. Dunn, J. M. and Hellman, G. (1986). Dualling: A critique of an argument of Popper and Miller, *The British Journal for the Philosophy of Science* **37**: 220–223.

31. Dunn, J. M. (1987a). Incompleteness of the bibinary semantics for R, *Bulletin of the Section of Logic of the Polish Academy of Sciences* **16**: 107–110.

32. Dunn, J. M. (1987b). Relevant predication 1: The formal theory, *Journal of Philosophical Logic* **16**: 347–381.

33. Dunn, J. M. (1988). The impossibility of certain higher-order non-classical logics with extensionality, *in* D. F. Austin (ed.), *Philosophical Analysis: A Defense by Example. (A Festschrift for Edmund Gettier)*, Kluwer, Dordrecht, pp. 261–279.

34. Dunn, J. M. and Meyer, R. K. (1989). Gentzen's cut and Ackermann's gamma, *in* J. Norman and R. Sylvan (eds.), *Directions in Relevant Logic*, Kluwer, Dordrecht, pp. 229–240.

35. Dunn, J. M. (1990a). The frame problem and relevant predication, *in* H. E. Kyburg, R. P. Loui and G. N. Carlson (eds.), *Knowledge Representation and Defeasible Reasoning*, Kluwer, Dordrecht, pp. 89–95.

36. Dunn, J. M. (1990b). Relevant predication 2: Intrinsic properties and internal relations, *Philosophical Studies* **60**: 177–206.

37. Dunn, J. M. (1990c). Relevant predication 3: Essential properties, *in* J. M. Dunn and A. Gupta (eds.), *Truth or Consequences: Essays in Honor of Nuel Belnap*, Kluwer, Amsterdam, pp. 77–95.

38. Dunn, J. M. (1991). Gaggle theory: An abstraction of Galois connections and residuation, with applications to negation, implication, and various logical operators, *in* J. van Eijck (ed.), *Logics in AI: European Workshop JELIA '90, Amsterdam, The Netherlands, September 10–14, 1990*, number 478 in *Lecture Notes in Computer Science*, Springer, Berlin, pp. 31–51.

39. Franco, J., Dunn, J. M. and Wheeler, W. (1992). Recent work at the interface of logic, combinatorics, and computer science, *Annals of Mathematics and Artificial Intelligence* **6**: 1–16.

40. Dunn, J. M. (1993a). Partial gaggles applied to logics with restricted structural rules, *in* K. Došen and P. Schroeder-Heister (eds.), *Substructural Logics*, Clarendon and Oxford University Press, Oxford, UK, pp. 63–108.

41. Dunn, J. M. (1993b). Star and perp: Two treatments of negation, *Philosophical Perspectives* **7**: 331–357. (Language and Logic, 1993, J. E. Tomberlin (ed.)).

42. Allwein, G. and Dunn, J. M. (1993). Kripke models for linear logic, *Journal of Symbolic Logic* **58**: 514–545.

43. Dunn, J. M. (1995a). Gaggle theory applied to intuitionistic, modal and relevance logics, *in* I. Max and W. Stelzner (eds.), *Logik und Mathematik. Frege-Kolloquium Jena 1993*, W. de Gruyter, Berlin, pp. 335–368.

44. Dunn, J. M. (1995b). Positive modal logic, *Studia Logica* **55**: 301–317.

45. Dunn, J. M. (1996a). Generalized ortho negation, *in* H. Wansing (ed.), *Negation: A Notion in Focus*, W. de Gruyter, New York, NY, pp. 3–26.

46. Dunn, J. M. (1996b). Is existence a (relevant) predicate?, *Philosophical Topics* **24**: 1–34.

47. Dunn, J. M. (1997). A logical framework for the notion of *natural property*, *in* J. Earman and J. D. Norton (eds.), *The Cosmos of Science: Essays of exploration*, number 6 in *Pittsburgh-Konstanz Series in the Philosophy and History of Science*, University of Pittsburgh Press and Universitäts-Verlag Konstanz, Pittsburgh and Konstanz, pp. 458–497.

48. Hartonas, C. and Dunn, J. M. (1997). Stone duality for lattices, *Algebra Universalis* **37**: 391–401. [Preliminary version: Duality theorems for partial orders, semilattices, Galois connections and lattices, *Indiana University Logic Group Preprint Series* IULG–93–26, 1993.]

49. Dunn, J. M. and Meyer, R. K. (1997). Combinators and structurally free logic, *Logic Journal of the IGPL* **5**: 505–537.

50. Bimbó, K. and Dunn, J. M. (1998). Two extensions of the structurally free logic LC, *Logic Journal of the IGPL* **6**: 403–424.

51. Dunn, J. M. (1999). A comparative study of various model-theoretic treatments of negation: A history of formal negation, *in* D. M. Gabbay and H. Wansing (eds.), *What is Negation?*, Kluwer, Dordrecht, pp. 23–51.

52. Dunn, J. M. (2000). Partiality and its dual, *Studia Logica* **65**: 5–40.

53. Dunn, J. M. (2001a). The concept of information and the development of modern logic, *in* W. Stelzner and M. Stöckler (eds.), *Zwischen traditioneller und moderner Logik: Nichtklassische Ansätze*, (Perspektiven der Analytischen Philosophie), Mentis-Verlag, Paderborn, pp. 423–447.

54. Dunn, J. M. (2001b). A representation of relation algebras using Routley–Meyer frames, *in* C. A. Anderson and M. Zelëny (eds.), *Logic, Meaning and Computation. Essays in Memory of Alonzo Church*, Kluwer, Dordrecht, pp. 77–108. [Preliminary version: A representation of relation algebras using Routley–Meyer frames, *Indiana University Logic Group Preprint Series* IULG–93–28, 1993.]

55. Dunn, J. M. (2001c). Ternary relational semantics and beyond: Programs as arguments (data) and programs as functions (programs), *Logical Studies* **7**: 1–20. Reprinted in А. С. Карпенко, (ред.), Логические исследования, Наука, Москва, 2001, cc. 282–301.

56. Bimbó, K. and Dunn, J. M. (2001). Four-valued logic, *Notre Dame Journal of Formal Logic* **42**: 171–192.

57. Shramko, Y., Dunn, J. M. and Takenaka, T. (2001). The trilattice of constructive truth values, *Journal of Logic and Computation* **11**: 761–788.

58. Dunn, J. M. and Restall, G. (2002). Relevance logic, *in* D. Gabbay and F. Guenthner (eds.), *Handbook of Philosophical Logic*, 2nd edn, Vol. 6, Kluwer, Amsterdam, pp. 1–128.

59. Bimbó, K. and Dunn, J. M. (2005). Relational semantics for Kleene logic and action logic, *Notre Dame Journal of Formal Logic* **46**: 461–490.

60. Dunn, J. M., Gehrke, M. and Palmigiano, A. (2005). Canonical extensions and relational completeness of some substructural logics, *Journal of Symbolic Logic* **70**: 713–740.

61. Dunn, J. M., Hagge, T. J., Moss, L. S. and Wang, Z. (2005). Quantum logic as motivated by quantum computing, *Journal of Symbolic Logic* **70**: 353–359.

62. Dunn, J. M. and Zhou, C. (2005). Negation in the context of gaggle theory, *Studia Logica* **80**: 235–264.

63. Dunn, J. M. (2008). Information in computer science, *in* P. Adriaans and J. van Benthem (eds.), *Philosophy of Information*, Vol. 8 of *Handbook of the Philosophy*

of Science, (D. M. Gabbay, P. Thagard, J. Woods (eds.)), Elsevier, Amsterdam, pp. 581–608.

64. Bimbó, K. and Dunn, J. M. (2009). Symmetric generalized Galois logics, *Logica Universalis* **3**: 125–152.

65. Bimbó, K., Dunn, J. M. and Maddux, R. D. (2009). Relevance logics and relation algebras, *Review of Symbolic Logic* **2**: 102–131.

66. Dunn, J. M. (2010). Contradictory information: Too much of a good thing, *Journal of Philosophical Logic* **39**: 425–452.

67. Bimbó, K. and Dunn, J. M. (2010). Calculi for symmetric generalized Galois logics, *in* J. van Benthem and M. Moortgat (eds.), *Festschrift for Joachim Lambek*, Vol. 36 of *Linguistic Analysis*, Linguistic Analysis, Vashon, WA, pp. 307–343.

68. Beall, J., Brady, R., Dunn, J. M., Hazen, A. P., Mares, E., Meyer, R. K., Priest, G., Restall, G., Ripley, D., Slaney, J. and Sylvan, R. (2012). On the ternary relation and conditionality, *Journal of Philosophical Logic* **41**: 595–612.

69. Bimbó, K. and Dunn, J. M. (2012). New consecution calculi for $R^t_\to$, *Notre Dame Journal of Formal Logic* **53**: 491–509.

70. Bimbó, K. and Dunn, J. M. (2013). On the decidability of implicational ticket entailment, *Journal of Symbolic Logic* **78**: 214–236.

71. Dunn, J. M., Moss, L. S. and Wang, Z. (2013). The third life of quantum logic: Quantum logic inspired by quantum computing, *Journal of Philosophical Logic* **42**: 443–459.

72. Dunn, J. M. (2014a). Some stories and theorems inspired by Raymond Smullyan, *in* J. Rosenhouse (ed.), *Four Lives: A Celebration of Raymond Smullyan*, Dover Publications, Mineola, NY, pp. 65–75.

73. Dunn, J. M. (2014b). Arrows pointing at arrows: Arrow logic, relevance logic, and relation algebras, *in* A. Baltag and S. Smets (eds.), *Johan van Benthem on Logic and Information Dynamics*, Vol. 5 of Outstanding Contributions to Logic, Springer, New York, NY, pp. 881–894.

74. Bimbó, K. and Dunn, J. M. (2014). Extracting BB'IW inhabitants of simple types from proofs in the sequent calculus $LT^t_\to$ for implicational ticket entailment, *Logica Universalis* **8**: 141–164.

75. Dunn, J. M. and Eisenberg, P. (2014). Chorus: Hector-Neri Castañeda. A conversation about Hector by two of his colleagues, *in* A. Palma (ed.), *Castañeda and his Guises: Essays on the Work of Hector-Neri Castañeda*, W. de Gruyter, Boston, MA, pp. 15–18.

76. Dunn, J. M. (2015). The relevance of relevance to relevance logic, *in* M. Banerjee and S. N. Krishna (eds.), *Logic and its Applications: 6th Indian Conference, ICLA 2015, Mumbai, India, January 8–10, 2015*, number 8923 in *Lecture Notes in Computer Science*, Springer, Heidelberg, pp. 11–29.

77. Dunn, J. M. (2016a). An engineer in philosopher's clothing, *in* K. Bimbó (ed.), *J. Michael Dunn on Information Based Logics*, Vol. 8 of *Outstanding Contributions to Logic*, Springer Nature, Switzerland, pp. xvii–xxxiii.

78. Dunn, J. M. (2016b). A "reply" to my "critics", *in* K. Bimbó (ed.), *J. Michael Dunn on Information Based Logics*, Vol. 8 of *Outstanding Contributions to Logic*, Springer Nature, Switzerland, pp. 417–434.

79. Dunn, J. M. (2016c). Too much of a good thing, (interview), [in Russian], *Date Palm Compote* **10**: 59–62. (DOI 10.24411/2587-9308-2016-00015) [У нас все слишком хорошо получается, (интервью), Беседовали и переводили Александр Беликов, Евгений Логинов и Андрей Мерцалов, Финиковый Компот, #10 февраль 2016, сс. 59–62.]

80. Bimbó, K. and Dunn, J. M. (eds.) (2017a). *Proceedings of the Third Workshop, May 16–17, 2016, Edmonton, Canada*, Vol. 4(3) of *IFCoLog Journal of Logics and Their Applications*, College Publications, London, UK.

81. Bimbó, K. and Dunn, J. M. (2017b). The emergence of set-theoretical semantics for relevance logics around 1970, *in* K. Bimbó and J. M. Dunn (eds.), *Proceedings of the Third Workshop, May 16–17, 2016, Edmonton, Canada*, Vol. 4(3) of *IFCoLog Journal of Logics and Their Applications*, College Publications, London, UK, pp. 557–589.

82. Dunn, J. M. (2018). Humans as rational toolmaking animals [in Russian], *in* D. V. Zaitsev (ed.), *Contemporary Logic: Its Foundations, Subject and Prospects of Development*, Forum, Moscow, pp. 128–160. [Люди—это разумные орудия труда создающие животные, *in* Д. В. Зайцев (ред.), *Современная логика: Основания, предмет и перспективы развития*, Форум, Москва, сс. 128–160.]

83. Bimbó, K. and Dunn, J. M. (2018). Larisa Maksimova's early contributions to relevance logic, *in* S. Odintsov (ed.), *L. Maksimova on Implication, Interpolation and Definability*, Vol. 15 of *Outstanding Contributions to Logic*, Springer Nature, Switzerland, pp. 33–60.

84. Bimbó, K., Dunn, J. M. and Ferenz, N. (2018). Two manuscripts, one by Routley, one by Meyer: The origins of the Routley–Meyer semantics for relevance logics, *Australian Journal of Logic* **15**: 171–209.

85. Dunn, J. M. (2019a). Natural language versus formal language, *in* H. Omori and H. Wansing (eds.), *New Essays on Belnap–Dunn Logic*, number 418 in *Synthese Library*, Springer Nature, Switzerland, pp. 13–19.

86. Dunn, J. M. (2019b). Two, three, four, infinity: The path to the four-valued logic and beyond, *in* H. Omori and H. Wansing (eds.), *New Essays on Belnap–Dunn Logic*, number 418 in *Synthese Library*, Springer International Publishing, Switzerland, pp. 67–86.

87. Dunn, J. M. and Kiefer, N. M. (2019). Contradictory information: Better than nothing? The paradox of the two firefighters, *in* C. Başkent and T. Ferguson

(eds.), *Graham Priest on Dialetheism and Paraconsistency*, Vol. 18 of *Outstanding Contributions to Logic*, Springer Nature, Switzerland, pp. 231–247.

88. Dunn, J. M. (2021). R-Mingle is nice, and so is Arnon Avron, *in* O. Arieli and A. Zamansky (eds.), *Arnon Avron on Semantics and Proof Theory of Nonclassical Logics*, Vol. 21 of *Outstanding Contributions to Logic*, Springer Nature, Switzerland, pp. 141–165.

89. Yang, E. and Dunn, J. M. (2021a). Implicational tonoid logics: Algebraic and relational semantics, *Logica Universalis* **15**: 435–456.

90. Yang, E. and Dunn, J. M. (2021b). Implicational partial Galois logics: Relational semantics, *Logica Universalis* **15**: 457–476.

91. Dunn, J. M. and Golan, A. (2021). Information and its value, *in* M. Chen, J. M. Dunn, A. Golan and A. Ullah (eds.), *Advances in Info-Metrics: Information and Information Processing across Disciplines*, Oxford University Press, New York, NY, pp. 3–31.

92. Bimbó, K. and Dunn, J. M. (2022a). St. Alasdair on lattices everywhere, *in* I. Düntsch and E. Mares (eds.), *Alasdair Urquhart on Nonclassical and Algebraic Logic and Complexity of Proofs*, Vol. 22 of *Outstanding Contributions to Logic*, Springer Nature, Switzerland, pp. 323–346.

93. Dunn, J. M. (2022). Kripke's argument for γ, *in* K. Bimbó (ed.), *Relevance Logics and other Tools for Reasoning. Essays in Honor of J. Michael Dunn*, Vol. 46 of *Tributes*, College Publications, London, UK, pp. 178–181.

94. Bimbó, K. and Dunn, J. M. (2022b). Modalities in lattice-*R*, *in* K. Bimbó (ed.), *Relevance Logics and other Tools for Reasoning. Essays in Honor of J. Michael Dunn*, Vol. 46 of *Tributes*, College Publications, London, UK, pp. 89–127.

95. Bimbó, K. and Dunn, J. M. (202+a). Fine's semantics for relevance logic and its relevance, *in* F. L. G. Faroldi and F. Van De Putte (eds.), *Kit Fine on Truthmakers, Relevance and Non-Classical Logic*, Vol. ## of *Outstanding Contributions to Logic*, Springer Nature, (2019, 24 pages, forthcoming).

96. Bimbó, K. and Dunn, J. M. (202+b). Entailment, mingle and binary accessibility, *in* Y. Weiss and R. Padró (eds.), *Saul A. Kripke on Modal Logic*, Vol. ## of *Outstanding Contributions to Logic*, Springer Nature, (2020, 29 pages, forthcoming).

A list of Dunn's publications up to 2015, which also includes smaller sundry pieces from abstracts to book reviews, may be found in *J. Michael Dunn on Information Based Logics*, K. Bimbó (ed.), (Outstanding Contributions to Logic, Vol. 8), Springer Nature, Switzerland, 2016, pp. xxxv–xliii.

www.ingramcontent.com/pod-product-compliance
Lightning Source LLC
Chambersburg PA
CBHW050118170426
43197CB00011B/1627